THE TIME MACHINE
AND OTHER STORIES

STORY THE FIRST

The Time Machine

§ 1

THE Time Traveller (for so it will be convenient to speak of him) was expounding a recondite matter to us. His grey eyes shone and twinkled, and his usually pale face was flushed and animated. The fire burned brightly, and the soft radiance of the incandescent lights in the lilies of silver caught the bubbles that flashed and passed in our glasses. Our chairs, being his patents, embraced and caressed us rather than submitted to be sat upon, and there was that luxurious after-dinner atmosphere when thought runs gracefully free of the trammels of precision. And he put it to us in this way—marking the points with a lean forefinger—as we sat and lazily admired his earnestness over this new paradox (as we thought it) and his fecundity.

"You must follow me carefully. I shall have to controvert one or two ideas that are almost universally accepted. The geometry, for instance, they taught you at school is founded on a misconception."

"Is not that rather a large thing to expect us to begin upon?" said Filby, an argumentative person with red hair.

"I do not mean to ask you to accept anything without reasonable ground for it. You will soon admit as much as I need from you. You know of course that a mathematical line, a line of thickness *nil*, has no real existence. They taught you that? Neither has a mathematical plane. These things are mere abstractions."

9

"That is all right," said the Psychologist.

"Nor, having only length, breadth, and thickness, can a cube have a real existence."

"There I object," said Filby. "Of course a solid body may exist. All real things——"

"So most people think. But wait a moment. Can an *instantaneous* cube exist?"

"Don't follow you," said Filby.

"Can a cube that does not last for any time at all, have a real existence?"

Filby became pensive. "Clearly," the Time Traveller proceeded, "any real body must have extension in *four* directions: it must have Length, Breadth, Thickness, and— Duration. But through a natural infirmity of the flesh, which I will explain to you in a moment, we incline to overlook this fact. There are really four dimensions, three which we call the three planes of Space, and a fourth, Time. There is, however, a tendency to draw an unreal distinction between the former three dimensions and the latter, because it happens that our consciousness moves intermittently in one direction along the latter from the beginning to the end of our lives."

"That," said a very young man, making spasmodic efforts to relight his cigar over the lamp; "that . . . very clear indeed."

"Now, it is very remarkable that this is so extensively overlooked," continued the Time Traveller, with a slight accession of cheerfulness. "Really this is what is meant by the Fourth Dimension, though some people who talk about the Fourth Dimension do not know they mean it. It is only another way of looking at Time. *There is no difference between Time and any of the three dimensions of Space except that our consciousness moves along it.* But some foolish people have got hold of the wrong side of that idea. You have all heard what they have to say about this Fourth Dimension?"

"I have not," said the Provincial Mayor.

"It is simply this. That Space, as our mathematicians have it, is spoken of as having three dimensions, which one may call Length, Breadth, and Thickness, and is always definable by reference to three planes, each at right angles to the others. But some philosophical people have been asking why *three* dimensions particularly—why not another direction at right angles to the other three?—and have even tried to construct a Four-Dimensional geometry. Professor Simon Newcomb was expounding this to the New York Mathematical Society only a month or so ago. You know how on a flat surface, which has only two dimensions, we can represent a figure of a three-dimensional solid, and similarly they think that by models of three dimensions they could represent one of four—if they could master the perspective of the thing. See?"

"I think so," murmured the Provincial Mayor; and, knitting his brows, he lapsed into an introspective state, his lips moving as one who repeats mystic words. "Yes, I think I see it now," he said after some time, brightening in a quite transitory manner.

"Well, I do not mind telling you I have been at work upon this geometry of Four Dimensions for some time. Some of my results are curious. For instance, here is a portrait of a man at eight years old, another at fifteen, another at seventeen, another at twenty-three, and so on. All these are evidently sections, as it were, Three-Dimensional representations of his Four-Dimensioned being, which is a fixed and unalterable thing.

"Scientific people," proceeded the Time Traveller, after the pause required for the proper assimilation of this, "know very well that Time is only a kind of Space. Here is a popular scientific diagram, a weather record. This line I trace with my finger shows the movement of the barometer. Yesterday it was so high, yesterday night it fell, then this morning it rose again, and so gently upward to here. Surely the mercury did not trace this line in any of the dimensions of Space generally recognised? But certainly it traced such a line, and that line, therefore, we must conclude was along the Time-Dimension."

"But," said the Medical Man, staring hard at a coal in the fire, "if Time is really only a fourth dimension of Space, why is it, and why has it always been, regarded as something different? And why cannot we move in Time as we move about in the other dimensions of Space?"

The Time Traveller smiled. "Are you so sure we can move freely in Space? Right and left we can go, backward and forward freely enough, and men always have done so. I admit we move freely in two dimensions. But how about up and down? Gravitation limits us there."

"Not exactly," said the Medical Man. "There are balloons."

"But before the balloons, save for spasmodic jumping and the inequalities of the surface, man had no freedom of vertical movement."

"Still they could move a little up and down," said the Medical Man.

"Easier, far easier down than up."

"And you cannot move at all in Time, you cannot get away from the present moment."

"My dear sir, that is just where you are wrong. That is just where the whole world has gone wrong. We are always getting away from the present moment. Our mental existences, which are immaterial and have no dimensions, are passing along the Time-Dimensions with a uniform velocity from the cradle to the grave. Just as we should travel *down* if we began our existence fifty miles above the earth's surface."

"But the great difficulty is this," interrupted the Psychologist. "You *can* move about in all directions of Space, but you cannot move about in Time."

"That is the germ of my great discovery. But you are wrong to say that we cannot move about in Time. For instance, if I am recalling an incident very vividly I go back to the instant of its occurrence: I become absent-minded, as you say. I jump back for a moment. Of course we have no means of staying back for any length of Time, any more than a savage or an animal has of staying six feet above the ground. But a civilised man is better off than the savage in this respect. He can go up against gravitation in a balloon, and why should he not hope that ultimately he may be able to stop or accelerate his drift along the Time-Dimension, or even turn about and travel the other way?"

"Oh, *this*," began Filby, "is all——"

"Why not?" said the Time Traveller.

"It's against reason," said Filby.

"What reason?" said the Time Traveller.

"You can show black is white by argument," said Filby, "but you will never convince me."

"Possibly not," said the Time Traveller. "But now you begin to see the object of my investigations into the geometry of Four Dimensions. Long ago I had a vague inkling of a machine——"

"To travel through Time!" exclaimed the Very Young Man.

"That shall travel indifferently in any direction of Space and Time, as the driver determines."

Filby contented himself with laughter.

"But I have experimental verification," said the Time Traveller.

"It would be remarkably convenient for the historian," the Psychologist suggested. "One might travel back and verify the accepted account of the Battle of Hastings, for instance!"

"Don't you think you would attract attention?" said the Medical Man. "Our ancestors had no great tolerance for anachronisms."

"One might get one's Greek from the very lips of Homer and Plato," the Very Young Man thought.

"In which case they would certainly plough you for the Little-go. The German scholars have improved Greek so much."

"Then there is the future," said the Very Young Man. "Just think! One might invest all one's money, leave it to accumulate at interest, and hurry on ahead!"

"To discover a society," said I, "erected on a strictly communistic basis."

"Of all the wild extravagant theories!" began the Psychologist.

"Yes, so it seemed to me, and so I never talked of it until——"

"Experimental verification!" cried I. "You are going to verify *that?*"

"The experiment!" cried Filby, who was getting brain-weary.

"Let's see your experiment anyhow," said the Psychologist, "though it's all humbug, you know."

The Time Traveller smiled round at us. Then, still smiling faintly, and with his hands deep in his trousers pockets, he walked slowly out of the room, and we heard his slippers shuffling down the long passage to his laboratory.

The Psychologist looked at us. "I wonder what he's got?"

"Some sleight-of-hand trick or other," said the Medical Man, and Filby tried to tell us about a conjurer he had seen at Burslem; but before he had finished his preface the Time Traveller came back, and Filby's anecdote collapsed.

The thing the Time Traveller held in his hand was a glittering metallic framework, scarcely larger than a small clock, and very delicately made. There was ivory in it, and some transparent crystalline substance. And now I must be explicit, for this that follows—unless his explanation is to be accepted—is an absolutely unaccountable thing. He took one of the small octagonal tables that were scattered about the room, and set it in front of the fire, with two legs on the hearth rug. On this table he placed the mechanism. Then he drew up a chair, and sat down. The only other object on the table was a small shaded lamp, the bright light of which fell full upon the model. There were also perhaps a dozen candles about, two in brass candlesticks upon the mantel and several in sconces, so that the room was brilliantly illuminated. I sat in a low armchair nearest the fire, and I drew this forward so as to be almost between the Time Traveller and the fireplace. Filby sat behind him, looking over his shoulder. The Medical Man and the Provincial Mayor watched him in profile from the right, the Psychologist from the left. The Very Young Man stood behind the Psychologist. We were all on the alert. It appears incredible to me that any kind of trick, however subtly conceived and however adroitly done, could have been played upon us under these conditions.

The Time Traveller looked at us, and then at the mechanism. "Well?" said the Psychologist.

"This little affair," said the Time Traveller, resting his

elbows upon the table and pressing his hands together above the apparatus, "is only a model. It is my plan for a machine to travel through time. You will notice that it looks singularly askew, and that there is an odd twinkling appearance about this bar, as though it was in some way unreal." He pointed to the part with his finger. "Also, here is one little white lever, and here is another."

The Medical Man got up out of his chair and peered into the thing. "It's beautifully made," he said.

"It took two years to make," retorted the Time Traveller. Then, when we had all imitated the action of the Medical Man, he said: "Now I want you clearly to understand that this lever, being pressed over, sends the machine gliding into the future, and this other reverses the motion. This saddle represents the seat of a time traveller. Presently I am going to press the lever, and off the machine will go. It will vanish, pass into future Time, and disappear. Have a good look at the thing. Look at the table too, and satisfy yourselves there is no trickery. I don't want to waste this model, and then be told I'm a quack."

There was a minute's pause perhaps. The Psychologist seemed about to speak to me, but changed his mind. Then the Time Traveller put forth his fingers towards the lever. "No," he said suddenly. "Lend me your hand." And turning to the Psychologist, he took that individual's hand in his own and told him to put out his forefinger. So that it was the Psychologist himself who sent forth the model Time Machine on its interminable voyage. We all saw the lever turn. I am absolutely certain there was no trickery. There was a breath of wind, and the lamp flame jumped. One of the candles on the mantel was blown out, and the little machine suddenly swung round, became indistinct, was seen as a ghost for a second perhaps, as an eddy of faintly glittering brass and ivory; and it was gone—vanished! Save for the lamp the table was bare.

Every one was silent for a minute. Then Filby said he was damned.

The Psychologist recovered from his stupor, and suddenly looked under the table. At that the Time Traveller laughed cheerfully. "Well?" he said, with a reminiscence of the

Psychologist. Then, getting up, he went to the tobacco jar on the mantel, and with his back to us began to fill his pipe.

We stared at each other. "Look here," said the Medical Man, "are you in earnest about this? Do you seriously believe that that machine has travelled into time."

"Certainly," said the Time Traveller, stooping to light a spill at the fire. Then he turned, lighting his pipe, to look at the Psychologist's face. (The Psychologist, to show that he was not unhinged, helped himself to a cigar and tried to light it uncut.) "What is more, I have a big machine nearly finished in there"—he indicated the laboratory—"and when that is put together I mean to have a journey on my own account."

"You mean to say that that machine has travelled into the future?" said Filby.

"Into the future or the past—I don't, for certain, know which."

After an interval the Psychologist had an inspiration. "It must have gone into the past if it has gone anywhere," he said.

"Why?" said the Time Traveller.

"Because I presume that it has not moved in space, and if it travelled into the future it would still be here all this time, since it must have travelled through this time."

"But," said I, "if it travelled into the past it would have been visible when we came first into this room; and last Thursday when we were here; and the Thursday before that; and so forth!"

"Serious objections," remarked the Provincial Mayor, with an air of impartiality, turning towards the Time Traveller.

"Not a bit," said the Time Traveller, and, to the Psychologist: "You think. *You* can explain that. It's presentation below the threshold, you know, diluted presentation."

"Of course," said the Psychologist, and reassured us. "That's a simple point of psychology. I should have thought of it. It's plain enough, and helps the paradox delightfully. We cannot see it, nor can we appreciate this machine, any more than we can the spoke of a wheel spinning, or a bullet flying through the air. If it is travelling through time fifty times or a hundred times faster than we are, if it gets through

a minute while we get through a second, the impression
it creates will of course be only one-fiftieth or one-hundredth
of what it would make if it were not travelling in time.
That's plain enough." He passed his hand through the
space in which the machine had been. "You see?" he said,
laughing.

We sat and stared at the vacant table for a minute or so.
Then the Time Traveller asked us what we thought of it all.

"It sounds plausible enough to-night," said the Medical
Man; "but wait until to-morrow. Wait for the common
sense of the morning."

"Would you like to see the Time Machine itself?" asked
the Time Traveller. And therewith, taking the lamp in his
hand, he led the way down the long, draughty corridor to his
laboratory. I remember vividly the flickering light, his
queer, broad head in silhouette, the dance of the shadows,
how we all followed him, puzzled but incredulous, and how
there in the laboratory we beheld a larger edition of the little
mechanism which we had seen vanish from before our eyes.
Parts were of nickel, parts of ivory, parts had certainly been
filed or sawn out of rock crystal. The thing was generally
complete, but the twisted crystalline bars lay unfinished upon
the bench beside some sheets of drawings, and I took one
up for a better look at it. Quartz it seemed to be.

"Look here," said the Medical Man, "are you perfectly
serious? Or is this a trick—like that ghost you showed us
last Christmas?"

"Upon that machine," said the Time Traveller, holding
the lamp aloft, "I intend to explore time. Is that plain?
I was never more serious in my life."

None of us quite knew how to take it.

I caught Filby's eye over the shoulder of the Medical Man,
and he winked at me solemnly.

§ 2

I think that at that time none of us quite believed in the
Time Machine. The fact is, the Time Traveller was one
of those men who are too clever to be believed: you never

felt that you saw all round him; you always suspected some subtle reserve, some ingenuity in ambush, behind his lucid frankness. Had Filby shown the model and explained the matter in the Time Traveller's words, we should have shown *him* far less scepticism. For we should have perceived his motives: a pork butcher could understand Filby. But the Time Traveller had more than a touch of whim among his elements, and we distrusted him. Things that would have made the fame of a less clever man seemed tricks in his hands. It is a mistake to do things too easily. The serious people who took him seriously never felt quite sure of his deportment: they were somehow aware that trusting their reputations for judgment with him was like furnishing a nursery with egg-shell china. So I don't think any of us said very much about time travelling in the interval between that Thursday and the next, though its odd potentialities ran, no doubt, in most of our minds: its plausibility, that is, its practical incredibleness, the curious possibilities of anachronism and of utter confusion it suggested. For my own part, I was particularly preoccupied with the trick of the model. That I remember discussing with the Medical Man, whom I met on Friday at the Linnæan. He said he had seen a similar thing at Tübingen, and laid considerable stress on the blowing out of the candle. But how the trick was done he could not explain.

The next Thursday I went again to Richmond—I suppose I was one of the Time Traveller's most constant guests—and, arriving late, found four or five men already assembled in his drawing-room. The Medical Man was standing before the fire with a sheet of paper in one hand and his watch in the other. I looked round for the Time Traveller, and—"It's half past seven now," said the Medical Man. "I suppose we'd better have dinner?"

"Where's ——?" said I, naming our host.

"You've just come? It's rather odd. He's unavoidably detained. He asks me in this note to lead off with dinner at seven if he's not back. Says he'll explain when he comes."

"It seems a pity to let the dinner spoil," said the Editor of a well-known daily paper; and thereupon the Doctor rang the bell.

The Psychologist was the only person besides the Doctor and myself who had attended the previous dinner. The other men were Blank, the Editor aforementioned, a certain journalist, and another—a quiet, shy man with a beard—whom I didn't know, and who, as far as my observation went, never opened his mouth all the evening. There was some speculation at the dinner table about the Time Traveller's absence, and I suggested time travelling, in a half jocular spirit. The Editor wanted that explained to him, and the Psychologist volunteered a wooden account of the "ingenious paradox and trick" we had witnessed that day week. He was in the midst of his exposition when the door from the corridor opened slowly and without noise. I was facing the door, and saw it first. "Hallo!" I said. "At last!" And the door opened wider, and the Time Traveller stood before us. I gave a cry of surprise. "Good heavens! man, what's the matter?" cried the Medical Man, who saw him next. And the whole tableful turned towards the door.

He was in an amazing plight. His coat was dusty and dirty, and smeared with green down the sleeves; his hair disordered, and as it seemed to me greyer—either with dust and dirt or because its colour had actually faded. His face was ghastly pale; his chin had a brown cut on it—a cut half healed; his expression was haggard and drawn, as by intense suffering. For a moment he hesitated in the doorway, as if he had been dazzled by the light. Then he came into the room. He walked with just such a limp as I have seen in footsore tramps. We stared at him in silence, expecting him to speak.

He said not a word, but came painfully to the table, and made a motion towards the wine. The Editor filled a glass of champagne, and pushed it towards him. He drained it, and it seemed to do him good: for he looked round the table, and the ghost of his old smile flickered across his face. "What on earth have you been up to, man?" said the Doctor. The Time Traveller did not seem to hear. "Don't let me disturb you," he said, with a certain faltering articulation. "I'm all right." He stopped, held out his glass for more, and took it off at a draught. "That's good," he said. His eyes grew brighter, and a faint colour came into his cheeks.

His glance flickered over our faces with a certain dull approval, and then went round the warm and comfortable room. Then he spoke again, still as it were feeling his way among his words. "I'm going to wash and dress, and then I'll come down and explain things. . . . Save me some of that mutton. I'm starving for a bit of meat."

He looked across at the Editor, who was a rare visitor, and hoped he was all right. The Editor began a question. "Tell you presently," said the Time Traveller. "I'm—funny! Be all right in a minute."

He put down his glass, and walked towards the staircase door. Again I remarked his lameness and the soft padding sound of his footfall, and standing in my place, I saw his feet as he went out. He had nothing on them but a pair of tattered, bloodstained socks. Then the door closed upon him. I had half a mind to follow, till I remembered how he detested any fuss about himself. For a minute, perhaps, my mind was wool gathering. Then, "Remarkable Behaviour of an Eminent Scientist," I heard the Editor say, thinking (after his wont) in head-lines. And this brought my attention back to the bright dinner table.

"What's the game?" said the Journalist. "Has he been doing the Amateur Cadger? I don't follow." I met the eye of the Psychologist, and read my own interpretation in his face. I thought of the Time Traveller limping painfully upstairs. I don't think any one else had noticed his lameness.

The first to recover completely from his surprise was the Medical Man, who rang the bell—the Time Traveller hated to have servants waiting at dinner—for a hot plate. At that the Editor turned to his knife and fork with a grunt, and the Silent Man followed suit. The dinner was resumed. Conversation was exclamatory for a little while, with gaps of wonderment; and then the Editor got fervent in his curiosity. "Does our friend eke out his modest income with a crossing? Or has he his Nebuchadnezzar phases?" he inquired. "I feel assured it's this business of the Time Machine," I said, and took up the Psychologist's account of our previous meeting. The new guests were frankly incredulous. The Editor raised objections. "What *was* this time travelling? A man couldn't cover himself with dust by rolling in a paradox, could he?" And then, as the idea came home to him, he resorted to caricature. Hadn't they any clothes-brushes in the Future? The Journalist, too, would not believe at any price, and joined the Editor in the easy work of heaping ridicule on the whole thing. They were both the new kind of journalist—very joyous, irreverent young men. "Our Special Correspondent in the Day after To-morrow reports," the Journalist was saying—or rather shouting—when the Time Traveller came back. He was dressed in ordinary evening clothes, and nothing save his haggard look remained of the change that had startled me.

"I say," said the Editor hilariously, "these chaps here say you have been travelling into the middle of next week!! Tell us all about little Rosebery, will you? What will you take for the lot?"

The Time Traveller came to the place reserved for him without a word. He smiled quietly, in his old way. "Where's my mutton?" he said. "What a treat it is to stick a fork into meat again!"

"Story!" cried the Editor.

"Story be damned!" said the Time Traveller. "I want something to eat. I won't say a word until I get some peptone into my arteries. Thanks. And the salt."

"One word," said I. "Have you been time travelling?"

"Yes," said the Time Traveller, with his mouth full, nodding his head.

"I'd give a shilling a line for a verbatim note," said the Editor. The Time Traveller pushed his glass towards the Silent Man and rang it with his finger nail; at which the Silent man, who had been staring at his face, started convulsively, and poured him wine. The rest of the dinner was uncomfortable. For my own part, sudden questions kept on rising to my lips, and I dare say it was the same with the others. The Journalist tried to relieve the tension by telling anecdotes of Hettie Potter. The Time Traveller devoted his attention to his dinner, and displayed the appetite of a tramp. The Medical Man smoked a cigarette, and watched the Time Traveller through his eyelashes. The Silent Man seemed even more clumsy than usual, and drank champagne with regularity and determination out of sheer nervousness. At last the Time Traveller pushed his plate away, and looked round us. "I suppose I must apologise," he said. "I was simply starving. I've had a most amazing time." He reached out his hand for a cigar, and cut the end. "But come into the smoking-room. It's too long a story to tell over greasy plates." And ringing the bell in passing, he led the way into the adjoining room.

"You have told Blank, and Dash, and Chose about the machine?" he said to me, leaning back in his easy chair and naming the three new guests.

"But the thing's a mere paradox," said the Editor.

"I can't argue to-night. I don't mind telling you the story, but I can't argue. I will," he went on, "tell you the story of what has happened to me, if you like, but you must refrain from interruptions. I want to tell it. Badly. Most of it will sound like lying. So be it! It's true—every word of it, all the same. I was in my laboratory at four o'clock, and since then . . . I've lived eight days . . . such days as no human being ever lived before! I'm nearly worn out, but I shan't sleep till I've told this thing over to you. Then I shall go to bed. But no interruptions! Is it agreed?"

"Agreed," said the Editor, and the rest of us echoed, "Agreed." And with that the Time Traveller began his story as I have set it forth. He sat back in his chair at first, and spoke like a weary man. Afterwards he got more animated. In writing it down I feel with only too much keenness the inadequacy of pen and ink—and, above all, my own inadequacy—to express its quality. You read, I will suppose, attentively enough; but you cannot see the speaker's white, sincere face in the bright circle of the little lamp, nor hear the intonation of his voice. You cannot know how his expression followed the turns of his story! Most of us hearers were in shadow, for the candles in the smoking-room had not been lighted, and only the face of the Journalist and the legs of the Silent Man from the knees downward were illuminated. At first we glanced now and again at each other. After a time we ceased to do that, and looked only at the Time Traveller's face.

§ 3

"I told some of you last Thursday of the principles of the Time Machine, and showed you the actual thing itself, incomplete in the workshop. There it is now, a little travel-worn, truly; and one of the ivory bars is cracked, and a brass rail bent; but the rest of it's sound enough. I expected to finish it on Friday; but on Friday, when the putting together was nearly done, I found that one of the nickel bars was exactly one inch too short, and this I had to get remade; so that the thing was not complete until this morning. It was at ten o'clock to-day that the first of all Time Machines began its career. I gave it a last tap, tried all the screws again, put one more drop of oil on the quartz rod, and sat myself in the saddle. I suppose a suicide who holds a pistol to his skull feels much the same wonder at what will come next as I felt then. I took the starting lever in one hand and the stopping one in the other, pressed the first, and almost immediately the second. I seemed to reel; I felt a nightmare sensation of falling; and, looking round, I saw the laboratory exactly as before; Had anything happened? For a moment I suspected that my intellect had tricked me. Then I noted the clock. A moment before, as it seemed, it had stood at a minute or so past ten; now it was nearly half-past three!

"I drew a breath, set my teeth, gripped the starting lever with both hands, and went off with a thud. The laboratory got hazy and went dark. Mrs. Watchett came in and walked, apparently without seeing me, towards the garden door. I suppose it took her a minute or so to traverse the place, but to me she seemed to shoot across the room like a rocket. I pressed the lever over to its extreme position. The night came like the turning out of a lamp, and in another moment came to-morrow. The laboratory grew faint and hazy, then fainter and ever fainter. To-morrow night came black, then day again, night again, day again, faster and faster still. An eddying murmur filled my ears, and a strange, dumb confusedness descended on my mind.

"I am afraid I cannot convey the peculiar sensations of

time travelling. They are excessively unpleasant. There is a feeling exactly like that one has upon a switchback—of a helpless headlong motion! I felt the same horrible anticipation, too, of an imminent smash. As I put on pace, night followed day like the flapping of a black wing. The dim suggestion of the laboratory seemed presently to fall away from me, and I saw the sun hopping swiftly across the sky, leaping it every minute, and every minute marking a day. I supposed the laboratory had been destroyed and I had come into the open air. I had a dim impression of scaffolding, but I was already going too fast to be conscious of any moving things. The slowest snail that ever crawled dashed by too fast for me. The twinkling succession of darkness and light was excessively painful to the eye. Then, in the intermittent darknesses, I saw the moon spinning swiftly through her quarters from new to full, and had a faint glimpse of the circling stars. Presently, as I went on, still gaining velocity, the palpitation of night and day merged into one continuous greyness; the sky took on a wonderful deepness of blue, a splendid luminous colour like that of early twilight; the jerking sun became a streak of fire, a brilliant arch, in space; the moon a fainter fluctuating band; and I could see nothing of the stars, save now and then a brighter circle flickering in the blue.

"The landscape was misty and vague. I was still on the hillside upon which this house now stands, and the shoulder rose above me grey and dim. I saw trees growing and changing like puffs of vapour, now brown, now green; they grew, spread, shivered, and passed away. I saw huge buildings rise up faint and fair, and pass like dreams. The whole surface of the earth seemed changed —melting and flowing under my eyes. The little hands upon the dials that registered my speed raced round faster and faster. Presently I noted that the sun belt swayed up and down, from solstice to solstice, in a minute or less, and that consequently my pace was over a year a minute; and minute by minute the white snow flashed across the world, and vanished, and was followed by the bright, brief green of spring.

"The unpleasant sensations of the start were less poignant now. They merged at last into a kind of hysterical exhilaration. I remarked indeed a clumsy swaying of the marchine, for which I was unable to account. But my mind was too confused to attend to it, so with a kind of madness growing upon me, I flung myself into futurity. At first I scarce thought of stopping, scarce thought of anything but these new sensations. But presently a fresh series of impressions grew up in my mind—a certain curiosity and therewith a certain dread—until at last they took complete possession of me. What strange developments of humanity, what wonderful advances upon our rudimentary civilisation, I thought, might not appear when I came to look nearly into the dim elusive world that raced and fluctuated before my eyes! I saw great and splendid architecture rising about me, more massive than any buildings of our own time, and yet, as it seemed, built of glimmer and mist. I saw a richer green flow up the hillside, and remain there without any wintry intermission. Even through the veil of my confusion the earth seemed very fair. And so my mind came round to the business of stopping.

"The peculiar risk lay in the possibility of my finding some substance in the space which I, or the machine, occupied. So long as I travelled at a high velocity through time, this scarcely mattered; I was, so to speak, attenuated—was slipping like a vapour through the interstices of intervening substances! But to come to a stop involved the jamming of myself, molecule by molecule, into whatever lay in my way; meant bringing my atoms into such intimate contact with those of the obstacle that a profound chemical reaction—possibly a far-reaching explosion—would result, and blow myself and my apparatus out of all possible dimensions—into the Unknown. This possibility had occurred to me again and again while I was making the machine; but then I had cheerfully accepted it as an unavoidable risk—one of the risks a man has got to take! Now the risk was inevitable, I no longer saw it in the same cheerful light. The fact is that, insensibly, the absolute strangeness of everything, the sickly jarring and swaying of the machine, above all, the feeling of prolonged falling, had absolutely upset my nerve. I told myself that I could never stop, and with a gust of petulance I resolved to stop forthwith. Like an impatient fool, I lugged over the lever, and incontinently the thing went reeling over, and I was flung headlong through the air.

"There was the sound of a clap of thunder in my ears. I may have been stunned for a moment. A pitiless hail was hissing round me, and I was sitting on soft turf in front of the overset machine. Everything still seemed grey, but presently I remarked that the confusion in my ears was gone. I looked round me. I was on what seemed to be a little lawn in a garden, surrounded by rhododendron bushes, and I noticed that their mauve and purple blossoms were dropping in a shower under the beating of the hailstones. The rebounding, dancing hail hung in a cloud over the machine, and drove along the ground like smoke. In a moment I was wet to the skin. 'Fine hospitality,' said I, 'to a man who has travelled innumerable years to see you.'

"Presently I thought what a fool I was to get wet. I stood up and looked round me. A colossal figure, carved apparently in some white stone, loomed indistinctly beyond the rhododendrons through the hazy downpour. But all else of the world was invisible.

"My sensations would be hard to describe. As the columns of hail grew thinner, I saw the white figure more distinctly. It was very large, for a silver birch-tree touched its shoulder. It was of white marble, in shape something like a winged sphinx, but the wings, instead of being carried vertically at the sides, were spread so that it seemed to hover. The pedestal, it appeared to me, was of bronze, and was thick with verdigris. It chanced that the face was towards me; the sightless eyes seemed to watch me; there was the faint shadow of a smile on the lips. It was greatly weather-worn, and that imparted an unpleasant suggestion of disease. I stood looking at it for a little space—half a minute, perhaps, or half an hour. It seemed to advance and to recede as the hail drove before it denser or thinner. At last I tore my eyes from it for a moment, and saw that the hail curtain had worn threadbare, and that the sky was lightening with the promise of the sun.

"I looked up again at the crouching white shape, and the

full temerity of my voyage came suddenly upon me. What
might appear when that hazy curtain was altogether withdrawn? What might not have happened to men? What if
cruelty had grown into a common passion? What if in this
interval the race had lost its manliness, and had developed
into something inhuman, unsympathetic, and overwhelmingly powerful? I might seem some old-world savage animal,
only the more dreadful and disgusting for our common likeness—a foul creature to be incontinently slain.

"Already I saw other vast shapes—huge buildings with
intricate parapets and tall columns, with a wooded hillside
dimly creeping in upon me through the lessening storm.
I was seized with a panic fear. I turned frantically to the
Time Machine, and strove hard to readjust it. As I did so
the shafts of the sun smote through the thunderstorm. The
grey downpour was swept aside and vanished like the trailing garments of a ghost. Above me, in the intense blue of
the summer sky, some faint brown shreds of cloud whirled
into nothingness. The great buildings about me stood out
clear and distinct, shining with the wet of the thunderstorm, and picked out in white by the unmelted hailstones
piled along their courses. I felt naked in a strange world.
I felt as perhaps a bird may feel in the clear air, knowing
the hawk wins above and will swoop. My fear grew to
frenzy. I took a breathing space, set my teeth, and again
grappled fiercely, wrist and knee, with the machine. It
gave under my desperate onset and turned over. It struck
my chin violently. One hand on the saddle, the other on
the lever, I stood panting heavily in attitude to mount again.

"But with this recovery of a prompt retreat my courage
recovered. I looked more curiously and less fearfully at
this world of the remote future. In a circular opening, high
up in the wall of the nearer house, I saw a group of figures
clad in rich soft robes. They had seen me, and their faces
were directed towards me.

"Then I heard voices approaching me. Coming through
the bushes by the White Sphinx were the heads and
shoulders of men running. One of these emerged in a
pathway leading straight to the little lawn upon which I
stood with my machine. He was a slight creature—perhaps

four feet high—clad in a purple tunic, girdled at the waist with a leather belt. Sandals or buskins—I could not clearly distinguish which—were on his feet; his legs were bare to the knees, and his head was bare. Noticing that, I noticed for the first time how warm the air was.

"He struck me as being a very beautiful and graceful creature, but indescribably frail. His flushed face reminded me of the more beautiful kind of consumptive—that hectic beauty of which we used to hear so much. At the sight of him I suddenly regained confidence. I took my hands from the machine.

§ 4

"In another moment we were standing face to face, I and this fragile thing out of futurity. He came straight up to me and laughed into my eyes. The absence from his bearing of any sign of fear struck me at once. Then he turned to the two others who were following him and spoke to them in a strange and very sweet and liquid tongue.

"There were others coming, and presently a little group of perhaps eight or ten of these exquisite creatures were about me. One of them addressed me. It came into my head, oddly enough, that my voice was too harsh and deep for them. So I shook my head, and, pointing to my ears, shook it again. He came a step forward, hesitated, and then touched my hand. Then I felt other soft little tentacles upon my back and shoulders. They wanted to make sure I was real. There was nothing in this at all alarming. Indeed, there was something in these pretty little people that inspired confidence—a graceful gentleness, a certain child-like ease. And besides, they looked so frail that I could fancy myself flinging the whole dozen of them about like nine-pins. But I made a sudden motion to warn them when I saw their little pink hands feeling at the Time Machine. Happily then, when it was not too late, I thought of a danger I had hitherto forgotten, and reaching over the bars of the machine I unscrewed the little levers that would set it in motion, and put these in my pocket. Then I turned again to see what I could do in the way of communication.

"And then, looking more nearly into their features, I saw some further peculiarities in their Dresden-china type of prettiness. Their hair, which was uniformly curly, came to a sharp end at the neck and cheek; there was not the faintest suggestion of it on the face, and their ears were singularly minute. The mouths were small, with bright red, rather thin lips, and the little chins ran to a point. The eyes were large and mild; and—this may seem egotism on my part—I fancied even then that there was a certain lack of the interest I might have expected in them.

"As they made no effort to communicate with me, but simply stood round me smiling and speaking in soft, cooing notes to each other, I began the conversation. I pointed to the Time Machine and to myself. Then hesitating for a moment how to express time, I pointed to the sun. At once a quaintly pretty little figure in chequered purple and white followed my gesture, and then astonished me by imitating the sound of thunder.

"For a moment I was staggered, though the import of his gesture was plain enough. The question had come into my mind abruptly: were these creatures fools? You may hardly understand how it took me. You see, I had always antici-pated that the people of the year Eight Hundred and Two Thousand odd would be incredibly in front of us in know-ledge, art, everything. Then one of them suddenly asked me a question that showed him to be on the intellectual level of one of our five-year-old children—asked me, in fact, if I had come from the sun in a thunderstorm! It let loose the judgment I had suspended upon their clothes, their frail light limbs, and fragile features. A flow of disappointment rushed across my mind. For a moment I felt that I had built the Time Machine in vain.

"I nodded, pointed to the sun, and gave them such a vivid rendering of a thunderclap as startled them. They all with-drew a pace or so and bowed. Then came one laughing towards me, carrying a chain of beautiful flowers altogether new to me, and put it about my neck. The idea was re-ceived with melodious applause; and presently they were all running to and fro for flowers, and laughingly flinging them upon me until I was almost smothered with blossom. You

who have never seen the like can scarcely imagine what
delicate and wonderful flowers countless years of culture had
created. Then someone suggested that their plaything
should be exhibited in the nearest building, and so I was led
past the sphinx of white marble, which had seemed to watch
me all the while with a smile at my astonishment, towards
a vast grey edifice of fretted stone. As I went with them the
memory of my confident anticipations of a profoundly grave
and intellectual posterity came, with irresistible merriment,
to my mind.

"The building had a huge entry, and was altogether of
colossal dimensions. I was naturally most occupied with the
growing crowd of little people, and with the big open portals
that yawned before me shadowy and mysterious. My
general impression of the world I saw over their heads was
of a tangled waste of beautiful bushes and flowers, a long-
neglected and yet weedless garden. I saw a number of tall
spikes of strange white flowers, measuring a foot perhaps
across the spread of the waxen petals. They grew scattered,
as if wild, among the variegated shrubs, but, as I say, I did
not examine them closely at this time. The Time Machine
was left deserted on the turf among the rhododendrons.

"The arch of the doorway was richly carved, but naturally
I did not observe the carving very narrowly, though I fancied
I saw suggestions of old Phœnician decorations as I passed
through, and it struck me that they were very badly broken
and weather-worn. Several more brightly clad people met
me in the doorway, and so we entered, I, dressed in dingy
nineteenth-century garments, looking grotesque enough,
garlanded with flowers, and surrounded by an eddying mass
of bright, soft-coloured robes and shining white limbs, in
a melodious whirl of laughter and laughing speech.

"The big doorway opened into a proportionately great hall
hung with brown. The roof was in shadow, and the windows, partially glazed with coloured glass and partially un-
glazed, admitted a tempered light. The floor was made up
of huge blocks of some very hard white metal, not plates
nor slabs,—blocks, and it was so much worn, as I judged by
the going to and fro of past generations, as to be deeply
channelled along the more frequented ways. Transverse to

the length were innumerable tables made of slabs of polished stone, raised perhaps a foot from the floor, and upon these were heaps of fruits. Some I recognised as a kind of hypertrophied raspberry and orange, but for the most part they were strange.

"Between the tables was scattered a great number of cushions. Upon these my conductors seated themselves, signing for me to do likewise. With a pretty absence of ceremony they began to eat the fruit with their hands, flinging peel and stalks and so forth, into the round openings in the sides of the tables. I was not loth to follow their example, for I felt thirsty and hungry. As I did so I surveyed the hall at my leisure.

"And perhaps the thing that struck me most was its dilapidated look. The stained-glass windows, which displayed only a geometrical pattern, were broken in many places, and the curtains that hung across the lower end were thick with dust. And it caught my eye that the corner of the marble table near me was fractured. Nevertheless, the general effect was extremely rich and picturesque. There were, perhaps, a couple of hundred people dining in the hall, and most of them, seated as near to me as they could come, were watching me with interest, their little eyes shining over the fruit they were eating. All were clad in the same soft, and yet strong, silky material.

"Fruit, by the bye, was all their diet. These people of the remote future were strict vegetarians, and while I was with them, in spite of some carnal cravings, I had to be frugivorous also. Indeed, I found afterwards that horses, cattle, sheep, dogs, had followed the Ichthyosaurus into extinction. But the fruits were very delightful; one, in particular, that seemed to be in season all the time I was there —a floury thing in a three-sided husk—was especially good, and I made it my staple. At first I was puzzled by all these strange fruits, and by the strange flowers I saw, but later I began to perceive their import.

"However, I am telling you of my fruit dinner in the distant future now. So soon as my appetite was a little checked, I determined to make a resolute attempt to learn the speech of these new men of mine. Clearly that was the next thing to do. The fruits seemed a convenient thing to begin upon, and holding one of these up I began a series of interrogative sounds and gestures. I had some considerable difficulty in conveying my meaning. At first my efforts met with a stare of surprise or inextinguishable laughter, but presently a fair-haired little creature seemed to grasp my intention and repeated a name. They had to chatter and explain the business at great length to each other, and my first attempts to make the exquisite little sounds of their language caused an immense amount of amusement. However, I felt like a schoolmaster amidst children, and persisted, and presently I had a score of noun substantives at least at my command; and then I got to demonstrative pronouns, and even the verb 'to eat.' But it was slow work, and the little people soon tired and wanted to get away from my interrogations, so I determined, rather of necessity, to let them give their lessons in little doses when they felt inclined. And very little doses I found they were before long, for I never met people more indolent or more easily fatigued.

"A queer thing I soon discovered about my little hosts, and that was their lack of interest. They would come to me with eager cries of astonishment, like children, but like children they would soon stop examining me and wander away after some other toy. The dinner and my conversational beginnings ended, I noted for the first time that almost all those who had surrounded me at first were gone. It is odd, too, how speedily I came to disregard these little people. I went out through the portal into the sunlit world again so soon as my hunger was satisfied. I was continually meeting more of these men of the future, who would follow me a little distance, chatter and laugh about me, and, having smiled and gesticulated in a friendly way, leave me again to my own devices.

The calm of evening was upon the world as I emerged from the great hall, and the scene was lit by the warm glow of the setting sun. At first things were very confusing. Everything was so entirely different from the world I had known—even the flowers. The big building I had left was situated on the slope of a broad river valley, but the Thames had shifted perhaps a mile from its present position. I resolved to mount to the summit of a crest, perhaps a mile and a half away, from which I could get a wider view of this our planet in the year Eight Hundred and Two Thousand Seven Hundred and One A.D. For that, I should explain, was the date the little dials of my machine recorded.

"As I walked I was watchful for every impression that could possibly help to explain the condition of ruinous splendour in which I found the world—for ruinous it was. A little way up the hill, for instance, was a great heap of granite; bound together by masses of aluminium, a vast labyrinth of precipitous walls and crumbled heaps, amidst which were thick heaps of very beautiful pagoda-like plants —nettles possibly—but wonderfully tinted with brown about the leaves, and incapable of stinging. It was evidently the derelict remains of some vast structure, to what end built I could not determine. It was here that I was destined, at a later date, to have a very strange experience—the first intimation of a still stranger discovery—but of that I will speak in its proper place.

"Looking round with a sudden thought, from a terrace on which I rested for a while, I realised that there were no small houses to be seen. Apparently the single house, and possibly even the household, had vanished. Here and there among the greenery were palace-like buildings, but the house and the cottage, which form such characteristic features of our own English landscape, had disappeared.

"'Communism,' said I to myself.

"And on the heels of that came another thought. I looked at the half-dozen little figures that were following me. Then, in a flash, I perceived that all had the same form of costume, the same soft hairless visage, and the same girlish rotundity of limb. It may seem strange, perhaps, that I had not noticed this before. But everything was so strange. Now, I saw the fact plainly enough. In costume, and in all the differences of texture and bearing that now mark off the sexes from each other, these people of the future were alike. And the children seemed to my eyes to be but the miniatures of their parents. I judged, then, that the children of that time were extremely precocious, physically at

B

least, and I found afterwards abundant verification of my opinion.

"Seeing the ease and security in which these people were living, I felt that this close resemblance of the sexes was, after all, what one would expect; for the strength of a man and the softness of a woman, the institution of the family, and the differentiation of occupations are mere militant necessities of an age of physical force. Where population is balanced and abundant, much child-bearing becomes an evil rather than a blessing to the State; where violence comes but rarely and offspring are secure, there is less necessity—indeed there is no necessity—for an efficient family, and the specialisation of the sexes with reference to their children's needs disappears. We see some beginnings of this even in our own time, and in this future age it was complete. This, I must remind you, was my speculation at the time. Later, I was to appreciate how far it fell short of the reality.

"While I was musing upon these things, my attention was attracted by a pretty little structure, like a well under a cupola. I thought in a transitory way of the oddness of wells still existing, and then resumed the thread of my speculations. There were no large buildings towards the top of the hill, and as my walking powers were evidently miraculous, I was presently left alone for the first time. With a strange sense of freedom and adventure I pushed on up to the crest.

"There I found a seat of some yellow metal that I did not recognise, corroded in places with a kind of pinkish rust and half smothered in soft moss, the arm rests cast and filed into the resemblance of griffins' heads. I sat down on it, and I surveyed the broad view of our old world under the sunset of that long day. It was as sweet and fair a view as I have ever seen. The sun had already gone below the horizon and the west was flaming gold, touched with some horizontal bars of purple and crimson. Below was the valley of the Thames, in which the river lay like a band of burnished steel. I have already spoken of the great palaces dotted about among the variegated greenery, some in ruins and some still occupied. Here and there rose a white or silvery figure in the waste garden of the earth, here and there came the sharp vertical line of some cupola or obelisk. There were no hedges, no signs of proprietary rights, no evidence of agriculture; the whole earth had become a garden.

"So watching, I began to put my interpretation upon the things I had seen, and as it shaped itself to me that evening, my interpretation was something in this way. (Afterwards I found I had got only a half-truth—or only a glimpse of one facet of the truth.)

"It seemed to me that I had happened upon humanity upon the wane. The ruddy sunset set me thinking of the sunset of mankind. For the first time I began to realise an odd consequence of the social effort in which we are at present engaged. And yet, come to think, it is a logical consequence enough. Strength is the outcome of need; security sets a premium on feebleness. The work of ameliorating the conditions of life—the true civilising process that makes life more and more secure—had gone steadily on to a climax. One triumph of a united humanity over Nature had followed another. Things that are now mere dreams had become projects deliberately put in hand and carried forward. And the harvest was what I saw!

"After all, the sanitation and the agriculture of to-day are still in the rudimentary stage. The science of our time has attacked but a little department of the field of human disease, but, even so, it spreads its operations very steadily and persistently. Our agriculture and horticulture destroy a weed just here and there and cultivate perhaps a score or so of wholesome plants, leaving the greater number to fight out a balance as they can. We improve our favourite plants and animals—and how few they are—gradually by selective breeding; now a new and better peach, now a seedless grape, now a sweeter and larger flower, now a more convenient breed of cattle. We improve them gradually, because our ideals are vague and tentative, and our knowledge is very limited; because Nature, too, is shy and slow in our clumsy hands. Some day all this will be better organised, and still better. That is the drift of the current in spite of the eddies. The whole world will be intelligent, educated, and co-operating; things will move faster and faster towards the subjugation of Nature. In the end, wisely and carefully we shall re-adjust the balance of animal and vegetable life to suit our human needs.

"This adjustment, I say, must have been done, and done well; done indeed for all Time, in the space of Time across which my machine had leaped. The air was free from gnats, the earth from weeds or fungi; everywhere were fruits and sweet and delightful flowers; brilliant butterflies flew hither and thither. The ideal of preventive medicine was attained. Diseases had been stamped out. I saw no evidence of any contagious diseases during all my stay. And I shall have to tell you later that even the processes of putre-faction and decay had been profoundly affected by these changes.

"Social triumphs, too, had been effected. I saw mankind housed in splendid shelters, gloriously clothed, and as yet I had found them engaged in no toil. There were no signs of struggle, neither social nor economical struggle. The shop, the advertisement, traffic, all that commerce which constitutes the body of our world, was gone. It was natural on that golden evening that I should jump at the idea of a social paradise. The difficulty of increasing population had been met, I guessed, and population had ceased to increase.

"But with this change in condition comes inevitably adaptations to the change. What, unless biological science is a mass of errors, is the cause of human intelligence and vigour? Hardship and freedom: conditions under which the active, strong, and subtle survive and the weaker go to the wall; conditions that put a premium upon the loyal alliance of capable men, upon self-restraint, patience, and decision. And the institution of the family, and the emotions that arise therein, the fierce jealousy, the tenderness for offspring, parental self-devotion, all found their justification and sup-port in the imminent dangers of the young. *Now,* where are these imminent dangers? There is a sentiment arising, and it will grow, against connubial jealousy, against fierce maternity, against passion of all sorts; unnecessary things now, and things that make us uncomfortable, savage sur-vivals, discords in a refined and pleasant life.

"I thought of the physical slightness of the people, their

lack of intelligence, and those big abundant ruins, and it strengthened my belief in a perfect conquest of Nature. For after the battle comes Quiet. Humanity had been strong, energetic, and intelligent, and had used all its abundant vitality to alter the conditions under which it lived. And now came the reaction of the altered conditions.

"Under the new conditions of perfect comfort and security, that restless energy, that with us is strength, would become weakness. Even in our own time certain tendencies and desires, once necessary to survival, are a constant source of failure. Physical courage and the love of battle, for instance, are no great help—may even be hindrances—to a civilised man. And in a state of physical balance and security, power, intellectual as well as physical, would be out of place. For countless years I judged there had been no danger of war or solitary violence, no danger from wild beasts, no wasting disease to require strength of constitution, no need of toil. For such a life, what we should call the weak are as well equipped as the strong, are indeed no longer weak. Better equipped indeed they are, for the strong would be fretted by an energy for which there was no outlet. No doubt the exquisite beauty of the buildings I saw was the outcome of the last surgings of the now purposeless energy of mankind before it settled down into perfect harmony with the conditions under which it lived—the flourish of that triumph which began the last great peace. This has ever been the fate of energy in security; it takes to art and to eroticism, and then come languor and decay.

"Even this artistic impetus would at last die away—had almost died in the Time I saw. To adorn themselves with flowers, to dance, to sing in the sunlight; so much was left of the artistic spirit, and no more. Even that would fade in the end into a contented inactivity. We are kept keen on the grindstone of pain and necessity, and, it seemed to me, that here was that hateful grindstone broken at last!

"As I stood there in the gathering dark I thought that in this simple explanation I had mastered the problem of the world—mastered the whole secret of these delicious people. Possibly the checks they had devised for the increase of population had succeeded too well, and their numbers had

rather diminished than kept stationary.   That would
account for the abandoned ruins. Very simple was my explanation, and plausible enough—as most wrong theories
are!

§ 5

"As I stood there musing over this too perfect triumph of
man, the full moon, yellow and gibbous, came up out of an
overflow of silver light in the north-east. The bright little
figures ceased to move about below, a noiseless owl flitted
by, and I shivered with the chill of the night. I determined
to descend and find where I could sleep.

"I looked for the building I knew. Then my eye travelled
along to the figure of the White Sphinx upon the pedestal
of bronze, growing distinct as the light of the rising moon
grew brighter. I could see the silver birch against it. There
was the tangle of rhododendron bushes, black in the pale
light, and there was the little lawn. I looked at the lawn
again. A queer doubt chilled my complacency. 'No,' said
I stoutly to myself, 'that was not the lawn.'

"But it *was* the lawn. For the white leprous face of the
sphinx was towards it. Can you image what I felt as this
conviction came home to me? But you cannot. The Time
Machine was gone!

"At once, like a lash across the face, came the possibility
of losing my own age, of being left helpless in this strange
new world. The bare thought of it was an actual physical
sensation. I could feel it grip me at the throat and stop my
breathing. In another moment I was in a passion of fear
and running with great leaping strides down the slope. Once
I fell headlong and cut my face; I lost no time in stanching
the blood, but jumped up and ran on, with a warm trickle
down by cheek and chin. All the time I ran I was saying
to myself, 'They have moved it a little, pushed it under the
bushes out of the way.' Nevertheless, I ran with all my
might. All the time, with the certainty that sometimes
comes with excessive dread, I knew that such assurance
was folly, knew instinctively that the machine was removed
out of my reach. My breath came with pain. I suppose

I covered the whole distance from the hill crest to the little lawn, two miles, perhaps, in ten minutes. And I am not a young man. I cursed aloud, as I ran, at my confident folly in leaving the machine, wasting good breath thereby. I cried aloud, and none answered. Not a creature seemed to be stirring in that moonlit world.

"When I reached the lawn my worst fears were realised. Not a trace of the thing was to be seen. I felt faint and cold when I faced the empty space among the black tangle of bushes. I ran round it furiously, as if the thing might be hidden in a corner, and then stopped abruptly, with my hands clutching my hair. Above me towered the sphinx, upon the bronze pedestal, white, shining, leprous, in the light of the rising moon. It seemed to smile in mockery of my dismay.

"I might have consoled myself by imagining the little people had put the mechanism in some shelter for me, had I not felt assured of their physical and intellectual inadequacy. That is what dismayed me: the sense of some hitherto unsuspected power, through whose intervention my invention had vanished. Yet, of one thing I felt assured: unless some other age had produced its exact duplicate, the machine could not have moved in time. The attachment of the levers—I will show you the method later—prevented anyone from tampering with it in that way when they were removed. It had moved, and was hid, only in space. But then, where could it be?

"I think I must have had a kind of frenzy. I remember running violently in and out among the moonlit bushes all round the sphinx, and startling some white animal that, in the dim light, I took for a small deer. I remember, too, late that night, beating the bushes with my clenched fists until my knuckles were gashed and bleeding from the broken twigs. Then, sobbing and raving in my anguish of mind, I went down to the great building of stone. The big hall was dark, silent, and deserted. I slipped on the uneven floor, and fell over one of the malachite tables, almost breaking my shin. I lit a match and went on past the dusty curtains, of which I have told you.

"There I found a second great hall covered with cushions,

upon which, perhaps, a score or so of the little people were sleeping. I have no doubt they found my second appearance strange enough, coming suddenly out of the quiet darkness with inarticulate noises and the splutter and flare of a match. For they had forgotten about matches. 'Where is my Time Machine?' I began, bawling like an angry child, laying hands upon them and shaking them up together. It must have been very queer to them. Some laughed, most of them looked sorely frightened. When I saw them standing round me, it came into my head that I was doing as foolish a thing as it was possible for me to do under the circumstances, in trying to revive the sensation of fear. For, reasoning from their daylight behaviour, I thought that fear must be forgotten.

"Abruptly, I dashed down the match, and, knocking one of the people over in my course, went blundering across the big dining-hall again, out under the moonlight. I heard cries of terror and their little feet running and stumbling this way and that. I do not remember all I did as the moon crept up the sky. I suppose it was the unexpected nature of my loss that maddened me. I felt hopelessly cut off from my own kind—a strange animal in an unknown world. I must have raved to and fro, screaming and crying upon God and Fate. I have a memory of horrible fatigue, as the long night of despair wore away; of looking in this impossible place and that; of groping among moonlit ruins and touching strange creatures in the black shadows; at last, of lying on the ground near the sphinx and weeping with absolute wretchedness. I had nothing left but misery. Then I slept, and when I woke again it was full day, and a couple of sparrows were hopping round me on the turf within reach of my arm.

"I sat up in the freshness of the morning, trying to remember how I had got there, and why I had such a profound sense of desertion and despair. Then things came clear in my mind. With the plain, reasonable daylight, I could look my circumstances fairly in the face. I saw the wild folly of my frenzy overnight, and I could reason with myself. 'Suppose the worst?' I said. 'Suppose the machine altogether lost—perhaps destroyed? It behoves me to be calm and patient, to learn the way of the people, to get a clear idea of the method of my loss, and the means of getting materials and tools; so that in the end, perhaps, I may make another.' That would be my only hope, a poor hope perhaps, but better than despair. And, after all, it was a beautiful and curious world.

"But probably the machine had only been taken away. Still, I must be calm and patient, find its hiding-place, and recover it by force or cunning. And with that I scrambled to my feet and looked about me, wondering where I could bathe. I felt weary, stiff, and travel-soiled. The freshness of the morning made me desire an equal freshness. I had exhausted my emotion. Indeed, as I went about my business, I found myself wondering at my intense excitement overnight. I made a careful examination of the ground about the little lawn. I wasted some time in futile questionings, conveyed, as well as I was able, to such of the little people as came by . They all failed to understand my gestures; some were simply stolid, some thought it was a jest and laughed at me. I had the hardest task in the world to keep my hands off their pretty laughing faces. It was a foolish impulse, but the devil begotten of fear and blind anger was ill curbed and still eager to take advantage of my perplexity. The turf gave better counsel. I found a groove ripped in it, about midway between the pedestal of the sphinx and the marks of my feet where, on arrival, I had struggled with the overturned machine. There were other signs of removal about, with queer narrow footprints like those I could imagine made by a sloth. This directed my closer attention to the pedestal. It was, as I think I have said, of bronze. It was not a mere block, but highly decorated with deep framed panels on either side. I went and rapped at these. The pedestal was hollow. Examining the panels with care I found them discontinuous with the frames. There were no handles or keyholes, but possibly the panels, if they were doors, as I supposed, opened from within. One thing was clear enough to my mind. It took no very great mental effort to infer that my Time Machine was inside that pedestal. But how it got there was a different problem.

"I saw the heads of two orange-clad people coming through the bushes and under some blossom-covered apple-trees towards me. I turned smiling to them and beckoned them to me. They came, and then, pointing to the bronze pedestal, I tried to intimate my wish to open it. But at my first gesture towards this they behaved very oddly. I don't know how to convey their expression to you. Suppose you were to use a grossly improper gesture to a delicate-minded woman —it is how she would look. They went off as if they had received the last possible insult. I tried a sweet-looking little chap in white next, with exactly the same result. Somehow, his manner made me feel ashamed of myself. But, as you know, I wanted the Time Machine, and I tried him once more. As he turned off, like the others, my temper got the better of me. In three strides I was after him, had him by the loose part of his robe round the neck, and began dragging him towards the sphinx. Then I saw the horror and repugnance of his face, and all of a sudden I let him go.

"But I was not beaten yet. I banged with my fist at the bronze panels. I thought I heard something stir inside—to be explicit, I thought I heard a sound like a chuckle—but I must have been mistaken. Then I got a big pebble from the river, and came and hammered till I had flattened a coil in the decorations, and the verdigris came off in powdery flakes. The delicate little people must have heard me hammering in gusty outbreaks a mile away on either hand, but nothing come of it. I saw a crowd of them upon the slopes, looking furtively at me. At last, hot and tired, I sat down to watch the place. But I was too restless to watch long; I am too Occidental for a long vigil. I could work at a problem for years, but to wait inactive for twenty-four hours—that is another matter.

"I got up after a time, and began walking aimlessly through the bushes towards the hill again. 'Patience,' said I to myself. 'If you want your machine again you must leave that sphinx alone. If they mean to take your machine away, it's little good your wrecking their bronze panels, and if they don't, you will get it back as soon as you can ask for it. To sit among all those unknown things before a puzzle like that is hopeless. That way lies monomania. Face this world. Learn its ways, watch it, be careful of too hasty guesses at its meaning. In the end you will find clues to it all. Then suddenly the humour of the situation came into my mind: the thought of the years I had spent in study and toil to get into the future age, and now my passion of anxiety to get out of it. I had made myself the most complicated and the most hopeless trap that ever a man devised. Although it was at my own expense, I could not help myself. I laughed aloud.

"Going through the big palace, it seemed to me that the little people avoided me. It may have been my fancy, or it may have had something to do with my hammering at the gates of bronze. Yet I felt tolerably sure of the avoidance. I was careful, however, to show no concern and to abstain from any pursuit of them, and in the course of a day or two things got back to the old footing. I made what progress I could in the language, and, in addition, I pushed my explorations here and there. Either I missed some subtle point, or their language was excessively simple—almost exclusively composed of concrete substantives and verbs. There seemed to be few, if any, abstract terms, or little use of figurative language. Their sentences were usually simple and of two words, and I failed to convey or understand any but the simplest propositions. I determined to put the thought of my Time Machine and the mystery of the bronze doors under the sphinx as much as possible in a corner of memory, until my growing knowledge would lead me back to them in a natural way. Yet a certain feeling, you may understand, tethered me in a circle of a few miles round the point of my arrival.

"So far as I could see, all the world displayed the same exuberant richness as the Thames valley. From every hill I climbed I saw the same abundance of splendid buildings, endlessly varied in material and style, the same clustering thickets of evergreens, the same blossom-laden trees and tree-ferns. Here and there water shone like silver, and beyond, the land rose into blue undulating hills, and so faded into the serenity of the sky. A peculiar feature, which presently attracted my attention, was the presence of certain circular wells, several, as it seemed to me, of a very great depth. One lay by the path up the hill, which I had followed during my first walk. Like the others, it was rimmed with bronze, curiously wrought, and protected by a little cupola from the rain. Sitting by the side of these wells, and peering down into the shafted darkness, I could see no gleam of water, nor could I start any reflection with a lighted match. But in all of them I heard a certain sound: a thud—thud—thud, like the beating of some big engine; and I discovered, from the flaring of my matches, that a steady current of air set down the shafts. Further, I threw a scrap of paper into the throat of one, and, instead of fluttering slowly down, it was at once sucked swiftly out of sight.

"After a time, too, I came to connect these wells with tall towers standing here and there upon the slopes; for above them there was often just such a flicker in the air as one sees on a hot day above a sun-scorched beach. Putting things together, I reached a strong suggestion of an extensive system of subterranean ventilation, whose true import it was difficult to imagine. I was at first inclined to associate it with the sanitary apparatus of these people. It was an obvious conclusion, but it was absolutely wrong.

"And here I must admit that I learned very little of drains and bells and modes of conveyance, and the like conveniences, during my time in this real future. In some of these visions of Utopias and coming times which I have read, there is a vast amount of detail about building, and social arrangements, and so forth. But while such details are easy enough to obtain when the whole world is contained in one's imagination, they are altogether inaccessible to a real traveller amid such realities as I found here. Conceive the tale of London which a negro, fresh from Central Africa, would take back to his tribe! What would he know of railway companies, of social movements, of telephone and telegraph wires, of the Parcels Delivery Company, and postal orders and the like? Yet we, at least, should be willing enough to explain these things to him! And even of what he knew, how much could he make his untravelled friend either apprehend or believe? Then, think how narrow the gap between a negro and a white man of our own times, and how wide the interval between myself and these of the

Golden Age! I was sensible of much which was unseen, and which contributed to my comfort; but save for a general impression of automatic organisation, I fear I can convey very little of the difference to your mind.

"In the matter of sepulture, for instance, I could see no signs of crematoria nor anything suggestive of tombs. But it occurred to me that, possibly, there might be cemeteries (or crematoria) somewhere beyond the range of my explorings. This, again, was a question I deliberately put to myself, and my curiosity was at first entirely defeated upon the point. The thing puzzled me, and I was led to make a further remark, which puzzled me still more: that aged and infirm among this people there were none.

"I must confess that my satisfaction with my first theories of an automatic civilisation and a decadent humanity did not long endure. Yet I could think of no other. Let me put my difficulties. The several big palaces I had explored were mere living places, great dining-halls and sleeping apartments. I could find no machinery, no appliances of any kind. Yet these people were clothed in pleasant fabrics that must at times need renewal, and their sandals, though undecorated, were fairly complex specimens of metal-work. Somehow such things must be made. And the little people displayed no vestige of a creative tendency. There were no shops, no workshops, no signs of importations among them. They spent all their time in playing gently, in bathing in the river, in making love in a half-playful fashion, in eating fruit and sleeping. I could not see how things were kept going.

"Then, again, about the Time Machine; something, I knew not what, had taken it into the hollow pedestal of the White Sphinx. *Why?* For the life of me I could not imagine. Those waterless wells, too, those flickering pillars. I felt I lacked a clue. I felt—how shall I put it? Suppose you found an inscription, with sentences here and there in excellent plain English, and interpolated therewith, others made up of words, of letters even, absolutely unknown to you? Well, on the third day of my visit, that was how the world of Eight Hundred and Two Thousand Seven Hundred and One presented itself to me!

"That day, too, I made a friend—of a sort. It happened that, as I was watching some of the little people bathing in a shallow, one of them was seized with cramp and began drifting downstream. The main current ran rather swiftly, but not too strongly for even a moderate swimmer. It will give you an idea, therefore, of the strange deficiency in these creatures, when I tell you that none made the slightest attempt to rescue the weakly crying little thing which was drowning before their eyes. When I realised this, I hurriedly slipped off my clothes and, wading in at a point lower down, I caught the poor mite and drew her safe to land. A little rubbing of the limbs soon brought her round, and I had the satisfaction of seeing she was all right before I left her. I had got to such a low estimate of her kind that I did not expect any gratitude from her. In that, however, I was wrong.

"This happened in the morning. In the afternoon I met my little woman, as I believe it was, as I was returning towards my centre from an exploration, and she received me with cries of delight and presented me with a big garland of flowers—evidently made for me and me alone. The thing took my imagination. Very possibly I had been feeling desolate. At any rate I did my best to display my appreciation of the gift. We were soon seated together in a little stone arbour, engaged in conversation, chiefly of smiles. The creature's friendliness affected me exactly as a child's might have done. We passed each other flowers, and she kissed my hands. I did the same to hers. Then I tried to talk, and found that her name was Weena, which, though I don't know what it meant, somehow seemed appropriate enough. That was the beginning of a queer friendship which lasted a week, and ended—as I will tell you!

"She was exactly like a child. She wanted to be with me always. She tried to follow me everywhere, and on my next journey out and about it went to my heart to tire her down, and leave her at last, exhausted and calling after me rather plaintively. But the problems of the world had to be mastered. I had not, I said to myself, come into the future to carry on a miniature flirtation. Yet her distress when I left her was very great, her expostulations at the parting were sometimes frantic, and I think, altogether, I had as much trouble as comfort from her devotion. Nevertheless she was, somehow, a very great comfort. I thought it was mere childish affection that made her cling to me. Until it was too late, I did not clearly know what I had inflicted upon her when I left her. Nor until it was too late did I clearly understand what she was to me. For, by merely seeming fond of me, and showing in her weak, futile way that she cared for me, the little doll of a creature presently gave my return to the neighbourhood of the White Sphinx almost the feeling of coming home; and I would watch for her tiny figure of white and gold so soon as I came over the hill.

"It was from her, too, that I learned that fear had not yet left the world. She was fearless enough in the daylight, and she had the oddest confidence in me; for once, in a foolish moment, I made threatening grimaces at her, and she simply laughed at them. But she dreaded the dark, dreaded shadows, dreaded black things. Darkness to her was the one thing dreadful. It was a singularly passionate emotion, and it set me thinking and observing. I discovered then, among other things, that these little people gathered into the great houses after dark, and slept in droves. To enter upon them without a light was to put them into a tumult of apprehension. I never found one out of doors, or one sleeping alone within doors, after dark. Yet I was still such a blockhead that I missed the lesson of that fear, and in spite of Weena's distress I insisted upon sleeping away from these slumbering multitudes.

"It troubled her greatly, but in the end her odd affection for me triumphed, and for five of the nights of our acquaintance, including the last night of all, she slept with her head pillowed on my arm. But my story slips away from me as I speak of her. It must have been the night before her rescue that I was awakened about dawn. I had been restless, dreaming most disagreeably that I was drowned, and that sea-anemones were feeling over my face with their soft palps. I woke with a start, and with an odd fancy that some greyish animal had just rushed out of the chamber. I tried to get to sleep again, but I felt restless and uncomfortable. It was that dim grey hour when things are just creeping out of darkness, when everything is colourless and clear cut, and yet unreal. I got up, and went down into the great hall, and so out upon the flagstones in front of the palace. I thought I would make a virtue of necessity, and see the sunrise.

"The moon was setting, and the dying moonlight and the first pallor of dawn were mingled in a ghastly half-light. The bushes were inky black, the ground a sombre grey, the sky colourless and cheerless. And up the hill I thought I could see ghosts. There several times, as I scanned the slope, I saw white figures. Twice I fancied I saw a solitary white, ape-like creature running rather quickly up the hill, and once near the ruins I saw a leash of them carrying some dark body. They moved hastily. I did not see what became of them. It seemed that they vanished among the bushes. The dawn was still indistinct, you must understand. I was feeling that chill, uncertain, early-morning feeling you may have known. I doubted my eyes.

"As the eastern sky grew brighter, and the light of the day came on and its vivid colouring returned upon the world once more, I scanned the view keenly. But I saw no vestige of my white figures. They were mere creatures of the half-light. 'They must have been ghosts,' I said; 'I wonder whence they dated.' For a queer notion of Grant Allen's came into my head, and amused me. If each generation die and leave ghosts, he argued, the world at last will get overcrowded with them. On that theory they would have grown innumerable some Eight Hundred Thousand Years hence, and it was no great wonder to see four at once. But the jest was unsatisfying, and I was thinking of these figures all the morning, until Weena's rescue drove them out of my head. I associated them in some indefinite way with the white animal I had startled in my first passionate search for the Time Machine. But Weena was a pleasant substitute. Yet all the same, they were soon destined to take far deadlier possession of my mind.

"I think I have said how much hotter than our own was the weather of this Golden Age. I cannot account for it. It may be that the sun was hotter, or the earth nearer the sun. It is usual to assume that the sun will go on cooling steadily

in the future. But people, unfamiliar with such speculations as those of the younger Darwin, forget that the planets
must ultimately fall back one by one into the parent body.
As these catastrophes occur, the sun will blaze with renewed
energy; and it may be that some inner planet had suffered
this fate. Whatever the reason, the fact remains that the
sun was very much hotter than we know it.

"Well, one very hot morning—my fourth, I think—as I
was seeking shelter from the heat and glare in a colossal
ruin near the great house where I slept and fed, there happened this strange thing: Clambering among these heaps
of masonry, I found a narrow gallery, whose end and side
windows were blocked by fallen masses of stone. By contrast with the brilliancy outside, it seemed at first impenetrably dark to me. I entered it groping, for the change
from light to blackness made spots of colour swim before
me. Suddenly I halted spellbound. A pair of eyes,
luminous by reflection against the daylight without, was
watching me out of the darkness.

"The old instinctive dread of wild beasts came upon me.
I clenched my hands and steadfastly looked into the glaring
eyeballs. I was afraid to turn. Then the thought of the
absolute security in which humanity appeared to be living
came to my mind. And then I remembered that strange
terror of the dark. Overcoming my fear to some extent, I
advanced a step and spoke. I will admit that my voice was
harsh and ill-controlled. I put out my hand and touched
something soft. At once the eyes darted sideways, and
something white ran past me. I turned with my heart in
my mouth, and saw a queer little ape-like figure, its head
held down in a peculiar manner, running across the sunlit
space behind me. It blundered against a block of granite,
staggered aside, and in a moment was hidden in a black
shadow beneath another pile of ruined masonry.

"My impression of it is, of course, imperfect; but I know
it was a dull white, and had strange large greyish-red eyes;
also that there was flaxen hair on its head and down its
back. But, as I say, it went too fast for me to see distinctly.
I cannot even say whether it ran on all-fours, or only with its
forearms held very low. After an instant's pause I followed

it into the second heap of ruins. I could not find it at first;
but, after a time in the profound obscurity, I came upon one
of those round well-like openings of which I have told you,
half closed by a fallen pillar. A sudden thought came to
me. Could this Thing have vanished down the shaft? I
lit a match, and, looking down, I saw a small, white, moving
creature, with large bright eyes which regarded me steadfastly as it retreated. It made me shudder. It was so like
a human spider! It was clambering down the wall, and now
I saw for the first time a number of metal foot and hand
rests forming a kind of ladder down the shaft. Then the
light burned my fingers and fell out of my hand, going out
as it dropped, and when I had lit another the little monster
had disappeared.

"I do not know how long I sat peering down that well.
It was not for some time that I could succeed in persuading
myself that the thing I had seen was human. But, gradually,
the truth dawned on me: that Man had not remained one
species, but had differentiated into two distinct animals:
that my graceful children of the Upper World were not the
sole descendants of our generation, but that this bleached,
obscene, nocturnal Thing, which had flashed before me,
was also heir to all the ages.

"I thought of the flickering pillars and of my theory of an
underground ventilation. I began to suspect their true
import. And what, I wondered, was this Lemur doing in
my scheme of a perfectly balanced organisation? How was
it related to the indolent serenity of the beautiful Upper-
worlders? And what was hidden down there, at the foot
of that shaft? I sat upon the edge of the well telling myself that, at any rate, there was nothing to fear, and that
there I must descend for the solution of my difficulties. And
withal I was absolutely afraid to go! As I hesitated, two
of the beautiful upper-world people came running in their
amorous sport across the daylight into the shadow. The
male pursued the female, flinging flowers at her as he ran.

"They seemed distressed to find me, my arm against the
overturned pillar, peering down the well. Apparently it
was considered bad form to remark these apertures; for
when I pointed to this one, and tried to frame a question

about it in their tongue, they were still more visibly distressed and turned away. But they were interested by my matches, and I struck some to amuse them. I tried them again about the well, and again I failed. So presently I left them, meaning to go back to Weena, and see what I could get from her. But my mind was already in revolution; my guesses and impressions were slipping and sliding to a new adjustment. I had now a clue to the import of these wells, to the ventilating towers, to the mystery of the ghosts; to say nothing of a hint at the meaning of the bronze gates and the fate of the Time Machine! And very vaguely there came a suggestion towards the solution of the economic problem that had puzzled me.

"Here was the new view. Plainly, this second species of Man was subterranean. There were three circumstances in particular which made me think that its rare emergence above ground was the outcome of a long-continued underground habit. In the first place, there was the bleached look common in most animals that live largely in the dark —the white fish of the Kentucky caves, for instance. Then, those large eyes, with that capacity for reflecting light, are common features of nocturnal things—witness the owl and the cat. And last of all, that evident confusion in the sunshine, that hasty yet fumbling and awkward flight towards dark shadow, and that peculiar carriage of the head while in the light—all reinforced the theory of an extreme sensitiveness of the retina.

"Beneath my feet, then, the earth must be tunnelled enormously, and these tunnellings were the habitat of the new race. The presence of ventilating-shafts and wells along the hill slopes—everywhere, in fact, except along the river valley—showed how universal were its ramifications. What so natural, then, as to assume that it was in this artificial underworld that such work as was necessary to the comfort of the daylight race was done? The notion was so plausible that I at once accepted it, and went on to assume the how of this splitting of the human species. I dare say you will anticipate the shape of my theory; though, for myself, I very soon felt that it fell far short of the truth.

"At first, proceeding from the problems of our own age,

it seemed clear as daylight to me that the gradual widening
of the present merely temporary and social difference between the Capitalist and the Labourer, was the key to the
whole position. No doubt it will seem grotesque enough to
you—and wildly incredible!—and yet even now there are
existing circumstances to point that way. There is a tendency to utilise underground space for the less ornamental
purposes of civilisation; there is the Metropolitan Railway
in London, for instance, there are new electric railways,
there are subways, there are underground workrooms and
restaurants, and they increase and multiply. Evidently, I
thought, this tendency had increased till Industry had
gradually lost its birthright in the sky. I mean that it had
gone deeper and deeper into larger and ever larger underground factories, spending a still-increasing amount of its
time therein, till, in the end—! Even now, does not an
East-end worker live in such artificial conditions as practically to be cut off from the natural surface of the earth?

"Again, the exclusive tendency of richer people—due, no
doubt, to the increasing refinement of their education, and
the widening gulf between them and the rude violence of
the poor—is already leading to the closing, in their interest,
of considerable portions of the surface of the land. About
London, for instance, perhaps half the prettier country is
shut in against intrusion. And this same widening gulf—
which is due to the length and expense of the higher educational process and the increased facilities for and tempta-
tions towards refined habits on the part of the rich—will
make that exchange between class and class, that promotion
by intermarriage which at present retards the splitting of
our species along lines of social stratification, less and less
frequent. So, in the end, above ground you must have the
Haves, pursuing pleasure and comfort, and beauty, and
below ground the Have-nots, the Workers getting continually adapted to the conditions of their labour. Once
they were there, they would, no doubt, have to pay rent,
and not a little of it, for the ventilation of their caverns;
and if they refused, they would starve or be suffocated for
arrears. Such of them as were so constituted as to be
miserable and rebellious would die; and, in the end, the

balance being permanent, the survivors would become as well adapted to the conditions of underground life, and as happy in their way, as the Upper-world people were to theirs. As it seemed to me, the refined beauty and the etiolated pallor, followed naturally enough.

"The great triumph of Humanity I had dreamed of took a different shape in my mind. It had been no such triumph of moral education and general co-operation as I had imagined. Instead, I saw a real aristocracy, armed with a perfected science and working to a logical conclusion the industrial system of to-day. Its triumph had not been simply a triumph over Nature, but a triumph over Nature and the fellow-man. This, I must warn you, was my theory at the time. I had no convenient cicerone in the pattern of the Utopian books. My explanation may be absolutely wrong. I still think it is the most plausible one. But even on this supposition the balanced civilisation that was at last attained must have long since passed its zenith, and was now far fallen into decay. The too-perfect security of the Upperworlders had led them to a slow movement of degeneration, to a general dwindling in size, strength, and intelligence. That I could see clearly enough already. What had happened to the Undergrounders I did not yet suspect; but from what I had seen of the Morlocks—that, by the bye, was the name by which these creatures were called—I could imagine that the modification of the human type was even far more profound than among the 'Eloi,' the beautiful race that I already knew.

"Then came troublesome doubts. Why had the Morlocks taken my Time Machine? For I felt sure it was they who had taken it. Why, too, if the Eloi were masters, could they not restore the machine to me? And why were they so terribly afraid of the dark? I proceeded, as I have said, to question Weena about this Underworld, but here again I was disappointed. At first she would not understand my questions, and presently she refused to answer them. She shivered as though the topic was unendurable. And when I pressed her, perhaps a little harshly, she burst into tears. They were the only tears, except my own, I ever saw in that Golden Age. When I saw them I ceased abruptly to trouble about the Morlocks, and was only concerned in banishing these signs of the human inheritance from Weena's eyes. And very soon she was smiling and clapping her hands, while I solemnly burned a match.

§ 6

"It may seem odd to you, but it was two days before I could follow up the new-found clue in what was manifestly the proper way. I felt a peculiar shrinking from those pallid bodies. They were just the half-bleached colour of the worms and things one sees preserved in spirit in a zoological museum. And they were filthily cold to the touch. Probably my shrinking was largely due to the sympathetic influence of the Eloi, whose disgust of the Morlocks I now began to appreciate.

"The next night I did not sleep well. Probably my health was a little disordered. I was oppressed with perplexity and doubt. Once or twice I had a feeling of intense fear for which I could perceive no definite reason. I remember creeping into the great hall where the little people were sleeping in the moonlight—that night Weena was among them—and feeling reassured by their presence. It occurred to me even then, that in the course of a few days the moon must pass through its last quarter, and the nights grow dark, when the appearances of these unpleasant creatures from below, these whitened Lemurs, this new vermin that had replaced the old, might be more abundant. And on both these days I had the restless feeling of one who shirks an inevitable duty. I felt assured that the Time Machine was only to be recovered by boldly penetrating these underground mysteries. Yet I could not face the mystery. If only I had had a companion it would have been different. But I was so horribly alone, and even to clamber down into the darkness of the well appalled me. I don't know if you will understand my feeling, but I never felt quite safe at my back.

"It was this restlessness, this insecurity, perhaps, that drove me further and further afield in my exploring expedi-

tions. Going to the south-westward towards the rising country that is now called Combe Wood, I observed far off,
in the direction of nineteenth-century Banstead, a vast
green structure, different in character from any I had
hitherto seen. It was larger than the largest of the palaces
or ruins I knew, and the façade had an Oriental look: the
face of it having the lustre, as well as the pale-green tint, a
kind of bluish-green, of a certain type of Chinese porcelain.
This difference in aspect suggested a difference in use, and
I was minded to push on and explore. But the day was
growing late, and I had come upon the sight of the place
after a long and tiring circuit; so I resolved to hold over the
adventure for the following day, and I returned to the welcome and the caresses of little Weena. But next morning
I perceived clearly enough that my curiosity regarding the
Palace of Green Porcelain was a piece of self-deception, to
enable me to shirk, by another day, an experience I dreaded.
I resolved I would make the descent without further waste
of time, and started out in the early morning towards a well
near the ruins of granite and aluminium.

"Little Weena ran with me. She danced beside me to
the well, but when she saw me lean over the mouth and
look downward, she seemed strangely disconcerted. 'Good-
bye, little Weena,' I said, kissing her; and then, putting her
down, I began to feel over the parapet for the climbing hooks.
Rather hastily, I may as well confess, for I feared my courage
might leak away! At first she watched me in amazement.
Then she gave a most piteous cry, and, running to me, she
began to pull at me with her little hands. I think her
opposition nerved me rather to proceed. I shook her off,
perhaps a little roughly, and in another moment I was in the
throat of the well. I saw her agonised face over the parapet,
and smiled to reassure her. Then I had to look down at the
unstable hooks to which I clung.

"I had to clamber down a shaft of perhaps two hundred
yards. The descent was effected by means of metallic bars
projecting from the sides of the well, and these being
adapted to the needs of a creature much smaller and lighter
than myself, I was speedily cramped and fatigued by the
descent. And not simply fatigued! One of the bars bent

suddenly under my weight, and almost swung me off into the blackness beneath. For a moment I hung by one hand, and after that experience I did not dare to rest again. Though my arms and back were presently acutely painful, I went on clambering down the sheer descent with as quick a motion as possible. Glancing upward, I saw the aperture, a small blue disc, in which a star was visible, while little Weena's head showed as a round black projection. The thudding sound of a machine below grew louder and more oppressive. Everything save that little disc above was profoundly dark, and when I looked up again Weena had disappeared.

"I was in an agony of discomfort. I had some thought of trying to go up the shaft again, and leave the Underworld alone. But even while I turned this over in my mind I continued to descend. At last, with intense relief, I saw dimly coming up, a foot to the right of me, a slender loophole in the wall. Swinging myself in, I found it was the aperture of a narrow horizontal tunnel in which I could lie down and rest. It was not too soon. My arms ached, my back was cramped, and I was trembling with the prolonged terror of a fall. Besides this, the unbroken darkness had had a distressing effect upon my eyes. The air was full of the throb and hum of machinery pumping air down the shaft.

"I do not know how long I lay. I was roused by a soft hand touching my face. Starting up in the darkness I snatched at my matches and, hastily striking one, I saw three stooping white creatures similar to the one I had seen above ground in the ruin, hastily retreating before the light. Living, as they did, in what appeared to me impenetrable darkness, their eyes were abnormally large and sensitive, just as are the pupils of the abysmal fishes, and they reflected the light in the same way. I have no doubt they could see me in that rayless obscurity, and they did not seem to have any fear of me apart from the light. But, so soon as I struck a match in order to see them, they fled incontinently, vanishing into dark gutters and tunnels, from which their eyes glared at me in the strangest fashion.

"I tried to call to them, but the language they had was apparently different from that of the Overworld people; so that I was needs left to my own unaided efforts, and the thought of flight before exploration was even then in my mind. But I said to myself, 'You are in for it now,' and, feeling my way along the tunnel, I found the noise of machinery grow louder. Presently the walls fell away from me, and I came to a large open space, and, striking another match, saw that I had entered a vast arched cavern, which stretched into utter darkness beyond the range of my light. The view I had of it was as much as one could see in the burning of a match.

"Necessarily my memory is vague. Great shapes like big machines rose out of the dimness, and cast grotesque black shadows, in which dim spectral Morlocks sheltered from the glare. The place, by the bye, was very stuffy and oppressive, and the faint halitus of freshly shed blood was in the air. Some way down the central vista was a little table of white metal, laid with what seemed a meal. The Morloks at any rate were carniverous! Even at the time, I remember wondering what large animal could have survived to furnish the red joint I saw. It was all very indistinct: the heavy smell, the big unmeaning shapes, the obscene figures lurking in the shadows, and only waiting for the darkness to come at me again! Then the match burned down, and stung my fingers, and fell, a wriggling red spot in the blackness.

"I have thought since how particularly ill equipped I was for such an experience. When I had started with the Time Machine, I had started with the absurd assumption that the men of the Future would certainly be infinitely ahead of ourselves in all their appliances. I had come without arms, without medicine, without anything to smoke—at times I missed tobacco frightfully—even without enough matches. If only I had thought of a Kodak! I could have flashed that glimpse of the underworld in a second, and examined it at leisure. But, as it was, I stood there with only the weapons and the powers that Nature had endowed me with—hands, feet, and teeth; these, and four safety-matches that still remained to me.

"I was afraid to push my way in among all this machinery in the dark, and it was only with my last glimpse of light I discovered that my store of matches had run low. It had never occurred to me until that moment that there was any need to economise them, and I had wasted almost half the box in astonishing the Upper-worlders, to whom fire was a novelty. Now, as I say, I had four left, and while I stood in the dark, a hand touched mine, lank fingers came feeling over my face, and I was sensible of a peculiar unpleasant odour. I fancied I heard the breathing of a crowd of those dreadful little beings about me. I felt the box of matches in my hand being gently disengaged, and other hands behind me plucking at my clothing. The sense of these unseen creatures examining me was indescribably unpleasant. The sudden realisation of my ignorance of their ways of thinking and doing came home to me very vividly in the darkness. I shouted at them as loudly as I could. They started away, and then I could feel them approaching me again. They clutched at me more boldly, whispering odd sounds to each other. I shivered violently, and shouted again—rather discordantly. This time they were not so seriously alarmed, and they made a queer laughing noise as they came back at me. I will confess I was horribly frightened. I determined to strike another match and escape under the protection of its glare. I did so, and eking out the flicker with a scrap of paper from my pocket, I made good my retreat to the narrow tunnel. But I had scarce entered this when my light was blown out, and in the blackness I could hear the Morlocks rustling like wind among leaves, and pattering like the rain, as they hurried after me.

"In a moment I was clutched by several hands, and there was no mistaking that they were trying to haul me back. I struck another light, and waved it in their dazzled faces. You can scarce imagine how nauseatingly inhuman they looked—those pale, chinless faces and great, lidless, pinkish-grey eyes!—as they stared in their blindness and bewilderment. But I did not stay to look, I promise you: I retreated again, and when my second match had ended, I struck my third. It had almost burned through when I reached the opening into the shaft. I lay down on the edge, for the throb of the great pump below made me giddy. Then I felt sideways for the projecting hooks, and, as I did so, my feet were grasped from behind, and I was violently tugged backward. I lit my last match . . . and it incontinently went out. But I had my hand on the climbing bars now, and, kicking violently, I disengaged myself from the clutches of the Morlocks and was speedily clambering up the shaft, while they stayed peering and blinking up at me: all but one little wretch who followed me for some way, and well-nigh secured my boot as a trophy.

"That climb seemed interminable to me. With the last twenty or thirty feet of it a deadly nausea came upon me. I had the greatest difficulty in keeping my hold. The last few yards was a frightful struggle against this faintness. Several times my head swam, and I felt all the sensations of falling. A last, however, I got over the well-mouth some-how, and staggered out of the ruin into the blinding sun-light. I fell upon my face. Even the soil smelt sweet and clean. Then I remember Weena kissing my hands and ears, and the voices of others among the Eloi. Then, for a time, I was insensible.

§ 7

"Now, indeed, I seemed in a worse case than before. Hitherto, except during my night's anguish at the loss of the Time Machine, I had felt a sustaining hope of ultimate escape, but that hope was staggered by these new discoveries. Hitherto I had merely thought myself impeded by the childish simplicity of the little people, and by some un-known forces which I had only to understand to overcome; but there was an altogether new element in the sickening quality of the Morlocks—a something inhuman and malign. Instinctively I loathed them. Before, I had felt as a man might feel who had fallen into a pit: my concern was with the pit and how to get out of it. Now I felt like a beast in a trap, whose enemy would come upon him soon.

"The enemy I dreaded may surprise you. It was the dark-ness of the new moon. Weena had put this into my head by some at first incomprehensible remarks about the Dark Nights. It was not now such a very difficult problem to

guess what the coming Dark Nights might mean. The moon
was on the wane: each night there was a longer interval of
darkness.   And I now understood to some slight degree at
least the reason of the fear of the little Upper-world people
for the dark.   I wondered vaguely what foul villainy it
might be that the Morlocks did under the new moon.   I felt
pretty sure now that my second hypothesis was all wrong.
The Upper-world people might once have been the
favoured aristocracy, and the Morlocks their mechanical
servants; but that had long since passed away.   The two
species that had resulted from the evolution of man were
sliding down towards, or had already arrived at, an altogether new relationship.   The Eloi, like the Carlovingian
kings, had decayed to a mere beautiful futility.   They still
possessed the earth on sufferance: since the Morlocks, subterranean for innumerable generations, had come at last to
find the daylit surface intolerable.   And the Morlocks made
their garments, I inferred, and maintained them in their
habitual needs, perhaps through the survival of an old habit
of service.   They did it as a standing horse paws with his
foot, or as a man enjoys killing animals in sport: because
ancient and departed necessities had impressed it on the
organism.   But, clearly, the old order was already in part
reversed.   The Nemesis of the delicate ones was creeping
on apace.   Ages ago, thousands of generations ago, man had
thrust his brother man out of the ease and the sunshine.
And now that brother was coming back—changed!   Already
the Eloi had begun to learn one old lesson anew.   They
were becoming reacquainted with Fear.   And suddenly there
came into my head the memory of the meat I had seen in
the Under-world.   It seemed odd how it floated into my
mind: not stirred up as it were by the current of my meditations, but coming in almost like a question from outside.   I
tried to recall the form of it.   I had a vague sense of something familiar, but I could not tell what it was at the time.

"Still, however helpless the little people in the presence
of their mysterious Fear, I was differently constituted.   I
came out of this age of ours, this ripe prime of the human
race, when Fear does not paralyse and mystery has lost its
terrors.   I at least would defend myself.   Without further

delay I determined to make myself arms and a fastness where
I might sleep. With that refuge as a base, I could face this
strange world with some of that confidence I had lost in
realising to what creatures night by night I lay exposed. I
felt I could never sleep again until my bed was secure from
them. I shuddered with horror to think how they must
already have examined me.

"I wandered during the afternoon along the valley of
the Thames, but found nothing that commended itself to my
mind as inaccessible. All the buildings and trees seemed
easily practicable to such dexterous climbers as the Morlocks, to judge by their wells, must be. Then the tall pinnacles of the Palace of Green Porcelain and the polished
gleam of its walls came back to my memory; and in the
evening, taking Weena like a child upon my shoulder, I
went up the hills towards the south-west. The distance, I
had reckoned, was seven or eight miles, but it must have
been nearer eighteen. I had first seen the place on a moist
afternoon when distances are deceptively diminished. In
addition, the heel of one of my shoes was loose, and a nail
was working through the sole—they were comfortable old
shoes I wore about indoors—so that I was lame. And it was
already long past sunset when I came in sight of the palace,
silhouetted black against the pale yellow of the sky.

"Weena had been hugely delighted when I began to carry
her, but after a time she desired me to let her down, and
ran along by the side of me, occasionally darting off on
either hand to pick flowers to stick in my pockets. My
pockets had always puzzled Weena, but at the last she had
concluded that they were an eccentric kind of vase for floral
decoration. At least she utilised them for that purpose.
And that reminds me! In changing my jacket I found . . ."

*The Time Traveller paused, put his hand into his pocket,
and silently placed two withered flowers, not unlike very
large white mallows, upon the little table. Then he resumed
his narrative.*

"As the hush of evening crept over the world and we
proceeded over the hill crest towards Wimbledon, Weena
grew tired and wanted to return to the house of grey stone.

But I pointed out the distant pinnacles of the Palace of Green Porcelain to her, and contrived to make her understand that we were seeking a refuge there from her Fear. You know that great pause that comes upon things before the dusk? Even the breeze stops in the trees. To me there is always an air of expectation about that evening stillness. The sky was clear, remote, and empty save for a few horizontal bars far down in the sunset. Well, that night the expectation took the colour of my fears. In that darkling calm my senses seemed preternaturally sharpened. I fancied I could even feel the hollowness of the ground beneath my feet: could, indeed, almost see through it the Morlocks on their ant-hill going hither and thither and waiting for the dark. In my excitement I fancied that they would receive my invasion of their burrows as a declaration of war. And why had they taken my Time Machine?

"So we went on in the quiet, and the twilight deepened into night. The clear blue of the distance faded, and one star after another came out. The ground grew dim and the trees black. Weena's fears and her fatigue grew upon her. I took her in my arms and talked to her and caressed her. Then, as the darkness grew deeper, she put her arms round my neck, and, closing her eyes, tightly pressed her face against my shoulder. So we went down a long slope into a valley, and there in the dimness I almost walked into a little river. This I waded, and went up the opposite side of the valley, past a number of sleeping houses, and by a statue—a Faun, or some such figure, *minus* the head. Here too, were acacias. So far I had seen nothing of the Morlocks, but it was yet early in the night, and the darker hours before the old moon rose were still to come.

"From the brow of the next hill I saw a thick wood spreading wide and black before me. I hesitated at this. I could see no end to it, either to the right or the left. Feeling tired—my feet in particular, were very sore—I carefully lowered Weena from my shoulder as I halted, and sat down upon the turf. I could no longer see the Palace of Green Porcelain, and I was in doubt of my direction. I looked into the thickness of the wood and thought of what it might hide. Under that dense tangle of branches one would be out of

sight of the stars. Even were there no other lurking danger
—a danger I did not care to let my imagination loose upon—
there would still be all the roots to stumble over and the
tree-boles to strike against.

I was very tired, too, after the excitements of the day; so
I decided that I would not face it, but would pass the night
upon the open hill.

"Weena, I was glad to find, was fast asleep. I carefully
wrapped her in my jacket, and sat down beside her to wait
for the moonrise. The hillside was quiet and deserted, but
from the black of the wood there came now and then a
stir of living things. Above me shone the stars, for the
night was very clear. I felt a certain sense of friendly comfort in their twinkling. All the old constellations had gone
from the sky, however: that slow movement which is imperceptible in a hundred human lifetimes, had long since rearranged them in unfamiliar groupings. But the Milky
Way, it seemed to me, was still the same tattered streamer of
star-dust as of yore. Southward (as I judged it) was a very
bright red star that was new to me; it was even more splendid than our own green Sirius. And amid all these scintil-
lating points of light one bright planet shone kindly and
steadily like the face of an old friend.

"Looking at these stars suddenly dwarfed my own troubles
and all the gravities of terrestrial life. I thought of their
unfathomable distance, and the slow inevitable drift of
their movements out of the unknown past into the unknown
future. I thought of the great precessional cycle that the
pole of the earth describes. Only forty times had that silent
revolution occurred during all the years that I had traversed. And during these few revolutions all the
activity, all the traditions, the complex organisations,
the nations, languages, literatures, aspirations, even the
mere memory of Man as I knew him, had been swept out of
existence. Instead were these frail creatures who had forgotten their high ancestry, and the white Things of which
I went in terror. Then I thought of the Great Fear that
was between the two species, and for the first time, with a
sudden shiver, came the clear knowledge of what the meat

I had seen might be. Yet it was too horrible! I looked at
little Weena sleeping beside me, her face white and starlike
under the stars, and forthwith dismissed the thought.

"Through that long night I held my mind off the Morlocks as well as I could, and whiled away the time by trying
to fancy I could find signs of the old constellations in the
new confusion. The sky kept very clear, except for a hazy
cloud or so. No doubt I dozed at times. Then, as my vigil
wore on, came a faintness in the eastward sky, like the reflection of some colourless fire, and the old moon rose, thin
and peaked and white. And close behind, and overtaking
it, and overflowing it, the dawn came, pale at first, and then
growing pink and warm. No Morlocks had approached us.
Indeed, I had seen none upon the hill that night. And in
the confidence of renewed day it almost seemed to me that
my fear had been unreasonable. I stood up and found my
foot with the loose heel swollen at the ankle and paintful
under the heel; so I sat down again, took off my shoes,
and flung them away.

"I awakened Weena, and we went down into the wood,
now green and pleasant instead of black and forbidding.
We found some fruit wherewith to break our fast. We
soon met others of the dainty ones, laughing and dancing in
the sunlight as though there was no such thing in nature
as the night. And then I thought once more of the meat
that I had seen. I felt assured now of what it was, and
from the bottom of my heart I pitied this last feeble rill
from the great flood of humanity. Clearly, at some time
in the Long-Ago of human decay the Morlocks' food had
run short. Possibly they had lived on rats and suchlike
vermin. Even now man is far less discriminating and exclusive in his food than he was—far less than any monkey. His
prejudice against human flesh is no deep-seated instinct.
And so these inhuman sons of men—! I tried to look at
the thing in a scientific spirit. After all, they were less
human and more remote than our cannibal ancestors of three
or four thousand years ago. And the intelligence that would
have made this state of things a torment had gone. Why
should I trouble myself? These Eloi were mere fatted
cattle, which the ant-like Morlocks preserved and preyed

upon—probably saw to the breeding of. And there was Weena dancing at my side!

"Then I tried to preserve myself from the horror that was coming upon me, by regarding it as a rigorous punishment of human selfishness. Man had been content to live in ease and delight upon the labours of his fellow man, had taken Necessity as his watchword and excuse, and in the fullness of time Necessity had come home to him. I even tried a Carlyle-like scorn of this wretched aristocracy in decay. But this attitude of mind was impossible. However great their intellectual degradation, the Eloi had kept too much of the human form not to claim my sympathy, and to make me perforce a sharer in their degradation and their Fear.

"I had at that time very vague ideas as to the course I should pursue. My first was to secure some safe place of refuge, and to make myself such arms of metal or stone as I could contrive. That necessity was immediate. In the next place, I hoped to procure some means of fire, so that I should have the weapon of a torch at hand, for nothing, I knew, would be more efficient against these Morlocks. Then I wanted to arrange some contrivance to break open the doors of bronze under the White Sphinx. I had in mind a battering-ram. I had a persuasion that if I could enter those doors and carry a blaze of light before me I should discover the Time Machine and escape. I could not imagine the Morlocks were strong enough to move it far away. Weena I had resolved to bring with me to our own time. And turning such schemes over in my mind I pursued our way towards the building which my fancy had chosen as our dwelling.

§ 8

"I found the Palace of Green Porcelain, when we approached it about noon, deserted and falling into ruin. Only ragged vestiges of glass remained in its windows, and great sheets of the green facing had fallen away from the corroded metallic framework. It lay very high upon a turfy down, and looking north-eastward before I entered it,

c

I was surprise to see a large estuary, or even creek, where I
judged Wandsworth and Battersea must once have been. I
thought then—though I never followed up the thought—of
what might have happened, or might be happening, to the
living things in the sea.

"The material of the Palace proved on examination to
be indeed porcelain, and along the face of it I saw an inscription in some unknown character. I thought, rather
foolishly, that Weena might help me to interpret this, but I
only learned that the bare idea of writing had never entered
her head. She always seemed to me, I fancy, more human
than she was, perhaps because her affection was so human.

"Within the big valves of the door—which were open and
broken—we found, instead of the customary hall, a long
gallery lit by many side windows. At the first glance I
was reminded of a museum. The tiled floor was thick with
dust, and a remarkable array of miscellaneous objects was
shrouded in the same grey covering. Then I perceived,
standing strange and gaunt in the centre of the hall, what
was clearly the lower part of a huge skeleton. I recognised
by the oblique feet that it was some extinct creature after
the fashion of the Megatherium. The skull and the upper
bones lay beside it in the thick dust, and in one place, where
rain-water had dropped through a leak in the roof, the thing
itself had been worn away. Further in the gallery was the
huge skeleton barrel of a Brontosaurus. My museum
hypothesis was confirmed. Going towards the side I found
what appeared to be sloping shelves, and, clearing away the
thick dust, I found the old familiar glass cases of our own
time. But they must have been air-tight to judge from the
fair preservation of some of their contents.

"Clearly we stood among the ruins of some latter-day
South Kensington! Here, apparently, was the Palæonto-
logical Section, and a very splendid array of fossils it must
have been, though the inevitable process of decay that had
been staved off for a time, and had, through the extinction
of bacteria and fungi, lost ninety-nine hundredths of its
force, was, nevertheless, with extreme sureness if with extreme slowness at work again upon all its treasures. Here
and there I found traces of the little people in the shape of

rare fossils broken to pieces or threaded in strings upon reeds. And the cases had in some instances been bodily removed— by the Morlocks as I judged. The place was very silent. The thick dust deadened our footsteps. Weena, who had been rolling a sea-urchin down the sloping glass of a case, presently came, as I stared about me, and very quietly took my hand and stood beside me.

"And at first I was so much surprised by this ancient monument of an intellectual age, that I gave no thought to the possibilities it presented. Even my preoccupation about the Time Machine receded a little from my mind.

"To judge from the size of the place, this Palace of Green Porcelain had a great deal more in it than a Gallery of Palæontology; possibly historical galleries; it might be, even a library! To me, at least in my present circumstances, these would be vastly more interesting than this spectacle of old-time geology in decay. Exploring, I found another short gallery running transversely to the first. This appeared to be devoted to minerals, and the sight of a block of sulphur set my mind running on gunpowder. But I could find no saltpetre; indeed, no nitrates of any kind. Doubtless they had deliquesced ages ago. Yet the sulphur hung in my mind, and set up a train of thinking. As for the rest of the contents of that gallery, though on the whole they were the best preserved of all I saw, I had little interest. I am no specialist in mineralogy, and I went on down a very ruinous aisle running parallel to the first hall I had entered. Apparently this section had been devoted to natural history, but everything had long since passed out of recognition. A few shrivelled and blackened vestiges of what had once been stuffed animals, desiccated mummies in jars that had once held spirit, a brown dust of departed plants; that was all! I was sorry for that, because I should have been glad to trace the patent readjustments by which the conquest of animated nature had been attained. Then we came to a gallery of simply colossal proportions, but singularly ill-lit, the floor of it running downward at a slight angle from the end at which I entered. At intervals white globes hung from the ceiling—many of them cracked and smashed—which suggested that originally the place had been artificially lit.

Here I was more in my element, for rising on either side
of me were the huge bulks of big machines, all greatly corroded and many broken down, but some still fairly complete.   You know I have a certain weakness for mechanism,
and I was inclined to linger among these; the more so as
for the most part they had the interest of puzzles, and I
could make only the vaguest guesses at what they were for.
I fancied that if I could solve their puzzles I should find
myself in possession of powers that might be of use against
the Morlocks.

"Suddenly Weena came very close to my side.   So suddenly that she startled me.   Had it not been for her I do
not think I should have noticed that the floor of the gallery
sloped at all.*   The end I had come in at was quite above
ground, and was lit by rare slit-like windows.   As you went
down the length the ground come up against these windows,
until at last there was a pit like the 'area' of a London house
before each, and only a narrow line of daylight at the top.
I went slowly along, puzzling about the machines, and had
been too intent upon them to notice the gradual diminution
of the light, until Weena's increasing apprehensions drew
my attention.   Then I saw that the gallery ran down at last
into a thick darkness.   I hesitated, and then, as I looked
round me, I saw that the dust was less abundant and its
surface less even.   Further away towards the dimness, it
appeared to be broken by a number of small narrow footprints.   My sense of the immediate presence of the Morlocks revived at that.   I felt that I was wasting my time in
this academic examination of machinery.   I called to mind
that it was already far advanced in the afternoon, and that
I had still no weapon, no refuge, and no means of making a
fire. And then, down in the remote blackness of the gallery,
I heard a peculiar pattering, and the same odd noises I had
heard down the well.

"I took Weena's hand.   Then, struck with a sudden idea,
I left her and turned to a machine from which projected a
lever not unlike those in a signal-box.   Clambering upon
the stand, and grasping this lever in my hands, I put all my

*It may be, of course, that the floor did not slope, but that the museum was built into
the side of a hill. ---Ed.

weight upon it sideways. Suddenly Weena, deserted in the central aisle began to whimper. I had judged the strength of the lever pretty correctly, for it snapped after a minute's strain, and I rejoined her with a mace in my hand more than sufficient, I judged, for any Morlock skull I might encounter. And I longed very much to kill a Morlock or so. Very inhuman, you may think, to want to go killing one's own descendants! But it was impossible, somehow, to feel any humanity in the things. Only my disinclination to leave Weena, and a persuasion that if I began to slake my thirst for murder my Time Machine might suffer, restrained me from going straight down the gallery and killing the brutes I heard.

"Well, mace in one hand and Weena in the other, I went out of that gallery and into another and still larger one, which at the first glance reminded me of a military chapel hung with tattered flags. The brown and charred rags that hung from the sides of it, I presently recognised as the decaying vestiges of books. They had long since dropped to pieces, and every semblance of print had left them. But here and there were warped boards and cracked metallic clasps that told the tale well enough. Had I been a literary man I might, perhaps, have moralised upon the futility of all ambition. But as it was, the thing that struck me with keenest force was the enormous waste of labour to which this sombre wilderness of rotting paper testified. At the time I will confess that I thought chiefly of the *Philosophical Transactions* and my own seventeen papers upon physical optics.

"Then, going up a broad staircase, we came to what may once have been a gallery of technical chemistry. And here I had not a little hope of useful discoveries. Except at one end where the roof had collapsed, this gallery was well preserved. I went eagerly to every unbroken case. And at last, in one of the really air-tight cases, I found a box of matches. Very eagerly I tried them. They were perfectly good. They were not even damp. I turned to Weena. 'Dance,' I cried to her in her own tongue. For now I had a weapon indeed against the horrible creatures we feared. And so, in that derelict museum, upon the thick soft carpeting of dust, to

Weena's huge delight, I solemnly performed a kind of composite dance, whistling *The Land of the Leal* as cheerfully as I could. In part it was a modest *cancan,* in part a step-dance, in part a skirt-dance (so far as my tail-coat permitted), and in part original. For I am naturally inventive, as you know.

"Now, I still think that for this box of matches to have escaped the wear of time for immemorial years was a most strange, as for me it was a most fortunate, thing. Yet oddly enough, I found a far unlikelier substance, and that was camphor. I found it in a sealed jar, that by chance, I suppose, had been really hermetically sealed. I fancied at first that it was paraffin wax, and smashed the glass accordingly. But the odour of camphor was unmistakable. In the universal decay this volatile substance had chanced to survive, perhaps through many thousands of centuries. It reminded me of a sepia painting I had once seen done from the ink of a fossil Belemnite that must have perished and become fossilised millions of years ago. I was about to throw it away, but I remembered that it was inflammable and burned with a good bright flame—was, in fact, an excellent candle—and I put it in my pocket. I found no explosives, however, nor any means of breaking down the bronze doors. As yet my iron crowbar was the most helpful thing I had chanced upon. Nevertheless I left that gallery greatly elated.

"I cannot tell you all the story of that long afternoon. It would require a great effort of memory to recall my explorations in at all the proper order. I remember a long gallery of rusting stands of arms, and how I hesitated between my crowbar and a hatchet or a sword. I could not carry both, however, and my bar of iron promised best against the bronze gates. There were numbers of guns, pistols, and rifles. The most were masses of rust, but many were of some new metal, and still fairly sound. But any cartridges or powder there may once have been had rotted into dust. One corner I saw was charred and shattered; perhaps, I thought, by an explosion among the specimens. In another place was a vast array of idols—Polynesian, Mexican, Grecian, Phœnician, every country on earth I should think. And here, yielding to an irresistible impulse, I wrote my name upon the nose of a steatite monster from South America that particularly took my fancy.

"As the evening drew on, my interest waned. I went through gallery after gallery, dusty, silent, often ruinous, the exhibits sometimes mere heaps of rust and lignite, sometimes fresher. In one place, I suddenly found myself near the model of a tin-mine, and then by the merest accident I discovered, in an air-tight case, two dynamite cartridges! I shouted 'Eureka,' and smashed the case with joy. Then came a doubt. I hesitated. Then, selecting a little side gallery, I made my essay. I never felt such a disappointment as I did in waiting five, ten, fifteen minutes for an explosion that never came. Of course the things were dummies, as I might have guessed from their presence. I really believe that, had they not been so, I should have rushed off incontinently and blown Sphinx, bronze doors, and (as it proved) my chances of finding the Time Machine, all together into non-existence.

"It was after that, I think, that we came to a little open court within the palace. It was turfed, and had three fruit-trees. So we rested and refreshed ourselves. Towards sunset I began to consider our position. Night was creeping upon us, and my inaccessible hiding-place had still to be found. But that troubled me very little now. I had in my possession a thing that was, perhaps, the best of all defences against the Morlocks—I had matches! I had the camphor in my pocket, too, if a blaze were needed. It seemed to me that the best thing we could do would be to pass the night in the open, protected by a fire. In the morning there was the getting of the Time Machine. Towards that, as yet, I had only my iron mace. But now, with my growing knowledge, I felt very differently towards those bronze doors. Up to this, I had refrained from forcing them, largely because of the mystery on the other side. They had never impressed me as being very strong, and I hoped to find my bar of iron not altogether inadequate for the work.

§9

"We emerged from the palace while the sun was still in part above the horizon. I was determined to reach the White Sphinx early the next morning, and ere the dusk I purposed pushing through the woods that had stopped me on the previous journey. My plan was to go as far as possible that night, and then, building a fire, to sleep in the protection of its glare. Accordingly, as we went along I gathered any sticks or dried grass I saw, and presently had my arms full of such litter. Thus loaded, our progress was slower than I had anticipated, and besides Weena was tired. And I began to suffer from sleepiness too; so that it was full night before we reached the wood. Upon the shrubby hill of its edge Weena would have stopped, fearing the darkness before us; but a singular sense of impending calamity, that should indeed have served me as a warning, drove me onward. I had been without sleep for a night and two days, and I was feverish and irritable. I felt sleep coming upon me, and the Morlocks with it.

"While we hesitated, among the black bushes behind us, and dim against their blackness, I saw three crouching figures. There was scrub and long grass all about us, and I did not feel safe from their insidious approach. The forest, I calculated, was rather less than a mile across. If we could get through it to the bare hillside, there, as it seemed to me, was an altogether safer resting-place; I thought that with my matches and my camphor I could contrive to keep my path illuminated through the woods. Yet it was evident that if I was to flourish matches with my hands I should have to abandon my firewood; so, rather reluctantly, I put it down. And then it came into my head that I would amaze our friends behind by lighting it. I was to discover the atrocious folly of this proceeding, but it came to my mind as an ingenious move for covering our retreat.

"I don't know if you have ever thought what a rare thing flame must be in the absence of man and in a temperate climate. The sun's heat is rarely strong enough to burn, even when it is focused by dewdrops, as is sometimes the case in more tropical districts. Lightning may blast and blacken, but it rarely gives rise to widespread fire. Decaying vegetation may occasionally smoulder with the heat of its fermentation, but this rarely results in flame. In this decadence, too, the art of fire-making had been forgotten on the earth. The red tongues that went licking up my heap of wood were an altogether new and strange thing to Weena.

"She wanted to run to it and play with it. I believe she would have cast herself into it had I not restrained her. But I caught her up, and, in spite of her struggles, plunged boldly before me into the wood. For a little way the glare of my fire lit the path. Looking back presently, I could see, through the crowded stems, that from my heap of sticks the blaze had spread to some bushes adjacent, and a curved line of fire was creeping up the grass of the hill. I laughed at that, and turned again to the dark trees before me. It was very black, and Weena clung to me convulsively, but there was still, as my eyes grew accustomed to the darkness, sufficient light for me to avoid the stems. Overhead it was simply black, except where a gap of remote blue sky shone down upon us here and there. I struck none of my matches because I had no hand free. Upon my left arm I carried my little one, in my right hand I had my iron bar.

"For some way I heard nothing but the crackling twigs under my feet, the faint rustle of the breeze above, and my own breathing and the throb of the blood-vessels in my ears. Then I seemed to know of a pattering about me. I pushed on grimly. The pattering grew more distinct, and then I caught the same queer sounds and voices I had heard in the Underworld. There were evidently several of the Morlocks, and they were closing in upon me. Indeed, in another minute I felt a tug at my coat, then something at my arm. And Weena shivered violently, and became very still.

"It was time for a match. But to get one I must put her down. I did so, and, as I fumbled with my pocket, a struggle began in the darkness about my knees, perfectly silent on her part and with the same peculiar cooing sounds from the Morlocks. Soft little hands, too, were creeping over my coat and back, touching even my neck. Then the match scratched and fizzed. I held it flaring, and saw the

white backs of the Morlocks in flight amid the trees. I
hastily took a lump of camphor from my pocket, and prepared to light it as soon as the match should wane. Then
I looked at Weena. She was lying clutching my feet and
quite motionless, with her face to the ground. With a
sudden fright I stooped to her. She seemed scarcely to
breathe. I lit the block of camphor and flung it to the
ground, and as it split and flared up and drove back the
Morlocks and the shadows, I knelt down and lifted her.
The wood behind seemed full of the stir and murmur of a
great company!

"She seemed to have fainted. I put her carefully upon
my shoulder and rose to push on, and then there came a
horrible realisation. In manœuvring with my matches and
Weena, I had turned myself about several times, and now
I had not the faintest idea in what direction lay my path.
For all I knew, I might be facing back towards the Palace
of Green Porcelain. I found myself in a cold sweat. I had
to think rapidly what to do. I determined to build a fire
and encamp where we were. I put Weena, still motionless,
down upon a turfy bole, and very hastily, as my first lump
of camphor waned, I began collecting sticks and leaves.
Here and there out of the darkness round me the Morlocks'
eyes shone like carbuncles.

"The camphor flickered and went out. I lit a match,
and as I did so, two white forms that had been approaching
Weena dashed hastily away. One was so blinded by the
light that he came straight for me, and I felt his bones grind
under the blow of my fist. He gave a whoop of dismay,
staggered a little way, and fell down. I lit another piece
of camphor, and went on gathering my bonfire. Presently
I noticed how dry was some of the foliage above me, for
since my arrival on the Time Machine, a matter of a week,
no rain had fallen. So, instead of casting about among the
trees for fallen twigs, I began leaping up and dragging down
branches. Very soon I had a choking smoky fire of green
wood and dry sticks, and could economise my camphor.
Then I turned to where Weena lay beside my iron mace.
I tried what I could to revive her, but she lay like one dead.
I could not even satisfy myself whether or not she breathed.

"Now, the smoke of the fire beat over towards me, and it must have made me heavy of a sudden. Moreover, the vapour of camphor was in the air. My fire would not need replenishing for an hour or so. I felt very weary after my exertion, and sat down. The wood, too, was full of slumbrous murmur that I did not understand. I seemed just to nod and open my eyes. But all was dark, and the Morlocks had their hands upon me. Flinging off their clinging fingers I hastily felt in my pocket for the match-box, and—it had gone! Then they gripped and closed with me again. In a moment I knew what had happened. I had slept, and my fire had gone out, and the bitterness of death came over my soul. The forest seemed full of the smell of burning wood. I was caught by the neck, by the hair, by the arms, and pulled down. It was indescribably horrible in the darkness to feel all these soft creatures heaped upon me. I felt as if I was in a monstrous spider's web. I was overpowered, and went down. I felt little teeth nipping at my neck. I rolled over, and as I did so my hand came against my iron lever. It gave me strength. I struggled up, shaking the human rats from me, and, holding the bar short, I thrust where I judged their faces might be. I could feel the succulent giving of flesh and bone under my blows, and for a moment I was free.

"The strange exultation that so often seems to accompany hard fighting came upon me. I knew that both I and Weena were lost, but I determined to make the Morlocks pay for their meat. I stood with my back to a tree, swinging the iron bar before me. The whole wood was full of the stir and cries of them. A minute passed. Their voices seemed to rise to a higher pitch of excitement, and their movements grew faster. Yet none came within reach. I stood glaring at the blackness. Then suddenly came hope. What if the Morlocks were afraid? And close on the heels of that came a strange thing. The darkness seemed to grow luminous. Very dimly I began to see the Morlocks about me—three battered at my feet—and then I recognised, with incredulous surprise, that the others were running, in an incessant stream, as it seemed, from behind me, and away through the wood in front. And their backs seemed no

longer white, but reddish. As I stood agape, I saw a little
red spark go drifting across a gap of starlight between the
branches, and vanish. And at that I understood the smell
of burning wood, the slumbrous murmur that was growing
now into a gusty roar, the red glow, and the Morlocks' flight.

"Stepping out from behind my tree and looking back, I
saw, through the black pillars of the nearer trees, the flames
of the burning forest. It was my first fire coming after me.
With that I looked for Weena, but she was gone. The hissing and crackling behind me, the explosive thud as each
fresh tree burst into flame, left little time for reflection.
My iron bar still gripped, I followed in the Morlocks' path.
It was a close race. Once the flames crept forward so swiftly
on my right as I ran that I was outflanked and had to strike
off to the left. But at last I emerged upon a small open
space, and as I did so, a Morlock came blundering towards
me, and past me, and went on straight into the fire!

"And now I was to see the most weird and horrible thing,
I think, of all that I beheld in that future age. This whole
space was as bright as day with the reflection of the fire.
In the centre was a hillock or tumulus, surmounted by a
scorched hawthorn. Beyond this was another arm of the
burning forest, with yellow tongues already writhing from
it, completely encircling the space with a fence of fire.
Upon the hillside were some thirty or forty Morlocks,
dazzled by the light and heat, and blundering hither and
thither against each other in their bewilderment. At first
I did not realise their blindness, and struck furiously at them
with my bar, in a frenzy of fear, as they approached me,
killing one and crippling several more. But when I had
watched the gestures of one of them groping under the
hawthorn against the red sky, and heard their moans, I was
assured of their absolute helplessness and misery in the
glare, and I struck no more of them.

"Yet every now and then one would come straight towards
me, setting loose a quivering horror that made me quick to
elude him. At one time the flames died down somewhat,
and I feared the foul creatures would presently be able to
see me. I was even thinking of beginning the fight by
killing some of them before this should happen; but the

fire burst out again brightly, and I stayed my hand. I walked about the hill among them and avoided them, looking for some trace of Weena. But Weena was gone.

"At last I sat down on the summit of the hillock, and watched this strange incredible company of blind things groping to and fro, and making uncanny noises to each other, as the glare of the fire beat on them. The coiling up-rush of smoke streamed across the sky, and through the rare tatters of that red canopy, remote as though they belonged to another universe, shone the little stars. Two or three Morlocks came blundering into me, and I drove them off with blows of my fists, trembling as I did so.

"For the most part of that night I was persuaded it was a nightmare. I bit myself and screamed in a passionate desire to awake. I beat the ground with my hands, and got up and sat down again, and wandered here and there, and again sat down. Then I would fall to rubbing my eyes and calling upon God to let me awake. Thrice I saw Mor-locks put their heads down in a kind of agony and rush into the flames. But, at last, above the subsiding red of the fire, above the streaming masses of black smoke and the whiten-ing and blackening tree stumps, and the diminishing numbers of these dim creatures, came the white light of the day.

"I searched again for traces of Weena, but there were none. It was plain that they had left her poor little body in the forest. I cannot describe how it relieved me to think that it had escaped the awful fate to which it seemed destined. As I thought of that, I was almost moved to begin a massacre of the helpless abominations about me, but I contained myself. The hillock, as I have said, was a kind of island in the forest. From its summit I could now make out through a haze of smoke the Palace of Green Porcelain, and from that I could get my bearings for the White Sphinx. And so, leaving the remnant of these damned souls still going hither and thither and moaning, as the day grew clearer, I tied some grass about my feet and limped on across smoking ashes and among black stems, that still pulsated internally with fire, towards the hiding-place of the Time Machine. I walked slowly, for I was almost

exhausted, as well as lame, and I felt the intensest wretched-
nes for the horrible death of little Weena. It seemed an
overwhelming calamity. Now, in this old familiar room,
it is more like the sorrow of a dream than an actual loss.
But that morning it left me absolutely lonely again—terribly
alone. I began to think of this house of mine, of this
fireside, of some of you, and with such thoughts came a
longing that was pain.

"But, as I walked over the smoking ashes under the bright
morning sky, I made a discovery. In my trouser pocket
were still some loose matches. The box must have leaked
before it was lost.

§ 10

"About eight or nine in the morning I came to the same
seat of yellow metal from which I had viewed the world
upon the evening of my arrival. I thought of my hasty
conclusions upon that evening and could not refrain from
laughing bitterly at my confidence. Here was the same
beautiful scene, the same abundant foliage, the same splendid palaces and magnificent ruins, the same silver river
running between its fertile banks. The gay robes of the
beautiful people moved hither and thither among the trees.
Some were bathing in exactly the place where I had saved
Weena, and that suddenly gave me a keen stab of pain.
And like blots upon the landscape rose the cupolas above
the ways to the under-world. I understood now what all
the beauty of the over-world people covered. Very pleasant
was their day, as pleasant as the day of the cattle in the
field. Like the cattle, they knew of no enemies and provided
against no needs. And their end was the same.

"I grieved to think how brief the dream of the human
intellect had been. It had committed suicide. It had set
itself steadfastly towards comfort and ease, a balanced society
with security and permanency as its watchword, it had
attained its hopes—to come to this at last. Once, life and
property must have reached almost absolute safety. The
rich had been assured of his wealth and comfort, the toiler

assured of his life and work. No doubt in that perfect world there had been no unemployed problem, no social question left unsolved. And a great quiet had followed.

"It is a law of nature we overlook, that intellectual versatility is the compensation for change, danger, and trouble. An animal perfectly in harmony with its environment is a perfect mechanism. Nature never appeals to intelligence until habit and instinct are useless. There is no intelligence where there is no change and no need of change. Only those animals partake of intelligence that have to meet a huge variety of needs and dangers.

"So, as I see it, the Upper-world man had drifted towards his feeble prettiness, and the Under-world to mere mechanical industry. But that perfect state had lacked one thing even for mechanical perfection—absolute permanency. Apparently as time went on, the feeding of the Under-world, however it was effected, had become disjointed. Mother Necessity, who had been staved off for a few thousand years, came back again, and she began below. The Under-world being in contact with machinery, which, however perfect, still needs some little thought outside habit, had probably retained perforce rather more initiative, if less of every other human character, than the upper. And when other meat failed them, they turned to what old habit had hitherto forbidden. So I say I saw it in my last view of the world of Eight Hundred and Two Thousand Seven Hundred and One. It may be as wrong an explanation as mortal wit could invent. It is how the thing shaped itself to me, and as that I give it to you.

"After the fatigues, excitements, and terrors of the past days, and in spite of my grief, this seat and the tranquil view of the warm sunlight were very pleasant. I was very tired and sleepy, and soon my theorising passed into dozing. Catching myself at that, I took my own hint, and spreading myself out upon the turf I had a long and refreshing sleep.

"I awoke a little before sunsetting. I now felt safe against being caught napping by the Morlocks, and, stretching myself, I came on down the hill towards the White Sphinx.

I had my crowbar in one hand, and the other hand played with the matches in my pocket.

"And now came a most unexpected thing. As I approached the pedestal of the sphinx I found the bronze valves were open. They had slid down into grooves.

"At that I stopped short before them, hesitating to enter.

"Within was a small apartment, and on a raised place in the corner of this was the Time Machine. I had the small levers in my pocket. So here, after all my elaborate preparations for the siege of the White Sphinx, was a meek surrender. I threw my iron bar away, almost sorry not to use it.

"A sudden thought came into my head as I stooped towards the portal. For once, at least, I grasped the mental operations of the Morlocks. Suppressing a strong inclination to laugh, I stepped through the bronze frame and up to the Time Machine. I was surprised to find it had been carefully oiled and cleaned. I have suspected since that the Morlocks had even partially taken it to pieces while trying in their dim way to grasp its purpose.

"Now as I stood and examined it, finding a pleasure in the mere touch of the contrivance, the thing I had expected happened. The bronze panels suddenly slid up and struck the frame with a clang. I was in the dark—trapped. So the Morlocks thought. At that I chuckled gleefully.

"I could already hear their murmuring laughter as they came towards me. Very calmly I tried to strike the match. I had only to fix on the levers and depart then like a ghost. But I had overlooked one little thing. The matches were of that abominable kind that light only on the box.

"You may imagine how all my calm vanished. The little brutes were close upon me. One touched me. I made a sweeping blow in the dark at them with the levers, and began to scramble into the saddle of the machine. Then came one hand upon me and then another. Then I had simply to fight against their persistent fingers for my levers, and at the same time feel for the studs over which these fitted. Once, indeed, they almost got away from me. As it slipped from my hand, I had to butt in the dark with my head—I could hear the Morlock's skull ring—to recover it. It was a nearer thing than the fight in the forest, I think, this last scramble.

"But at last the lever was fixed and pulled over. The clinging hands slipped from me. The darkness presently fell from my eyes. I found myself in the same grey light and tumult I have already described.

§ 11

"I have already told you of the sickness and confusion that comes with time travelling. And this time I was not seated properly in the saddle, but sideways and in an unstable fashion. For an indefinite time I clung to the machine as it swayed and vibrated, quite unheeding how I went, and when I brought myself to look at the dials again I was amazed to find where I had arrived. One dial records days, another thousands of days, another millions of days, and another thousands of millions. Now, instead of reversing the levers, I had pulled them over so as to go forward with them, and when I came to look at these indicators I found that the thousands hand was sweeping round as fast as the seconds hand of a watch—into futurity.

"As I drove on, a peculiar change crept over the appearance of things. The palpitating greyness drew darker; then —though I was still travelling with prodigious velocity—the blinking succession of day and night, which was usually indicative of a slower pace, returned, and grew more and more marked. This puzzled me very much at first. The alternations of night and day grew slower and slower, and so did the passage of the sun across the sky, until they seemed to stretch through the centuries. At last a steady twilight brooded over the earth, a twilight only broken now and then when a comet glared across the darkling sky. The band of light that had indicated the sun had long since disappeared; for the sun had ceased to set—it simply rose and fell in the west, and grew ever broader and more red. All trace of the moon had vanished. The circling of the stars, growing slower and slower, had given place to creeping points of light. At last, some time before I stopped, the sun, red and very large, halted motionless upon the horizon, a vast dome glowing with a dull heat, and now and then suffering a momentary extinction. At one time it had for a little while glowed more brilliantly again, but it speedily reverted to its sullen red heat. I perceived by this slowing down of its rising and setting that the work of the tidal drag was done. The earth had come to rest with one face to the sun, even as in our own time the moon faces the earth. Very cautiously, for I remembered my former head-long fall, I began to reverse my motion. Slower and slower went the circling hands until the thousands one seemed motionless and the daily one was no longer a mere mist upon its scale. Still slower, until the dim outlines of a desolate beach grew visible.

"I stopped very gently and sat upon the Time Machine, looking round. The sky was no longer blue. North-eastward it was inky black, and out of the blackness shone brightly and steadily the pale white stars. Overhead it was a deep Indian red and starless, and south-eastward it grew brighter to a glowing scarlet where, cut by the horizon, lay the huge hull of the sun, red and motionless. The rocks about me were of a harsh reddish colour, and all the trace of life that I could see at first was the intensely green vegetation that covered every projecting point on their south-eastern face. It was the same rich green that one sees on forest moss or on the lichen in caves: plants which like these grow in a perpetual twilight.

"The machine was standing on a sloping beach. The sea stretched away to the south-west, to rise into a sharp bright horizon against the wan sky. There were no breakers and no waves, for not a breath of wind was stirring. Only a slight oily swell rose and fell like a gentle breathing, and showed that the eternal sea was still moving and living. And along the margin where the water sometimes broke was a thick incrustation of salt—pink under the lurid sky. There was a sense of oppression in my head, and I noticed that I was breathing very fast. The sensation reminded me of my only experience of mountaineering, and from that I judged the air to be more rarefied than it is now.

"Far away up the desolate slope I heard a harsh scream,

and saw a thing like a huge white butterfly go slanting and fluttering up into the sky and, circling, disappear over some low hillocks beyond. The sound of its voice was so dismal that I shivered and seated myself more firmly upon the machine. Looking round me again, I saw that, quite near, what I had taken to be a reddish mass of rock was moving slowly towards me. Then I saw the thing was really a monstrous crab-like creature. Can you imagine a crab as large as yonder table, with its many legs moving slowly and uncertainly, its big claws swaying, its long antennæ, like carters' whips, waving and feeling, and its stalked eyes gleaming at you on either side of its metallic front? Its back was corrugated and ornamented with ungainly bosses, and a greenish incrustation blotched it here and there. I could see the many palps of its complicated mouth flickering and feeling as it moved.

"As I stared at this sinister apparition crawling towards me, I felt a tickling on my cheek as though a fly had lighted there. I tried to brush it away with my hand, but in a moment it returned, and almost immediately came another by my ear. I struck at this, and caught something threadlike. It was drawn swiftly out of my hand. With a frightful qualm, I turned, and saw that I had grasped the antenna of another monster crab that stood just behind me. Its evil eyes were wriggling on their stalks, its mouth was all alive with appetite, and its vast ungainly claws, smeared with an algal slime, were descending upon me. In a moment my hand was on the lever, and I had placed a month between myself and these monsters. But I was still on the same beach, and I saw them distinctly now as soon as I stopped. Dozens of them seemed to be crawling here and there, in the sombre light, among the foliated sheets of intense green.

"I cannot convey the sense of abominable desolation that hung over the world. The red eastern sky, the northward blackness, the salt Dead Sea, the stony beach crawling with these foul, slow-stirring monsters, the uniform poisonous-looking green of the lichenous plants, the thin air that hurts one's lungs; all contributed to an appalling effect. I moved on a hundred years, and there was the same red sun—a little larger, a little duller—the same dying sea, the same chill air, and the same crowd of earthy crustacea creeping in and out among the green weed and the red rocks. And in the westward sky I saw a curved pale line like a vast new moon.

"So I travelled, stopping ever and again, in great strides of a thousand years or more, drawn on by the mystery of the earth's fate, watching with a strange fascination the sun grow larger and duller in the westward sky, and the life of the old earth ebb away. At last, more than thirty million years hence, the huge red-hot dome of the sun had come to obscure nearly a tenth part of the darkling heavens. Then I stopped once more, for the crawling multitude of crabs had disappeared, and the red beach, save for its livid green liverworts and lichens, seemed lifeless. And now it was flecked with white. A bitter cold assailed me. Rare white flakes ever and again came eddying down. To the north-eastward, the glare of snow lay under the starlight of the sable sky, and I could see an undulating crest of hillocks pinkish white. There were fringes of ice along the sea margin, with drifting masses further out; but the main expanse of that salt ocean, all bloody under the eternal sunset, was still unfrozen.

I looked about me to see if any traces of animal life remained. A certain indefinable apprehension still kept me in the saddle of the machine. But I saw nothing moving, in earth or sky or sea. The green slime on the rocks alone testified that life was not extinct. A shallow sand-bank had appeared in the sea and the water had receded from the beach. I fancied I saw some black object flopping about upon this bank, but it became motionless as I looked at it, and I judged that my eye had been deceived, and that the black object was merely a rock. The stars in the sky were intensely bright and seemed to me to twinkle very little.

"Suddenly I noticed that the circular westward outline of the sun had changed; that a concavity, a bay, had appeared in the curve. I saw this grow larger. For a minute perhaps I stared aghast at this blackness that was creeping over the day, and then I realised that an eclipse was beginning. Either the moon or the planet Mercury was passing across the sun's disk. Naturally, at first I took it to be the moon, but there is much to incline me to believe that what I really saw was the transit of an inner planet passing very near to the earth.

"The darkness grew apace; a cold wind began to blow in freshening gusts from the east, and the showering white flakes in the air increased in number. From the edge of the sea came a ripple and whisper. Beyond these lifeless sounds the world was silent. Silent? It would be hard to convey the stillness of it. All the sounds of man, the bleating of sheep, the cries of birds, the hum of insects, the stir that makes the background of our lives—all that was over. As the darkness thickened, the eddying flakes grew more abundant, dancing before my eyes; and the cold of the air more intense. At last, one by one, swiftly, one after the other, the white peaks of the distant hills vanished into blackness. The breeze rose to a moaning wind. I saw the black central shadow of the eclipse sweeping towards me. In another moment the pale stars alone were visible. All else was rayless obscurity. The sky was absolutely black.

"A horror of this great darkness came on me. The cold, that smote to my marrow, and the pain I felt in breathing, overcame me. I shivered, and a deadly nausea seized me. Then like a red-hot bow in the sky appeared the edge of the sun. I got off the machine to recover myself. I felt giddy and incapable of facing the return journey. As I stood sick and confused I saw again the moving thing upon the shoal —there was no mistake now that it was a moving thing— against the red water of the sea. It was a round thing, the size of a football perhaps, or, it may be, bigger, and tentacles trailed down from it; it seemed black against the weltering blood-red water, and it was hopping fitfully about. Then I felt I was fainting. But a terrible dread of lying helpless in that remote and awful twilight sustained me while I clambered upon the saddle.

§ 12

"So I came back. For a long time I must have been insensible upon the machine. The blinking succession of the days and nights was resumed, the sun got golden again,

the sky blue. I breathed with greater freedom. The fluctuating contours of the land ebbed and flowed. The hands
spun backward upon the dials. At last I saw again the
dim shadows of houses, the evidences of decadent humanity.
These, too, changed and passed, and others came. Presently,
when the million dial was at zero, I slackened speed. I
began to recognise our own petty and familiar architecture,
the thousands hand ran back to the starting-point, the night
and day flapped slower and slower. Then the old walls of
the laboratory came round me. Very gently, now, I slowed
the mechanism down.

"I saw one little thing that seemed odd to me. I think
I have told you that when I set out, before my velocity
became very high, Mrs. Watchett had walked across the
room, travelling, as it seemed to me, like a rocket. As I
returned, I passed again across that minute when she
traversed the laboratory. But now her every motion appeared
to be the exact inversion of her previous ones. The door
at the lower end opened, and she glided quietly up the
laboratory, back foremost, and disappeared behind the door
by which she had previously entered. Just before that I
seemed to see Hillyer for a moment; but he passed like a
flash.

"Then I stopped the machine, and saw about me again
the old familiar laboratory, my tools, my appliances, just as
I had left them. I got off the thing very shakily, and sat
down upon my bench. For several minutes I trembled
violently. Then I became calmer. Around me was my old
workshop again, exactly as it had been. I might have slept
there, and the whole thing have been a dream.

"And yet, not exactly! The thing had started from the
south-east corner of the laboratory. It had come to rest
again in the north-west, against the wall where you saw it.
That gives you the exact distance from my little lawn to
the pedestal of the White Sphinx, into which the Morlocks
had carried my machine.

"For a time my brain went stagnant. Presently I got up
and came through the passage here, limping, because my
heel was still painful, and feeling sorely begrimed. I saw
the *Pall Mall Gazette* on the table by the door. I found

the date was indeed to-day, and looking at the timepiece, saw the hour was amost eight o'clock. I heard your voices and the clatter of plates. I hesitated—I felt so sick and weak. Then I sniffed good wholesome meat, and opened the door on you. You know the rest. I washed, and dined, and now I am telling you the story."

"I know," he said, after a pause, "that all this will be absolutely incredible to you. To me the one incredible thing is that I am here to-night in this old familiar room looking into your friendly faces and telling you these strange adventures."

He looked at the Medical Man. "No. I cannot expect you to believe it. Take it as a lie—or a prophecy. Say I dreamed it in the workshop. Consider I have been speculating upon the destinies of our race until I have hatched this fiction. Treat my assertion of its truth as a mere stroke of art to enhance its interest. And taking it as a story, what do you think of it?"

He took up his pipe, and began, in his old accustomed manner, to tap with it nervously upon the bars of the grate. There was a momentary stillness. Then chairs began to creak and shoes to scrape upon the carpet. I took my eyes off the Time Traveller's face, and looked round at his audience. They were in the dark, and little spots of colour swam before them. The Medical Man seemed absorbed in the contemplation of our host. The Editor was looking hard at the end of his cigar—the sixth. The Journalist fumbled for his watch. The others, as far as I remember, were motionless.

The Editor stood up with a sigh. "What a pity it is you're not a writer of stories!" he said, putting his hand on the Time Traveller's shoulder.

"You don't believe it?"

"Well——"

"I thought not."

The Time Traveller turned to us. "Where are the matches?" he said. He lit one and spoke over his pipe, puffing. "To tell you the truth . . . I hardly believe it myself. . . . And yet . . . ."

His eye fell with a mute inquiry upon the withered white flowers upon the little table. Then he turned over the hand holding his pipe, and I saw he was looking at some half-healed scars on his knuckles.

The Medical Man rose, came to the lamp, and examined the flowers. "The gynæceum's odd," he said. The Psychologist leant forward to see, holding out his hand for a specimen.

"I'm hanged if it isn't a quarter to one," said the Journalist. "How shall we get home?"

"Plenty of cabs at the station," said the Psychologist.

"It's a curious thing," said the Medical Man; "but I certainly don't know the natural order of these flowers. May I have them?"

The Time Traveller hesitated. Then suddenly: "Certainly not."

"Where did you really get them?" said the Medical Man.

The Time Traveller put his hand to his head. He spoke like one who was trying to keep hold of on idea that eluded him. "They were put into my pocket by Weena, when I travelled into Time." He stared round the room. "I'm damned if it isn't all going. This room and you and the atmosphere of every day is too much for my memory. Did I ever make a Time Machine, or a model of a Time Machine? Or is it all only a dream? They say life is a dream, a precious poor dream at times—but I can't stand another that won't fit. It's madness. And where did the dream come from? . . . I must look at that machine. If there is one!"

He caught up the lamp swiftly, and carried it, flaring red, through the door into the corridor. We followed him. There in the flickering light of the lamp was the machine sure enough, squat, ugly, and askew; a thing of brass, ebony, ivory, and translucent glimmering quartz. Solid to the touch—for I put out my hand and felt the rail of it—and with brown spots and smears upon the ivory, and bits of grass and moss upon the lower parts, and one rail bent awry.

The Time Traveller put the lamp down on the bench, and ran his hand along the damaged rail. "It's all right now," he said. "The story I told you was true. I'm sorry to have brought you out here in the cold." He took up the lamp, and, in an absolute silence, we returned to the smoking-room."

He came into the hall with us and helped the Editor on with his coat. The Medical Man looked into his face and, with a certain hesitation, told him he was suffering from overwork, at which he laughed hugely. I remember him standing in the open doorway, bawling good-night.

I shared a cab with the Editor. He thought the tale a "gaudy lie." For my own part I was unable to come to a conclusion. The story was so fantastic and incredible, the telling so credible and sober. I lay awake most of the night thinking about it. I determined to go next day and see the Time Traveller again. I was told he was in the laboratory, and, being on easy terms in the house, I went up to him. The laboratory, however, was empty. I stared for a minute at the Time Machine and put out my hand and touched the lever. At that the squat substantial-looking mass swayed like a bough shaken by the wind. Its instability startled me extremely, and I had a queer reminiscence of the childish days when I used to be forbidden to meddle. I came back through the corridor. The Time Traveller met me in the smoking-room. He was coming from the house. He had a small camera under one arm and a knapsack under the other. He laughed when he saw me, and gave me an elbow to shake. "I'm frightfully busy," said he, "with that thing in there."

"But is it not some hoax?" I said. "Do you really travel through time?"

"Really and truly I do." And he looked frankly into my eyes. He hesitated. His eye wandered about the room. "I only want half an hour," he said. "I know why you came, and it's awfully good of you. There's some magazines here. If you'll stop to lunch I'll prove you this time travelling up to the hilt, specimen and all. If you'll forgive my leaving you now?"

I consented, hardly comprehending then the full import of his words, and he nodded and went on down the corridor. I heard the door of the laboratory slam, seated myself in a chair, and took up a daily paper. What was he going to do before lunch-time? Then suddenly I was reminded by an advertisement that I had promised to meet Richardson, the publisher, at two. I looked at my watch, and saw that I could barely save that engagement. I got up and went down the passage to tell the Time Traveller.

As I took hold of the handle of the door I heard an exclamation, oddly truncated at the end, and a click and a thud. A gust of air whirled round me as I opened the door, and from within came the sound of broken glass falling on the floor. The Time Traveller was not there. I seemed to see a ghostly, indistinct figure sitting in a whirling mass of black and brass for a moment—a figure so transparent that the bench behind, with its sheets of drawings, was absolutely distinct; but this phantasm vanished as I rubbed my eyes. The Time Machine had gone. Save for a subsiding stir of dust, the further end of the laboratory was empty. A pane of the skylight had, apparently, just been blown in.

I felt an unreasonable amazement. I knew that something strange had happened, and for the moment could not distinguish what the strange thing might be. As I stood staring, the door into the garden opened, and the man-servant appeared.

We looked at each other. Then ideas began to come. "Has Mr. —— gone out that way?" said I.

"No, sir. No one has come out this way. I was expecting to find him here."

At that I understood. At the risk of disappointing Richardson I stayed on, waiting for the Time Traveller; waiting for the second, perhaps still stranger story, and the specimens and photographs he would bring with him. But I am beginning now to fear that I must wait a lifetime. The Time Traveller vanished three years ago. And, as everybody knows now, he has never returned.

EPILOGUE

One cannot choose but wonder. Will he ever return? It may be that he swept back into the past, and fell among the blood drinking, hairy savages of the Age of Unpolished Stone; into the abysses of the Cretaceous Sea; or among the grotesque saurians, the huge reptilian brutes of the Jurassic times. He may even now—if I may use the phrase—be wandering on some plesoisaurus-haunted Oolitic coral reef, or beside the lonely saline lakes of the Triassic Age. Or did he go forward, into one of the nearer ages, in which men are still men, but with the riddles of our own time answered and its wearisome problems solved? Into the manhood of the race: for I, for my own part, cannot think that these latter days of weak experiment, fragmentary theory, and mutual discord are indeed man's culminating time! I say, for my own part. He, I know—for the question had been discussed among us long before the Time Machine was made—thought but cheerlessly of the Advancement of Mankind, and saw in the growing pile of civilisation only a foolish heaping that must inevitably fall back upon and destroy its makers in the end. If that is so, it remains for us to live as though it were not so. But to me the future is still black and blank—is a vast ignorance, lit at a few casual places by the memory of his story. And I have by me, for my comfort, two strange white flowers—shrivelled now, and brown and flat and brittle —to witness that even when mind and strength had gone, gratitude and a mutual tenderness still lived on in the heart of man.

STORY THE SECOND

The Empire of the Ants

§ 1

WHEN Captain Gerilleau received instructions to take his new gunboat, the *Benjamin Constant,* to Badama on the Batemo arm of the Guaramadema and there assist the inhabitants against a plague of ants, he suspected the authorities of mockery. His promotion had been romantic and irregular, the affections of a prominent Brazilian lady and the captain's liquid eyes had played a part in the process, and the *Diario* and *O Futuro* had been lamentably disrespectful in their comments. He felt he was to give further occasion for disrespect.

He was a Creole, his conceptions of etiquette and discipline were pure-blooded Portuguese, and it was only to Holroyd, the Lancashire engineer, who had come over with the boat, and as an exercise in the use of English—his "th" sounds were very uncertain—that he opened his heart.

"It is in effect," he said, "to make me absurd! What can a man do against ants? Dey come, dey go."

"They say," said Holroyd, "that these don't go. That chap you said was a Sambo——"

"Zambo—it is a sort of mixture of blood."

"Sambo. He said the people are going!"

The captain smoked fretfully for a time. "Dese tings 'ave to happen," he said at last. "What is it? Plagues of ants and suchlike as God wills. Dere was a plague in Trinidad —the little ants that carry leaves. Orl der orange-trees, all

92

der mangoes! What does it matter? Sometimes ant armies come into your houses—fighting ants; a different sort. You go and they clean the house. Then you come back again;—the house is clean, like new! No cockroaches, no fleas, no jiggers in the floor."

"That Sambo chap," said Holroyd, "says thesc are a different sort of ant."

The captain shrugged his shoulders, fumed, and gave his attention to a cigarette.

Afterwards he reopened the subject. "My dear 'Olroyd, what am I to do about dese infernal ants?"

The captain reflected. "It is ridiculous," he said. But in the afternoon he put on his full uniform and went ashore, and jars and boxes came back to the ship and subsequently he did. And Holroyd sat on deck in the evening coolness and smoked profoundly and marvelled at Brazil. They were six days up the Amazon, some hundreds of miles from the ocean, and east and west of him there was a horizon like the sea, and to the south nothing but a sand-bank island with some tufts of scrub. The water was always running like a sluice, thick with dirt, animated with crocodiles and hovering birds, and fed by some inexhaustible source of tree trunks; and the waste of it, the headlong waste of it, filled his soul. The town of Alemquer, with its meagre church, its thatched sheds for houses, its discoloured ruins of ampler days, seemed a little thing lost in this wilderness of Nature, a sixpence dropped on Sahara. He was a young man, this was his first sight of the tropics, he came straight from England, where Nature is hedged, ditched, and drained into the perfection of submission, and he had suddenly discovered the insignificance of man. For six days they had been steaming up from the sea by unfrequented channels, and man had been as rare as a rare butterfly. One saw one day a canoe, another day a distant station, the next no men at all. He began to perceive that man is indeed a rare animal, having but a precarious hold upon this land.

He perceived it more clearly as the days passed, and he made his devious way to the Batemo, in the company of this remarkable commander, who ruled over one big gun, and was forbidden to waste his ammunition. Holroyd was learn-

ing Spanish industriously, but he was still in the present
tense and substantive stage of speech, and the only other
person who had any words of English was a negro stoker,
who had them all wrong. The second in command was a
Portuguese, da Cunha, who spoke French, but it was a different sort of French from the French Holroyd had learned
in Southport, and their intercourse was confined to polite-
nesses and simple propositions about the weather. And the
weather, like everything else in this amazing new world, the
weather had no human aspect, and was hot by night and hot
by day, and the air steam, even the wind was hot steam,
smelling of vegetation in decay: and the alligators and the
strange birds, the flies of many sorts and sizes, the beetles,
the ants, the snakes and monkeys seemed to wonder what
man was doing in an atmosphere that had no gladness in
its sunshine and no coolness in its night. To wear clothing
was intolerable, but to cast it aside was to scorch by day, and
expose an ampler area to the mosquitoes by night; to go on
deck by day was to be blinded by glare and to stay below
was to suffocate. And in the daytime came certain flies, extremely clever and noxious about one's wrist and ankle.
Captain Gerilleau, who was Holroyd's sole distraction from
these physical distresses, developed into a formidable bore,
telling the simple story of his heart's affections day by day,
a string of anonymous women, as if he was telling beads.
Sometimes he suggested sport, and they shot at alligators,
and at rare intervals they came to human aggregations in the
waste of trees, and stayed for a day or so, and drank and sat
about; and, one night, danced with Creole girls, who found
Holroyd's poor elements of Spanish, without either past
tense or future, amply sufficient for their purposes. But
these were mere luminous chinks in the long grey passage of
the streaming river, up which the throbbing engines beat.
A certain liberal heathen deity, in the shape of a demi-john,
held seductive court aft, and, it is probable, forward.

But Gerilleau learned things about the ants, more things
and more, at this stopping-place and that, and became interested in his mission.

"Dey are a new sort of ant," he said. "We have got to be
—what do you call it?—entomologie Big. Five centimetres!

Some bigger! It is ridiculous. We are like the monkeys—sent to pick insects. . . . But dey are eating up the country."

He burst out indignantly. "Suppose—suddenly, there are complications with Europe. Here am I—soon we shall be above the Rio Negro—and my gun, useless!"

He nursed his knee and mused.

"Dose people who were dere at de dancing place, dey 'ave come down. Dey 'ave lost all they got. De ants come to deir house one afternoon. Everyone run out. You know when de ants come one must—everyone runs out and they go over the house. If you stayed they'd eat you. See? Well, presently dey go back; dey say, 'The ants 'ave gone.' . . . De ants 'aven't gone. Dey try to go in—de son, 'e goes in. De ants fight."

"Swarm over him?"

"Bite 'im. Presently he comes out again—screaming and running. He runs past them to the river. See? He get into de water and drowns de ants—yes." Gerilleau paused, brought his liquid eyes close to Holroyd's face, tapped Holroyd's knee with his knuckle. "That night he dies, just as if he was stung by a snake."

"Poisoned—by the ants?"

"Who knows?" Gerilleau shrugged his shoulders, "Perhaps they bit him badly. . . . When I joined dis service I joined to fight men. Dese things, dese ants, dey come and go. It is no business for men."

After that he talked frequently of the ants to Holroyd, and whenever they chanced to drift against any speck of humanity in that waste of water and sunshine and distant trees, Holroyd's improving knowledge of the language enabled him to recognise the ascendant word Saüba, more and more completely dominating the whole.

He perceived the ants were becoming interesting, and the nearer he drew to them the more interesting they became. Gerilleau abandoned his old themes almost suddenly, and the Portuguese lieutenant became a conversational figure; he knew something about the leaf-cutting ant, and expanded his knowledge. Gerilleau sometimes rendered what he had to tell to Holroyd. He told of the little workers that swarm and fight, and the big workers that command and rule, and

how these latter always crawled to the neck and how their
bites drew blood. He told how they cut leaves and made
fungus beds, and how their nests in Caracas are sometimes
a hundred yards across. Two days the three men spent disputing whether ants have eyes. The discussion grew dangerously heated on the second afternoon, and Holroyd saved
the situation by going ashore in a boat to catch ants and see.
He captured various specimens and returned and some had
eyes and some hadn't. Also, they argued, do ants bite or
sting?

"Dese ants," said Gerilleau, after collecting information
at a rancho, "have big eyes. They don't run about blind—
not as most ants do. No! Dey get in corners and watch
what you do."

"And they sting?" asked Holroyd.

"Yes. Dey sting. Dere is poison in the sting." He
meditated. "I do not see what men can do against ants.
Dey come and go."

"But these don't go."

"They will," said Gerilleau.

Past Tamandu there is a long low coast of eighty miles
without any population, and then one comes to the con-
fluence of the main river and the Batemo arm like a great
lake, and then the forest came nearer, came at last intimately
near. The character of the channel changes, snags abound,
and the *Benjamin Constant* moored by a cable that night,
under the very shadow of dark trees. For the first time for
many days came a spell of coolness, and Holroyd and Gerilleau sat late, smoking cigars and enjoying this delicious sensation. Gerilleau's mind was full of ants and what they
could do. He decided to sleep at last, and lay down on a
mattress on deck, a man hopelessly perplexed; his last words,
when he already seemed asleep, were to ask, with a flourish
of despair: "What can one do with ants? . . . De whole thing
is absurd."

Holroyd was left to scratch his bitten wrists, and meditate
alone.

He sat on the bulwark and listened to the changes in
Gerilleau's breathing until he was fast asleep, and then the
ripple and lap of the stream took his mind, and brought back

that sense of immensity that had been growing upon him since first he had left Para and come up the river. The monitor showed but one small light, and there was first a little talking forward and then stillness. His eyes went from the dim black outlines of the middle works of the gunboat towards the bank, to the black overwhelming mysteries of forest, lit now and then by a fire-fly, and never still from the murmur of alien and mysterious activities. . . .

It was the inhuman immensity of this land that astonished and oppressed him. He knew the skies were empty of men, the stars were specks in an incredible vastness of space; he knew the ocean was enormous and untamable, but in England he had come to think of the land as man's. In England it is indeed man's, the wild things live by sufferance, grow on lease, everywhere the roads, the fences, and absolute security runs. In an atlas, too, the land is man's, and all coloured to show his claim to it—in vivid contrast to the universal independent blueness of the sea. He had taken it for granted that a day would come when everywhere about the earth, plough and culture, light tramways, and good roads, an ordered security, would prevail. But now, he doubted.

This forest was interminable, it had an air of being invincible, and Man seemed at best an infrequent precarious intruder. One travelled for miles amidst the still, silent struggle of giant trees, of strangulating creepers, of assertive flowers, everywhere the alligator, the turtle, and endless varieties of birds and insects seemed at home, dwelt irreplaceably—but man, man at most held a footing upon resentful clearings, fought weeds, fought beasts and insects for the barest foothold, fell a prey to snake and beast, insect and fever, and was presently carried away. In many places down the river he had been manifestly driven back, this deserted creek or that preserved the name of a *casa,* and here and there ruinous white walls and a shattered tower enforced the lesson. The puma, the jaguar, were more the masters here. . . .

Who were the real masters?

In a few miles of this forest there must be more ants than there are men in the whole world! This seemed to Holroyd

D

a perfectly new idea. In a few thousand years men had emerged from barbarism to a stage of civilisation that made them feel lords of the future and masters of the earth! But what was to prevent the ants evolving also? Such ants as one knew lived in little communities of a few thousand individuals, made no concerted efforts against the greater world. But they had a language, they had an intelligence! Why should things stop at that any more than men had stopped at the barbaric stage? Suppose presently the ants began to store knowledge, just as men had done by means of books and records, use weapons, form great empires, sustain a planned and organised war?

Things came back to him that Gerilleau had gathered about these ants they were approaching. They used a poison like the poison of snakes. They obeyed greater leaders even as the leaf-cutting ants do. They were carnivorous, and where they came they stayed. . . .

The forest was very still. The water lapped incessantly against the side. About the lantern overhead there eddied a noiseless whirl of phantom moths.

Gerilleau stirred in the darkness and sighed. "What can one *do*?" he murmured, and turned over and was still again.

Holroyd was roused from meditations that were becoming sinister by the hum of a mosquito.

§ 2

The next morning Holroyd learned they were within forty kilometres of Badama, and his interest in the banks intensified. He came up whenever an opportunity offered to examine his surroundings. He could see no signs of human occupation whatever, save for a weedy ruin of a house and the green-stained façade of the long-deserted monastery at Mojû, with a forest tree growing out of a vacant window space, and great creepers netted across its vacant portals. Several flights of strange yellow butterflies with semi-transparent wings crossed the river that morning, and many alighted on the monitor and were killed by the men. It was towards afternoon that they came upon the derelict cuberta.

She did not at first appear to be derelict; both her sails were set and hanging slack in the afternoon calm, and there was the figure of a man sitting on the fore planking beside the shipped sweeps. Another man appeared to be sleeping face downwards on the sort of longitudinal bridge these big canoes have in the waist. But it was presently apparent, from the sway of her rudder and the way she drifted into the course of the gunboat, that something was out of order with her. Gerilleau surveyed her through a field-glass, and became interested in the queer darkness of the face of the sitting man, a red-faced man he seemed, without a nose— crouching he was rather than sitting, and the longer the captain looked the less he liked to look at him, and the less able he was to take his glasses away.

But he did so at last, and went a little way to call up Holroyd. Then he went back to hail the cuberta. He hailed her again, and so she drove past him. *Santa Rosa* stood out clearly as her name.

As she came by and into the wake of the monitor, she pitched a little, and suddenly the figure of the crouching man collapsed as though all its joints had given way. His hat fell off, his head was not nice to look at, and his body flopped lax and rolled out of sight behind the bulwarks.

"Caramba!" cried Gerilleau, and resorted to Holroyd forthwith.

Holroyd was half-way up the companion. "Did you see dat?" said the captain.

"Dead!" said Holroyd. "Yes. You'd better send a boat aboard. There's something wrong."

"Did you—by any chance—see his face?"

"What was it like?"

"It was—ugh!—I have no words." And the captain suddenly turned his back on Holroyd and became an active and strident commander.

The gunboat came about, steamed parallel to the erratic course of the canoe, and dropped the boat with Lieutenant da Cunha and three sailors to board her. Then the curiosity of the captain made him draw up almost alongside as the lieutenant got aboard, so that the whole of the *Santa Rosa,* deck and hold, was visible to Holroyd.

He saw now clearly that the sole crew of the vessel was these two dead men, and though he could not see their faces, he saw by their outstretched hands, which were all of ragged flesh, that they had been subjected to some strange exceptional process of decay. For a moment his attention, concentrated on these two enigmatical bundles of dirty clothes and laxly flung limbs, and then his eyes went forward to discover the open hold piled high with trunks and cases, and aft, to where the little cabin gaped inexplicably empty. Then he became aware that the planks of the middle decking were dotted with moving black specks.

His attention was riveted by these specks. They were all walking in directions radiating from the fallen man in a manner—the image came unsought to his mind—like the crowd dispersing from a bull-fight.

He became aware of Gerilleau beside him. "Capo," he said, "have you your glasses? Can you focus as closely as those planks there?"

Gerilleau made an effort, grunted, and handed him the glasses.

There followed a moment of scrutiny. "It's ants," said the Englishman, and handed the focused field-glasses back to Gerilleau.

His impression of them was of a crowd of large black ants, very like ordinary ants except for their size, and for the fact that some of the larger of them bore a sort of clothing of grey. But at the time his inspection was too brief for particulars. The head of Lieutenant da Cunha appeared over the side of the cuberta, and a brief colloquy ensued.

"You must go aboard," said Gerilleau.

The lieutenant objected that the boat was full of ants.

"You have your boots," said Gerilleau.

The lieutenant changed the subject. "How did these men die?" he asked.

Captain Gerilleau embarked upon speculations that Holroyd could not follow, and the two men disputed with a certain increasing vehemence. Holroyd took up the field-glass and resumed his scrutiny, first of the ants and then of the dead man amidships.

He has described these ants to me very particularly.

He says they were as large as any ants he has ever seen, black and moving with a steady deliberation very different from the mechanical fussiness of the common ant. About one in twenty was much larger than its fellows, and with an exceptionally large head. These reminded him at once of the master workers who are said to rule over the leaf-cutter ants; like them they seemed to be directing and co-ordinating the general movements. They tilted their bodies back in a manner altogether singular as if they made some use of the fore feet. And he had a curious fancy that he was too far off to verify, that most of these ants of both kinds were wearing accoutrements, had things strapped about their bodies by bright white bands like white metal threads. . . .

He put down the glasses abruptly, realising that the question of discipline between the captain and his subordinate had become acute.

"It is your duty," said the captain, "to go aboard. It is my instructions."

The lieutenant seemed on the verge of refusing. The head of one of the mulatto sailors appeared beside him.

"I believe these men were killed by the ants," said Holroyd abruptly in English.

The captain burst into a rage. He made no answer to Holroyd. "I have commanded you to go aboard," he screamed to his subordinate in Portuguese. "If you do not go aboard forthwith it is mutiny—rank mutiny. Mutiny and cowardice! Where is the courage that should animate us? I will have you in irons, I will have you shot like a dog." He began a torrent of abuse and curses, he danced to and fro. He shook his fists, he behaved as if beside himself with rage, and the lieutenant, white and still, stood looking at him. The crew appeared forward, with amazed faces.

Suddenly, in a pause of this outbreak, the lieutenant came to some heroic decision, saluted, drew himself together and clambered upon the deck of the cuberta.

"Ah!" said Gerilleau, and his mouth shut like a trap. Holroyd saw the ants retreating before da Cunha's boots. The

Portuguese walked slowly to the fallen man, stooped down, hesitated, clutched his coat and turned him over. A black swarm of ants rushed out of the clothes, and da Cunha stepped back very quickly and trod two or three times on the deck.

Holroyd put up the glasses. He saw the scattered ants about the invader's feet, and doing what he had never seen ants doing before. They had nothing of the blind movements of the common ant; they were looking at him—as a rallying crowd of men might look at some gigantic monster that had dispersed it.

"How did he die?" the captain shouted.

Holroyd understood the Portuguese to say the body was too much eaten to tell.

"What is there forward?" asked Gerilleau.

The lieutenant walked a few paces, and began his answer in Portuguese. He stopped abruptly and beat off something from his leg. He made some peculiar steps as if he was trying to stamp on something invisible, and went quickly towards the side. Then he controlled himself, turned about, walked deliberately forward to the hold, clambered up to the fore decking, from which the sweeps are worked, stooped for a time over the second man, groaned audibly, and made his way back and aft to the cabin; moving very rigidly. He turned and began a conversation with his captain, cold and respectful in tone on either side, contrasting vividly with the wrath and insult of a few moments before. Holroyd gathered only fragments of its purport.

He reverted to the field-glass, and was surprised to find the ants had vanished from all the exposed surfaces of the deck. He turned towards the shadows beneath the decking, and it seemed to him they were full of watching eyes.

The cuberta, it was agreed, was derelict, but too full of ants to put men aboard to sit and sleep: it must be towed. The lieutenant went forward to take in and adjust the cable, and the men in the boat stood up to be ready to help him. Holroyd's glasses searched the canoe.

He became more and more impressed by the fact that a great if minute and furtive activity was going on. He perceived that a number of gigantic ants—they seemed nearly a couple of inches in length—carrying oddly-shaped burthens for which he could imagine no use—were moving in rushes from one point of obscurity to another. They did not move in columns across the exposed places, but in open, spaced-out lines, oddly suggestive of the rushes of modern infantry advancing under fire. A number were taking cover under the dead man's clothes, and a perfect swarm was gathering along the side over which da Cunha must presently go.

He did not see them actually rush for the lieutenant as he returned, but he has no doubt they did make a concerted rush. Suddenly the lieutenant was shouting and cursing and beating at his legs. "I'm stung!" he shouted, with a face of hate and accusation towards Gerilleau.

Then he vanished over the side, dropped into his boat, and plunged at once into the water. Holroyd heard the splash.

The three men in the boat pulled him out and brought him aboard, and that night he died.

§ 3

Holroyd and the captain came out of the cabin in which the swollen and contorted body of the lieutenant lay, and stood together at the stern of the monitor, staring at the sinister vessel they trailed behind them. It was a close, dark night that had only phantom flickerings of sheet lightning to illuminate it. The cuberta, a vague black triangle, rocked about in the steamer's wake, her sails bobbing and flapping, and the black smoke from the funnels, spark-lit ever and again, streamed over her swaying masts.

Gerilleau's mind was inclined to run on the unkind things the lieutenant had said in the heat of his last fever. "He says I murdered 'im," he protested. "It is simply absurd. Someone 'ad to go aboard. Are we to run away from these confounded ants whenever they show up?"

Holroyd said nothing. He was thinking of a disciplined rush of little black shapes across bare sunlit planking.

"It was his place to go," harped Gerilleau. "He died in the execution of his duty. What has he to complain of?

Murdered! . . . But the poor fellow was—what is it?—demented. He was not in his right mind. The poison swelled him. . . . U'm."

They came to a long silence.

"We will sink that canoe—burn it."

"And then?"

The inquiry irritated Gerilleau. His shoulders went up, his hands flew out at right angles from his body. "What is one to *do*?" he said, his voice going up to an angry squeak.

"Anyhow," he broke out vindictively, "every ant in dat cuberta!—I will burn dem alive!"

Holroyd was not moved to conversation. A distant ululation of howling monkeys filled the sultry night with foreboding sounds, and as the gunboat drew near the black mysterious banks this was reinforced by a depressing clamour of frogs.

"What is one to *do*?" the captain repeated after a vast interval, and suddenly becoming active and savage and blasphemous, decided to burn the *Santa Rosa* without further delay. Everyone aboard was pleased by that idea, everyone helped with zest; they pulled in the cable, cut it, and dropped the boat and fired her with tow and kerosene, and soon the cuberta was crackling and flaring merrily amidst the immensities of the tropical night. Holroyd watched the mounting yellow flare against the blackness, and the livid flashes of sheet lightning that came and went above the forest summits, throwing them into momentary silhouette, and his stoker stood behind him watching also.

The stoker was stirred to the depths of his linguistics. "*Saüba* go pop, pop," he said. "Wahaw!" and laughed richly.

But Holroyd was thinking that these little creatures on the decked canoe had also eyes and brains.

The whole thing impressed him as incredibly foolish and wrong, but—what was one to *do*? This question came back enormously reinforced on the morrow, when at last the gunboat reached Badama.

This place, with its leaf-thatch-covered houses and sheds, its creeper-invaded sugar-mill, its little jetty of timber and canes, was very still in the morning heat, and showed never a sign of living men. Whatever ants there were at that distance were too small to see.

"All the people have gone," said Gerilleau, "but we will do one thing anyhow. We will 'oot and vissel."

So Holroyd hooted and whistled.

Then the captain fell into a doubting fit of the worst kind. "Dere is one thing we can do," he said presently.

"What's that?" said Holroyd.

" 'Oot and vissel again."

So they did.

The captain walked his deck and gesticulated to himself. He seemed to have many things on his mind. Fragments of speeches came from his lips. He appeared to be addressing some imaginary public tribunal either in Spanish or Portuguese. Holroyd's improving ear detected something about ammunition. He came out of these preoccupations suddenly into English. "My dear 'Olroyd!" he cried, and broke off with "But what *can* one do?"

They took the boat and the field-glasses, and went close in to examine the place. They made out a number of big ants, whose still postures had a certain effect of watching them, dotted about the edge of the rude embarkation jetty. Gerilleau tried ineffectual pistol shots at these. Holroyd thinks he distinguished curious earthworks running between the nearer houses, that may have been the work of the insect conquerors of those human habitations. The explorers pulled past the jetty, and became aware of a human skeleton wearing a loin cloth, and very bright and clean and shining, lying beyond. They came to a pause regarding this. . . .

"I 'ave all dose lives to consider," said Gerilleou suddenly.

Holroyd turned and stared at the captain, realising slowly that he referred to the unappetising mixture of races that constituted his crew.

"To send a landing party—it is impossible—impossible. They will be poisoned, they will swell, they will swell up and abuse me and die. It is totally impossible. . . . If we land, I must land alone, alone, in thick boots and with my life in my hand. Perhaps I should live. Or again—I might not land. I do not know. I do not know."

Holroyd thought he did, but he said nothing.

"De whole thing," said Gerilleau suddenly, "'as been got up to make me ridiculous. De whole thing!"

They paddled about and regarded the clean white skeleton from various points of view, and then they returned to the gunboat. Then Gerilleau's indecisions became terrible. Steam was got up, and in the afternoon the monitor went on up the river with an air of going to ask somebody something, and by sunset came back again, and anchored. A thunderstorm gathered and broke furiously, and then the night became beautifully cool and quiet and everyone slept on deck. Except Gerilleau, who tossed about and muttered. In the dawn he awakened Holroyd.

"Lord!" said Holroyd, "what now?"

"I have decided," said the captain.

"What—to land?" said Holroyd, sitting up brightly.

"No!" said the captain, and was for a time very reserved. "I have decided," he repeated, and Holroyd manifested symptoms of impatience.

"Well—yes," said the captain. "*I shall fire de big gun!*"

And he did! Heaven knows what the ants thought of it, but he did. He fired it twice with great sternness and ceremony. All the crew had wadding in their ears, and there was an effect of going into action about the whole affair, and first they hit and wrecked the old sugar-mill, and then they smashed the abandoned store behind the jetty. And then Gerilleau experienced the inevitable reaction.

"It is no good," he said to Holroyd; "no good at all. No sort of bally good. We must go back—for instructions. Dere will be de devil of a row about dis ammunition—oh! de *devil* of a row! You don't know, 'Olroyd. . . ."

He stood regarding the world in infinite perplexity for a space.

"But what else was there to *do?*" he cried.

In the afternoon the monitor started down stream again, and in the evening a landing party took the body of the lieutenant and buried it on the bank upon which the new ants have so far not appeared. . . .

§ 4

I heard this story in a fragmentary state from Holroyd not three weeks ago.

These new ants have got into his brain, and he has come back to England with the idea, as he says, of "exciting people" about them "before it is too late." He says they threaten British Guiana, which cannot be much over a trifle of a thousand miles from their present sphere of activity, and that the Colonial Office ought to get to work upon them at once. He declaims with great passion: "These are intelligent ants. Just think what that means!"

There can be no doubt they are a serious pest, and that the Brazilian Government is well advised in offering a prize of five hundred pounds for some effectual method of extirpation. It is certain, too, that since they first appeared in the hills beyond Badama, about three years ago, they have achieved extraordinary conquests. The whole of the south bank of the Batemo River, for nearly sixty miles, they have in their effectual occupation; they have driven men out completely, occupied plantations and settlements, and boarded and captured at least one ship. It is even said they have in some inexplicable way bridged the very considerable Capuarana arm and pushed many miles towards the Amazon itself. There can be little doubt that they are far more reasonable and with a far better social organisation than any previously known ant species; instead of being in dispersed societies they are organised into what is in effect a single nation; but their peculiar and immediate formidableness lies not so much in this as in the intelligent use they make of poison against their larger enemies. It would seem this poison of theirs is closely akin to snake poison, and it is highly probable they actually manufacture it, and that the larger individuals among them carry the needle-like crystals of it in their attacks upon men.

Of course it is extremely difficult to get any detailed information about these new competitors for the sovereignty of the globe. No eye-witnesses of their activity, except for such glimpses as Holroyd's, have survived the encounter.

The most extraordinary legends of their prowess and capacity are in circulation in the region of the Upper Amazon, and grow daily as the steady advance of the invader stimulates men's imaginations through their fears. These strange little creatures are credited not only with the use of implements and a knowledge of fire and metals and with organised feats of engineering that stagger our Northern minds—unused as we are to such feats as that of the Saübus of Rio de Janeiro, who, in 1841, drove a tunnel under Parahyba, where it is as wide as the Thames at London Bridge—but with an organised and detailed method of record and communication analogous to our books. So far their action has been a steady progressive settlement, involving the flight or slaughter of every human being in the new areas they invade. They are increasing rapidly in numbers, and Holroyd at least is firmly convinced that they will finally dispossess man over the whole of tropical South America.

And why should they stop at tropical South America?

"Well, there they are, anyhow. By 1911 or thereabouts, if they go on as they are going, they ought to strike the Capuarana Extension Railway, and force themselves upon the attention of the European capitalist.

By 1920 they will be half-way down the Amazon. I fix 1950 or '60 at the latest for the discovery of Europe.

STORY THE THIRD

A Vision of Judgment

§ 1

Bʀᴜ-ᴀ-ᴀ-ᴀ.

I listened, not understanding.

Wa-ra-ra-ra.

"Good Lord!" said I, still only half awake. "What an infernal shindy!"

Ra-ra-ra-ra-ra-ra-ra-ra-ra Ta-ra-rra-ra.

"It's enough," said I, "to wake——" and stopped short. Where was I?

Ta-rra-rara—louder and louder.

"It's either some new invention——"

Toora-toora-toora! Deafening!

"No," said I, speaking loud in order to hear myself. "That's the Last Trump."

Tooo-rraa!

§ 2

The last note jerked me out of my grave like a hooked minnow.

I saw my monument (rather a mean little affair, and I wished I knew who'd done it), and the old elm tree and the sea view vanished like a puff of steam, and then all about me —a multitude no man could number, nations, tongues, kingdoms, peoples—children of all the ages, in an amphitheatral space as vast as the sky. And over against us, seated on a throne of dazzling white cloud, the Lord God and all the

109

host of his angels. I recognised Azreal by his darkness and
Michael by his sword, and the great angel who had blown
the trumpet stood with the trumpet still half raised.

§ 3

"Prompt," said the little man beside me. "Very prompt.
Do you see the angel with the book?"

He was ducking and craning his head about to see over
and under and between the souls that crowded round us.
"Everybody's here," he said. "Everybody. And now we
shall know——"

"There's Darwin," he said, going off at a tangent. "*He'll*
catch it! And there—you see?—that tall, important-looking
man trying to catch the eye of the Lord God, that's the
Duke. But there's a lot of people one doesn't know.

"Oh! there's Priggles, the publisher. I have always wondered about printers' overs. Priggles was a clever man . . .
But we shall know now—even about him.

"I shall hear all that. I shall get most of the fun before
. . . *My* letter's S."

He drew the air in between his teeth.

"Historical characters, too. See? That's Henry the
Eighth. There'll be a good bit of evidence. Oh, damn!
He's Tudor."

He lowered his voice. "Notice this chap, just in front of
us, all covered with hair. Paleolithic, you know. And
there again——"

But I did not heed him, because I was looking at the Lord
God.

§ 4

"Is this *all*?" asked the Lord God.

The angel at the book—it was one of countless volumes,
like the British Museum Reading-room Catalogue, glanced
at us and seemed to count us in the instant.

"That's all," he said, and added: "It was, O God, a very
little planet."

The eyes of God surveyed us.

"Let us begin," said the Lord God.

§ 5

The angel opened the book and read a name. It was a name full of A's, and the echoes of it came back out of the uttermost parts of space. I did not catch it clearly, because the little man beside me said, in a sharp jerk, *"What's* that?" It sounded like "Ahab" to me; but it could not have been the Ahab of Scripture.

Instantly a small black figure was lifted up to a puffy cloud at the very feet of God. It was a stiff little figure, dressed in rich outlandish robes and crowned, and it folded its arms and scowled.

"Well?" said God, looking down at him.

We were privileged to hear the reply, and indeed the acoustic properties of the place were marvellous.

"I plead guilty," said the little figure,

"Tell them what you have done," said the Lord God.

"I was a king," said the little figure, "a great king, and I was lustful and proud and cruel. I made wars, I devastated countries, I built palaces, and the mortar was the blood of men. Hear, O God, the witnesses against me, calling to you for vengeance. Hundreds and thousands of witnesses." He waved his hands towards us. "And worse! I took a prophet —one of your prophets——"

"One of my prophets," said the Lord God.

"And because he would not bow to me, I tortured him for four days and nights, and in the end he died. I did more, O God, I blasphemed. I robbed you of your honours——"

"Robbed me of my honours," said the Lord God.

"I caused myself to be worshipped in your stead. No evil was there but I practised it; no cruelty where-with I did not stain my soul. And at last you smote me, O God!"

God raised his eyebrows slightly.

"And I was slain in battle. And so I stand before you, meet for your nethermost Hell! Out of your greatness daring no lies, daring no pleas, but telling the truth of my iniquities before all mankind."

He ceased. His face I saw distinctly, and it seemed to me white and terrible and proud and strangely noble. I thought of Milton's Satan.

"Most of that is from the Obelisk," said the Recording Angel, finger on page.

"It is," said the Tyrannous Man, with a faint touch of surprise.

Then suddenly God bent forward and took this man in his hand, and held him up on his palm as if to see him better. He was just a little dark stroke in the middle of God's palm.

"*Did* he do all this?" said the Lord God.

The Recording Angel flattened his book with his hand.

"In a way," said the Recording Angel, carelessly.

Now when I looked again at the little man his face had changed in a very curious manner. He was looking at the Recording Angel with a strange apprehension in his eyes, and one hand fluttered to his mouth. Just the movement of a muscle or so, and all that dignity of defiance was gone.

"Read," said the Lord God.

And the angel read, explaining very carefully and fully all the wickedness of the Wicked Man. It was quite an intellectual treat.—A little "daring" in places, I thought, but of course Heaven has its privileges. . . .

§ 6

Everybody was laughing. Even the prophet of the Lord whom the Wicked Man had tortured had a smile on his face. The Wicked Man was really such a preposterous little fellow.

"And then," read the Recording Angel, with a smile that set us all agog, "one day, when he was a little irascible from over-eating, he——"

"Oh, not *that*," cried the Wicked Man, "nobody knew of *that*.

"It didn't happen," screamed the Wicked Man. "I was

bad—I was really bad. Frequently bad, but there was nothing so silly—so absolutely silly——"

The angel went on reading.

"O God!" cried the Wicked Man. "Don't let them know that! I'll repent! I'll apologise. . . .

The Wicked Man on God's hand began to dance and weep. Suddenly shame overcame him. He made a wild rush to jump off the ball of God's little finger, but God stopped him by a dexterous turn of the wrist. Then he made a rush for the gap between hand and thumb, but thumb closed. And all the while the angel went on reading —reading. The Wicked Man rushed to and fro across God's palm, and then suddenly turned about and fled up the sleeve of God.

I expected God would turn him out, but the mercy of God is infinite.

The Recording Angel paused.

"Eh?" said the Recording Angel.

"Next," said God, and before the Recording Angel could call upon the name a hairy creature in filthy rags stood upon God's palm.

§ 7

"Has God got Hell up his sleeve then?" said the little man beside me.

"Is there a Hell?" I asked.

"If you notice," he said—he peered between the feet of the great angels—"there's no particular indication of the Celestial City."

" 'Ssh!" said a little woman near us, scowling. "Hear this blessed Saint!"

§ 8

"He was Lord of the Earth, but I was the prophet of the God of Heaven," cried the Saint, "and all the people marvelled at the sign. For I, O God, knew of the glories of thy Paradise. No pain, no hardship, gashing with knives, splinters thrust under my nails, strips of flesh flayed off, all for the glory and honour of God."

God smiled.

"And at last I went, I in my rags and sores, smelling of my holy discomforts——"

Gabriel laughed abruptly.

"And lay outside his gates, as a sign, as a wonder——"

"As a perfect nuisance," said the Recording Angel, and began to read, heedless of the fact that the Saint was still speaking of the gloriously unpleasant things he had done that Paradise might be his.

And behold, in that book the record of the Saint also was a revelation, a marvel.

It seemed not ten seconds before the Saint, also, was rushing to and fro over the great palm of God. Not ten seconds! And at last he also shrieked beneath that pitiless and cynical exposition, and fled also, even as the Wicked Man had fled, into the shadow of the sleeve. And it was permitted us to see into the shadow of the sleeve. And the two sat side by side, stark of all delusions, in the shadow of the robe of God's charity, like brothers.

And thither also I fled in my turn.

§ 9

"And now," said God, as he shook us out of his sleeve upon the planet he had given us to live upon, the planet that whirled about green Sirius for a sun, "now that you understand me and each other a little better, . . . try again."

Then he and his great angels turned themselves about and suddenly had vanished.

The Throne had vanished.

All about me was a beautiful land, more beautiful than any I had even seen before—waste, austere, and wonderful; and all about me were the enlightened souls of men in new clean bodies. . . .

STORY THE FOURTH

The Land Ironclads*

§ 1

THE young lieutenant lay beside the war correspondent and admired the idyllic calm of the enemy's lines through his field-glass.

"So far as I can see," he said at last, "one man."

"What's he doing?" asked the war correspondent.

"Field-glass at us," said the young lieutenant.

"And this is war!"

"No," said the young lieutenant; "it's Bloch."

"The game's a draw."

"No! They've got to win or else they lose. A draw's a win for our side."

They had discussed the political situation fifty times or so, and the war correspondent was weary of it. He stretched out his limbs. "Aaaai s'pose it *is!*" he yawned.

Flut!

What was that?"

"Shot at us."

The war correspondent shifted to a slightly lower position. "No one shot at him," he complained.

"I wonder if they think we shall get so bored we shall go home?"

The war correspondent made no reply.

"There's the harvest, of course. . . ."

They had been there a month. Since the first brisk move-

*First Published in December, 1903.

ments after the declaration of war things had gone slower and slower, until it seemed as though the whole machine of events must have run down. To begin with, they had had almost a scampering time; the invader had come across the frontier on the very dawn of the war in half-a-dozen parallel columns behind a cloud of cyclists and cavalry, with a general air of coming straight on the capital, and the defender horsemen had held him up, and peppered him and forced him to open out to outflank, and had then bolted to the next position in the most approved style, for a couple of days, until in the afternoon, bump! they had the invader against their prepared lines of defence. He did not suffer so much as had been hoped and expected: he was coming on, it seemed, with his eyes open, his scouts winded the guns, and down he sat at once without the shadow of an attack and began grubbing trenches for himself, as though he meant to sit down there to the very end of time. He was slow, but much more wary than the world had been led to expect, and he kept convoys tucked in and shielded his slow-marching infantry sufficiently well to prevent any heavy adverse scoring.

"But he ought to attack," the young lieutenant had insisted.

"He'll attack us at dawn, somewhere along the lines. You'll get the bayonets coming into the trenches just about when you can see," the war correspondent had held until a week ago.

The young lieutenant winked when he said that.

When one early morning the men the defenders sent to lie out five hundred yards before the trenches, with a view to the unexpected emptying of magazines into any night attack, gave way to causeless panic and blazed away at nothing for ten minutes, the war correspondent understood the meaning of that wink.

"What would you do if you were the enemy?" said the war correspondent, suddenly.

"If I had men like I've got now?"

"Yes."

"Take those trenches."

"How?"

"Oh—dodges! Crawl out half-way at night before moon-rise and get into touch with the chaps we send out. Blaze at 'em if they tried to shift, and so bag some of 'em in the daylight. Learn that patch of ground by heart, lie all day in squatty holes, and come on nearer next night. There's a bit over there, lumpy ground, where they could get across to rushing distance—easy. In a night or so. It would be a mere game for our fellows; it's what they're made for. . . . Guns? Shrapnel and stuff wouldn't stop good men who meant business."

"Why don't *they* do that?"

"Their men aren't brutes enough; that's the trouble. They're a crowd of devitalised townsmen, and that's the truth of the matter. They're clerks, they're factory hands, they're students, they're civilised men. They can write, they can talk, they can make and do all sorts of things, but they're poor amateurs at war. They've got no physical staying power, and that's the whole thing. They've never slept in the open one night in their lives; they've never drunk anything but the purest water-company water; they've never gone short of three meals a day since they left their feeding-bottles. Half their cavalry never cocked leg over horse till it enlisted six months ago. They ride their horses as though they were bicycles—you watch 'em! They're fools at the game, and they know it. Our boys of fourteen can give their grown men points. . . . Very well——"

The war correspondent mused on his face with his nose between his knuckles.

"If a decent civilisation," he said, "cannot produce better men for war than——"

He stopped with belated politeness. "I mean——"

"Than our open-air life," said the young lieutenant.

"Exactly," said the war correspondent. "Then civilisation has to stop."

"It looks like it," the young lieutenant admitted.

"Civilisation has science, you know," said the war correspondent. "It invented and it made the rifles and guns and things you use."

"Which our nice healthy hunters and stockmen and so

on, rowdy-dowdy cowpunchers and nigger-whackers, can use
ten times better than—— *What's that?*"

"What?" said the war correspondent, and then seeing his
companion busy with his field-glass he produced his own:
"Where?" said the war correspondent, sweeping the enemy's
lines.

"It's nothing," said the young lieutenant, still looking.

"What's nothing?"

The young lieutenant put down his glass and pointed.
"I thought I saw something there, behind the stems of those
trees. Something black. What it was I don't know."

The war correspondent tried to get even by intense scrutiny.

"It wasn't anything," said the young lieutenant, rolling
over to regard the darkling evening sky, and generalised:
"There never will be anything any more for ever. Unless——"

The war correspondent looked inquiry.

"They may get their stomachs wrong, or something—
living without proper drains."

A sound of bugles came from the tents behind. The war
correspondent slid backward down the sand and stood up.
"Boom!" came from somewhere far away to the left.
"Halloa!" he said, hesitated, and crawled back to peer again.
"Firing at this time is jolly bad manners."

The young lieutenant was uncommunicative for a space.

Then he pointed to the distant clump of trees again. "One
of our big guns. They were firing at that," he said.

"The thing that wasn't anything?"

"Something over there, anyhow."

Both men were silent, peering through their glasses for a
space. "Just when it's twilight," the lieutenant complained.
He stood up.

"I might stay here a bit," said the war correspondent.

The lieutenant shook his head. "There's nothing to see,"
he apologised, and then went down to where his little squad
of sun-brown, loose-limbed men had been yarning in the
trench. The war correspondent stood up also, glanced for a
moment at the businesslike bustle below him, gave perhaps

twenty seconds to those enigmatical trees again, then turned his face toward the camp.

He found himself wondering whether his editor would consider the story of how somebody thought he saw something black behind a clump of trees, and how a gun was fired at this illusion by somebody else, too trivial for public consumption.

"It's the only gleam of a shadow of interest," said the war correspondent, "for ten whole days.

"No," he said presently; "I'll write that other article, 'Is War Played Out?'"

He surveyed the darkling lines in perspective, the tangle of trenches one behind another, one commanding another, which the defender had made ready. The shadows and mists swallowed up their receding contours, and here and there a lantern gleamed, and here and there knots of men were busy about small fires. "No troops on earth could do it," he said. . . .

He was depressed. He believed that there were other things in life better worth having than proficiency in war; he believed that in the heart of civilisation, for all its stresses, its crushing concentrations of forces, its injustice and suffering, there lay something that might be the hope of the world; and the idea that any people, by living in the open air, hunting perpetually, losing touch with books and art and all the things that intensify life, might hope to resist and break that great development to the end of time, jarred on his civilised soul.

Apt to his thought came a file of the defender soldiers, and passed him in the gleam of a swinging lamp that marked the way.

He glanced at their red-lit faces, and one shone out for a moment, a common type of face in the defender's ranks: ill-shaped nose, sensuous lips, bright clear eyes full of alert cunning, slouch hat cocked on one side and adorned with the peacock's plume of the rustic Don Juan turned soldier, a hard brown skin, a sinewy frame, an open, tireless stride, and a master's grip on the rifle.

The war correspondent returned their salutations and went on his way.

"Louts," he whispered. "Cunning, elementary louts. And they are going to beat the townsmen at the game of war!"

From the red glow among the nearer tents came first one and then half-a-dozen hearty voices, bawling in a drawling unison the words of a particularly slab and sentimental patriotic song.

"Oh, *go* it!" muttered the war correspondent, bitterly.

§ 2

It was opposite the trenches called after Hackbone's Hut that the battle began. There the ground stretched broad and level between the lines, with scarcely shelter for a lizard, and it seemed to the startled, just-awakened men who came crowding into the trenches that this was one more proof of that inexperience of the enemy of which they had heard so much. The war correspondent would not believe his ears at first, and swore that he and the war artist, who, still imperfectly roused, was trying to put on his boots by the light of a match held in his hand, were the victims of a common illusion. Then, after putting his head in a bucket of cold water, his intelligence came back as he towelled. He listened. "Gollys!" he said; "that's something more than scare firing this time. It's like ten thousand carts on a bridge of tin."

There came a sort of enrichment to that steady uproar. "Machine-guns!"

Then, "Guns!"

The artist, with one boot on, thought to look at his watch, and went to it hopping.

"Half an hour from dawn," he said. "You were right about their attacking, after all. . . ."

The war correspondent came out of the tent, verifying the presence of chocolate in his pocket as he did so. He had to halt for a moment or so until his eyes were toned down to the night a little. "Pitch!" he said. He stood for a space to season his eyes before he felt justified in striking out for a black gap among the adjacent tents. The artist coming out behind him fell over a tent-rope. It was half-past two o'clock in the morning of the darkest night in time, and against a sky of dull black silk the enemy was talking search-lights, a wild jabber of search-lights. "He's trying to blind our riflemen," said the war correspondent with a flash, and waited for the artist and then set off with a sort of discreet haste again. "Whoa!" he said, presently. "Ditches!"

They stopped.

"It's the confounded search-lights," said the war correspondent.

They saw lanterns going to and fro, near by, and men falling in to march down to the trenches. They were for following them, and then the artist began to get his night eyes. "If we scramble this," he said, "and it's only a drain, there's a clear run up to the ridge." And that way they took. Lights came and went in the tents behind, as the men turned out, and ever and again they came to broken ground and staggered and stumbled. But in a little while they drew near the crest. Something that sounded like the impact of a tremendous railway accident happened in the air above them, and the shrapnel bullets seethed about them like a sudden handful of hail. "Right-oh!" said the war correspondent, and soon they judged they had come to the crest and stood in the midst of a world of great darkness and frantic glares, whose principal fact was sound.

Right and left of them and all about them was the uproar, an army-full of magazine fire, at first chaotic and monstrous, and then, eked out by little flashes and gleams and suggestions, taking the beginnings of a shape. It looked to the war correspondent as though the enemy must have attacked in line and with his whole force—in which case he was either being or was already annihilated.

"Dawn and the dead," he said, with his instinct for head-lines. He said this to himself, but afterwards by means of shouting he conveyed an idea to the artist "They must have meant it for a surprise," he said.

It was remarkable how the firing kept on. After a time he began to perceive a sort of rhythm in this inferno of noise. It would decline—decline perceptibly, droop towards something that was comparatively a pause—a pause of inquiry. "Aren't you all dead yet?" this pause seemed to say.

The flickering fringe of rifle-flashes would become attenuated and broken, and the whack-bang of the enemy's big guns two miles away there would come up out of the deeps. Then suddenly, east or west of them, something would startle the rifles to a frantic outbreak again.

The war correspondent taxed his brain for some theory of conflict that would account for this, and was suddenly aware that the artist and he were vividly illuminated. He could see the ridge on which they stood, and before them in black outline a file of riflemen hurrying down towards the nearer trenches. It became visible that a light rain was falling, and farther away towards the enemy was a clear space with men—"our men?"—running across it in disorder. He saw one of those men throw up his hands and drop. And something else black and shining loomed up on the edge of the beam-coruscating flashes; and behind it and far away a calm, white eye regarded the world. "Whit, whit, whit," sang something in the air, and then the artist was running for cover, with the war correspondent behind him. Bang came shrapnel, bursting close at hand as it seemed, and our two men were lying flat in a dip in the ground, and the light and everything had gone again, leaving a vast note of interrogation upon the light.

The war correspondent came within bawling range. "What the deuce was it? Shooting our men down!"

"Black," said the artist, "and like a fort. Not two hundred yards from the first trench."

He sought for comparisons in his mind. "Something between a big blockhouse and a giant's dish-cover," he said.

"And they were running!" said the war correspondent.

"You'd run if a thing like that, with a search-light to help it, turned up like a prowling nightmare in the middle of the night."

They crawled to what they judged the edge of the dip and lay regarding the unfathomable dark. For a space they could distinguish nothing, and then a sudden convergence of the search-lights of both sides brought the strange thing out again.

In that flickering pallor it had the effect of a large and clumsy black insect, an insect the size of an iron-clad cruiser, crawling obliquely to the first line of trenches and firing shots out of port-holes in its side. And on its carcass the bullets must have been battering with more than the passionate violence of hail on a roof of tin.

Then in the twinkling of an eye the curtain of the dark had fallen again and the monster had vanished, but the crescendo of musketry marked its approach to the trenches.

They were beginning to talk about the thing to each other, when a flying bullet kicked dirt into the artist's face, and they decided abruptly to crawl down into the cover of the trenches. They had got down with an unobtrusive persistence into the second line, before the dawn had grown clear enough for anything to be seen. They found themselves in a crowd of expectant riflemen, all noisily arguing about what would happen next. The enemy's contrivance had done execution upon the outlying men, it seemed, but they did not believe it would do any more. "Come the day and we'll capture the lot of them," said a burly soldier.

"Them?" said the war correspondent.

"They say there's a regular string of 'em, crawling along the front of our lines. . . . Who cares?"

The darkness filtered away so imperceptibly that at no moment could one declare decisively that one could see. The search-lights ceased to sweep hither and thither. The enemy's monsters were dubious patches of darkness upon the dark, and then no longer dubious, and so they crept out into distinctness. The war correspondent, munching chocolate absent-mindedly, beheld at last a spacious picture of battle under the cheerless sky, whose central focus was an array of fourteen or fifteen huge clumsy shapes lying in perspective on the very edge of the first line of trenches, at intervals of perhaps three hundred yards, and evidently firing down upon the crowded riflemen. They were so close in that the defender's guns had ceased, and only the first line of trenches was in action.

The second line commanded the first, and as the light grew, the war correspondent could make out the riflemen who were fighting these monsters, crouched in knots and crowds behind the transverse banks that crossed the trenches against the eventuality of an enfilade. The trenches close to the big machines were empty save for the crumpled suggestions of dead and wounded men; the defenders had been driven right and left as soon as the prow of a land ironclad had loomed up over the front of the trench. The war correspondent produced his field-glass, and was immediately a centre of inquiry from the soldiers about him.

They wanted to look, they asked questions, and after he had announced that the men across the traverses seemed unable to advance or retreat, and were crouching under cover rather than fighting, he found it advisable to loan his glasses to a burly and incredulous corporal. He heard a strident voice, and found a lean and sallow soldier at his back talking to the artist.

"There's chaps down there caught," the man was saying. "If they retreat they got to expose themselves, and the fire's too straight. . . ."

"They aren't firing much, but every shot's a hit."

"Who?"

"The chaps in that thing. The men who're coming up——"

"Coming up where?"

"We're evacuating them trenches where we can. Our chaps are coming back up the zigzags. . . . No end of 'em hit. . . . But when we get clear our turn'll come. Rather! Those things won't be able to cross a trench or get into it; and before they can get back our guns'll smash 'em up. Smash 'em right up. See?" A brightness came into his eyes. "Then we'll have a go at the beggars inside," he said. . . .

The war correspondent thought for a moment, trying to realise the idea. Then he set himself to recover his field-glasses from the burly corporal. . . .

The daylight was getting clearer now. The clouds were lifting, and a gleam of lemon-yellow amidst the level masses to the east portended sunrise. He looked again at the land ironclad. As he saw it in the bleak, grey dawn, lying obliquely upon the slope and on the very lip of the foremost trench, the suggestion of a stranded vessel was very strong indeed. It might have been from eighty to a hundred

feet long—it was about two hundred and fifty yards away
—its vertical side was ten feet high or so, smooth for that
height, and then with a complex patterning under the eaves
of its flattish turtle cover. This patterning was a close interlacing of port-holes, rifle barrels, and telescope tubes—sham
and real—indistinguishable one from the other. The thing
had come into such a position as to enfilade the trench,
which was empty now, so far as he could see, except for
two or three crouching knots of men and the tumbled dead.
Behind it, across the plain, it had scored the grass with a
train of linked impressions, like the dotted tracings sea-
things leave in sand. Left and right of that track dead
men and wounded men were scattered—men it had picked
off as they fled back from their advanced positions in the
search-light glare from the invader's lines. And now it lay
with its head projecting a little over the trench it had won,
as if it were a single sentient thing planning the next phase
of its attack. . . .

He lowered his glasses and took a more comprehensive
view of the situation. These creatures of the night had
evidently won the first line of trenches and the fight had
come to a pause. In the increasing light he could make out
by a stray shot or a chance exposure that the defender's
marksmen were lying thick in the second and third line
of trenches up towards the low crest of the position, and
in such of the zigzags as gave them a chance of a converging
fire. The men about him were talking of guns. "We're
in the line of the big guns at the crest, but they'll soon shift
one to pepper them," the lean man said, reassuringly.

"Whup," said the corporal.

"Bang! bang! bang! Whir-r-r-r!" it was a sort of nervous
jump, and all the rifles were going off by themselves. The war
correspondent found himself and the artist, two idle men
crouching behind a line of preoccupied backs, of industrious
men discharging magazines. The monster had moved. It
continued to move regardless of the hail that splashed its skin
with bright new specks of lead. It was singing a mechanical
little ditty to itself, "Tuf-tuf, tuf-tuf, tuf-tuf," and squirting
out little jets of steam behind. It had humped itself up, as a
limpet does before it crawls; it had lifted its skirt and dis-

played along the length of it—*feet!* They were thick, stumpy feet, between knobs and buttons in shape—flat, broad things, reminding one of the feet of elephants or the legs of caterpillars; and then, as the skirt rose higher, the war correspondent, scrutinising the thing through his glasses again, saw that these feet hung, as it were, on the rims of wheels. His thoughts whirled back to Victoria Street, Westminster, and he saw himself in the piping times of peace, seeking matter for an interview.

"Mr.—Mr. Diplock," he said; "and he called them Pedrails. . . . Fancy meeting them here!"

The marksman beside him raised his head and shoulders in a speculative mood to fire more certainly—it seemed so natural to assume the attention of the monster must be distracted by this trench before it—and was suddenly knocked backwards by a bullet through his neck. His feet flew up, and he vanished out of the margin of the watcher's field of vision. The war correspondent grovelled tighter, but after a glance behind him at a painful little confusion, he resumed his field-glass, for the thing was putting down its feet one after the other, and hoisting itself farther and farther over the trench. Only a bullet in the head could have stopped him looking just then.

The lean man with the strident voice ceased firing to turn and reiterate his point. "They can't possibly cross," he bawled. "They——"

"Bang! Bang! Bang! Bang!"—drowned everything.

The lean man continued speaking for a word or so, then gave it up, shook his head to enforce the impossibility of anything crossing a trench like the one below, and resumed business once more.

And all the while that great bulk was crossing. When the war correspondent turned his glass on it again it had bridged the trench, and its queer feet were rasping away at the farther bank, in the attempt to get a hold there. It got its hold. It continued to crawl until the greater bulk of it was over the trench—until it was all over. Then it paused for a moment, adjusted its skirt a little nearer the ground, gave an unnerving "toot, toot," and came on abruptly at a

pace of, perhaps, six miles an hour straight up the gentle
slope towards our observer.

The war correspondent raised himself on his elbow and
looked a natural inquiry at the artist.

For a moment the men about him stuck to their position
and fired furiously. Then the lean man in a mood of pre-
cipitancy slid backwards, and the war correspondent said
"Come along" to the artist, and led the movement along the
trench.

As they dropped down, the vision of a hillside of trench
being rushed by a dozen vast cockroaches disappeared for
a space, and instead was one of a narrow passage, crowded
with men, for the most part receding, though one or two
turned or halted. He never turned back to see the nose of
the monster creep over the brow of the trench; he never
even troubled to keep in touch with the artist. He heard
the "whit" of bullets about him soon enough, and saw a
man before him stumble and drop, and then he was one
of a furious crowd fighting to get into a transverse zigzag
ditch that enabled the defenders to get under cover up and
down the hill. It was like a theatre panic. He gathered
from signs and fragmentary words that on ahead another
of these monsters had also won to the second trench.

He lost his interest in the general course of the battle
for a space altogether; he became simply a modest egotist,
in a mood of hasty circumspection, seeking the farthest
rear, amidst a dispersed multitude of disconcerted riflemen
similarly employed. He scrambled down through trenches,
he took his courage in both hands and sprinted across the
open, he had moments of panic when it seemed madness
not to be quadrupedal, and moments of shame when he
stood up and faced about to see how the fight was going.
And he was one of many thousand very similar men that
morning. On the ridge he halted in a knot of scrub, and
was for a few minutes almost minded to stop and see things
out.

The day was now fully come. The grey sky had changed
to blue, and of all the cloudy masses of the dawn there
remained only a few patches of dissolving fleeciness. The
world below was bright and singularly clear. The ridge

was not, perhaps, more than a hundred feet or so above the general plain, but in this flat region it sufficed to give the effect of extensive view. Away on the north side of the ridge, little and far, were the camps, the ordered wagons, all the gear of a big army; with officers galloping about and men doing aimless things. Here and there men were falling in, however, and the cavalry was forming up on the plain beyond the tents. The bulk of men who had been in the trenches were still on the move to the rear, scattered like sheep without a shepherd over the farther slopes. Here and there were little rallies and attempts to wait and do—something vague; but the general drift was away from any concentration. There on the southern side was the elaborate lacework of trenches and defences, across which these iron turtles, fourteen of them spread out over a line of perhaps three miles, were now advancing as fast as a man could trot, and methodically shooting down and breaking up any persistent knots of resistance. Here and there stood little clumps of men, outflanked and unable to get away, showing the white flag, and the invader's cyclist infantry was advancing now across the open, in open order, but unmolested, to complete the work of the machines. Surveyed at large, the defenders already looked a beaten army. A mechanism that was effectually ironclad against bullets, that could at a pinch cross a thirty-foot trench, and that seemed able to shoot out rifle-bullets with unerring precision, was clearly an inevitable victor against anything but rivers, precipices, and guns.

He looked at his watch. "Half-past four! Lord! What things can happen in two hours. Here's the whole blessed army being walked over, and at half-past two——

"And even now our blessed louts haven't done a thing with their guns!"

He scanned the ridge right and left of him with his glasses. He turned again to the nearest land ironclad, advancing now obliquely to him and not three hundred yards away, and then scanned the ground over which he must retreat if he was not to be captured.

"They'll do nothing," he said, and glanced again at the enemy.

And then from far away to the left came the thud of a gun, followed very rapidly by a rolling gun-fire.

He hesitated and decided to stay.

§ 3

The defender had relied chiefly upon his rifles in the event of an assault. His guns he kept concealed at various points upon and behind the ridge ready to bring them into action against any artillery preparations for an attack on the part of his antagonist. The situation had rushed upon him with the dawn, and by the time the gunners had their guns ready for motion, the land ironclads were already in among the foremost trenches. There is a natural reluctance to fire into one's own broken men, and many of the guns, being intended simply to fight an advance of the enemy's artillery, were not in positions to hit anything in the second line of trenches. After that the advance of the land ironclads was swift. The defender-general found himself suddenly called upon to invent a new sort of warfare, in which guns were to fight alone amidst broken and retreating infantry. He had scarcely thirty minutes in which to think it out. He did not respond to the call, and what happened that morning was that the advance of the land ironclads forced the fight, and each gun and battery made what play its circumstances dictated. For the most part it was poor play.

Some of the guns got in two or three shots, some one or two, and the percentage of misses was unusually high. The howitzers, of course, did nothing. The land ironclads in each case followed much the same tactics. As soon as a gun came into play the monster turned itself almost end-on, so as to minimise the chances of a square hit, and made not for the gun, but for the nearest point on its flank from which the gunners could be shot down. Few of the hits scored were very effectual; only one of the things was disabled, and that was the one that fought the three batteries attached to the brigade on the left wing. Three that were hit when close upon the guns were clean shot through without being put out of action. Our war correspondent did not see that one momentary arrest of the tide of victory on the left; he saw only the very ineffectual fight of half-battery 96B close at hand upon his right. This he watched some time beyond the margin of safety.

Just after he heard the three batteries opening up upon his left he became aware of the thud of horses' hoofs from the sheltered side of the slope, and presently saw first one and then two other guns galloping into position along the north side of the ridge, well out of sight of the great bulk that was now creeping obliquely towards the crest and cutting up the lingering infantry beside it and below, as it came.

The half-battery swung round into line—each gun describing its curve—halted, unlimbered, and prepared for action. . . .

"Bang!"

The land ironclad had become visible over the brow of the hill, and just visible as a long black back to the gunners. It halted, as though it hesitated.

The two remaining guns fired, and then their big antagonist had swung round and was in full view, end-on, against the sky, coming at a rush.

The gunners became frantic in their haste to fire again. They were so near the war correspondent could see the expression of their excited faces through his field-glass. As he looked he saw a man drop, and realised for the first time that the ironclad was shooting.

For a moment the big black monster crawled with an accelerated pace towards the furiously active gunners. Then, as if moved by a generous impulse, it turned its full broadside to their attack, and scarcely forty yards away from them. The war correspondent turned his field-glass back to the gunners and perceived it was now shooting down the men about the guns with the most deadly rapidity.

Just for a moment it seemed splendid, and then it seemed horrible. The gunners were dropping in heaps about their guns. To lay a hand on a gun was death. "Bang!" went the gun on the left, a hopeless miss, and that was the only second shot the half-battery fired. In another moment half-a-dozen surviving artillerymen were holding up their hands amidst a scattered muddle of dead and wounded men, and the fight was done.

The war correspondent hesitated between stopping in his scrub and waiting for an opportunity to surrender decently, or taking to an adjacent gully he had discovered. If he surrendered it was certain he would get no copy off; while, if he escaped, there were all sorts of chances. He decided to follow the gully, and take the first offer in the confusion beyond the camp of picking up a horse.

§ 4

Subsequent authorities have found fault with the first land ironclads in many particulars, but assuredly they served their purpose on the day of their appearance. They were essentially long, narrow, and very strong steel frameworks carrying the engines, and borne upon eight pairs of big pedrail wheels, each about ten feet in diameter, each a driving wheel and set upon long axles free to swivel round a common axis. This arrangement gave them the maximum of adaptability to the contours of the ground. They crawled level along the ground with one foot high upon a hillock and another deep in a depression, and they could hold themselves erect and steady sideways upon even a steep hillside. The engineers directed the engines under the command of the captain, who had look-out points at small ports all round the upper edge of the adjustable skirt of twelve-inch iron-plating which protected the whole affair, and who could also raise or depress a conning-tower set about the port-holes through the centre of the iron top cover. The riflemen each occupied a small cabin of peculiar construction, and these cabins were slung along the sides of and before and behind the great main framework, in a manner suggestive of the slinging of the seats of an Irish jaunting-car. Their rifles, however, were very different pieces of apparatus from the simple mechanisms in the hands of their adversaries.

These were in the first place automatic, ejected their cartridges and loaded again from a magazine each time they fired, until the ammunition store was at an end, and they had the most remarkable sights imaginable, sights which threw a bright little camera-obscura picture into the light-tight box in which the riflemen sat below. This camera-obscura picture was marked with two crossed lines, and whatever was covered by the intersection of these two lines, that the rifle hit. The sighting was ingeniously contrived. The rifleman stood at the table with a thing like an elaboration of a draughtman's dividers in his hand, and he opened and closed these dividers, so that they were always at the apparent height—if it was an ordinary-sized man—of the man he wanted to kill. A little twisted strand of wire like an electric-light wire ran from this implement up to the gun, and as the dividers opened and shut the sights went up or down. Changes in the clearness of the atmosphere, due to changes of moisture, were met by an ingenious use of that meteorologically sensitive substance, catgut, and when the land ironclad moved forward the sights got a compensatory deflection in the direction of its motion. The rifleman stood up in his pitch-dark chamber and watched the little picture before him. One hand held the dividers for judging distance, and the other grasped a big knob like a door-handle. As he pushed this knob about the rifle above swung to correspond, and the picture passed to and fro like an agitated panorama. When he saw a man he wanted to shoot he brought him up to the cross-lines, and then pressed a finger upon a little push like an electric bell-push, conveniently placed in the centre of the knob. Then the man was shot. If by any chance the rifleman missed his target he moved the knob a trifle, or readjusted his dividers, pressed the push, and got him the second time.

This rifle and its sights protruded from a port-hole, exactly like a great number of other port-holes that ran in a triple row under the eaves of the cover of the land ironclad. Each port-hole displayed a rifle and sight in dummy, so that the real ones could only be hit by a chance shot, and if one was, then the young man below said "Pshaw!" turned on an electric light, lowered the injured instrument into his camera, replaced the injured part, or put up a new rifle if the injury was considerable.

You must conceive these cabins as hung clear above the

swing of the axles, and inside the big wheels upon which the great elephant-like feet were hung, and behind these cabins along the centre of the monster ran a central gallery into which they opened, and along which worked the big compact engines. It was like a long passage into which this throbbing machinery had been packed, and the captain stood about the middle, close to the ladder that led to his conning-tower, and directed the silent, alert engineers—for the most part by signs. The throb and noise of the engines mingled with the reports of the rifles and the intermittent clangour of the bullet hail upon the armour. Ever and again he would touch the wheel that raised his conning-tower, step up his ladder until his engineers could see nothing of him above the waist, and then come down again with orders. Two small electric lights were all the illumination of this space—they were placed to make him most clearly visible to his subordinates; the air was thick with the smell of oil and petrol, and had the war correspondent been suddenly transferred from the spacious dawn outside to the bowels of this apparatus he would have thought himself fallen into another world.

The captain, of course, saw both sides of the battle. When he raised his head into his conning-tower there were the dewy sunrise, the amazed and disordered trenches, the flying and falling soldiers, the depressed-looking groups of prisoners, the beaten guns; when he bent down again to signal "half speed," "quarter speed," "half circle round toward the right," or what not, he was in the oil-smelling twilight of the ill-lit engine-room. Close beside him on either side was the mouth-piece of a speaking-tube, and ever and again he would direct one side or other of his strange craft to "concentrate fire forward on gunners," or to "clear out trench about a hundred yards on our right front."

He was a young man, healthy enough but by no means sun-tanned, and of a type of feature and expression that prevails in His Majesty's Navy: alert, intelligent, quiet. He and his engineers and riflemen all went about their work, calm and reasonable men. They had none of that flapping strenuousness of the half-wit in a hurry, that excessive strain upon the blood-vessels, that hysteria of effort which is so frequently regarded as the proper state of mind for heroic deeds.

For the enemy these young engineers were defeating they felt a certain qualified pity and a quite unqualified contempt. They regarded these big, healthy men they were shooting down precisely as these same big, healthy men might regard some inferior kind of nigger. They despised them for making war; despised their bawling patriotisms and their emotionality profoundly; despised them, above all, for the petty cunning and the almost brutish want of imagination their method of fighting displayed. "If they *must* make war," these young men thought, "why in thunder don't they do it like sensible men?" They resented the assumption that their own side was too stupid to do anything more than play their enemy's game, that they were going to play this costly folly according to the rules of unimaginative men. They resented being forced to the trouble of making man-killing machinery; resented the alternative of having to massacre these people or endure their truculent yappings; resented the whole unfathomable imbecility of war.

Meanwhile, with something of the mechanical precision of a good clerk posting a ledger, the riflemen moved their knobs and pressed their buttons. . . .

The captain of Land Ironclad Number Three had halted on the crest close to his captured half-battery. His lined-up prisoners stood hard by and waited for the cyclists behind to come for them. He surveyed the victorious morning through his conning-tower.

He read the general's signals. "Five and Four are to keep among the guns to the left and prevent any attempt to recover them. Seven and Eleven and Twelve, stick to the guns you have got; Seven, get into position to command the guns taken by Three. Then we're to do something else, are we? Six and One, quicken up to about ten miles an hour and walk round behind that camp to the levels near the river—we shall bag the whole crowd of them," interjected the young man. "Ah, here we are! Two and Three, Eight and Nine, Thirteen and Fourteen, space out to a thousand yards, wait for the word, and then go slowly to cover the advance of the cyclist infantry against any charge of mounted

troops. That's all right. But where's Ten? Halloa! Ten
to repair and get movable as soon as possible. They've
broken up Ten!"

The discipline of the new war machines was business-like
rather than pedantic, and the head of the captain came down
out of the conning-tower to tell his men: "I say, you chaps
there. They've broken up Ten. Not badly, I think; but
anyhow, he's stuck."

But that still left thirteen of the monsters in action to
finish up the broken army.

The war correspondent stealing down his gully looked
back and saw them all lying along the crest and talking
fluttering congratulatory flags to one another. Their iron
sides were shining golden in the light of the rising sun.

§ 5

The private adventures of the war correspondent terminated in surrender about one o'clock in the afternoon,
and by that time he had stolen a horse, pitched off it, and
narrowly escaped being rolled upon; found the brute had
broken its leg, and shot it with his revolver. He had spent
some hours in the company of a squad of dispirited riflemen,
had quarrelled with them about topography at last, and gone
off by himself in a direction that should have brought him
to the banks of the river and didn't. Moreover, he had eaten
all his chocolate and found nothing in the whole world to
drink. Also, it had become extremely hot. From behind a
broken, but attractive, stone wall he had seen far away in
the distance the defender-horsemen trying to charge cyclists
in open order, with land ironclads outflanking them on
either side. He had discovered that cyclists could retreat
over open turf before horsemen with a sufficient margin of
speed to allow of frequent dismounts and much terribly effective sharp-shooting, and he had a sufficient persuasion that
those horsemen, having charged their hearts out, had halted
just beyond his range of vision and surrendered. He had been
urged to sudden activity by a forward movement of one of

those machines that had threatened to enfilade his wall. He had discovered a fearful blister on his heel.

He was now in a scrubby gravelly place, sitting down and meditating on his pocket-handkerchief, which had in some extraordinary way become in the last twenty-four hours extremely ambiguous in hue. "It's the whitest thing I've got," he said.

He had known all along that the enemy was east, west, and south of him, but when he heard land ironclads Number One and Six talking in their measured, deadly way not half a mile to the north he decided to make his own little unconditional peace without any further risks. He was for hoisting his white flag to a bush and taking up a position of modest obscurity near it until some one came along. He became aware of voices, clatter, and the distinctive noises of a body of horses, quite near, and he put his handkerchief in his pocket again and went to see what was going forward.

The sound of firing ceased, and then as he drew near he heard the deep sounds of many simple, coarse, but hearty and noble-hearted soldiers of the old school swearing with vigour.

He emerged from his scrub upon a big level plain, and far away a fringe of trees marked the banks of the river.

In the centre of the picture was a still-intact road bridge, and a big railway bridge a little to the right. Two land ironclads rested, with a general air of being long, harmless sheds, in a pose of anticipatory peacefulness right and left of the picture, completely commanding two miles and more of the river levels. Emerged and halted a few yards from the scrub was the remainder of the defender's cavalry, dusty, a little disordered and obviously annoyed, but still a very fine show of men. In the middle distance three or four men and horses were receiving medical attendance, and nearer a knot of officers regarded the distant novelties in mechanism with profound distaste. Every one was very distinctly aware of the twelve other ironclads, and of the multitude of townsmen soldiers, on bicycles or afoot, encumbered now by prisoners and captured war-gear, but otherwise thoroughly effective, who were sweeping like a great net in their rear.

"Checkmate," said the war correspondent, walking out

into the open. "But I surrender in the best of company. Twenty-four hours ago I thought war was impossible—and these beggars have captured the whole blessed army! Well! Well!" He thought of his talk with the young lieutenant. "If there's no end to the surprises of science, the civilised people have it, of course. As long as their science keeps going they will necessarily be ahead of open-country men. Still. . . ." He wondered for a space what might have happened to the young lieutenant.

The war correspondent was one of those inconsistent people who always want the beaten side to win. When he saw all these burly, sun-tanned horsemen, disarmed and lined up; when he saw their horses unskilfully led away by the singularly not equestrian cyclists to whom they had surrendered; when he saw these truncated Paladins watching this scandalous sight, he forgot altogether that he had called these men "cunning louts" and wished them beaten not four-and-twenty hours ago. A month ago he had seen that regiment in its pride going forth to war, and had been told of its terrible prowess, how it could charge in open order with each man firing from his saddle, and sweep before it anything else that ever came out to battle in any sort of order, foot or horse. And it had had to fight a few score of young men in atrociously unfair machines!

"Manhood *versus* Machinery" occurred to him as a suitable headline. Journalism curdles all one's mind to phrases.

He strolled as near the lined-up prisoners as the sentinels seemed disposed to permit, and surveyed them and compared their sturdy proportions with those of their lightly built captors.

"Smart degenerates," he muttered. "Anæmic cockneydom."

The surrendered officers came quite close to him presently, and he could hear the colonel's high-pitched tenor. The poor gentleman had spent three years of arduous toil upon the best material in the world perfecting that shooting from the saddle charge, and he was inquiring with phrases of blasphemy, natural in the circumstances, what one could be expected to do against this suitably consigned ironmongery.

"Guns," said some one.

"Big guns they can walk round. You can't shift big guns to keep pace with them, and little guns in the open they rush. I saw 'em rushed. You might do a surprise now and then—assassinate the brutes, perhaps——"

"You might make things like 'em."

"What? *More* ironmongery? Us? . . ."

"I'll call my article," meditated the war correspondent, "'Mankind *versus* Ironmongery,' and quote the old boy at the beginning."

And he was much too good a journalist to spoil his contrast by remarking that the half-dozen comparatively slender young men in blue pyjamas who were standing about their victorious land ironclad, drinking coffee and eating biscuits, had also in their eyes and carriage something not altogether degraded below the level of a man.

STORY THE FIFTH

The Beautiful Suit

THERE was once a little man whose mother made him a beautiful suit of clothes. It was green and gold, and woven so that I cannot describe how delicate and fine it was, and there was a tie of orange fluffiness that tied up under his chin. And the buttons in their newness shone like stars. He was proud and pleased by his suit beyond measure, and stood before the long looking-glass when first he put it on, so astonished and delighted with it that he could hardly turn himself away.

He wanted to wear it everywhere, and show it to all sorts of people. He thought over all the places he had ever visited, and all the scenes he had ever heard described, and tried to imagine what the feel of it would be if he were to go now to those scenes and places wearing his shining suit, and he wanted to go out forthwith into the long grass and the hot sunshine of the meadow wearing it. Just to wear it! But his mother told him "No." She told him he must take great care of his suit, for never would he have another nearly so fine; he must save it and save it, and only wear it on rare and great occasions. It was his wedding-suit, she said. And she took the buttons and twisted them up with tissue paper for fear their bright newness should be tarnished, and she tacked little guards over the cuffs and elbows, and wherever the suit was most likely to come to harm. He hated and resisted these things, but what could he do? And at last her warnings and persuasions had effect, and he consented to take off his beautiful suit and fold it into its

139

proper creases, and put it away. It was almost as though he
gave it up again. But he was always thinking of wearing
it, and of the supreme occasions when some day it might
be worn without the guards, without the tissue paper on
the buttons, utterly and delightfully, never caring, beautiful
beyond measure.

One night, when he was dreaming of it after his habit,
he dreamt he took the tissue paper from one of the buttons,
and found its brightness a little faded, and that distressed
him mightily in his dream. He polished the poor faded
button and polished it, and, if anything, it grew duller.
He woke up and lay awake, thinking of the brightness
slightly dulled, and wondering how he would feel if perhaps
when the great occasions (whatever it might be) should
arrive, one button should chance to be ever so little short
of its first glittering freshness, and for days and days that
thought remained with him distressingly. And when next
his mother let him wear his suit, he was tempted and nearly
gave way to the temptation just to fumble off a bit of
tissue paper and see if indeed the buttons were keeping as
bright as ever.

He went trimly along on his way to church, full of this
wild desire. For you must know his mother did, with
repeated and careful warnings, let him wear his suit at
times, on Sundays, for example, to and fro from church,
when there was no threatening of rain, no dust blowing,
nor anything to injure it, with its buttons covered and its
protections tacked upon it, and a sunshade in his hand to
shadow it if there seemed too strong a sunlight for its colours.
And always, after such occasions, he brushed it over and
folded it exquisitely as she had taught him, and put it away
again.

Now all these restrictions his mother set to the wearing
of his suit he obeyed, always he obeyed them, until one
strange night he woke up and saw the moonlight shining
outside his window. It seemed to him the moonlight was
not common moonlight, nor the night a common night, and
for a while he lay quite drowsily, with this odd persuasion
in his mind. Thought joined on to thought like things that
whisper warmly in the shadows. Then he sat up in his

little bed suddenly very alert, with his heart beating very fast, and a quiver in his body from top to toe. He had made up his mind. He knew that now he was going to wear his suit as it should be worn. He had no doubt in the matter. He was afraid, terribly afraid, but glad, glad.

He got out of his bed and stood for a moment by the window looking at the moonshine-flooded garden, and trembling at the thing he meant to do. The air was full of a minute clamour of crickets and murmurings, of the infinitesimal shoutings of little living things. He went very gently across the creaking boards, for fear that he might wake the sleeping house, to the big dark clothes-press wherein his beautiful suit lay folded, and he took it out garment by garment, and softly and very eagerly tore off its tissue-paper covering and its tacked protections until there it was, perfect and delightful as he had seen it when first his mother had given it to him—a long time it seemed ago. Not a button had tarnished, not a thread had faded on this dear suit of his; he was glad enough for weeping as in a noiseless hurry he put it on. And then back he went, soft and quick, to the window that looked out upon the garden, and stood there for a minute, shining in the moonlight, with his buttons twinkling like stars, before he got out on the sill, and, making as little of a rustling as he could, clambered down to the garden path below. He stood before his mother's house, and it was white and nearly as plain as by day, with every window-blind but his own shut like an eye that sleeps. The trees cast still shadows like intricate black lace upon the wall.

The garden in the moonlight was very different from the garden by day; moonshine was tangled in the hedges and stretched in phantom cobwebs from spray to spray. Every flower was gleaming white or crimson black, and the air was a-quiver with the thridding of small crickets and nightingales singing unseen in the depths of the trees.

There was no darkness in the world, but only warm, mysterious shadows, and all the leaves and spikes were edged and lined with iridescent jewels of dew. The night was warmer than any night had ever been; the heavens by some miracle at once vaster and nearer, and, spite of the great ivory-tinted moon that ruled the world, the sky was full of stars.

The little man did not shout nor sing for all his infinite gladness. He stood for a time like one awe-stricken, and then, with a queer small cry and holding out his arms, he ran out as if he would embrace at once the whole round immensity of the world. He did not follow the neat set paths that cut the garden squarely, but thrust across the beds and through the wet, tall, scented herbs, through the night-stock and the nicotine and the clusters of phantom white mallow flowers and through the thickets of southern-wood and lavender, and knee-deep across a wide space of mignonette. He came to the great hedge, and he thrust his way through it; and though the thorns of the brambles scored him deeply and tore threads from his wonderful suit, and though burrs and goose-grass and havers caught and clung to him, he did not care. He did not care, for he knew it was all part of the wearing for which he had longed. "I am glad I put on my suit," he said; "I am glad I wore my suit."

Beyond the hedge he came to the duck-pond, or at least to what was the duck-pond by day. But by night it was a great bowl of silver moonshine all noisy with singing frogs, of wonderful silver moonshine twisted and clotted with strange patternings, and the little man ran down into its waters between the thin black rushes, knee-deep and waist-deep and to his shoulders, smiting the water to black and shining wavelets with either hand, swaying and shivering wavelets, amidst which the stars were netted in the tangled reflections of the brooding trees upon the bank. He waded until he swam, and so he crossed the pond and came out upon the other side, trailing, as it seemed to him, not duckweed, but very silver in long, clinging, dripping masses. And up he went through the transfigured tangles of the willow-herb and the uncut seeding grasses of the farther bank. He came glad and breathless into the high-road. "I am glad," he said, "beyond measure, that I had clothes that fitted this occasion."

The high-road ran straight as an arrow flies, straight into the deep-blue pit of sky beneath the moon, a white and shin-

ing road between the singing nightingales, and along it he
went, running now and leaping, and now walking and
rejoicing, in the clothes his mother had made for him with
tireless, loving hands. The road was deep in dust, but that
for him was only soft whiteness; and as he went a great
dim moth came fluttering round his wet and shimmering and
hastening figure. At first he did not heed the moth, and
then he waved his hands at it, and made a sort of dance with
it as it circled round his head. "Soft moth!" he cried, "dear
moth! And wonderful night, wonderful night of the world!
Do you think my clothes are beautiful, dear moth? As
beautiful as your scales and all this silver vesture of the
earth and sky?"

And the moth circled closer and closer until at last its
velvet wings just brushed his lips. . . .

And next morning they found him dead, with his neck
broken, in the bottom of the stone pit, with his beautiful
clothes a little bloody, and foul and stained with the duckweed from the pond. But his face was a face of such happiness that, had you seen it, you would have understood indeed
how that he had died happy, never knowing that cool and
streaming silver for the duckweed in the pond.

STORY THE SIXTH

The Door in the Wall

§ 1

ONE confidential evening, not three months ago, Lionel Wallace told me this story of the Door in the Wall. And at the time I thought that so far as he was concerned it was a true story.

He told it me with such a direct simplicity of conviction that I could not do otherwise than believe in him. But in the morning, in my own flat, I woke to a different atmosphere; and as I lay in bed and recalled the things he had told me, stripped of the glamour of his earnest slow voice, denuded of the focused, shaded table light, the shadowy atmosphere that wrapped about him and me, and the pleasant bright things, the dessert and glasses and napery of the dinner we had shared, making them for the time a bright little world quite cut off from everyday realities, I saw it all as frankly incredible. "He was mystifying!" I said, and then: "How well he did it! . . . It isn't quite the thing I should have expected him, of all people, to do well."

Afterwards as I sat up in bed and sipped my morning tea, I found myself trying to account for the flavour of reality that perplexed me in his impossible reminiscences, by supposing they did in some way suggest, present, convey—I hardly know which word to use—experiences it was otherwise impossible to tell.

Well, I don't resort to that explanation now. I have got over my intervening doubts. I believe now, as I believed

144

at the moment of telling, that Wallace did to the very best
of his ability strip the truth of his secret for me. But
whether he himself saw, or only thought he saw, whether
he himself was the possessor of an inestimable privilege or
the victim of a fantastic dream, I cannot pretend to guess.
Even the facts of his death, which ended my doubts for
ever, throw no light on that.

That much the reader must judge for himself.

I forget now what chance comment or criticism of mine
moved so reticent a man to confide in me. He was, I think,
defending himself against an imputation of slackness and
unreliability I had made in relation to a great public
movement, in which he had disappointed me. But he
plunged suddenly. "I have," he said, "a preoccupation——

"I know," he went on, after a pause, "I have been negli-
gent. The fact is—it isn't a case of ghosts or apparitions—
but—it's an odd thing to tell of, Redmond—I am haunted.
I am haunted by something—that rather takes the light out
of things, that fills me with longings . . ."

He paused, checked by that English shyness that so often
overcomes us when we would speak of moving or grave
or beautiful things. "You were at Saint Athelstan's all
through," he said, and for a moment that seemed to me
quite irrelevant. "Well"—and he paused. Then very halt-
ingly at first, but afterwards more easily, he began to tell of
the thing that was hidden in his life, the haunting memory
of a beauty and a happiness that filled his heart with insatiable longings, that made all the interests and spectacle
of worldly life seem dull and tedious and vain to him.

Now that I have the clue to it, the thing seems written
visibly in his face. I have a photograph in which that look
of detachment has been caught and intensified. It reminds
me of what a woman once said of him—a woman who had
loved him greatly. "Suddenly," she said, "the interest goes
out of him. He forgets you. He doesn't care a rap for you
—under his very nose . . . "

Yet the interest was not always out of him, and when he
was holding his attention to a thing Wallace could contrive
to be an extremely successful man. His career, indeed, is
set with successes. He left me behind him long ago; he soared

up over my head, and cut a figure in the world that I couldn't cut—anyhow. He was still a year short of forty, and they say now that he would have been in office and very probably in the new Cabinet if he had lived. At school he always beat me without effort—as it were by nature. We were at school together at Saint Althelstan's College in West Kensington for almost all our school-time. He came into the school as my co-equal, but he left far above me, in a blaze of scholarships and brilliant performance. Yet I think I made a fair average running. And it was at school I heard first of the "Door in the Wall"—that I was to hear of a second time only a month before his death.

To him at least the Door in the Wall was a real door, leading through a real wall to immortal realities. Of that I am now quite assured.

And it came into his life quite early, when he was a little fellow between five and six. I remember how, as he sat making his confession to me with a slow gravity, he reasoned and reckoned the date of it. "There was," he said, "a crimson Virginia creeper in it—all one bright uniform crimson, in a clear amber sunshine against a white wall. That came into the impression somehow, though I don't clearly remember how, and there were horse-chesnut leaves upon the clean pavement outside the green door. They were blotched yellow and green, you know, not brown nor dirty, so that they must have been new fallen. I take it that means October. I look out for horse-chestnut leaves every year and I ought to know.

"If I'm right in that, I was about five years and four months old."

He was, he said, rather a precocious little boy—he learned to talk at an abnormally early age, and he was so sane and "old-fashioned," as people say, that he was permitted an amount of initiative that most children scarcely attain by seven or eight. His mother died when he was two, and he was under the less vigilant and authoritative care of a nursery governess. His father was a stern, preoccupied lawyer, who gave him little attention and expected great things of him. For all his brightness he found life grey and dull, I think. And one day he wandered.

He could not recall the particular neglect that enabled him to get away, nor the course he took among the West Kensington roads. All that had faded among the incurable blurs of memory. But the white wall and the green door stood out quite distinctly.

As his memory of that childish experience ran, he did at the very first sight of that door experience a peculiar emotion, an attraction, a desire to get to the door and open it and walk in. And at the same time he had the clearest conviction that either it was unwise or it was wrong of him—he could not tell which—to yield to this attraction. He insisted upon it as a curious thing that he knew from the very beginning—unless memory has played him the queerest trick—that the door was unfastened, and that he could go in as he chose.

I seem to see the figure of that little boy, drawn and repelled. And it was very clear in his mind, too, though why it should be so was never explained, that his father would be very angry if he went in through that door.

Wallace described all these moments of hesitation to me with the utmost particularity. He went right past the door, and then, with his hands in his pockets and making an infantile attempt to whistle, strolled right along beyond the end of the wall. There he recalls a number of mean dirty shops, and particularly that of a plumber and decorator with a dusty disorder of earthenware pipes, sheet lead, ball taps, pattern books of wall paper, and tins of enamel. He stood pretending to examine these things, and *coveting*, passionately desiring, the green door.

Then, he said, he had a gust of emotion. He made a run for it, lest hesitation should grip him again; he went plump with outstretched hand through the green door and let it slam behind him. And so, in a trice, he came into the garden that has haunted all his life.

It was very difficult for Wallace to give me his full sense of that garden into which he came.

There was something in the very air of it that exhilarated, that gave one a sense of lightness and good happening and well-being; there was something in the sight of it that made all its colour clean and perfect and subtly luminous. In the instant of coming into it one was exquisitely glad—as only in rare moments, and when one is young and joyful one can be glad in this world. And everything was beautiful there. . . .

Wallace mused before he went on telling me. "You see," he said, with the doubtful inflection of a man who pauses at incredible things, "there were two great panthers there. . . . Yes, spotted panthers. And I was not afraid. There was a long wide path with marble-edged flower borders on either side, and these two huge velvety beasts were playing there with a ball. One looked up and came towards me, a little curious as it seemed. It came right up to me, rubbed its soft round ear very gently against the small hand I held out, and purred. It was, I tell you, an enchanted garden. I know. And the size? Oh! it stretched far and wide, this way and that. I believe there were hills far away. Heaven knows where West Kensington had suddenly got to. And somehow it was just like coming home.

"You know, in the very moment the door swung to behind me, I forgot the road with its fallen chestnut leaves, its cabs and tradesmen's carts, I forgot the sort of gravitational pull back to the discipline and obedience of home, I forgot all hesitations and fear, forgot discretion, forgot all the intimate realities of this life. I became in a moment a very glad and wonder-happy little boy—in another world. It was a world with a different quality, a warmer, more penetrating and mellower light, with a faint clear gladness in its air, and wisps of sun-touched cloud in the blueness of its sky. And before me ran this long wide path, invitingly, with weedless beds on either side, rich with untended flowers, and these two great panthers. I put my little hands fearlessly on their soft fur, and caressed their round ears and the sensitive corners under their ears, and played with them, and it was as though they welcomed me home. There was a keen sense of homecoming in my mind, and when presently a tall, fair girl appeared in the pathway and came to meet me, smiling, and said 'Well?' to me, and lifted me and kissed me, and put me down and led me by the hand, there was no amazement, but only an impression of delightful rightness, of being reminded of happy things that had in some strange way been overlooked. There were broad red steps, I remember, that came into view between spikes of delphinium, and up these we went to a great avenue between very old and shady dark trees. All down this avenue, you know, between the red chapped stems, were marble seats of honour and statuary, and very tame and friendly white doves.

"Along this cool avenue my girl-friend led me, looking down—I recall the pleasant lines, the finely-modelled chin of her sweet kind face—asking me questions in a soft, agreeable voice, and telling me things, pleasant things I know, though what they were I was never able to recall. . . . Presently a Capuchin monkey, very clean, with a fur of ruddy brown and kindly hazel eyes, came down a tree to us and ran beside me, looking up at me and grinning, and presently leaped to my shoulder. So we two went on our way in great happiness."

He paused.

"Go on," I said.

"I remember little things. We passed an old man musing among laurels, I remember, and a place gay with paroquets, and came through a broad shaded colonnade to a spacious cool palace, full of pleasant fountains, full of beautiful things, full of the quality and promise of heart's desire. And there were many things and many people, some that still seem to stand out clearly and some that are vaguer; but all these people were beautiful and kind. In some way —I don't know how—it was conveyed to me that they all were kind to me, glad to have me there, and filling me with gladness by their gestures, by the touch of their hands, by the welcome and love in their eyes. Yes——"

He mused for a while. "Playmates I found there. That was very much to me, because I was a lonely little boy. They played delightful games in a grass-covered court where there was a sun-dial set about with flowers. And as one played one loved. . . .

"But—it's odd—there's a gap in my memory. I don't remember the games we played. I never remembered. Afterwards, as a child, I spent long hours trying, even with tears, to recall the form of that happiness. I wanted to

play it all over again—in my nursery—by myself. No! All
I remember is the happiness and two dear playfellows who
were most with me. . . . Then presently came a sombre
dark woman, with a grave, pale face and dreamy eyes, a
sombre woman, wearing a soft long robe of pale purple,
who carried a book, and beckoned and took me aside with
her into a gallery above a hall—though my playmates were
loth to have me go, and ceased their game and stood watching as I was carried away. 'Come back to us!' they cried.
'Come back to us soon!' I looked up at her face, but she
heeded them not at all. Her face was very gentle and grave.
She took me to a seat in the gallery, and I stood beside her,
ready to look at her book as she opened it upon her knee.
The pages fell open. She pointed, and I looked, marvelling, for in the living pages of that book I saw myself;
it was a story about myself, and in it were all the things that
had happened to me since ever I was born. . . .

"It was wonderful to me, because the pages of that book
were not pictures, you understand, but realities."

Wallace paused gravely—looked at me doubtfully.

"Go on," I said. "I understand."

"They were realities—yes, they must have been; people
moved and things came and went in them; my dear mother,
whom I had near forgotten; then my father, stern and upright, the servants, the nursery, all the familiar things of
home. Then the front door and the busy streets, with traffic
to and fro. I looked and marvelled, and looked half doubtfully again into the woman's face and turned the pages
over, skipping this and that, to see more of this book and
more, and so at last I came to myself hovering and hesitating
outside the green door in the long white wall, and felt
again the conflict and the fear.

" 'And next?' I cried, and would have turned on, but the
cool hand of the grave woman delayed me.

" 'Next?' I insisted, and struggled gently with her hand,
pulling up her fingers with all my childish strength, and as
she yielded and the page came over she bent down upon
me like a shadow and kissed my brow.

"But the page did not show the enchanted garden, nor
the panthers, nor the girl who had led me by the hand, nor

the playfellows who had been so loth to let me go. It showed a long grey street in West Kensington, in that chill hour of afternoon before the lamps are lit; and I was there, a wretched little figure, weeping aloud, for all that I could do to restrain myself, and I was weeping because I could not return to my dear playfellows who had called after me, 'Come back to us! Come back to us soon!' I was there. This was no page in a book, but harsh reality; that enchanted place and the restraining hand of the grave mother at whose knee I stood had gone—whither had they gone?"

He halted again, and remained for a time staring into the fire.

"Oh! the woefulness of that return!" he murmured.

"Well?" I said, after a minute or so.

"Poor little wretch I was!—brought back to this grey world again! As I realised the fullness of what had happened to me, I gave way to quite ungovernable grief. And the shame and humiliation of that public weeping and my disgraceful home-coming remain with me still. I see again the benevolent-looking old gentleman in gold spectacles who stopped and spoke to me—prodding me first with his umbrella. 'Poor little chap,' said he; 'and are you lost then?'—and me a London boy of five and more! And he must needs bring in a kindly young policeman and make a crowd of me, and so march me home. Sobbing, conspicuous, and frightened, I came back from the enchanted garden to the steps of my father's house.

"That is as well as I can remember my vision of that garden—the garden that haunts me still. Of course, I can convey nothing of that indescribable quality of translucent unreality, that difference from the common things of experience that hung about it all; but that—that is what happened. If it was a dream, I am sure it was a day-time and altogether extraordinary dream. . . . H'm!—naturally there followed a terrible questioning, by my aunt, my father, the nurse, the governess—everyone. . . .

"I tried to tell them, and my father gave me my first thrashing for telling lies. When afterwards I tried to tell my aunt, she punished me again for my wicked persistence. Then, as I said, everyone was forbidden to listen to me, to hear a word about it. Even my fairy-tale books were taken away from me for a time—because I was too 'imaginative.' Eh? Yes, they did that! My father belonged to the old school. . . . And my story was driven back upon myself. I whispered it to my pillow—my pillow that was often damp and salt to my whispering lips with childish tears. And I added always to my official and less fervent prayers this one heartfelt request: 'Please God I may dream of the garden. Oh! take me back to my garden.' Take me back to my garden! I dreamt often of the garden. I may have added to it, I may have changed it; I do not know. . . . All this, you understand, is an attempt to reconstruct from fragmentary memories a very early experience. Between that and the other consecutive memories of my boyhood there is a gulf. A time came when it seemed impossible I should ever speak of that wonder glimpse again."

I asked an obvious question.

"No," he said. "I don't remember that I ever attempted to find my way back to the garden in those early years. This seems odd to me now, but I think that very probably a closer watch was kept on my movements after this misadventure to prevent my going astray. No, it wasn't till you knew me that I tried for the garden again. And I believe there was a period—incredible as it seems now—when I forgot the garden altogether—when I was about eight or nine it may have been. Do you remember me as a kid at Saint Althelstan's?"

"Rather!"

"I didn't show any signs, did I, in those days of having a secret dream?"

§ 2

He looked up with a sudden smile.

"Did you ever play North-West Passage with me? . . . No, of course you didn't come my way!"

"It was the sort of game," he went on, "that every imaginative child plays all day. The idea was the discovery of a North-West Passage to school. The way to school was plain enough; the game consisted in finding some way that wasn't plain, starting off ten minutes early in some almost hopeless direction, and working my way round through unaccustomed streets to my goal. And one day I got entangled among some rather low-class streets on the other side of Campden Hill, and I began to think that for once the game would be against me and that I should get to school late. I tried rather desperately a street that seemed a *cul-de-sac*, and found a passage at the end. I hurried through that with renewed hope. 'I shall do it yet,' I said, and passed a row of frowsy little shops that were inexplicably familiar to me, and behold! there was my long white wall and the green door that led to the enchanted garden!

"The thing whacked upon me suddenly. Then, after all, that garden, that wonderful garden, wasn't a dream!"

He paused.

"I suppose my second experience with the green door marks the world of difference there is between the busy life of a schoolboy and the infinite leisure of a child. Anyhow, this second time I didn't for a moment think of going in straight away. You see——. For one thing, my mind was full of the idea of getting to school in time—set on not breaking my record for punctuality. I must surely have felt *some* little desire at least to try the door—yes. I must have felt that. . . . But I seem to remember the attraction of the door mainly as another obstacle to my overmastering determination to get to school. I was immensely interested by this discovery I had made, of course—I went on with my mind full of it—but I went on. It didn't check me. I ran past, tugging out my watch, found I had ten minutes still to spare, and then I was going downhill into familiar surroundings. I got to school, breathless, it is true, and wet with perspiration, but in time. I can remember hanging up my coat and hat. . . . Went right by it and left it behind me. Odd, eh?"

He looked at me thoughtfully. "Of course I didn't know then that it wouldn't always be there. Schoolboys have limited imaginations. I suppose I thought it was an awfully jolly thing to have it there, to know my way back to it; but there was the school tugging at me. I expect I was a good deal distraught and inattentive that morning, recalling what I could of the beautiful strange people I should presently see again. Oddly enough I had no doubt in my mind that they would be glad to see me. . . . Yes, I must have thought of the garden that morning just as a jolly sort of place to which one might resort in the interludes of a strenuous scholastic career.

"I didn't go that day at all. The next day was a half-holiday, and that may have weighed with me. Perhaps, too, my state of inattention brought down impositions upon me, and docked the margin of time necessary for the *détour*. I don't know. What I do know is that in the meantime the enchanted garden was so much upon my mind that I could not keep it to myself.

"I told—what was his name?—a ferrety-looking youngster we used to call Squiff."

"Young Hopkins," said I.

"Hopkins it was. I did not like telling him. I had a feeling that in some way it was against the rules to tell him, but I did. He was walking part of the way home with me; he was talkative, and if we had not talked about the enchanted garden we should have talked of something else, and it was intolerable to me to think about any other subject. So I blabbed.

"Well, he told my secret. The next day in the play interval I found myself surrounded by half a dozen bigger boys, half teasing, and wholly curious to hear more of the enchanted garden. There was that big Fawcett—you remember him?—and Carnaby and Morley Reynolds. You weren't there by any chance? No, I think I should have remembered if you were. . . .

"A boy is a creature of odd feelings. I was, I really believe, in spite of my secret self-disgust, a little flattered to have the attention of these big fellows. I remember particularly a moment of pleasure caused by the praise of Crawshaw—you remember Crawshaw major, the son of Crawshaw the composer?—who said it was the best lie he had ever heard. But at the same time there was a really painful undertow of shame at telling what I felt was indeed a sacred secret. That beast Fawcett made a joke about the girl in green——"

Wallace's voice sank with the keen memory of that shame. "I pretended not to hear," he said. "Well, then Carnaby suddenly called me a young liar, and disputed with me when I said the thing was true. I said I knew where to find the green door, could lead them all there in ten minutes. Carnaby became outrageously virtuous, and said I'd have to—and bear out my words or suffer. Did you ever have Carnaby twist your arm? Then perhaps you'll understand how it went with me. I swore my story was true. There was nobody in the school then to save a chap from Carnaby, though Crawshaw put in a word or so. Carnaby had got his game. I grew excited and red-eared, and a little frightened. I behaved altogether like a silly little chap, and the outcome of it all was that instead of starting alone for my enchanted garden, I led the way presently—cheeks flushed, ears hot, eyes smarting, and my soul one burning misery and shame—for a party of six mocking, curious, and threatening schoolfellows.

"We never found the white wall and the green door. . . ."

"You mean——?"

"I mean I couldn't find it. I would have found it if I could.

"And afterwards when I could go alone I couldn't find it. I never found it. I seem now to have been always looking for it through my school-boy days, but I never came upon it—never."

"Did the fellows—make it disagreeable?"

"Beastly. . . . Carnaby held a council over me for wanton lying. I remember how I sneaked home and upstairs to hide the marks of my blubbering. But when I cried myself to sleep at last it wasn't for Carnaby, but for the garden, for the beautiful afternoon I had hoped for, for the sweet friendly women and the waiting playfellows, and the game I had hoped to learn again, that beautiful forgotten game. . . .

"I believed firmly that if I had not told—— . . . I had bad times after that—crying at night and wool-gathering by day. For two terms I slacked and had bad reports. Do you remember? Of course you would! It was *you*—your beating me in mathematics that brought me back to the grind again."

§ 3

For a time my friend stared silently into the red heart of the fire. Then he said: "I never saw it again until I was seventeen.

"It leaped upon me for the third time—as I was driving to Paddington on my way to Oxford and a scholarship. I had just one momentary glimpse. I was leaning over the apron of my hanson smoking a cigarette, and no doubt thinking myself no end of a man of the world, and suddenly there was the door, the wall, the dear sense of unforgettable and still attainable things.

"We clattered by—I too taken by surprise to stop my cab until we were well past and round a corner. Then I had a queer moment, a double and divergent movement of my will: I tapped the little door in the roof of the cab, and brought my arm down to pull out my watch. 'Yes, sir!' said the cabman, smartly. 'Er—well—it's nothing,' I cried. '*My* mistake! We haven't much time! Go on!' And he went on. . . .

"I got my scholarship. And the night after I was told of that I sat over my fire in my little upper room, my study, in my father's house, with his praise—his rare praise—and his sound counsels ringing in my ears, and I smoked my favourite pipe—the formidable bulldog of adolescence—and thought of that door in the long white wall. 'If I had stopped,' I thought, 'I should have missed my scholarship, I should have missed Oxford—muddled all the fine career before me! I begin to see things better!' I fell musing deeply, but I did not doubt then this career of mine was a thing that merited sacrifice.

"Those dear friends and that clear atmosphere seemed very sweet to me, very fine but remote. My grip was fixing now upon the world. I saw another door opening—the door of my career."

He stared again into the fire. Its red light picked out a

stubborn strength in his face for just one flickering moment, and then it vanished again.

"Well," he said and sighed, "I have served that career. I have done—much work, much hard work. But I have dreamt of the enchanted garden a thousand dreams, and seen its door, or at least glimpsed its door, four times since then. Yes—four times. For a while this world was so bright and interesting, seemed so full of meaning and opportunity, that the half-effaced charm of the garden was by comparison gentle and remote. Who wants to pat panthers on the way to dinner with pretty women and distinguished men? I came down to London from Oxford, a man of bold promise that I have done something to redeem. Something —and yet there have been disappointments. . . .

"Twice I have been in love—I will not dwell on that—but once, as I went to someone who, I knew, doubted whether I dared to come, I took a short cut at a venture through an unfrequented road near Earl's Court, and so happened on a white wall and a familiar green door. 'Odd!' said I to myself, 'but I thought this place was on Campden Hill. It's the place I never could find somehow—like counting Stonehenge—the place of that queer daydream of mine.' And I went by it intent upon my purpose. It had no appeal to me that afternoon.

"I had just a moment's impulse to try the door, three steps aside were needed at the most—though I was sure enough in my heart that it would open to me—and then I thought that doing so might delay me on the way to that appointment in which my honour was involved. Afterwards I was sorry for my punctuality—I might at least have peeped in and waved a hand to those panthers, but I knew enough by this time not to seek again belatedly that which is not found by seeking. Yes, that time made me very sorry. . . .

"Years of hard work after that, and never a sight of the door. It's only recently it has come back to me. With it there has come a sense as though some thin tarnish had spread itself over my world. I began to think of it as a sorrowful and bitter thing that I should never see that door again. Perhaps I was suffering a little from overwork— perhaps it was what I've heard spoken of as the feeling of

forty. I don't know. But certainly the keen brightness that
makes effort easy has gone out of things recently, and that
just at a time—with all these new political developments—
when I ought to be working. Odd, isn't it? But I do begin
to find life toilsome, its rewards, as I come near them, cheap.
I began a little while ago to want the garden quite badly.
Yes—and I've seen it three times."

"The garden?"

"No—the door! And I haven't gone in!"

He leaned over the table to me, with an enormous sorrow
in his voice as he spoke. "Thrice I have had my chance—
thrice! If ever that door offers itself to me again, I swore,
I will go in, out of this dust and heat, out of this dry glitter
of vanity, out of these toilsome futilities. I will go and
never return. This time I will stay. . . . I swore it, and
when the time came—*I didn't go.*

"Three times in one year have I passed that door and
failed to enter. Three times in the last year.

"The first time was on the night of the snatch division
on the Tenants' Redemption Bill, on which the Government
was saved by a majority of three. You remember? No one
on our side—perhaps very few on the opposite side—expected
the end that night. Then the debate collapsed like egg-
shells. I and Hotchkiss were dining with his cousin at
Brentford; we were both unpaired, and we were called up
by telephone, and set off at once in his cousin's motor. We
got in barely in time, and on the way we passed my wall
and door—livid in the moonlight, blotched with hot yellow
as the glare of our lamps lit it, but unmistakable. 'My God!'
cried I. 'What?' said Hotchkiss. 'Nothing!' I answered,
and the moment passed.

" 'I've made a great sacrifice,' I told the whip as I got in.
'They all have,' he said, and hurried by.

"I do not see how I could have done otherwise then. And
the next occasion was as I rushed to my father's bedside to
bid that stern old man farewell. Then, too, the claims of
life were imperative. But the third time was different; it
happened a week ago. It fills me with hot remorse to recall
it. I was with Gurker and Ralphs—it's no secret now, you
know, that I've had my talk with Gurker. We had been

dining at Frobisher's, and the talk had become intimate between us. The question of my place in the reconstructed Ministry lay always just over the boundary of the discussion. Yes—yes. That's all settled. It needn't be talked about yet, but there's no reason to keep a secret from you . . . Yes —thanks! thanks! But let me tell you my story.

"Then, on that night things were very much in the air. My position was a very delicate one. I was keenly anxious to get some definite word from Gurker, but was hampered by Ralphs' presence. I was using the best power of my brain to keep that light and careless talk not too obviously directed to the point that concerned me. I had to. Ralphs' behaviour since has more than justified my caution. . . . Ralphs, I knew, would leave us beyond the Kensington High Street, and then I could surprise Gurker by a sudden frankness. One has sometimes to resort to these little devices. . . . And then it was that in the margin of my field of vision I became aware once more of the white wall, the green door before us down the road.

"We passed it talking. I passed it. I can still see the shadow of Gurker's marked profile, his opera hat tilted forward over his prominent nose, the many folds of his neck wrap going before my shadow and Ralphs' as we sauntered past.

"I passed within twenty inches of the door. 'If I say good-night to them, and go in,' I asked myself, 'what will happen?' And I was all a-tingle for that word with Gurker.

"I could not answer that question in the tangle of my other problems. 'They will think me mad,' I thought. 'And suppose I vanish now!—Amazing disappearance of a prominent politician!' That weighed with me. A thousand inconceivable petty worldlinesses weighed with me in that crisis."

Then he turned on me with a sorrowful smile, and, speaking slowly, "Here I am!" he said.

"Here I am!" he repeated, "and my chance has gone from me. Three times in one year the door has been offered me— the door that goes into peace, into delight, into a beauty beyond dreaming, a kindness no man on earth can know. And I have rejected it, Redmond, and it has gone——"

"How do you know?"

"I know. I know. I am left now to work it out, to stick to the tasks that held me so strongly when my moments came. You say I have success—this vulgar, tawdry, irksome, envied thing. I have it." He had a walnut in his big hand. "If that was my success," he said, and crushed it, and held it out for me to see.

"Let me tell you something, Redmond. This loss is destroying me. For two months, for ten weeks nearly now, I have done no work at all, except the most necessary and urgent duties. My soul is full of inappeasable regrets. At nights—when it is less likely I shall be recognised—I go out. I wander. Yes. I wonder what people would think of that if they knew. A Cabinet Minister, the responsible head of that most vital of all departments, wandering alone—grieving —sometimes near audibly lamenting—for a door, for a garden!"

§ 4

I can see now his rather pallid face, and the unfamiliar sombre fire that had come into his eyes. I see him very vividly to-night. I sit recalling his words, his tones, and last evening's *Westminster Gazette* still lies on my sofa, containing the notice of his death. At lunch to-day the club was busy with his death. We talked of nothing else.

They found his body very early yesterday morning in a deep excavation near East Kensington Station. It is one of two shafts that have been made in connection with an extension of the railway southward. It is protected from the intrusion of the public by a hoarding upon the high road, in which a small doorway has been cut for the convenience of some of the workmen who live in that direction. The doorway was left unfastened through a misunderstanding between two gangers, and through it he made his way.

My mind is darkened with questions and riddles.

It would seem he walked all the way from the House that

night—he has frequently walked home during the past Session—and so it is I figure his dark form coming along the late and empty streets, wrapped up, intent. And then did the pale electric lights near the station cheat the rough planking into a semblance of white? Did that fatal unfastened door awaken some memory?

Was there, after all, ever any green door in the wall at all?

I do not know. I have told his story as he told it to me. There are times when I believe that Wallace was no more than the victim of the coincidence between a rare but not unprecedented type of hallucination and a careless trap, but that indeed is not my profoundest belief. You may think me superstitious, if you will, and foolish; but, indeed, I am more than half convinced that he had, in truth, an abnormal gift, and a sense, something—I know not what—that in the guise of wall and door offered him an outlet, a secret and peculiar passage of escape into another and altogether more beautiful world. At any rate, you will say, it betrayed him in the end. But did it betray him? There you touch the inmost mystery of these dreamers, these men of vision and the imagination. We see our world fair and common, the hoarding and the pit. By our daylight standard he walked out of security into darkness, danger, and death.

But did he see like that?

STORY THE SEVENTH

The Pearl of Love

THE pearl is lovelier than the most brilliant of crystalline stones, the moralist declares, because it is made through the suffering of a living creature. About that I can say nothing because I feel none of the fascination of pearls. Their cloudy lustre moves me not at all. Nor can I decide for myself upon that age-long dispute whether The Pearl of Love is the cruellest of stories or only a gracious fable of the immortality of beauty.

Both the story and the controversy will be familiar to students of mediæval Persian prose. The story is a short one, though the commentary upon it is a respectable part of the literature of that period. They have treated it as a poetic invention and they have treated it as an allegory meaning this, that, or the other thing. Theologians have had their copious way with it, dealing with it particularly as concerning the restoration of the body after death, and it has been greatly used as a parable by those who write about æsthetics. And many have held it to be the statement of a fact, simply and baldly true.

The story is laid in North India, which is the most fruitful soil for sublime love stories of all the lands in the world. It was in a country of sunshine and lakes and rich forests and hills and fertile valleys; and far away the great mountains hung in the sky, peaks, crests, and ridges of inaccessible and eternal snow. There was a young prince, lord of all the land; and he found a maiden of indescribable beauty and delightfulness and he made her his queen and laid his heart at her feet. Love was theirs, full of joys and sweetness, full of hope, exquisite, brave and marvellous love, beyond anything you have ever dreamt of love. It was theirs for a year and a part of a year, and then suddenly, because of some venomous sting that came to her in a thicket, she died.

She died and for a while the prince was utterly prostrated. He was silent and motionless with grief. They feared he might kill himself, and he had neither sons nor brothers to succeed him. For two days and nights he lay upon his face, fasting, across the foot of the couch which bore her calm and lovely body. Then he arose and ate, and went about very quietly like one who has taken a great resolution. He caused her body to be put in a coffin of lead mixed with silver, and for that he had an outer coffin made of the most precious and scented woods wrought with gold, and about that there was to be a sarcophagus of alabaster, inlaid with precious stones. And while these things were being done he spent his time for the most part by the pools and in the garden-houses and pavilions and groves and in those chambers in the palace where they two had been most together, brooding upon her loveliness. He did not rend his garments nor defile himself with ashes and sackcloth as the custom was, for his love was too great for such extravagances. At last he came forth again among his councillors and before the people, and told them what he had a mind to do.

He said he could never more touch woman, he could never more think of them, and so he would find a seemly youth to adopt for his heir and train him to his task, and that he would do his princely duties as became him; but that for the rest of it, he would give himself with all his power and all his strength and all his wealth, all that he could command, to make a monument worthy of his incomparable, dear, lost mistress. A building it should be of perfect grace and beauty, more marvellous than any other building had ever been or could ever be, so that to the end of time it should be a wonder, and men would treasure it and speak of it and desire to see it and come from all the lands of the earth to visit and recall the name and the memory of

his queen. And this building he said was to be called the
Pearl of Love.

And this his councillors and people permitted him to do,
and so he did.

Year followed year and all the years he devoted himself
to building and adorning the Pearl of Love. A great foundation was hewn out of the living rock in a place whence one
seemed to be looking at the snowy wilderness of the great
mountain across the valley of the world. Villages and hills
there were, a winding river, and very far away three great
cities. Here they put the sarcophagus of alabaster beneath
a pavilion of cunning workmanship; and about it there were
set pillars of strange and lovely stone and wrought and
fretted walls, and a great casket of masonry bearing a dome
and pinnacles and cupolas, as exquisite as a jewel. At first
the design of the Pearl of Love was less bold and subtle
than it became later. At first it was smaller and more
wrought and encrusted; there were many pierced screens and
delicate clusters of rosy hued pillars, and the sarcophagus lay
like a child that sleeps among flowers. The first dome was
covered with green tiles, framed and held together by silver,
but this was taken away again because it seemed close, because it did not soar grandly enough for the broadening
imagination of the prince.

For by this time he was no longer the graceful youth
who had loved the girl queen. He was now a man, grave
and intent, wholly set upon the building of the Pearl of
Love. With every year of effort he had learnt new possibilities in arch and wall and buttress; he had acquired
greater power over the material he had to use and he had
learnt of a hundred stones and hues and effects that he could
never have thought of in the beginning. His sense of colour
had grown finer and colder; he cared no more for the
enamelled gold-lined brightness that had pleased him first,
the brightness of an illuminated missal; he sought now for
blue colourings like the sky and for the subtle hues of great
distances, for recondite shadows and sudden broad floods of
purple opalescence and for grandeur and space. He wearied
altogether of carvings and pictures and inlaid ornamenta-

tion and all the little careful work of men. "Those were pretty things," he said of his earlier decorations; and had them put aside into subordinate buildings where they would not hamper his main design. Greater and greater grew his artistry. With awe and amazement people saw the Pearl of Love sweeping up from its first beginnings to a superhuman breadth and height and magnificence. They did not know clearly what they had expected, but never had they expected so sublime a thing as this. "Wonderful are the miracles," they whispered, "that love can do," and all the women in the world, whatever other loves they had, loved the prince for the splendour of his devotion.

Through the middle of the building ran a great aisle, a vista, that the prince came to care for more and more. From the inner entrance of the building he looked along the length of an immense pillared gallery and across the central area from which the rose-hued columns had long since vanished, over the top of the pavilion under which lay the sarcophagus, through a marvellously designed opening, to the snowy wildernesses of the great mountain, the lord of all mountains, two hundred miles away. The pillars and arches and buttresses and galleries soared and floated on either side, perfect yet unobtrusive, like great archangels waiting in the shadows about the presence of God. When men saw that austere beauty for the first time they were exalted, and then they shivered and their hearts bowed down. Very often would the prince come to stand there and look at that vista, deeply moved and not yet fully satisfied. The Pearl of Love had still something for him to do, he felt, before his task was done. Always he would order some little alteration to be made or some recent alteration to be put back again. And one day he said that the sarcophagus would be clearer and simpler without the pavilion; and after regarding it very steadfastly for a long time, he had the pavilion dismantled and removed.

The next day he came and said nothing, and the next day and the next. Then for two days he stayed away altogether. Then he returned, bringing with him an architect and two master craftsmen and a small retinue.

All looked, standing together silently in a little group, amidst the serene vastness of their achievement. No trace of toil remained in its perfection. It was as if the God of nature's beauty had taken over their offspring to himself.

Only one thing there was to mar the absolute harmony. There was a certain disproportion about the sarcophagus. It had never been enlarged, and indeed how could it have been enlarged since the early days? It challenged the eye; it nicked the streaming lines. In that sarcophagus was the casket of lead and silver, and in the casket of lead and silver was the queen, the dear immortal cause of all this beauty. But now that sarcophagus seemed no more than a little dark oblong that lay incongruously in the great vista of the Pearl of Love. It was as if someone had dropped a small valise upon the crystal sea of heaven.

Long the prince mused, but no one knew the thoughts that passed through his mind.

At last he spoke. He pointed.

"Take that thing away," he said.

STORY THE EIGHTH

The Country of the Blind

THREE hundred miles and more from Chimborazo, one hundred from the snows of Cotopaxi, in the wildest wastes of Ecuador's Andes, there lies that mysterious mountain valley, cut off from the world of men, the Country of the Blind. Long years ago that valley lay so far open to the world that men might come at last through frightful gorges and over an icy pass into its equable meadows; and thither indeed men came, a family or so of Peruvian half-breeds fleeing from the lust and tyranny of an evil Spanish ruler. Then came the stupendous outbreak of Mindobamba, when it was night in Quito for seventeen days, and the water was boiling at Yaguachi and all the fish floating dying even as far as Guayaquil; everywhere along the Pacific slopes there were landslips and swift thawings and sudden floods, and one whole side of the old Arauca crest slipped and came down in thunder, and cut off the Country of the Blind for ever from the exploring feet of men. But one of these early settlers had chanced to be on the hither side of the gorges when the world had so terribly shaken itself, and he perforce had to forget his wife and his child and all the friends and possessions he had left up there, and start life over again in the lower world. He started it again but ill, blindness overtook him, and he died of punishment in the mines; but the story he told begot a legend that lingers along the length of the Cordilleras of the Andes to this day.

He told of his reason for venturing back from that fastness, into which he had first been carried lashed to a llama, beside a vast bale of gear, when he was a child. The valley, he said, had in it all that the heart of man could desire—sweet water, pasture, and even climate, slopes of rich brown soil with tangles of a shrub that bore an excellent fruit, and on one side great hanging forests of pine that held the avalanches high. Far overhead, on three sides, vast cliffs of grey-green rock were capped by cliffs of ice; but the glacier stream came not to them but flowed away by the farther slopes, and only now and then huge ice masses fell on the valley side. In this valley it neither rained nor snowed, but the abundant springs gave a rich green pasture, that irrigation would spread over all the valley space. The settlers did well indeed there. Their beasts did well and multiplied, and but one thing marred their happiness. Yet it was enough to mar it greatly. A strange disease had come upon them, and had made all the children born to them there—and, indeed, several older children also—blind. It was to seek some charm or antidote against this plague of blindness that he had with fatigue and danger and difficulty returned down the gorge. In those days, in such cases, men did not think of germs and infections but of sins; and it seemed to him that the reason of this affliction must lie in the negligence of these priestless immigrants to set up a shrine so soon as they entered the valley. He wanted a shrine—a handsome, cheap, effectual shrine—to be erected in the valley; he wanted relics and such-like potent things of faith, blessed objects and mysterious medals and prayers. In his wallet he had a bar of native silver for which he would not account; he insisted there was none in the valley with something of the insistence of an inexpert liar. They had all clubbed their money and ornaments together, having little need for such treasure up there, he said, to buy them holy help against their ill. I figure this dim-eyed young mountaineer, sunburnt, gaunt, and anxious, hat-brim clutched feverishly, a man all unused to the ways of the lower world, telling this story to some keen-eyed, attentive priest before the great convulsion; I can picture him presently seeking to return with pious and infallible remedies against that trouble, and the infinite dismay with which he must have faced the tumbled vastness where the gorge had once come out. But the rest of his story of mischances is lost to me, save that I know of his evil death after several years. Poor stray from that remoteness! The stream that had once made the gorge now bursts from the mouth of a rocky cave, and the legend his poor, ill-told story set going developed into the legend of a race of blind men somewhere "over there" one may still hear to-day.

And amidst the little population of that now isolated and forgotten valley the disease ran its course. The old became groping and purblind, the young saw but dimly, and the children that were born to them saw never at all. But life was very easy in that snow-rimmed basin, lost to all the world, with neither thorns nor briars, with no evil insects nor any beasts save the gentle breed of llamas they had lugged and thrust and followed up the beds of the shrunken rivers in the gorges up which they had come. The seeing had become purblind so gradually that they scarcely noted their loss. They guided the sightless youngsters hither and thither until they knew the whole valley marvellously, and when at last sight died out among them the race lived on. They had even time to adapt themselves to the blind control of fire, which they made carefully in stoves of stone. They were a simple strain of people at the first, unlettered, only slightly touched with the Spanish civilisation, but with something of a tradition of the arts of old Peru and of its lost philosophy. Generation followed generation. They forgot many things; they devised many things. Their tradition of the greater world they came from became mythical in colour and uncertain. In all things save sight they were strong and able; and presently the chance of birth and heredity sent one who had an original mind and who could talk and persuade among them, and then afterwards another. These two passed, leaving their effects, and the little community grew in numbers and in understanding, and met and settled social and economic problems that arose. Generation followed generation. Generation followed generation. There came a time when a child was born who was fifteen generations from that ancestor who went out of the valley with a bar of silver to seek God's aid, and who never returned. Thereabouts it chanced that a man came into this community from the outer world. And this is the story of that man.

He was a mountaineer from the country near Quito, a man who had been down to the sea and had seen the world, a reader of books in an original way, an acute and enterprising man, and he was taken on by a party of Englishmen who had come out to Ecuador to climb mountains, to replace one of their three Swiss guides who had fallen ill. He climbed here and he climbed there, and then came the attempt on Parascotopetl, the Matterhorn of the Andes, in which he was lost to the outer world. The story of the accident has been written a dozen times. Pointer's narrative is the best. He tells how the party worked their difficult and almost vertical way up to the very foot of the last and greatest precipice, and how they built a night shelter amidst the snow upon a little shelf of rock, and, with a touch of real dramatic power, how presently they found Nunez had gone from them. They shouted, and there was no reply; shouted and whistled, and for the rest of that night they slept no more.

As the morning broke they saw the traces of his fall. It seems impossible he could have uttered a sound. He had slipped eastward towards the unknown side of the mountain; far below he had struck a steep slope of snow, and ploughed his way down it in the midst of a snow avalanche. His track went straight to the edge of a frightful precipice, and beyond that everything was hidden. Far, far below, and hazy with distance, they could see trees rising out of a narrow, shut-in valley—the lost Country of the Blind. But they did not know it was the lost Country of the Blind, nor distinguish it in any way from any other narrow streak of upland valley. Unnerved by the disaster, they abandoned their attempt in the afternoon, and Pointer was called away to the war before he could make another attack. To this day Parascotopetl lifts an unconquered crest, and Pointer's shelter crumbles unvisited amidst the snows.

And the man who fell survived.

At the end of the slope he fell a thousand feet, and came down in the midst of a cloud of snow upon a snow slope even steeper than the one above. Down this he was whirled, stunned and insensible, but without a bone broken in his body; and then at last came to gentler slopes, and at last rolled out and lay still, buried amidst a softening heap of the white masses that had accompanied and saved him. He came to himself with a dim fancy that he was ill in bed; then realised his position with a mountaineer's intelligence, and worked himself loose and, after a rest or so, out until he saw the stars. He rested flat upon his chest for a space, wondering where he was and what had happened to him. He explored his limbs, and discovered that several of his buttons were gone and his coat turned over his head. His knife had gone from his pocket, and his hat was lost, though he had tied it under his chin. He recalled that he had been looking for loose stones to raise his piece of the shelter wall. His ice-axe had disappeared.

He decided he must have fallen, and looked up to see, exaggerated by the ghastly light of the rising moon, the tremendous flight he had taken. For a while he lay, gazing blankly at that vast pale cliff towering above, rising moment by moment out of a subsiding tide of darkness. Its phantasmal, mysterious beauty held him for a space, and then he was seized with a paroxysm of sobbing laughter. . . .

After a great interval of time he became aware that he was near the lower edge of the snow. Below, down what was now a moonlit and practicable slope, he saw the dark and broken appearance of rock-strewn turf. He struggled to his feet, aching in every joint and limb, got down painfully from the heaped loose snow about him, went downward until he was on the turf, and there dropped rather than lay beside a boulder, drank deep from the flask in his inner pocket, and instantly fell asleep. . . .

He was awakened by the singing of birds in the trees far below.

He sat up and perceived he was on a little alp at the foot of a vast precipice, that was grooved by the gully down which he and his snow had come. Over against him another wall of rock reared itself against the sky. The gorge between these precipices ran east and west and was full of the morning sunlight, which lit to the westward the mass of fallen mountain that closed the descending gorge. Below him it seemed there was a precipice equally steep, but behind the snow in the gully he found a sort of chimney-cleft dripping with snow-water down which a desperate man might venture. He found it easier than it seemed, and came at last to another desolate alp, and then after a rock climb of no particular difficulty to a steep slope of trees. He took his bearings and turned his face up the gorge, for he saw it opened out above upon green meadows, among which he now glimpsed quite distinctly a cluster of stone huts of unfamiliar fashion. At times his progress was like clambering along the face of a wall, and after a time the rising sun ceased to strike along the gorge, the voices of the singing birds died away, and the air grew cold and dark about him. But the distant valley with its houses was all the brighter for that. He came presently to talus, and among the rocks he noted—for he was an observant man—an unfamiliar fern that seemed to clutch out of the crevices with intense green hands. He picked a frond or so and gnawed its stalk and found it helpful.

About midday he came at last out of the throat of the gorge into the plain and the sunlight. He was stiff and weary; he sat down in the shadow of a rock, filled up his flask with water from a spring and drank it down, and remained for a time resting before he went on to the houses.

They were very strange to his eyes, and indeed the whole aspect of that valley became, as he regarded it, queerer and more unfamiliar. The greater part of its surface was lush green meadow, starred with many beautiful flowers, irrigated with extraordinary care, and bearing evidence of systematic cropping piece by piece. High up and ringing the valley about was a wall, and what appeared to be a circumferential water-channel, from which the little trickles of water that fed the meadow plants came, and on the higher slopes above this flocks of llamas cropped the scanty herbage. Sheds, apparently shelters or feeding-places for the llamas, stood against the boundary wall here and there. The irrigation streams ran together into a main channel down the centre of the valley, and this was enclosed on either side by a wall breast high. This gave a singularly urban quality to this secluded place, a quality that was greatly enhanced by the fact that a number of paths paved with black and white stones, and each with a curious little kerb at the side, ran hither and thither in an orderly manner. The houses of the central village were quite unlike the casual and higgledy-piggledy agglomeration of the mountain villages he knew; they stood in a continuous row on either side of a central street of astonishing cleanness; here and there their parti-coloured façade was pierced by a door, and not a solitary window broke their even frontage. They were parti-coloured with extraordinary irregularity; smeared with a sort of plaster that was sometimes grey, sometimes drab, sometimes slate-coloured or dark brown; and it was the sight of this wild plastering first brought the word "blind" into the thoughts of the explorer. "The good man who did that," he thought, "must have been as blind as a bat."

He descended a steep place, and so came to the wall and channel that ran about the valley, near where the latter spouted out its surplus contents into the deeps of the gorge in a thin and wavering thread of cascade. He could now see a number of men and women resting on piled heaps of grass, as if taking a siesta, in the remoter part of the meadow, and nearer the village a number of recumbent children, and then nearer at hand three men carrying pails on yokes along a little path that ran from the encircling wall towards the houses. These latter were clad in garments of llama cloth and boots and belts of leather, and they wore caps of cloth with back and ear flaps. They followed one another in single file, walking slowly and yawning as they walked, like men who have been up all night. There was something so reassuringly prosperous and respectable in their bearing that after a moment's hesitation Nunez stood forward as conspicuously as possible upon his rock, and gave vent to a mighty shout that echoed round the valley.

The three men stopped, and moved their heads as though

they were looking about them. They turned their faces this way and that, and Nunez gesticulated with freedom. But they did not appear to see him for all his gestures, and after a time, directing themselves towards the mountains far away to the right, they shouted as if in answer. Nunez bawled again, and then once more, and as he gestured ineffectually the word "blind" came up to the top of his thoughts. "The fools must be blind," he said.

When at last, after much shouting and wrath, Nunez crossed the stream by a little bridge, came through a gate in the wall, and approached them, he was sure that they were blind. He was sure that this was the Country of the Blind of which the legends told. Conviction had sprung upon him, and a sense of great and rather enviable adventure. The three stood side by side, not looking at him, but with their ears directed towards him, judging him by his unfamiliar steps. They stood close together like men a little afraid, and he could see their eyelids closed and sunken, as though the very balls beneath had shrunk away. There was an expression near awe on their faces.

"A man," one said, in hardly recognisable Spanish—"a man it is—a man or a spirit—coming down from the rocks."

But Nunez advanced with the confident steps of a youth who enters upon life. All the old stories of the lost valley and the Country of the Blind had come back to his mind, and through his thoughts ran this old proverb, as if it were a refrain—

"In the Country of the Blind the One-eyed Man is King."

"In the Country of the Blind the One-eyed Man is King."

And very civilly he gave them greeting. He talked to them and used his eyes.

"Where does he come from, brother Pedro?" asked one.

"Down out of the rocks."

"Over the mountains I come," said Nunez, "out of the country beyond there—where men can see. From near Bogota, where there are a hundred thousands of people, and where the city passes out of sight."

"Sight?" muttered Pedro. "Sight?"

"He comes," said the second blind man, "out of the rocks."

The cloth of their coats Nunez saw was curiously fashioned, each with a different sort of stitching.

They startled him by a simultaneous movement towards him, each with a hand outstretched. He stepped back from the advance of these spread fingers.

"Come hither," said the third blind man, following his motion and clutching him neatly.

And they held Nunez and felt him over, saying no word further until they had done so.

"Carefully," he cried, with a finger in his eye, and found they thought that organ, with its fluttering lids, a queer thing in him. They went over it again.

"A strange creature, Correa," said the one called Pedro. "Feel the coarseness of his hair. Like a llama's hair."

"Rough he is as the rocks that begot him," said Correa, investigating Nunez's unshaven chin with a soft and slightly moist hand. "Perhaps he will grow finer." Nunez struggled a little under their examination, but they gripped him firm.

"Carefully," he said again.

"He speaks," said the third man. "Certainly he is a man."

"Ugh!" said Pedro, at the roughness of his coat.

"And you have come into the world?" asked Pedro.

"*Out* of the world. Over mountains and glaciers; right over above there, half-way to the sun. Out of the great big world that goes down, twelve days' journey to the sea."

They scarcely seemed to heed him. "Our fathers have told us men may be made by the forces of Nature," said Correa. "It is the warmth of things and moisture, and rottenness—rottenness."

"Let us lead him to the elders," said Pedro.

"Shout first," said Correa, "lest the children be afraid. This is a marvellous occasion."

So they shouted, and Pedro went first and took Nunez by the hand to lead him to the houses.

He drew his hand away. "I can see," he said.

"See?" said Correa.

"Yes, see," said Nunez, turning towards him, and stumbled against Pedro's pail.

"His senses are still imperfect," said the third blind man.

"He stumbles, and talks unmeaning words. Lead him by the hand."

"As you will," said Nunez, and was led along, laughing. It seemed they knew nothing of sight.

Well, all in good time he would teach them.

He heard people shouting, and saw a number of figures gathering together in the middle roadway of the village.

He found it taxed his nerve and patience more than he had anticipated, that first encounter with the population of the Country of the Blind. The place seemed larger as he drew near to it, and the smeared plasterings queerer, and a crowd of children and men and women (the women and girls, he was pleased to note, had some of them quite sweet faces, for all that their eyes were shut and sunken) came about him, holding on to him, touching him with soft, sensitive hands, smelling at him, and listening at every word he spoke. Some of the maidens and children, however, kept aloof as if afraid, and indeed his voice seemed coarse and rude beside their softer notes. They mobbed him. His three guides kept close to him with an effect of proprietorship, and said again and again, "A wild man out of the rocks."

"Bogota," he said. "Bogota. Over the mountain crests."

"A wild man—using wild words," said Pedro. "Did you hear that—*Bogota*? His mind is hardly formed yet. He has only the beginnings of speech."

A little boy nipped his hand. "Bogota!" he said mockingly.

"Ay! A city to your village. I come from the great world—where men have eyes and see."

"His name's Bogota," they said.

"He stumbled," said Correa; "stumbled twice as we came hither."

"Bring him to the elders."

And they thrust him suddenly through a doorway into a room as black as pitch, save at the end there faintly glowed a fire. The crowd closed in behind him and shut out all but the faintest glimmer of day, and before he could arrest himself he had fallen headlong over the feet of a seated man. His arm, outflung, struck the face of someone else as he went down; he felt the soft impact of features and heard a cry of anger, and for a moment he struggled against a number of hands that clutched him. It was a one-sided fight. An inkling of the situation came to him, and he lay quiet.

"I fell down," he said; "I couldn't see in this pitchy darkness."

There was a pause as if the unseen persons about him tried to understand his words. Then the voice of Correa said: "He is but newly formed. He stumbles as he walks and mingles words that mean nothing with his speech."

Others also said things about him that he heard or understood imperfectly.

"May I sit up?" he asked, in a pause. "I will not struggle against you again."

They consulted and let him rise.

The voice of an older man began to question him, and Nunez found himself trying to explain the great world out of which he had fallen, and the sky and mountains and sight and such-like marvels, to these elders who sat in darkness in the Country of the Blind. And they would believe and understand nothing whatever he told them, a thing quite outside his expectation. They would not even understand many of his words. For fourteen generations these people had been blind and cut off from all the seeing world; the names for all the things of sight had faded and changed; the story of the outer world was faded and changed to a child's story; and they had ceased to concern themselves with anything beyond the rocky slopes above their circling wall. Blind men of genius had arisen among them and questioned the shreds of belief and tradition they had brought with them from their seeing days, and had dismissed all these things as idle fancies, and replaced them with new and saner explanations. Much of their imagination had shrivelled with their eyes, and they had made for themselves new imaginations with their ever more sensitive ears and finger-tips. Slowly Nunez realised this; that his expectation of wonder and reverence at his origin and his gifts was not to be borne out; and after his poor attempt to explain sight to them had been set aside as the confused version of a new-made being describing the marvels of his incoherent sensations, he subsided, a little dashed, into listening to their instruction. And the eldest of the blind men explained to him life and philosophy and religion, how that the world (meaning their valley) had been first an empty hollow in the rocks, and then had come, first, inanimate things without the gift of touch, and llamas and a few other creatures that had little sense, and then men, and at last angels, whom one could hear singing and making fluttering sounds, but whom no one could touch at all, which puzzled Nunez greatly until he thought of the birds.

He went on to tell Nunez how this time had been divided into the warm and the cold, which are the blind equivalents of day and night, and how it was good to sleep in the warm and work during the cold, so that now, but for his advent, the whole town of the blind would have been asleep. He said Nunez must have been specially created to learn and serve the wisdom they had acquired, and for that all his mental incoherency and stumbling behaviour he must have courage, and do his best to learn, and at that all the people in the doorway murmured encouragingly. He said the night—for the blind call their day night—was now far gone, and it behoved every one to go back to sleep. He asked Nunez if he knew how to sleep, and Nunez said he did, but that before sleep he wanted food.

They brought him food—llama's milk in a bowl, and rough salted bread—and led him into a lonely place to eat out of their hearing, and afterwards to slumber until the chill of the mountain evening roused them to begin their day again. But Nunez slumbered not at all.

Instead, he sat up in the place where they had left him, resting his limbs and turning the unanticipated circumstances of his arrival over and over in his mind.

Every now and then he laughed, sometimes with amusement, and sometimes with indignation.

"Unformed mind!" he said. "Got no senses yet! They little know they've been insulting their heaven-sent king and master. I see I must bring them to reason. Let me think—let me think."

He was still thinking when the sun set.

Nunez had an eye for all beautiful things, and it seemed to him that the glow upon the snowfields and glaciers that rose about the valley on every side was the most beautiful thing he had ever seen. His eyes went from that inaccessible glory to the village and irrigated fields, fast sinking into the twilight, and suddenly a wave of emotion took him, and he thanked God from the bottom of his heart that the power of sight had been given him.

He heard a voice calling to him from out of the village.

"Ya ho there, Bogota! Come hither!"

At that he stood up smiling. He would show these people once and for all what sight would do for a man. They would seek him, but not find him.

"You move not, Bogota," said the voice.

He laughed noiselessly, and made two stealthy steps aside from the path.

"Trample not on the grass, Bogota; that is not allowed."

Nunez had scarcely heard the sound he made himself. He stopped amazed.

The owner of the voice came running up the piebald path towards him.

He stepped back into the pathway. "Here I am," he said.

"Why did you not come when I called you?" said the blind man. "Must you be led like a child? Cannot you hear the path as you walk?"

Nunez laughed. "I can see it," he said.

"There is no such word as *see*," said the blind man, after a pause. "Cease this folly, and follow the sound of my feet."

Nunez followed, a little annoyed.

"My time will come," he said.

"You'll learn," the blind man answered. "There is much to learn in the world."

"Has no one told you, 'In the Country of the Blind the One-eyed Man is King'?"

"What is blind?" asked the blind man carelessly over his shoulder.

Four days passed, and the fifth found the King of the Blind still incognito, as a clumsy and useless stranger among his subjects.

It was, he found, much more difficult to proclaim himself than he had supposed, and in the meantime, while he meditated his *coup d'état*, he did what he was told and learned the manners and customs of the Country of the Blind. He found working and going about at night a particularly irksome thing, and he decided that that should be the first thing he would change.

They led a simple, laborious life, these people, with all the elements of virtue and happiness, as these things can be understood by men. They toiled, but not oppressively; they had food and clothing sufficient for their needs; they had days and seasons of rest; they made much of music and singing, and there was love among them, and little children.

It was marvellous with what confidence and precision they went about their ordered world. Everything, you see, had been made to fit their needs; each of the radiating paths of the valley area had a constant angle to the others, and was distinguished by a special notch upon its kerbing; all obstacles and irregularities of path or meadow had long since been cleared away; all their methods and procedure arose naturally from their special needs. Their senses had become marvellously acute; they could hear and judge the slightest gesture of a man a dozen paces away—could hear the very beating of his heart. Intonation had long replaced expression with them, and touches gesture, and their work with hoe and spade and fork was as free and confident as garden work can be. Their sense of smell was extraordinarily fine; they could distinguish individual differences as readily as a dog can, and they went about the tending of the llamas, who lived among the rocks above and came to the wall for food and shelter, with ease and confidence. It was only when at last Nunez sought to assert himself that he found how easy and confident their movements could be.

He rebelled only after he had tried persuasion.

He tried at first on several occasions to tell them of sight. "Look you here, you people," he said. "There are things you do not understand in me."

Once or twice one or two of them attended to him; they sat with faces downcast and ears turned intelligently towards him, and he did his best to tell them what it was to see. Among his hearers was a girl, with eyelids less red and sunken than the others, so that one could almost fancy she was hiding eyes, whom especially he hoped to persuade. He spoke of the beauties of sight, of watching the mountains, of the sky and the sunrise, and they heard him with amused incredulity that presently became condemnatory. They told him there were indeed no mountains at all, but that the end of the rocks where the llamas grazed was indeed the end of the world; thence sprang a cavernous roof of the universe, from which the dew and the avalanches fell; and when he maintained stoutly the world had neither end nor roof such as they supposed, they said his thoughts were wicked. So far as he could describe sky and clouds and stars to them it seemed to them a hideous void, a terrible blankness in the place of the smooth roof to things in which they believed—it was an article of faith with them that the cavern roof was exquisitely smooth to the touch. He saw that in some manner he shocked them, and gave up that aspect of the matter altogether, and tried to show them the practical value of sight. One morning he saw Pedro in the path called Seventeen and coming towards the central houses, but still too far off for hearing or scent, and he told them as much. "In a little while," he prophesied, "Pedro will be here." An old man remarked that Pedro had no business on path Seventeen, and then, as if in confirmation, that individual as he drew near turned and went transversely into path Ten, and so back with nimble paces towards the outer wall. They mocked Nunez when Pedro did not arrive, and afterwards, when he asked Pedro questions to clear his character, Pedro denied and outfaced him, and was afterwards hostile to him.

Then he induced them to let him go a long way up the sloping meadows towards the wall with one complacent

individual, and to him he promised to describe all that
happened among the houses. He noted certain goings and
comings, but the things that really seemed to signify to
these people happened inside of or behind the windowless
houses—the only things they took note of to test him by—
and of these he could see or tell nothing; and it was after
the failure of this attempt, and the ridicule they could not
repress, that he resorted to force. He thought of seizing
a spade and suddenly smiting one or two of them to earth,
and so in fair combat showing the advantage of eyes. He
went so far with that resolution as to seize his spade, and
then he discovered a new thing about himself, and that was
that it was impossible for him to hit a blind man in cold
blood.

He hesitated, and found them all aware that he snatched
up the spade. They stood alert, with their heads on one
side, and bent ears towards him for what he would do next.

"Put that spade down," said one, and he felt a sort of
helpless horror. He came near obedience.

Then he thrust one backwards against a house wall, and
fled past him and out of the village.

He went athwart one of their meadows, leaving a track
of trampled grass behind his feet, and presently sat down
by the side of one of their ways. He felt something of
the buoyancy that comes to all men in the beginning of
a fight, but more perplexity. He began to realise that you
cannot even fight happily with creatures who stand upon
a different mental basis to yourself. Far away he saw a
number of men carrying spades and sticks come out of the
street of houses, and advance in a spreading line along
the several paths towards him. They advanced slowly,
speaking frequently to one another, and ever and again
the whole cordon would halt and sniff the air and listen.

The first time they did this Nunez laughed. But afterwards he did not laugh.

One struck his trail in the meadow grass, and came stooping and feeling his way along it.

For five minutes he watched the slow extension of the
cordon, and then his vague disposition to do something forth-

with became frantic. He stood up, went a pace or so towards the circumferential wall, turned, and went back a little way. There they all stood in a crescent, still and listening.

He also stood still, gripping his spade very tightly in both hands. Should he charge them?

The pulse in his ears ran into the rhythm of "In the Country of the Blind the One-eyed Man is King!"

Should he charge them?

He looked back at the high and unclimbable wall behind —unclimbable because of its smooth plastering, but withal pierced with many little doors, and at the approaching line of seekers. Behind these, others were now coming out of the street of houses.

Should he charge them?

"Bogota!" called one. "Bogota! where are you?"

He gripped his spade still tighter, and advanced down the meadows towards the place of habitations, and directly he moved they converged upon him. "I'll hit them if they touch me," he swore; "by Heaven, I will. I'll hit." He called aloud, "Look here, I'm going to do what I like in this valley. Do you hear? I'm going to do what I like and go where I like!"

They were moving in upon him quickly, groping, yet moving rapidly. It was like playing blind man's buff, with everyone blindfolded except one. "Get hold of him!" cried one. He found himself in the arc of a loose curve of pursuers. He felt suddenly he must be active and resolute.

"You don't understand," he cried in a voice that was meant to be great and resolute, and which broke. "You are blind, and I can see. Leave me alone!"

"Bogota! Put down that spade, and come off the grass!"

The last order, grotesque in its urban familiarity, produced a gust of anger.

"I'll hurt you," he said, sobbing with emotion. "By Heaven, I'll hurt you. Leave me alone!"

He began to run, not knowing clearly where to run. He ran from the nearest blind man, because it was a horror to hit him. He stopped, and then made a dash to escape from their closing ranks. He made for where a gap was wide, and the men on either side, with a quick perception of the approach of his paces, rushed in on one another. He sprang forward, and then saw he must be caught, and *swish!* the spade had struck. He felt the soft thud of hand and arm, and the man was down with a yell of pain, and he was through.

Through!' And then he was close to the street of houses again, and blind men, whirling spades and stakes, were running with a sort of reasoned swiftness hither and thither.

He heard steps behind him just in time, and found a tall man rushing forward and swiping at the sound of him. He lost his nerve, hurled his spade a yard wide at his antagonist, and whirled about and fled, fairly yelling as he dodged another.

He was panic-stricken. He ran furiously to and fro, dodging when there was no need to dodge, and in his anxiety to see on every side of him at once, stumbling. For a moment, he was down and they heard his fall. Far away in the circumferential wall a little doorway looked like heaven, and he set off in a wild rush for it. He did not even look round at his pursuers until it was gained, and he had stumbled across the bridge, clambered a little way among the rocks, to the surprise and dismay of a young llama, who went leaping out of sight, and lay down sobbing for breath.

And so his *coup d'état* came to an end.

He stayed outside the wall of the valley of the Blind for two nights and days without food or shelter, and meditated upon the unexpected. During these meditations he repeated very frequently and always with a profounder note of derision the exploded proverb: "In the Country of the Blind the One-eyed Man is King." He thought chiefly of ways of fighting and conquering these people, and it grew clear that for him no practicable way was possible. He had no weapons, and now it would be hard to get one.

The canker of civilisation had got to him even in Bogota, and he could not find it in himself to go down and assassinate a blind man. Of course, if he did that, he might then

dictate terms on the threat of assassinating them all. But
—sooner or later he must sleep! . . .

He tried also to find food among the pine trees, to be
comfortable under pine boughs while the frost fell at night,
and—with less confidence—to catch a llama by artifice in
order to try to kill it—perhaps by hammering it with a stone
—and so finally, perhaps, to eat some of it. But the llamas
had a doubt of him and regarded him with distrustful brown
eyes, and spat when he drew near. Fear came on him the
second day and fits of shivering. Finally he crawled down
to the wall of the Country of the Blind and tried to make
terms. He crawled along by the stream, shouting, until
two blind men came out to the gate and talked to him.

"I was mad," he said. "But I was only newly made."

They said that was better.

He told them he was wiser now, and repented of all he
had done.

Then he wept without intention, for he was very weak
and ill now, and they took that as a favourable sign.

They asked him if he still thought he could *"see."*

"No," he said. "That was folly. The word means nothing—less than nothing!"

They asked him what was overhead.

"About ten times ten the height of a man there is a roof
above the world—of rock—and very, very smooth." . . . He
burst again into hysterical tears. "Before you ask me any
more, give me some food or I shall die."

He expected dire punishments, but these blind people
were capable of toleration. They regarded his rebellion as
but one more proof of his general idiocy and inferiority;
and after they had whipped him they appointed him to do
the simplest and heaviest work they had for anyone to do,
and he, seeing no other way of living, did submissively
what he was told.

He was ill for some days, and they nursed him kindly.
That refined his submission. But they insisted on his lying
in the dark, and that was a great misery. And blind
philosophers came and talked to him of the wicked levity
of his mind, and reproved him so impressively for his doubts

about the lid of rock that covered their cosmic casserole that he almost doubted whether indeed he was not the victim of hallucination in not seeing it overhead.

So Nunez became a citizen of the Country of the Blind, and these people ceased to be a generalised people and became individualities and familiar to him, while the world beyond the mountains became more and more remote and unreal. There was Yacob, his master, a kindly man when not annoyed; there was Pedro, Yacob's nephew; and there was Medina-saroté, who was the youngest daughter of Yacob. She was little esteemed in the world of the blind, because she had a clear-cut face, and lacked that satisfying, glossy smoothness that is the blind man's ideal of feminine beauty; but Nunez thought her beautiful at first, and presently the most beautiful thing in the whole creation. Her closed eyelids were not sunken and red after the common way of the valley, but lay as though they might open again at any moment; and she had long eyelashes, which were considered a grave disfigurement. And her voice was strong, and did not satisfy the acute hearing of the valley swains. So that she had no lover.

There came a time when Nunez thought that, could he win her, he would be resigned to live in the valley for all the rest of his days.

He watched her; he sought opportunities of doing her little services, and presently he found that she observed him. Once at a rest-day gathering they sat side by side in the dim starlight, and the music was sweet. His hand came upon hers and he dared to clasp it. Then very tenderly she returned his pressure. And one day, as they were at their meal in the darkness, he felt her hand very softly seeking him, and as it chanced the fire leaped then and he saw the tenderness of her face.

He sought to speak to her.

He went to her one day when she was sitting in the summer moonlight spinning. The light made her a thing of silver and mystery. He sat down at her feet and told her he loved her, and told her how beautiful she seemed to him. He had a lover's voice, he spoke with a tender reverence that came near to awe, and she had never before been touched by adoration. She made him no definite answer, but it was clear his words pleased her.

After that he talked to her whenever he could take an opportunity. The valley became the world for him, and the world beyond the mountains where men lived in sunlight seemed no more than a fairy tale he would some day pour into her ears. Very tentatively and timidly he spoke to her of sight.

Sight seemed to her the most poetical of fancies, and she listened to his description of the stars and the mountains and her own sweet white-lit beauty as though it was a guilty indulgence. She did not believe, she could only half understand, but she was mysteriously delighted, and it seemed to him that she completely understood.

His love lost its awe and took courage. Presently he was for demanding her of Yacob and the elders in marriage, but she became fearful and delayed. And it was one of her elder sisters who first told Yacob that Medina-saroté and Nunez were in love.

There was from the first very great opposition to the marriage of Nunez and Medina-saroté; not so much because they valued her as because they held him as a being apart, an idiot, incompetent thing below the permissible level of a man. Her sisters opposed it bitterly as bringing discredit on them all; and old Yacob, though he had formed a sort of liking for his clumsy, obedient serf, shook his head and said the thing could not be. The young men were all angry at the idea of corrupting the race, and one went so far as to revile and strike Nunez. He struck back. Then for the first time he found an advantage in seeing, even by twilight, and after that fight was over no one was disposed to raise a hand against him. But they still found his marriage impossible.

Old Yacob had a tenderness for his last little daughter, and was grieved to have her weep upon his shoulder.

"You see, my dear, he's an idiot. He has delusions; he can't do anything right."

"I know," wept Medina-saroté. "But he's better than he

was. He's getting better. And he's strong, dear father, and kind—stronger and kinder than any other man in the world. And he loves me—and, father, I love him."

Old Yacob was greatly distressed to find her inconsolable, and, besides—what made it more distressing—he liked Nunez for many things. So he went and sat in the windowless council-chamber with the other elders and watched the trend of the talk, and said, at the proper time, "He's better than he was. Very likely, some day, we shall find him as sane as ourselves."

Then afterwards one of the elders, who thought deeply, had an idea. He was the great doctor among these people, their medicine-man, and he had a very philosophical and inventive mind, and the idea of curing Nunez of his peculiarities appealed to him. One day when Yacob was present he returned to the topic of Nunez.

"I have examined Bogota," he said, "and the case is clearer to me. I think very probably he might be cured."

"That is what I have always hoped," said old Yacob.

"His brain is affected," said the blind doctor.

The elders murmured assent.

"Now, *what* affects it?"

"Ah!" said old Yacob.

"*This,*" said the doctor, answering his own question. "Those queer things that are called the eyes, and which exist to make an agreeable soft depression in the face, are diseased, in the case of Bogota, in such a way as to affect his brain. They are greatly distended, he has eyelashes, and his eyelids move, and consequently his brain is in a state of constant irritation and destruction."

"Yes?" said old Yacob. "Yes?"

"And I think I may say with reasonable certainty that, in order to cure him completely, all that we need do is a simple and easy surgical operation—namely, to remove these irritant bodies."

"And then he will be sane?"

"Then he will be perfectly sane, and a quite admirable citizen."

"Thank Heaven for science!" said old Yacob, and went forth at once to tell Nunez of his happy hopes.

But Nunez's manner of receiving the good news struck him as being cold and disappointing.

"One might think," he said, "from the tone you take, that you did not care for my daughter."

It was Medina-saroté who persuaded Nunez to face the blind surgeons.

"*You* do not want me," he said, "to lose my gift of sight?"

She shook her head.

"My world is sight."

Her head drooped lower.

"There are the beautiful things, the beautiful little things —the flowers, the lichens among the rocks, the lightness and softness on a piece of fur, the far sky with its drifting down of clouds, the sunsets and the stars. And there is *you*. For you alone it is good to have sight, to see your sweet, serene face, your kindly lips, your dear, beautiful hands folded together. . . . It is these eyes of mine you won, these eyes that hold me to you, that these idiots seek. Instead, I must touch you, hear you, and never see you again. I must come under that roof of rock and stone and darkness, that horrible roof under which your imagination stoops. . . . No; you would not have me do that?"

A disagreeable doubt had arisen in him. He stopped, and left the thing a question.

"I wish," she said, "sometimes——" She paused.

"Yes?" said he, a little apprehensively.

"I wish sometimes—you would not talk like that."

"Like what?"

"I know it's pretty—it's your imagination. I love it, but now——"

He felt cold. "*Now?*" he said faintly.

She sat quite still.

"You mean—you think—I should be better, better perhaps——"

He was realising things very swiftly. He felt anger, indeed, anger at the dull course of fate, but also sympathy for her lack of understanding—a sympathy near akin to pity.

"*Dear,*" he said, and he could see by her whiteness how intensely her spirit pressed against the things she could not say. He put his arms about her, he kissed her ear, and they sat for a time in silence.

"If I were to consent to this?" he said at last, in a voice that was very gentle.

She flung her arms about him, weeping wildly. "Oh, if you would," she sobbed, "if only you would!"

For a week before the operation that was to raise him from his servitude and inferiority to the level of a blind citizen, Nunez knew nothing of sleep, and all through the warm sunlit hours, while the others slumbered happily, he sat brooding or wandered aimlessly, trying to bring his mind to bear on his dilemma. He had given his answer, he had given his consent, and still he was not sure. And at last work-time was over, the sun rose in splendour over the golden crests, and his last day of vision began for him. He had a few minutes with Medina-saroté before she went apart to sleep.

"To-morrow," he said, "I shall see no more."

"Dear heart!" she answered, and pressed his hands with all her strength.

"They will hurt you but little," she said; "and you are going through this pain—you are going through it, dear lover, for *me.* . . . Dear, if a woman's heart and life can do it, I will repay you. My dearest one, my dearest with the tender voice, I will repay."

He was drenched in pity for himself and her.

He held her in his arms, and pressed his lips to hers, and looked on her sweet face for the last time. "Good-bye!" he whispered at that dear sight, "good-bye!"

And then in silence he turned away from her.

She could hear his slow retreating footsteps, and something in the rhythm of them threw her into a passion of weeping.

He had fully meant to go to a lonely place where the

meadows were beautiful with white narcissus, and there remain until the hour of his sacrifice should come, but as he went he lifted up his eyes and saw the morning, the morning like an angel in golden armour, marching down the steeps. . . .

It seemed to him that before this splendour he, and this blind world in the valley, and his love, and all, were no more than a pit of sin.

He did not turn aside as he had meant to do, but went on, and passed through the wall of the circumference and out upon the rocks, and his eyes were always upon the sunlit ice and snow.

He saw their infinite beauty, and his imagination soared over them to the things beyond he was now to resign for ever. .

He thought of that great free world he was parted from, the world that was his own, and he had a vision of those further slopes, distance beyond distance, with Bogota, a place of multitudinous stirring beauty, a glory by day, a luminous mystery by night, a place of palaces and fountains and statues and white houses, lying beautifully in the middle distance. He thought how for a day or so one might come down through passes, drawing ever nearer and nearer to its busy streets and ways. He thought of the river journey, day by day, from great Bogota to the still vaster world beyond, through towns and villages, forest and desert places, the rushing river day by day, until its banks receded and the big steamers came splashing by, and one had reached the sea—the limitless sea, with its thousand islands, its thousands of islands, and its ships seen dimly far away in their incessant journeyings round and about that greater world. And there, unpent by mountains, one saw the sky—the sky, not such a disc as one saw it here, but an arch of immeasurable blue, a deep of deeps in which the circling stars were floating. . . .

His eyes scrutinised the great curtain of the mountains with a keener inquiry.

For example, if one went so, up that gully and to that chimney there, then one might come out high among those

stunted pines that ran round in a sort of shelf and rose
still higher and higher as it passed above the gorge. And
then? That talus might be managed. Thence perhaps a
climb might be found to take him up to the precipice that
came below the snow; and if that chimney failed, then
another farther to the east might serve his purpose better.
And then? Then one would be out upon the amber-lit
snow there, and half-way up to the crest of those beautiful
desolations.

He glanced back at the village, then turned right round
and regarded it steadfastly.

He thought of Medina-saroté, and she had become small
and remote.

He turned again towards the mountain wall, down which
the day had come to him.

Then very circumspectly he began to climb.

When sunset came he was no longer climbing, but he
was far and high. He had been higher, but he was still very
high. His clothes were torn, his limbs were blood-stained,
he was bruised in many places, but he lay as if he were at
his ease, and there was a smile on his face.

From where he rested the valley seemed as if it were in
a pit and nearly a mile below. Already it was dim with
haze and shadow, though the mountain summits around
him were things of light and fire, and the little details
of the rocks near at hand were drenched with subtle
beauty—a vein of green mineral piercing the grey, the
flash of crystal faces here and there, a minute, minutely
beautiful orange lichen close beside his face. There were
deep mysterious shadows in the gorge, blue deepening
into purple, and purple into a luminous darkness, and
overhead was the illimitable vastness of the sky. But he
heeded these things no longer, but lay quite inactive
there, smiling as if he were satisfied merely to have escaped
from the valley of the Blind in which he had thought to be
King.

The glow of the sunset passed, and the night came, and
still he lay peacefully contented under the cold stars.

THE STOLEN BACILLUS
AND OTHER INCIDENTS

STORY THE FIRST

The Stolen Bacillus

"THIS again," said the Bacteriologist, slipping a glass slide under the microscope, "is a preparation of the celebrated Bacillus of cholera—the cholera germ."

The pale-faced man peered down the microscope. He was evidently not accustomed to that kind of thing, and held a limp white hand over his disengaged eye. "I see very little," he said.

"Touch this screw," said the Bacteriologist; "perhaps the microscope is out of focus for you. Eyes vary so much. Just the fraction of a turn this way or that."

"Ah! now I see," said the visitor. "Not so very much to see after all. Little streaks and shreds of pink. And yet those little particles, those mere atomies, might multiply and devastate a city! Wonderful!"

He stood up, and releasing the glass slip from the microscope, held it in his hand towards the window. "Scarcely visible," he said, scrutinising the preparation. He hesitated. "Are these—alive? Are they dangerous now?"

"Those have been stained and killed," said the Bacteriologist. "I wish, for my own part, we could kill and stain every one of them in the universe."

"I suppose," the pale man said with a slight smile, "that you scarcely care to have such things about you in the living —in the active state?"

"On the contrary, we are obliged to," said the Bacteriologist. "Here, for instance——" He walked across the room and took up one of several sealed tubes. "Here is the living thing. This is a cultivation of the actual living disease bacteria." He hesitated. "Bottled cholera, so to speak."

A slight gleam of satisfaction appeared momentarily in the face of the pale man. "It's a deadly thing to have in your possession," he said, devouring the little tube with his eyes. The Bacteriologist watched the morbid pleasure in his visitor's expression. This man, who had visited him that afternoon with a note of introduction from an old friend, interested him from the very contrast of their dispositions. The lank black hair and deep grey eyes, the haggard expression and nervous manner, the fitful yet keen interest of his visitor were a novel change from the phlegmatic deliberations of the ordinary scientific worker with whom the Bacteriologist chiefly associated. It was perhaps natural, with a hearer evidently so impressionable to the lethal nature of his topic, to take the most effective aspect of the matter.

He held the tube in his hand thoughtfully. "Yes, here is the pestilence imprisoned. Only break such a little tube as this into a supply of drinking-water, say to these minute particles of life that one must needs stain and examine with the highest powers of the microscope even to see, and that one can neither smell nor taste—say to them, 'Go forth, increase and multiply, and replenish the cisterns,' and death—mysterious, untraceable death, death swift and terrible, death full of pain and indignity—would be released upon this city, and go hither and thither seeking his victims. Here he would take the husband from the wife, here the child from its mother, here the statesman from his duty, and here the toiler from his trouble. He would follow the water-mains, creeping along streets, picking out and punishing a house here and a house there where they did not boil their drinking-water, creeping into the wells of the mineral-water makers, getting washed into salad, and lying dormant in ices. He would wait ready to be drunk in the horse-troughs, and by unwary children in the public fountains. He would soak into the soil, to reappear in springs and wells at a thousand unexpected places. Once start him at the water supply, and before we could ring him in, and catch him again, he would have decimated the metropolis."

He stopped abruptly. He had been told rhetoric was his weakness.

"But he is quite safe here, you know—quite safe."

The pale-faced man nodded. His eyes shone. He cleared his throat. "These Anarchist—rascals," said he, "are fools, blind fools—to use bombs when this kind of thing is attainable. I think——"

A gentle rap, a mere light touch of the finger-nails was heard at the door. The Bacteriologist opened it. "Just a minute, dear," whispered his wife.

When he re-entered the laboratory his visitor was looking at his watch. "I had no idea I had wasted an hour of your time," he said. "Twelve minutes to four. I ought to have left here by half-past three. But your things were really too interesting. No, positively I cannot stop a moment longer. I have an engagement at four."

He passed out of the room reiterating his thanks, and the Bacteriologist accompanied him to the door, and then returned thoughtfully along the passage to his laboratory. He was musing on the ethnology of his visitor. Certainly the man was not a Teutonic type nor a common Latin one. "A morbid product, anyhow, I am afraid," said the Bacteriologist to himself. "How he gloated on those cultivations of disease-germs!" A disturbing thought struck him. He turned to the bench by the vapour-bath, and then very quickly to his writing-table. Then he felt hastily in his pockets, and then rushed to the door. "I may have put it down on the hall table," he said.

"Minnie!" he shouted hoarsely in the hall.

"Yes, dear," came a remote voice.

"Had I anything in my hand when I spoke to you, dear, just now?"

Pause.

"Nothing, dear, because I remember——"

"Blue ruin!" cried the Bacteriologist, and incontinently ran to the front door and down the steps of his house to the street.

Minnie, hearing the door slam violently, ran in alarm to the window. Down the street a slender man was getting into a cab. The Bacteriologist, hatless, and in his carpet slippers, was running and gesticulating wildly towards this group. One slipper came off, but he did not wait for it. "He has gone *mad!*" said Minnie; "it's that horrid science of his"; and, opening the window, would have called after him. The slender man, suddenly glancing round, seemed struck with the same idea of mental disorder. He pointed hastily to the Bacteriologist, said something to the cabman, the apron of the cab slammed, the whip swished, the horses's feet clattered, and in a moment cab, and Bacteriologist hotly in pursuit, had receded up the vista of the roadway and disappeared round the corner.

Minnie remained straining out of the window for a minute. Then she drew her head back into the room again. She was dumbfounded. "Of course he is eccentric," she meditated. "But running about London—in the height of the season, too—in his socks!" A happy thought struck her. She hastily put her bonnet on, seized his shoes, went into the hall, took down his hat and light overcoat from the pegs, emerged upon the doorstep, and hailed a cab that opportunely crawled by. "Drive me up the road and round Havelock Crescent, and see if we can find a gentleman running about in a velveteen coat and no hat."

"Velveteen coat, ma'am, and no 'at. Very good, ma'am." And the cabman whipped up at once in the most matter-of-fact way, as if he drove to this address every day in his life.

Some few minutes later the little group of cabmen and loafers that collects round the cabmen's shelter at Haverstock Hill were startled by the passing of a cab with a ginger-coloured screw of a horse, driven furiously.

They were silent as it went by, and then as it receded— "That's 'Arry 'Icks. Wot's *he* got?" said the stout gentleman known as Old Tootles.

"He's a-using his whip, he is, *to* rights," said the ostler boy.

"Hullo!" said poor old Tommy Byles; "here's another bloomin' loonatic. Blowed if there ain't."

"It's old George," said old Tootles, "and he's drivin' a loonatic, *as* you say. Ain't he a-clawin' out of the keb? Wonder if he's after 'Arry 'Icks?"

The group round the cabmen's shelter became animated. Chorus: "Go it, George!" "It's a race!" "You'll ketch 'em!" "Whip up!"

"She's a goer, she is!" said the ostler boy.

"Strike me giddy!" cried old Tootles. "Here! *I'm* a-goin' to begin in a minute. Here's another comin'. If all the kebs in Hampstead ain't gone mad this morning!"

"It's a fieldmale this time," said the ostler boy.

"She's a followin' *him*," said old Tootles. "Usually the other way about."

"What's she got in her 'and?"

"Looks like a 'igh 'at."

"What a bloomin' lark it is! Three to one on old George," said the ostler boy. "Next!"

Minnie went by in a perfect roar of applause. She did not like it but she felt that she was doing her duty, and whirled on down Haverstock Hill and Camden Town High Street with her eyes ever intent on the animated back view of old George, who was driving her vagrant husband so incomprehensively away from her.

The man in the foremost cab sat crouched in the corner, his arms tightly folded, and the little tube that contained such vast possibilities of destruction gripped in his hand. His mood was a singular mixture of fear and exultation. Chiefly he was afraid of being caught before he could accomplish his purpose, but behind this was a vaguer but larger fear of the awfulness of his crime. But his exultation far exceeded his fear. No Anarchist before him had ever approached this conception of his. Ravachol, Vaillant, all those distinguished persons whose fame he had envied dwindled into insignificance beside him. He had only to make sure of the water supply, and break the little tube into a reservoir. How brilliantly he had planned it, forged the letter of introduction and got into the laboratory, and how brilliantly he had seized his opportunity! The world should hear of him at last. All those people who had

sneered at him, neglected him, preferred other people to
him, found his company undesirable, should consider him at
last. Death, death, death! They had always treated him as
a man of no importance. All the world had been in a con-
spiracy to keep him under. He would teach them yet what
it is to isolate a man. What was this familiar street? Great
Saint Andrew's Street, of course! How fared the chase? He
craned out of the cab. The Bacteriologist was scarcely
fifty yards behind. That was bad. He would be caught
and stopped yet. He felt in his pocket for money, and
found half-a-sovereign. This he thrust up through the trap
in the top of the cab into the man's face. "More," he
shouted, "if only we get away."

The money was snatched out of his hand. "Right you
are," said the cabman, and the trap slammed, and the lash
lay along the glistening side of the horse. The cab swayed,
and the Anarchist, half-standing under the trap, put the
hand containing the little glass tube upon the apron to preserve his balance. He felt the brittle thing crack, and the
broken half of it rang upon the floor of the cab. He fell
back into the seat with a curse, and stared dismally at the
two or three drops of moisture on the apron.

He shuddered.

"Well! I suppose I shall be the first. *Phew!* Anyhow,
I shall be a Martyr. That's something. But it is a filthy
death, nevertheless. I wonder if it hurts as much as they
say."

Presently a thought occurred to him—he groped between
his feet. A little drop was still in the broken end of the
tube, and he drank that to make sure. It was better to make
sure. At any rate, he would not fail.

Then it dawned upon him that there was no further need
to escape the Bacteriologist. In Wellington Street he told
the cabman to stop, and got out. He slipped on the step,
and his head felt queer. It was rapid stuff this cholera
poison. He waved his cabman out of existence, so to speak,
and stood on the pavement with his arm folded upon his
breast awaiting the arrival of the Bacteriologist. There was
something tragic in his pose. The sense of imminent death

gave him a certain dignity. He greeted his pursuer with a defiant laugh.

"Vive l'Anarchie! You are too late, my friend. I have drunk it. The cholera is abroad!"

The Bacteriologist from his cab beamed curiously at him through his spectacles. "You have drunk it! An Anarchist! I see now." He was about to say something more, and then checked himself. A smile hung in the corner of his mouth. He opened the apron of his cab as if to descend, at which the Anarchist waved him a dramatic farewell and strode off towards Waterloo Bridge, carefully jostling his infected body against as many people as possible. The Bacteriologist was so preoccupied with the vision of him that he scarcely manifested the slightest surprise at the appearance of Minnie upon the pavement with his hat and shoes and overcoat. "Very good of you to bring my things," he said, and remained lost in contemplation of the receding figure of the Anarchist.

"You had better get in," he said, still staring. Minnie felt absolutely convinced now that he was mad, and directed the cabman home on her own responsibility. "Put on my shoes? Certainly, dear," said he, as the cab began to turn, and hid the strutting black figure, now small in the distance, from his eyes. Then suddenly something grotesque struck him, and he laughed. Then he remarked, "It is really very serious, though."

"You see, that man came to my house to see me, and he is an Anarchist. No—don't faint, or I cannot possibly tell you the rest. And I wanted to astonish him, not knowing he was an Anarchist, and took up a cultivation of that new species of Bacterium I was telling you of, that infest, and I think cause, the blue patches upon various monkeys; and like a fool, I said it was Asiatic cholera. And he ran away with it to poison the water of London, and he certainly might have made things look blue for this civilised city. And now he has swallowed it. Of course, I cannot say what will happen, but you know it turned that kitten blue, and the three puppies—in patches, and the sparrow—bright blue.

But the bother is, I shall have all the trouble and expense of preparing some more.

"Put on my coat on this hot day! Why? Because we might meet Mrs. Jabber. My dear, Mrs. Jabber is not a draught. But why should I wear a coat on a hot day because of Mrs. ——? Oh! *very* well."

STORY THE SECOND

The Flowering of the Strange Orchid

THE buying of orchids always has in it a certain speculative flavour. You have before you the brown shrivelled lump of tissue, and for the rest you must trust your judgment, or the auctioneer, or your good-luck, as your taste may incline. The plant may be moribund or dead, or it may be just a respectable purchase, fair value for your money, or perhaps—for the thing has happened again and again—there slowly unfolds before the delighted eyes of the happy purchaser, day after day, some new variety, some novel richness, a strange twist of the labellum, or some subtler colouration or unexpected mimicry. Pride, beauty, and profit blossom together on one delicate green spike, and, it may be, even immortality. For the new miracle of Nature may stand in need of a new specific name, and what so convenient as that of its discoverer? "Johnsmithia"! There have been worse names.

It was perhaps the hope of some such happy discovery that made Winter-Wedderburn such a frequent attendant at these sales—that hope, and also, maybe, the fact that he had nothing else of the slightest interest to do in the world. He was a shy, lonely, rather ineffectual man, provided with just enough income to keep off the spur of necessity, and not enough nervous energy to make him seek any exacting employment. He might have collected stamps or coins, or translated Horace, or bound books, or invented new species of diatoms. But, as it happened, he grew orchids, and had one ambitious little hothouse.

"I have a fancy," he said over his coffee, "that something is going to happen to me to-day." He spoke—as he moved and thought—slowly.

"Oh, don't say *that!*" said his housekeeper—who was also his remote cousin. For "something happening" was a euphemism that meant only one thing to her.

"You misunderstand me. I mean nothing unpleasant . . . though what I do mean I scarcely know.

"To-day," he continued, after a pause, "Peters' are going to sell a batch of plants from the Andamans and the Indies. I shall go up and see what they have. It may be I shall buy something good, unawares. That may be it."

He passed his cup for his second cupful of coffee.

"Are those the things collected by that poor young fellow you told me of the other day?" asked his cousin as she filled his cup.

"Yes," he said, and became meditative over a piece of toast.

"Nothing ever does happen to me," he remarked presently, beginning to think aloud. "I wonder why? Things enough happen to other people. There is Harvey. Only the other week—on Monday he picked up sixpence, on Wednesday his chicks all had the staggers, on Friday his cousin came home from Australia, and on Saturday he broke his ankle. What a whirl of excitement!—compared to me."

"I think I would rather be without so much excitement," said his housekeeper. "It can't be good for you."

"I suppose it's troublesome. Still . . . you see, nothing ever happens to me. When I was a little boy I never had accidents. I never fell in love as I grew up. Never married. . . . I wonder how it feels to have something happen to you, something really remarkable.

"That orchid-collector was only thirty-six—twenty years younger than myself—when he died. And he had been married twice and divorced once; he had had malarial fever four times, and once he broke his thigh. He killed a Malay once, and once he was wounded by a poisoned dart. And in the end he was killed by jungle-leeches. It must have all been very troublesome, but then it must have been very interesting, you know—except, perhaps, the leeches."

"I am sure it was not good for him," said the lady, with conviction.

"Perhaps not." And then Wedderburn looked at his watch. "Twenty-three minutes past eight. I am going up by the quarter to twelve train, so that there is plenty of time. I think I shall wear my alpaca jacket—it is quite warm enough—and my grey felt hat and brown shoes. I suppose——"

He glanced out of the window at the serene sky and sunlit garden, and then nervously at his cousin's face.

"I think you had better take an umbrella if you are going to London," she said in a voice that admitted of no denial. "There's all between here and the station coming back."

When he returned he was in a state of mild excitement. He had made a purchase. It was rare that he could make up his mind quickly enough to buy, but this time he had done so.

"These are Vandas," he said, "and a Dendrobe and some Palæonophis." He surveyed his purchases lovingly as he consumed his soup. They were laid out on the spotless tablecloth before him, and he was telling his cousin all about them as he slowly meandered through his dinner. It was his custom to live all his visit to London over again in the evening for her and his own entertainment.

"I knew something would happen to-day. And I have bought all these. Some of them—some of them—I feel sure, do you know, that some of them will be remarkable. I don't know how it is, but I feel just as sure as if someone had told me that some of these will turn out remarkable.

"That one"—he pointed to a shrivelled rhizome—"was not identified. It may be a Palæonophis—or it may not. It may be a new species, or even a new genus. And it was the last that poor Batten ever collected."

"I don't like the look of it," said his housekeeper. "It's such an ugly shape."

"To me it scarcely seems to have a shape."

"I don't like those things that stick out," said his house-keeper.

"It shall be put away in a pot to-morrow."

"It looks," said the housekeeper, "like a spider shamming dead."

Wedderburn smiled and surveyed the root with his head on one side. "It is certainly not a pretty lump of stuff. But you can never judge of these things from their dry appearance. It may turn out to be a very beautiful orchid indeed. How busy I shall be to-morrow! I must see to-night just exactly what to do with these things, and to-morrow I shall set to work.

"They found poor Batten lying dead, or dying, in a man-grove swamp—I forget which," he began again presently, "with one of these very orchids crushed up under his body. He had been unwell for some days with some kind of native fever, and I suppose he fainted. These mangrove swamps are very unwholesome. Every drop of blood, they say, was taken out of him by the jungle-leeches. It may be that very plant that cost him his life to obtain."

"I think none the better of it for that."

"Men must work though women may weep," said Wed-derburn with profound gravity.

"Fancy dying away from every comfort in a nasty swamp! Fancy being ill of fever with nothing to take but chloro-dyne and quinine—if men were left to themselves they would live on chlorodyne and quinine—and no one round you but horrible natives! They say the Andaman islanders are most disgusting wretches—and, anyhow, they can scarcely make good nurses, not having the necessary train-ing. And just for people in England to have orchids!"

"I don't suppose it was comfortable, but some men seem to enjoy that kind of thing," said Wedderburn. "Anyhow, the natives of his party were sufficiently civilised to take care of all his collection until his colleague, who was an orni-thologist, came back again from the interior; though they could not tell the species of the orchid and had let it wither. And it makes these things more interesting."

"It makes them disgusting. I should be afraid of some of

the malaria clinging to them. And just think, there has been a dead body lying across that ugly thing! I never thought of that before. There! I declare I cannot eat another mouthful of dinner."

"I will take them off the table if you like, and put them in the window-seat. I can see them just as well there."

The next few days he was indeed singularly busy in his steamy little hothouse, fussing about with charcoal, lumps of teak, moss, and all the other mysteries of the orchid cultivator. He considered he was having a wonderfully eventful time. In the evening he would talk about these new orchids to his friends, and over and over again he reverted to his expectation of something strange.

Several of the Vandas and the Dendrobium died under his care, but presently the strange orchid began to show signs of life. He was delighted and took his housekeeper right away from jam-making to see it at once, directly he made the discovery.

"That is a bud," he said, "and presently there will be a lot of leaves there, and those little things coming out here are aërial rootlets."

"They look to me like little white fingers poking out of the brown," said his housekeeper. "I don't like them."

"Why not?"

"I don't know. They look like fingers trying to get at you. I can't help my likes and dislikes."

"I don't know for certain, but I don't *think* there are any orchids I know that have aërial rootlets quite like that. It may be my fancy, of course. You see they are a little flattened at the ends."

"I don't like 'em," said his housekeeper, suddenly shivering and turning away. "I know it's very silly of me—and I'm very sorry, particularly as you like the thing so much. But I can't help thinking of that corpse."

"But it may not be that particular plant. That was merely a guess of mine."

His housekeeper shrugged her shoulders. "Anyhow I don't like it," she said.

Wedderburn felt a little hurt at her dislike to the plant.

But that did not prevent his talking to her about orchids generally, and this orchid in particular, whenever he felt inclined.

"There are such queer things about orchids," he said one day; "such possibilities of surprises. You know, Darwin studied their fertilisation, and showed that the whole structure of an ordinary orchid-flower was contrived in order that moths might carry the pollen from plant to plant. Well, it seems that there are lots of orchids known the flower of which cannot possibly be used for fertilisation in that way. Some of the Cypripediums, for instance; there are no insects known that can possibly fertilise them, and some of them have never been found with seed."

"But how do they form new plants?"

"By runners and tubers, and that kind of outgrowth. That is easily explained. The puzzle is, what are the flowers for?

"Very likely," he added, "*my* orchid may be something extraordinary in that way. If so I shall study it. I have often thought of making researches as Darwin did. But hitherto I have not found the time, or something else has happened to prevent it. The leaves are beginning to unfold now. I do wish you would come and see them!"

But she said that the orchid-house was so hot it gave her the headache. She had seen the plant once again, and the aërial rootlets, which were now some of them more than a foot long, had unfortunately reminded her of tentacles reaching out after something; and they got into her dreams, growing after her with incredible rapidity. So that she had settled to her entire satisfaction that she would not see that plant again, and Wedderburn had to admire its leaves alone. They were of the ordinary broad form, and a deep glossy green, with splashes and dots of deep red towards the base. He knew of no other leaves quite like them. The plant was placed on a low bench near the thermometer, and close by was a simple arrangement by which a tap dripped on the hot-water pipes and kept the air steamy. And he spent his afternoons now with some regularity meditating on the approaching flowering of this strange plant.

And at last the great thing happened. Directly he entered the little glass house he knew that the spike had burst out, although his great *Palæonophis Lowii* hid the corner where his new darling stood. There was a new odour in the air, a rich, intensely sweet scent, that overpowered every other in that crowded, steaming little greenhouse.

Directly he noticed this he hurried down to the strange orchid. And, behold! the trailing green spikes bore now three great splashes of blossom, from which this overpowering sweetness proceeded. He stopped before them in an ecstasy of admiration.

The flowers were white, with streaks of golden orange upon the petals; the heavy labellum was coiled into an intricate projection, and a wonderful bluish purple mingled there with the gold. He could see at once that the genus was altogether a new one. And the insufferable scent! How hot the place was! The blossoms swam before his eyes.

He would see if the temperature was right. He made a step towards the thermometer. Suddenly everything appeared unsteady. The bricks on the floor were dancing up and down. Then the white blossoms, the green leaves behind them, the whole greenhouse, seemed to sweep sideways, and then in a curve upward.

*   *   *   *   *

At half-past four his cousin made the tea, according to their invariable custom. But Wedderburn did not come in for his tea. "He is worshipping that horrid orchid," she told herself, and waited ten minutes. "His watch must have stopped. I will go and call him."

She went straight to the hothouse, and, opening the door, called his name. There was no reply. She noticed that the air was very close, and loaded with an intense perfume. Then she saw something lying on the bricks between the hot-water pipes.

For a minute, perhaps, she stood motionless.

He was lying, face upward, at the foot of the strange orchid. The tentacle-like aërial rootlets no longer swayed freely in the air, but were crowded together, a tangle of grey ropes, and stretched tight with their ends closely applied to his chin and neck and hands.

She did not understand. Then she saw from under one of the exultant tentacles upon his cheek there trickled a little thread of blood.

With an inarticulate cry she ran towards him, and tried to pull him away from the leech-like suckers. She snapped two of these tentacles, and their sap dripped red.

Then the overpowering scent of the blossom began to make her head reel. How they clung to him! She tore at the tough ropes, and he and the white inflorescence swam about her. She felt she was fainting, knew she must not. She left him and hastily opened the nearest door, and, after she had panted for a moment in the fresh air, she had a brilliant inspiration. She caught up a flower-pot and smashed in the windows at the end of the greenhouse. Then she re-entered. She tugged now with renewed strength at Wedderburn's motionless body, and brought the strange orchid crashing to the floor. It still clung with the grimmest tenacity to its victim. In a frenzy, she lugged it and him into the open air.

Then she thought of tearing through the sucker rootlets one by one, and in another minute she had released him and was dragging him away from the horror.

He was white and bleeding from a dozen circular patches.

The odd-job man was coming up the garden, amazed at the smashing of glass, and saw her emerge, hauling the inanimate body with red-stained hands. For a moment he thought impossible things.

"Bring some water!" she cried, and her voice dispelled his fancies. When, with unnatural alacrity, he returned with the water, he found her weeping with excitement, and with Wedderburn's head upon her knee, wiping the blood from his face.

"What's the matter?" said Wedderburn, opening his eyes feebly, and closing them again at once.

"Go and tell Annie to come out here to me, and then go for Doctor Haddon at once," she said to the odd-job man so soon as he brought the water; and added, seeing he hesitated, "I will tell you all about it when you come back."

Presently Wedderburn opened his eyes again, and seeing that he was troubled by the puzzle of his position, she explained to him, "You fainted in the hot-house."

"And the orchid?"

"I will see to that," she said.

Wedderburn had lost a good deal of blood, but beyond that he had suffered no very great injury. They gave him brandy mixed with some pink extract of meat, and carried him upstairs to bed. His housekeeper told her incredible story in fragments to Dr. Haddon. "Come to the orchid-house and see," she said.

The cold outer air was blowing in through the open door, and the sickly perfume was almost dispelled. Most of the torn aërial rootlets lay already withered amidst a number of dark stains upon the bricks. The stem of the inflorescence was broken by the fall of the plant, and the flowers were growing limp and brown at the edges of the petals. The doctor stooped towards it, then saw that one of the aërial rootlets still stirred feebly, and hesitated.

The next morning the strange orchid still lay there, black now and putrescent. The door banged intermittently in the morning breeze, and all the array of Wedderburn's orchids was shrivelled and prostrate. But Wedderburn himself was bright and garrulous upstairs in the glory of his strange adventure.

STORY THE THIRD

In the Avu Observatory

THE observatory at Avu, in Borneo, stands on the spur of the mountain. To the north rises the old crater, black at night against the unfathomable blue of the sky. From the little circular building, with its mushroom dome, the slopes plunge steeply downward into the black mysteries of the tropical forest beneath. The little house in which the observer and his assistant live is about fifty yards from the observatory, and beyond this are the huts of their native attendants.

Thaddy, the chief observer, was down with a slight fever. His assistant, Woodhouse, paused for a moment in silent contemplation of the tropical night before commencing his solitary vigil. The night was very still. Now and then voices and laughter came from the native huts, or the cry of some strange animal was heard from the midst of the mystery of the forest. Nocturnal insects appeared in ghostly fashion out of the darkness, and fluttered round his light. He thought, perhaps, of all the possibilities of discovery that still lay in the black tangle beneath him; for to the naturalist the virgin forests of Borneo are still a wonderland full of strange questions and half-suspected discoveries. Woodhouse carried a small lantern in his hand, and its yellow glow contrasted vividly with the infinite series of tints between lavender-blue and black in which the landscape was painted. His hands and face were smeared with ointment against the attacks of the mosquitoes.

Even in these days of celestial photography, work done

in a purely temporary erection, and with only the most primitive appliances in addition to the telescope, still involves a very large amount of cramped and motionless watching. He sighed as he thought of the physical fatigues before him, stretched himself, and entered the observatory.

The reader is probably familiar with the structure of an ordinary astronomical observatory. The building is usually cylindrical in shape, with a very light hemispherical roof capable of being turned round from the interior. The telescope is supported upon a stone pillar in the centre, and a clockwork arrangement compensates for the earth's rotation, and allows a star once found to be continuously observed. Besides this, there is a compact tracery of wheels and screws about its point of support, by which the astronomer adjusts it. There is, of course, a slit in the movable roof which follows the eye of the telescope in its survey of the heavens. The observer sits or lies on a sloping wooden arrangement, which he can wheel to any part of the observatory as the position of the telescope may require. Within it is advisable to have things as dark as possible, in order to enhance the brilliance of the stars observed.

The lantern flared as Woodhouse entered his circular den, and the general darkness fled into black shadows behind the big machine, from which it presently seemed to creep back over the whole place again as the light waned. The slit was a profound transparent blue, in which six stars shone with tropical brilliance, and their light lay, a pallid gleam, along the black tube of the instrument. Woodhouse shifted the roof, and then proceeding to the telescope, turned first one wheel and then another, the great cylinder slowly swinging into a new position. Then he glanced through the finder, the little companion telescope, moved the roof a little more, made some further adjustments, and set the clockwork in motion. He took off his jacket, for the night was very hot, and pushed into position the uncomfortable seat to which he was condemned for the next four hours. Then with a sigh he resigned himself to his watch upon the mysteries of space.

There was no sound now in the observatory, and the

lantern waned steadily. Outside there was the occasional
cry of some animal in alarm or pain, or calling to its mate,
and the intermittent sounds of the Malay and Dyak servants. Presently one of the men began a queer chanting
song, in which the others joined at intervals. After this it
would seem that they turned in for the night, for no further
sound came from their direction, and the whispering stillness became more and more profound.

The clockwork ticked steadily. The shrill hum of a
mosquito explored the place and grew shriller in indignation
at Woodhouse's ointment. Then the lantern went out and
all the observatory was black.

Woodhouse shifted his position presently, when the slow
movement of the telescope had carried it beyond the limits
of his comfort.

He was watching a little group of stars in the Milky Way,
in one of which his chief had seen or fancied a remarkable
colour variability. It was not a part of the regular work
for which the establishment existed, and for that reason
perhaps Woodhouse was deeply interested. He must have
forgotten things terrestrial. All his attention was concentrated upon the great blue circle of the telescope field—a
circle powdered, so it seemed, with an innumerable multitude of stars, and all luminous against the blackness of its
setting. As he watched he seemed to himself to become
incorporeal, as if he too were floating in the ether of space.
Infinitely remote was the faint red spot he was observing.

Suddenly the stars were blotted out. A flash of blackness passed, and they were visible again.

"Queer," said Woodhouse. "Must have been a bird."

The thing happened again, and immediately after the
great tube shivered as though it had been struck. Then
the dome of the observatory resounded with a series of
thundering blows. The stars seemed to sweep aside as the
telescope—which had been unclamped—swung round and
away from the slit in the roof.

"Great Scott!" cried Woodhouse. "What's this?"

Some huge vague black shape, with a flapping something
like a wing, seemed to be struggling in the aperture of the

roof. In another moment the slit was clear again, and the luminous haze of the Milky Way shone warm and bright.

The interior of the roof was perfectly black, and only a scraping sound marked the whereabouts of the unknown creature.

Woodhouse had scrambled from the seat to his feet. He was trembling violently and in a perspiration with the suddenness of the occurrence. Was the thing, whatever it was, inside or out? It was big, whatever else it might be. Something shot across the skylight, and the telescope swayed. He started violently and put his arm up. It was in the observatory, then, with him. It was clinging to the roof, apparently. What the devil was it? Could it see him?

He stood for perhaps a minute in a state of stupefaction. The beast, whatever it was, clawed at the interior of the dome, and then something flapped almost into his face, and he saw the momentary gleam of starlight on a skin like oiled leather. His water-bottle was knocked off his little table with a smash.

The sense of some strange bird-creature hovering a few yards from his face in the darkness was indescribably unpleasant to Woodhouse. As his thought returned he concluded that it must be some night-bird or large bat. At any risk he would see what it was, and pulling a match from his pocket, he tried to strike it on the telescope seat. There was a smoking streak of phosphorescent light, the match flared for a moment, and he saw a vast wing sweeping towards him, a gleam of grey-brown fur, and then he was struck in the face and the match knocked out of his hand. The blow was aimed at his temple, and a claw tore sideways down to his cheek. He reeled and fell, and he heard the extinguished lantern smash. Another blow followed as he fell. He was partly stunned, he felt his own warm blood stream out upon his face. Instinctively he felt his eyes had been struck at, and, turning over on his face to protect them, tried to crawl under the protection of the telescope.

He was struck again upon the back, and he heard his jacket rip, and then the thing hit the roof of the observatory. He edged as far as he could between the wooden seat and the eyepiece of the instrument, and turned his body round so that it was chiefly his feet that were exposed. With these he could at least kick. He was still in a mystified state. The strange beast banged about in the darkness, and presently clung to the telescope, making it sway and the gear rattle. Once it flapped near him, and he kicked out madly and felt a soft body with his feet. He was horribly scared now. It must be a big thing to swing the telescope like that. He saw for a moment the outline of a head black against the starlight, with sharply-pointed upstanding ears and a crest between them. It seemed to him to be as big as a mastiff's. Then he began to bawl out as loudly as he could for help.

At that the thing came down upon him again. As it did so his hand touched something beside him on the floor. He kicked out, and the next moment his ankle was gripped and held by a row of keen teeth. He yelled again, and tried to free his leg by kicking with the other. Then he realised he had the broken water-bottle at his hand, and, snatching it, he struggled into a sitting posture, and feeling in the darkness towards his foot, gripped a velvety ear, like the ear of a big cat. He had seized the water-bottle by its neck and brought it down with a shivering crash upon the head of the strange beast. He repeated the blow, and then stabbed and jabbed with the jagged end of it, in the darkness, where he judged the face might be.

The small teeth relaxed their hold, and at once Woodhouse pulled his leg free and kicked hard. He felt the sickening feel of fur and bone giving under his boot. There was a tearing bite at his arm, and he struck over it at the face, as he judged, and hit damp fur.

There was a pause; then he heard the sound of claws and the dragging of a heavy body away from him over the observatory floor. Then there was silence, broken only by his own sobbing breathing, and a sound like licking. Everything was black except the parallelogram of the blue skylight with the luminous dust of stars, against which the end of the telescope now appeared in silhouette. He waited, as it seemed, an interminable time.

Was the thing coming on again? He felt in his trouser-pocket for some matches, and found one remaining. He tried to strike this, but the floor was wet, and it spat and went out. He cursed. He could not see where the door was situated. In his struggle he had quite lost his bearings. The strange beast, disturbed by the splutter of the match, began to move again. "Time!" called Woodhouse, with a sudden gleam of mirth, but the thing was not coming at him again. He must have hurt it, he thought, with the broken bottle. He felt a dull pain in his ankle. Probably he was bleeding there. He wondered if it would support him if he tried to stand up. The night outside was very still. There was no sound of anyone moving. The sleepy fools had not heard those wings battering upon the dome, nor his shouts. It was no good wasting strength in shouting. The monster flapped its wings and startled him into a defensive attitude. He hit his elbow against the seat, and it fell over with a crash. He cursed this, and then he cursed the darkness.

Suddenly the oblong patch of starlight seemed to sway to and fro. Was he going to faint? It would never do to faint. He clenched his fists and set his teeth to hold himself together. Where had the door got to? It occurred to him he could get his bearings by the stars visible through the skylight. The patch of stars he saw was in Sagittarius and south-eastward; the door was north—or was it north by west? He tried to think. If he could get the door open he might retreat. It might be the thing was wounded. The suspense was beastly. "Look here!" he said, "if you don't come on, I shall come at you."

Then the thing began clambering up the side of the observatory, and he saw its black outline gradually blot out the skylight. Was it in retreat? He forgot about the door, and watched as the dome shifted and creaked. Somehow he did not feel very frightened or excited now. He felt a curious sinking sensation inside him. The sharply-defined patch of light, with the black form moving across it, seemed to be growing smaller and smaller. That was curious. He began to feel very thirsty, and yet he did not feel inclined to get anything to drink. He seemed to be sliding down a long funnel.

He felt a burning sensation in his throat, and then he perceived it was broad daylight, and that one of the Dyak servants was looking at him with a curious expression. Then there was the top of Thaddy's face upside down. Funny fellow, Thaddy, to go about like that! Then he grasped the situation better, and perceived that his head was on Thaddy's knee, and Thaddy was giving him brandy. And then he saw the eyepiece of the telescope with a lot of red smears on it. He began to remember.

"You've made this observatory in a pretty mess," said Thaddy.

The Dyak boy was beating up an egg in brandy. Woodhouse took this and sat up. He felt a sharp twinge of pain. His ankle was tied up, so were his arm and the side of his face. The smashed glass, red-stained, lay about the floor, the telescope seat was overturned, and by the opposite wall was a dark pool. The door was open, and he saw the grey summit of the mountain against a brilliant background of blue sky.

"Pah!" said Woodhouse. "Who's been killing calves here? Take me out of it."

Then he remembered the Thing, and the fight he had had with it.

"What *was* it?" he said to Thaddy—"the Thing I fought with?"

"*You* know that best," said Thaddy. "But, anyhow, don't worry yourself now about it. Have some more to drink."

Thaddy, however, was curious enough, and it was a hard struggle between duty and inclination to keep Woodhouse quiet until he was decently put away in bed, and had slept upon the copious dose of meat-extract Thaddy considered advisable. They then talked it over together.

"It was," said Woodhouse, "more like a big bat than anything else in the world. It had sharp, short ears, and soft fur, and its wings were leathery. Its teeth were little, but devilish sharp, and its jaw could not have been very strong or else it would have bitten through my ankle."

"It has pretty nearly," said Thaddy.

"It seemed to me to hit out with its claws pretty freely. That is about as much as I know about the beast. Our conversation was intimate, so to speak, and yet not confidential."

"The Dyak chaps talk about a Big Colugo, a Klangutang —whatever that may be. It does not often attack man, but I suppose you made it nervous. They say there is a Big Colugo and a Little Colugo, and a something else that sounds like gobble. They all fly about at night. For my own part I know there are flying foxes and flying lemurs about here, but they are none of them very big beasts."

"There are more things in heaven and earth," said Woodhouse—and Thaddy groaned at the quotation—"and more particularly in the forests of Borneo, than are dreamt of in our philosophies. On the whole, if the Borneo fauna is going to disgorge any more of its novelties upon me, I should prefer that it did so when I was not occupied in the observatory at night and alone."

STORY THE FOURTH

The Triumphs of a Taxidermist

HERE are some of the secrets of taxidermy. They were told me by the taxidermist in a mood of elation. He told me them in the time between the first glass of whisky and the fourth, when a man is no longer cautious and yet not drunk. We sat in his den together; his library it was, his sitting and his eating-room—separated by a bead curtain, so far as the sense of sight went, from the noisome den where he plied his trade.

He sat on a deck chair, and when he was not tapping refractory bits of coal with them, he kept his feet—on which he wore, after the manner of sandals, the holey relics of a pair of carpet slippers—out of the way upon the mantelpiece, among the glass eyes. And his trousers, by-the-by—though they have nothing to do with his triumphs—were a most horrible yellow plaid, such as they made when our fathers wore side-whiskers and there were crinolines in the land. Further, his hair was black, his face rosy, and his eye a fiery brown; and his coat was chiefly of grease upon a basis of velveteen. And his pipe had a bowl of china showing the Graces, and his spectacles were always askew, the left eye glaring nakedly at you, small and penetrating; the right, seen through a glass darkly, magnified and mild. Thus his discourse ran: "There never was a man who could stuff like me, Bellows, never. I have stuffed elephants and I have stuffed moths, and the things have looked all the livelier and better for it. And I have stuffed human beings —chiefly amateur ornithologists. But I stuffed a nigger once.

220

"No, there is no law against it. I made him with all his fingers out, and used him as a hat-rack, but that fool Homersby got up a quarrel with him late one night and spoilt him. That was before your time. It is hard to get skins, or I would have another.

"Unpleasant? I don't see it. Seems to me taxidermy is a promising third course to burial or cremation. You could keep all your dear ones by you. Bric-à-brac of that sort stuck about the house would be as good as most company, and much less expensive. You might have them fitted up with clockwork to do things.

"Of course they would have to be varnished, but they need not shine more than lots of people do naturally. Old Manningtree's bald head. . . . Anyhow, you could talk to them without interruption. Even aunts. There is a great future before taxidermy, depend upon it. There is fossils again . . ."

He suddenly became silent.

"No, I don't think I ought to tell you that." He sucked at his pipe thoughtfully. "Thanks, yes. Not too much water.

"Of course, what I tell you now will go no further. You know I have made some dodos and a great auk? No! Evidently you are an amateur at taxidermy. My dear fellow, half the great auks in the world are about as genuine as the handkerchief of Saint Veronica, as the Holy Coat of Treves. We make 'em of grebes' feathers and the like. And the great auk's eggs too!"

"Good heavens!"

"Yes, we make them out of fine porcelain. I tell you it is worth while. They fetch—one fetched £300 only the other day. That one was really genuine, I believe, but of course one is never certain. It is very fine work, and afterwards you have to get them dusty, for no one who owns one of these precious eggs has ever the temerity to clean the thing. That's the beauty of the business. Even if they suspect an egg they do not like to examine it too closely. It's such brittle capital at the best.

"You did not know that taxidermy rose to heights like

that. My boy, it has risen higher. I have rivalled the hands of Nature herself. One of the *genuine* great auks" —his voice fell to a whisper—"one of the *genuine* great auks *was made by me.*"

"No. You must study ornithology, and find out which it is yourself. And what is more, I have been approached by a syndicate of dealers to stock one of the unexplored skerries to the north of Iceland with specimens. I may—some day. But I have another little thing in hand just now. Ever heard of the dinornis?

"It is one of those big birds recently extinct in New Zealand. 'Moa' is its common name, so called because extinct: there is no moa now. See? Well, they have got bones of it, and from some of the marshes even feathers and dried bits of skin. Now, I am going to—well, there is no need to make any bones about it—going to *forge* a complete stuffed moa. I know a chap out there who will pretend to make the find in a kind of antiseptic swamp, and say he stuffed it at once, as it threatened to fall to pieces. The feathers are peculiar, but I have got a simply lovely way of dodging up singed bits of ostrich plume. Yes, that is the new smell you noticed. They can only discover the fraud with a microscope, and they will hardly care to pull a nice specimen to bits for that.

"In this way, you see, I give my little push in the advancement of science.

"But all this is merely imitating Nature. I have done more than that in my time. I have—beaten her."

He took his feet down from the mantel-board, and leant over confidentially towards me. "I have *created* birds," he said in a low voice. "*New* birds. Improvements. Like no birds that were ever seen before."

He resumed his attitude during an impressive silence.

"Enrich the universe; *rath*-er. Some of the birds I made were new kinds of humming birds, and very beautiful little things, but some of them were simply rum. The rummest, I think, was the *Anomalopteryx Jejuna. Jejunus-a-um*— empty—so called because there was really nothing in it; a thoroughly empty bird—except for stuffing. Old Javvers has the thing now, and I suppose he is almost as proud of it as I am. It is a masterpiece, Bellows. It has all the silly clumsiness of your pelican, all the solemn want of dignity of your parrot, all the gaunt ungainliness of a flamingo, with all the extravagant chromatic conflict of a mandarin duck. *Such* a bird. I made it out of the skeletons of a stork and a toucan and a job lot of feathers. Taxidermy of that kind is just pure joy, Bellows, to a real artist in the art.

"How did I come to make it? Simple enough, as all great inventions are. One of those young genii who write us Science Notes in the papers got hold of a German pamphlet about the birds of New Zealand, and translated some of it by means of a dictionary and his mother-wit—he must have been one of a very large family with a small mother—and he got mixed between the living apteryx and the extinct anomalopteryx; talked about a bird five feet high, living in the jungles of the North Island, rare, shy, specimens difficult to obtain, and so on. Javvers, who even for a collector, is a miraculously ignorant man, read these paragraphs, and swore he would have the thing at any price. Raided the dealers with enquiries. It shows what a man can do by persistence—will-power. Here was a bird-collector swearing he would have a specimen of a bird that did not exist, that never had existed, and which for very shame of its own profane ungainliness, probably would not exist now if it could help itself. And he got it. *He got it.*"

"Have some more whisky, Bellows?" said the taxidermist, rousing himself from a transient contemplation of the mysteries of will-power and the collecting turn of mind. And, replenished, he proceeded to tell me of how he concocted a most attractive mermaid, and how an itinerant preacher, who could not get an audience because of it, smashed it because it was idolatry, or worse, at Burslem Wakes. But as the conversation of all the parties to this transaction, creator, would-be preserver, and destroyer, was uniformly unfit for publication, this cheerful incident must still remain unprinted.

The reader unacquainted with the dark ways of the collector may perhaps be inclined to doubt my taxidermist, but so far as great auks' eggs, and the bogus stuffed birds are concerned, I find that he has the confirmation of distinguished ornithological writers. And the note about the New Zealand bird certainly appeared in a morning paper of unblemished reputation, for the Taxidermist keeps a copy and has shown it to me.

STORY THE FIFTH

A Deal in Ostriches

"TALKING of the prices of birds, I've seen an ostrich that cost three hundred pounds," said the Taxidermist, recalling his youth of travel. "Three hundred pounds!"

He looked at me over his spectacles. "I've seen another that was refused at four."

"No," he said, "it wasn't any fancy points. They was just plain ostriches. A little off colour, too—owing to dietary. And there wasn't any particular restriction of the demand either. You'd have thought five ostriches would have ruled cheap on an East Indiaman. But the point was, one of 'em had swallowed a diamond.

"The chap it got it off was Sir Mohini Padishah, a tremendous swell, a Piccadilly swell you might say up to the neck of him, and then an ugly black head and a whopping turban, with this diamond in it. The blessed bird pecked suddenly and had it, and when the chap made a fuss it realised it had done wrong, I suppose, and went and mixed itself with the others to preserve its *incog*. It all happened in a minute. I was among the first to arrive, and there was this heathen going over his gods, and two sailors and the man who had charge of the birds laughing fit to split. It was a rummy way of losing a jewel, come to think of it. The man in charge hadn't been about just at the moment, so that he didn't know which bird it was. Clean lost, you see. I didn't feel half sorry, to tell you the truth. The beggar had been swaggering over his blessed diamond ever since he came aboard.

225

H

"A thing like that goes from stem to stern of a ship in no time. Everyone was talking about it. Padishah went below to hide his feelings. At dinner—he pigged at a table by himself, him and two other Hindoos—the captain kind of jeered at him about it, and he got very excited. He turned round and talked into my ear. He would not buy the birds; he would have his diamond. He demanded his rights as a British subject. His diamond must be found. He was firm upon that. He would appeal to the House of Lords. The man in charge of the birds was one of those wooden-headed chaps you can't get a new idea into anyhow. He refused any proposal to interfere with the birds by way of medicine. His instructions were to feed them so-and-so and treat them so-and-so, and it was as much as his place was worth not to feed them so-and-so and treat them so-and-so. Padishah had wanted a stomach-pump—though you can't do that to a bird, you know. This Padishah was full of bad law, like most of these blessed Bengalis, and talked of having a lien on the birds, and so forth. But an old boy, who said his son was a London barrister, argued that what a bird swallowed became *ipso facto* part of the bird, and that Padishah's only remedy lay in an action for damages, and even then it might be possible to show contributory negligence. He hadn't any right of way about an ostrich that didn't belong to him. That upset Padishah extremely, the more so as most of us expressed an opinion that that was the reasonable view. There wasn't any lawyer aboard to settle the matter, so we all talked pretty free. At last, after Aden, it appears that he came round to the general opinion, and went privately to the man in charge and made an offer for all five ostriches.

"The next morning there was a fine shindy at breakfast. The man hadn't any authority to deal with the birds, and nothing on earth would induce him to sell; but it seems he told Padishah that a Eurasian named Potter had already made him an offer, and on that Padishah denounced Potter before us all. But I think the most of us thought it rather smart of Potter, and I know that when Potter said that he'd wired at Aden to London to buy the birds, and would have an answer at Suez, I cursed pretty richly at a lost opportunity.

"At Suez, Padishah gave way to tears—actual wet tears—when Potter became the owner of the birds, and offered him two hundred and fifty right off for the five, being more than two hundred per cent. on what Potter had given. Potter said he'd be hanged if he parted with a feather of them—that he meant to kill them off one by one and find the diamond; but afterwards, thinking it over, he relented a little. He was a gambling hound, was this Potter, a little queer at cards, and this kind of prize-packet business must have suited him down to the ground. Anyhow, he offered, for a lark, to sell the birds separately to separate people by auction at a starting price of £80 for a bird. But one of them, he said, he meant to keep for luck.

"You must understand this diamond was a valuable one —a little Jew chap, a diamond merchant, who was with us, had put it at three or four thousand when Padishah had shown it to him—and this idea of an ostrich gamble caught on. Now it happened that I'd been having a few talks on general subjects with the man who looked after these ostriches, and quite incidentally he'd said one of the birds was ailing, and he fancied it had indigestion. It had one feather in its tail almost all white, by which I knew it, and so when, next day, the auction started with it, I capped Padishah's eighty-five by ninety. I fancy I was a bit too sure and eager with my bid, and some of the others spotted the fact that I was in the know. And Padishah went for that particular bird like an irresponsible lunatic. At last the Jew diamond merchant got it for £175, and Padishah said £180 just after the hammer came down—so Potter declared. At any rate the Jew merchant secured it, and there and then he got a gun and shot it. Potter made a Hades of a fuss because he said it would injure the sale of the other three, and Padishah, of course, behaved like an idiot; but all of us were very much excited. I can tell you I was precious glad when that dissection was over, and no diamond had turned up—precious glad. I'd gone to one-forty on that particular bird myself.

"The little Jew was like most Jews—he didn't make any great fuss over bad luck; but Potter declined to go on with the auction until it was understood that the goods could not be delivered until the sale was over. The little Jew wanted to argue that the case was exceptional, and as the discussion ran pretty even, the thing was postponed until the next morning. We had a lively dinner-table that evening, I can tell you, but in the end Potter got his way, since it would stand to reason he would be safer if he stuck to all the birds, and that we owed him some consideration for his sportsmanlike behaviour. And the old gentleman whose son was a lawyer said he'd been thinking the thing over and that it was very doubtful if, when a bird had been opened and the diamond recovered, it ought not to be handed back to the proper owner. I remember I suggested it came under the laws of treasure-trove—which was really the truth of the matter. There was a hot argument, and we settled it was certainly foolish to kill the bird on board the ship. Then the old gentleman, going at large through his legal talk, tried to make out the sale was a lottery and illegal, and appealed to the captain; but Potter said he sold the birds *as* ostriches. He didn't want to sell any diamonds, he said, and didn't offer that as an inducement. The three birds he put up, to the best of his knowledge and belief, did *not* contain a diamond. It was in the one he kept—so he hoped.

"Prices ruled high next day all the same. The fact that now there were four chances instead of five of course caused a rise. The blessed birds averaged £227, and, oddly enough, this Padishah didn't secure one of 'em—not one. He made too much shindy, and when he ought to have been bidding he was talking about liens, and, besides, Potter was a bit down on him. One fell to a quiet little officer chap, another to the little Jew, and the third was syndicated by the engineers. And then Potter seemed suddenly sorry for having sold them, and said he'd flung away a clear thousand pounds, and that very likely he'd draw a blank and that he always had been a fool, but when I went and had a bit of a talk to him, with the idea of getting him to hedge on his last chance, I found he'd already sold the bird he'd reserved to a political chap that was on board, a chap who'd been studying Indian morals and social questions in his vacation. That last was the three hundred pounds bird. Well, they landed three of the blessed creatures at Brindisi —though the old gentleman said it was a breach of the Customs regulations—and Potter and Padishah landed too. The Hindoo seemed half mad as he saw his blessed diamond going this way and that, so to speak. He kept on saying he'd get an injunction—he had injunction on the brain— and giving his name and address to the chaps who'd bought the birds, so that they'd know where to send the diamond. None of them wanted his name and address, and none of them would give their own. It was a fine row I can tell you—on the platform. They all went off by different trains. I came on to Southampton, and there I saw the last of the birds, as I came ashore; it was the one the engineers bought, and it was standing up near the bridge, in a kind of crate, and looking as leggy and silly a setting for a valuable diamond as ever you saw—if it *was* a setting for a valuable diamond.

"*How did it end?* Oh! like that. Well—perhaps. Yes, there's one more thing that may throw light on it. A week or so after landing I was down Regent Street doing a bit of shopping, and who should I see arm-in-arm and having a purple time of it but Padishah and Potter. If you come to think of it——

"Yes. *I've* thought that. Only, you see, there's no doubt the diamond was real. And Padishah was an eminent Hindoo. I've seen his name in the papers—often. But whether the bird swallowed the diamond certainly is another matter, as you say."

STORY THE SIXTH

Through a Window

AFTER his legs were set, they carried Bailey into the study and put him on a couch before the open window. There he lay, a live—even a feverish man down to the loins, and below that a double-barrelled mummy swathed in white wrappings. He tried to read, even tried to write a little, but most of the time he looked out of the window.

He had thought the window cheerful to begin with, but now he thanked God for it many times a day. Within, the room was dim and grey, and in the reflected light the wear of the furniture showed plainly. His medicine and drink stood on the little table, with such litter as the bare branches of a bunch of grapes or the ashes of a cigar upon a green plate, or a day old evening paper. The view outside was flooded with light, and across the corner of it came the head of the acacia, and at the foot the top of the balcony-railing of hammered iron. In the foreground was the weltering silver of the river, never quiet and yet never tiresome. Beyond was the reedy bank, a broad stretch of meadow land, and then a dark line of trees ending in a group of poplars at the distant bend of the river, and, upstanding behind them, a square church tower.

Up and down the river, all day long, things were passing. Now a string of barges drifting down to London, piled with lime or barrels of beer; then a steam-launch, disengaging heavy masses of black smoke, and disturbing the whole width of the river with long rolling waves; then an impetuous electric launch, and then a boatload of pleasure-seekers, a solitary sculler, or a four from some rowing club. Perhaps the river was quietest of a morning or late at night. One moonlight night some people drifted down singing, and with a zither playing—it sounded very pleasantly across the water.

In a few days Bailey began to recognise some of the craft; in a week he knew the intimate history of half-a-dozen. The launch *Luzon*, from Fitzgibbon's, two miles up, would go fretting by, sometimes three or four times a day, conspicuous with its colouring of Indian-red and yellow, and its two Oriental attendants; and one day, to Bailey's vast amusement, the house-boat *Purple Emperor* came to a stop outside, and breakfasted in the most shameless domesticity. Then one afternoon, the captain of a slow-moving barge began a quarrel with his wife as they came into sight from the left, and had carried it to personal violence before he vanished behind the window-frame to the right. Bailey regarded all this as an entertainment got up to while away his illness, and applauded all the more moving incidents. Mrs. Green, coming in at rare intervals with his meals, would catch him clapping his hands or softly crying, "Encore!" But the river players had other engagements, and his encore went unheeded.

"I should never have thought I could take such an interest in things that did not concern me," said Bailey to Wilderspin, who used to come in in his nervous, friendly way and try to comfort the sufferer by being talked to. "I thought this idle capacity was distinctive of little children and old maids. But it's just circumstances. I simply can't work, and things have to drift; it's no good to fret and struggle. And so I lie here and am as amused as a baby with a rattle, at this river and its affairs.

"Sometimes, of course, it gets a bit dull, but not often.

"I would give anything, Wilderspin, for a swamp—just one swamp—once. Heads swimming and a steam launch to the rescue, and a chap or so hauled out with a boat-hook. . . . There goes Fitzgibbon's launch! They have a new boat-hook, I see, and the little blackie is still in the dumps. I don't think he's very well, Wilderspin. He's been like that for two or three days, squatting sulky-fashion and meditating over the churning of the water. Unwholesome for him to be always staring at the frothy water running away from the stern."

They watched the little steamer fuss across the patch of sunlit river, suffer momentary occultation from the acacia, and glide out of sight behind the dark window-frame.

"I'm getting a wonderful eye for details," said Bailey: "I spotted that new boat-hook at once. The other nigger is a funny little chap. He never used to swagger with the old boat-hook like that."

"Malays, aren't they?" said Wilderspin.

"Don't know," said Bailey. "I thought one called all that sort of mariner Lascar."

Then he began to tell Wilderspin what he knew of the private affairs of the house-boat, *Purple Emperor*. "Funny," he said, "how these people come from all points of the compass—from Oxford and Windsor, from Asia and Africa— and gather and pass opposite the window just to entertain me. One man floated out of the infinite the day before yesterday, caught one perfect crab opposite, lost and recovered a scull, and passed on again. Probably he will never come into my life again. So far as I am concerned, he has lived and had his little troubles, perhaps thirty— perhaps forty—years on the earth, merely to make an ass of himself for three minutes in front of my window. Wonderful thing, Wilderspin, if you come to think of it."

"Yes," said Wilderspin; "*isn't* it?"

A day or two after this Bailey had a brilliant morning. Indeed, towards the end of the affair, it became almost as exciting as any window show very well could be. We will, however, begin at the beginning.

Bailey was all alone in the house, for his housekeeper had gone into the town three miles away to pay bills, and the servant had her holiday. The morning began dull. A canoe went up about half-past nine, and later a boat-load of camping men came down. But this was mere margin. Things became cheerful about ten o'clock.

It began with something white fluttering in the remote distance where the three poplars marked the river bend.

"Pocket-handkerchief," said Bailey, when he saw it. "No. Too big! Flag perhaps."

However, it was not a flag, for it jumped about. "Man in whites running fast, and this way," said Bailey. "That's luck! But his whites are precious loose!"

Then a singular thing happened. There was a minute pink gleam among the dark trees in the distance, and a little puff of pale grey that began to drift and vanish eastward. The man in white jumped and continued running. Presently the report of the shot arrived.

"What the devil!" said Bailey. "Looks as if someone was shooting at him."

He sat up stiffly and stared hard. The white figure was coming along the pathway through the corn. "It's one of those niggers from the Fitzgibbon's," said Bailey; "or may I be hanged! I wonder why he keeps sawing with his arm."

Then three other figures became indistinctly visible against the dark background of the trees.

Abruptly on the opposite bank a man walked into the picture. He was black-bearded, dressed in flannels, had a red belt, and a vast grey felt hat. He walked, leaning very much forward and with his hands swinging before him. Behind him one could see the grass swept by the towing-rope of the boat he was dragging. He was steadfastly regarding the white figure that was hurrying through the corn. Suddenly he stopped. Then, with a peculiar gesture, Bailey could see that he began pulling in the tow-rope hand over hand. Over the water could be heard the voices of the people in the still invisible boat.

"What are you after, Hagshot?" said someone.

The individual with the red belt shouted something that was inaudible, and went on lugging in the rope, looking over his shoulder at the advancing white figure as he did so. He came down the bank, and the rope bent a lane among the reeds and lashed the water between his pulls.

Then just the bows of the boat came into view, with the towing-mast and a tall, fair-haired man standing up and trying to see over the bank. The boat bumped unexpectedly among the reeds, and the tall, fair-haired man disappeared

suddenly, having apparently fallen back into the invisible
part of the boat. There was a curse and some indistinct
laughter. Hagshot did not laugh, but hastily clambered
into the boat and pushed off. Abruptly the boat passed
out of Bailey's sight.

But it was still audible. The melody of voices suggested
that its occupants were busy telling each other what to do.

The running figure was drawing near the bank. Bailey
could now see clearly that it was one of Fitzgibbon's
Orientals, and began to realise what the sinuous thing
the man carried in his hand might be. Three other men
followed one another through the corn, and the foremost
carried what was probably the gun. They were perhaps
two hundred yards or more behind the Malay.

"It's a man hunt, by all that's holy!" said Bailey.

The Malay stopped for a moment and surveyed the bank
to the right. Then he left the path, and, breaking through
the corn, vanished in that direction. The three pursuers
followed suit, and their heads and gesticulating arms above
the corn, after a brief interval, also went out of Bailey's
field of vision.

Bailey so far forgot himself as to swear. "Just as things
were getting lively!" he said. Something like a woman's
shriek came through the air. Then shouts, a howl, a dull
whack upon the balcony outside that made Bailey jump, and
then the report of a gun.

"This is precious hard on an invalid," said Bailey.

But more was to happen yet in his picture. In fact, a
great deal more. The Malay appeared again, running now
along the bank up stream. His stride had more swing and
less pace in it than before. He was threatening someone
ahead with the ugly krees he carried. The blade, Bailey
noticed, was dull—it did not shine as steel should.

Then came the tall, fair man, brandishing a boat-hook,
and after him three other men in boating costume running
clumsily with oars. The man with the grey hat and red
belt was not with them. After an interval the three men
with the gun reappeared, still in the corn, but now near
the river bank. They emerged upon the towing-path, and

hurried after the others. The opposite bank was left blank and desolate again.

The sick-room was disgraced by more profanity. "I would give my life to see the end of this," said Bailey. There were indistinct shouts up stream. Once they seemed to be coming nearer, but they disappointed him.

Bailey sat and grumbled. He was still grumbling when his eye caught something black and round among the waves. "Hullo!" he said. He looked narrowly and saw two triangular black bodies frothing every now and then about a yard in front of this.

He was still doubtful when the little band of pursuers came into sight again, and began to point to this floating object. They were talking eagerly. Then the man with the gun took aim.

"He's swimming the river, by George!" said Bailey.

The Malay looked round, saw the gun, and went under. He came up so close to Bailey's bank of the river that one of the bars of the balcony hid him for a moment. As he emerged the man with the gun fired. The Malay kept steadily onward—Bailey could see the wet hair on his forehead now and the krees between his teeth—and was presently hidden by the balcony.

This seemed to Bailey an unendurable wrong. The man was lost to him for ever now, so he thought. Why couldn't the brute have got himself decently caught on the opposite bank, or shot in the water?

"It's worse than Edwin Drood," said Bailey.

Over the river, too, things had become an absolute blank. All seven men had gone down stream again, probably to get the boat and follow across. Bailey listened and waited. There was silence. "Surely it's not over like this," said Bailey.

Five minutes passed—ten minutes. Then a tug with two barges went up stream. The attitudes of the men upon these were the attitudes of those who see nothing remarkable in earth, water, or sky. Clearly the whole affair had passed out of sight of the river. Probably the hunt had gone into the beech woods behind the house.

"Confound it!" said Bailey. "To be continued again, and no chance this time of the sequel. But this is hard on a sick man."

He heard a step on the staircase behind him, and looking round saw the door open. Mrs. Green came in and sat down, panting. She still had her bonnet on, her purse in her hand, and her little brown basket upon her arm. "Oh, there!" she said, and left Bailey to imagine the rest.

"Have a little whisky and water, Mrs. Green, and tell me about it," said Bailey.

Sipping a little, the lady began to recover her powers of explanation.

One of those black creatures at the Fitzgibbon's had gone mad, and was running about with a big knife, stabbing people. He had killed a groom, and stabbed the under-butler, and almost cut the arm off a boating gentleman.

"Running amuck with a krees," said Bailey. "I thought that was it."

And he was hiding in the wood when she came through it from the town.

"What! Did he run after you?" asked Bailey, with a certain touch of glee in his voice.

"No, that was the horrible part of it," Mrs. Green explained. She had been right through the woods and had *never known he was there*. It was only when she met young Mr. Fitzgibbon carrying his gun in the shrubbery that she heard anything about it. Apparently, what upset Mrs. Green was the lost opportunity for emotion. She was determined, however, to make the most of what was left her.

"To think he was there all the time!" she said, over and over again.

Bailey endured this patiently enough for perhaps ten minutes. At last he thought it advisable to assert himself. "It's twenty past one, Mrs. Green," he said. "Don't you think it time you got me something to eat?"

This brought Mrs. Green suddenly to her knees.

"Oh Lord, sir!" she said. "Oh! don't go making me go out of this room, sir, till I know he's caught. He might have got into the house, sir. He might be creeping, creeping, with that knife of his, along the passage this very——"

She broke off suddenly and glared over him at the window. Her lower jaw dropped. Bailey turned his head sharply.

For the space of half a second things seemed just as they were. There was the tree, the balcony, the shining river, the distant church tower. Then he noticed that the acacia was displaced about a foot to the right, and that it was quivering, and the leaves were rustling. The tree was shaken violently, and a heavy panting was audible.

In another moment a hairy brown hand had appeared and clutched the balcony railings, and in another the face of the Malay was peering through these at the man on the couch. His expression was an unpleasant grin, by reason of the krees he held between his teeth, and he was bleeding from an ugly wound in his cheek. His hair wet to drying stuck out like horns from his head. His body was bare save for the wet trousers that clung to him. Bailey's first impulse was to spring from the couch, but his legs reminded him that this was impossible.

By means of the balcony and tree the man slowly raised himself until he was visible to Mrs. Green. With a choking cry she made for the door and fumbled with the handle.

Bailey thought swiftly and clutched a medicine bottle in either hand. One he flung, and it smashed against the acacia. Silently and deliberately, and keeping his bright eyes fixed on Bailey, the Malay clambered into the balcony. Bailey, still clutching his second bottle, but with a sickening, sinking feeling about his heart, watched first one leg come over the railing and then the other.

It was Bailey's impression that the Malay took about an hour to get his second leg over the rail. The period that elapsed before the sitting position was changed to a standing one seemed enormous—days, weeks, possibly a year or so. Yet Bailey had no clear impression of anything going on in his mind during that vast period, except a vague wonder at his inability to throw the second medicine bottle. Suddenly the Malay glanced over his shoulder. There was the crack of a rifle. He flung up his arms and came down upon the couch. Mrs. Green began a dismal shriek that seemed likely to last until Doomsday. Bailey stared at the brown body with its shoulder blade driven in, that writhed painfully across his legs and rapidly staining and soaking the spotless bandages. Then he looked at the long krees, with the reddish streaks upon its blade, that lay an inch beyond the trembling brown fingers upon the floor. Then at Mrs. Green, who had backed hard against the door and was staring at the body and shrieking in gusty outbursts as if she would wake the dead. And then the body was shaken by one last convulsive effort.

The Malay gripped the krees, tried to raise himself with his left hand, and collapsed. Then he raised his head, stared for a moment at Mrs. Green, and twisting his face round looked at Bailey. With a gasping groan the dying man succeeded in clutching the bed clothes with his disabled hand, and by a violent effort, which hurt Bailey's legs exceedingly, writhed sideways towards what must be his last victim. Then something seemed released in Bailey's mind and he brought down the second bottle with all his strength on to the Malay's face. The krees fell heavily upon the floor.

"Easy with those legs," said Bailey, as young Fitzgibbon and one of the boating party lifted the body off him.

Young Fitzgibbon was very white in the face. "I didn't mean to kill him," he said.

"It's just as well," said Bailey.

STORY THE SEVENTH

The Temptation of Harringay

IT is quite impossible to say whether this thing really happened. It depends entirely on the word of R. M. Harringay, who is an artist.

Following his version of the affair, the narrative deposes that Harringay went into his studio about ten o'clock to see what he could make of the head that he had been working at the day before. The head in question was that of an Italian organ-grinder, and Harringay thought—but was not quite sure—that the title would be the "Vigil." So far he is frank, and his narrative bears the stamp of truth. He had seen the man expectant for pennies, and with a promptness that suggested genius, had had him in at once.

"Kneel. Look up at that bracket," said Harringay. "As if you expected pennies."

"Don't *grin!*" said Harringay. "I don't want to paint your gums. Look as though you were unhappy."

Now, after a night's rest, the picture proved decidedly unsatisfactory. "It's good work," said Harringay. "That little bit in the neck. . . But."

He walked about the studio, and looked at the thing from this point and from that. Then he said a wicked word. In the original the word is given.

"Painting," he says he said. "Just a painting of an organ-grinder—a mere portrait. If it was a live organ-grinder I wouldn't mind. But somehow I never make things alive. I wonder if my imagination is wrong." This, too, has a truthful air. His imagination *is* wrong.

239

"That creative touch! To take canvas and pigment and make a man—as Adam was made of red ochre! But this thing! If you met it walking about the streets you would know it was only a studio production. The little boys would tell it to 'Garnome and git frimed.' Some little touch. . . Well—it won't do as it is."

He went to the blinds and began to pull them down. They were made of blue holland with the rollers at the bottom of the window, so that you pull them down to get more light. He gathered his palette, brushes, and mahl stick from his table. Then he turned to the picture and put a speck of brown in the corner of the mouth; and shifted his attention thence to the pupil of the eye. Then he decided that the chin was a trifle too impassive for a vigil.

Presently he put down his impedimenta, and lighting a pipe surveyed the progress of his work. "I'm hanged if the thing isn't sneering at me," said Harringay, and he still believes it sneered.

The animation of the figure had certainly increased, but scarcely in the direction he wished. There was no mistake about the sneer. "Vigil of the Unbeliever," said Harringay. "Rather subtle and clever that! But the left eyebrow isn't cynical enough."

He went and dabbed at the eyebrow, and added a little to the lobe of the ear to suggest materialism. Further consideration ensued. "Vigil's off. I'm afraid," said Harringay. "Why not Mephistopheles? But that's a bit too common. 'A Friend of the Doge'—not so seedy. The armour won't do, though. Too Camelot. How about a scarlet robe and call him 'One of the Sacred College'? Humour in that, and an appreciation of Middle Italian History."

"There's always Benvenuto Cellini," said Harringay; "with a clever suggestion of a gold cup in one corner. But that would scarcely suit the complexion."

He describes himself as babbling in this way in order to keep down an unaccountably unpleasant sensation of fear. The thing was certainly acquiring anything but a pleasing expression. Yet it was as certainly becoming far

to talk studio to me?" He filled his number twelve hoghair with red paint.

"The true artist," said the picture, "is always an ignorant man. An artist who theorises about his work is no longer artist but critic. Wagner . . . I say!—What's that red paint for?"

"I'm going to paint you out," said Harringay. "I don't want to hear all that Tommy Rot. If you think just because I'm an artist by trade I'm going to talk studio to you, you make a precious mistake."

"One minute," said the picture, evidently alarmed. "I want to make you an offer—a genuine offer. It's right what I'm saying. You lack inspirations. Well. No doubt you've heard of the Cathedral of Cologne, and the Devil's Bridge, and——"

"Rubbish," said Harringay. "Do you think I want to go to perdition simply for the pleasure of painting a good picture, and getting it slated. Take that."

His blood was up. His danger only nerved him to action, so he says. So he planted a dab of vermilion in his creature's mouth. The Italian spluttered and tried to wipe it off—evidently horribly surprised. And then—according to Harringay—there began a very remarkable struggle, Harringay splashing away with the red paint, and the picture wriggling about and wiping it off as fast as he put it on. "*Two* masterpieces," said the demon. "Two indubitable masterpieces for a Chelsea artist's soul. It's a bargain?" Harringay replied with the paint brush.

For a few minutes nothing could be heard but the brush going and the spluttering and ejaculations of the Italian. A lot of the strokes he caught on his arm and hand, though Harringay got over his guard often enough. Presently the paint on the palette gave out and the two antagonists stood breathless, regarding each other. The picture was so smeared with red that it looked as if it had been rolling about a slaughterhouse, and it was painfully out of breath and very uncomfortable with the wet paint trickling down its neck. Still, the first round was in its favour on the whole. "Think," it said, sticking pluckily to its point, "two supreme masterpieces—in different styles. Each equivalent to the Cathedral . . ."

"I know," said Harringay, and rushed out of the studio and along the passage towards his wife's boudoir.

In another minute he was back with a large tin of enamel —Hedge Sparrow's Egg Tint, it was, and a brush. At the sight of that the artistic devil with the red eye began to scream. "*Three* masterpieces—culminating masterpieces."

Harringay delivered cut two across the demon, and followed with a thrust in the eye. There was an indistinct rumbling. "*Four* masterpieces," and a spitting sound.

But Harringay had the upper hand now and meant to keep it. With rapid, bold strokes he continued to paint over the writhing canvas, until at last it was a uniform field of shining Hedge Sparrow tint. Once the mouth reappeared and got as far as "Five master——" before he filled it with enamel; and near the end the red eye opened and glared at him indignantly. But at last nothing remained save a gleaming panel of drying enamel. For a little while a faint stirring beneath the surface puckered it slightly here and there, but presently even that died away and the thing was perfectly still.

Then Harringay—according to Harringay's account—lit his pipe and sat down and stared at the enamelled canvas, and tried to make out clearly what had happened. Then he walked round behind it, to see if the back of it was at all remarkable. Then it was he began to regret that he had not photographed the Devil before he painted him out.

This is Harringay's story—not mine. He supports it by a small canvas (24 by 20) enamelled a pale green, and by violent asseverations. It is also true that he never has produced a masterpiece, and in the opinion of his intimate friends probably never will.

STORY THE EIGHTH

The Flying Man

THE Ethnologist looked at the *bhimraj* feather thoughtfully. "They seemed loth to part with it," he said.

"It is sacred to the Chiefs," said the lieutenant; "just as yellow silk, you know, is sacred to the Chinese Emperor."

The Ethnologist did not answer. He hesitated. Then opening the topic abruptly, "What on earth is this cock-and-bull story they have of a flying man?"

The lieutenant smiled faintly. "What did they tell you?"

"I see," said the Ethnologist, "that you know of your fame."

The lieutenant rolled himself a cigarette. "I don't mind hearing about it once more. How does it stand at present?"

"It's so confoundedly childish," said the Ethnologist, becoming irritated. "How did you play it off upon them?"

The lieutenant made no answer, but lounged back in his folding chair, still smiling.

"Here am I, come four hundred miles out of my way to get what is left of the folk-lore of these people, before they are utterly demoralised by missionaries and the military, and all I find are a lot of impossible legends about a sandy-haired scrub of an infantry lieutenant. How he is invulnerable—how he can jump over elephants—how he can fly. That's the toughest nut. One old gentleman described your wings, said they had black plumage and were not quite as long as a mule. Said he often saw you by moonlight hovering over the crests out towards the Shendu country. Confound it, man!"

The lieutenant laughed cheerfully. "Go on," he said. "Go on."

The Ethnologist did. At last he wearied. "To trade so," he said, "on these unsophisticated children of the mountains. How could you bring yourself to do it, man?"

"I'm sorry," said the lieutenant, "but truly the thing was forced upon me. I can assure you I was driven to it. And at the time I had not the faintest idea of how the Chin imagination would take it. Or curiosity. I can only plead it was an indiscretion and not malice that made me replace the folk-lore by a new legend. But as you seem aggrieved, I will try and explain the business to you.

"It was in the time of the last Lushai expedition but one, and Walters thought these people you have been visiting were friendly. So, with an airy confidence in my capacity for taking care of myself, he sent me up the gorge —fourteen miles of it—with three of the Derbyshire men and half a dozen Sepoys, two mules, and his blessing, to see what popular feeling was like at that village you visited. A force of ten—not counting the mules—fourteen miles, and during a war! You saw the road?"

"*Road!*" said the Ethnologist.

"It's better now than it was. When we went up we had to wade in the river for a mile where the valley narrows, with a smart stream frothing round our knees and the stones as slippery as ice. There it was I dropped my rifle. Afterwards the Sappers blasted the cliff with dynamite and made the convenient way you came by. Then below, where those very high cliffs come, we had to keep on dodging across the river—I should say we crossed it a dozen times in a couple of miles.

"We got in sight of the place early the next morning. You know how it lies, on a spur halfway between the big hills, and as we began to appreciate how wickedly quiet the village lay under the sunlight, we came to a stop to consider.

"At that they fired a lump of filed brass idol at us, just by way of a welcome. It came twanging down the slope to the right of us where the boulders are, missed my shoulder by an inch or so, and plugged the mule that carried all the provisions and utensils. I never heard such a death-rattle before or since. And at that we became aware of a number of gentlemen carrying matchlocks, and dressed in things like plaid dusters, dodging about along the neck between the village and the crest to the east.

"'Right about face,' I said. 'Not too close together.'

"And with that encouragement my expedition of ten men came round and set off at a smart trot down the valley again hitherward. We did not wait to save anything our dead had carried, but we kept the second mule with us—he carried my tent and some other rubbish—out of a feeling of friendship.

"So ended the battle—ingloriously. Glancing back, I saw the valley dotted with the victors, shouting and firing at us. But no one was hit. These Chins and their guns are very little good except at a sitting shot. They will sit and finick over a boulder for hours taking aim, and when they fire running it is chiefly for stage effect. Hooker, one of the Derbyshire men, fancied himself rather with the rifle, and stopped behind for half a minute to try his luck as we turned the bend. But he got nothing.

"I'm not a Xenophon to spin much of a yarn about my retreating army. We had to pull the enemy up twice in the next two miles when he became a bit pressing, by exchanging shots with him, but it was a fairly monotonous affair—hard breathing chiefly—until we got near the place where the hills run in towards the river and pinch the valley into a gorge. And there we very luckily caught a glimpse of half a dozen round black heads coming slanting-ways over the hill to the left of us—the east that is—and almost parallel with us.

"At that I called a halt. 'Look here,' says I to Hooker and the other Englishmen; 'what are we to do now?' and I pointed to the heads.

"'Headed orf, or I'm a nigger,' said one of the men.

"'We shall be,' said another. 'You know the Chin way, George?'

"'They can pot everyone of us at fifty yards,' says Hooker, 'in the place where the river is narrow. It's just suicide to go on down.'

"I looked at the hill to the right of us. It grew steeper

lower down the valley, but it still seemed climbable. And all the Chins we had seen hitherto had been on the other side of the stream.

" 'It's that or stopping,' says one of the Sepoys.

"So we started slanting up the hill. There was something faintly suggestive of a road running obliquely up the face of it, and that we followed. Some Chins presently came into view up the valley, and I heard some shots. Then I saw one of the Sepoys was sitting down about thirty yards below us. He had simply sat down without a word, apparently not wishing to give trouble. At that I called a halt again; I told Hooker to try another shot, and went back and found the man was hit in the leg. I took him up, carried him along to put him on the mule—already pretty well laden with the tent and other things which we had no time to take off. When I got up to the rest with him, Hooker had his empty Martini in his hand, and was grinning and pointing to a motionless black spot up the valley. All the rest of the Chins were behind boulders or back round the bend. 'Five hundred yards,' says Hooker, 'if an inch. And I'll swear I hit him in the head.'

"I told him to go and do it again, and with that we went on again.

"Now the hillside kept getting steeper as we pushed on, and the road we were following more and more of a shelf. At last it was mere cliff above and below us. 'It's the best road I have seen yet in Chin Lushai land,' said I to encourage the men, though I had a fear of what was coming.

"And in a few minutes the way bent round a corner of the cliff. Then, finis! the ledge came to an end.

"As soon as he grasped the position one of the Derbyshire men fell a-swearing at the trap we had fallen into. The Sepoys halted quietly. Hooker grunted and reloaded, and went back to the bend.

"Then two of the Sepoy chaps helped their comrade down and began to unload the mule.

"Now, when I came to look about me, I began to think we had not been so very unfortunate after all. We were on a shelf perhaps ten yards across it at widest. Above it the cliff projected so that we could not be shot down upon, and below was an almost sheer precipice of perhaps two or three hundred feet. Lying down we were invisible to anyone across the ravine. The only approach was along the ledge, and on that one man was as good as a host. We were in a natural stronghold, with only one disadvantage, our sole provision against hunger and thirst was one live mule. Still we were at most eight or nine miles from the main expedition, and no doubt, after a day or so, they would send up after us if we did not return.

"After a day or so . . ."

The lieutenant paused. "Ever been thirsty, Graham?"

"Not that kind," said the Ethnologist.

"H'm. We had the whole of that day, the night, and the next day of it, and only a trifle of dew we wrung out of our clothes and the tent. And below us was the river going giggle, giggle, round a rock in mid stream. I never knew such a barrenness of incident, or such a quantity of sensation. The sun might have had Joshua's command still upon it for all the motion one could see; and it blazed like a near furnace. Towards the evening of the first day one of the Derbyshire men said something—nobody heard what—and went off round the bend of the cliff. We heard shots, and when Hooker looked round the corner he was gone. And in the morning the Sepoy whose leg was shot was in delirium, and jumped or fell over the cliff. Then we took the mule and shot it, and that must needs go over the cliff too in its last struggles, leaving eight of us.

"We could see the body of the Sepoy down below, with the head in the water. He was lying face downwards, and so far as I could make out was scarcely smashed at all. Badly as the Chins might covet his head, they had the sense to leave it alone until the darkness came.

"At first we talked of all the chances there were of the main body hearing the firing, and reckoned whether they would begin to miss us, and all that kind of thing, but we dried up as the evening came on. The Sepoys played games with bits of stone among themselves, and afterwards told stories. The night was rather chilly. The second day nobody spoke. Our lips were black and our throats afire, and we lay about on the ledge and glared at one another. Perhaps it's as well we kept our thoughts to ourselves. One of the British soldiers began writing some blasphemous rot on the rock with a bit of pipeclay, about his last dying will, until I stopped it. As I looked over the edge down into the valley and saw the river rippling I was nearly tempted to go after the Sepoy. It seemed a pleasant and desirable thing to go rushing down through the air with something to drink—or no more thirst at any rate—at the bottom. I remembered in time, though, that I was the officer in command, and my duty to set a good example, and that kept me from any such foolishness.

"Yet, thinking of that, put an idea into my head. I got up and looked at the tent and tent ropes, and wondered why I had not thought of it before. Then I came and peered over the cliff again. This time the height seemed greater and the pose of the Sepoy rather more painful. But it was that or nothing. And to cut it short, I parachuted.

"I got a big circle of canvas out of the tent, about three times the size of that table-cover, and plugged the hole in the centre, and I tied eight ropes round it to meet in the middle and make a parachute. The other chaps lay about and watched me as though they thought it was a new kind of delirium. Then I explained my notion to the two British soldiers and how I meant to do it, and as soon as the short dusk had darkened into night, I risked it. They held the thing high up, and I took a run the whole length of the ledge. The thing filled with air like a sail, but at the edge I will confess I funked and pulled up.

"As soon as I stopped I was ashamed of myself—as well I might be in front of privates—and went back and started again. Off I jumped this time—with a kind of sob, I remember—clean into the air, with the big white sail bellying out above me.

"I must have thought at a frightful pace. It seemed a long time before I was sure that the thing meant to keep steady. At first it heeled sideways. Then I noticed the face of the rock which seemed to be streaming up past me, and me motionless. Then I looked down and saw in the darkness the river and the dead Sepoy rushing up towards me. But in the indistinct light I also saw three Chins, seemingly aghast at the sight of me, and that the Sepoy was decapitated. At that I wanted to go back again.

"Then my boot was in the mouth of one, and in a moment he and I were in a heap with the canvas fluttering down on the top of us. I fancy I dashed out his brains with my foot. I expected nothing more than to be brained myself by the other two, but the poor heathen had never heard of Baldwin, and incontinently bolted.

"I struggled out of the tangle of dead Chin and canvas, and looked round. About ten paces off lay the head of the Sepoy staring in the moonlight. Then I saw the water and went and drank. There wasn't a sound in the world but the footsteps of the departing Chins, a faint shout from above, and the gluck of the water. So soon as I had drunk my full I started off down the river.

"That about ends the explanation of the flying man story. I never met a soul the whole eight miles of the way. I got to Walters' camp by ten o'clock, and a born idiot of a sentinel had the cheek to fire at me as I came trotting out of the darkness. So soon as I had hammered my story into Winter's thick skull, about fifty men started up the valley to clear the Chins out and get our men down. But for my own part I had too good a thirst to provoke it by going with them.

"You have heard what kind of a yarn the Chins made of it. Wings as long as a mule, eh?—And black feathers! The gay lieutenant bird! Well, well."

The lieutenant meditated cheerfully for a moment. Then he added, "You would scarcely credit it, but when they got to the ridge at last, they found two more of the Sepoys had jumped over."

"The rest were all right?" asked the Ethnologist.

"Yes," said the lieutenant; "the rest were all right, barring a certain thirst, you know."

And at the memory he helped himself to soda and whisky again.

STORY THE NINTH

The Diamond Maker

SOME business had detained me in Chancery Lane until nine in the evening, and thereafter, having some inkling of a headache, I was disinclined either for entertainment or further work. So much of the sky as the high cliffs of that narrow cañon of traffic left visible spoke of a serene night, and I determined to make my way down to the Embankment, and rest my eyes and cool my head by watching the variegated lights upon the river. Beyond comparison the night is the best time for this place; a merciful darkness hides the dirt of the waters, and the lights of this transition age, red, glaring orange, gas-yellow, and electric white, are set in shadowy outlines of every possible shade between grey and deep purple. Through the arches of Waterloo Bridge a hundred points of light mark the sweep of the Embankment, and above its parapet rise the towers of Westminster, warm grey against the starlight. The black river goes by with only a rare ripple breaking its silence, and disturbing the reflections of the lights that swim upon its surface.

"A warm night," said a voice at my side.

I turned my head, and saw the profile of a man who was leaning over the parapet beside me. It was a refined face, not unhandsome, though pinched and pale enough, and the coat collar turned up and pinned round the throat marked his status in life as sharply as a uniform. I felt I was committed to the price of a bed and breakfast if I answered him.

I looked at him curiously. Would he have anything to

tell me worth the money, or was he the common incapable —incapable even of telling his own story? There was a quality of intelligence in his forehead and eyes, and a certain tremulousness in his nether lip that decided me.

"Very warm," said I; " but not too warm for us here."

"No," he said, still looking across the water, "it is pleasant enough here . . .just now."

"It is good," he continued after a pause, "to find anything so restful as this in London. After one has been fretting about business all day, about getting on, meeting obligations, and parrying dangers, I do not know what one would do if it were not for such pacific corners." He spoke with long pauses between the sentences. "You must know a little of the irksome labour of the world, or you would not be here. But I doubt if you can be so brain-weary and footsore as I am . . . Bah! Sometimes I doubt if the game is worth the candle. I feel inclined to throw the whole thing over—name, wealth, and position—and take to some modest trade. But I know if I abandoned my ambition— hardly as she uses me—I should have nothing but remorse left for the rest of my days."

He became silent. I looked at him in astonishment. If ever I saw a man hopelessly hard-up it was the man in front of me. He was ragged and he was dirty, unshaven and unkempt; he looked as though he had been left in a dust-bin for a week. And he was talking to *me* of the irksome worries of a large business. I almost laughed outright. Either he was mad or playing a sorry jest on his own poverty.

"If high aims and high positions," said I, "have their drawbacks of hard work and anxiety, they have their compensations. Influence, the power of doing good, of assisting those weaker and poorer than ourselves; and there is even a certain gratification in display. . . ."

My banter under the circumstances was in very vile taste. I spoke on the spur of the contrast of his appearance and speech. I was sorry even while I was speaking.

He turned a haggard but very composed face upon me.

Said he: "I forget myself. Of course you would not under stand."

He measured me for a moment. "No doubt it is very absurd. You will not believe me even when I tell you so that it is fairly safe to tell you. And it will be a comfort to tell someone. I really have a big business in hand, a very big business. But there are troubles just now. The fact is . . . I make diamonds."

"I suppose," said I, "you are out of work just at present?"

"I am sick of being disbelieved," he said impatiently, and suddenly unbuttoning his wretched coat he pulled out a little canvas bag that was hanging by a cord round his neck. From this he produced a brown pebble. "I wonder if you know enough to know what that is?" He handed it to me.

Now, a year or so ago, I had occupied my leisure in taking a London science degree, so that I have a smattering of physics and mineralogy. The thing was not unlike an uncut diamond of the darker sort, though far too large, being almost as big as the top of my thumb. I took it, and saw it had the form of a regular octahedron, with the carved faces peculiar to the most precious of minerals. I took out my penknife and tried to scratch it—vainly. Leaning forward towards the gas-lamp, I tried the thing on my watch-glass, and scored a white line across that with the greatest ease.

I looked at my interlocutor with rising curiosity. "It certainly is rather like a diamond. But, if so, it is a Behemoth of diamonds. Where did you get it?"

"I tell you I made it," he said. "Give it back to me."

He replaced it hastily and buttoned his jacket. "I will sell it you for one hundred pounds," he suddenly whispered eagerly. With that my suspicions returned. The thing might, after all, be merely a lump of that almost equally hard substance, corundum, with an accidental resemblance in shape to the diamond. Or if it was a diamond, how came he by it, and why should he offer it at a hundred pounds?

We looked into one another's eyes. He seemed eager, but honestly eager. At that moment I believed it was a diamond he was trying to sell. Yet I am a poor man, a hundred pounds would leave a visible gap in my fortunes and no sane man would buy a diamond by gaslight from a ragged tramp on his personal warranty only. Still, a diamond that size conjured up a vision of many thousands of pounds. Then, thought I, such a stone could scarcely exist without being mentioned in every book on gems, and again I called to mind the stories of contraband and light-fingered Kaffirs at the Cape. I put the question of purchase on one side.

"How did you get it?" said I.

"I made it."

I had heard something of Moissan, but I knew his artificial diamonds were very small. I shook my head.

"You seem to know something of this kind of thing. I will tell you a little about myself. Perhaps then you may think better of the purchase." He turned round with his back to the river, and put his hands in his pockets. He sighed. "I know you will not believe me."

"Diamonds," he began—and as he spoke his voice lost its faint flavour of the tramp and assumed something of the easy tone of an educated man—"are to be made by throwing carbon out of combination in a suitable flux and under a suitable pressure; the carbon crystallises out, not as black-lead or charcoal-powder, but as small diamonds. So much has been known to chemists for years, but no one yet has hit upon exactly the right flux in which to melt up the carbon, or exactly the right pressure for the best results. Consequently the diamonds made by chemists are small and dark, and worthless as jewels. Now I, you know, have given up my life to this problem—given my life to it.

"I began to work at the conditions of diamond making when I was seventeen, and now I am thirty-two. It seemed to me that it might take all the thought and energies of a man for ten years, or twenty years, but, even if it did, the game was still worth the candle. Suppose one to have at last just hit the right trick, before the secret got out and diamonds became as common as coal, one might realise millions. Millions!"

He paused and looked for my sympathy. His eyes shone

hungrily. "To think," said he, "that I am on the verge of it all, and here!

"I had," he proceeded, "about a thousand pounds when I was twenty-one, and this, I thought, eked out by a little teaching, would keep my researches going. A year or two was spent in study, at Berlin chiefly, and then I continued on my own account. The trouble was the secrecy. You see, if once I had let out what I was doing, other men might have been spurred on by my belief in the practicability of the idea; and I do not pretend to be such a genius as to have been sure of coming in first, in the case of a race for the discovery. And you see it was important that if I really meant to make a pile, people should not know it was an artificial process and capable of turning out diamonds by the ton. So I had to work all alone. At first I had a little laboratory, but as my resources began to run out I had to conduct my experiments in a wretched unfurnished room in Kentish Town, where I slept at last on a straw mattress on the floor among all my apparatus. The money simply flowed away. I grudged myself everything except scientific appliances. I tried to keep things going by a little teaching, but I am not a very good teacher, and I have no university degree, nor very much education except in chemistry, and I found I had to give a lot of time and labour for precious little money. But I got nearer and nearer the thing. Three years ago I settled the problem of the composition of the flux, and got near the pressure by putting this flux of mine and a certain carbon composition into a closed-up gun-barrel, filling up with water, sealing tightly, and heating."

He paused.

"Rather risky," said I.

"Yes. It burst, and smashed all my windows and a lot of my apparatus; but I got a kind of diamond powder nevertheless. Following out the problem of getting a big pressure upon the molten mixture from which the things were to crystallise, I hit upon some researches of Daubré's at the Paris Laboratorie des Poudres et Salpêtres. He exploded dynamite in a tightly screwed steel cylinder, too strong to burst, and I found he could crush rocks into a muck not unlike the South African bed in which diamonds are found. It was a tremendous strain on my resources, but I got a steel cylinder made for my purpose after his pattern. I put in all my stuff and my explosives, built up a fire in my furnace, put the whole concern in, and—went out for a walk."

I could not help laughing at his matter-of-fact manner. "Did you not think it would blow up the house? Were there other people in the place?"

"It was in the interest of science," he said ultimately. "There was a costermonger family on the floor below, a begging-letter writer in the room behind mine, and two flower-women were upstairs. Perhaps it was a bit thoughtless. But possibly some of them were out.

"When I came back the thing was just where I left it, among the white-hot coals. The explosive hadn't burst the case. And then I had a problem to face. You know time is an important element in crystallisation. If you hurry the process the crystals are small—it is only by prolonged standing that they grow to any size. I resolved to let this apparatus cool for two years, letting the temperature go down slowly during that time. And I was now quite out of money; and with a big fire and the rent of my room, as well as my hunger to satisfy, I had scarcely a penny in the world.

"I can hardly tell you all the shifts I was put to while I was making the diamonds. I have sold newspapers, held horses, opened cab-doors. For many weeks I addressed envelopes. I had a place as assistant to a man who owned a barrow, and used to call down one side of the road while he called down the other. Once for a week I had absolutely nothing to do, and I begged. What a week that was! One day the fire was going out and I had eaten nothing all day, and a little chap taking his girl out, gave me sixpence—to show-off. Thank heaven for vanity! How the fish-shops smelt! But I went and spent it all on coals, and had the furnace bright red again, and then—— Well, hunger makes a fool of a man.

"At last, three weeks ago, I let the fire out. I took my cylinder and unscrewed it while it was still so hot that it punished my hands, and I scraped out the crumbling lava-like mass with a chisel, and hammered it into a powder upon an iron plate. And I found three big diamonds and five small ones. As I sat on the floor hammering, my door opened, and my neighbour, the begging-letter writer, came in. He was drunk—as he usually is. ''Nerchist,' said he. 'You're drunk,' said I. ''Structive scoundrel,' said he. 'Go to your father,' said I, meaning the Father of Lies. 'Never you mind,' said he, and gave me a cunning wink, and hic-cupped, and leaning up against the door, with his other eye against the door-post, began to babble of how he had been prying in my room, and how he had gone to the police that morning, and how they had taken down everything he had to say—''siffiwas a ge'm,' said he. Then I suddenly realised I was in a hole. Either I should have to tell these police my little secret, and get the whole thing blown upon, or be lagged as an Anarchist. So I went up to my neigh-bour and took him by the collar, and rolled him about a bit, and then I gathered up my diamonds and cleared out. The evening newspapers called my den the Kentish-Town Bomb Factory. And now I cannot part with the things for love or money.

"If I go in to respectable jewellers they ask me to wait, and go and whisper to a clerk to fetch a policeman, and then I say I cannot wait. And I found out a receiver of stolen goods, and he simply stuck to the one I gave him and told me to prosecute if I wanted it back. I am going about now with several hundred thousand pounds-worth of diamonds round my neck, and without either food or shelter. You are the first person I have taken into my con-fidence. But I like your face and I am hard-driven."

He looked into my eyes.

"It would be madness," said I, "for me to buy a diamond under the circumstances. Besides, I do not carry hundreds of pounds about in my pocket. Yet I more than half believe your story. I will, if you like, do this: come to my office to-morrow. . . ."

"You think I am a thief!" said he keenly. "You will tell the police. I am not coming into a trap."

"Somehow I am assured you are no thief. Here is my card. Take that, anyhow. You need not come to any appointment. Come when you will."

He took the card, and an earnest of my good-will.

"Think better of it and come," said I.

He shook his head doubtfully. "I will pay back your half-crown with interest some day—such interest as will amaze you," said he. "Anyhow, you will keep the secret? . . . Don't follow me."

He crossed the road and went into the darkness towards the little steps under the archway leading into Essex Street, and I let him go. And that was the last I ever saw of him.

Afterwards I had two letters from him asking me to send bank-notes—not cheques—to certain addresses. I weighed the matter over, and took what I conceived to be the wisest course. Once he called upon me when I was out. My urchin described him as a very thin, dirty, and ragged man, with a dreadful cough. He left no message. That was the finish of him so far as my story goes. I wonder sometimes what has become of him. Was he an ingenious monomaniac, or a fraudulent dealer in pebbles, or has he really made diamonds as he asserted? The latter is just sufficiently credible to make me think at times that I have missed the most brilliant opportunity of my life. He may of course be dead, and his diamonds carelessly thrown aside—one, I repeat, was almost as big as my thumb. Or he may be still wandering about trying to sell the things. It is just possible he may yet emerge upon society, and, passing athwart my heavens in the serene altitude sacred to the wealthy and the well-advertised, reproach me silently for my want of enterprise. I sometimes think I might at least have risked five pounds.

STORY THE TENTH

Æpyornis Island

THE man with the scarred face leant over the table and looked at my bundle.

"Orchids?" he asked.

"A few," I said.

"Cypripediums," he said.

"Chiefly," said I.

"Anything new? I thought not. *I* did these islands twenty-five—twenty-seven years ago. If you find anything new here—well, it's brand new. I didn't leave much."

"I'm not a collector," said I.

"I was young then," he went on. "Lord! how I used to fly round." He seemed to take my measure. "I was in the East Indies two years and in Brazil seven. Then I went to Madagascar."

"I know a few explorers by name," I said, anticipating a yarn. "Whom did you collect for?"

"Dawsons'. I wonder if you've heard the name of Butcher ever?"

"Butcher—Butcher?" The name seemed vaguely present in my memory; then I recalled *Butcher* v. *Dawson*. "Why!" said I, "you are the man who sued them for four years' salary—got cast away on a desert island. . . ."

"Your servant," said the man with the scar, bowing. "Funny case, wasn't it? Here was me, making a little fortune on that island, doing nothing for it neither, and them quite unable to give me notice. It often used to amuse me thinking over it while I was there. I did calcula-

260

tions of it—big—all over the blessed atoll in ornamental figuring."

"How did it happen?" said I. "I don't rightly remember the case."

"Well. . . . You've heard of the Æpyornis?"

"Rather. Andrews was telling me of a new species he was working on only a month or so ago. Just before I sailed. They've got a thigh-bone, it seems, nearly a yard long. Monster the thing must have been!"

"I believe you," said the man with the scar. "It *was* a monster. Sindbad's roc was just a legend of 'em. But when did they find these bones?"

"Three or four years ago—'91, I fancy. Why?"

"Why? Because I found them—Lord!—it's nearly twenty years ago. If Dawsons' hadn't been silly about that salary they might have made a perfect ring in 'em. . . . I couldn't help the infernal boat going adrift."

He paused. "I suppose it's the same place. A kind of swamp about ninety miles north of Antananarivo. Do you happen to know? You have to go to it along the coast by boats. You don't happen to remember, perhaps?"

"I don't. I fancy Andrews said something about a swamp."

"It must be the same. It's on the east coast. And somehow there's something in the water that keeps things from decaying. Like creosote it smells. It reminded me of Trinidad. Did they get any more eggs? Some of the eggs I found were a foot and a half long. The swamp goes circling round, you know, and cuts off this bit. It's mostly salt, too. Well. . . . What a time I had of it! I found the things quite by accident. We went for eggs, me and two native chaps, in one of those rum canoes all tied together, and found the bones at the same time. We had a tent and provisions for four days, and we pitched on one of the firmer places. To think of it brings that odd tarry smell back even now. It's funny work. You go probing into the mud with iron rods, you know. Usually the egg gets smashed. I wonder how long it is since these Æpyornises really lived. The missionaries say the natives have legends about when they were alive, but I never heard any such stories myself.* But certainly those eggs we got were as fresh as if they had been new laid. Fresh! Carrying them down to the boat one of my nigger chaps dropped one on a rock and it smashed. How I lammed into the beggar! But sweet it was, as if it was new laid, not even smelly, and its mother dead these four hundred years, perhaps. Said a centipede had bit him. However, I'm getting off the straight with the story. It had taken us all day to dig into the slush and get these eggs out unbroken, and we were all covered with beastly black mud, and naturally I was cross. So far as I knew they were the only eggs that have ever been got out not even cracked. I went afterwards to see the ones they have at the Natural History Museum in London; all of them were cracked and just stuck together like a mosaic, and bits missing. Mine were perfect, and I meant to blow them when I got back. Naturally I was annoyed at the silly duffer dropping three hours' work just on account of a centipede. I hit him about rather."

The man with the scar took out a clay pipe. I placed my pouch before him. He filled up absent-mindedly.

"How about the others? Did you get those home? I don't remember——"

"That's the queer part of the story. I had three others. Perfectly fresh eggs. Well, we put 'em in the boat, and then I went up to the tent to make some coffee, leaving my two heathens down by the beach—the one fooling about with his sting and the other helping him. It never occurred to me that the beggar would take advantage of the peculiar position I was in to pick a quarrel. But I suppose the centipede poison and the kicking I had given him had upset the one—he was always a cantankerous sort—and he persuaded the other.

"I remember I was sitting and smoking and boiling up the water over a spirit-lamp business I used to take on these expeditions. Incidentally I was admiring the swamp under the sunset. All black and blood-red it was, in streaks—a

*No European is known to have seen a live Æpyornis, with the doubtful exception of Macer, who visited Madagascar in 1745.--- H. G. W.

beautiful sight. And up beyond the land rose grey and hazy to the hills, and the sky behind them red, like a furnace mouth. And fifty yards behind the back of me was these blessed heathen—quite regardless of the tranquil air of things—plotting to cut off with the boat and leave me all alone with three days' provisions and a canvas tent, and nothing to drink whatsoever beyond a little keg of water. I heard a kind of yelp behind me, and there they were in this canoe affair—it wasn't properly a boat—and, perhaps, twenty yards from land. I realised what was up in a moment. My gun was in the tent, and, besides, I had no bullets—only duck shot. They knew that. But I had a little revolver in my pocket, and I pulled that out as I ran down to the beach.

" 'Come back!' says I, flourishing it.

"They jabbered something at me, and the man that broke the egg jeered. I aimed at the other—because he was unwounded and had the paddle, and I missed. They laughed. However, I wasn't beat. I knew I had to keep cool, and I tried him again and made him jump with the whang of it. He didn't laugh that time. The third time I got his head, and over he went, and the paddle with him. It was a precious lucky shot for a revolver. I reckon it was fifty yards. He went right under. I don't know if he was shot, or simply stunned and drowned. Then I began to shout to the other chap to come back, but he huddled up in the canoe and refused to answer. So I fired out my revolver at him and never got near him.

"I felt a precious fool, I can tell you. There I was on this rotten black beach, flat swamp all behind me, and the flat sea, cold after the sun set, and just this black canoe drifting steadily out to sea. I tell you I damned Dawsons' and Jamrach's and Museums and all the rest of it just to rights. I bawled to this nigger to come back, until my voice went up into a scream.

"There was nothing for it but to swim after him and take my luck with the sharks. So I opened my clasp-knife and put it in my mouth, and took off my clothes and waded in. As soon as I was in the water I lost sight of the canoe, but I aimed, as I judged, to head it off. I hoped the man in it was too bad to navigate it, and that it would keep on drifting in the same direction. Presently it came up over the horizon again to the south-westward about. The afterglow of sunset was well over now and the dim of night creeping up. The stars were coming through the blue. I swum like a champion, though my legs and arms were soon aching.

"However, I came up to him by the time the stars were fairly out. As it got darker I began to see all manner of glowing things in the water—phosphorescence, you know. At times it made me giddy. I hardly knew which was stars and which was phosphorescence, and whether I was swimming on my head or my heels. The canoe was as black as sin, and the ripple under the bows like liquid fire. I was naturally chary of clambering up into it. I was anxious to see what he was up to first. He seemed to be lying cuddled up in a lump in the bows, and the stern was all out of water. The thing kept turning round slowly as it drifted—kind of waltzing, don't you know. I went to the stern and pulled it down, expecting him to wake up. Then I began to clamber in with my knife in my hand, and ready for a rush. But he never stirred. So there I sat in the stern of the little canoe, drifting away over the calm phosphorescent sea and with all the host of the stars above me, waiting for something to happen.

"After a long time I called him by name, but he never answered. I was too tired to take any risks by going along to him. So we sat there. I fancy I dozed once or twice. When the dawn came I saw he was as dead as a door-nail and all puffed up and purple. My three eggs and the bones were lying in the middle of the canoe, and the keg of water and some coffee and biscuits wrapped in a Cape *Argus* by his feet, and a tin of methylated spirit underneath him. There was no paddle, nor, in fact, anything except the spirit tin that I could use as one, so I settled to drift until I was picked up. I held an inquest on him, brought in a verdict against some snake, scorpion, or centipede unknown, and sent him overboard.

"After that I had a drink of water and a few biscuits, and took a look round. I suppose a man low down as I was don't see very far; leastways, Madagascar was clean out of sight, and any trace of land at all. I saw a sail going south-westward—looked like a schooner but her hull never came up. Presently the sun got high in the sky and began to beat down upon me. Lord! it pretty near made my brains boil. I tried dipping my head in the sea, but after a while my eye fell on the Cape *Argus*, and I lay down flat in the canoe and spread this over me. Wonderful things these newspapers! I never read one through thoroughly before, but it's odd what you get up to when you're alone, as I was. I suppose I read that blessed old Cape *Argus* twenty times. The pitch in the canoe simply reeked with the heat and rose up into big blisters.

"I drifted ten days," said the man with the scar. "It's a little thing in the telling, isn't it? Every day was like the last. Except in the morning and the evening I never kept a lookout even—the blaze was so infernal. I didn't see a sail after the first three days, and those I saw took no notice of me. About the sixth night a ship went by scarcely half a mile away from me, with all its lights ablaze and its ports open, looking like a big firefly. There was music aboard. I stood up and shouted and screamed at it. The second day I broached one of the Æpyornis eggs, scraped the shell away at the end bit by bit, and tried it, and I was glad to find it was good enough to eat. A bit flavoury—not bad, I mean—but with something of the taste of a duck's egg. There was a kind of circular patch, about six inches across, on one side of the yolk, and with streaks of blood and a white mark like a ladder in it that I thought queer, but I did not understand what this meant at the time, and I wasn't inclined to be particular. The egg lasted me three days, with biscuits and a drink of water. I chewed coffee-berries too—invigorating stuff. The second egg I opened about the eighth day, and it scared me."

The man with the scar paused. "Yes," he said, "developing."

"I dare say you find it hard to believe. I did, with the

thing before me. There the egg had been, sunk in that cold
black mud, perhaps three hundred years. But there was no
mistaking it. There was the—what is it?—embryo, with its
big head and curved back, and its heart beating under its
throat, and the yolk shrivelled up and great membranes
spreading inside of the shell and all over the yolk. Here
was I hatching out the eggs of the biggest of all extinct
birds, in a little canoe in the midst of the Indian Ocean. If
old Dawson had known that!   It was worth four years'
salary. What do *you* think?

"However, I had to eat that precious thing up, every bit
of it, before I sighted the reef, and some of the mouthfuls
were beastly unpleasant. I left the third one alone. I held
it up to the light, but the shell was too thick for me to get
any notion of what might be happening inside; and though
I fancied I heard blood pulsing, it might have been the
rustle in my own ears, like what you listen to in a seashell.

"Then came the atoll. Came out of the sunrise, as it
were, suddenly, close to me. I drifted straight towards it
until I was about half a mile from shore, not more, and
then the current took a turn, and I had to paddle as hard as I
could with my hands and bits of the Æpyornis shell to make
the place. However, I got there. It was just a common
atoll about four miles round, with a few trees growing and a
spring in one place, and the lagoon full of parrot-fish. I
took the egg ashore and put it in a good place, well above
the tide lines and in the sun, to give it all the chance I
could, and pulled the canoe up safe, and loafed about
prospecting. It's rum how dull an atoll is. As soon as I had
found a spring all the interest seemed to vanish. When I
was a kid I thought nothing could be finer or more adventurous than the Robinson Crusoe business, but that place
was as monotonous as a book of sermons. I went round
finding eatable things and generally thinking; but I tell you
I was bored to death before the first day was out. It shows
my luck—the very day I landed the weather changed. A
thunderstorm went by to the north and flicked its wing over
the island, and in the night there came a drencher and a

howling wind slap over us It wouldn't have taken much,
you know, to upset that canoe.

"I was sleeping under the canoe, and the egg was luckily
among the sand higher up the beach, and the first thing I
remember was a sound like a hundred pebbles hitting the
boat at once, and a rush of water over my body. I'd been
dreaming of Antananarivo, and I sat up and halloaed to
Intoshi to ask her what the devil was up, and clawed out at
the chair where the matches used to be Then I remembered where I was. There were phosphorescent waves
rolling up as if they meant to eat me, and all the rest of the
night as black as pitch. The air was simply yelling. The
clouds seemed down on your head almost, and the rain fell
as if heaven was sinking and they were baling out the
waters above the firmament. One great roller came writhing
at me, like a fiery serpent, and I bolted. Then I thought
of the canoe, and ran down to it as the water went hissing
back again; but the thing had gone. I wondered about the
egg, then, and felt my way to it. It was all right and well
out of reach of the maddest waves, so I sat down beside it
and cuddled it for company. Lord! what a night that
was!

"The storm was over before the morning. There wasn't
a rag of cloud left in the sky when the dawn came, and all
along the beach there were bits of plank scattered—which
was the disarticulated skeleton, so to speak, of my canoe.
However, that gave me something to do, for taking advantage of two of the trees being together, I rigged up a kind of
storm-shelter with these vestiges. And that day the egg
hatched.

"Hatched, sir, when my head was pillowed on it and I
was asleep. I heard a whack and felt a jar and sat up, and
there was the end of the egg pecked out and a rum little
brown head looking out at me. 'Lord!' I said, 'You're welcome,' and with a little difficulty he came out.

"He was a nice friendly little chap at first, about the size
of a small hen—very much like most other young birds, only
bigger. His plumage was a dirty brown to begin with, with
a sort of grey scab that fell off it very soon, and scarcely

feathers—a kind of downy hair. I can hardly express how pleased I was to see him. I tell you, Robinson Crusoe don't make near enough of his loneliness. But here was interesting company. He looked at me and winked his eye from the front backward, like a hen, and gave a chirp and began to peck about at once, as though being hatched three hundred years too late was just nothing. 'Glad to see you, Man Friday!' says I, for I had naturally settled he was to be called Man Friday if ever he was hatched, as soon as ever I found the egg in the canoe had developed. I was a bit anxious about his feed, so I gave him a lump of raw parrot-fish at once. He took it, and opened his beak for more. I was glad of that, for, under the circumstances, if he'd been at all fanciful, I should have had to eat him after all.

"You'd be surprised what an interesting bird that Æpyornis chick was. He followed me about from the very beginning. He used to stand by me and watch while I fished in the lagoon, and go shares in anything I caught. And he was sensible, too. There were nasty green warty things, like pickled gherkins, used to lie about on the beach, and he tried one of these and it upset him. He never even looked at any of them again.

"And he grew. You could almost see him grow. And as I was never much of a society man, his quiet friendly ways suited me to a T. For nearly two years we were as happy as we could be on that island. I had no business worries, for I knew my salary was mounting up at Dawsons'. We would see a sail now and then, but nothing ever came near us. I amused myself, too, by decorating the island with designs worked in sea-urchins and fancy shells of various kinds. I put ÆPYORNIS ISLAND all around the place very nearly, in big letters, like what you see done with coloured stones at railway stations in the old country, and mathematical calculations and drawings of various sorts. And I used to lie watching the blessed bird stalking round and growing, growing; and think how I could make a living out of him by showing him about if I ever got taken off. After his first moult he began to get handsome, with a crest and a blue wattle, and a lot of green feathers at the behind of him. And then I used to puzzle whether Dawsons' had any right to claim him or not. Stormy weather and in the rainy season we lay snug under the shelter I had made out of the old canoe, and I used to tell him lies about my friends at home. And after a storm we would go round the island together to see if there was any drift. It was a kind of idyll, you might say. If only I had had some tobacco it would have been simply just like heaven.

"It was about the end of the second year our little paradise went wrong. Friday was then about fourteen feet high to the bill of him, with a big, broad head like the end of a pickaxe, and two huge brown eyes with yellow rims, set together like a man's—not out of sight of each other like a hen's. His plumage was fine—none of the half-mourning style of your ostrich—more like a cassowary as far as colour and texture go. And then it was he began to cock his comb at me and give himself airs, and show signs of a nasty temper. . . .

"At last came a time when my fishing had been rather unlucky, and he began to hang about me in a queer, meditative way. I thought he might have been eating sea-cucumbers or something, but it was really just discontent on his part. I was hungry, too, and when at last I landed a fish I wanted it for myself. Tempers were short that morning on both sides. He pecked at it and grabbed it, and I gave him a whack on the head to make him leave go. And at that he went for me. Lord! . . .

"He gave me this in the face." The man indicated his scar. "Then he kicked me. It was like a cart-horse. I got up, and, seeing he hadn't finished, I started off full tilt with my arms doubled up over my face. But he ran on those gawky legs of his faster than a race-horse, and kept landing out at me with sledge-hammer kicks and bringing his pick-axe down on the back of my head. I made for the lagoon, and went in up to my neck. He stopped at the water, for he hated getting his feet wet, and began to make a shindy, something like a peacock's, only hoarser. He started strutting up and down the beach. I'll admit I felt small to see this blessed fossil lording it there. And my head and face were all bleeding, and—well, my body just one jelly of bruises.

"I decided to swim across the lagoon and leave him alone for a bit, until the affair blew over. I shinned up the tallest palm-tree, and sat there thinking of it all. I don't suppose I ever felt so hurt by anything before or since. It was the brutal ingratitude of the creature. I'd been more than a brother to him. I'd hatched him, educated him. A great gawky, out-of-date bird! And me a human being—heir of the ages and all that.

"I thought after a time he'd begin to see things in that light himself, and feel a little sorry for his behaviour. I thought if I was to catch some nice little bits of fish, perhaps, and go to him presently in a casual kind of way, and offer them to him, he might do the sensible thing. It took me some time to learn how unforgiving and cantankerous an extinct bird can be. Malice!

"I won't tell you all the little devices I tried to get that bird round again. I simply can't. It makes my cheek burn with shame even now to think of the snubs and buffets I had from this infernal curiosity. I tried violence. I chucked lumps of coral at him from a safe distance, but he only swallowed them. I shied my open knife at him and almost lost it, though it was too big for him to swallow. I tried starving him out and struck fishing, but he took to picking along the beach at low water after worms, and rubbed along on that. Half my time I spent up to my neck in the lagoon, and the rest up the palm-trees. One of them was scarcely high enough, and when he caught me up it he had a regular Bank Holiday with the calves of my legs. It got unbearable. I don't know if you have ever tried sleeping up a palm-tree. It gave me the most horrible night-mares. Think of the shame of it, too! Here was this extinct animal mooning about my island like a sulky duke, and me not allowed to rest the sole of my foot on the place. I used to cry with weariness and vexation. I told him straight that I didn't mean to be chased about a desert island by any damned anachronisms. I told him to go and peck a navigator of his own age. But he only snapped his beak at me. Great ugly bird, all legs and neck!

"I shouldn't like to say how long that went on altogether. I'd have killed him sooner if I'd known how. However, I hit on a way of settling him at last. It is a South American dodge. I joined all my fishing-lines together with stems of seaweed and things, and made a stoutish string, perhaps twelve yards in length or more, and I fastened two lumps of coral rock to the ends of this. It took me some time to do, because every now and then I had to go into the lagoon or up a tree as the fancy took me. This I whirled rapidly round my head, and then let it go at him. The first time I missed, but the next time the string caught his legs beauti- fully, and wrapped round them again and again. Over he went. I threw it standing waist-deep in the lagoon, and as soon as he went down I was out of the water and sawing at his neck with my knife . . .

"I don't like to think of that even now. I felt like a mur- derer while I did it, though my anger was hot against him. When I stood over him and saw him bleeding on the white sand, and his beautiful great legs and neck writhing in his last agony . . . Pah!

"With that tragedy loneliness came upon me like a curse. Good Lord! you can't imagine how I missed that bird. I sat by his corpse and sorrowed over him, and shivered as I looked round the desolate, silent reef. I thought of what a jolly little bird he had been when he was hatched, and of a thousand pleasant tricks he had played before he went wrong. I thought if I'd only wounded him I might have nursed him round into a better understanding. If I'd had any means of digging into the coral rock I'd have buried him. I felt exactly as if he was human. As it was, I couldn't think of eating him, so I put him in the lagoon, and the little fishes picked him clean. I didn't even save the feathers. Then one day a chap cruising about in a yacht had a fancy to see if my atoll still existed.

"He didn't come a moment too soon, for I was about sick enough of the desolation of it, and only hesitating whether

I should walk out into the sea and finish up the business that way, or fall back on the green things. . . .

"I sold the bones to a man named Winslow—a dealer near the British Museum, and he says he sold them to old Havers. It seems Havers didn't understand they were extra large, and it was only after his death they attracted attention. They called 'em Æpyornis—what was it?"

"*Æpyornis vastus*," said I. "It's funny, the very thing was mentioned to me by a friend of mine. When they found an Æpyornis with a thigh a yard long, they thought they had reached the top of the scale, and called him *Æpyornis maximus*. Then someone turned up another thigh-bone four feet six or more, and that they called *Æpyornis titan*. Then your *vastus* was found after old Havers died, in his collection, and then a *vastissimus* turned up."

"Winslow was telling me as much," said the man with the scar. "If they get any more Æpyornises, he reckons some scientific swell will go and burst a blood-vessel. But it was a queer thing to happen to a man, wasn't it—altogether?"

STORY THE ELEVENTH

The Remarkable Case of Davidson's Eyes

THE transitory mental aberration of Sidney Davidson, re-markable enough in itself, is still more remarkable if Wade's explanation is to be credited. It sets one dreaming of the oddest possibilities of intercommunication in the future, of spending an intercalary five minutes on the other side of the world, or being watched in our most secret operations by unsuspected eyes. It happened that I was the immediate witness of Davidson's seizure, and so it falls naturally to me to put the story upon paper.

When I say that I was the immediate witness of his seizure, I mean that I was the first on the scene. The thing happened at the Harlow Technical College, just beyond the Highgate Archway. He was alone in the larger laboratory when the thing happened. I was in a smaller room, where the balances are, writing up some notes. The thunderstorm had completely upset my work, of course. It was just after one of the louder peals that I thought I heard some glass smash in the other room. I stopped writing, and turned round to listen. For a moment I heard nothing; the hail was playing the devil's tattoo on the corrugated zinc of the roof. Then came another sound, a smash—no doubt of it this time. Something heavy had been knocked off the bench. I jumped up at once and went and opened the door leading into the big laboratory.

I was surprised to hear a queer sort of laugh, and saw Davidson standing unsteadily in the middle of the room, with a dazzled look on his face. My first impression was that he was drunk. He did not notice me. He was clawing out at something invisible a yard in front of his face. He put out his hand slowly, rather hesitatingly, and then clutched nothing. "What's come to it?" he said. He held up his hands to his face, fingers spread out. "Great Scott!" he said. The thing happened three or four years ago, when everyone swore by that personage. Then he began raising his feet clumsily, as though he had expected to find them glued to the floor.

"Davidson!" cried I. "What's the matter with you?" He turned round in my direction and looked about for me. He looked over me and at me and on either side of me, without the slightest sign of seeing me. "Waves," he said; "and a remarkably neat schooner. I'd swear that was Bellow's voice. *Hullo!*" He shouted suddenly at the top of his voice.

I thought he was up to some foolery. Then I saw littered about his feet the shattered remains of the best of our electrometers. "What's up, man?" said I. "You've smashed the electrometer!"

"Bellows again!" said he. "Friends left, if my hands are gone. Something about electrometers. Which way *are* you, Bellows?" He suddenly came staggering towards me. "The damned stuff cuts like butter," he said. He walked straight into the bench and recoiled. "None so buttery that!" he said, and stood swaying.

I felt scared. "Davidson," said I, "what on earth's come over you?"

He looked round him in every direction. "I could swear that was Bellows. Why don't you show yourself like a man, Bellows?"

It occurred to me that he must be suddenly struck blind. I walked round the table and laid my hand upon his arm. I never saw a man more startled in my life. He jumped away from me, and came round into an attitude of self-defence, his face fairly distorted with terror. "Good God!" he cried. "What was that?"

"It's I—Bellows. Confound it, Davidson!"

He jumped when I answered him and stared—how can I express it?—right through me. He began talking, not to me, but to himself. "Here in broad daylight on a clear beach. Not a place to hide in." He looked about him wildly. "Here! I'm *off*." He suddenly turned and ran headlong into the big electromagnet—so violently that, as we found afterwards, he bruised his shoulder and jawbone cruelly. At that he stepped back a pace, and cried out with almost a whimper: "What, in Heaven's name, has come over me?" He stood, blanched with terror and trembling violently, with his right arm clutching his left, where that had collided with the magnet.

By that time I was excited and fairly scared. "Davidson," said I, "don't be afraid."

He was startled at my voice, but not so excessively as before. I repeated my words in as clear and as firm a tone as I could assume. "Bellows," he said, "is that you?"

"Can't you see it's me?"

He laughed. "I can't even see it's myself. Where the devil are we?"

"Here," said I, "in the laboratory."

"The laboratory!" he answered in a puzzled tone, and put his hand to his forehead. "I *was* in the laboratory—till that flash came, but I'm hanged if I'm there now. What ship is that?"

"There's no ship," said I. "Do be sensible, old chap."

"No ship!" he repeated, and seemed to forget my denial forthwith. "I suppose," said he slowly, "we're both dead. But the rummy part is I feel just as though I still had a body. Don't get used to it all at once, I suppose. The old ship was struck by lightning, I suppose. Jolly quick thing, Bellows—eigh?"

"Don't talk nonsense. You're very much alive. You are in the laboratory, blundering about. You've just smashed a new electrometer. I don't envy you when Boyce arrives."

He stared away from me towards the diagrams of cryohy-drates. "I must be deaf," said he. "They've fired a gun, for there goes the puff of smoke, and I never heard a sound."

I put my hand on his arm again, and this time he was less

alarmed. "We seem to have a sort of invisible bodies," said he. "By Jove! there's a boat coming round the headland. It's very much like the old life after all—in a different climate."

I shook his arm. "Davidson," I cried, "wake up!"

It was just then that Boyce came in. So soon as he spoke Davidson exclaimed: "Old Boyce! Dead too! What a lark!" I hastened to explain that Davidson was in a kind of somnambulistic trance. Boyce was interested at once. We both did all we could to rouse the fellow out of his extraordinary state. He answered our questions, and asked us some of his own, but his attention seemed distracted by his hallucination about a beach and a ship. He kept interpolating observations concerning some boat and the davits, and sails filling with the wind. It made one feel queer, in the dusky laboratory, to hear him saying such things.

He was blind and helpless. We had to walk him down the passage, one at each elbow, to Boyce's private room, and while Boyce talked to him there, and humoured him about this ship idea, I went along the corridor and asked old Wade to come and look at him. The voice of our Dean sobered him a little, but not very much. He asked where his hands were, and why he had to walk about up to his waist in the ground. Wade thought over him a long time—you know how he knits his brows—and then made him feel the couch, guiding his hands to it. "That's a couch," said Wade. "The couch in the private room of Prof. Boyce. Horsehair stuffing."

Davidson felt about, and puzzled over it, and answered presently that he could feel it all right, but he couldn't see it.

"What *do* you see?" asked Wade. Davidson said he could see nothing but a lot of sand and broken-up shells. Wade gave him some other things to feel, telling him what they were, and watching him keenly.

"The ship is almost hull down," said Davidson presently, apropos of nothing.

"Never mind the ship," said Wade. "Listen to me, Davidson. Do you know what hallucination means?"

"Rather," said Davidson.

"Well, everything you see is hallucinatory."

"Bishop Berkeley," said Davidson.

"Don't mistake me," said Wade. "You are alive and in this room of Boyce's. But something has happened to your eyes. You cannot see; you can feel and hear, but not see. Do you follow me?"

"It seems to me that I see too much." Davidson rubbed his knuckles into his eyes. "Well?" he said.

"That's all. Don't let it perplex you. Bellows here and I will take you home in a cab."

"Wait a bit." Davidson thought. "Help me to sit down," said he presently; "and now—I'm sorry to trouble you—but will you tell me all that over again?"

Wade repeated it very patiently. Davidson shut his eyes, and pressed his hands upon his forehead. "Yes," said he. "It's quite right. Now my eyes are shut I know you're right. That's you, Bellows, sitting by me on the couch. I'm in England again. And we're in the dark."

Then he opened his eyes. "And there," said he, "is the sun just rising, and the yards of the ship, and a tumbled sea, and a couple of birds flying. I never saw anything so real. And I'm sitting up to my neck in a bank of sand."

He bent forward and covered his face with his hands. Then he opened his eyes again. "Dark sea and sunrise! And yet I'm sitting on a sofa in old Boyce's room! . . . God help me!"

That was the beginning. For three weeks this strange affection of Davidson's eyes continued unabated. It was far worse than being blind. He was absolutely helpless, and had to be fed like a newly hatched bird, and led about and undressed. If he attempted to move, he fell over things or struck himself against walls or doors. After a day or so he got used to hearing our voices without seeing us, and willingly admitted he was at home, and that Wade was right in what he told him. My sister, to whom he was engaged, insisted on coming to see him, and would sit for hours every day while he talked about this beach of his. Holding her hand seemed to comfort him immensely. He explained that when we left the College and drove home—he lived in Hampstead village—it appeared to him as if we drove right through a sandhill—it was perfectly black until he emerged again—and through rocks and trees and solid obstacles, and when he was taken to his own room it made him giddy and almost frantic with the fear of falling, because going upstairs seemed to lift him thirty or forty feet above the rocks of his imaginary island. He kept saying he should smash all the eggs. The end was that he had to be taken down into his father's consulting-room and laid upon a couch that stood there.

He described the island as being a bleak kind of place on the whole, with very little vegetation, except some peaty stuff, and a lot of bare rock. There were multitudes of penguins, and they made the rocks white and disagreeable to see. The sea was often rough, and once there was a thunderstorm, and he lay and shouted at the silent flashes. Once or twice seals pulled up on the beach, but only on the first two or three days. He said it was very funny the way in which the penguins used to waddle right through him, and how he seemed to lie among them without disturbing them.

I remember one odd thing, and that was when he wanted very badly to smoke. We put a pipe in his hands—he almost poked his eye out with it—and lit it. But he couldn't taste anything. I've since found it's the same with me—I don't know if it's the usual case—that I cannot enjoy tobacco at all unless I can see the smoke.

But the queerest part of his vision came when Wade sent him out in a Bath-chair to get fresh air. The Davidsons hired a chair, and got that deaf and obstinate dependant of theirs, Widgery, to attend to it. Widgery's ideas of healthy expeditions were peculiar. My sister, who had been to the Dogs' Home, met them in Camden Town, towards King's Cross, Widgery trotting along complacently, and Davidson, evidently most distressed, trying in his feeble, blind way to attract Widgery's attention.

He positively wept when my sister spoke to him. "Oh, get me out of this horrible darkness!" he said, feeling for her hand. "I must get out of it, or I shall die." He was quite incapable of explaining what was the matter, but my sister decided he must go home, and presently, as they went uphill towards Hampstead, the horror seemed to drop from him. He said it was good to see the stars again, though it was then about noon and a blazing sky.

"It seemed," he told me afterwards, "as if I was being carried irresistibly towards the water. I was not very much alarmed at first. Of course it was night there—a lovely night."

"Of course?" I asked, for that struck me as odd.

"Of course," said he. "It's always night there when it is day here. . . . Well, we went right into the water, which was calm and shining under the moonlight—just a broad swell that seemed to grow broader and flatter as I came down into it. The surface glistened just like a skin—it might have been empty space underneath for all I could tell to the contrary. Very slowly, for I rode slanting into it, the water crept up to my eyes. Then I went under and the skin seemed to break and heal again about my eyes. The moon gave a jump up in the sky and grew green and dim, and fish, faintly glowing, came darting round me—and things that seemed made of luminous glass; and I passed through a tangle of seaweeds that shone with an oily lustre. And so I drove down into the sea, and the stars went out one by one, and the moon grew greener and darker, and the seaweed became a luminous purple-red. It was all very faint and mysterious, and everything seemed to quiver. And all the while I could hear the wheels of the Bath-chair creaking, and the footsteps of people going by, and a man in the distance selling the special *Pall Mall*.

"I kept sinking down deeper and deeper into the water. It became inky black about me, not a ray from above came down into that darkness, and the phosphorescent things grew brighter and brighter. The snaky branches of the deeper weeds flickered like the flames of spirit-lamps; but, after a time, there were no more weeds. The fishes came staring and gaping towards me, and into me and through me. I never imagined such fishes before. They had lines of fire along the sides of them as though they had been outlined with a luminous pencil. And there was a ghastly thing swimming backward with a lot of twining arms. And then I saw, coming very slowly towards me through the gloom, a hazy mass of light that resolved itself as it drew nearer into multitudes of fishes, struggling and darting round something that drifted. I drove on straight towards it, and presently I saw in the midst of the tumult, and by the light of the fish, a bit of splintered spar looming over me, and a dark hull tilting over, and some glowing phosphorescent forms that were shaken and writhed as the fish bit at them. Then it was I began to try to attract Widgery's attention. A horror came upon me. Ugh! I should have driven right into those half-eaten —— things. If your sister had not come! They had great holes in them, Bellows, and . . . Never mind. But it was ghastly!"

For three weeks Davidson remained in this singular state, seeing what at the time we imagined was an altogether phantasmal world, and stone blind to the world around him. Then, one Tuesday, when I called I met old Davidson in the passage. "He can see his thumb!" the old gentleman said, in a perfect transport. He was struggling into his overcoat. "He can see his thumb, Bellows!" he said, with the tears in his eyes. "The lad will be all right yet."

I rushed in to Davidson. He was holding up a little book before his face, and looking at it and laughing in a weak kind of way.

"It's amazing," said he. "There's a kind of patch come there." He pointed with his finger. "I'm on the rocks as usual, and the penguins are staggering and flapping about as usual, and there's been a whale showing every now and then, but it's got too dark now to make him out. But put something *there*, and I see it—I do see it. It's very dim and broken in places, but I see it all the same, like a faint spectre of itself. I found it out this morning while they were dressing me. It's like a hole in this infernal phantom world. Just put your hand by mine. No—not there. Ah! Yes! I see it. The base of your thumb and a bit of cuff! It looks like the ghost of a bit of your hand sticking out of the darkling sky. Just by it there's a group of stars like a cross coming out."

From that time Davidson began to mend. His account of the change, like his account of the vision, was oddly convincing. Over patches of his field of vision, the phantom world grew fainter, grew transparent, as it were, and through these translucent gaps he began to see dimly the real world about him. The patches grew in size and number, ran together and spread until only here and there were blind spots left upon his eyes. He was able to get up and steer himself about, feed himself once more, read, smoke, and behave like an ordinary citizen again. At first it was very confusing to him to have these two pictures overlapping each other like the changing views of a lantern, but in a little while he began to distinguish the real from the illusory.

At first he was unfeignedly glad, and seemed only too anxious to complete his cure by taking exercise and tonics. But as that odd island of his began to fade away from him, he became queerly interested in it. He wanted particularly to go down into the deep sea again, and would spend half his time wandering about the low-lying parts of London, trying to find the water-logged wreck he had seen drifting. The glare of real daylight very soon impressed him so vividly as to blot out everything of his shadowy world, but of a night time, in a darkened room, he could still see the white-splashed rocks of the island, and the clumsy penguins staggering to and fro. But even these grew fainter and fainter, and, at last, soon after he married my sister, he saw them for the last time.

And now to tell of the queerest thing of all. About two years after his cure I dined with the Davidsons, and after dinner a man named Atkins called in. He is a lieutenant in the Royal Navy, and a pleasant, talkative man. He was on friendly terms with my brother-in-law, and was soon on friendly terms with me. It came out that he was engaged to Davidson's cousin, and incidentally he took out a kind of pocket photograph case to show us a new render-

ing of his *fiancée*. "And, by the bye," said he, "here's the
old *Fulmar*."

Davidson looked at it casually. Then suddenly his face
lit up. "Good heavens!" said he. "I could almost
swear——"

"What?" said Atkins.

"That I had seen that ship before."

"Don't see how you can have. She hasn't been out of the
South Seas for six years, and before then——"

"But," began Davidson, and then: "Yes—that's the ship I
dreamt of; I'm sure that's the ship I dreamt of. She was
standing off an island that swarmed with penguins, and she
fired a gun."

"Good Lord!" said Atkins, who had now heard the particulars of the seizure. "How the deuce could you dream
that?"

And then, bit by bit, it came out that on the very day
Davidson was seized, H.M.S. *Fulmar* had actually been off a
little rock to the south of Antipodes Island. A boat had
landed overnight to get penguins' eggs, had been delayed,
and a thunderstorm drifting up, the boat's crew had waited
until the morning before rejoining the ship. Atkins had
been one of them, and he corroborated, word for word, the
descriptions Davidson had given of the island and the boat.
There is not the slightest doubt in any of our minds that
Davidson has really seen the place. In some unaccountable
way, while he moved hither and thither in London, his
sight moved hither and thither in a manner that corresponded, about this distant island. *How* is absolutely a
mystery.

That completes the remarkable story of Davidson's eyes.
It's perhaps the best authenticated case in existence of real
vision at a distance. Explanation there is none forthcoming, except what Prof. Wade has thrown out. But his
explanation involves the Fourth Dimension, and a disserta-
tion on theoretical kinds of space. To talk of there being
"a kink in space" seems mere nonsense to me; it may be
because I am no mathematician. When I said that nothing
would alter the fact that the place is eight thousand miles

away, he answered that two points might be a yard away on a sheet of paper, and yet be brought together by bending the paper round. The reader may grasp his argument, but I certainly do not. His idea seems to be that Davidson, stooping between the poles of the big electromagnet, had some extraordinary twist given to his retinal elements through the sudden change in the field of force due to the lightning.

He thinks, as a consequence of this, that it may be possible to live visually in one part of the world, while one lives bodily in another. He has even made some experiments in support of his views; but, so far, he has simply succeeded in blinding a few dogs. I believe that is the net result of his work, though I have not seen him for some weeks. Latterly I have been so busy with my work in connection with the Saint Pancras installation that I have had little opportunity of calling to see him. But the whole of his theory seems fantastic to me. The facts concerning Davidson stand on an altogether different footing, and I can testify personally to the accuracy of every detail I have given.

STORY THE
TWELFTH

The Lord of the Dynamos

THE chief attendant of the three dynamos that buzzed and rattled at Camberwell and kept the electric railway going, came out of Yorkshire, and his name was James Holroyd. He was a practical electrician but fond of whisky, a heavy, red-haired brute with irregular teeth. He doubted the existence of the Deity but accepted Carnot's cycle, and he had read Shakespeare and found him weak in chemistry. His helper came out of the mysterious East, and his name was Azuma-zi. But Holroyd called him Pooh-bah. Holroyd liked a nigger help because he would stand kicking—a habit with Holroyd—and did not pry into the machinery and try to learn the ways of it. Certain odd possibilities of the negro mind brought into abrupt contact with the crown of our civilisation Holroyd never fully realised, though just at the end he got some inkling of them.

To define Azuma-zi was beyond ethnology. He was, perhaps, more negroid than anything else, though his hair was curly rather than frizzy, and his nose had a bridge. Moreover, his skin was brown rather than black, and the whites of his eyes were yellow. His broad cheek-bones and narrow chin gave his face something of the viperine V. His head, too, was broad behind, and low and narrow at the forehead, as if his brain had been twisted round in the reverse way to a European's. He was short of stature and still shorter of English. In conversation he made numerous odd noises of no known marketable value, and his infrequent words were carved and wrought into heraldic grotesqueness. Holroyd tried to elucidate his religious beliefs, and—especially after whisky—lectured to him against superstition and missionaries. Azuma-zi, however, shirked the discussion of his gods, even though he was kicked for it.

Azuma-zi had come, clad in white but insufficient raiment, out of the stoke-hole of the *Lord Clive,* from the Straits Settlements and beyond, into London. He had heard even in his youth of the greatness and riches of London, where all the women are white and fair and even the beggars in the streets are white; and he had arrived, with newly-earned gold coins in his pocket, to worship at the shrine of civilisation. The day of his landing was a dismal one; the sky was dun, and a wind-worried drizzle filtered down to the greasy streets, but he plunged boldly into the delights of Shadwell, and was presently cast up, shattered in health, civilised in costume, penniless, and, except in matters of the direst necessity, practically a dumb animal, to toil for James Holroyd, and to be bullied by him in the dynamo shed at Camberwell. And to James Holroyd bullying was a labour of love.

There were three dynamos with their engines at Camberwell. The two that have been there since the beginning are small machines; the larger one was new. The smaller machines made a reasonable noise; their straps hummed over the drums, every now and then the brushes buzzed and fizzled, and the air churned steadily, whoo! whoo! whoo! between their poles. One was loose in its foundations and kept the shed vibrating. But the big dynamo drowned these little noises altogether with the sustained drone of its iron core, which somehow set part of the ironwork humming. The place made the visitor's head reel with the throb, throb, throb of the engines, the rotation of the big wheels, the spinning ball-valves, the occasional spittings of the steam, and over all the deep, unceasing, surging note of the big dynamo. This last noise was from an engineering point of view a defect, but Azuma-zi accounted it unto the monster for mightiness and pride.

If it were possible we would have the noises of that shed always about the reader as he reads, we would tell all our story to such an accompaniment. It was a steady stream of din, from which the ear picked out first one thread and then another; there was the intermittent snorting, panting, and seething of the steam-engines, the suck and thud of their pistons, the dull beat on the air as the spokes of the great driving wheels came round, a note the leather straps made as they ran tighter and looser, and a fretful tumult from the dynamos; and, over all, sometimes inaudible, as the ear tired of it, and then creeping back upon the senses again, was this trombone note of the big machine. The floor never felt steady and quiet beneath one's feet, but quivered and jarred. It was a confusing, unsteady place, and enough to send anyone's thoughts jerking into odd zigzags. And for three months, while the big strike of the engineers was in progress, Holroyd who was a blackleg, and Azuma-zi who was a mere black, were never out of the stir and eddy of it, but slept and fed in the little wooden shanty between the shed and the gates.

Holroyd delivered a theological lecture on the text of his big machine soon after Azuma-zi came. He had to shout to be heard in the din. "Look at that," said Holroyd; "where's your 'eathen idol to match 'im?" And Azuma-zi looked. For a moment Holroyd was inaudible, and then Azuma-zi heard: "Kill a hundred men. Twelve per cent. on the ordinary shares," said Holroyd, "and that's something like a Gord."

Holroyd was proud of his big dynamo, and expatiated upon its size and power to Azuma-zi until heaven knows what odd currents of thought that and the incessant whirling and shindy set up within the curly black cranium. He would explain in the most graphic manner the dozen or so ways in which a man might be killed by it, and once he gave Azuma-zi a shock as a sample of its quality. After that, in the breathing times of his labour—it was heavy labour, being not only his own, but most of Holroyd's—Azuma-zi would sit and watch the big machine. Now and then the brushes would sparkle and spit blue flashes, at which Holroyd would swear, but all the rest was as smooth and rhythmic as breathing. The band ran shouting over the shaft, and ever behind one as one watched was the complacent thud of the piston. So it lived all day in this big airy shed, with him and Holroyd to wait upon it; not prisoned up and slaving to drive a ship as the other engines he knew—mere captive devils of the British Solomon—had been, but a machine enthroned. Those two smaller dynamos Azuma-zi by force of contrast despised; the large one he privately christened the Lord of the Dynamos. They were fretful and irregular, but the big dynamo was steady. How great it was! How serene and easy in its working! Greater and calmer even than the Buddhas he had seen at Rangoon, and yet not motionless, but living! The great black coils spun, spun, spun, the rings ran round under the brushes, and the deep note of its coil steadied the whole. It affected Azuma-zi queerly.

Azuma-zi was not fond of labour. He would sit about and watch the Lord of the Dynamos while Holroyd went away to persuade the yard porter to get whisky, although his proper place was not in the dynamo shed but behind the engines, and, moreover, if Holroyd caught him skulking he got hit for it with a rod of stout copper wire. He would go and stand close to the colossus, and look up at the great leather band running overhead. There was a black patch on the band that came round, and it pleased him somehow among all the clatter to watch this return again and again. Odd thoughts spun with the whirl of it. Scientific people tell us that savages give souls to rocks and trees—and a machine is a thousand times more alive than a rock or a tree. And Azuma-zi was practically a savage still; the veneer of civilisation lay no deeper than his slop suit, his bruises, and the coal grime on his face and hands. His father before him had worshipped a meteoric stone; kindred blood, it may be, had splashed the broad wheels of Juggernaut.

He took every opportunity Holroyd gave him of touching and handling the great dynamo that was fascinating him. He polished and cleaned it until the metal parts were blinding in the sun. He felt a mysterious sense of service in doing this. He would go up to it and touch its spinning coils gently. The gods he had worshipped were all far away. The people in London hid their gods.

At last his dim feelings grew more distinct and took shape in thoughts, and at last in acts. When he came into the roaring shed one morning he salaamed to the Lord of the Dynamos, and then, when Holroyd was away, he went and whispered to the thundering machine that he was its servant, and prayed it to have pity on him and save him from Holroyd. As he did so a rare gleam of light came in through the open archway of the throbbing machine-shed, and the Lord of the Dynamos, as he whirled and roared, was radiant with pale gold. Then Azuma-zi knew that his service was acceptable to his Lord. After that he did not feel so lonely as he had done, and he had indeed been very much alone in London. Even when his work-time was over, which was rare, he loitered about the shed.

The next time Holroyd maltreated him, Azuma-zi went presently to the Lord of the Dynamos and whispered, "Thou seest, O my Lord!" and the angry whirr of the machinery seemed to answer him. Thereafter it appeared to him that whenever Holroyd came into the shed a different note mingled with the sound of the dynamo. "My Lord bides his time," said Azuma-zi to himself. "The iniquity of the fool is not yet ripe." And he waited and watched for the reckoning. One day there was evidence of short circuiting, and Holroyd, making an unwary examination—it was in the afternoon—got a rather severe shock. Azuma-zi from behind the engine saw him jump off and curse at the peccant coil.

"He is warned," said Azuma-zi to himself. "Surely my Lord is very patient."

Holroyd had at first initiated his "nigger" into such elementary conceptions of the dynamo's working as would enable him to take temporary charge of the shed in his absence. But when he noticed the manner in which Azuma-zi hung about the monster he became suspicious. He dimly perceived his assistant was "up to something," and connecting him with the anointing of the coils with oil that had rotted the varnish in one place, he issued an edict, shouted above the confusion of the machinery, "Don't 'ee go nigh that big dynamo any more, Pooh-bah, or a'll take thy skin off!" Besides, if it pleased Azuma-zi to be near the big machine it was plain sense and decency to keep him away from it.

Azuma-zi obeyed at the time, but later he was caught bowing before the Lord of the Dynamos. At which Holroyd twisted his arm and kicked him as he turned to go away. As Azuma-zi presently stood behind the engine and glared at the back of the hated Holroyd, the noises of the machinery took a new rhythm and sounded like four words in his native tongue.

It is hard to say exactly what madness is. I fancy Azuma-zi was mad. The incessant din and whirl of the dynamo shed may have churned up his little store of knowledge and big store of superstitious fancy, at last, into something akin to frenzy. At any rate, when the idea of making Holroyd a sacrifice to the Dynamo Fetich was thus suggested to him, it filled him with a strange tumult of exultant emotion.

That night the two men and their black shadows were alone in the shed together. The shed was lit with one big arc-light and winked and flickered purple. The shadows lay black behind the dynamos, the ball governors of the engines whirled from light to darkness, and their pistons beat loud and steadily. The world outside seen through the open end of the shed seemed incredibly dim and remote. It seemed absolutely silent, too, since the riot of the machinery drowned every external sound. Far away was the black fence of the yard with grey shadowy houses behind, and above was the deep blue sky and the pale little stars. Azuma-zi suddenly walked across the centre of the shed above which the leather bands were running, and went into the shadow by the big dynamo. Holroyd heard a click, and the spin of the armature changed.

"What are you dewin' with that switch?" he bawled in surprise. "Han't I told you——"

Then he saw the set expression of Azuma-zi's eyes as the Asiatic came out of the shadow towards him.

κ

In another moment the two men were grappling fiercely in front of the great dynamo.

"You coffee-headed fool!" gasped Holroyd, with a brown hand at his throat. "Keep off those contact rings." In another moment he was tripped and reeling back upon the Lord of the Dynamos. He instinctively loosened his grip upon his antagonist to save himself from the machine.

The messenger, sent in furious haste from the station to find out what had happened in the dynamo shed, met Azuma-zi at the porter's lodge by the gate. Azuma-zi tried to explain something, but the messenger could make nothing of the black's incoherent English, and hurried on to the shed. The machines were all noisily at work, and nothing seemed to be disarranged. There was, however, a queer smell of singed hair. Then he saw an odd-looking crumpled mass clinging to the front of the big dynamo, and, approaching, recognised the distorted remains of Holroyd.

The man stared and hesitated a moment. Then he saw the face, and shut his eyes convulsively. He turned on his heel before he opened them, so that he should not see Holroyd again, and went out of the shed to get advice and help.

When Azuma-zi saw Holroyd die in the grip of the Great Dynamo he had been a little scared about the consequences of his act. Yet he felt strangely elated, and knew that the favour of the Lord Dynamo was upon him. His plan was already settled when he met the man coming from the station, and the scientific manager who speedily arrived on the scene jumped at the obvious conclusion of suicide. This expert scarcely noticed Azuma-zi, except to ask a few questions. Did he see Holroyd kill himself? Azuma-zi explained he had been out of sight at the engine furnace until he heard a difference in the noise from the dynamo. It was not a difficult examination, being untinctured by suspicion.

The distorted remains of Holroyd, which the electrician removed from the machine, were hastily covered by the porter with a coffee-stained tablecloth. Somebody, by a happy inspiration, fetched a medical man. The expert was chiefly anxious to get the machine at work again, for seven or eight trains had stopped midway in the stuffy tunnels of the electric railway. Azuma-zi, answering or misunderstanding the questions of the people who had by authority or impudence come into the shed, was presently sent back to the stoke-hole by the scientific manager. Of course a crowd collected outside the gates of the yard—a crowd, for no known reason, always hovers for a day or two near the scene of a sudden death in London—two or three reporters percolated somehow into the engine shed, and one even got to Azuma-zi; but the scientific expert cleared them out again, being himself an amateur journalist.

Presently the body was carried away, and public interest departed with it. Azuma-zi remained very quietly at his furnace, seeing over and over again in the coals a figure that wriggled violently and became still. An hour after the murder, to anyone coming into the shed things would have looked exactly as if nothing remarkable had ever happened there. Peeping presently from his engine-room the black saw the Lord Dynamo spin and whirl beside his little brothers, and the driving-wheels were beating round and the steam in the pistons went thud, thud, exactly as it had been earlier in the evening. After all, from the mechanical point of view it had been a most insignificant incident—the mere temporary deflection of a current. But now the slender form and slender shadow of the scientific manager replaced the sturdy outline of Holroyd travelling up and down the lane of light upon the vibrating floor under the straps between the engines and the dynamos.

"Have I not served my Lord?" said Azuma-zi inaudibly from his shadow, and the note of the great dynamo rang out full and clear. As he looked at the big whirling mechanism the strange fascination of it that had been a little in abeyance since Holroyd's death resumed its sway.

Never had Azuma-zi seen a man killed so swiftly and pitilessly. The big humming machine had slain its victim without wavering for a second from its steady beating. It was indeed a mighty god.

The unconscious scientific manager stood with his back to him, scribbling on a piece of paper. His shadow lay at the foot of the monster.

Was the Lord Dynamo still hungry? His servant was ready.

Azuma-zi made a stealthy step forward; then stopped. The scientific manager suddenly ceased his writing, walked down the shed to the endmost of the dynamos, and began to examine the brushes.

Azuma-zi hesitated, and then slipped across noiselessly into the shadow by the switch. There he waited. Presently the manager's footsteps could be heard returning. He stopped in his old position, unconscious of the stoker crouching ten feet away from him. Then the big dynamo suddenly fizzled, and in another moment Azuma-zi had sprung out of the darkness upon him.

The scientific manager was gripped round the body and swung towards the big dynamo. Kicking with his knee and forcing his antagonist's head down with his hands, he loosened the grip on his waist and swung round away from the machine. Then the black grasped him again, putting a curly head against his chest, and they swayed and panted as it seemed for an age or so. Then the scientific manager was impelled to catch a black ear in his teeth and bite furiously. The black yelled hideously.

They rolled over on the floor, and the black, who had apparently slipped from the vice of the teeth or parted with some ear—the scientific manager wondered which at the time—tried to throttle him. The scientific manager was making some ineffectual efforts to claw something with his hands and to kick, when the welcome sound of quick foot-steps sounded on the floor. The next moment Azuma-zi had left him and darted towards the big dynamo. There was a splutter amid the roar.

The officer of the company who had entered stood staring as Azuma-zi caught the naked terminals in his hands, gave one horrible convulsion, and then hung motionless from the machine, his face violently distorted.

"I'm jolly glad you came in when you did," said the scientific manager, still sitting on the floor.

He looked at the still quivering figure. "It is not a nice death to die, apparently—but it is quick."

The official was still staring at the body. He was a man of slow apprehension.

There was a pause.

The scientific manager got up on his feet rather awkwardly. He ran his fingers along his collar thoughtfully, and moved his head to and fro several times.

"Poor Holroyd! I see now." Then almost mechanically he went towards the switch in the shadow and turned the current into the railway circuit again. As he did so the singed body loosened its grip upon the machine and fell forward on its face. The core of the dynamo roared out loud and clear, and the armature beat the air.

So ended prematurely the worship of the Dynamo Deity, perhaps the most short-lived of all religions. Yet withal it could at least boast a Martyrdom and a Human Sacrifice.

STORY THE
THIRTEENTH

The Hammerpond Park Burglary

IT is a moot point whether burglary is to be considered as a sport, a trade, or an art. For a trade, the technique is scarcely rigid enough, and its claims to be considered an art are vitiated by the mercenary element that qualifies its triumphs. On the whole it seems to be most justly ranked as sport, a sport for which no rules are at present formulated, and of which the prizes are distributed in an extremely informal manner. It was this informality of burglary that led to the regrettable extinction of two promising beginners at Hammerpond Park.

The stakes offered in this affair consisted chiefly of diamonds and other personal *bric-à-brac* belonging to the newly married Lady Aveling. Lady Aveling, as the reader will remember, was the only daughter of Mrs. Montague Pangs, the well-known hostess. Her marriage to Lord Aveling was extensively advertised in the papers, the quantity and quality of her wedding presents, and the fact that the honeymoon was to be spent at Hammerpond. The announcement of these valuable prizes created a considerable sensation in the small circle in which Mr. Teddy Watkins was the undisputed leader, and it was decided that, accompanied by a duly qualified assistant, he should visit the village of Hammerpond in his professional capacity.

Being a man of naturally retiring and modest disposition, Mr. Watkins determined to make this visit *incog.*, and after due consideration of the conditions of his enterprise, he selected the rôle of a landscape artist and the unassuming surname of Smith. He preceded his assistant, who, it was decided, should join him only on the last afternoon of his stay at Hammerpond. Now the village of Hammerpond is perhaps one of the prettiest little corners in Sussex; many thatched houses still survive, the flint-built church with its tall spire nestling under the down is one of the finest and least restored in the county, and the beech-woods and bracken jungles through which the road runs to the great house are singularly rich in what the vulgar artist and photographer call "bits." So that Mr. Watkins, on his arrival with two virgin canvases, a brand-new easel, a paint-box, portmanteau, an ingenious little ladder made in sections (after the pattern of the late lamented master Charles Peace), crowbar, and wire coils, found himself welcomed with effusion and some curiosity by half-a-dozen other brethren of the brush. It rendered the disguise he had chosen unexpectedly plausible, but it inflicted upon him a considerable amount of æsthetic conversation for which he was very imperfectly prepared.

"Have you exhibited very much?" said Young Porson in the bar-parlour of the "Coach and Horses," where Mr. Watkins was skilfully accumulating local information on the night of his arrival.

"Very little," said Mr. Watkins, "just a snack here and there."

"Academy?"

"Of course. *And* at the Crystal Palace."

"Did they hang you well?" said Porson.

"Don't talk rot," said Mr. Watkins; "I don't like it."

"I mean did they put you in a good place?"

"Whadyer mean?" said Mr. Watkins suspiciously. "One 'ud think you were trying to make out I'd been put away."

Porson had been brought up by aunts, and was a gentlemanly young man even for an artist; he did not know what being "put away" meant, but he thought it best to explain that he intended nothing of the sort. As the question of hanging seemed a sore point with Mr. Watkins, he tried to divert the conversation a little.

"Do you do figure-work at all?"

"No, never had a head for figures," said Mr. Watkins, "my miss—Mrs. Smith, I mean, does all that."

"She paints too!" said Porson. "That's rather jolly."

"Very," said Mr. Watkins, though he really did not think so, and, feeling the conversation was drifting a little beyond his grasp, added, "I came down here to paint Hammerpond House by moonlight."

"Really!" said Porson. "That's rather a novel idea."

"Yes," said Mr. Watkins, "I thought it rather a good notion when it occurred to me. I expect to begin to-morrow night."

"What! You don't mean to paint in the open, by night."

"I do, though."

"But how will you see your canvas?"

"Have a bloomin' cop's——" began Mr. Watkins, rising too quickly to the question, and then realising this, bawled to Miss Durgan for another glass of beer. "I'm goin' to have a thing called a dark lantern," he said to Porson.

"But it's about new moon now," objected Porson. "There won't be any moon."

"There'll be the house," said Watkins, "at any rate I'm goin', you see, to paint the house first and the moon afterwards."

"Oh!" said Porson, too staggered to continue the conversation.

"They doo say," said old Durgan, the landlord, who had maintained a respectful silence during the technical conversation, "as there's no less than three p'licemen from 'Azleworth on dewty every night in the house—'count of this Lady Aveling 'n her jewellery. One'm won fower-and-six last night, off second footman—tossin'."

Towards sunset next day Mr. Watkins, virgin canvas, easel, and a very considerable case of other appliances in hand, strolled up the pleasant pathway through the beech-woods to Hammerpond Park, and pitched his apparatus in a strategic position commanding the house. Here he was observed by Mr. Raphael Sant, who was returning across the park from a study of the chalk-pits. His curiosity having been fired by Porson's account of the new arrival, he turned aside with the idea of discussing nocturnal art.

Mr. Watkins was apparently unaware of his approach. A friendly conversation with Lady Hammerpond's butler had just terminated, and that individual, surrounded by the three pet dogs which it was his duty to take for an airing after dinner had been served, was receding in the distance. Mr. Watkins was mixing colour with an air of great industry. Sant, approaching more nearly, was surprised to see the colour in question was as harsh and brilliant an emerald green as it is possible to imagine. Having cultivated an extreme sensibility to colour from his earliest years, he drew the air in sharply between his teeth at the very first glimpse of this brew. Mr. Watkins turned round. He looked annoyed.

"What on earth are you going to do with that *beastly* green?" said Sant.

Mr. Watkins realised that his zeal to appear busy in the eyes of the butler had evidently betrayed him into some technical error. He looked at Sant and hesitated.

"Pardon my rudeness," said Sant; "but really, that green is altogether too amazing. It came as a shock. What *do* you mean to do with it?"

Mr. Watkins was collecting his resources. Nothing could save the situation but decision. "If you come here interrupting my work," he said, "I'm a-goin' to paint your face with it."

Sant retired, for he was a humorist and a peaceful man. Going down the hill he met Porson and Wainwright. "Either that man is a genius or he is a dangerous lunatic," said he. "Just go up and look at his green." And he continued his way, his countenance brightened by a pleasant anticipation of a cheerful affray round an easel in the gloaming, and the shedding of much green paint.

But to Porson and Wainwright Mr. Watkins was less aggressive, and explained that the green was intended to be the first coating of his picture. It was, he admitted in response to a remark, an absolutely new method, invented by himself. But subsequently he became more reticent; he explained he was not going to tell every passer-by the secret of his own particular style, and added some scathing remarks upon the meanness of people "hanging about" to pick up such tricks of the masters as they could, which immediately relieved him of their company.

Twilight deepened, first one, then another star appeared. The rooks amid the tall trees to the left of the house had long since lapsed into slumbrous silence, the house itself lost all the details of its architecture and became a dark grey outline, and then the windows of the salon shone out brilliantly, the conservatory was lighted up, and here and there a bedroom window burnt yellow. Had anyone approached the easel in the park it would have been found deserted. One brief uncivil word in brilliant green sullied the purity of its canvas. Mr. Watkins was busy in the shrubbery with his assistant, who had discreetly joined him from the carriage-drive.

Mr. Watkins was inclined to be self-congratulatory upon the ingenious device by which he had carried all his apparatus boldly, and in the sight of all men, right up to the scene of operations. "That's the dressing-room," he said to his assistant, "and, as soon as the maid takes the candle away and goes down to supper, we'll call in. My! how nice the house do look, to be sure, against the starlight, and with all its windows and lights! Swopme, Jim, I almost wish I *was* a painter-chap. Have you fixed that there wire across the path from the laundry?"

He cautiously approached the house until he stood below the dressing-room window, and began to put together his folding ladder. He was much too experienced a practitioner to feel any unusual excitement. Jim was reconnoitring the smoking-room. Suddenly, close beside Mr. Watkins in the bushes, there was a violent crash and a stifled curse. Someone had tumbled over the wire which his assistant had just arranged. He heard feet running on the gravel pathway beyond. Mr. Watkins, like all true artists, was a singularly shy man, and he incontinently dropped his folding ladder and began running circumspectly through the shrubbery. He was indistinctly aware of two people hot upon his heels, and he fancied that he distinguished the outline of his assistant in front of him. In another moment he had vaulted the low stone wall bounding the shrubbery, and was in the open park. Two thuds on the turf followed his own leap.

It was a close chase in the darkness through the trees. Mr. Watkins was a loosely-built man and in good training, and he gained hand-over-hand upon the hoarsely panting figure in front. Neither spoke, but, as Mr. Watkins pulled up alongside, a qualm of awful doubt came over him. The other man turned his head at the same moment and gave an exclamation of surprise. "It's not Jim," thought Mr. Watkins, and simultaneously the stranger flung himself, as it were, at Watkins' knees, and they were forthwith grappling on the ground together. "Lend a hand, Bill," cried the stranger as the third man came up. And Bill did—two hands in fact, and some accentuated feet. The fourth man, presumably Jim, had apparently turned aside and made off in a different direction. At any rate, he did not join the trio.

Mr. Watkins' memory of the incidents of the next two minutes is extremely vague. He has a dim recollection of having his thumb in the corner of the mouth of the first man, and feeling anxious about its safety, and for some seconds at least he held the head of the gentleman answering to the name of Bill, to the ground by the hair. He was also kicked in a great number of different places, apparently by a vast multitude of people. Then the gentleman who was not Bill got his knee below Mr. Watkins' diaphragm, and tried to curl him up upon it.

When his sensations became less entangled he was sitting upon the turf, and eight or ten men—the night was dark, and he was rather too confused to count—standing round him, apparently waiting for him to recover. He mournfully assumed that he was captured, and would probably have made some philosophical reflections on the fickleness of fortune, had not his internal sensations disinclined him for speech.

He noticed very quickly that his wrists were not hand-cuffed, and then a flask of brandy was put in his hands. This touched him a little—it was such unexpected kindness.

"He's a-comin' round,' said a voice which he fancied as belonging to the Hammerpond second footman.

"We've got 'em, sir, both of 'em," said the Hammerpond butler, the man who had handed him the flask. "Thanks to *you*."

No one answered this remark. Yet he failed to see how it applied to him.

"He's fair dazed," said a strange voice; "the villains half-murdered him."

Mr. Teddy Watkins decided to remain fair dazed until he had a better grasp of the situation. He perceived that two of the black figures round him stood side-by-side with a dejected air, and there was something in the carriage of their shoulders that suggested to his experienced eye hands that were bound together. Two! In a flash he rose to his position. He emptied the little flask and staggered—obsequious hands assisting him—to his feet. There was a sympathetic murmur.

"Shake hands, sir, shake hands," said one of the figures near him. "Permit me to introduce myself. I am very greatly indebted to you. It was the jewels of my wife, Lady Aveling, which attracted these scoundrels to the house."

"Very glad to make your lordship's acquaintance," said Teddy Watkins.

"I presume you saw the rascals making for the shrubbery, and dropped down on them?"

"That's exactly how it happened," said Mr. Watkins.

"You should have waited till they got in at the window," said Lord Aveling; "they would get it hotter if they had actually committed the burglary. And it was lucky for you two of the policemen were out by the gates, and followed up the three of you. I doubt if you could have secured the two of them—though it was confoundedly plucky of you all the same."

"Yes, I ought to have thought of all that," said Mr. Watkins; "but one can't think of everything."

"Certainly not," said Lord Aveling. "I am afraid they have mauled you a little," he added. The party was now moving towards the house. "You walk rather lame. May I offer you my arm?"

And instead of entering Hammerpond House by the dressing-room window, Mr. Watkins entered it—slightly intoxicated, and inclined now to cheerfulness again—on the arm of a real live peer, and by the front door. "This," thought Mr. Watkins, "is burgling in style!" The "scoundrels," seen by the gaslight, proved to be mere local amateurs unknown to Mr. Watkins, and they were taken down into the pantry and there watched over by the three policemen, two gamekeepers with loaded guns, the butler, an ostler, and a carman, until the dawn allowed of their removal to Hazelhurst police-station. Mr. Watkins was made much of in the saloon. They devoted a sofa to him, and would not hear of a return to the village that night. Lady Aveling was sure he was brilliantly original, and said her idea of Turner was just such another rough, half-inebriated, deep-eyed, brave, and clever man. Someone brought up a remarkable little folding-ladder that had been picked up in the shrubbery, and showed him how it was put together. They also described how wires had been found in the shrubbery, evidently placed there to trip up unwary pursuers. It was lucky he had escaped these snares. And they showed him the jewels.

Mr. Watkins had the sense not to talk too much, and in any conversational difficulty fell back on his internal pains. At last he was seized with stiffness in the back, and yawning. Everyone suddenly awoke to the fact that it was a shame to keep him talking after his affray, so he retired early to his room, the little red room next to Lord Aveling's suite.

*   *   *   *   *

The dawn found a deserted easel bearing a canvas with a green inscription in the Hammerpond Park, and it found Hammerpond House in commotion. But if the dawn found Mr. Teddy Watkins and the Aveling diamonds, it did not communicate the information to the police.

STORY THE
FOURTEENTH

The Moth

PROBABLY you have heard of Hapley—not W. T. Hapley, the son, but the celebrated Hapley, the Hapley of *Periplaneta Hapliia*, Hapley the entomologist.

If so you know at least of the great feud between Hapley and Professor Pawkins, though certain of its consequences may be new to you. For those who have not, a word or two of explanation is necessary, which the idle reader may go over with a glancing eye if his indolence so incline him.

It is amazing how very widely diffused is the ignorance of such really important matters as this Hapley-Hawkins feud. Those epoch-making controversies, again, that have convulsed the Geological Society are, I verily believe, almost entirely unknown outside the fellowship of that body. I have heard men of fair general education even refer to the great scenes at these meetings as vestry-meeting squabbles. Yet the great hate of the English and Scotch geologists has lasted now half a century, and has "left deep and abundant marks upon the body of the science." And this Hapley-Pawkins business, though perhaps a more personal affair, stirred passions as profound, if not profounder. Your common man has no conception of the zeal that animates a scientific investigator, the fury of contradiction you can arouse in him. It is the *odium theologicum* in a new form. There are men, for instance, who would gladly burn Sir Ray Lankester at Smithfield for his treatment of the Mollusca in

302

the Encyclopædia. That fantastic extension of the Cephalopods to cover the Pteropods. . . . But I wander from
Hapley and Pawkins.

It began years and years ago with a revision of Microlepidoptera (whatever these may be) by Pawkins, in which
he extinguished a new species created by Hapley. Hapley,
who was always quarrelsome, replied by a stinging impeach-
ment of the entire classification of Pawkins.* Pawkins in
his "Rejoinder"† suggested that Hapley's microscope was
as defective as his power of observation, and called him an
"irresponsible meddler"—Hapley was not a professor at that
time. Hapley in his retort,‡ spoke of "blundering col-
lectors," and described, as if inadvertently, Pawkins's revision as a "miracle of ineptitude." It was war to the knife.
However, it would scarcely interest the reader to detail how
these two great men quarrelled, and how the split between
them widened until from the Microlepidoptera they were
at war upon every open question in entomology. There
were memorable occasions. At times the Royal Entomological Society meetings resembled nothing so much as the
Chamber of Deputies. On the whole, I fancy Pawkins was
nearer the truth than Hapley. But Hapley was skilful
with his rhetoric, had a turn for ridicule rare in a scientific
man, was endowed with vast energy, and had a fine sense of
injury in the matter of the extinguished species; while
Pawkins was a man of dull presence, prosy of speech , in
shape not unlike a water-barrel, over-conscientious with
testimonials, and suspected of jobbing museum appointments. So the young men gathered round Hapley and
applauded him. It was a long struggle, vicious from the
beginning and growing at last to pitiless antagonism. The
successive turns of fortune, now an advantage to one side
and now to another—now Hapley tormented by some success
of Pawkins, and now Pawkins outshone by Hapley, belong
rather to the history of entomology than to this story.

*"Remarks on a Recent Revision of Microlepidoptera." *Quart. Journ.
Entomological Soc.,* 1863.

†"Rejoinder to Certain Remarks." etc. *Ibid.* 1864.

‡"Further Remarks," etc. *Ibid.*

But in 1891 Pawkins, whose health had been bad for some time, published some work upon the "mesoblast" of the Death's-Head Moth. What the mesoblast of the Death's-Head Moth may be does not matter a rap in this story. But the work was far below his usual standard, and gave Hapley an opening he had coveted for years. He must have worked night and day to make the most of his advantage.

In an elaborate critique he rent Pawkins to tatters—one can fancy the man's disordered black hair, and his queer dark eye flashing as he went for his antagonist—and Pawkins made a reply, halting, ineffectual, with painful gaps of silence, and yet malignant. There was no mistaking his will to wound Hapley, nor his incapacity to do it. But few of those who heard him—I was absent from that meeting—realised how ill the man was.

Hapley got his opponent down, and meant to finish him. He followed with a brutal attack upon Pawkins, in the form of a paper upon the development of moths in general, a paper showing evidence of an extraordinary amount of labour, couched in a violently controversial tone. Violent as it was, an editorial note witnesses that it was modified. It must have covered Pawkins with shame and confusion of face. It left no loophole; it was murderous in argument, and utterly contemptuous in tone; an awful thing for the declining years of a man's career.

The world of entomologists waited breathlessly for the rejoinder from Pawkins. He would try one, for Pawkins had always been game. But when it came it surprised them. For the rejoinder of Pawkins was to catch influenza, proceed to pneumonia, and die.

It was perhaps as effectual a reply as he could make under the circumstances, and largely turned the current of feeling against Hapley. The very people who had most gleefully cheered on those gladiators became serious at the consequence. There could be no reasonable doubt the fret of the defeat had contributed to the death of Pawkins. There was a limit even to scientific controversy, said serious people. Another crushing attack was already in the press and appeared on the day before the funeral. I don't think Hapley exerted himself to stop it. People remembered how Hapley had hounded down his rival and forgot that rival's defects. Scathing satire reads ill over fresh mould. The thing provoked comment in the daily papers. It was that made me think you had probably heard of Hapley and this controversy. But, as I have already remarked, scientific workers live very much in a world of their own; half the people, I dare say, who go along Piccadilly to the Academy every year could not tell you where the learned societies abide. Many even think that research is a kind of happy-family cage in which all kinds of men lie down together in peace.

In his private thoughts Hapley could not forgive Pawkins for dying. In the first place, it was a mean dodge to escape the absolute pulverisation Hapley had in hand for him, and in the second, it left Hapley's mind with a queer gap in it. For twenty years he had worked hard, sometimes far into the night, and seven days a week, with microscope, scalpel, collecting-net, and pen, and almost entirely with reference to Pawkins. The European reputation he had won had come as an incident in that great antipathy. He had gradually worked up to a climax in this last controversy. It had killed Pawkins, but it had also thrown Hapley out of gear, so to speak, and his doctor advised him to give up work for a time, and rest. So Hapley went down into a quiet village in Kent, and thought day and night of Pawkins and good things it was now impossible to say about him.

At last Hapley began to realise in what direction the preoccupation tended. He determined to make a fight for it, and started by trying to read novels. But he could not get his mind off Pawkins, white in the face and making his last speech—every sentence a beautiful opening for Hapley. He turned to fiction—and found it had no grip on him. He read the "Island Nights' Entertainments" until his "sense of causation" was shocked beyond endurance by the Bottle Imp. Then he went to Kipling, and found he "proved nothing" besides being irreverent and vulgar. These scientific people have their limitations. Then unhappily he tried Besant's "Inner House," and the opening chapter set his mind upon learned societies and Pawkins at once.

So Hapley turned to chess, and found it a little more soothing. He soon mastered the moves and the chief gambits and commoner closing positions; and began to beat the Vicar. But then the cylindrical contours of the opposite king began to resemble Pawkins standing up and gasping ineffectually against checkmate, and Hapley decided to give up chess.

Perhaps the study of some new branch of science would after all be better diversion. The best rest is change of occupation. Hapley determined to plunge at diatoms, and had one of his smaller microscopes and Halibut's monograph sent down from London. He thought that perhaps if he could get up a vigorous quarrel with Halibut, he might be able to begin life afresh and forget Pawkins. And very soon he was hard at work in his habitual strenuous fashion at these microscopic denizens of the wayside pool.

It was on the third day of the diatoms that Hapley became aware of a novel addition to the local fauna. He was working late at the microscope, and the only light in the room was the brilliant little lamp with the special form of green shade. Like all experienced microscopists, he kept both eyes open. It is the only way to avoid excessive fatigue. One eye was over the instrument, and bright and distinct before that was the circular field of the microscope, across which a brown diatom was slowly moving. With the other eye Hapley saw, as it were, without seeing. He was only dimly conscious of the brass side of the instrument, the illuminated part of the tablecloth, a sheet of note-paper, the foot of the lamp, and the darkened room beyond.

Suddenly his attention drifted from one eye to the other. The tablecloth was of the material called tapestry by shopmen, and rather brightly coloured. The pattern was in gold, with a small amount of crimson and pale blue upon a greyish ground. At one point the pattern seemed displaced, and there was a vibrating movement of the colours at this point.

Hapley suddenly moved his head back and looked with both eyes. His mouth fell open with astonishment.

It was a large moth or butterfly; its wings spread in butterfly fashion!

It was strange it should be in the room at all, for the windows were closed. Strange that it should not have attracted his attention when fluttering to its present position. Strange that it should match the tablecloth. Stranger far that to him, Hapley, the great entomologist, it was altogether unknown. There was no delusion. It was crawling slowly towards the foot of the lamp.

"New Genus, by heavens! And in England!" said Hapley, staring.

Then he suddenly thought of Pawkins. Nothing would have maddened Pawkins more. . . . And Pawkins was dead!

Something about the head and body of the insect became singularly suggestive of Pawkins, just as the chess king had been.

"Confound Pawkins!" said Hapley. "But I must catch this." And looking round him for some means of capturing the moth, he rose slowly out of his chair. Suddenly the insect rose, struck the edge of the lampshade—Hapley heard the "ping"—and vanished into the shadow.

In a moment Hapley had whipped off the shade, so that the whole room was illuminated. The thing had disappeared, but soon his practised eye detected it upon the wallpaper near the door. He went towards it poising the lampshade for capture. Before he was within striking distance, however, it had risen and was fluttering round the room. After the fashion of its kind, it flew with sudden starts and turns, seeming to vanish here and reappear there. Once Hapley struck, and missed; then again.

The third time he hit his microscope. The instrument swayed, struck and overturned the lamp, and fell noisily upon the floor. The lamp turned over on the table, and very luckily, went out. Hapley was left in the dark. With a start he felt the strange moth blunder into his face.

It was maddening. He had no lights. If he opened the door of the room the thing would get away. In the darkness he saw Pawkins quite distinctly laughing at him.

Pawkins had ever an oily laugh. He swore furiously and stamped his foot on the floor.

There was a timid rapping at the door.

Then it opened, perhaps a foot, and very slowly. The alarmed face of the landlady appeared behind a pink candle flame; she wore a nightcap over her grey hair and had some purple garment over her shoulders. "What *was* that fearful smash?" she said. "Has anything——" The strange moth appeared fluttering about the chink of the door. "Shut that door!" said Hapley, and suddenly rushed at her.

The door slammed hastily. Hapley was left alone in the dark. Then in the pause he heard his landlady scuttle upstairs, lock her door, and drag something heavy across the room and put against it.

It became evident to Hapley that his conduct and appearance had been strange and alarming. Confound the moth! and Pawkins! However, it was a pity to lose the moth now. He felt his way into the hall and found the matches, after sending his hat down upon the floor with a noise like a drum. With the lighted candle he returned to the sitting-room. No moth was to be seen. Yet once for a moment it seemed that the thing was fluttering round his head. Hapley very suddenly decided to give up the moth and go to bed. But he was excited. All night long his sleep was broken by dreams of the moth, Pawkins, and his landlady. Twice in the night he turned out and soused his head in cold water.

One thing was very clear to him. His landlady could not possibly understand about the strange moth, especially as he had failed to catch it. No one but an entomologist would understand quite how he felt. She was probably frightened at his behaviour, and yet he failed to see how he could explain it. He decided to say nothing further about the events of last night. After breakfast he saw her in her garden, and decided to go out and talk to reassure her. He talked to her about beans and potatoes, bees, caterpillars, and the price of fruit. She replied in her usual manner, but she looked at him a little suspiciously, and kept walking as he walked, so that there was always a bed of flowers, or a

ow of beans, or something of the sort, between them. After
a while he began to feel singularly irritated at this, and, to
conceal his vexation, went indoors and presently went out
for a walk.

The moth, or butterfly, trailing an odd flavour of Pawkins
with it, kept coming into that walk though he did his best
to keep his mind off it. Once he saw it quite distinctly,
with its wings flattened out, upon the old stone wall that
runs along the west edge of the park, but going up to it he
found it was only two lumps of grey and yellow lichen.
"This," said Hapley, "is the reverse of mimicry. Instead of
a butterfly looking like a stone, here is a stone looking like a
butterfly!" Once something hovered and fluttered round
his head, but by an effort of will he drove that impression
out of his mind again.

In the afternoon Hapley called upon the Vicar, and
argued with him upon theological questions. They sat in
the little arbour covered with brier, and smoked as they
wrangled. "Look at that moth!" said Hapley, suddenly,
pointing to the edge of the wooden table.

"Where?" said the Vicar.

"You don't see a moth on the edge of the table there?"
said Hapley.

"Certainly not," said the Vicar.

Hapley was thunderstruck. He gasped. The Vicar was
staring at him. Clearly the man saw nothing. "The eye
of faith is no better than the eye of science," said Hapley
awkwardly.

"I don't see your point," said the Vicar, thinking it was
part of the argument.

That night Hapley found the moth crawling over his
counterpane. He sat on the edge of the bed in his shirt-
sleeves and reasoned with himself. Was it pure hallucination? He knew he was slipping, and he battled for his
sanity with the same silent energy he had formerly displayed against Pawkins. So persistent is mental habit that
he felt as if it were still a struggle with Pawkins. He was
well versed in psychology. He knew that such visual illusions do come as a result of mental strain. But the point

was, he did not only *see* the moth, he had heard it when i
touched the edge of the lamp-shade and afterwards when i
hit against the wall, and he had felt it strike his face in the
dark.

He looked at it. It was not at all dream-like but perfectl
clear and solid-looking in the candle-light. He saw the
hairy body and the short feathery antennæ, the jointed leg
even a place where the down was rubbed from the wing
He suddenly felt angry with himself for being afraid of a
little insect.

His landlady had got the servant to sleep with her tha
night, because she was afraid to be alone. In addition she
had locked the door and put the chest of drawers against it
They listened and talked in whispers after they had gone
to bed, but nothing occurred to alarm them. About eleven
they had ventured to put the candle out and had both
dozed off to sleep. They woke with a start, and sat up in
bed, listening in the darkness.

Then they heard slippered feet going to and fro in
Hapley's room. A chair was overturned and there was a
violent dab at the wall. Then a china mantel ornamen
smashed upon the fender. Suddenly the door of the room
opened, and they heard him upon the landing. They clung
to one another, listening. He seemed to be dancing upon
the staircase. Now he would go down three or four step
quickly, then up again, then hurry down into the hall. They
heard the umbrella-stand go over, and the fanlight break
Then the bolt shot and the chain rattled. He was opening
the door.

They hurried to the window. It was a dim grey night
an almost unbroken sheet of watery cloud was sweeping
across the moon, and the hedge and trees in front of the
house were black against the pale roadway. They saw
Hapley, looking like a ghost in his shirt and white trousers
running to and fro in the road and beating the air. Now he
would stop, now he would dart very rapidly at something in
visible, now he would move upon it with stealthy strides
At last he went out of sight up the road towards the down
Then while they argued who should go down and lock the

door, he returned. He was walking very fast, and he came straight into the house, closed the door carefully, and went quietly up to his bedroom. Then everything was silent.

"Mrs. Colville," said Hapley, calling down the staircase next morning, "I hope I did not alarm you last night."

"You may well ask that!" said Mrs. Colville.

"The fact is, I am a sleep-walker, and the last two nights, I have been without my sleeping mixture. There is nothing to be alarmed about, really. I am sorry I made such an ass of myself. I will go over the down to Shoreham, and get some stuff to make me sleep soundly. I ought to have done that yesterday."

But half-way over the down, by the chalk pits, the moth came upon Hapley again. He went on, trying to keep his mind upon chess problems, but it was no good. The thing fluttered into his face, and he struck at it with his hat in self-defence. Then rage, the old rage—the rage he had so often felt against Pawkins—came upon him again. He went on, leaping and striking at the eddying insect. Suddenly he trod on nothing, and fell headlong.

There was a gap in his sensations, and Hapley found himself sitting on the heap of flints in front of the opening of the chalk pits, with a leg twisted back under him. The strange moth was still fluttering round his head. He struck at it with his hand, and turning his head saw two men approaching him. One was the village doctor. It occurred to Hapley that this was lucky. Then it came into his mind with extraordinary vividness, that no one would ever be able to see the strange moth except himself, and that it behoved him to keep silent about it.

Late that night, however, after his broken leg was set, he was feverish and forgot his self-restraint. He was lying flat on his bed, and he began to run his eyes round the room to see if the moth was still about. He tried not to do this, but it was no good. He soon caught sight of the thing resting close to his hand, by the night-light, on the green tablecloth. The wings quivered. With a sudden wave of anger he smote at it with his fist, and the nurse woke up with a shriek. He had missed it.

"That moth!" he said; and then: "It was fancy. Nothing!"

All the time he could see quite clearly the insect going round the cornice and darting across the room, and he could also see that the nurse saw nothing of it and looked at him strangely. He must keep himself in hand. He knew he was a lost man if he did not keep himself in hand. But as the night waned the fever grew upon him, and the very dread he had of seeing the moth made him see it. About five, just as the dawn was grey, he tried to get out of bed and catch it, though his leg was afire with pain. The nurse had to struggle with him.

On account of this, they tied him down to the bed. At this the moth grew bolder, and once he felt it settle in his hair. Then, because he struck out violently with his arms, they tied these also. At this the moth came and crawled over his face, and Hapley wept, swore, screamed, prayed for them to take it off him, unavailingly.

The doctor was a blockhead, a just-qualified general practitioner, and quite ignorant of mental science. He simply said there was no moth. Had he possessed the wit, he might still perhaps have saved Hapley from his fate by entering into his delusion, and covering his face with gauze as he prayed might be done. But, as I say, the doctor was a blockhead; and until the leg was healed Hapley was kept tied to his bed, with the imaginary moth crawling over him. It never left him while he was awake and it grew to a monster in his dreams. While he was awake he longed for sleep, and from sleep he awoke screaming.

So now, Hapley is spending the remainder of his days in a padded room, worried by a moth that no one else can see. The asylum doctor calls it hallucination; but Hapley, when he is in his easier mood and can talk, says it is the ghost of Pawkins, and consequently a unique specimen and well worth the trouble of catching.

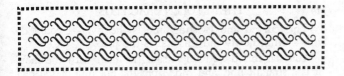

STORY THE FIFTEENTH

The Treasure in the Forest

THE canoe was now approaching the land. The bay opened out, and a gap in the white surf of the reef marked where the little river ran out to the sea; the thicker and deeper green of the virgin forest showed its course down the distant hill slope. The forest here came close to the beach. Far beyond, dim and almost cloudlike in texture, rose the mountains, like suddenly frozen waves. The sea was still save for an almost imperceptible swell. The sky blazed.

The man with the carved paddle stopped. "It should be somewhere here," he said. He shipped the paddle and held his arms out straight before him.

The other man had been in the fore part of the canoe, closely scrutinising the land. He had a sheet of yellow paper on his knee.

"Come and look at this, Evans," he said.

Both men spoke in low tones, and their lips were hard and dry.

The man called Evans came swaying along the canoe until he could look over his companion's shoulder.

The paper had the appearance of a rough map. By much folding it was creased and worn to the pitch of separation, and the second man held the discoloured fragments together where they had parted. On it one could dimly make out, in almost obliterated pencil, the outline of the bay.

"Here," said Evans, "is the reef and here is the gap." He ran his thumb-nail over the chart.

"This curved and twisting line is the river—I could do with a drink now!—and this star is the place."

"You see this dotted line," said the man with the map; "it is a straight line, and runs from the opening of the reef to a clump of palm-trees. The star comes just where it cuts the river. We must mark the place as we go into the lagoon."

"It's queer," said Evans, after a pause, "what these little marks down here are for. It looks like the plan of a house or something; but what all these little dashes, pointing this way and that, may mean I can't get a notion. And what's the writing?"

"Chinese," said the man with the map.

"Of course! *He* was a Chinee," said Evans.

"They all were," said the man with the map.

They both sat for some minutes staring at the land, while the canoe drifted slowly. Then Evans looked towards the paddle.

"Your turn with the paddle now, Hooker," said he.

And his companion quietly folded up his map, put it in his pocket, passed Evans carefully, and began to paddle. His movements were languid, like those of a man whose strength was nearly exhausted.

Evans sat with his eyes half closed, watching the frothy breakwater of the coral creep nearer and nearer. The sky was like a furnace now, for the sun was near the zenith. Though they were so near the Treasure he did not feel the exaltation he had anticipated. The intense excitement of the struggle for the plan, and the long night voyage from the mainland in the unprovisioned canoe had, to use his own expression, "taken it out of him." He tried to arouse himself by directing his mind to the ingots the Chinamen had spoken of, but it would not rest there; it came back headlong to the thought of sweet water rippling in the river, and to the almost unendurable dryness of his lips and throat. The rhythmic wash of the sea upon the reef was becoming audible now, and it had a pleasant sound in his ears; the water washed along the side of the canoe, and the paddle dripped between each stroke. Presently he began to doze.

He was still dimly conscious of the island, but a queer dream texture interwove with his sensations. Once again it was the night when he and Hooker had hit upon the Chinamen's secret; he saw the moonlit trees, the little fire burning, and the black figures of the three Chinamen—silvered on one side by moonlight, and on the other glowing from the firelight—and heard them talking together in pigeon-English —for they came from different provinces. Hooker had caught the drift of their talk first, and had motioned to him to listen. Fragments of the conversation were inaudible and fragments incomprehensible. A Spanish galleon from the Philippines hopelessly aground, and its treasure buried against the day of return, lay in the background of the story; a shipwrecked crew thinned by disease, a quarrel or so, and the needs of discipline, and at last taking to their boats never to be heard of again. Then Chang-hi, only a year since, wandering ashore, had happened upon the ingots hidden for two hundred years, had deserted his junk, and reburied them with infinite toil, single-handed but very safe. He laid great stress on the safety—it was a secret of his. Now he wanted help to return and exhume them. Presently the little map fluttered and the voices sank. A fine story for two stranded British wastrels to hear! Evans' dream shifted to the moment when he had Chang-hi's pigtail in his hand. The life of a Chinaman is scarcely sacred like a European's. The cunning little face of Chang-hi, first keen and furious like a startled snake, and then fearful, treacherous and piti- ful, became overwhelmingly prominent in the dream. At the end Chang-hi had grinned, a most incomprehensible and startling grin.   Abruptly things became very un- pleasant, as they will do at times in dreams. Chang-hi gib- bered and threatened him. He saw in his dream heaps and heaps of gold, and Chang-hi intervening and struggling to hold him back from it. He took Chang-hi by the pigtail— how big the yellow brute was, and how he struggled and grinned! He kept growing bigger, too. Then the bright heaps of gold turned to a roaring furnace, and a vast devil, surprisingly like Chang-hi, but with a huge black tail, began to feed him with coals. They burnt his mouth horribly.

Another devil was shouting his name: "Evans, Evans, you sleepy fool!"—or was it Hooker?

He woke up. They were in the mouth of the lagoon.

"There are the three palm-trees. It must be in a line with that clump of bushes," said his companion. "Mark that. If we go to those bushes and then strike into the bush in a straight line from here, we shall come to it when we come to the stream."

They could see now where the mouth of the stream opened out. At the sight of it Evans revived. "Hurry up, man," he said, "Or by heaven I shall have to drink sea water!" He gnawed his hand and stared at the gleam of silver among the rocks and green tangle.

Presently he turned almost fiercely upon Hooker. "Give *me* the paddle," he said.

So they reached the river mouth. A little way up Hooker took some water in the hollow of his hand, tasted it, and spat it out. A little further he tried again. "This will do," he said, and they began drinking eagerly.

"Curse this!" said Evans, suddenly. "It's too slow." And, leaning dangerously over the fore part of the canoe, he began to suck up the water with his lips.

Presently they made an end of drinking, and, running the canoe into a little creek, were about to land among the thick growth that overhung the water.

"We shall have to scramble through this to the beach to find our bushes and get the line to the place," said Evans.

"We had better paddle round," said Hooker.

So they pushed out again into the river and paddled back down it to the sea, and along the shore to the place where the clump of bushes grew. Here they landed, pulled the light canoe far up the beach, and then went up towards the edge of the jungle until they could see the opening of the reef and the bushes in a straight line. Evans had taken a native implement out of the canoe. It was L-shaped, and the transverse piece was armed with polished stone. Hooker carried the paddle. "It is straight now in this direction," said he; "we must push through this till we strike the stream. Then we must prospect."

They pushed through a close tangle of reeds, broad fronds, and young trees, and at first it was toilsome going, but very speedily the trees became larger and the ground beneath them opened out. The blaze of the sunlight was replaced by insensible degrees by cool shadow. The trees became at last vast pillars that rose up to a canopy of greenery far overhead. Dim white flowers hung from their stems, and ropy creepers swung from tree to tree. The shadow deepened. On the ground, blotched fungi and a red-brown incrustation became frequent.

Evans shivered. "It seems almost cold here after the blaze outside."

"I hope we are keeping to the straight," said Hooker.

Presently they saw, far ahead, a gap in the sombre darkness where white shafts of hot sunlight smote into the forest. There also was brilliant green undergrowth, and coloured flowers. Then they heard the rush of water.

"Here is the river. We should be close to it now," said Hooker.

The vegetation was thick by the river bank. Great plants, as yet unnamed, grew among the roots of the big trees, and spread rosettes of huge green fans towards the strip of sky. Many flowers and a creeper with shiny foliage clung to the exposed stems. On the water of the broad, quiet pool which the treasure seekers now overlooked there floated big oval leaves and a waxen, pinkish-white flower not unlike a water-lily. Further, as the river bent away from them, the water suddenly frothed and became noisy in a rapid.

"Well?" said Evans.

"We have swerved a little from the straight," said Hooker. "That was to be expected."

He turned and looked into the dim cool shadows of the silent forest behind them. "If we beat a little way up and down the stream we should come to something."

"You said——" began Evans.

"*He* said there was a heap of stones," said Hooker. The two men looked at each other for a moment.

"Let us try a little down-stream first," said Evans.

They advanced slowly, looking curiously about them. Suddenly Evans stopped. "What the devil's that?" he said.

Hooker followed his finger. "Something blue," he said. It had come into view as they topped a gentle swell of the ground. Then he began to distinguish what it was.

He advanced suddenly with hasty steps, until the body that belonged to the limp hand and arm had become visible. His grip tightened on the implement he carried. The thing was the figure of a Chinaman lying on his face. The abandon of the pose was unmistakable.

The two men drew closer together, and stood staring silently at this ominous dead body. It lay in a clear space among the trees. Near by was a spade after the Chinese pattern, and further off lay a scattered heap of stones, close to a freshly dug hole.

"Somebody has been here before," said Hooker, clearing his throat.

Then suddenly Evans began to swear and rave, and stamp upon the ground.

Hooker turned white but said nothing. He advanced towards the prostrate body. He saw the neck was puffed and purple, and the hands and ankles swollen. "Pah!" he said, and suddenly turned away and went towards the excavation. He gave a cry of surprise. He shouted to Evans, who was following him slowly.

"You fool! It's all right. It's here still." Then he turned again and looked at the dead Chinaman, and then again at the hole.

Evans hurried to the hole. Already half exposed by the ill-fated wretch beside them lay a number of dull yellow bars. He bent down in the hole, and, clearing off the soil with his bare hands, hastily pulled one of the heavy masses out. As he did so a little thorn pricked his hand. He pulled the delicate spike out with his fingers and lifted the ingot.

"Only gold or lead could weigh like this," he said exultantly.

Hooker was still looking at the dead Chinaman. He was puzzled.

"He stole a march on his friends," he said at last. "He came here alone, and some poisonous snake has killed him. . . . I wonder how he found the place."

Evans stood with the ingot in his hands. What did a dead Chinaman signify? "We shall have to take this stuff to the mainland piecemeal, and bury it there for a while. How shall we get it to the canoe?"

He took his jacket off and spread it on the ground, and flung two or three ingots into it. Presently he found that another little thorn had punctured his skin.

"This is as much as we can carry," said he. Then suddenly, with a queer rush of irritation, "What are you staring at?"

Hooker turned to him. "I can't stand . . . him." He nodded towards the corpse. "It's so like——"

"Rubbish!" said Evans. "All Chinamen are alike."

Hooker looked into his face. "I'm going to bury *that*, anyhow, before I lend a hand with this stuff."

"Don't be a fool, Hooker," said Evans. "Let that mass of corruption bide."

Hooker hesitated, and then his eye went carefully over the brown soil about them. "It scares me somehow," he said.

"The thing is," said Evans, "what to do with these ingots. Shall we re-bury them over here, or take them across the strait in the canoe?"

Hooker thought. His puzzled gaze wandered among the tall tree-trunks, and up into the remote sunlit greenery overhead. He shivered again as his eye rested upon the blue figure of the Chinaman. He stared searchingly among the grey depths between the trees.

"What's come to you, Hooker?" said Evans. "Have you lost your wits?"

"Let's get the gold out of this place, anyhow," said Hooker.

He took the ends of the collar of the coat in his hands, and Evans took the opposite corners, and they lifted the mass. "Which way?" said Evans. "To the canoe?"

"It's queer," said Evans, when they had advanced only a few steps, "but my arms ache still with that paddling. . . ."

"Curse it!" he said. "But they ache! I must rest."

They let the coat down. Evans' face was white, and little drops of sweat stood out upon his forehead. "It's stuffy, somehow, in this forest."

Then with an abrupt transition to unreasonable anger: "What is the good of waiting here all the day? Lend a hand, I say! You have done nothing but moon since we saw the dead Chinaman."

Hooker was looking steadfastly at his companion's face. He helped raise the coat bearing the ingots, and they went forward perhaps a hundred yards in silence. Evans began to breathe heavily. "Can't you speak?" he said.

"What's the matter with you?" said Hooker.

Evans stumbled, and then with a sudden curse flung the coat from him. He stood for a moment staring at Hooker, and then with a groan clutched at his own throat.

"Don't come near me," he said, and went and leant against a tree. Then in a steadier voice, "I'll be better in a minute."

Presently his grip upon the trunk loosened, and he slipped slowly down the stem of the tree until he was a crumpled heap at its foot. His hands were clenched convulsively. His face became distorted with pain. Hooker approached him.

"Don't touch me! Don't touch me!" said Evans in a stifled voice. "Put the gold back on the coat."

"Can't I do anything for you?" said Hooker.

"Put the gold back on the coat."

As Hooker handled the ingots he felt a little prick on the ball of his thumb. He looked at his hand and saw a slender thorn, perhaps two inches in length.

Evans gave an inarticulate cry and rolled over.

Hooker's jaw dropped. He stared at the thorn for a moment with dilated eyes. Then he looked at Evans, who was now crumpled together on the ground, his back bending and straightening spasmodically. Then he looked through the pillars of the trees and net-work of creeper stems, to where in the dim grey shadow the blue-clad body of the

Chinaman was still indistinctly visible. He thought of the little dashes in the corner of the plan, and in a moment he understood.

"God help me!" he said. For the thorns were similar to those the Dyaks poison and use in their blowing-tubes. He understood now what Chang-hi's assurance of the safety of his treasure meant. He understood that grin now.

"Evans!" he cried.

But Evans was silent and motionless now, save for a horrible spasmodic twitching of his limbs. A profound silence brooded over the forest.

Then Hooker began to suck furiously at the little pink spot on the ball of his thumb—sucking for dear life. Presently he felt a strange aching pain in his arms and shoulders, and his fingers seemed difficult to bend. Then he knew that sucking was no good.

Abruptly he stopped, and sitting down by the pile of ingots, and resting his chin upon his hands and his elbows upon his knees, stared at the distorted but still stirring body of his companion. Chang-hi's grin came in his mind again. The dull pain spread towards his throat and grew slowly in intensity. Far above him a faint breeze stirred the greenery, and the white petals of some unknown flower came floating down through the gloom.

THE PLATTNER STORY
AND OTHERS

STORY THE FIRST

The Plattner Story

WHETHER the story of Gottfried Plattner is to be credited or not, is a pretty question in the value of evidence. On the one hand, we have seven witnesses—to be perfectly exact, we have six and a half pairs of eyes, and one undeniable fact; and on the other we have—what is it?—prejudice, common sense, the inertia of opinion. Never were there seven more honest-seeming witnesses; never was there a more undeniable fact than the inversion of Gottfried Plattner's anatomical structure, and—never was there a more preposterous story than the one they have to tell! The most preposterous part of the story is the worthy Gottfried's contribution (for I count him as one of the seven). Heaven forbid that I should be led into giving countenance to superstition by a passion for impartiality, and so come to share the fate of Eusapia's patrons! Frankly, I believe there is something crooked about this business of Gottfried Plattner; but what that crooked factor is, I will admit as frankly, I do not know. I have been surprised at the credit accorded to the story in the most unexpected and authoritative quarters. The fairest way to the reader, however, will be for me to tell it without further comment.

Gottfried Plattner is, in spite of his name, a free-born Englishman. His father was an Alsatian who came to England in the Sixties, married a respectable English girl of unexceptionable antecedents, and died, after a wholesome and uneventful life (devoted, I understand, chiefly to the laying of parquet flooring), in 1887. Gottfried's age is seven-and-twenty. He is, by virtue of his heritage of three lan-

325

guages, Modern Languages Master in a small private schoo
in the South of England. To the casual observer he i
singularly like any other Modern Languages Master in an
other small private school. His costume is neither very costl
nor very fashionable, but, on the other hand, it is no
markedly cheap or shabby; his complexion, like his heigh
and his bearing, is inconspicuous. You would notic
perhaps that, like the majority of people, his face was no
absolutely symmetrical, his right eye a little larger tha
the left, and his jaw a trifle heavier on the right side. I
you, as an ordinary careless person, were to bare his ches
and feel his heart beating, you would probably find it quit
like the heart of anyone else. But here you and the traine
observer would part company. If you found his heart quit
ordinary, the trained observer would find it quite otherwise
And once the thing was pointed out to you, you too woul
perceive the peculiarity easily enough. It is that Gottfried'
heart beats on the right side of his body.

Now that is not the only singularity of Gottfried's struc
ture, although it is the only one that would appeal to th
untrained mind. Careful sounding of Gottfried's interna
arrangements, by a well-known surgeon, seems to poin
to the fact that all the other unsymmetrical parts of hi
body are similarly misplaced. The right lobe of his live
is on the left side, the left on his right; while his lung:
too, are similarly contraposed. What is still more singula
unless Gottfried is a consummate actor we must believ
that his right hand has recently become his left. Sinc
the occurrences we are about to consider (as impartiall
as possible), he has found the utmost difficulty in writin
except from right to left across the paper with his left hanc
He cannot throw with his right hand, he is perplexed a
meal times between knife and fork, and his ideas of th
rule of the road—he is a cyclist—are still a dangerous cor
fusion. And there is not a scrap of evidence to show tha
before these occurrences Gottfried was at all left-handed.

There is yet another wonderful fact in this preposterou
business. Gottfried produces three photographs of himsel
You have him at the age of five or six, thrusting fat leg

at you from under a plaid frock, and scowling. In that
photograph his left eye is a little larger than his right, and
his jaw is a trifle heavier on the left side. This is the
reverse of his present living conditions. The photograph
of Gottfried at fourteen seems to contradict these facts, but
that is because it is one of those cheap "Gem" photographs
that were then in vogue, taken direct upon metal, and
therefore reversing things just as a looking-glass would.
The third photograph represents him at one-and-twenty, and
confirms the record of the others. There seems here evidence of the strongest confirmatory character that Gottfried
has exchanged his left side for his right. Yet how a human
being can be so changed, short of a fantastic and pointless
miracle, it is exceedingly hard to suggest.

In one way, of course, these facts might be explicable on
the supposition that Plattner has undertaken an elaborate
mystification on the strength of his heart's displacement.
Photographs may be fudged, and left-handedness imitated.
But the character of the man does not lend itself to any
such theory. He is quiet, practical, unobtrusive, and
thoroughly sane from the Nordau standpoint. He likes
beer and smokes moderately, takes walking exercise daily,
and has a healthily high estimate of the value of his teaching. He has a good but untrained tenor voice, and takes a
pleasure in singing airs of a popular and cheerful character.
He is fond, but not morbidly fond, of reading—chiefly fiction
pervaded with a vaguely pious optimism,—sleeps well, and
rarely dreams. He is, in fact, the very last person to evolve
a fantastic fable. Indeed, so far from forcing this story
upon the world, he has been singularly reticent on the
matter. He meets inquirers with a certain engaging—bash-
fulness is almost the word, that disarms the most suspicious.
He seems genuinely ashamed that anything so unusual has
occurred to him.

It is to be regretted that Plattner's aversion to the idea
of post-mortem dissection may postpone, perhaps for ever,
the positive proof that his entire body has had its left and
right sides transposed. Upon that fact mainly the credibility
of his story hangs. There is no way of taking a man and

moving him about *in space,* as ordinary people understand space, that will result in our changing his sides. Whatever you do, his right is still his right, his left his left. You can do that with a perfectly thin and flat thing, of course. If you were to cut a figure out of paper, any figure with a right and left side, you could change its sides simply by lifting it up and turning it over. But with a solid it is different. Mathematical theorists tell us that the only way in which the right and left sides of a solid body can be changed is by taking that body clean out of space as we know it,—taking it out of ordinary existence, that is, and turning it somewhere outside space. This is a little abstruse, no doubt, but anyone with a slight knowledge of mathematical theory will assure the reader of its truth. To put the thing in technical language, the curious inversion of Plattner's right and left sides is proof that he has moved out of our space into what is called the Fourth Dimension, and that he has returned again to our world. Unless we choose to consider ourselves the victims of an elaborate and motiveless fabrication, we are almost bound to believe that this has occurred.

So much for the tangible facts. We come now to the account of the phenomena that attended his temporary disappearance from the world. It appears that in the Sussexville Proprietary School, Plattner not only discharged the duties of Modern Languages Master, but also taught chemistry, commercial geography, bookkeeping, shorthand, drawing, and any other additional subject to which the changing fancies of the boy's parents might direct attention. He knew little or nothing of these various subjects, but in secondary as distinguished from Board or elementary schools, knowledge in the teacher is, very properly, by no means so necessary as high moral character and gentlemanly tone. In chemistry he was particularly deficient, knowing, he says, nothing beyond the Three Gases (whatever the three gases may be). As, however, his pupils began by knowing nothing, and derived all their information from him, this caused him (or anyone) but little inconvenience for several terms. Then a little boy named Whibble joined the school, who had been educated, it seems, by some mischievous relative into an inquiring habit of mind. This little boy followed Plattner's lessons with marked and sustained interest, and in order to exhibit his zeal on the subject, brought at various times substances for Plattner to analyse. Plattner, flattered by this evidence of his power to awaken interest and trusting to the boy's ignorance, analysed these and even made general statements as to their composition. Indeed he was so far stimulated by his pupil as to obtain a work upon analytical chemistry, and study it during his supervision of the evening's preparation. He was surprised to find chemistry quite an interesting subject.

So far the story is absolutely commonplace. But now the greenish powder comes upon the scene. The source of that greenish powder seems, unfortunately, lost. Master Whibble tells a tortuous story of finding it done up in a packet in a disused limekiln near the Downs. It would have been an excellent thing for Plattner, and possibly for Master Whibble's family, if a match could have been applied to that powder there and then. The young gentleman certainly did not bring it to school in a packet, but in a common eight-ounce graduated medicine bottle, plugged with masticated newspaper. He gave it to Plattner at the end of the afternoon school. Four boys had been detained after school prayers in order to complete some neglected tasks, and Plattner was supervising these in the small classroom in which the chemical teaching was conducted. The appliances for the practical teaching of chemistry in the Sussexville Proprietary School, as in most private schools in this country, are characterised by a severe simplicity. They are kept in a cupboard standing in a recess and having about the same capacity as a common travelling trunk. Plattner, being bored with his passive superintendence, seems to have welcomed the intervention of Whibble with his green powder as an agreeable diversion, and, unlocking this cupboard, proceeded at once with his analytical experiments. Whibble sat, luckily for himself, at a safe distance, regarding him. The four malefactors, feigning a profound absorption in their work, watched him furtively with the

keenest interest. For even within the limits of the Three
Gases, Plattner's practical chemistry was, I understand,
temerarious.

They are practically unanimous in their account of Plattner's proceedings. He poured a little of the green powder
into a test-tube, and tried the substance with water, hydro-
chloric acid, nitric acid, and sulphuric acid in succession.
Getting no result, he emptied out a little heap—nearly half
the bottleful, in fact—upon a slate and tried a match. He
held the medicine bottle in his left hand. The stuff began
to smoke and melt, and then—exploded with deafening
violence and a blinding flash.

The five boys, seeing the flash and being prepared for
catastrophes, ducked below their desks, and were none of
them seriously hurt. The window was blown out into the
playground, and the blackboard on its easel was upset. The
slate was smashed to atoms. Some plaster fell from the
ceiling. No other damage was done to the school edifice
or appliances, and the boys at first, seeing nothing of Plattner, fancied he was knocked down and lying out of their
sight below the desks. They jumped out of their places
to go to his assistance, and were amazed to find the space
empty. Being still confused by the sudden violence of the
report, they hurried to the open door, under the impression
that he must have been hurt, and have rushed out of the
room. But Carson, the foremost, nearly collided in the
doorway with the principal, Mr. Lidgett.

Mr. Lidgett is a corpulent, excitable man with one eye.
The boys describe him as stumbling into the room mouthing
some of those tempered expletives irritable schoolmasters
accustom themselves to use—lest worse befall. "Wretched
mumchancer!" he said. "Where's Mr. Plattner?" The boys
are agreed on the very words. ("Wobbler," "snivelling
puppy," and "mumchancer" are, it seems, among the ordinary
small change of Mr. Lidgett's scholastic commerce.)

Where's Mr. Plattner? That was a question that was
to be repeated many times in the next few days. It really
seemed as though that frantic hyperbole, "blown to atoms,"
had for once realised itself. There was not a visible particle

of Plattner to be seen; not a drop of blood nor a stitch of clothing to be found. Apparently he had been blown clean out of existence and left not a wrack behind. Not so much as would cover a sixpenny piece, to quote a proverbial expression! The evidence of his absolute disappearance, as a consequence of that explosion, is indubitable.

It is not necessary to enlarge here upon the commotion excited in the Sussexville Proprietary School, and in Sussexville and elsewhere, by this event. It is quite possible, indeed, that some of the readers of these pages may recall the hearing of some remote and dying version of that excitement during the last summer holidays. Lidgett, it would seem, did everything in his power to suppress and minimise the story. He instituted a penalty of twenty-five lines for any mention of Plattner's name among the boys, and stated in the schoolroom that he was clearly aware of his assistant's whereabouts. He was afraid, he explains, that the possibility of an explosion happening, in spite of the elaborate precautions taken to minimise the practical teaching of chemistry, might injure the reputation of the school; and so might any mysterious quality in Plattner's departure. Indeed, he did everything in his power to make the occurrence seem as ordinary as possible. In particular, he cross-examined the five eye-witnesses of the occurrence so searchingly that they began to doubt the plain evidence of their senses. But, in spite of these efforts, the tale, in a magnified and distorted state, made a nine days' wonder in the district, and several parents withdrew their sons on colourable pretexts. Not the least remarkable point in the matter is the fact that a large number of people in the neighbourhood dreamed singularly vivid dreams of Plattner during the period of excitement before his return, and that these dreams had a curious uniformity. In almost all of them Plattner was seen, sometimes singly, sometimes in company, wandering about through a coruscating iridescence. In all cases his face was pale and distressed, and in some he gesticulated towards the dreamer. One or two of the boys, evidently under the influence of nightmare, fancied that Plattner approached them with remarkable swiftness, and seemed to look closely into their very eyes. Others fled with Plattner from the pursuit of vague and extraordinary creatures of a globular shape. But all these fancies were forgotten in inquiries and speculations when, on the Wednesday next but one after the Monday of the explosion, Plattner returned.

The circumstances of his return were as singular as those of his departure. So far as Mr. Lidgett's somewhat choleric outline can be filled in from Plattner's hesitating statements, it would appear that on Wednesday evening, towards the hour of sunset, the former gentleman, having dismissed evening preparation, was engaged in his garden, picking and eating strawberries, a fruit of which he is inordinately fond. It is a large old-fashioned garden, secured from observation, fortunately, by a high and ivy-covered red-brick wall. Just as he was stooping over a particularly prolific plant, there was a flash in the air and a heavy thud, and before he could look round, some heavy body struck him violently from behind. He was pitched forward, crushing the strawberries he held in his hand, and with such force that his silk hat—Mr. Lidgett adheres to the older ideas of scholastic costume—was driven violently down upon his forehead, and almost over one eye. This heavy missile, which slid over him sideways and collapsed into a sitting posture among the strawberry plants, proved to be our long-lost Mr. Gottfried Plattner, in an extremely dishevelled condition. He was collarless and hatless, his linen was dirty, and there was blood upon his hands. Mr. Lidgett was so indignant and surprised that he remained on all-fours, and with his hat jammed down on his eye, while he expostulated vehemently with Plattner for his disrespectful and unaccountable conduct.

This scarcely idyllic scene completes what I may call the exterior version of the Plattner story—its exoteric aspect. It is quite unnecessary to enter here into all the details of his dismissal by Mr. Lidgett. Such details, with the full names and dates and references, will be found in the larger report of these occurrences that was laid before the Society for the Investigation of Abnormal Phenomena. The singular transposition of Plattner's right and left sides was scarcely observed for the first day or so, and then first in connection with his disposition to write from right to left across the blackboard. He concealed rather than ostended this curious confirmatory circumstance, as he considered it would unfavourably affect his prospects in a new situation. The displacement of his heart was discovered some months after, when he was having a tooth extracted under anæsthetics. He then, very unwillingly, allowed a cursory surgical examination to be made of himself, with a view to a brief account in the *Journal of Anatomy*. That exhausts the statement of the material facts; and we may now go on to consider Plattner's account of the matter.

But first let us clearly differentiate between the preceding portion of this story and what is to follow. All I have told thus far is established by such evidence as even a criminal lawyer would approve. Every one of the winesses is still alive; the reader, if he have the leisure, may hunt the lads out to-morrow, or even brave the terrors of the redoubtable Lidgett, and cross-examine and trap and test to his heart's content; Gottfried Plattner, himself, and his twisted heart and his three photographs are producible. It may be taken as proved that he did disappear for nine days as the consequence of an explosion; that he returned almost as violently, under circumstances in their nature annoying to Mr. Lidgett, whatever the details of those circumstances may be; and that he returned inverted, just as a reflection returns from a mirror. From the last fact, as I have already stated, it follows almost inevitably that Plattner, during those nine days, must have been in some state of existence altogether out of space. The evidence to these statements is, indeed, far stronger than that upon which most murderers are hanged. But for his own particular account of where he had been, with its confused explanations and well-nigh self-contradictory details, we have only Mr. Gottfried Plattner's word. I do not wish to discredit that, but I must point out —what so many writers upon obscure psychic phenomena fail to do—that we are passing here from the practically undeniable to that kind of matter which any reasonable man is entitled to believe or reject as he thinks proper. The previous statements render it plausible; its discordance with common experience tilts it towards the incredible. I would prefer not to sway the beam of the reader's judgment either way, but simply to tell the story as Plattner told it me.

He gave me his narrative, I may state, at my house at Chislehurst; and so soon as he had left me that evening, I went into my study and wrote down everything as I remembered it. Subsequently he was good enough to read over a type-written copy, so that its substantial correctness is undeniable.

He states that at the moment of the explosion he distinctly thought he was killed. He felt lifted off his feet and driven forcibly backward. It is a curious fact for psychologists that he thought clearly during his backward flight, and wondered whether he should hit the chemistry cupboard or the blackboard easel. His heels struck ground, and he staggered and fell heavily into a sitting position on something soft and firm. For a moment the concussion stunned him. He became aware at once of a vivid scent of singed hair, and he seemed to hear the voice of Lidgett asking for him. You will understand that for a time his mind was greatly confused.

At first he was distinctly under the impression that he was still in the classroom. He perceived quite distinctly the surprise of the boys and the entry of Mr. Lidgett. He is quite positive upon that score. He did not hear their remarks, but that he ascribed to the deafening effect of the experiment. Things about him seemed curiously dark and faint, but his mind explained that on the obvious but mistaken idea that the explosion had engendered a huge volume of dark smoke. Through the dimness the figures of Lidgett and the boys moved, as faint and silent as ghosts. Plattner's face still tingled with the stinging heat of the flash. He was, he says, "all muddled." His first definite thoughts seem to have been of his personal safety. He thought he was perhaps blinded and deafened. He felt his limbs and face in a gingerly manner. Then his perceptions grew clearer, and he was astonished to miss the old familiar desks and other schoolroom furniture about him. Only dim, uncertain, grey shapes stood in the place of these. Then came a thing that made him shout aloud, and awoke his stunned faculties to instant activity. *Two of the boys, gesticulating, walked one after the other clean through him!* Neither manifested the slightest consciousness of his presence. It is difficult to imagine the sensation he felt. They came against him, he says, with no more force than a wisp of mist.

Plattner's first thought after that was that he was dead. Having been brought up with thoroughly sound views in these matters, however, he was a little surprised to find his body still about him. His second conclusion was that he was not dead, but that the others were: that the explosion had destroyed the Sussexville Proprietary School and every soul in it except himself. But that, too, was scarcely satisfactory. He was thrown back upon astonished observation.

Everything about him was extraordinarily dark: at first it seemed to have an altogether ebony blackness. Overhead was a black firmament. The only touch of light in the scene was a faint greenish glow at the edge of the sky in one direction, which threw into prominence a horizon of undulating black hills. This, I say, was his impression at first. As his eye grew accustomed to the darkness, he began to distinguish a faint quality of differentiating greenish colour in the circumambient night. Against this background the furniture and occupants of the classroom, it seems, stood out like phosphorescent spectres, faint and impalpable. He extended his hand, and thrust it without an effort through the wall of the room by the fireplace.

He describes himself as making a strenuous effort to attract attention. He shouted to Lidgett, and tried to seize the boys as they went to and fro. He only desisted from these attempts when Mrs. Lidgett, whom he as an Assistant Master naturally disliked, entered the room. He says the sensation of being in the world, and yet not a part of it, was an extraordinarily disagreeable one. He compared his feelings not inaptly to those of a cat watching a mouse through a window. Whenever he made a motion to communicate with the dim, familiar world about him, he found an invisible, incomprehensible barrier preventing intercourse.

He then turned his attention to his solid environment. He found the medicine bottle still unbroken in his hand, with the remainder of the green powder therein. He put this in his pocket, and began to feel about him. Apparently, he was sitting on a boulder of rock covered with a velvety moss. The dark country about him he was unable to see, the faint, misty picture of the schoolroom blotting it out, but he had a feeling (due perhaps to a cold wind) that he was near the crest of a hill, and that a steep valley fell away beneath his feet. The green glow along the edge of the sky seemed to be growing in extent and intensity. He stood up, rubbing his eyes.

It would seem that he made a few steps, going steeply downhill, and then stumbled, nearly fell, and sat down again upon a jagged mass of rock to watch the dawn. He became aware that the world about him was absolutely silent. It was as still as it was dark, and though there was a cold wind blowing up the hill-face, the rustle of grass, the sighing of the boughs that should have accompanied it, were absent. He could hear, therefore, if he could not see, that the hillside upon which he stood was rocky and desolate. The green grew brighter every moment, and as it did so a faint, transparent blood-red mingled with, but did not mitigate, the blackness of the sky overhead and the rocky desolations about him. Having regard to what follows, I am inclined to think that that redness may have been an optical effect due to contrast. Something black fluttered momentarily against the livid yellow-green of the lower sky, and then the thin and penetrating voice of a bell rose out of the black gulf below him. An oppressive expectation grew with the growing light.

It is probable that an hour or more elapsed while he sat there, the strange green light growing brighter every moment, and spreading slowly, in flamboyant fingers, upward towards the zenith. As it grew, the spectral vision of *our* world became relatively or absolutely fainter. Probably both, for the time must have been about that of our earthly sunset. So far as his vision of our world went, Plattner by his few steps downhill, had passed through the floor of the classroom, and was now, it seemed, sitting in mid-air in the larger schoolroom downstairs. He saw the boarders distinctly, but much more faintly than he had seen Lidgett. They were preparing their evening tasks, and he noticed with interest that several were cheating with their Euclid riders by means of a crib, a compilation whose existence he had hitherto never suspected. As the time passed they faded steadily, as steadily as the light of the green dawn increased.

Looking down into the valley, he saw that the light had crept far down its rocky sides, and that the profound blackness of the abyss was now broken by a minute green glow, like the light of a glow-worm. And almost immediately the limb of a huge heavenly body of blazing green rose over the basaltic undulations of the distant hills, and the monstrous hill-masses about him came out gaunt and desolate, in green light and deep, ruddy black shadows. He became aware of a vast number of ball-shaped objects drifting as thistledown drifts over the high ground. There were none of these nearer to him than the opposite side of the gorge. The bell below twanged quicker and quicker, with something like impatient insistence, and several lights moved hither and thither. The boys at work at their desks were now almost imperceptibly faint.

This extinction of our world, when the green sun of this other universe rose, is a curious point upon which Plattner insists. During the Other-World night it is difficult to move about, on account of the vividness with which the things of this world are visible. It becomes a riddle to explain why, if this is the case, we in this world catch no glimpse of the Other-World. It is due, perhaps, to the comparatively vivid illumination of this world of ours. Plattner describes the midday of the Other-World, at its brightest, as not being nearly so bright as this world at full moon, while its night is profoundly black. Consequently, the amount of light, even in an ordinary dark room, is sufficient to render the things of the Other-World invisible, on the same principle that faint phosphorescence is only visible in the profoundest darkness. I have tried, since he told me his story, to see something of the Other-World by sitting for a long space in a photographer's dark room at night. I have certainly seen indistinctly the form of greenish slopes and rocks, but only, I must admit, very indistinctly indeed. The reader may possibly be more successful. Plattner tells me that since his return he has seen and recognised places in the Other-World in his dreams, but this is probably due to his memory of these scenes. It seems quite possible that people with unusually keen eyesight may occasionally catch a glimpse of this strange Other-World about us.

However, this is a digression. As the green sun rose, a long street of black buildings became perceptible, though only darkly and indistinctly, in the gorge, and, after some hesitation, Plattner began to clamber down the precipitous descent towards them. The descent was long and exceedingly tedious, being so not only by the extraordinary steepness, but also by reason of the looseness of the boulders with which the whole face of the hill was strewn. The noise of his descent—now and then his heels struck fire from the rocks—seemed now the only sound in the universe, for the beating of the bell had ceased. As he drew nearer he perceived that the various edifices had a singular resemblance to tombs and mausoleums and monuments, saving only that they were all uniformly black instead of being white as most sepulchres are. And then he saw, crowding out of the largest building very much as people disperse from church, a number of pallid, rounded, pale-green figures. These scattered in several directions about the broad street of the place, some going through side alleys and reappearing upon the steepness of the hill, others entering some of the small black buildings which lined the way.

At the sight of these things drifting up towards him, Plattner stopped, staring. They were not walking, they were indeed limbless; and they had the appearance of human heads beneath which a tadpole-like body swung. He was too astonished at their strangeness, too full indeed of strangeness, to be seriously alarmed by them. They drove towards him, in front of the chill wind that was blowing uphill, much as soap-bubbles drive before a draught. And as he looked at the nearest of those approaching, he saw it was indeed a human head, albeit with singularly large eyes, and wearing such an expression of distress and anguish as he had never seen before upon mortal countenance. He was surprised to find that it did not turn to regard him, but seemed to be watching and following some unseen moving thing. For a moment he was puzzled, and then it occurred to him that this creature was watching with its enormous eyes something that was happening in the world he had just left. Nearer it came, and nearer, and he was too astonished to cry out. It made a very faint fretting sound as it came close to him. Then it struck his face with a gentle pat—its touch was very cold—and drove past him, and upward towards the crest of the hill.

An extraordinary conviction flashed across Plattner's mind that this head had a strong likeness to Lidgett. Then he turned his attention to the other heads that were now swarming thickly up the hillside. None made the slightest sign of recognition. One or two, indeed, came close to his head and almost followed the example of the first, but he dodged convulsively out of the way. Upon most of them he saw the same expression of unavailing regret he had seen upon the first, and heard the same faint sounds of wretchedness from them. One or two wept, and one rolling swiftly uphill wore an expression of diabolical rage. But others were cold, and several had a look of gratified interest in their eyes. One, at least, was almost in an ecstasy of happiness. Plattner does not remember that he recognised any more likenesses in those he saw at this time.

For several hours, perhaps, Plattner watched these strange things dispersing themselves over the hills, and not till long after they had ceased to issue from the clustering black buildings in the gorge did he resume his downward climb. The darkness about him increased so much that he had a difficulty in stepping true. Overhead the sky was now a bright pale green. He felt neither hunger nor thirst. Later, when he did, he found a chilly stream running down the centre of the gorge, and the rare moss upon the boulders, when he tried it at last in desperation, was good to eat.

He groped about among the tombs that ran down the gorge, seeking vaguely for some clue to these inexplicable things. After a long time he came to the entrance of the big mausoleum-like building from which the heads had issued. In this he found a group of green lights burning upon a kind of basaltic altar, and a bell-rope from a belfry overhead hanging down into the centre of the place. Round the wall ran a lettering of fire in a character unknown to him. While he was still wondering at the purport of these things, he heard the receding tramp of heavy feet echoing far down the street. He ran out into the darkness again, but he could see nothing. He had a mind to pull the bell-rope, and finally decided to follow the footsteps. But although he ran far, he never overtook them; and his shouting was of no avail. The gorge seemed to extend an interminable distance. It was as dark as earthly starlight throughout its length, while the ghastly green day lay along the upper edge of its precipices. There were none of the heads, now, below. They were all, it seemed, busily occupied along the upper slopes. Looking up, he saw them drifting hither and thither, some hovering stationary, some flying swiftly through the air. It reminded him, he said, of "big snowflakes"; only these were black and pale green.

In pursuing the firm, undeviating footsteps that he never overtook, in groping into new regions of this endless devil's dyke, in clambering up and down the pitiless heights, in wandering about the summits, and in watching the drifting faces, Plattner states that he spent the better part of seven or eight days. He did not keep count, he says. Though once or twice he found eyes watching him, he had word with no living soul. He slept among the rocks on the hillside. In the gorge things earthly were invisible, because, from the earthly standpoint, it was far underground. On the altitudes, so soon as the earthly day began, the world became visible to him. He found himself sometimes stumbling over the dark green rocks, or arresting himself on a precipitous brink, while all about him the green branches of the Sussexville lanes were swaying; or, again, he seemed to be walking through the Sussexville streets, or watching unseen the private business of some household. And then it was he discovered, that to almost every human being in our world there pertained some of these drifting heads; that everyone in the world is watched intermittently by these helpless disembodiments.

What are they—these Watchers of the Living? Plattner never learned. But two that presently found and followed him, were like his childhood's memory of his father and mother. Now and then other faces turned their eyes upon him: eyes like those of dead people who had swayed him, or injured him, or helped him in his youth and manhood. Whenever they looked at him, Plattner was overcome with a strange sense of responsibility. To his mother he ventured to speak; but she made no answer. She looked sadly, stead-fastly, and tenderly—a little reproachfully, too, it seemed—into his eyes.

He simply tells this story: he does not endeavour to explain. We are left to surmise who these Watchers of the Living may be, or if they are indeed the Dead, why they should so closely and passionately watch a world they have left for ever. It may be—indeed to my mind it seems just—that, when our life has closed, when evil or good is no longer a choice for us, we may still have to witness the working out of the train of consequences we have laid. If human souls continue after death, then surely human interests continue after death. But that is merely my own guess at the meaning of the things seen. Plattner offers no interpretation, for none was given him. It is well the reader should understand this clearly. Day after day, with his head reeling, he wandered about this green-lit world out-side the world, weary and, towards the end, weak and hungry. By day—by our earthly day, that is—the ghostly vision of the old familiar scenery of Sussexville, all about him, irked and worried him. He could not see where to put his feet, and ever and again with a chilly touch one of these Watching Souls would come against his face.

And after dark the multitude of these Watchers about him, and their intent distress, confused his mind beyond describing. A great longing to return to the earthly life that was so near and yet so remote consumed him. The unearthliness of things about him produced a positively painful mental distress. He was worried beyond describing by his own particular followers. He would shout at them to desist from staring at him, scold at them, hurry away from them. They were always mute and intent. Run as he might over the uneven ground, they followed his destinies.

On the ninth day, towards evening, Plattner heard the invisible footsteps approaching, far away down the gorge. He was then wandering over the broad crest of the same hill upon which he had fallen in his entry into this strange Other-World of his. He turned to hurry down into the gorge, feeling his way hastily, and was arrested by the sight of the thing that was happening in a room in a back street near the school. Both of the people in the room he knew by sight. The windows were open, the blinds up, and the setting sun shone clearly into it, so that it came out quite brightly at first, a vivid oblong of room, lying like a magic-lantern picture upon the black landscape and the livid green dawn. In addition to the sunlight, a candle had just been lit in the room.

On the bed lay a lank man, his ghastly white face terrible upon the tumbled pillow. His clenched hands were raised above his head. A little table beside the bed carried a few medicine bottles, some toast and water, and an empty glass. Every now and then the lank man's lips fell apart, to indicate a word he could not articulate. But the woman did not notice that he wanted anything, because she was busy turning out papers from an old-fashioned bureau in the opposite corner of the room. At first the picture was very vivid indeed, but as the green dawn behind it grew brighter and brighter, so it became fainter and more and more transparent.

As the echoing footsteps paced nearer and nearer, those footsteps that sound so loud in that Other-World and come so silently in this, Plattner perceived about him a great multitude of dim faces gathering together out of the darkness and watching the two people in the room. Never before had he seen so many of the Watchers of the Living. A multitude had eyes only for the sufferer in the room, another multitude, in infinite anguish, watched the woman as she hunted with greedy eyes for something she could not find. They crowded about Plattner, they came across his sight and buffeted his face, the noise of their unavailing regrets was all about him. He saw clearly only now and then. At other times the pictures quivered dimly, through the veil of green reflections upon their movements. In the room it must have been very still, and Plattner says the candle flame streamed up into a perfectly vertical line of smoke, but in his ears each footfall and its echoes beat like a clap of thunder. And the faces! Two more particularly, near the woman's: one a woman's also, white and clear-featured, a face which might have once been cold and hard but which was now softened by the touch of a wisdom strange to earth. The other might have been the woman's father. Both were evidently absorbed in the contemplation of some act of hateful meanness, so it seemed, which they could no longer guard against and prevent. Behind were others, teachers it may be who had taught ill, friends whose influence had failed. And over the man, too—a multitude, but none that seemed to be parents or teachers! Faces that might once have been coarse, now purged to strength by sorrow! And in the forefront one face, a girlish one, neither angry nor remorseful but merely patient and weary, and, as it seemed to Plattner, waiting for relief. His powers of description fail him at the memory of this multitude of ghastly countenances. They gathered on the stroke of the bell. He saw them all in the space of a second. It would seem that he was so worked upon by his excitement that quite involuntarily his restless fingers took the bottle of green powder out of his pocket and held it before him. But he does not remember that.

Abruptly the footsteps ceased. He waited for the next and there was silence, and then suddenly, cutting through the unexpected stillness like a keen, thin blade, came the first

stroke of the bell. At that the multitudinous faces swayed
to and fro, and a louder crying began all about him. The
woman did not hear; she was burning something now in
the candle flame. At the second stroke everything grew
dim, and a breath of wind, icy cold, blew through the host
of watchers. They swirled about him like an eddy of dead
leaves in the spring, and at the third stroke something was
extended through them to the bed. You have heard of a
beam of light. This was like a beam of darkness, and
looking again at it, Plattner saw that it was a shadowy arm
and hand.

The green sun was now topping the black desolations of
the horizon, and the vision of the room was very faint.
Plattner could see that the white of the bed struggled, and
was convulsed; and that the woman looked round over her
shoulder at it, startled.

The cloud of watchers lifted high like a puff of green
dust before the wind, and swept swiftly downward towards
the temple in the gorge. Then suddenly Plattner understood the meaning of the shadowy black arm that stretched
across his shoulder and clutched its prey. He did not dare
turn his head to see the Shadow behind the arm. With a
violent effort, and covering his eyes, he set himself to run,
made perhaps twenty strides ,then slipped on a boulder and
fell. He fell forward on his hands; and the bottle smashed
and exploded as he touched the ground.

In another moment he found himself, stunned and bleeding, sitting face to face with Lidgett in the old walled
garden behind the school.

There the story of Plattner's experiences ends. I have
resisted, I believe successfully, the natural disposition of a
writer of fiction to dress up incidents of this sort. I have
told the thing as far as possible in the order in which
Plattner told it to me. I have carefully avoided any attempt
at style, effect, or construction. It would have been easy,
for instance, to have worked the scene of the death-bed into
a kind of plot in which Plattner might have been involved.
But quite apart from the objectionableness of falsifying a

most extraordinary true story, any such trite devices would spoil, to my mind, the peculiar effect of this dark world, with its livid green illumination and its drifting Watchers of the Living, which, unseen and unapproachable to us, is yet lying all about us.

It remains to add, that a death did actually occur in Vincent Terrace, just beyond the school garden, and, so far as can be proved, at the moment of Plattner's return. Deceased was a rate-collector and insurance agent. His widow, who was much younger than himself, married last month a Mr. Whymper, a veterinary surgeon of Allbeeding. As the portion of this story given here has in various forms circulated orally in Sussexville, she has consented to my use of her name, on condition that I make it distinctly known that she emphatically contradicts every detail of Plattner's account of her husband's last moments. She burnt no will, she says, although Plattner never accused her of doing so: her husband made but one will, and that just after their marriage. Certainly, from a man who had never seen it, Plattner's account of the furniture of the room was curiously accurate.

One other thing, even at the risk of an irksome repetition, I must insist upon lest I seem to favour the credulous superstitious view. Plattner's absence from the world for nine days is, I think, proved. But that does not prove his story. It is quite conceivable that even outside space hallucinations may be possible. That, at least, the reader must bear distinctly in mind.

STORY THE SECOND

The Argonauts of the Air

ONE saw Monson's Flying Machine from the windows of the trains passing either along the South-Western main line or along the line between Wimbledon and Worcester Park, —to be more exact, one saw the huge scaffoldings which limited the flight of the apparatus. They rose over the tree-tops, a massive alley of interlacing iron and timber, and an enormous web of ropes and tackle, extending the best part of two miles. From the Leatherhead branch this alley was foreshortened and in part hidden by a hill with villas; but from the main line one had it in profile, a complex tangle of girders and curving bars, very impressive to the excursionists from Portsmouth and Southampton and the West. Monson had taken up the work where Maxim had left it, had gone on at first with an utter contempt for the journalistic wit and ignorance that had irritated and hampered his predecessor, and had spent (it was said) rather more than half his immense fortune upon his experiments. The results, to an impatient generation, seemed inconsiderable. When some five years had passed after the growth of the colossal iron groves at Worcester Park, and Monson still failed to put in a fluttering appearance over Trafalgar Square, even the Isle of Wight trippers felt their liberty to smile. And such intelligent people as did not consider Monson a fool stricken with the mania for invention, denounced him as being (for no particular reason) a self-advertising quack.

Yet now and again a morning trainload of season-ticket holders would see a white monster rush headlong through

346

the airy tracery of guides and bars, and hear the further
stays, nettings, and buffers snap, creak, and groan with the
impact of the blow. Then there would be an efflorescence
of black-set white-rimmed faces along the sides of the train,
and the morning papers would be neglected for a vigorous
discussion of the possibility of flying (in which nothing new
was ever said by any chance), until the train reached
Waterloo, and its cargo of season-ticket holders dispersed
themselves over London. Or the fathers and mothers in
some multitudinous train of weary excursionists returning
exhausted from a day of rest by the sea, would find the dark
fabric, standing out against the evening sky, useful in divert-
ing some bilious child from its introspection, and be suddenly startled by the swift transit of a huge black flapping
shape that strained upward against the guides. It was a
great and forcible thing beyond dispute, and excellent for
conversation; yet, all the same, it was but flying in leading-
strings, and most of those who witnessed it scarcely counted
its flight as flying. More of a switchback it seemed to the
run of the folk.

Monson, I say, did not trouble himself very keenly about
the opinions of the press at first. But possibly he, even, had
formed but a poor idea of the time it would take before the
tactics of flying were mastered, the swift assured adjustment
of the big soaring shape to every gust and chance movement
of the air; nor had he clearly reckoned the money this prolonged struggle against gravitation would cost him. And he
was not so pachydermatous as he seemed. Secretly he had
his periodical bundles of cuttings sent him by Romeike, he
had his periodical reminders from his banker; and if he did
not mind the initial ridicule and scepticism, he felt the
growing neglect as the months went by and the money
dribbled away. Time was when Monson had sent the enterprising journalist, keen after readable matter, empty from his
gates. But when the enterprising journalist ceased from
troubling, Monson was anything but satisfied in his heart
of hearts. Still day by day the work went on, and the multitudinous subtle difficulties of the steering diminished in
number. Day by day, too, the money trickled away, until

his balance was no longer a matter of hundreds of thousands, but of tens. And at last came an anniversary.

Monson, sitting in the little drawing-shed, suddenly noticed the date on Woodhouse's calendar.

"It was five years ago to-day that we began," he said to Woodhouse suddenly.

"Is it?" said Woodhouse.

"It's the alterations play the devil with us," said Monson, biting a paper-fastener.

The drawings for the new vans to the hinder screw lay on the table before him as he spoke. He pitched the mutilated brass paper-fastener into the waste-paper basket and drummed with his fingers. "These alterations! Will the mathematicians ever be clever enough to save us all this patching and experimenting? Five years—learning by rule of thumb, when one might think that it was possible to calculate the whole thing out beforehand. The cost of it! I might have hired three senior wranglers for life. But they'd only have developed some beautifully useless theorems in pneumatics. What a time it has been, Woodhouse!"

"These mouldings will take three weeks," said Woodhouse. "At special prices."

"Three weeks!" said Monson, and sat drumming.

"Three weeks certain," said Woodhouse, an excellent engineer, but no good as a comforter. He drew the sheets towards him and began shading a bar.

Monson stopped drumming, and began to bite his finger-nails, staring the while at Woodhouse's head.

"How long have they been calling this Monson's Folly?" he said suddenly.

"Oh! Year or so," said Woodhouse carelessly, without looking up.

Monson sucked the air in between his teeth, and went to the window. The stout iron columns carrying the elevated rails upon which the start of the machine was made rose up close by, and the machine was hidden by the upper edge of the window. Through the grove of iron pillars, red painted and ornate with rows of bolts, one had a glimpse of the pretty scenery towards Esher. A train went gliding noiselessly across the middle distance, its rattle drowned by the hammering of the workmen overhead. Monson could imagine the grinning faces at the windows of the carriages. He swore savagely under his breath, and dabbed viciously at a blowfly that suddenly became noisy on the window-pane.

"What's up?" said Woodhouse, staring in surprise at his employer.

"I'm about sick of this."

Woodhouse scratched his cheek. "Oh!" he said, after an assimilating pause. He pushed the drawing away from him.

"Here these fools . . . I'm trying to conquer a new element —trying to do a thing that will revolutionise life. And instead of taking an intelligent interest, they grin and make their stupid jokes, and call me and my appliances names."

"Asses!" said Woodhouse, letting his eye fall again on the drawing.

The epithet, curiously enough, made Monson wince. "I'm about sick of it, Woodhouse, anyhow," he said, after a pause.

Woodhouse shrugged his shoulders.

"There's nothing for it but patience, I suppose," said Monson, sticking his hands in his pockets. "I've started. I've made my bed, and I've got to lie on it. I can't go back. I'll see it through, and spend every penny I have and every penny I can borrow. But I tell you, Woodhouse, I'm infernally sick of it, all the same. If I'd paid a tenth part of the money towards some political greaser's expenses—I'd have been a baronet before this."

Monson paused. Woodhouse stared in front of him with a blank expression he always employed to indicate sympathy, and tapped his pencil-case on the table. Monson stared at him for a minute.

"Oh, *damn!*" said Monson suddenly, and abruptly rushed out of the room.

Woodhouse continued his sympathetic rigour for perhaps half a minute. Then he sighed and resumed the shading of the drawings. Something had evidently upset Monson. Nice chap, and generous, but difficult to get on with. It was the way with every amateur who had anything to do with engineering—wanted everything finished at once. But Monson had usually the patience of the expert. Odd he was so irritable. Nice and round that aluminium rod did look now! Woodhouse threw back his head, and put it, first this side and then that, to appreciate his bit of shading better.

"Mr. Woodhouse," said Hooper, the foreman of the labourers, putting his head in at the door.

"Hullo!" said Woodhouse, without turning round.

"Nothing happened, sir?" said Hooper.

"Happened?" said Woodhouse.

"The governor just been up the rails swearing like a tornader."

"*Oh!*" said Woodhouse.

"It ain't like him, sir."

"No?"

"And I was thinking perhaps"——

"Don't think," said Woodhouse, still admiring the drawings.

Hooper knew Woodhouse, and he shut the door suddenly with a vicious slam. Woodhouse stared stonily before him for some further minutes, and then made an ineffectual effort to pick his teeth with his pencil. Abruptly he desisted, pitched that old, tried, and stumpy servitor across the room, got up, stretched himself, and followed Hooper.

He looked ruffled—it was visible to every workman he met. When a millionaire who has been spending thousands on experiments that employ quite a little army of people suddenly indicates that he is sick of the undertaking, there is almost invariably a certain amount of mental friction in the ranks of the little army he employs. And even before he indicates his intentions there are speculations and murmurs, a watching of faces and a study of straws. Hundreds of people knew before the day was out that Monson was ruffled, Woodhouse ruffled, Hooper ruffled. A workman's wife, for instance (whom Monson had never seen), decided to keep her money in the savings-bank instead of buying a velveteen dress. So far-reaching are even the casual curses of a millionaire.

Monson found a certain satisfaction in going on the works and behaving disagreeably to as many people as possible. After a time even that palled upon him, and he rode off the grounds, to every one's relief there, and through the lanes south-eastward, to the infinite tribulation of his house steward at Cheam.

And the immediate cause of it all, the little grain of annoyance that had suddenly precipitated all this discontent with his life-work was—these trivial things that direct all our great decisions!—half a dozen ill-considered remarks made by a pretty girl, prettily dressed, with a beautiful voice and something more than prettiness in her soft grey eyes. And of these half-dozen remarks, two words especially— "Monson's Folly." She had felt she was behaving charmingly to Monson; she reflected the next day how exceptionally effective she had been, and no one would have been more amazed than she, had she learned the effect she had left on Monson's mind. I hope, considering everything, that she never knew.

"How are you getting on with your flying-machine?" she asked. ("I wonder if I shall ever meet any one with the sense not to ask that," thought Monson.) "It will be very dangerous at first, will it not?" ("Thinks I'm afraid.") "Jorgon is going to play presently; have you heard him before?" ("My mania being attended to, we turn to rational conversation.") Gush about Jorgon; gradual decline of conversation, ending with—"You must let me know when your flying-machine is finished, Mr. Monson, and then I will consider the advisability of taking a ticket." ("One would think I was still playing inventions in the nursery.") But the bitterest thing she said was not meant for Monson's ears. To Phlox, the novelist, she was always conscientiously brilliant. "I have been talking to Mr. Monson, and he can think of nothing, positively nothing, but that flying-machine of his. Do you know, all his workmen call that place of his 'Monson's Folly'? He is quite impossible. It is really very, very sad. I always regard him myself in the light of sunken treasure—the Lost Millionaire, you know."

She was pretty and well educated,—indeed, she had written an epigrammatic novelette; but the bitterness was that she was typical. She summarised what the world thought of the man who was working sanely, steadily, and surely towards a more tremendous revolution in the appliances of civilisation, a more far-reaching alteration in the ways of humanity than has ever been effected since history began. They did not even take him seriously. In a little while he would be proverbial. "I *must* fly now," he said on his way home, smarting with a sense of absolute social failure. "I must fly soon. If it doesn't come off soon, by God! I shall run amuck."

He said that before he had gone through his pass-book and his litter of papers. Inadequate as the cause seems, it was that girl's voice and the expression of her eyes that precipitated his discontent. But certainly the discovery that he had no longer even one hundred thousand pounds' worth of realisable property behind him was the poison that made the wound deadly.

It was the next day after this that he exploded upon Woodhouse and his workmen, and thereafter his bearing was consistently grim for three weeks, and anxiety dwelt in Cheam and Ewell, Maldon, Morden, and Worcester Park, places that had thriven mightily on his experiments.

Four weeks after that first swearing of his, he stood with Woodhouse by the reconstructed machine as it lay across the elevated railway, by means of which it gained its initial impetus. The new propeller glittered a brighter white than the rest of the machine, and a gilder, obedient to a whim of Monson's, was picking out the aluminium bars with gold. And looking down the long avenue between the ropes (gilded now with the sunset), one saw red signals, and two miles away an ant-hill of workmen busy altering the last falls of the run into a rising slope.

"I'll *come*," said Woodhouse. "I'll come right enough. But I tell you it's infernally foolhardy. If only you would give another year——"

"I tell you I won't. I tell you the thing works. I've given years enough"——

"It's not that," said Woodhouse. "We're all right with the machine. But it's the steering"——

"Haven't I been rushing, night and morning, backwards and forwards, through the squirrel's cage? If the thing steers true here, it will steer true all across England. It's just funk, I tell you, Woodhouse. We could have gone a year ago. And besides"——

"Well?" said Woodhouse.

"The money!" snapped Monson over his shoulder.

"Hang it! I never thought of the money," said Woodhouse, and then, speaking now in a very different tone to that with which he had said the words before, he repeated, "I'll come. Trust me."

Monson turned suddenly, and saw all that Woodhouse had not the dexterity to say, shining on his sunset-lit face. He looked for a moment, then impulsively extended his hand. "Thanks," he said.

"All right," said Woodhouse, gripping the hand, and with a queer softening of his features. "Trust me."

Then both men turned to the big apparatus that lay with its flat wings extended upon the carrier, and stared at it meditatively. Monson, guided perhaps by a photographic study of the flight of birds, and by Lilienthal's methods, had gradually drifted from Maxim's shapes towards the bird form again. The thing, however, was driven by a huge screw behind in the place of the tail; and so hovering, which needs an almost vertical adjustment of a flat tail, was rendered impossible. The body of the machine was small, almost cylindrical, and pointed. Forward and aft on the pointed ends were two small petroleum engines for the screw, and the navigators sat deep in a canoe-like recess, the foremost one steering, and being protected by a low screen, with two plate-glass windows, from the blinding rush of air. On either side a monstrous flat framework with a curved front border could be adjusted so as either to lie horizontally, or to be tilted upward or down. These wings

M

worked rigidly together, or, by releasing a pin, one could be tilted through a small angle independently of its fellow. The front edge of either wing could also be shifted back so as to diminish the wing-area about one-sixth. The machine was not only not designed to hover, but it was also incapable of fluttering. Monson's idea was to get into the air with the initial rush of the apparatus, and then to skim, much as a playing-card may be skimmed, keeping up the rush by means of the screw at the stern. Rooks and gulls fly enormous distances in that way with scarcely a perceptible movement of the wings. The bird really drives along on an aërial switchback. It glides slanting downward for a space, until it has gained considerable momentum, and then altering the inclination of its wings, glides up again almost to its original altitude. Even a Londoner who has watched the birds in the aviary in Regent's Park knows that.

But the bird is practising this art from the moment it leaves its nest. It has not only the perfect apparatus, but the perfect instinct to use it. A man off his feet has the poorest skill in balancing. Even the simple trick of the bicycle costs him some hours of labour. The instantaneous adjustments of the wings, the quick response to a passing breeze, the swift recovery of equilibrium, the giddy, eddying movements that require such absolute precision—all that he must learn, learn with infinite labour and infinite danger, if ever he is to conquer flying. The flying-machine that will start off some fine day, driven by neat "little levers," with a nice open deck like a liner, and all loaded up with bombshells and guns, is the easy dreaming of a literary man. In lives and in treasure the cost of the conquest of the empire of the air may even exceed all that has been spent in man's great conquest of the sea. Certainly it will be costlier than the greatest war that has ever devastated the world.

No one knew these things better than these two practical men. And they knew they were in the front rank of the coming army. Yet there is hope even in a forlorn hope. Men are killed outright in the reserves sometimes, while others who have been left for dead in the thickest corner crawl out and survive.

"If we miss these meadows"—said Woodhouse presently in his slow way.

"My dear chap," said Monson, whose spirits had been rising fitfully during the last few days, "we mustn't miss these meadows. There's a quarter of a square mile for us to hit, fences removed, ditches levelled. We shall come down all right—rest assured. And if we don't"——

"Ah!" said Woodhouse. "If we don't!"

Before the day of the start, the newspaper people got wind of the alterations at the northward end of the framework, and Monson was cheered by a decided change in the comments Romeike forwarded him. "He will be off some day," said the papers. "He will be off some day," said the South-Western season-ticket holders one to another; the seaside excursionists, the Saturday-to-Monday trippers from Sussex and Hampshire and Dorset and Devon, the eminent literary people from Haslemere, all remarked eagerly one to another, "He will be off some day," as the familiar scaffolding came in sight. And actually, one bright morning, in full view of the ten-past-ten train from Basingstoke, Monson's flying-machine started on its journey.

They saw the carrier running swiftly along its rail, and the white and gold screw spinning in the air. They heard the rapid rumble of wheels, and thud as the carrier reached the buffers at the end of its run. Then a whirr as the Flying-Machine was shot forward into the networks. All that the majority of them had seen and heard before. The thing went with a drooping flight through the framework and rose again, and then every beholder shouted, or screamed, or yelled, or shrieked after his kind. For instead of the customary concussion and stoppage, the Flying-Machine flew out of its five years' cage like a bolt from a crossbow, and drove slantingly upward into the air, curved round a little, so as to cross the line, and soared in the direction of Wimbledon Common.

It seemed to hang momentarily in the air and grow smaller, then it ducked and vanished over the clustering blue tree-tops to the east of Coombe Hill, and no one stopped staring and gasping until long after it had disappeared.

That was what the people in the train from Basingstoke saw. If you had drawn a line down the middle of that train, from engine to guard's van, you would not have found a living soul on the opposite side to the flying-machine. It was a mad rush from window to window as the thing crossed the line. And the engine-driver and stoker never took their eyes off the low hills about Wimbledon, and never noticed that they had run clean through Coombe and Malden and Raynes Park, until, with returning animation, they found themselves pelting, at the most indecent pace, into Wimbledon station.

From the moment when Monson had started the carrier with a *"Now!"* neither he nor Woodhouse said a word. Both men sat with clenched teeth. Monson had crossed the line with a curve that was too sharp, and Woodhouse had opened and shut his white lips; but neither spoke. Woodhouse simply gripped his seat, and breathed sharply through his teeth, watching the blue country to the west rushing past, and down, and away from him. Monson knelt at his post forward, and his hands trembled on the spoked wheel that moved the wings. He could see nothing before him but a mass of white clouds in the sky.

The machine went slanting upward, travelling with an enormous speed still, but losing momentum every moment. The land ran away underneath with diminishing speed.

"Now!" said Woodhouse at last, and with a violent effort Monson wrenched over the wheel and altered the angle of the wings. The machine seemed to hang for half a minute motionless in mid-air, and then he saw the hazy blue house-covered hills of Kilburn and Hampstead jump up before his eyes and rise steadily, until the little sunlit dome of the Albert Hall appeared through his windows. For a moment he scarcely understood the meaning of this upward rush of the horizon, but as the nearer and nearer houses came into view, he realised what he had done. He had turned the wings over too far, and they were swooping steeply downward towards the Thames.

The thought, the question, the realisation were all the business of a second of time. "Too much!" gasped Woodhouse. Monson brought the wheel half-way back with a jerk, and forthwith the Kilburn and Hampstead ridge dropped again to the lower edge of his windows. They had been a thousand feet above Coombe and Malden station; fifty seconds after they whizzed, at a frightful pace, not eighty feet above the East Putney station, on the Metropolitan District line, to the screaming astonishment of a platformful of people. Monson flung up the vans against the air, and over Fulham they rushed up their atmospheric switchback again, steeply—too steeply. The 'buses went floundering across the Fulham Road, the people yelled.

Then down again, too steeply still, and the distant trees and houses about Primrose Hill leapt up across Monson's window, and then suddenly he saw straight before him the greenery of Kensington Gardens and the towers of the Imperial Institute. They were driving straight down upon South Kensington. The pinnacles of the Natural History Museum rushed up into view. There came one fatal second of swift thought, a moment of hesitation. Should he try and clear the towers, or swerve eastward?

He made a hesitating attempt to release the right wing, left the catch half released, and gave a frantic clutch at the wheel.

The nose of the machine seemed to leap up before him. The wheel pressed his hand with irresistible force, and jerked itself out of his control.

Woodhouse, sitting crouched together, gave a hoarse cry, and sprang up towards Monson. "Too far!" he cried, and then he was clinging to the gunwale for dear life, and Monson had been jerked clean overhead, and was falling backwards upon him.

So swiftly had the thing happened that barely a quarter of the people going to and fro in Hyde Park, and Brompton Road, and the Exhibition Road saw anything of the aerial catastrophe. A distant winged shape had appeared above the clustering houses to the south, had fallen and risen, growing larger as it did so; had swooped swiftly down towards the Imperial Institute, a broad spread of flying wings, had swept round in a quarter circle, dashed eastward,

and then suddenly sprang vertically into the air. A black
object shot out of it, and came spinning downward. A
man! Two men clutching each other! They came whirling down, separated as they struck the roof of the Students'
Club, and bounded off into the green bushes on its southward side.

For perhaps half a minute, the pointed stem of the big
machine still pierced vertically upward, the screw spinning
desperately. For one brief instant, that yet seemed an age
to all who watched, it had hung motionless in mid-air. Then
a spout of yellow flame licked up its length from the stern
engine, and swift, swifter, swifter, and flaring like a rocket, it
rushed down upon the solid mass of masonry which was formerly the Royal College of Science. The big screw of white
and gold touched the parapet, and crumpled up like wet
linen. Then the blazing spindle-shaped body smashed and
splintered, smashing and splintering in its fall, upon the
north-westward angle of the building.

But the crash, the flame of blazing paraffin that shot
heavenward from the shattered engines of the machine, the
crushed horrors that were found in the garden beyond the
Students' Club, the masses of yellow parapet and red brick
that fell headlong into the roadway, the running to and fro
of people like ants in a broken ant-hill, the galloping of
fire-engines, the gathering of crowds—all these things do not
belong to this story, which was written only to tell how the
first of all successful flying-machines was launched and flew.
Though he failed, and failed disastrously, the record of
Monson's work remains—a sufficient monument—to guide
the next of that band of gallant experimentalists who will
sooner or later master this great problem of flying. And
between Worcester Park and Malden there still stands that
portentous avenue of iron-work, rusting now, and dangerous
here and there, to witness to the first desperate struggle for
man's right of way through the air.

STORY THE THIRD

The Story of the late Mr. Elvesham

I SET this story down, not expecting it will be believed, but, if possible, to prepare a way of escape for the next victim. He perhaps may profit by my misfortune. My own case, I know, is hopeless, and I am now in some measure prepared to meet my fate.

My name is Edward George Eden. I was born at Trentham, in Staffordshire, my father being employed in the gardens there. I lost my mother when I was three years old and my father when I was five, my uncle, George Eden, then adopting me as his own son. He was a single man, self-educated, and well-known in Birmingham as an enterprising journalist; he educated me generously, fired my ambition to succeed in the world, and at his death, which happened four years ago, left me his entire fortune, a matter of about five hundred pounds after all outgoing charges were paid. I was then eighteen. He advised me in his will to expend the money in completing my education. I had already chosen the profession of medicine, and through his posthumous generosity, and my good fortune in a scholarship competition, I became a medical student at University College, London. At the time of the beginning of my story I lodged at 11A University Street, in a little upper room, very shabbily furnished, and draughty, overlooking the back of Shoolbred's premises. I used this little room both to live in and sleep in, because I was anxious to eke out my means to the very last shillingsworth.

I was taking a pair of shoes to be mended at a shop in the Tottenham Court Road when I first encountered the little old man with the yellow face, with whom my life has now become so inextricably entangled. He was standing on the kerb, and staring at the number on the door in a doubtful way, as I opened it. His eyes—they were dull grey eyes, and reddish under the rims—fell to my face, and his countenance immediately assumed an expression of corrugated amiability.

"You come," he said, "apt to the moment. I had forgotten the number of your house. How do you do, Mr. Eden?"

I was a little astonished at his familiar address, for I had never set eyes on the man before. I was annoyed, too, at his catching me with my boots under my arm. He noticed my lack of cordiality.

"Wonder who the deuce I am, eh? A friend, let me assure you. I have seen you before, though you haven't seen me. Is there anywhere where I can talk to you?"

I hesitated. The shabbiness of my room upstairs was not a matter for every stranger. "Perhaps," said I, "We might walk down the street. I'm unfortunately prevented"—— My gesture explained the sentence before I had spoken it.

"The very thing," he said, and faced this way and then that. "The street? Which way shall we go?" I slipped my boots down in the passage. "Look here!" he said abruptly; "this business of mine is a rigmarole. Come and lunch with me, Mr. Eden. I'm an old man, a very old man, and not good at explanations, and what with my piping voice and the clatter of the traffic"——

He laid a persuasive skinny hand that trembled a little upon my arm.

I was not so old that an old man might not treat me to a lunch. Yet at the same time I was not altogether pleased by this abrupt invitation. "I had rather"——I began. "But I had rather," he said, catching me up, "and a certain civility is surely due to my grey hairs." And so I consented, and went away with him.

He took me to Blavitski's; I had to walk slowly to accommodate myself to his paces; and over such a lunch as I had never tasted before, he fended off my leading questions, and

I took a better note of his appearance. His clean-shaven face was lean and wrinkled, his shrivelled lips fell over a set of false teeth, and his white hair was thin and rather long; he seemed small to me—though, indeed, most people seemed small to me—and his shoulders were rounded and bent. And, watching him, I could not help but observe that he too was taking note of me, running his eyes, with a curious touch of greed in them, over me from my broad shoulders to my sun-tanned hands and up to my freckled face again. "And now," said he, as we lit our cigarettes, "I must tell you of the business in hand.

"I must tell you, then, that I am an old man, a very old man." He paused momentarily. "And it happens that I have money that I must presently be leaving, and never a child have I to leave it to." I thought of the confidence trick, and resolved I would be on the alert for the vestiges of my five hundred pounds. He proceeded to enlarge on his loneliness, and the trouble he had to find a proper disposition of his money. "I have weighed this plan and that plan, charities, institutions, and scholarships, and libraries, and I have come to this conclusion at last,"—he fixed his eyes on my face,—"that I will find some young fellow, ambitious, pure-minded, and poor, healthy in body and healthy in mind, and, in short, make him my heir, give him all that I have." He repeated, "Give him all that I have. So that he will suddenly be lifted out of all the trouble and struggle in which his sympathies have been educated, to freedom and influence."

I tried to seem disinterested. With a transparent hypocrisy, I said, "And you want my help, my professional services maybe, to find that person."

He smiled and looked at me over his cigarette, and I laughed at his quiet exposure of my modest pretence.

"What a career such a man might have!" he said. "It fills me with envy to think how I have accumulated that another man may spend——

"But there are conditions, of course, burdens to be imposed. He must, for instance, take my name. You cannot expect everything without some return. And I must go into

all the circumstances of his life before I can accept him.
He *must* be sound. I must know his heredity, how his
parents and grandparents died, have the strictest inquiries
made into his private morals"——

This modified my secret congratulations a little. "And do
I understand," said I, "that I——?"

"Yes," he said, almost fiercely. "You. *You.*"

I answered never a word. My imagination was dancing
wildly, my innate scepticism was useless to modify its trans-
ports. There was not a particle of gratitude in my mind—
I did not know what to say nor how to say it. "But why
me in particular?" I said at last.

He had chanced to hear of me from Professor Haslar, he
said, as a typically sound and sane young man, and he
wished, as far as possible, to leave his money where health
and integrity were assured.

That was my first meeting with the little old man. He
was mysterious about himself; he would not give his name
yet, he said, and after I had answered some questions of his,
he left me at the Blavitski portal. I noticed that he drew a
handful of gold coins from his pocket when it came to
paying for the lunch. His insistence upon bodily health
was curious. In accordance with an arrangement we had
made I applied that day for a life policy in the Loyal Insurance Company for a large sum, and I was exhaustively
overhauled by the medical advisers of that company in the
subsequent week. Even that did not satisfy him, and he
insisted I must be re-examined by the great Doctor Hender-
son. It was Friday in Whitsun week before he came to a
decision. He called me down quite late in the evening,—
nearly nine it was,—from cramming chemical equations for
my Preliminary Scientific examination. He was standing
in the passage under the feeble gas-lamp, and his face was
a grotesque interplay of shadows. He seemed more bowed
than when I had first seen him, and his cheeks had sunk in
a little.

His voice shook with emotion. "Everything is satisfactory,
Mr. Eden," he said. "Everything is quite, quite satisfactory. And this night of all nights, you must dine with me

and celebrate your—accession." He was interrupted by a
cough. "You won't have long to wait, either," he said,
wiping his handkerchief across his lips, and gripping my
hand with his long bony claw that was disengaged. "Certainly not very long to wait."

We went into the street and called a cab. I remember
every incident of that drive vividly, the swift, easy motion,
the contrast of gas and oil and electric light, the crowds
of people in the streets, the place in Regent Street to which
we went, and the sumptuous dinner we were served with
there. I was disconcerted at first by the well-dressed waiter's
glances at my rough clothes, bothered by the stones of the
olives, but as the champagne warmed my blood, my confidence revived. At first the old man talked of himself.
He had already told me his name in the cab; he was Egbert
Elvesham, the great philosopher, whose name I had known
since I was a lad at school. It seemed incredible to me that
this man, whose intelligence had so early dominated mine,
this great abstraction, should suddenly realise itself as this
decrepit, familiar figure. I dare say every young fellow who
has suddenly fallen among celebrities has felt something of
my disappointment. He told me now of the future that the
feeble streams of his life would presently leave dry for me,
houses, copyrights, investments; I had never suspected that
philosophers were so rich. He watched me drink and eat
with a touch of envy. "What a capacity for living you
have!" he said; and then, with a sigh, a sigh of relief I could
have thought it, "It will not be long."

"Ay," said I, my head swimming now with champagne;
"I have a future perhaps—of a fairly agreeable sort, thanks to
you. I shall now have the honour of your name. But you
have a past. Such a past as is worth all my future."

He shook his head and smiled, as I thought with half-sad
appreciation of my flattering admiration. "That future," he
said; "would you in truth change it?" The waiter came with
liqueurs. "You will not perhaps mind taking my name,
taking my position, but would you indeed—willingly—take
my years?"

"With your achievements," said I, gallantly.

He smiled again. "Kümmel—both," he said to the waiter, and turned his attention to a little paper packet he had taken from his pocket. "This hour," said he, "this after-dinner hour is the hour of small things. Here is a scrap of my unpublished wisdom." He opened the packet with his shaking yellow fingers, and showed a little pinkish powder on the paper. "This," said he—"well, you must guess what it is. But Kümmel—put but a dash of this powder in it—is Himmel." His large greyish eyes watched mine with an inscrutable expression.

It was a bit of a shock to me to find this great teacher gave his mind to the flavour of liqueurs. However, I feigned a great interest in his weakness, for I was drunk enough for such small sycophancy.

He parted the powder between the little glasses, and rising suddenly with a strange unexpected dignity, held out his hand towards me. I imitated his action, and the glasses rang. "To a quick succession," said he, and raised his glass towards his lips.

"Not that," I said hastily. "Not that."

He paused, with the liqueur at the level of his chin, and his eyes blazing into mine.

"To a long life," said I.

He hesitated. "To a long life," said he, with a sudden bark of laughter, and with eyes fixed on one another we tilted the little glasses. His eyes looked straight into mine, and as I drained the stuff off, I felt a curiously intense sensation. The first touch of it set my brain in a furious tumult; I seemed to feel an actual physical stirring in my skull, and a seething humming filled my ears. I did not notice the flavour in my mouth, the aroma that filled my throat; I saw only the grey intensity of his gaze that burnt into mine. The draught, the mental confusion, the noise and stirring in my head, seemed to last an interminable time. Curious vague impressions of half-forgotten things danced and vanished on the edge of my consciousness. At last he broke the spell. With a sudden explosive sigh he put down his glass.

"Well?" he said.

"It's glorious," said I, though I had not tasted the stuff.

My head was spinning. I sat down. My brain was chaos. Then my perception grew clear and minute as though I saw things in a concave mirror. His manner seemed to have changed into something nervous and hasty. He pulled out his watch and grimaced at it. "Eleven-seven! And to-night I must—Seven—twenty-five. Waterloo! I must go at once." He called for the bill, and struggled with his coat. Officious waiters came to our assistance. In another moment I was wishing him good-bye, over the apron of a cab, and still with an absurd feeling of minute distinctness, as though—how can I express it?—I not only saw but *felt* through an inverted opera-glass.

"That stuff," he said. He put his hand to his forehead. "I ought not to have given it to you. It will make your head split to-morrow. Wait a minute. Here." He handed me out a little flat thing like a seidlitz-powder. "Take that in water as you are going to bed. The other thing was a drug. Not till you're ready to go to bed, mind. It will clear your head. That's all. One more shake—Futurus!"

I gripped his shrivelled claw. "Good-bye," he said, and by the droop of his eyelids I judged he too was a little under the influence of that brain-twisting cordial.

He recollected something else with a start, felt in his breast-pocket, and produced another packet, this time a cylinder the size and shape of a shaving-stick. "Here," said he, "I'd almost forgotten. Don't open this until I come to-morrow—but take it now."

It was so heavy that I well-nigh dropped it. "All ri'!" said I, and he grinned at me through the cab window as the cabman flicked his horse into wakefulness. It was a white packet he had given me, with red seals at either end and along its edge. "If this isn't money," said I, "it's platinum or lead."

I stuck it with elaborate care into my pocket, and with a whirling brain walked home through the Regent Street loiterers and the dark back streets beyond Portland Road. I remember the sensations of that walk very vividly, strange as they were. I was still so far myself that I could notice

my strange mental state, and wonder whether this stuff I
had had was opium—a drug beyond my experience. It is
hard now to describe the peculiarity of my mental strangeness—mental doubling vaguely expresses it. As I was walking up Regent Street I found in my mind a queer persuasion
that it was Waterloo station, and had an odd impulse to get
into the Polytechnic as a man might get into a train. I put
a knuckle in my eye, and it was Regent Street. How can I
express it? You see a skilful actor looking quietly at you, he
pulls a grimace, and lo!—another person. Is it too extravagant if I tell you that it seemed to me as if Regent Street
had, for the moment, done that? Then, being persuaded it
was Regent Street again, I was oddly muddled about some
fantastic reminiscences that cropped up. "Thirty years
ago," thought I, "it was here that I quarrelled with my
brother." Then I burst out laughing, to the astonishment
and encouragement of a group of night prowlers. Thirty
years ago I did not exist, and never in my life had I boasted a
brother. The stuff was surely liquid folly, for the poignant
regret for that lost brother still clung to me. Along Portland
Road the madness took another turn. I began to recall
vanished shops, and to compare the street with what it used
to be. Confused, troubled thinking was comprehensible
enough after the drink I had taken, but what puzzled me
were these curiously vivid phantasmal memories that had
crept into my mind; and not only the memories that had
crept in, but also the memories that had slipped out. I stopped
opposite Stevens', the natural history dealer's, and cudgelled
my brains to think what he had to do with me. A 'bus
went by, and sounded exactly like the rumbling of a train.
I seemed to be dipped into some dark, remote pit for the
recollection. "Of course," said I, at last, "he has promised
me three frogs to-morrow. Odd I should have forgotten."

Do they still show children dissolving views? In those I
remember one view would begin like a faint ghost, and grow
and oust another. In just that way it seemed to me that a
ghostly set of new sensations was struggling with those of
my ordinary self.

I went on through Euston Road to Tottenham Court

Road, puzzled, and a little frightened, and scarcely noticed the unusual way I was taking, for commonly I used to cut through the intervening network of back streets. I turned into University Street, to discover that I had forgotten my number. Only by a strong effort did I recall 11A, and even then it seemed to me that it was a thing some forgotten person had told me. I tried to steady my mind by recalling the incidents of the dinner, and for the life of me I could conjure up no picture of my host's face; I saw him only as a shadowy outline, as one might see oneself reflected in a window through which one was looking. In his place, however, I had a curious exterior vision of myself sitting at a table, flushed, bright-eyed, and talkative.

"I must take this other powder," said I. "This is getting impossible."

I tried the wrong side of the hall for my candle and the matches, and had a doubt of which landing my room might be on. "I'm drunk," I said, "that's certain," and blundered needlessly on the staircase to sustain the proposition.

At the first glance my room seemed unfamiliar. "What rot!" I said, and stared about me. I seemed to bring myself back by the effort and the odd phantasmal quality passed into the concrete familiar. There was the old looking-glass, with my notes on the albumens stuck in the corner of the frame, my old everyday suit of clothes pitched about the floor. And yet it was not so real after all. I felt an idiotic persuasion trying to creep into my mind, as it were, that I was in a railway carriage in a train just stopping, that I was peering out of the window at some unknown station. I gripped the bed-rail firmly to reassure myself. "It's clairvoyance, perhaps," I said. "I must write to the Psychical Research Society."

I put the rouleau on my dressing-table, sat on my bed and began to take off my boots. It was as if the picture of my present sensations was painted over some other picture that was trying to show through. "Curse it!" said I, "my wits are going, or am I in two places at once?" Half-undressed, I tossed the powder into a glass and drank it off. It effervesced, and became a fluorescent amber colour. Before I was in bed my mind was already tranquilised. I felt the pillow at my cheek, and thereupon I must have fallen asleep.

I awoke abruptly out of a dream of strange beasts, and found myself lying on my back. Probably everyone knows that dismal emotional dream from which one escapes, awake indeed but strangely cowed. There was a curious taste in my mouth, a tired feeling in my limbs, a sense of cutaneous discomfort. I lay with my head motionless on my pillow, expecting that my feeling of strangeness and terror would probably pass away, and that I should then doze off again to sleep. But instead of that, my uncanny sensations increased. At first I could perceive nothing wrong about me. There was a faint light in the room, so faint that it was the very next thing to darkness, and the furniture stood out in it as vague blots of absolute darkness. I stared with my eyes just over the bedclothes.

It came into my mind that someone had entered the room to rob me of my rouleau of money, but after lying for some moments, breathing regularly to simulate sleep, I realised this was mere fancy. Nevertheless, the uneasy assurance of something wrong kept fast hold of me. With an effort I raised my head from the pillow, and peered about me at the dark. What it was I could not conceive. I looked at the dim shapes around me, the greater and lesser darknesses that indicated curtains, table, fireplace, bookshelves, and so forth. Then I began to perceive something unfamiliar in the forms of the darkness. Had the bed turned round? Yonder should be the bookshelves, and something shrouded and pallid rose there, something that would not answer to the bookshelves, however I looked at it. It was far too big to be my shirt thrown on a chair.

Overcoming a childish terror, I threw back the bedclothes and thrust my leg out of bed. Instead of coming out of my truckle-bed upon the floor, I found my foot scarcely reached the edge of the mattress. I made another step, as it were, and sat up on the edge of the bed. By the side of my bed should be the candle, and the matches upon the broken chair. I put out my hand and touched—nothing. I waved my hand in the darkness, and it came against some heavy hanging, soft and thick in texture, which gave a rustling noise at my touch. I grasped this and pulled it; it appeared to be a curtain suspended over the head of my bed.

I was now thoroughly awake, and beginning to realise that I was in a strange room. I was puzzled. I tried to recall the overnight circumstances, and I found them now, curiously enough, vivid in my memory: the supper, my reception of the little packages, my wonder whether I was intoxicated, my slow undressing, the coolness to my flushed face of my pillow. I felt a sudden distrust. Was that last night, or the night before? At anyrate, this room was strange to me, and I could not imagine how I had got into it. The dim, pallid outline was growing paler, and I perceived it was a window, with the dark shape of an oval toilet-glass against the weak intimation of the dawn that filtered through the blind. I stood up, and was surprised by a curious feeling of weakness and unsteadiness. With trembling hands outstretched, I walked slowly towards the window, getting, nevertheless, a bruise on the knee from a chair by the way. I fumbled round the glass, which was large, with handsome brass sconces, to find the blind-cord. I could not find any. By chance I took hold of the tassel, and with the click of a spring the blind ran up.

I found myself looking out upon a scene that was altogether strange to me. The night was overcast, and through the flocculent grey of the heaped clouds there filtered a faint half-light of dawn. Just at the edge of the sky, the cloud canopy had a blood-red rim. Below, everything was dark and indistinct, dim hills in the distance, a vague mass of buildings, running up into pinnacles, trees like spilt ink, and below the window a tracery of black bushes and pale grey paths. It was so unfamiliar that for the moment I thought myself still dreaming. I felt the toilet-table; it appeared to be made of some polished wood, and was rather elaborately furnished—there were little cut-glass bottles and a brush upon it. There was also a queer little object, horse-shoe-shaped it felt, with smooth, hard projections, lying in a saucer. I could find no matches nor candle-stick.

I turned my eyes to the room again. Now the blind was up, faint spectres of its furnishing came out of the darkness. There was a huge curtained bed, and the fireplace at its foot had a large white mantel with something of the shimmer of marble.

I leant against the toilet-table, shut my eyes and opened them again, and tried to think. The whole thing was far too real for dreaming. I was inclined to imagine there was still some hiatus in my memory as a consequence of my draught of that strange liqueur; that I had come into my inheritance perhaps, and suddenly lost my recollection of everything since my good fortune had been announced. Perhaps if I waited a little, things would be clearer to me again. Yet my dinner with old Elvesham was now singularly vivid and recent. The champagne, the observant waiters, the powder, and the liqueurs—I could have staked my soul it all happened a few hours ago.

And then occurred a thing so trivial and yet so terrible to me that I shiver now to think of that moment. I spoke aloud. I said, "How the devil did I get here?" . . . *And the voice was not my own.*

It was not my own, it was thin, the articulation was slurred, the resonance of my facial bones was different. Then to reassure myself I ran one hand over the other, and felt loose folds of skin, the bony laxity of age. "Surely," I said in that horrible voice that had somehow established itself in my throat, "surely this thing is a dream!" Almost as quickly as if I did it involuntarily, I thrust my fingers into my mouth. My teeth had gone. My finger-tips ran on the flaccid surface of an even row of shrivelled gums. I was sick with dismay and disgust.

I felt then a passionate desire to see myself, to realise at once in its full horror the ghastly change that had come upon me. I tottered to the mantel, and felt along it for matches. As I did so, a barking cough sprang up in my throat, and I clutched the thick flannel nightdress I found about me. There were no matches there, and I suddenly realised that my extremities were cold. Sniffing and coughing, whimpering a little perhaps, I fumbled back to bed.

"It is surely a dream," I whimpered to myself as I clambered back, "surely a dream." It was a senile repetition. I pulled the bedclothes over my shoulders, over my ears, I thrust my withered hand under the pillow, and determined to compose myself to sleep. Of course it was a dream. In the morning the dream would be over, and I should wake up strong and vigorous again to my youth and studies. I shut my eyes, breathed regularly, and, finding myself wakeful, began to count slowly through the powers of three.

But the thing I desired would not come. I could not get to sleep. And the persuasion of the inexorable reality of the change that had happened to me grew steadily. Presently I found myself with my eyes wide open, the powers of three forgotten, and my skinny fingers upon my shrivelled gums. I was indeed, suddenly and abruptly, an old man. I had in some unaccountable manner fallen through my life and come to old age, in some way I had been cheated of all the best of my life, of love, of struggle, of strength and hope. I grovelled into the pillow and tried to persuade myself that such hallucination was possible. Imperceptibly, steadily, the dawn grew clearer.

At last, despairing of further sleep, I sat up in bed and looked about me. A chill twilight rendered the whole chamber visible. It was spacious and well-furnished, better furnished than any room I had ever slept in before. A candle and matches became dimly visible upon a little pedestal in a recess. I threw back the bedclothes, and shivering with the rawness of the early morning, albeit it was summer-time, I got out and lit the candle. Then, trembling horribly so that the extinguisher rattled on its spike, I tottered to the glass and saw—*Elvesham's face!* It was none the less horrible because I had already dimly feared as much. He had already seemed physically weak and pitiful to me, but seen now, dressed only in a coarse flannel nightdress that fell apart and showed the stringy neck, seen now as my own body, I cannot describe its desolate decrepitude. The hollow cheeks, the straggling tail of dirty grey hair, the rheumy bleared eyes, the quivering, shrivelled lips, the lower displaying a gleam of the pink interior lining, and those horrible dark gums showing. You who are mind and body together at your natural years, cannot imagine what this fiendish imprisonment meant to me. To be young and full of the desire and energy of youth, and to be caught, and presently to be crushed in this tottering ruin of a body. . . .

But I wander from the course of my story. For some time I must have been stunned at this change that had come upon me. It was daylight when I did so far gather myself together as to think. In some inexplicable way I had been changed, though how, short of magic, the thing had been done, I could not say. And as I thought, the diabolical ingenuity of Elvesham came home to me. It seemed plain to me that as I found myself in his, so he must be in possession of *my* body, of my strength that is, and my future. But how to prove it? Then as I thought, the thing became so incredible even to me, that my mind reeled, and I had to pinch myself, to feel my toothless gums, to see myself in the glass, and touch the things about me before I could steady myself to face the facts again. Was all life hallucination? Was I indeed Elvesham, and he me? Had I been dreaming of Eden overnight? Was there any Eden? But if I was Elvesham, I should remember where I was on the previous morning, the name of the town in which I lived, what happened before the dream began. I struggled with my thoughts. I recalled the queer doubleness of my memories overnight. But now my mind was clear. Not the ghost of any memories but those proper to Eden could I raise.

"This way lies insanity!" I cried in my piping voice. I staggered to my feet, dragged my feeble, heavy limbs to the washhand-stand, and plunged my grey head into a basin of cold water. Then, towelling myself, I tried again. It was no good. I felt beyond all question that I was indeed Eden, not Elvesham. But Eden in Elvesham's body!

Had I been a man of any other age, I might have given myself up to my fate as one enchanted. But in these sceptical days miracles do not pass current. Here was some trick of psychology. What a drug and a steady stare could do, a drug and a steady stare, or some similar treatment, could surely undo. Men have lost their memories before. But to exchange memories as one does umbrellas! I laughed. Alas! not a healthy laugh, but wheezing, senile titter. I could have fancied old Elvesham laughing at my plight, and a gust of petulant anger, unusual to me, swept across my feelings. I began dressing eagerly in the clothes I found lying about on the floor, and only realised when I was dressed that it was an evening suit I had assumed. I opened the wardrobe and found some ordinary clothes, a pair of plaid trousers, and an old-fashioned dressing-gown. I put a venerable smoking-cap on my venerable head, and, coughing a little from my exertions, tottered out upon the landing.

It was then perhaps a quarter to six, and the blinds were closely drawn and the house quite silent. The landing was a spacious one, a broad, richly carpeted staircase went down into the darkness of the hall below, and before me a door ajar showed me a writing-desk, a revolving bookcase, the back of a study chair, and a fine array of bound books, shelf upon shelf.

"My study," I mumbled, and walked across the landing. Then, at the sound of my voice a thought struck me, and I went back to the bedroom and put in the set of false teeth. They slipped in with the ease of old habit. "That's better," said I, gnashing them, and so returned to the study.

The drawers of the writing-desk were locked. Its revolving top was also locked. I could see no indications of the keys, and there were none in the pockets of my trousers. I shuffled back at once to the bedroom, and went through the dress suit, and afterwards the pockets of all the garments I could find. I was very eager; and one might have imagined that burglars had been at work, to see my room when I had done. Not only were there no keys to be found, but not a coin, nor a scrap of paper—save only the receipted bill of the overnight dinner.

A curious weariness asserted itself. I sat down and stared at the garments flung here and there, their pockets turned inside out. My first frenzy had already flickered out. Every moment I was beginning to realise the immense intelligence of the plans of my enemy, to see more and more clearly the hopelessness of my position. With an effort I rose and hurried into the study again. On the staircase was a house-maid pulling up the blinds. She stared, I think, at the expression of my face. I shut the door of the study behind me, and seizing a poker, began an attack upon the desk. That is how they found me. The cover of the desk was split, the lock smashed, the letters torn out of the pigeon-holes and tossed about the room. In my senile rage I had flung about the pens and other such light stationery, and overturned the ink. Moreover, a large vase upon the mantel had got broken—I do not know how. I could find no cheque-book, no money, no indications of the slightest use for the recovery of my body. I was battering madly at the drawers, when the butler, backed by two women-servants, intruded upon me.

That simply is the story of my change. No one will be-lieve my frantic assertions. I am treated as one demented, and even at this moment I am under restraint. But I am sane, absolutely sane, and to prove it I have sat down to write this story minutely as the thing happened to me. I appeal to the reader, whether there is any trace of insanity in the style or method of the story he has been reading. I am a young man locked away in an old man's body. But the clear fact is incredible to everyone. Naturally I appear demented to those who will not believe this, naturally I do not know the names of my secretaries, of the doctors who come to see me, of my servants and neighbours, of this town (wherever it is) where I find myself. Naturally I lose myself in my own house, and suffer inconveniences of every sort. Naturally I ask the oddest questions. Naturally I weep and cry out, and have paroxysms of despair. I have no money and no cheque-book. The bank will not recognise my signature, for I suppose that, allowing for the feeble muscles I now have, my handwriting is still Eden's. These people about me will not let me go to the bank personally. It seems, indeed, that there is no bank in this town, and that I have taken an account in some part of London. It seems that Elvesham kept the name of his solicitor secret from all his household—I can ascertain nothing. Elvesham was, of course, a profound student of mental science, and all my declarations of the facts of the case merely confirm the theory that my insanity is the outcome of overmuch brooding upon psychology. Dreams of the personal identity indeed! Two days ago I was a healthy youngster, with all life before me; now I am a furious old man, unkempt and desperate and miserable, prowling about a great luxurious strange house, watched, feared, and avoided as a lunatic by everyone about me. And in London is Elvesham beginning life again in a vigorous body, and with all the accumulated knowledge and wisdom of threescore and ten. He has stolen my life.

What has happened I do not clearly know. In the study are volumes of manuscript notes referring chiefly to the psychology of memory, and parts of what may be either calculations or ciphers in symbols absolutely strange to me. In some passages there are indications that he was also occupied with the philosophy of mathematics. I take it he has transferred the whole of his memories, the accumulation that makes up his personality, from this old withered brain of his to mine, and, similarly, that he has transferred mine to his discarded tenement. Practically, that is, he has changed bodies. But how such a change may be possible is without the range of my philosophy. I have been a materialist for all my thinking life, but here, suddenly, is a clear case of man's detachability from matter.

One desperate experiment I am about to try. I sit writing here before putting the matter to issue. This morning, with the help of a table-knife that I had secreted at breakfast, I succeeded in breaking open a fairly obvious secret drawer in this wrecked writing-desk. I discovered nothing save a little green glass phial containing a white powder. Round the neck of the phial was a label, and thereon was written this one word, "Release." This may be—is most probably, poison. I can understand Elvesham placing poison in my way, and I should be sure that it was his intention so to get rid of the only living witness against him, were it not for this careful concealment. The man has practically solved the problem of immortality. Save for the spite of chance, he will live

in my body, until it has aged, and then, again, throwing
that aside, he will assume some other victim's youth and
strength. When one remembers his heartlessness, it is
terrible to think of the ever-growing experience, that . . .
How long has he been leaping from body to body? . . .
But I tire of writing. The powder appears to be soluble in
water. The taste is not unpleasant.

There the narrative found upon Mr. Elvesham's desk
ends. His dead body lay between the desk and the chair.
The latter had been pushed back, probably by his last convulsions. The story was written in pencil, and in a crazy
hand quite unlike his usual minute characters. There
remain only two curious facts to record. Indisputably there
was some connection between Eden and Elvesham, since
the whole of Elvesham's property was bequeathed to the
young man. But he never inherited. When Elvesham committed suicide, Eden was, strangely enough, already dead.
Twenty-four hours before, he had been knocked down by a
cab and killed instantly, at the crowded crossing at the intersection of Gower Street and Euston Road. So that the only
human being who could have thrown light upon this fantastic narrative is beyond the reach of questions.

STORY THE FOURTH

In the Abyss

THE lieutenant stood in front of the steel sphere and gnawed a piece of pine splinter. "What do you think of it, Steevens?" he asked.

"It's an idea," said Steevens, in the tone of one who keeps an open mind.

"I believe it will smash—flat," said the lieutenant.

"He seems to have calculated it all out pretty well," said Steevens, still impartial.

"But think of the pressure," said the lieutenant. "At the surface of the water it's fourteen pounds to the inch, thirty feet down it's double that; sixty, treble; ninety, four times; nine hundred, forty times; five thousand, three hundred— that's a mile—it's two hundred and forty times fourteen pounds; that—let's see—thirty hundredweight—a ton and a half, Steevens; *a ton and a half* to the square inch. And the ocean where he's going is five miles deep. That's seven and a half"——

"Sounds a lot," said Steevens, "but it's jolly thick steel."

The lieutenant made no answer, but resumed his pine splinter. The object of their conversation was a huge ball of steel, having an exterior diameter of perhaps nine feet. It looked like the shot for some Titanic piece of artillery. It was elaborately nested in a monstrous scaffolding built into the framework of the vessel, and the gigantic spars that were presently to sling it overboard gave the stern of the ship an appearance that had raised the curiosity of every decent sailor who had sighted it, from the Pool of London to the Tropic of Capricorn. In two places, one above the other,

the steel gave place to a couple of circular windows of enormously thick glass, and one of these, set in a steel frame of
great solidity, was now partially unscrewed. Both the men
had seen the interior of this globe for the first time that
morning. It was elaborately padded with air cushions, with
little studs sunk between bulging pillows to work the simple
mechanism of the affair.   Everything was elaborately
padded, even the Myers apparatus which was to absorb
carbonic acid and replace the oxygen inspired by its tenant,
when he had crept in by the glass manhole, and had been
screwed in. It was so elaborately padded that a man might
have been fired from a gun in it with perfect safety. And it
had need to be, for presently a man was to crawl in through
that glass manhole, to be screwed up tightly, and to be
flung overboard, and to sink down—down—down, for five
miles, even as the lieutenant said. It had taken the strongest
hold of his imagination; it made him a bore at mess; and he
found Steevens, the new arrival aboard, a godsend to talk to
about it, over and over again.

"It's my opinion," said the lieutenant, "that that glass will
simply bend in and bulge and smash, under a pressure of
that sort. Daubrée has made rocks run like water under big
pressures—and, you mark my words"——

"If the glass did break in," said Steevens, "what then?"

"The water would shoot in like a jet of iron. Have you
ever felt a straight jet of high pressure water? It would hit
as hard as a bullet. It would simply smash him and flatten
him. It would tear down his throat, and into his lungs; it
would blow in his ears"——

"What a detailed imagination you have!" protested
Steevens, who saw things vividly.

"It's a simple statement of the inevitable," said the lieutenant.

"And the globe?"

"Would just give out a few little bubbles, and it would
settle down comfortably against the day of judgment,
among the oozes and the bottom clay—with poor Elstead
spread over his own smashed cushions like butter over
bread."

He repeated this sentence as though he liked it very much. "Like butter over bread," he said.

"Having a look at the jigger?" said a voice, and Elstead stood behind them, spick and span in white, with a cigarette between his teeth, and his eyes smiling out of the shadow of his ample hat-brim. "What's that about bread and butter, Weybridge? Grumbling as usual about the insufficient pay of naval officers? It won't be more than a day now before I start. We are to get the slings ready to-day. This clean sky and gentle swell is just the kind of thing for swinging off a dozen tons of lead and iron, isn't it?"

"It won't affect you much," said Weybridge.

"No. Seventy or eighty feet down, and I shall be there in a dozen seconds, there's not a particle moving, though the wind shriek itself hoarse up above, and the water lifts half-way to the clouds. No. Down there"—— He moved to the side of the ship and the other two followed him. All three leant forward on their elbows and stared down into the yellow-green water.

"Peace," said Elstead, finishing his thought aloud.

"Are you dead certain that clockwork will act?" asked Weybridge presently.

"It has worked thirty-five times," said Elstead. "It's bound to work."

"But if it doesn't?"

"Why shouldn't it?"

"I wouldn't go down in that confounded thing," said Weybridge, "for twenty thousand pounds."

"Cheerful chap you are," said Elstead, and spat sociably at a bubble below.

"I don't understand yet how you mean to work the thing," said Steevens.

"In the first place, I'm screwed into the sphere," said Elstead, "and when I've turned the electric light off and on three times to show I'm cheerful, I'm swung out over the stern by that crane, with all those big lead sinkers slung below me. The top lead weight has a roller carrying a hundred fathoms of strong cord rolled up, and that's all that joins the sinkers to the sphere, except the slings that will be cut when the affair is dropped. We use cord rather than wire rope because it's easier to cut and more buoyant—necessary points, as you will see.

"Through each of these lead weights you notice there is a hole, and an iron rod will be run through that and will project six feet on the lower side. If that rod is rammed up from below, it knocks up a lever and sets the clockwork in motion at the side of the cylinder on which the cord winds.

"Very well. The whole affair is lowered gently into the water, and the slings are cut. The sphere floats,—with the air in it, it's lighter than water,—but the lead weights go down straight and the cord runs out. When the cord is all paid out, the sphere will go down too, pulled down by the cord."

"But why the cord?" asked Steevens. "Why not fasten the weights directly to the sphere?"

"Because of the smash down below. The whole affair will go rushing down, mile after mile, at a headlong pace at last. It would be knocked to pieces on the bottom if it wasn't for that cord. But the weights will hit the bottom, and directly they do the buoyancy of the sphere will come into play. It will go on sinking slower and slower; come to a stop at last, and then begin to float upward again.

"That's where the clockwork comes in. Directly the weights smash against the sea bottom, the rod will be knocked through and will kick up the clockwork, and the cord will be rewound on the reel. I shall be lugged down to the sea bottom. There I shall stay for half an hour, with the electric light on, looking about me. Then the clockwork will release a spring knife, the cord will be cut, and up I shall rush again, like a soda-water bubble. The cord itself will help the flotation."

"And if you should chance to hit a ship?" said Weybridge.

"I should come up at such a pace, I should go clean through it," said Elstead, "like a cannon ball. You needn't worry about that."

"And suppose some nimble crustacean should wriggle into your clockwork"——

"It would be a pressing sort of invitation for me to stop,"

said Elstead, turning his back on the water and staring at
the sphere.

They had swung Elstead overboard by eleven o'clock.
The day was serenely bright and calm, with the horizon lost
in haze. The electric glare in the little upper compartment
beamed cheerfully three times. Then they let him down
slowly to the surface of the water, and a sailor in the stern
chains hung ready to cut the tackle that held the lead
weights and the sphere together. The globe, which had
looked so large on deck, looked the smallest thing conceivable under the stern of the ship. It rolled a little, and
its two dark windows, which floated uppermost, seemed like
eyes turned up in round wonderment at the people who
crowded the rail. A voice wondered how Elstead liked the
rolling. "Are you ready?" sang out the commander. "Ay,
ay, sir!" "Then let her go!"

The rope of the tackle tightened against the blade and
was cut, and an eddy rolled over the globe in a grotesquely
helpless fashion. Someone waved a handkerchief, someone
else tried an ineffectual cheer, a middy was counting slowly,
"Eight, nine, ten!" Another roll, then with a jerk and a
splash the thing righted itself.

It seemed to be stationary for a moment, to grow rapidly
smaller, and then the water closed over it, and it became
visible, enlarged by refraction and dimmer, below the surface. Before one could count three it had disappeared. There
was a flicker of white light far down in the water, that
diminished to a speck and vanished. Then there was
nothing but a depth of water going down into blackness,
through which a shark was swimming.

Then suddenly the screw of the cruiser began to rotate,
the water was crickled, the shark disappeared in a wrinkled
confusion, and a torrent of foam rushed across the crystalline clearness that had swallowed up Elstead. "What's the
idea?" said one A.B. to another.

"We're going to lay off about a couple of miles, 'fear he
should hit us when he comes up," said his mate.

The ship steamed slowly to her new position. Aboard her

almost everyone who was unoccupied remained watching the breathing swell into which the sphere had sunk. For the next half-hour it is doubtful if a word was spoken that did not bear directly or indirectly on Elstead. The December sun was now high in the sky, and the heat very considerable.

"He'll be cold enough down there," said Weybridge. "They say that below a certain depth sea water's always just about freezing."

"Where'll he come up?" asked Steevens. "I've lost my bearings."

"That's the spot," said the commander, who prided himself on his omniscience. He extended a precise finger southeastward. "And this, I reckon, is pretty nearly the moment," he said. "He's been thirty-five minutes."

"How long does it take to reach the bottom of the ocean?" asked Steevens.

"For a depth of five miles, and reckoning—as we did—an acceleration of two feet per second, both ways, is just about three-quarters of a minute."

"Then he's overdue," said Weybridge.

"Pretty nearly," said the commander. "I suppose it takes a few minutes for that cord of his to wind in."

"I forgot that," said Weybridge, evidently relieved.

And then began the suspense. A minute slowly dragged itself out, and no sphere shot out of the water. Another followed, and nothing broke the low oily swell. The sailors explained to one another that little point about the winding-in of the cord. The rigging was dotted with expectant faces. "Come up, Elstead!" called one hairy-chested salt impatiently, and the others caught it up, and shouted as though they were waiting for the curtain of a theatre to rise.

The commander glanced irritably at them.

"Of course, if the acceleration's less than two," he said, "he'll be all the longer. We aren't absolutely certain that was the proper figure. I'm no slavish believer in calculations."

Steevens agreed concisely. No one on the quarter-deck

spoke for a couple of minutes. Then Steevens' watchcase clicked.

When, twenty-one minutes after the sun reached the zenith, they were still waiting for the globe to reappear, and not a man aboard had dared to whisper that hope was dead. It was Weybridge who first gave expression to that realisation. He spoke while the sound of eight bells still hung in the air. "I always distrusted that window," he said quite suddenly to Steevens.

"Good God!" said Steevens; "you don't think——?"

"Well!" said Weybridge, and left the rest to his imagination.

"I'm no great believer in calculations myself," said the commander dubiously, "so that I'm not altogether hopeless yet." And at midnight the gunboat was steaming slowly in a spiral round the spot where the globe had sunk, and the white beam of the electric light fled and halted and swept discontentedly onward again over the waste of phosphorescent waters under the little stars.

"If his window hasn't burst and smashed him," said Weybridge, "then it's a cursed sight worse, for his clockwork has gone wrong, and he's alive now, five miles under our feet, down there in the cold and dark, anchored in that little bubble of his, where never a ray of light has shone or a human being lived, since the waters were gathered together. He's there without food, feeling hungry and thirsty and scared, wondering whether he'll starve or stifle. Which will it be? The Myers apparatus is running out, I suppose. How long do they last?"

"Good heavens!" he exclaimed; "what little things we are! What daring little devils! Down there, miles and miles of water—all water, and all this empty water about us and this sky. Gulfs!" He threw his hands out, and as he did so, a little white streak swept noiselessly up the sky, travelled more slowly, stopped, became a motionless dot, as though a new star had fallen up into the sky. Then it went sliding back again and lost itself amidst the reflections of the stars and the white haze of the sea's phosphorescence.

At the sight he stopped, arm extended and mouth open.

He shut his mouth, opened it again, and waved his arms with an impatient gesture. Then he turned, shouted "El-stead ahoy!" to the first watch, and went at a run to Lindley and the search-light. "I saw him," he said. "Starboard there! His light's on, and he's just shot out of the water. Bring the light round. We ought to see him drifting, when he lifts on the swell."

But they never picked up the explorer until dawn. Then they almost ran him down. The crane was swung out and a boat's crew hooked the chain to the sphere. When they had shipped the sphere, they unscrewed the manhole and peered into the darkness of the interior (for the electric light chamber was intended to illuminate the water about the sphere, and was shut off entirely from its general cavity).

The air was very hot within the cavity, and the india-rubber at the lip of the manhole was soft. There was no answer to to their eager questions and no sound of movement within. Elstead seemed to be lying motionless, crumpled up in the bottom of the globe. The ship's doctor crawled in and lifted him out to the men outside. For a moment or so they did not know whether Elstead was alive or dead. His face, in the yellow light of the ship's lamps, glistened with perspiration. They carried him down to his own cabin.

He was not dead, they found, but in a state of absolute nervous collapse, and besides cruelly bruised. For some days he had to lie perfectly still. It was a week before he could tell his experiences.

Almost his first words were that he was going down again. The sphere would have to be altered, he said, in order to allow him to throw off the cord if need be, and that was all. He had had the most marvellous experience. "You thought I should find nothing but ooze," he said. "You laughed at my explorations, and I've discovered a new world!" He told his story in disconnected fragments, and chiefly from the wrong end, so that it is impossible to re-tell it in his words. But what follows is the narrative of his experience.

It began atrociously, he said. Before the cord ran out, the thing kept rolling over. He felt like a frog in a football.

He could see nothing but the crane and the sky overhead, with an occasional glimpse of the people on the ship's rail. He couldn't tell a bit which way the thing would roll next. Suddenly he would find his feet going up, and try to step, and over he went rolling, head over heels, and just anyhow, on the padding. Any other shape would have been more comfortable, but no other shape was to be relied upon under the huge pressure of the nethermost abyss.

Suddenly the swaying ceased; the globe righted, and when he had picked himself up, he saw the water all about him greeny-blue, with an attenuated light filtering down from above, and a shoal of little floating things went rushing up past him, as it seemed to him, towards the light. And even as he looked, it grew darker and darker, until the water above was as dark as the midnight sky, albeit of a greener shade, and the water below black. And little transparent things in the water developed a faint glint of luminosity, and shot past him in faint greenish streaks.

And the feeling of falling! It was just like the start of a lift, he said, only it kept on. One has to imagine what that means, that keeping on. It was then of all times that Elstead repented of his adventure. He saw the chances against him in an altogether new light. He thought of the big cuttle-fish people knew to exist in the middle waters, the kind of things they find half digested in whales at times, or floating dead and rotten and half eaten by fish. Suppose one caught hold and wouldn't let go. And had the clock-work really been sufficiently tested? But whether he wanted to go on or go back mattered not the slightest now.

In fifty seconds everything was as black as night outside, except where the beam from his light struck through the waters, and picked out every now and then some fish or scrap of sinking matter. They flashed by too fast for him to see what they were. Once he thinks he passed a shark. And then the sphere began to get hot by friction against the water. They had under-estimated this, it seems.

The first thing he noticed was that he was perspiring, and then he heard a hissing growing louder under his feet, and saw a lot of little bubbles—very little bubbles they were

N

—rushing upward like a fan through the water outside. Steam! He felt the window, and it was hot. He turned on the minute glow-lamp that lit his own cavity, looked at the padded watch by the studs, and saw he had been travelling now for two minutes. It came into his head that the window would crack through the conflict of temperatures, for he knew the bottom water is very near freezing.

Then suddenly the floor of the sphere seemed to press against his feet, the rush of bubbles outside grew slower and slower, and the hissing diminished. The sphere rolled a little. The window had not cracked, nothing had given, and he knew that the dangers of sinking, at any rate, were over.

In another minute or so he would be on the floor of the abyss. He thought, he said, of Steevens and Weybridge and the rest of them five miles overhead, higher to him than the very highest clouds that ever floated over land are to us, steaming slowly and staring down and wondering what had happened to him.

He peered out of the window. There were no more bubbles now, and the hissing had stopped. Outside there was a heavy blackness—as black as black velvet—except where the electric light pierced the empty water and showed the colour of it—a yellow-green. Then three things like shapes of fire swam into sight, following each other through the water. Whether they were little and near or big and far off he could not tell.

Each was outlined in a bluish light almost as bright as the lights of a fishing smack, a light which seemed to be smoking greatly, and all along the sides of them were specks of this, like the lighter portholes of a ship. Their phosphorescence seemed to go out as they came into the radiance of his lamp, and he saw then that they were little fish of some strange sort, with huge heads, vast eyes, and dwindling bodies and tails. Their eyes were turned towards him, and he judged they were following him down. He supposed they were attracted by his glare.

Presently others of the same sort joined them. As he went on down, he noticed that the water became of a pallid colour, and that little specks twinkled in his ray like motes in a sunbeam. This was probably due to the clouds of ooze and mud that the impact of his leaden sinkers had disturbed.

By the time he was drawn down to the lead weights he was in a dense fog of white that his electric light failed altogether to pierce for more than a few yards, and many minutes elapsed before the hanging sheets of sediment subsided to any extent. Then, lit by his light and by the transient phosphorescence of a distant shoal of fishes, he was able to see under the huge blackness of the super-incumbent water an undulating expanse of greyish-white ooze, broken here and there by tangled thickets of a growth of sea lilies, waving hungry tentacles in the air.

Farther away were the graceful, translucent outlines of a group of gigantic sponges. About this floor there were scattered a number of bristling flattish tufts of rich purple and black, which he decided must be some sort of sea-urchin, and small, large-eyed or blind things having a curious resemblance, some to woodlice, and others to lobsters, crawled sluggishly across the track of the light and vanished into the obscurity again, leaving furrowed trails behind them.

Then suddenly the hovering swarm of little fishes veered about and came towards him as a flight of starlings might do. They passed over him like a phosphorescent snow, and then he saw behind them some larger creature advancing towards the sphere.

At first he could see it only dimly, a faintly moving figure remotely suggestive of a walking man, and then it came into the spray of light that the lamp shot out. As the glare struck it, it shut its eyes, dazzled. He stared in rigid astonishment.

It was a strange vertebrated animal. Its dark purple head was dimly suggestive of a chameleon, but it had such a high forehead and such a braincase as no reptile ever displayed before; the vertical pitch of its face gave it a most extraordinary resemblance to a human being.

Two large and protruding eyes projected from sockets in

chameleon fashion, and it had a broad reptilian mouth with horny lips beneath its little nostrils. In the position of the ears were two huge gill-covers, and out of these floated a branching tree of coralline filaments, almost like the tree-like gills that very young rays and sharks possess.

But the humanity of the face was not the most extra-ordinary thing about the creature. It was a biped; its almost globular body was poised on a tripod of two frog-like legs and a long thick tail, and its fore limbs, which grotes-quely caricatured the human hand, much as a frog's do, carried a long shaft of bone, tipped with copper. The colour of the creature was variegated; its head, hands, and legs were purple; but its skin, which hung loosely upon it, even as clothes might do, was a phosphorescent grey. And it stood there blinded by the light.

At last this unknown creature of the abyss blinked its eyes open, and, shading them with its disengaged hand, opened its mouth and gave vent to a shouting noise, articulate almost as speech might be, that penetrated even the steel case and padded jacket of the sphere. How a shouting may be accomplished without lungs Elstead does not profess to explain. It then moved sideways out of the glare into the mystery of shadow that bordered it on either side, and Elstead felt rather than saw that it was coming towards him. Fancying the light had attracted it, he turned the switch that cut off the current. In another moment some-thing soft dabbed upon the steel, and the globe swayed.

Then the shouting was repeated, and it seemed to him that a distant echo answered it. The dabbing recurred, and the whole globe swayed and ground against the spindle over which the wire was rolled. He stood in the blackness and peered out into the everlasting night of the abyss. And presently he saw, very faint and remote, other phosphorescent quasi-human forms hurrying towards him.

Hardly knowing what he did, he felt about in his swaying prison for the stud of the exterior electric light, and came by accident against his own small glow-lamp in its padded recess. The sphere twisted, and then threw him down; he heard shouts like shouts of surprise, and when he rose to his feet, he saw two pairs of stalked eyes peering into the lower window and reflecting his light.

In another moment hands were dabbing vigorously at his steel casing, and there was a sound, horrible enough in his position, of the metal protection of the clockwork being vigorously hammered. That, indeed, sent his heart into his mouth, for if these strange creatures succeeded in stopping that, his release would never occur. Scarcely had he thought as much when he felt the sphere sway violently, and the floor of it press hard against his feet. He turned off the small glow-lamp that lit the interior, and sent the ray of the large light in the separate compartment out into the water. The sea-floor and the man-like creatures had disappeared, and a couple of fish chasing each other dropped suddenly by the window.

He thought at once that these strange denizens of the deep sea had broken the rope, and that he had escaped. He drove up faster and faster, and then stopped with a jerk that sent him flying against the padded roof of his prison. For half a minute, perhaps, he was too astonished to think.

Then he felt that the sphere was spinning slowly, and rocking, and it seemed to him that it was also being drawn through the water. By crouching close to the window, he managed to make his weight effective and roll that part of the sphere downward, but he could see nothing save the pale ray of his light striking down ineffectively into the darkness. It occured to him that he would see more if he turned the lamp off, and allowed his eyes to grow accustomed to the profound obscurity.

In this he was wise. After some minutes the velvety blackness became a translucent blackness, and then, far away, and as faint as the zodiacal light of an English summer evening, he saw shapes moving below. He judged these creatures had detached his cable, and were towing him along the sea bottom.

And then he saw something faint and remote across the undulations of the submarine plain, a broad horizon of pale luminosity that extended this way and that way as far

as the range of his little window permitted him to see.
To this he was being towed, as a balloon might be towed
by men out of the open country into a town. He approached
it very slowly, and very slowly the dim irradiation was
gathered together into more definite shapes.

It was nearly five o'clock before he came over this
luminous area, and by that time he could make out an
arrangement suggestive of streets and houses grouped about
a vast roofless erection that was grotesquely suggestive of a
ruined abbey. It was spread out like a map below him.
The houses were all roofless enclosures of walls, and their
substance being, as he afterwards saw, of phosphorescent
bones, gave the place an appearance as if it were built of
drowned moonshine.

Among the inner caves of the place waving trees of
crinoid stretched their tentacles, and tall, slender, glassy
sponges shot like shining minarets and lilies of filmy light
out of the general glow of the city. In the open spaces of
the place he could see a stirring movement as of crowds of
people, but he was too many fathoms above them to distinguish the individuals in those crowds.

Then slowly they pulled him down, and as they did so,
the details of the place crept slowly upon his apprehension.
He saw that the courses of the cloudy buildings were
marked out with beaded lines of round objects, and then he
perceived that at several points below him, in broad open
spaces, were forms like the encrusted shapes of ships.

Slowly and surely he was drawn down, and the forms
below him became brighter, clearer, more distinct. He was
being pulled down, he perceived, towards the large building
in the centre of the town, and he could catch a glimpse
ever and again of the multitudinous forms that were lugging
at his cord. He was astonished to see that the rigging of
one of the ships, which formed such a prominent feature
of the place, was crowded with a host of gesticulating figures
regarding him, and then the walls of the great building
rose about him silently, and hid the city from his eyes.

And such walls they were, of water-logged wood, and
twisted wire-rope, and iron spars, and copper, and the bones

and skulls of dead men. The skulls ran in zigzag lines and spirals and fantastic curves over the buildings; and in and out of their eye-sockets, and over the whole surface of the place, lurked and played a multitude of silvery little fishes.

Suddenly his ears were filled with a low shouting and a noise like the violent blowing of horns, and this gave place to a fantastic chant. Down the sphere sank, past the huge pointed windows, through which he saw vaguely a great number of these strange, ghostlike people regarding him, and at last he came to rest, as it seemed, on a kind of altar that stood in the centre of the place.

And now he was at such a level that he could see these strange people of the abyss plainly once more. To his astonishment, he perceived that they were prostrating themselves before him, all save one, dressed as it seemed in a robe of placoid scales, and crowned with a luminous diadem, who stood with his reptilian mouth opening and shutting, as though he led the chanting of the worshippers.

A curious impulse made Elstead turn on his small glow-lamp again, so that he became visible to these creatures of the abyss, albeit the glare made them disappear forthwith into night. At this sudden sight of him, the chanting gave place to a tumult of exultant shouts; and Elstead, being anxious to watch them, turned his light off again, and vanished from before their eyes. But for a time he was too blind to make out what they were doing, and when at last he could distinguish them, they were kneeling again. And thus they continued worshipping him, without rest or intermission, for a space of three hours.

Most circumstantial was Elstead's account of this astounding city and its people, these people of perpetual night, who have never seen sun or moon or stars, green vegetation, nor any living, air-breathing creatures, who know nothing of fire, nor any light but the phosphorescent light of living things.

Startling as is his story, it is yet more startling to find that scientific men, of such eminence as Adams and Jenkins, find nothing incredible in it. They tell me they see no reason why intelligent, water-breathing, vertebrated creatures, inured to a low temperature and enormous pressure, and of such a heavy structure, that neither alive nor dead would they float, might not live upon the bottom of the deep sea, and quite unsuspected by us, descendants like ourselves of the great Theriomorpha of the New Red Sandstone age.

We should be known to them, however, as strange, meteoric creatures, wont to fall catastrophically dead out of the mysterious blackness of their watery sky. And not only we ourselves, but our ships, our metals, our appliances, would come raining down out of the night. Sometimes sinking things would smite down and crush them, as if it were the judgment of some unseen power above, and sometimes would come things of the utmost rarity or utility, or shapes of inspiring suggestion. One can understand, perhaps, something of their behaviour at the descent of a living man, if one thinks what a barbaric people might do, to whom an enhaloed, shining creature came suddenly out of the sky.

At one time or another Elstead probably told the officers of the *Ptarmigan* every detail of his strange twelve hours in the abyss. That he also intended to write them down is certain, but he never did, and so unhappily we have to piece together the discrepant fragments of his story from the reminiscences of Commander Simmons, Weybridge, Steevens, Lindley, and the others.

We see the thing darkly in fragmentary glimpses—the huge ghostly building, the bowing, chanting people, with their dark chameleon-like heads and faintly luminous clothing, and Elstead, with his light turned on again, vainly trying to convey to their minds that the cord by which the sphere was held was to be severed. Minute after minute slipped away, and Elstead, looking at his watch, was horrified to find that he had oxygen only for four hours more. But the chant in his honour kept on as remorselessly as if it was the marching song of his approaching death.

The manner of his release he does not understand, but to judge by the end of cord that hung from the sphere, it had been cut through by rubbing against the edge of the altar. Abruptly the sphere rolled over, and he swept up, out of their world, as an ethereal creature clothed in a vacuum would sweep through our own atmosphere back to its native ether again. He must have torn out of their sight as a hydrogen bubble hastens upward from our air. A strange ascension it must have seemed to them.

The sphere rushed up with even greater velocity than, when weighted with the lead sinkers, it had rushed down. It became exceedingly hot. It drove up with the windows uppermost, and he remembers the torrent of bubbles frothing against the glass. Every moment he expected this to fly. Then suddenly something like a huge wheel seemed to be released in his head, the padded compartment began spinning about him, and he fainted. His next recollection was of his cabin, and of the doctor's voice.

But that is the substance of the extraordinary story that Elstead related in fragments to the officers of the *Ptarmigan*. He promised to write it all down at a later date. His mind was chiefly occupied with the improvement of his apparatus, which was effected at Rio.

It remains only to tell that on February 2, 1896, he made his second descent into the ocean abyss, with the improvements his first experience suggested. What happened we shall probably never know. He never returned. The *Ptarmigan* beat about over the point of his submersion, seeking him in vain for thirteen days. Then she returned to Rio, and the news was telegraphed to his friends. So the matter remains for the present. But it is hardly probable that no further attempt will be made to verify his strange story of these hitherto unsuspected cities of the deep sea.

STORY THE FIFTH

The Apple

"I MUST get rid of it," said the man in the corner of the carriage, abruptly breaking the silence.

Mr. Hinchcliff looked up, hearing imperfectly. He had been lost in the rapt contemplation of the college cap tied by a string to his portmanteau handles—the outward and visible sign of his newly-gained pedagogic position—in the rapt appreciation of the college cap and the pleasant anticipations it excited. For Mr. Hinchcliff had just matriculated at London University, and was going to be junior assistant at the Holmwood Grammar School—a very enviable position. He stared across the carriage at his fellow-traveller.

"Why not give it away?" said this person. "Give it away! Why not?"

He was a tall, dark, sunburnt man with a pale face. His arms were folded tightly, and his feet were on the seat in front of him. He was pulling at a lank black moustache. He stared hard at his toes.

"Why not?" he said.

Mr. Hinchcliff coughed.

The stranger lifted his eyes—they were curious, dark-grey eyes—and stared blankly at Mr. Hinchcliff for the best part of a minute, perhaps. His expression grew to interest.

"Yes," he said slowly. "Why not? And end it."

"I don't quite follow you, I'm afraid," said Mr. Hinchcliff, with another cough.

"You don't quite follow me?" said the stranger quite mechanically, his singular eyes wandering from Mr. Hinchcliff to the bag with its ostentatiously displayed cap, and back to Mr. Hinchcliff's downy face.

"You're so abrupt, you know," apologised Mr. Hinchcliff.

"Why shouldn't I?" said the stranger, following his thoughts. "You are a student?" he said, addressing Mr. Hinchcliff.

"I am—by Correspondence—of the London University," said Mr. Hinchcliff, with irrepressible pride, and feeling nervously at his tie.

"In pursuit of knowledge," said the stranger, and suddenly took his feet off the seat, put his fist on his knees, and stared at Mr. Hinchcliff as though he had never seen a student before. "Yes," he said, and flung out an index finger. Then he rose, took a bag from the hat-rack, and unlocked it. Quite silently he drew out something round and wrapped in a quantity of silver-paper, and unfolded this carefully. He held it out towards Mr. Hinchcliff—a small, very smooth, golden-yellow fruit.

Mr. Hinchcliff's eyes and mouth were open. He did not offer to take this object—if he was intended to take it.

"That," said this fantastic stranger, speaking very slowly, "is the Apple of the Tree of Knowledge. Look at it—small, and bright, and wonderful—Knowledge—and I am going to give it to you."

Mr. Hinchcliff's mind worked painfully for a minute, and then the sufficient explanation, "Mad!" flashed across his brain, and illuminated the whole situation. One humoured madmen. He put his head a little on one side.

"The Apple of the Tree of Knowledge, eigh!" said Mr. Hinchcliff, regarding it with a finely assumed air of interest, and then looking at the interlocutor. "But don't you want to eat it yourself? And besides—how did you come by it?"

"It never fades. I have had it now three months. And it is ever bright and smooth and ripe and desirable, as you see it." He laid his hand on his knee and regarded the fruit musingly. Then he began to wrap it again in the papers, as though he had abandoned his intention of giving it away.

"But how did you come by it?" said Mr. Hinchcliff, who

had his argumentative side. "And how do you know that
it *is* the Fruit of the Tree?"

"I bought this fruit," said the stranger, "three months ago
—for a drink of water and a crust of bread. The man who
gave it to me—because I kept the life in him—was an
Armenian. Armenia! that wonderful country, the first of
all countries, where the ark of the Flood remains to this day,
buried in the glaciers of Mount Ararat. This man, I say,
fleeing with others from the Kurds who had come upon
them, went up into desolate places among the mountains—
places beyond the common knowledge of men. And fleeing from imminent pursuit, they came to a slope high
among the mountain-peaks, green with a grass like knife-
blades, that cut and slashed most pitilessly at anyone who
went into it. The Kurds were close behind, and there was
nothing for it but to plunge in, and the worst of it was that
the paths they made through it at the price of their blood
served for the Kurds to follow. Every one of the fugitives
was killed save this Armenian and another. He heard the
screams and cries of his friends, and the swish of the grass
about those who were pursuing them—it was tall grass
rising overhead. And then a shouting and answers, and
when presently he paused, everything was still. He pushed
out again, not understanding, cut and bleeding, until he
came out on a steep slope of rocks below a precipice, and
then he saw the grass was all on fire, and the smoke of it rose
like a veil between him and his enemies."

The stranger paused. "Yes?" said Mr. Hinchcliff. "Yes?"

"There he was, all torn and bloody from the knife-blades
of the grass, the rocks blazing under the afternoon sun—
the sky molten brass—and the smoke of the fire driving
towards him. He dared not stay there. Death he did not
mind, but torture! Far away beyond the smoke he heard
shouts and cries. Women screaming. So he went clambering up a gorge in the rocks—everywhere were bushes with
dry branches that stuck like thorns among the leaves—until
he clambered over the brow of a ridge that hid him. And
then he met his companion, a shepherd, who had also
escaped. And, counting cold and famine and thirst as

nothing against the Kurds, they went on into the heights, and among the snow and ice. They wandered three whole days.

"The third day came the vision. I suppose hungry men often do see visions, but then there is this fruit." He lifted the wrapped globe in his hand. "And I have heard it, too, from other mountaineers who have known something of the legend. It was in the evening time, when the stars were increasing, that they came down a slope of polished rock into a huge dark valley all set about with strange, contorted trees, and in these trees hung little globes like glow-worm spheres, strange round yellow lights.

"Suddenly this valley was lit far away, many miles away, far down it, with a golden flame marching slowly athwart it, that made the stunted trees against it black as night, and turned the slopes all about them and their figures to the likeness of fiery gold. And at the vision they, knowing the legends of the mountains, instantly knew that it was Eden they saw, or the sentinel of Eden, and they fell upon their faces like men struck dead.

"When they dared to look again the valley was dark for a space, and then the light came again—returning, a burning amber.

"At that the shepherd sprang to his feet, and with a shout began to run down towards the light, but the other man was too fearful to follow him. He stood stunned, amazed, and terrified, watching his companion recede towards the marching glare. And hardly had the shepherd set out when there came a noise like thunder, the beating of invisible wings hurrying up the valley, and a great and terrible fear; and at that the man who gave me the fruit turned—if he might still escape. And hurrying headlong up the slope again, with that tumult sweeping after him, he stumbled against one of these stunted bushes, and a ripe fruit came off it into his hand. This fruit. Forthwith, the wings and the thunder rolled all about him. He fell and fainted, and when he came to his senses, he was back among the blackened ruins of his own village, and I and the others were attending to the wounded. A vision? But the golden fruit of the tree was still clutched in his hand. There were others there who knew the legend, knew what that strange fruit might be." He paused. "And this is it," he said.

It was a most extraordinary story to be told in a third-class carriage on a Sussex railway. It was as if the real was a mere veil to the fantastic, and here was the fantastic poking through. "Is it?" was all Mr. Hinchcliff could say.

"The legend," said the stranger, "tells that those thickets of dwarfed trees growing about the garden sprang from the apple that Adam carried in his hand when he and Eve were driven forth. He felt something in his hand, saw the half-eaten apple, and flung it petulantly aside. And there they grow, in that desolate valley, girdled round with the everlasting snows, and there the fiery swords keep ward against the Judgment Day."

"But I thought these things were"—Mr. Hinchcliff paused —"fables—parables rather. Do you mean to tell me that there in Armenia"—

The stranger answered the unfinished question with the fruit in his open hand.

"But you don't know," said Mr. Hinchcliff, "that that *is* the fruit of the Tree of Knowledge. The man may have had—a sort of mirage, say. Suppose"—

"Look at it," said the stranger.

It was certainly a strange-looking globe, not really an apple, Mr. Hinchcliff saw, and a curious glowing golden colour, almost as though light itself was wrought into its substance. As he looked at it, he began to see more vividly the desolate valley among the mountains, the guarding swords of fire, the strange antiquities of the story he had just heard. He rubbed a knuckle into his eye. "But"—said he.

"It has kept like that, smooth and full, three months. Longer than that it is now by some days. No drying, no withering, no decay."

"And you yourself," said Mr. Hinchcliff, "really believe that"—

"Is the Forbidden Fruit."

There was no mistaking the earnestness of the man's manner and his perfect sanity. "The Fruit of Knowledge," he said.

"Suppose it was?" said Mr. Hinchcliff, after a pause, still staring at it. "But after all," said Mr. Hinchcliff, "it's not my kind of knowledge—not the sort of knowledge. I mean, Adam and Eve have eaten it already."

"We inherit their sins—not their knowledge," said the stranger. "That would make it all clear and bright again. We should see into everything, through everything, into the deepest meaning of everything"—

"Why don't you eat it, then?" said Mr. Hinchcliff, with an inspiration.

"I took it intending to eat it," said the stranger. "Man has fallen. Merely to eat again could scarcely"—

"Knowledge is power," said Mr. Hinchcliff.

"But is it happiness? I am older than you—more than twice as old. Time after time I have held this in my hand, and my heart has failed me at the thought of all that one might know, that terrible lucidity— Suppose suddenly all the world became pitilessly clear?"

"That, I think, would be a great advantage," said Mr. Hinchcliff, "on the whole."

"Suppose you saw into the hearts and minds of everyone about you, into their most secret recesses—people you loved, whose love you valued?"

"You'd soon find out the humbugs," said Mr. Hinchcliff, greatly struck by the idea.

"And worse—to know yourself, bare of your most intimate illusions. To see yourself in your place. All that your lusts and weaknesses prevented your doing. No merciful perspective."

"That might be an excellent thing too. 'Know thyself,' you know."

"You are young," said the stranger.

"If you don't care to eat it, and it bothers you, why don't you throw it away?"

"There again, perhaps, you will not understand me. To me, how could one throw away a thing like that, glowing, wonderful? Once one has it, one is bound. But, on the other hand, to *give* it away! To give it away to someone who thirsted after knowledge, who found no terror in the thought of that clear perception"—

"Of course," said Mr. Hinchcliff thoughtfully, "it might be some sort of poisonous fruit."

And then his eye caught something motionless, the end of a white board black-lettered outside the carriage window. "—MWOOD," he saw. He started convulsively. "Gracious!" said Mr. Hinchcliff. "Holmwood!"—and the practical present blotted out the mystic realisations that had been stealing upon him.

In another moment he was opening the carriage-door, portmanteau in hand. The guard was already fluttering his green flag. Mr. Hinchcliff jumped out. "Here!" said a voice behind him, and he saw the dark eyes of the stranger shining and the golden fruit, bright and bare, held out of the open carriage-door. He took it instinctively, the train was already moving.

"*No!*" shouted the stranger, and made a snatch at it as if to take it back.

"Stand away," cried a country porter, thrusting forward to close the door. The stranger shouted something Mr. Hinchcliff did not catch, head and arm thrust excitedly out of the window, and then the shadow of the bridge fell on him, and in a trice he was hidden. Mr. Hinchcliff stood astonished, staring at the end of the last waggon receding round the bend, and with the wonderful fruit in his hand. For the fraction of a minute his mind was confused, and then he became aware that two or three people on the platform were regarding him with interest. Was he not the new Grammar School master making his début? It occured to him that, so far as they could tell, the fruit might very well be the naïve refreshment of an orange. He flushed at the thought, and thrust the fruit into his side pocket, where it bulged undesirably. But there was no help for it, so he went towards them, awkwardly concealing his sense of awkwardness, to ask the way to the Grammar School, and the means of getting his portmanteau and the two tin boxes which lay up the platform thither. Of all the odd and fantastic yarns to tell a fellow!

His luggage could be taken on a truck for sixpence, he found, and he could precede it on foot. He fancied an ironical note in the voices. He was painfully aware of his contour.

The curious earnestness of the man in the train, and the glamour of the story he told, had, for a time, diverted the current of Mr. Hinchcliff's thoughts. It drove like a mist before his immediate concerns. Fires that went to and fro! But the preoccupation of his new position, and the impression he was to produce upon Holmwood generally, and the school people in particular, returned upon him with reinvigorating power before he left the station and cleared his mental atmosphere. But it is extraordinary what an inconvenient thing the addition of a soft and rather brightly-golden fruit, not three inches in diameter, may prove to a sensitive youth on his best appearance. In the pocket of his black jacket it bulged dreadfully, spoilt the lines altogether. He passed a little old lady in black, and he felt her eye drop upon the excrescence at once. He was wearing one glove and carrying the other, together with his stick, so that to bear the fruit openly was impossible. In one place, where the road into the town seemed suitably secluded, he took his encumbrance out of his pocket and tried it in his hat. It was just too large, the hat wobbled ludicrously, and just as he was taking it out again, a butcher's boy came driving round the corner.

"Confound it!" said Mr. Hinchcliff.

He would have eaten the thing, and attained omniscience there and then, but it would seem so silly to go into the town sucking a juicy fruit—and it certainly felt juicy. If one of the boys should come by, it might do him a serious injury with his discipline so to be seen. And the juice might make his face sticky and get upon his cuffs—or it might be an acid juice as potent as lemon, and take all the colour out of his clothes.

Then round a bend in the lane came two pleasant sunlit girlish figures. They were walking slowly towards the town

and chattering—at any moment they might look round and
see a hot-faced young man behind them carrying a kind of
phosphorescent yellow tomato!   They would be sure to
laugh.

"*Hang!*" said Mr Hinchcliff, and with a swift jerk sent
the encumbrance flying over the stone wall of an orchard
that there abutted on the road.   As it vanished, he felt a
faint twinge of loss that lasted scarcely a moment.   He
adjusted the stick and glove in his hand, and walked on,
erect and self-conscious, to pass the girl.

But in the darkness of the night Mr. Hinchcliff had a
dream, and saw the valley, and the flaming swords, and
the contorted trees, and knew that it really was the Apple
of the Tree of Knowledge that he had thrown regardlessly
away.   And he awoke very unhappy.

In the morning his regret had passed, but afterwards it
returned and troubled him; never, however, when he was
happy or busily occupied.   At last, one moonlight night
about eleven, when all Holmwood was quiet, his regrets
returned with redoubled force, and therewith an impulse
to adventure.   He slipped out of the house and over the
playground wall, went through the silent town to Station
Lane, and climbed into the orchard where he had thrown
the fruit.   But nothing was to be found of it there among
the dewy grass and the faint intangible globes of dandelion
down.

STORY THE SIXTH

Under the Knife

"What if I die under it?" The thought recurred again and again as I walked home from Haddon's. It was a purely personal question. I was spared the deep anxieties of a married man, and I knew there were few of my intimate friends but would find my death troublesome chiefly on account of their duty of regret. I was surprised indeed and perhaps a little humiliated, as I turned the matter over, to think how few could possibly exceed the conventional requirement. Things came before me stripped of glamour, in a clear dry light, during that walk from Haddon's house over Primrose Hill. There were the friends of my youth; I perceived now that our affection was a tradition which we foregathered rather laboriously to maintain. There were the rivals and helpers of my later career: I suppose I had been cold-blooded or undemonstrative—one perhaps implies the other. It may be that even the capacity for friendship is a question of physique. There had been a time in my own life when I had grieved bitterly enough at the loss of a friend; but as I walked home that afternoon the emotional side of my imagination was dormant. I could not pity myself, nor feel sorry for my friends, nor conceive of them as grieving for me.

I was interested in this deadness of my emotional nature —no doubt a concomitant of my stagnating physiology; and my thoughts wandered off along the line it suggested. Once before, in my hot youth, I had suffered a sudden loss of blood and had been within an ace of death. I remembered

403

now that my affections as well as my passions had drained out of me, leaving scarcely anything but a tranquil resignation, a dreg of self-pity. It had been weeks before the old ambitions, and tendernesses, and all the complex moral interplay of a man, had reasserted themselves. Now again I was bloodless; I had been feeding down for a week or more. I was not even hungry. It occurred to me that the real meaning of this numbness might be a gradual slipping away from the pleasure-pain guidance of the animal man. It has been proven, I take it, as thoroughly as anything can be proven in this world, that the higher emotions, the moral feelings, even the subtle tendernesses of love, are evolved from the elemental desires and fears of the simple animal: they are the harness in which man's mental freedom goes. And it may be that, as death overshadows us, as our possibility of acting diminishes, this complex growth of balanced impulse, propensity, and aversion whose interplay inspires our acts, goes with it. Leaving what?

I was suddenly brought back to reality by an imminent collision with a butcher-boy's tray. I found that I was crossing the bridge over the Regent's Park Canal which runs parallel with that in the Zoological Gardens. The boy in blue had been looking over his shoulder at a black barge advancing slowly, towed by a gaunt white horse. In the Gardens a nurse was leading three happy little children over the bridge. The trees were bright green; the spring hopefulness was still unstained by the dusts of summer; the sky in the water was bright and clear, but broken by long waves, by quivering bands of black, as the barge drove through. The breeze was stirring; but it did not stir me as the spring breeze used to do.

Was this dullness of feeling in itself an anticipation? It was curious that I could reason and follow out a network of suggestion as clearly as ever: so, at least, it seemed to me. It was calmness rather than dullness that was coming upon me. Was there any ground for the belief in the presentiment of death? Did a man near to death begin instinctively to withdraw himself from the meshes of matter and sense, even before the cold hand was laid upon his?

I felt strangely isolated—isolated without regret—from the life and existence about me. The children playing in the sun and gathering strength and experience for the business of life, the park-keeper gossiping with a nursemaid, the nursing mother, the young couple intent upon each other as they passed me, the trees by the wayside spreading new pleading leaves to the sunlight, the stir in their branches—I had been part of it all, but I had nearly done with it now.

Some way down the Broad Walk I perceived that I was tired, and that my feet were heavy. It was hot that afternoon, and I turned aside and sat down on one of the green chairs that line the way. In a minute I had dozed into a dream, and the tide of my thoughts washed up a vision of the resurrection. I was still sitting in the chair, but I thought myself actually dead, withered, tattered, dried, one eye (I saw) pecked out by birds. "Awake!" cried a voice; and incontinently the dust of the path and the mould under the grass became insurgent. I had never before thought of Regent's Park as a cemetery, but now through the trees, stretching as far as eye could see, I beheld a flat plain of writhing graves and heeling tombstones. There seemed to be some trouble: the rising dead appeared to stifle as they struggled upward, they bled in their struggles, the red flesh was tattered away from the white bones. "Awake!" cried a voice: but I determined I would not rise to such horrors. "Awake!" They would not let me alone. "Wike up!" said an angry voice. A cockney angel! The man who sells the tickets was shaking me, demanding my penny.

I paid my penny, pocketed my ticket, yawned, stretched my legs, and, feeling now rather less torpid, got up and walked on towards Langham Place. I speedily lost myself again in a shifting maze of thoughts about death. Going across Marylebone Road into that crescent at the end of Langham Place, I had the narrowest escape from the shaft of a cab, and went on my way with a palpitating heart and a bruised shoulder. It struck me that it would have been curious if my meditations on my death on the morrow had led to my death that day.

But I will not weary you with more of my experiences that day and the next. I knew more and more certainly that I should die under the operation; at times I think I was inclined to pose to myself. At home I found everything prepared; my room cleared of needless objects and hung with white sheets; a nurse installed and already at loggerheads with my housekeeper. They wanted me to go to bed early, and after a little resistance I obeyed.

In the morning I was very indolent, and though I read my newspapers and the letters that came by the first post, I did not find them very interesting. There was a friendly note from Addison, my old school friend, calling my attention to two discrepancies and a printer's error in my new book, with one from Langridge venting some vexation over Minton. The rest were business communications. I had a cup of tea but nothing to eat. The glow of pain at my side seemed more massive. I knew it was pain, and yet, if you can understand, I did not find it very painful. I had been awake and hot and thirsty in the night, but in the morning bed felt comfortable. In the night-time I had lain thinking of things that were past; in the morning I dozed over the question of immortality. Haddon came, punctual to the minute, with a neat black bag; and Mowbray soon followed. Their arrival stirred me up a little. I began to take a more personal interest in the proceedings. Haddon moved the little octagonal table close to the bedside, and, with his broad black back to me, began taking things out of his bag. I heard the light click of steel upon steel. My imagination, I found, was not altogether stagnant. "Will you hurt me much?" I said in an off-hand tone.

"Not a bit," Haddon answered over his shoulder. "We shall chloroform you. Your heart's as sound as a bell." And as he spoke, I had a whiff of the pungent sweetness of the anæsthetic.

They stretched me out, with a convenient exposure of my side, and, almost before I realised what was happening, the chloroform was being administered. It stings the nostrils, and there is a suffocating sensation, at first. I knew I should die—that this was the end of consciousness for me. And suddenly I felt that I was not prepared for death: I had a vague sense of a duty overlooked—I knew not what. What was it I had not done? I could think of nothing more to do, nothing desirable left in life; and yet I had the strangest disinclination for death. And the physical sensation was painfully oppressive. Of course the doctors did not know they were going to kill me. Possibly I struggled. Then I fell motionless, and a great silence, a monstrous silence, and an impenetrable blackness came upon me.

There must have been an interval of absolute unconsciousness, seconds or minutes. Then, with a chilly, unemotional clearness, I perceived that I was not yet dead. I was still in my body; but all the multitudinous sensations that come sweeping from it to make up the background of consciousness had gone, leaving me free of it all. No, not free of it all; for as yet something still held me to the poor stark flesh upon the bed—held me, yet not so closely that I did not feel myself external to it, independent of it, straining away from it. I do not think I saw, I do not think I heard; but I perceived all that was going on, and it was as if I both heard and saw. Haddon was bending over me, Mowbray behind me; the scalpel—it was a large scalpel—was cutting my flesh at the side under the flying ribs. It was interesting to see myself cut like cheese, without a pang, without even a qualm. The interest was much of a quality with that one might feel in a game of chess between strangers. Haddon's face was firm and his hand steady; but I was surprised to perceive (*how* I know not) that he was feeling the gravest doubt as to his own wisdom in the conduct of the operation.

Mowbray's thoughts, too, I could see. He was thinking that Haddon's manner showed too much of the specialist. New suggestions came up like bubbles through a stream of frothing meditation, and burst one after another in the little bright spot of his consciousness. He could not help noticing and admiring Haddon's swift dexterity, in spite of his envious quality and his disposition to detract. I saw my liver exposed. I was puzzled at my own condition. I did not feel that I was dead, but I was different in some way from my living self. The grey depression that had weighed on me for a year or more and coloured all my thoughts, was gone. I perceived and thought without any emotional tint at all. I wondered if everyone perceived things in this way under chloroform, and forgot it again when he came out of it. It would be inconvenient to look into some heads, and not forget.

Although I did not think that I was dead, I still perceived quite clearly that I was soon to die. This brought me back to the consideration of Haddon's proceedings. I looked into his mind, and saw that he was afraid of cutting a branch of the portal vein. My attention was distracted from details by the curious changes going on in his mind. His consciousness was like the quivering little spot of light which is thrown by the mirror of a galvanometer. His thoughts ran under it like a stream, some through the focus bright and distinct, some shadowy in the half-light of the edge. Just now the little glow was steady; but the least movement on Mowbray's part, the slightest sound from outside, even a faint difference in the slow movement of the living flesh he was cutting, set the light-spot shivering and spinning. A new sense-impression came rushing up through the flow of thoughts, and lo! the light-spot jerked away towards it, swifter than a frightened fish. It was wonderful to think that upon that unstable, fitful thing depended all the complex motions of the man; that for the next five minutes, therefore, my life hung upon its movements. And he was growing more and more nervous in his work. It was as if a little picture of a cut vein grew brighter, and struggled to oust from his brain another picture of a cut falling short of the mark. He was afraid: his dread of cutting too little was battling with his dread of cutting too far.

Then, suddenly, like an escape of water from under a lock-gate, a great uprush of horrible realisation set all his thoughts swirling, and simultaneously I perceived that the vein was cut. He started back with a hoarse exclamation, and I saw the brown-purple blood gather in a swift bead, and run trickling. He was horrified. He pitched the red-stained scalpel on to the octagonal table; and instantly both doctors flung themselves upon me, making hasty and ill-conceived efforts to remedy the disaster. "Ice!" said Mowbray, gasping. But I knew that I was killed, though my body still clung to me.

I will not describe their belated endeavours to save me, though I perceived every detail. My perceptions were sharper and swifter than they had ever been in life; my thoughts rushed through my mind with incredible swiftness, but with perfect definition. I can only compare their crowded clarity to the effects of a reasonable dose of opium. In a moment it would all be over, and I should be free. I knew I was immortal, but what would happen I did not know. Should I drift off presently, like a puff of smoke from a gun, in some kind of half-material body, an attenuated version of my material self? Should I find myself suddenly among the innumerable hosts of the dead, and know the world about me for the phantasmagoria it had always seemed? Should I drift to some spiritualistic *séance,* and there make foolish, incomprehensible attempts to affect a purblind medium? It was a state of unemotional curiosity, of colourless expectation. And then I realised a growing stress upon me, a feeling as though some huge human magnet was drawing me upward out of my body. The stress grew and grew. I seemed an atom for which monstrous forces were fighting. For one brief, terrible moment sensation came back to me. That feeling of falling headlong which comes in nightmares, that feeling a thousand times intensified, that and a black horror swept across my thoughts in a torrent. Then the two doctors, the naked body with its cut side, the little room, swept away from under me and vanished as a speck of foam vanishes down an eddy.

I was in mid-air. Far below was the West End of London, receding rapidly,—for I seemed to be flying swiftly upward,—and, as it receded, passing westward, like a panorama. I could see, through the faint haze of smoke the innumerable roofs chimney-set, the narrow roadways stippled with people and conveyances, the little specks of squares, and the church steeples like thorns sticking out of the fabric.

But it spun away as the earth rotated on its axis, and in a few seconds (as it seemed) I was over the scattered clumps of town about Ealing, the little Thames a thread of blue to the south, and the Chiltern Hills and the North Downs coming up like the rim of a basin, far away and faint with haze. Up I rushed. And at first I had not the faintest conception what this headlong rush upward could mean.

Every moment the circle of scenery beneath me grew wider and wider, and the details of town and field, of hill and valley, got more and more hazy and pale and indistinct, a luminous grey was mingled more and more with the blue of the hills and the green of the open meadows; and a little patch of cloud, low and far to the west, shone ever more dazzlingly white. Above, as the veil of atmosphere between myself and outer space grew thinner, the sky, which had been a fair springtime blue at first, grew deeper and richer in colour, passing steadily through the intervening shades until presently it was as dark as the blue sky of midnight, and presently as black as the blackness of a frosty starlight, and at last as black as no blackness I had ever beheld. And first one star and then many, and at last an innumerable host broke out upon the sky: more stars than anyone has ever seen from the face of the earth. For the blueness of the sky is the light of the sun and stars sifted and spread abroad blindingly: there is diffused light even in the darkest skies of winter, and we do not see the stars by day only because of the dazzling irradiation of the sun. But now I saw things—I know not how; assuredly with no mortal eyes—and that defect of bedazzlement blinded me no longer. The sun was incredibly strange and wonderful. The body of it was a disc of blinding white light: not yellowish as it seems to those who live upon the earth, but livid white, all streaked with scarlet streaks and rimmed about with a fringe of writhing tongues of red fire. And, shooting half-way across the heavens from either side of it, and brighter than the Milky Way, were two pinions of silver-white, making it look more like those winged globes I have seen in Egyptian sculpture, than anything else I can remember upon earth. These I knew for the solar corona, though I had never seen anything of it but a picture during the days of my earthly life.

When my attention came back to the earth again, I saw that it had fallen very far away from me. Field and town were long since indistinguishable, and all the varied hues of the country were merging into a uniform bright grey, broken only by the brilliant white of the clouds that lay scattered in flocculent masses over Ireland and the west of England. For now I could see the outlines of the north of France and Ireland, and all this island of Britain save where Scotland passed over the horizon to the north, or where the coast was blurred or obliterated by cloud. The sea was a dull grey, and darker than the land; and the whole panorama was rotating slowly towards the east.

All this had happened so swiftly that, until I was some thousand miles or so from the earth, I had no thought for myself. But now I perceived I had neither hands nor feet, neither parts nor organs, and that I felt neither alarm nor pain. All about me I perceived that the vacancy (for I had already left the air behind) was cold beyond the imagination of man; but it troubled me not. The sun's rays shot through the void, powerless to light or heat until they should strike on matter in their course. I saw things with a serene self-forgetfulness, even as if I were God. And down below there, rushing away from me,—countless miles in a second,— where a little dark spot on the grey marked the position of London, two doctors were struggling to restore life to the poor hacked and outworn shell I had abandoned. I felt then such release, such serenity as I can compare to no mortal delight I have ever known.

It was only after I had perceived all these things that the meaning of that headlong rush of the earth grew into comprehension. Yet it was so simple, so obvious, that I was amazed at my never anticipating the thing that was happening to me. I had suddenly been cut adrift from matter: all that was material of me was there upon earth, whirling away through space, held to the earth by gravitation, partaking of the earth's inertia, moving in its wreath of epicycles round the sun, and with the sun and the planets on their vast march through space. But the immaterial has no inertia, feels nothing of the pull of matter for matter: where it parts from its garment of flesh, there it remains (so far as space concerns it any longer) immovable in space. I was not leaving the earth: the earth was leaving *me*, and not only the earth, but the whole solar system was streaming past. And about me in space, invisible to me, scattered in the wake of the earth upon its journey, there must be an innumerable multitude of souls, stripped like myself of the material, stripped like myself of the passions of the individual and the generous emotions of the gregarious brute, naked intelligences, things of newborn wonder and thought, marvelling at the strange release that had suddenly come on them!

As I receded faster and faster from the strange white sun in the black heavens, and from the broad and shining earth upon which my being had begun, I seemed to grow, in some incredible manner, vast: vast as regards this world I had left, vast as regards the moments and periods of a human life. Very soon I saw the full circle of the earth, slightly gibbous, like the moon when she nears her full, but very large; and the silvery shape of America was now in the noon-day blaze wherein (as it seemed) little England had been basking but a few minutes ago. At first the earth was large and shone in the heavens, filling a great part of them; but every moment she grew smaller and more distant. As she shrunk, the broad moon in its third quarter crept into view over the rim of her disc. I looked for the constellations. Only that part of Aries directly behind the sun, and the Lion, which the earth covered, were hidden. I recognised the tortuous, tattered band of the Milky Way, with Vega very bright between sun and earth; and Sirius and Orion shone splendid against the unfathomable blackness in the opposite quarter of the heavens. The Pole Star was overhead, and the Great Bear hung over the circle of the earth. And away beneath and beyond the shining corona of the sun were strange groupings of stars I had never seen in my life—notably, a dagger-shaped group that I knew for the Southern Cross. All these were no larger than when they had shone on earth; but the little stars that one scarcely sees shone now against the setting of black vacancy as brightly as the first-magnitudes had done, while the larger worlds were points of indescribable glory and colour. Aldebaran was a spot of blood-red fire, and Sirius condensed to one point the light of a world of sapphires. And they shone steadily: they did not scintillate, they were calmly glorious. My impressions had an adamantine hardness and bright-ness: there was no blurring softness, no atmosphere, nothing but infinite darkness set with the myriads of these acute and brilliant points and specks of light. Presently, when I looked again, the little earth seemed no bigger than the sun, and it dwindled and turned as I looked until, in a second's space (as it seemed to me), it was halved; and so it went on swiftly dwindling. Far away in the opposite direction, a little pinkish pin's head of light, shining steadily, was the planet Mars. I swam motionless in vacancy, and, without a trace of terror or astonishment, watched the speck of cosmic dust we call the world fall away from me.

Presently it dawned upon me that my sense of duration had changed: that my mind was moving not faster but in-finitely slower, that between each separate impression there was a period of many days. The moon spun once round the earth as I noted this; and I perceived clearly the motion of Mars in his orbit. Moreover, it appeared as if the time between thought and thought grew steadily greater, until at last a thousand years was but a moment in my perception.

At first the constellations had shone motionless against the black background of infinite space; but presently it seemed as though the group of stars about Hercules and the Scorpion was contracting, while Orion and Aldebaran and their neighbours were scattering apart. Flashing suddenly out of the darkness there came a flying multitude of par-ticles of rock, glittering like dust-specks in a sunbeam, and encompassed in a faintly luminous haze. They swirled all about me, and vanished again in a twinkling far behind. And then I saw that a bright spot of light, that shone a little to one side of my path, was growing very rapidly larger, and perceived that it was the planet Saturn rushing towards me. Larger and larger it grew, swallowing up the heavens behind it, and hiding every moment a fresh multitude of stars. I perceived its flattened, whirling body, its disc-like belt, and seven of its little satellites. It grew and grew, till it towered enormous; and then I plunged amid a streaming multitude of clashing stones and dancing dust-particles and gas-eddies, and saw for a moment the mighty triple belt like three concentric arches of moonlight above me, its shadow black on the boiling tumult below. These things happened in one-tenth of the time it takes to tell of them. The planet went by like a flash of lightning; for a few seconds it blotted out the sun, and there and then became a mere black, dwindling, winged patch against the light. The earth, the mother mote of my being, I could no longer see.

So with a stately swiftness, in the profoundest silence, the solar system fell from me, as it had been a garment, until the sun was a mere star amid the multitude of stars, with its eddy of planet-specks, lost in the confused glittering of the remoter light. I was no longer a denizen of the solar system: I had come to the Outer Universe, I seemed to grasp and comprehend the whole world of matter. Ever more swiftly the stars closed in about the spot where Antares and Vega had vanished in a luminous haze, until that part of the sky had the semblance of a whirling mass of nebulæ, and ever before me yawned vaster gaps of vacant blackness and the stars shone fewer and fewer. It seemed as if I moved towards a point between Orion's belt and sword; and the void about that region opened vaster and vaster every second, an incredible gulf of nothingness, into which I was falling. Faster and ever faster the universe rushed by, a hurry of whirling motes at last, speeding silently into the void. Stars glowing brighter and brighter, with their circling planets catching the light in a ghostly fashion as I neared them, shone out and vanished again into inexistence; faint comets, clusters of meteorites, winking specks of matter, eddying light-points, whizzed past, some perhaps a hundred millions of miles or so from me at most, few nearer, travelling with unimaginable rapidity, shooting constellations, momentary darts of fire, through that black, enormous night. More than anything else it was like a dusty draught, sunbeam-lit. Broader, and wider, and deeper grew the starless space, the vacant Beyond, into which I was being drawn. At last a quarter of the heavens was black and blank, and the whole headlong rush of stellar universe closed in behind me like a veil of light that is gathered together. It drove away from me like a monstrous jack-o'-lantern driven by the wind. I had come out into the wilderness of space. Ever the vacant blackness grew broader, until the hosts of the stars seemed only like a swarm of fiery specks hurrying away from me, inconceivably remote, and the darkness, the nothingness and emptiness, was about me on every side. Soon the little universe of matter, the cage of points in which I had begun to be, was dwindling, now to a whirling disc of luminous glittering, and now to one minute disc of hazy light. In a little while it would shrink to a point, and at last would vanish altogether.

Suddenly feeling came back to me—feeling in the shape of overwhelming terror: such a dread of those dark vastitudes as no words can describe, a passionate resurgence of sympathy and social desire. Were there other souls, invisible to me as I to them, about me in the blackness? Or was I indeed, even as I felt, alone? Had I passed out of being into something that was neither being nor not-being? The covering of the body, the covering of matter, had been torn from me, and the hallucinations of companionship and security. Everything was black and silent. I had ceased to be. I was nothing. There was nothing, save only that infinitesimal dot of light that dwindled in the gulf. I strained myself to hear and see, and for a while there was naught but infinite silence, intolerable darkness, horror, and despair.

Then I saw that about the spot of light into which the whole world of matter had shrunk there was a faint glow. And in a band on either side of that the darkness was not absolute. I watched it for ages, as it seemed to me, and through the long waiting the haze grew imperceptibly more distinct. And then about the band appeared an irregular

cloud of the faintest, palest brown. I felt a passionate impatience; but the things grew brighter so slowly that they
scarcely seemed to change. What was unfolding itself?
What was this strange reddish dawn in the interminable
night of space?

The cloud's shape was grotesque. It seemed to be looped
along its lower side into four projecting masses, and, above,
it ended in a straight line. What phantom was it? I felt
assured I had seen that figure before; but I could not think
what, nor where, nor when it was. Then the realisation
rushed upon me. *It was a clenched Hand.* I was alone in
space, alone with his huge, shadowy Hand, upon which the
whole Universe of Matter lay like an unconsidered speck of
dust. It seemed as though I watched it through vast periods
of time. On the forefinger glittered a ring; and the universe
from which I had come was but a spot of light upon the
ring's curvature. And the thing that the hand gripped
had the likeness of a black rod. Through a long eternity I
watched this Hand, with the ring and the rod, marvelling
and fearing and waiting helplessly on what might follow.
It seemed as though nothing could follow: that I should
watch for ever, seeing only the Hand and the thing it held,
and understanding nothing of its import. Was the whole
universe but a refracting speck upon some greater Being?
Were our worlds but the atoms of another universe, and
those again of another, and so on through an endless pro-
gression? And what was I? Was I indeed immaterial? A
vague persuasion of a body gathering about me came into
my suspense. The abysmal darkness about the Hand filled
with impalpable suggestion, with uncertain, fluctuating
shapes.

Came a sound, like the sound of a tolling bell; faint, as if
infinitely far, muffled as though heard through thick swath-
ings of darkness: a deep, vibrating resonance, with vast
gulfs of silence between each stroke. And the Hand appeared to tighten on the rod. And I saw far above the
Hand, towards the apex of the darkness, a circle of dim
phosphorescence, a ghostly sphere whence these sounds
came throbbing; and at the last stroke the Hand vanished,

for the hour had come, and I heard a noise of many waters. But the black rod remained as a great band across the sky. And then a voice, which seemed to run to the uttermost parts of space, spoke, saying, "There will be no more pain."

At that an almost intolerable gladness and radiance rushed upon me, and I saw the circle shining white and bright, and the rod black and shining, and many things else distinct and clear. And the circle was the face of the clock, and the rod the rail of my bed. Haddon was standing at the foot, against the rail, with a small pair of scissors on his fingers; and the hands of my clock on the mantel over his shoulder were clasped together over the hour of twelve. Mowbray was washing something in a basin at the octagonal table, and at my side I felt a subdued feeling that could scarce be spoken of as pain.

The operation had not killed me. And I perceived, suddenly, that the dull melancholy of half a year was lifted from my mind.

STORY THE SEVENTH

The Sea-Raiders

§ 1

Until the extraordinary affair at Sidmouth, the peculiar species *Haploteuthis ferox* was known to science only generically, on the strength of a half-digested tentacle obtained near the Azores, and a decaying body pecked by birds and nibbled by fish, found early in 1896 by Mr. Jennings, near Land's End.

In no department of zoological science, indeed, are we quite so much in the dark as with regard to the deep-sea cephalopods. A mere accident, for instance, it was that led to the Prince of Monaco's discovery of nearly a dozen new forms in the summer of 1895, a discovery in which the before-mentioned tentacle was included. It chanced that a cachalot was killed off Terceira by some sperm whalers, and in its last struggles charged almost to the Prince's yacht, missed it, rolled under, and died within twenty yards of his rudder. And in its agony it threw up a number of large objects, which the Prince, dimly perceiving they were strange and important, was, by a happy expedient, able to secure before they sank. He set his screws in motion, and kept them circling in the vortices thus created until a boat could be lowered. And these specimens were whole cephalopods and fragments of cephalopods, some of gigantic proportions, and almost all of them unknown to science!

It would seem, indeed, that these large and agile creatures, living in the middle depths of the sea, must, to a large extent, for ever remain unknown to us, since under water they are too nimble for nets, and it is only by such rare un-looked-for accidents that specimens can be obtained. In the case of *Haploteuthis ferox,* for instance, we are still altogether ignorant of its habitat, as ignorant as we are of the breeding-ground of the herring or the sea-ways of the salmon. And zoologists are altogether at a loss to account for its sudden appearance on our coast. Possibly it was the stress of a hunger migration that drove it hither out of the deep. But it will be, perhaps, better to avoid necessarily inconclusive discussion, and to proceed at once with our narrative.

The first human being to set eyes upon a living *Haploteuthis*—the first human being to survive, that is, for there can be little doubt now that the wave of bathing fatalities and boating accidents that travelled along the coast of Cornwall and Devon in early May was due to this cause—was a retired tea-dealer of the name of Fison, who was stopping at a Sidmouth boarding-house. It was in the afternoon, and he was walking along the cliff path between Sidmouth and Ladram Bay. The cliffs in this direction are very high, but down the red face of them in one place a kind of ladder staircase has been made. He was near this when his attention was attracted by what at first he thought to be a cluster of birds struggling over a fragment of food that caught the sunlight, and glistened pinkish-white. The tide was right out, and this object was not only far below him, but remote across a broad waste of rock reefs covered with dark sea-weed and interspersed with silvery shining tidal pools. And he was, moreover, dazzled by the brightness of the further water.

In a minute, regarding this again, he perceived that his judgment was in fault, for over this struggle circled a number of birds, jackdaws and gulls for the most part, the latter gleaming blindingly when the sunlight smote their wings, and they seemed minute in comparison with it. And his curiosity was, perhaps, aroused all the more strongly because of his first insufficient explanations.

As he had nothing better to do than amuse himself, he decided to make this object, whatever it was, the goal of his afternoon walk, instead of Ladram Bay, conceiving it might perhaps be a great fish of some sort, stranded by some chance, and flapping about in its distress. And so he hurried down the long steep ladder, stopping at intervals of thirty feet or so to take breath and scan the mysterious movement.

At the foot of the cliff he was, of course, nearer his object than he had been; but, on the other hand, it now came up against the incandescent sky, beneath the sun, so as to seem dark and indistinct. Whatever was pinkish of it was now hidden by a skerry of weedy boulders. But he perceived that it was made up of seven rounded bodies, distinct or connected, and that the birds kept up a constant croaking and screaming, but seemed afraid to approach it too closely.

Mr. Fison, torn by curiosity, began picking his way across the wave-worn rocks, and, finding the wet seaweed that covered them thickly rendered them extremely slippery, he stopped, removed his shoes and socks, and coiled his trousers above his knees. His object was, of course, merely to avoid stumbling into the rocky pools about him, and perhaps he was rather glad, as all men are, of an excuse to resume, even for a moment, the sensations of his boyhood. At any rate, it is to this, no doubt, that he owes his life.

He approached his mark with all the assurance which the absolute security of this country against all forms of animal life gives its inhabitants. The round bodies moved to and fro, but it was only when he surmounted the skerry of boulders I have mentioned that he realised the horrible nature of the discovery. It came upon him with some suddenness.

The rounded bodies fell apart as he came into sight over the ridge, and displayed the pinkish object to be the partially devoured body of a human being, but whether of a man or woman he was unable to say. And the rounded bodies were new and ghastly-looking creatures, in shape somewhat resembling an octopus, and with huge and very long and flexible tentacles, coiled copiously on the ground. The skin had a glistening texture, unpleasant to see, like shiny leather. The downward bend of the tentacle-surrounded mouth, the curious excrescence at the bend, the tentacles, and the large intelligent eyes, gave the creatures a grotesque suggestion of a face. They were the size of a fair-sized swine about the body, and the tentacles seemed to him to be many feet in length. There were, he thinks, seven or eight at least of the creatures. Twenty yards beyond them, amid the surf of the now returning tide, two others were emerging from the sea.

Their bodies lay flatly on the rocks, and their eyes regarded him with evil interest; but it does not appear that Mr. Fison was afraid, or that he realised that he was in any danger. Possibly his confidence is to be ascribed to the limpness of their attitudes. But he was horrified, of course, and intensely excited and indignant at such revolting creatures preying upon human flesh. He thought they had chanced upon a drowned body. He shouted to them, with the idea of driving them off, and, finding they did not budge, cast about him, picked up a big rounded lump of rock, and flung it at one.

And then, slowly uncoiling their tentacles, they all began moving towards him—creeping at first deliberately, and making a soft purring sound to each other.

In a moment Mr. Fison realised that he was in danger. He shouted again, threw both his boots and started off, with a leap, forthwith. Twenty yards off he stopped and faced about, judging them slow, and behold! the tentacles of their leader were already pouring over the rocky ridge on which he had just been standing!

At that he shouted again, but this time not threatening, but a cry of dismay, and began jumping, striding, slipping, wading across the uneven expanse between him and the beach. The tall red cliffs seemed suddenly at a vast distance, and he saw, as though they were creatures in another world, two minute workmen engaged in the repair of the ladder-way, and little suspecting the race for life that was beginning below them. At one time he could hear the creatures splashing in the pools not a dozen feet behind him, and once he slipped and almost fell.

They chased him to the very foot of the cliffs, and desisted only when he had been joined by the workmen at the foot of the ladder-way up the cliff. All three of the men pelted them with stones for a time, and then hurried to the cliff top and along the path towards Sidmouth, to secure assistance and a boat, and to rescue the desecrated body from the clutches of these abominable creatures.

§ 2

And, as if he had not already been in sufficient peril that day, Mr. Fison went with the boat to point out the exact spot of his adventure.

As the tide was down, it required a considerable detour to reach the spot, and when at last they came off the ladder-way, the mangled body had disappeared. The water was now running, submerging first one slab of slimy rock and then another, and the four men in the boat—the workmen, that is, the boatman, and Mr. Fison—now turned their attention from the bearings off shore to the water beneath the keel.

At first they could see little below them, save a dark jungle of laminaria, with an occasional darting fish. Their minds were set on adventure, and they expressed their disappointment freely. But presently they saw one of the monsters swimming through the water seaward, with a curious rolling motion that suggested to Mr. Fison the spinning roll of a captive balloon. Almost immediately after, the waving streamers of laminaria were extraordinarily perturbed, parted for a moment, and three of these beasts became darkly visible, struggling for what was probably some fragment of the drowned man. In a moment the copious olive-green ribbons had poured again over this writhing group.

At that all four men, greatly excited, began beating the

water with oars and shouting, and immediately they saw a
tumultuous movement among the weeds. They desisted to
see more clearly, and as soon as the water was smooth, they
saw, as it seemed to them, the whole sea bottom among the
weeds set with eyes.

"Ugly swine!" cried one of the men. "Why, there's
dozens!"

And forthwith the things began to rise through the water
about them. Mr. Fison has since described to the writer
this startling eruption out of the waving laminaria meadows.
To him it seemed to occupy a considerable time, but it is
probable that really it was an affair of a few seconds only.
For a time nothing but eyes, and then he speaks of tentacles
streaming out and parting the weed fronds this way and
that. Then these things, growing larger, until at last the
bottom was hidden by their intercoiling forms, and the tips
of tentacles rose darkly here and there into the air above the
swell of the waters.

One came up boldly to the side of the boat, and, clinging
to this with three of its sucker-set tentacles threw four others
over the gunwale, as if with an intention either of over-
setting the boat or of clambering into it. Mr. Fison at once
caught up the boathook, and, jabbing furiously at the soft
tentacles, forced it to desist. He was struck in the back and
almost pitched overboard by the boatman, who was using
his oar to resist a similar attack on the other side of the boat.
But the tentacles on either side at once relaxed their hold
at this, slid out of sight, and splashed into the water.

"We'd better get out of this," said Mr. Fison, who was
trembling violently. He went to the tiller, while the boatman and one of the workmen seated themselves, and began
rowing. The other workman stood up in the fore part of
the boat, with the boathook, ready to strike any more tentacles that might appear. Nothing else seems to have been
said. Mr. Fison had expressed the common feeling beyond
amendment. In a hushed, scared mood, with faces white
and drawn, they set about escaping from the position into
which they had so recklessly blundered.

But the oars had scarcely dropped into the water before

dark, tapering, serpentine ropes had bound them, and were about the rudder; and creeping up the sides of the boat with a looping motion came the suckers again. The men gripped their oars and pulled, but it was like trying to move a boat in a floating raft of weeds. "Help here!" cried the boat-man, and Mr. Fison and the second workman rushed to help lug at the oar.

Then the man with the boathook—his name was Ewan, or Ewen—sprang up with a curse, and began striking down-ward over the side, as far as he could reach, at the bank of tentacles that now clustered along the boat's bottom. And, at the same time, the two rowers stood up to get a better purchase for the recovery of their oars. The boatman handed his to Mr. Fison, who lugged desperately, and, meanwhile, the boatman opened a big clasp-knife, and, lean-ing over the side of the boat, began hacking at the spiring arms upon the oar shaft.

Mr. Fison, staggering with the quivering rocking of the boat, his teeth set, his breath coming short, and the veins starting on his hands as he pulled at his oar, suddenly cast his eyes seaward. And there, not fifty yards off, across the long rollers of the incoming tide, was a large boat standing in towards them, with three women and a little child in it. A boatman was rowing, and a little man in a pink-ribboned straw hat and whites stood in the stern, hailing them. For a moment, of course, Mr. Fison thought of help, and then he thought of the child. He abandoned his oar forthwith, threw up his arms in a frantic gesture, and screamed to the party in the boat to keep away "for God's sake!" It says much for the modesty and courage of Mr. Fison that he does not seem to be aware that there was any quality of heroism in his action at this juncture. The oar he had abandoned was at once drawn under, and presently reap-peared floating about twenty yards away.

At the same moment Mr. Fison felt the boat under him lurch violently, and a hoarse scream, a prolonged cry of terror from Hill, the boatman, caused him to forget the party of excursionists altogether. He turned, and saw Hill crouching by the forward rowlock, his face convulsed with terror, and his right arm over the side and drawn tightly down. He gave now a succession of short sharp cries, "Oh! oh! oh!—oh!" Mr. Fison believes that he must have been hacking at the tentacles below the water-line, and have been grasped by them, but, of course, it is quite impossible to say now certainly what had happened. The boat was heeling over, so that the gunwale was within ten inches of the water, and both Ewan and the other labourer were striking down into the water, with oar and boathook, on either side of Hill's arm. Mr. Fison instinctively placed himself to counterpoise them.

Then Hill, who was a burly, powerful man, made a strenuous effort, and rose almost to a standing position. He lifted his arm, indeed, clean out of the water. Hanging to it was a complicated tangle of brown ropes; and the eyes of one of the brutes that had hold of him, glaring straight and resolute, showed momentarily above the surface. The boat heeled more and more, and the green-brown water came pouring in a cascade over the side. Then Hill slipped and fell with his ribs across the side, and his arm and the mass of tentacles about it splashed back into the water. He rolled over; his boot kicked Mr. Fison's knee as that gentleman rushed forward to seize him, and in another moment fresh tentacles had whipped about his waist and neck, and after a brief, convulsive struggle, in which the boat was nearly capsized, Hill was lugged overboard. The boat righted with a violent jerk that all but sent Mr. Fison over the other side, and hid the struggle in the water from his eyes.

He stood staggering to recover his balance for a moment, and as he did so, he became aware that the struggle and the inflowing tide had carried them close upon the weedy rocks again. Not four yards off a table of rock still rose in rhythmic movements above the inwash of the tide. In a moment Mr. Fison seized the oar from Ewan, gave one vigorous stroke, then, dropping it, ran to the bows and leapt. He felt his feet slide over the rock, and, by a frantic effort, leapt again towards a further mass. He stumbled over this, came to his knees and rose again.

"Look out!" cried someone, and a large drab body struck him. He was knocked flat into a tidal pool by one of the workmen, and as he went down he heard smothered, choking cries, that he believed at the time came from Hill. Then he found himself marvelling at the shrillness and variety of Hill's voice. Someone jumped over him, and a curving rush of foamy water poured over him, and passed. He scrambled to his feet, dripping, and, without looking seaward, ran as fast as his terror would let him shoreward. Before him, over the flat space of scattered rocks, stumbled the two workmen—one a dozen yards in front of the other.

He looked over his shoulder at last, and, seeing that he was not pursued, faced about. He was astonished. From the moment of the rising of the cephalopods out of the water, he had been acting too swiftly to fully comprehend his actions. Now it seemed to him as if he had suddenly jumped out of an evil dream.

For there were the sky, cloudless and blazing with the afternoon sun, the sea weltering under its pitiless brightness, the soft creamy foam of the breaking water, and the low, long, dark ridges of rock. The righted boat floated, rising and falling gently on the swell about a dozen yards from shore. Hill and the monsters, all the stress and tumult of that fierce fight for life, had vanished as though they had never been.

Mr. Fison's heart was beating violently; he was throbbing to the finger-tips, and his breath came deep.

There was something missing. For some seconds he could not think clearly enough what this might be. Sun, sky, sea, rocks—what was it? Then he remembered the boatload of excursionists. It had vanished. He wondered whether he had imagined it. He turned, and saw the two workmen standing side by side under the projecting masses of the tall pink cliffs. He hesitated whether he should make one last attempt to save the man Hill. His physical excitement seemed to desert him suddenly, and leave him aimless and helpless. He turned shoreward, stumbling and wading towards his two companions.

He looked back again, and there were now two boats float-

ing, and the one farthest out at sea pitched clumsily, bottom
upward.

§ 3

So it was *Haploteuthis ferox* made its appearance upon
the Devonshire coast. So far, this has been its most serious
aggression. Mr. Fison's account, taken together with the
wave of boating and bathing casualties to which I have
already alluded, and the absence of fish from the Cornish
coasts that year, points clearly to a shoal of these voracious
deep-sea monsters prowling slowly along the sub-tidal coast-
line. Hunger migration has, I know, been suggested as the
force that drove them hither; but, for my own part, I prefer
to believe the alternative theory of Hemsley. Hemsley holds
that a pack or shoal of these creatures may have become
enamoured of human flesh by the accident of a foundered
ship sinking among them, and have wandered in search of
it out of their accustomed zone; first waylaying and following ships, and so coming to our shores in the wake of the
Atlantic traffic. But to discuss Hemsley's cogent and ad-
mirably-stated arguments would be out of place here.

It would seem that the appetites of the shoal were satisfied
by the catch of eleven people—for so far as can be ascertained, there were ten people in the second boat, and certainly these creatures gave no further signs of their presence
off Sidmouth that day. The coast between Seaton and
Budleigh Salterton was patrolled all that evening and night
by four Preventive Service boats, the men in which were
armed with harpoons and cutlasses, and as the evening advanced, a number of more or less similarly equipped expeditions, organised by private individuals, joined them. Mr.
Fison took no part in any of these expeditions.

About midnight excited hails were heard from a boat
about a couple of miles out at sea to the south-east of Sidmouth, and a lantern was seen waving in a strange manner
to and fro and up and down. The nearer boats at once hurried towards the alarm. The venturesome occupants of the
boat, a seaman, a curate, and two schoolboys, had actually

seen the monsters passing under their boat. The creatures, it seems, like most deep-sea organisms, were phosphorescent, and they had been floating, five fathoms deep or so, like creatures of moonshine through the blackness of the water, their tentacles retracted and as if asleep, rolling over and over, and moving slowly in a wedge-like formation towards the south-east.

These people told their story in gesticulated fragments, as first one boat drew alongside and then another. At last there was a little fleet of eight or nine boats collected together, and from them a tumult, like the chatter of a marketplace, rose into the stillness of the night. There was little or no disposition to pursue the shoal, the people had neither weapons nor experience for such a dubious chase, and presently—even with a certain relief, it may be—the boats turned shoreward.

And now to tell what is perhaps the most astonishing fact in this whole astonishing raid. We have not the slightest knowledge of the subsequent movements of the shoal, although the whole south-west coast was now alert for it. But it may, perhaps, be significant that a cachalot was stranded off Sark on June 3. Two weeks and three days after this Sidmouth affair, a living *Haploteuthis* came ashore on Calais sands. It was alive, because several witnesses saw its tentacles moving in a convulsive way. But it is probable that it was dying. A gentleman named Pouchet obtained a rifle and shot it.

That was the last appearance of a living *Haploteuthis*. No others were seen on the French coast. On the 15th of June a dead body, almost complete, was washed ashore near Torquay, and a few days later a boat from the Marine Biological station, engaged in dredging off Plymouth, picked up a rotting specimen, slashed deeply with a cutlass wound. How the former specimen had come by its death it is impossible to say. And on the last day of June, Mr. Egbert Caine, an artist, bathing near Newlyn, threw up his arms, shrieked, and was drawn under. A friend bathing with him made no attempt to save him, but swam at once for the shore. This is the last fact to tell of this extraordinary raid from the deeper sea. Whether it is really the last of these horrible creatures it is, as yet, premature to say. But it is believed, and certainly it is to be hoped, that they have returned now, and returned for good, to the sunless depths of the middle seas, out of which they have so strangely and so mysteriously arisen.

STORY THE EIGHTH

Pollock and the Porroh Man

It was in a swampy village on the lagoon river behind the Turner Peninsula that Pollock's first encounter with the Porroh man occurred. The women of that country are famous for their good looks—they are Gallinas with a dash of European blood that dates from the days of Vasco de Gama and the English slave-traders, and the Porroh man, too, was possibly inspired by a faint Caucasian taint in his composition. (It's a curious thing to think that some of us may have distant cousins eating men on Sherboro Island or raiding with the Sofas.) At any rate, the Porroh man stabbed the woman to the heart as though he had been a mere low-class Italian, and very narrowly missed Pollock. But Pollock, using his revolver to parry the lightning stab which was aimed at his deltoid muscle, sent the iron dagger flying, and, firing, hit the man in the hand.

He fired again and missed, knocking a sudden window out of the wall of the hut. The Porroh man stooped in the doorway, glancing under his arm at Pollock. Pollock caught a glimpse of his inverted face in the sunlight, and then the Englishman was alone, sick and trembling with the excitement of the affair, in the twilight of the place. It had all happened in less time than it takes to read about it.

The woman was quite dead, and having ascertained this, Pollock went to the entrance of the hut and looked out. Things outside were dazzling bright. Half a dozen of the porters of the expedition were standing up in a group near the green huts they occupied, and staring towards him, wondering what the shots might signify. Behind the little

430

group of men was the broad stretch of black fetid mud by the river, a green carpet of rafts of papyrus and water-grass, and then the leaden water. The mangroves beyond the stream loomed indistinctly through the blue haze. There were no signs of excitement in the squat village, whose fence was just visible above the cane-grass.

Pollock came out of the hut cautiously and walked towards the river, looking over his shoulder at intervals. But the Porroh man had vanished. Pollock clutched his revolver nervously in his hand.

One of his men came to meet him, and as he came, pointed to the bushes behind the hut in which the Porroh man had disappeared. Pollock had an irritating persuasion of having made an absolute fool of himself; he felt bitter, savage, at the turn things had taken. At the same time, he would have to tell Waterhouse—the moral, exemplary, cautious Waterhouse—who would inevitably take the matter seriously. Pollock cursed bitterly at his luck, at Waterhouse, and especially at the West Coast of Africa. He felt consummately sick of the expedition. And in the back of his mind all the time was a speculative doubt where precisely within the visible horizon the Porroh man might be.

It is perhaps rather shocking, but he was not at all upset by the murder that had just happened. He had seen so much brutality during the last three months, so many dead women, burnt huts, drying skeletons, up the Kittam River in the wake of the Sofa cavalry, that his senses were blunted. What disturbed him was the persuasion that this business was only beginning.

He swore savagely at the black, who ventured to ask a question, and went on into the tent under the orange-trees where Waterhouse was lying, feeling exasperatingly like a boy going into the headmaster's study.

Waterhouse was still sleeping off the effects of his last dose of chlorodyne, and Pollock sat down on a packing-case beside him, and, lighting his pipe, waited for him to awake. About him were scattered the pots and weapons Waterhouse had collected from the Mendi people, and which he had been repacking for the canoe voyage to Sulyma.

Presently Waterhouse woke up, and after judicial stretching, decided he was all right again. Pollock got him some tea. Over the tea the incidents of the afternoon were described by Pollock, after some preliminary beating about the bush. Waterhouse took the matter even more seriously than Pollock had anticipated. He did not simply disapprove, he scolded, he insulted.

"You're one of those infernal fools who think a black man isn't a human being," he said. "I can't be ill a day without you must get into some dirty scrape or other. This is the third time in a month that you have come crossways-on with a native, and this time you're in for it with a vengeance. Porroh, too! They're down upon you enough as it is, about that idol you wrote your silly name on. And they're the most vindictive devils on earth! You make a man ashamed of civilisation. To think you come of a decent family! If ever I cumber myself up with a vicious, stupid young lout like you again"——

"Steady on, now," snarled Pollock, in the tone that always exasperated Waterhouse; "steady on."

At that Waterhouse became speechless. He jumped to his feet.

"Look here, Pollock," he said, after a struggle to control his breath. "You must go home. I won't have you any longer. I'm ill enough as it is through you"——

"Keep your hair on," said Pollock, staring in front of him. "I'm ready enough to go."

Waterhouse became calmer again. He sat down on the camp-stool. "Very well," he said. "I don't want a row, Pollock, you know, but it's confoundedly annoying to have one's plans put out by this kind of thing. I'll come to Sulyma with you, and see you safe aboard"——

"You needn't," said Pollock. "I can go alone. From here."

"Not far," said Waterhouse. "You don't understand this Porroh business."

"How should *I* know she belonged to a Porroh man?" said Pollock bitterly.

"Well, she did," said Waterhouse; "and you can't undo the thing. Go alone, indeed! I wonder what they'd do to you. You don't seem to understand that this Porroh hokey-pokey rules this country, is its law, religion, constitution, medicine, magic. . . . They appoint the chiefs. The Inquisition, at its best, couldn't hold a candle to these chaps. He will probably set Awajale, the chief here, on to us. It's lucky our porters are Mendis. We shall have to shift this little settlement of ours. . . . Confound you, Pollock! And, of course, you must go and miss him."

He thought, and his thoughts seemed disagreeable. Presently he stood up and took his rifle. "I'd keep close for a bit, if I were you," he said, over his shoulder, as he went out. "I'm going out to see what I can find out about it."

Pollock remained sitting in the tent, meditating. "I was meant for a civilised life," he said to himself, regretfully, as he filled his pipe. "The sooner I get back to London or Paris the better for me."

His eye fell on the sealed case in which Waterhouse had put the featherless poisoned arrows they had bought in the Mendi country. "I wish I had hit the beggar somewhere vital," said Pollock viciously.

Waterhouse came back after a long interval. He was not communicative, though Pollock asked him questions enough. The Porroh man, it seems, was a prominent member of that mystical society. The village was interested, but not threatening. No doubt the witch-doctor had gone into the bush. He was a great witch-doctor. "Of course, he's up to something," said Waterhouse, and became silent.

"But what can he do?" asked Pollock, unheeded.

"I must get you out of this. There's something brewing, or things would not be so quiet," said Waterhouse, after a gap of silence. Pollock wanted to know what the brew might be. "Dancing in a circle of skulls," said Waterhouse; "brewing a stink in a copper pot."_ Pollock wanted particulars. Waterhouse was vague, Pollock pressing.   At last

Waterhouse lost his temper. "How the devil should I know?" he said to Pollock's twentieth inquiry what the Porroh man would do. "He tried to kill you off-hand in the hut. *Now,* I fancy he will try something more elaborate. But you'll see fast enough. I don't want to help unnerve you. It's probably all nonsense."

That night, as they were sitting at their fire, Pollock again tried to draw Waterhouse out on the subject of Porroh methods. "Better get to sleep," said Waterhouse, when Pollock's bent became apparent; "we start early to-morrow. You may want all your nerve about you."

"But what line will he take?"

"Can't say. They're versatile people. They know a lot of rum dodges. You'd better get that copper-devil, Shakespear, to talk."

There was a flash and a heavy bang out of the darkness behind the huts, and a clay bullet came whistling close to Pollock's head. This, at least, was crude enough. The blacks and half-breeds sitting and yarning round their own fire jumped up, and someone fired into the dark.

"Better go into one of the huts," said Waterhouse quietly, still sitting unmoved.

Pollock stood up by the fire and drew his revolver. Fighting, at least, he was not afraid of. But a man in the dark is in the best of armour. Realising the wisdom of Waterhouse's advice, Pollock went into the tent and lay down there.

What little sleep he had was disturbed by dreams, variegated dreams, but chiefly of the Porroh man's face, upside down, as he went out of the hut, and looked up under his arm. It was odd that this transitory impression should have stuck so firmly in Pollock's memory. Moreover, he was troubled by queer pains in his limbs.

In the white haze of the early morning, as they were loading the canoes, a barbed arrow suddenly appeared quivering in the ground close to Pollock's foot. The boys made a perfunctory effort to clear out the thicket, but it led to no capture.

After these two occurrences, there was a disposition on the part of the expedition to leave Pollock to himself, and Pollock became, for the first time in his life, anxious to mingle with blacks. Waterhouse took one canoe, and Pollock, in spite of a friendly desire to chat with Waterhouse, had to take the other. He was left all alone in the front part of the canoe, and he had the greatest trouble to make the men—who did not love him—keep to the middle of the river, a clear hundred yards or more from either shore. However, he made Shakespear, the Freetown half-breed, come up to his own end of the canoe and tell him about Porroh, which Shakespear, failing in his attempts to leave Pollock alone, presently did with considerable freedom and gusto.

The day passed. The canoe glided swiftly along the ribbon of lagoon water, between the drift of water-figs, fallen trees, papyrus, and palm-wine palms, and with the dark mangrove swamp to the left, through which one could hear now and then the roar of the Atlantic surf. Shakespear told in his soft, blurred English of how the Porroh could cast spells; how men withered up under their malice; how they could send dreams and devils; how they tormented and killed the sons of Ijibu; how they kidnapped a white trader from Sulyma who had maltreated one of the sect, and how his body looked when it was found. And Pollock after each narrative cursed under his breath at the want of missionary enterprise that allowed such things to be, and at the inert British Government that ruled over this dark heathendom of Sierra Leone. In the evening they came to the Kasi Lake, and sent a score of crocodiles lumbering off the island on which the expedition camped for the night.

The next day they reached Sulyma, and smelt the sea breeze, but Pollock had to put up there for five days before he could get on to Freetown. Waterhouse, considering him to be comparatively safe here, and within the pale of Freetown influence, left him and went back with the expedition to Gbemma, and Pollock became very friendly with Perera, the only resident white trader at Sulyma—so friendly, indeed, that he went about with him everywhere. Perera was a little Portuguese Jew, who had lived in England, and he appreciated the Englishman's friendliness as a great compliment.

For two days nothing happened out of the ordinary; for the most part Pollock and Perera played Nap—the only game they had in common—and Pollock got into debt. Then, on the second evening, Pollock had a disagreeable intimation of the arrival of the Porroh man in Sulyma by getting a flesh-wound in the shoulder from a lump of filed iron. It was a long shot, and the missile had nearly spent its force when it hit him. Still it conveyed its message plainly enough. Pollock sat up in his hammock, revolver in hand, all that night, and next morning confided, to some extent, in the Anglo-Portuguese.

Perera took the matter seriously. He knew the local customs pretty thoroughly. "It is a personal question, you must know. It is revenge. And of course he is hurried by your leaving de country. None of de natives or half-breeds will interfere wid him very much—unless you make it wort deir while. If you come upon him suddenly, you might shoot him. But den he might shoot you.

"Den dere's dis—infernal magic," said Perera. "Of course, I don't believe in it—superstition—but still its's not nice to tink dat wherever you are, dere is a black man, who spends a moonlight night now and den a-dancing about a fire to send you bad dreams. . . . Had any bad dreams?"

"Rather," said Pollock. "I keep on seeing the beggar's head upside down grinning at me and showing all his teeth as he did in the hut, and coming close up to me, and then going ever so far off, and coming back. It's nothing to be afraid of, but somehow it simply paralyses me with terror in my sleep. Queer things—dreams. I know it's a dream all the time, and I can't wake up from it."

"It's probably only fancy," said Perera. "Den my niggers say Porroh men can send snakes. Seen any snakes lately?"

"Only one. I killed him this morning, on the floor near my hammock. Almost trod on him as I got up."

"Ah!" said Perera, and then, reassuringly, "Of course it is a—coincidence. Still I would keep my eyes open. Den dere's pains in de bones."

"I thought they were due to miasma," said Pollock.

"Probably dey are. When did dey begin?"

Then Pollock remembered that he first noticed them the night after the fight in the hut. "It's my opinion he don't want to kill you," said Perera—"at least not yet. I've heard deir idea is to scare and worry a man wid deir spells, and narrow misses, and rheumatic pains, and bad dreams, and all dat, until he's sick of life. Of course, it's all talk, you know. You mustn't worry about it. . . . But I wonder what he'll be up to next."

"I shall have to be up to something first," said Pollock, staring gloomily at the greasy cards that Perera was putting on the table. "It don't suit my dignity to be followed about, and shot at, and blighted in this way. I wonder if Porroh hokey-pokey upsets your luck at cards."

He looked at Perera suspiciously.

"Very likely it does," said Perera warmly, shuffling. "Dey are wonderful people."

That afternoon Pollock killed two snakes in his hammock, and there was also an extraordinary increase in the number of red ants that swarmed over the place; and these annoyances put him in a fit temper to talk over business with a certain Mendi rough he had interviewed before. The Mendi rough showed Pollock a little iron dagger, and demonstrated where one struck in the neck, in a way that made Pollock shiver, and in return for certain considerations Pollock promised him a double-barrelled gun with an ornamental lock.

In the evening, as Pollock and Perera were playing cards, the Mendi rough came in through the doorway, carrying something in a blood-soaked piece of native cloth.

"Not here!" said Pollock very hurriedly. "Not here!"

But he was not quick enough to prevent the man, who was anxious to get to Pollock's side of the bargain, from opening the cloth and throwing the head of the Porroh man upon the table. It bounded from there on to the floor, leaving a red trail on the cards, and rolled into a corner, where it came to rest upside down, but glaring hard at Pollock.

Perera jumped up as the thing fell among the cards, and began in his excitement to gabble in Portuguese. The Mendi was bowing, with the red cloth in his hand. "De gun!" he cried. Pollock stared back at the head in the corner. It bore exactly the expression it had in his dreams. something seemed to snap in his own brain as he looked at it.

Then Perera found his English again.

"You got him killed?" he said. "You did not kill him yourself?"

"Why should I?" said Pollock.

"But he will not be able to take it off now!"

"Take *what* off?" said Pollock.

"And all dese cards are spoiled!"

"*What* do you mean by taking off?" said Pollock.

"You must send me a new pack from Freetown. You can buy dem dere."

"But—'take it off'?"

"It is only superstition. I forgot. De niggers say dat if de witches—he was a witch— But it is rubbish. . . . You must make de Porroh man take it off, or kill him yourself. . . . It is very silly."

Pollock swore under his breath, still staring hard at the head in the corner.

"I can't stand that glare," he said. Then suddenly he rushed at the thing and kicked it. It rolled some yards or so, and came to rest in the same position as before, upside down, and looking at him.

"He is ugly," said the Anglo-Portuguese. "Very ugly. Dey do it on deir faces with little knives."

Pollock would have kicked the head again, but the Mendi man touched him on the arm. "De gun?" he said, looking nervously at the head.

"Two—if you will take that beastly thing away," said Pollock.

Then Mendi shook his head, and intimated that he only wanted one gun now due to him, and for which he would be obliged. Pollock found neither cajolery nor bullying any good with him. Perera had a gun to sell (at a profit of three hundred per cent.), and with that the man presently departed. Then Pollock's eyes, against his will, were recalled to the thing on the floor.

"It is funny dat his head keeps upside down," said Perera, with an uneasy laugh. "His brains must be heavy, like de weight in de little images one sees dat keep always upright wid lead in dem. You will take him wiv you when you go presently. You might take him now. De cards are all spoilt. Dere is a man sell dem in Freetown. De room is in a filthy mess as it is. You should have killed him yourself."

Pollock pulled himself together, and went and picked up the head. He would hang it up by the lamp-hook in the middle of the ceiling of his room, and dig a grave for it at once. He was under the impression that he hung it up by the hair, but that must have been wrong, for when he returned for it, it was hanging by the neck upside down.

He buried it before sunset on the north side of the shed he occupied, so that he should not have to pass the grave after dark when he was returning from Perera's. He killed two snakes before he went to sleep. In the darkest part of the night he awoke with a start, and heard a pattering sound and something scraping on the floor. He sat up noiselessly, and felt under his pillow for his revolver. A mumbling growl followed, and Pollock fired at the sound. There was a yelp, and something dark passed for a moment across the hazy blue of the doorway. "A dog!" said Pollock, lying down again.

In the early dawn he awoke again with a peculiar sense of unrest. The vague pain in his bones had returned. For some time he lay watching the red ants that were swarming over the ceiling, and then, as the light grew brighter, he looked over the edge of his hammock and saw something dark on the floor. He gave such a violent start that the hammock overset and flung him out.

He found himself lying, perhaps, a yard away from the head of the Porroh man. It had been disinterred by the dog, and the nose was grievously battered. Ants and flies swarmed over it. By an odd coincidence, it was still upside

down, and with the same diabolical expression in the
inverted eyes.

Pollock sat paralysed, and stared at the horror for some
time. Then he got up and walked round it—giving it a
wide berth—and out of the shed. The clear light of the
sunrise, the living stir of vegetation before the breath of the
dying land-breeze, and the empty grave with the marks
of the dog's paws, lightened the weight upon his mind a
little.

He told Perera of the business as though it was a jest—
a jest to be told with white lips. "You should not have
frighten de dog," said Perera, with poorly simulated hilarity.

The next two days, until the steamer came, were spent
by Pollock in making a more effectual disposition of his
possession. Overcoming his aversion to handling the thing,
he went down to the river mouth and threw it into the
sea-water, but by some miracle it escaped the crocodiles,
and was cast up by the tide on mud a little way up the
river, to be found by an intelligent Arab half-breed, and
offered for sale to Pollock and Perera as a curiosity, just
on the edge of night. The native hung about in the brief
twilight, making lower and lower offers, and at last, getting
scared in some way by the evident dread these wise white
men had for the thing, went off, and, passing Pollock's shed,
threw his burden in there for Pollock to discover in the
morning.

At this Pollock got into a kind of frenzy. He would burn
the thing. He went out straightway into the dawn, and
had constructed a big pyre of brushwood before the heat
of the day. He was interrupted by the hooter of the little
paddle steamer from Monrovia to Bathurst, which was coming through the gap in the bar. "Thank Heaven!" said
Pollock, with infinite piety, when the meaning of the sound
dawned upon him. With trembling hands he lit his pile
of wood hastily, threw the head upon it, and went away
to pack his portmanteau and make his adieux to Perera.

That afternoon, with a sense of infinite relief, Pollock
watched the flat swampy foreshore of Sulyma grow small
in the distance. The gap in the long line of white surge

became narrower and narrower. It seemed to be closing in and cutting him off from his trouble. The feeling of dread and worry began to slip from him bit by bit. At Sulyma belief in Porroh malignity and Porroh magic had been in the air, his sense of Porroh had been vast, pervading, threatening, dreadful. Now manifestly the domain of Porroh was only a little place, a little black band between the sea and the blue cloudy Mendi uplands.

"Good-bye, Porroh!" said Pollock. "Good-bye—certainly not *au revoir.*"

The captain of the steamer came and leant over the rail beside him, and wished him good-evening, and spat at the froth of the wake in token of friendly ease.

"I picked up a rummy curio on the beach this go," said the captain. "It's a thing I never saw done this side of Indy before."

"What might that be?" said Pollock.

"Pickled 'ed," said the captain.

"*What!*" said Pollock.

" ''Ed—smoked. 'Ed of one of these Porroh chaps, all ornamented with knife-cuts. Why! What's up? Nothing? I shouldn't have took you for a nervous chap. Green in the face. By gosh! you're a bad sailor. All right, eh? Lord, how funny you went! . . . Well, this 'ed I was telling you of is a bit rum in a way. I've got it, along with some snakes, in a jar of spirit in my cabin what I keeps for such curios, and I'm hanged if it don't float upsy down. Hullo!"

Pollock had given an incoherent cry, and had his hands in his hair. He ran towards the paddle-boxes with a half-formed idea of jumping into the sea, and then he realised his position and turned back towards the captain.

"Here!" said the captain. "Jack Philips, just keep him off me! Stand off! No nearer, mister! What's the matter with you? Are you mad?"

Pollock put his hand to his head. It was no good explaining. "I believe I am pretty nearly mad at times," he said. "It's a pain I have here. Comes suddenly. You'll excuse me, I hope."

He was white and in a perspiration. He saw suddenly

very clearly all the danger he ran of having his sanity doubted. He forced himself to restore the captain's confidence, by answering his sympathetic inquiries, noting his suggestions, even trying a spoonful of neat brandy in his cheek, and, that matter settled, asking a number of questions about the captain's private trade in curiosities. The captain described the head in detail. All the while Pollock was struggling to keep under a preposterous persuasion that the ship was as transparent as glass, and that he could distinctly see the inverted face looking at him from the cabin beneath his feet.

Pollock had a worse time almost on the steamer than he had at Sulyma. All day he had to control himself in spite of his intense perception of the imminent presence of that horrible head that was overshadowing his mind. At night his old nightmare returned, until, with a violent effort, he would force himself awake, rigid with the horror of it, and with the ghost of a hoarse scream in his throat.

He left the actual head behind at Bathurst, where he changed ship for Teneriffe, but not his dreams nor the dull ache in his bones. At Teneriffe Pollock transferred to a Cape liner, but the head followed him. He gambled, he tried chess, he even read books, but he knew the danger of drink. Yet whenever a round black shadow, a round black object came into his range, there he looked for the head, and—saw it. He knew clearly enough that his imagination was growing traitor to him, and yet at times it seemed the ship he sailed in, his fellow-passengers, the sailors, the wide sea, was all part of a filmy phantasmagoria that hung, scarcely veiling it, between him and a horrible real world. Then the Porroh man, thrusting his diabolical face through that curtain, was the one real and undeniable thing. At that he would get up and touch things, taste something, gnaw something, burn his hand with a match, or run a needle into himself.

So, struggling grimly and silently with his excited imagination, Pollock reached England. He landed at Southampton, and went on straight from Waterloo to his banker's in Cornhill in a cab. There he transacted some business with the manager in a private room, and all the while the head hung like an ornament under the black marble mantel and dripped upon the fender. He could hear the drops fall, and see the red on the fender.

"A pretty fern," said the manager, following his eyes. "But it makes the fender rusty."

"Very," said Pollock; "a *very* pretty fern. And that reminds me. Can you recommend me a physician for mind troubles? I've got a little—what is it?—hallucination."

The head laughed savagely, wildly. Pollock was surprised the manager did not notice it. But the manager only stared at his face.

With the address of a doctor, Pollock presently emerged in Cornhill. There was no cab in sight, and so he went on down to the western end of the street, and essayed the crossing opposite the Mansion House. The crossing is hardly easy even for the expert Londoner; cabs, vans, carriages, mail-carts, omnibuses go by in one incessant stream; to anyone fresh from the malarious solitudes of Sierra Leone it is a boiling, maddening confusion. But when an inverted head suddenly comes bouncing, like an indiarubber ball, between your legs, leaving distinct smears of blood every time it touches the ground, you can scarcely hope to avoid an accident. Pollock lifted his feet convulsively to avoid it, and then kicked at the thing furiously. Then something hit him violently in the back, and a hot pain ran up his arm.

He had been hit by the pole of an omnibus, and three of the fingers of his left hand smashed by the hoof of one of the horses—the very fingers, as it happened, that he shot from the Porroh man. They pulled him out from between the horses' legs, and found the address of the physician in his crushed hand.

For a couple of days Pollock's sensations were full of the sweet, pungent smell of chloroform, of painful operations that caused him no pain, of lying still and being given food and drink. Then he had a slight fever, and was very thirsty, and his old nightmare came back. It was only when it returned that he noticed it had left him for a day.

"If my skull had been smashed instead of my fingers, it might have gone altogether," said Pollock, staring thoughtfully at the dark cushion that had taken on for the time the shape of the head.

Pollock at the first opportunity told the physician of his mind trouble. He knew clearly that he must go mad unless something should intervene to save him. He explained that he had witnessed a decapitation in Dahomey, and was haunted by one of the heads. Naturally, he did not care to state the actual facts. The physician looked grave.

Presently he spoke hesitatingly. "As a child, did you get very much religious training?"

"Very little," said Pollock.

A shade passed over the physician's face. "I don't know if you have heard of the miraculous cures—it may be, of course, they are not miraculous—at Lourdes."

"Faith-healing will hardly suit me, I am afraid," said Pollock, with his eye on the dark cushion.

The head distorted its scarred features in an abominable grimace. The physician went upon a new track. "It's all imagination," he said, speaking with sudden briskness. "A fair case for faith-healing, anyhow. Your nervous system has run down, you're in that twilight state of health when the bogles come easiest. The strong impression was too much for you. I must make you up a little mixture that will strengthen your nervous system—especially your brain. And you must take exercise."

"I'm no good for faith-healing," said Pollock.

"And therefore we must restore tone. Go in search of stimulating air—Scotland, Norway, the Alps"—

"Jericho, if you like," said Pollock—"where Naaman went."

However, so soon as his fingers would let him, Pollock made a gallant attempt to follow out the doctor's suggestion. It was now November. He tried football, but to Pollock the game consisted in kicking a furious inverted head about a field. He was no good at the game. He kicked blindly, with a kind of horror, and when they put him back into goal, and the ball same swooping down upon him, he suddenly yelled and got out of its way. The discreditable stories that had driven him from England to wander in the tropics shut him off from any but men's society, and now his increasingly strange behaviour made even his man friends avoid him. The thing was no longer a thing of the eye merely; it gibbered at him, spoke to him. A horrible fear came upon him that presently, when he took hold of the apparition, it would no longer become some mere article of furniture, but would *feel* like a real dissevered head. Alone, he would curse at the thing, defy it, entreat it; once or twice, in spite of his grim self-control, he addressed it in the presence of others. He felt the growing suspicion in the eyes of the people that watched him—his landlady, the servant, his man.

One day early in December his cousin Arnold—his next of kin—came to see him and draw him out, and watch his sunken yellow face with narrow eager eyes. And it seemed to Pollock that the hat his cousin carried in his hand was no hat at all, but a Gorgon head that glared at him upside down, and fought with its eyes against his reason. However, he was still resolute to see the matter out. He got a bicycle, and, riding over the frosty road from Wandsworth to Kingston, found the thing rolling along at his side, and leaving a dark trail behind it. He set his teeth and rode faster. Then suddenly, as he came down the hill towards Richmond Park, the apparition rolled in front of him and under his wheel, so quickly that he had no time for thought, and, turning quickly to avoid it, was flung violently against a heap of stones and broke his left wrist.

The end came on Christmas morning. All night he had been in a fever, the bandages encircling his wrist like a band of fire, his dreams more vivid and terrible than ever. In the cold, colourless, uncertain light that came before the sunrise, he sat up in his bed, and saw the head upon the bracket in the place of the bronze jar that had stood there overnight.

"I know that is a bronze jar," he said, with a chill doubt

in his heart. Presently the doubt was irresistible. He got out of bed slowly, shivering, and advanced to the jar with his hand raised. Surely he would see now his imagination had deceived him, recognise the distinctive sheen of bronze. At last, after an age of hesitation, his fingers came down on the patterned cheek of the head. He withdrew them spasmodically. The last stage was reached. His sense of touch had betrayed him.

Trembling, stumbling against the bed, kicking against his shoes with his bare feet, a dark confusion eddying round him, he groped his way to the dressing-table, took his razor from the drawer, and sat down on the bed with this in his hand. In the looking-glass he saw his own face, colourless, haggard, full of the ultimate bitterness of despair.

He beheld in swift succession the incidents in the brief tale of his experience. His wretched home, his still more wretched schooldays, the years of vicious life he had led since then, one act of selfish dishonour leading to another; it was all clear and pitiless now, all its squalid folly, in the cold light of the dawn. He came to the hut, to the fight with the Porroh man, to the retreat down the river to Sulyma, to the Mendi assassin and his red parcel, to his frantic endeavours to destroy the head, to the growth of his hallucination. It was a hallucination! He *knew* it was. A hallucination merely. For a moment he snatched at hope. He looked away from the glass, and on the bracket, the inverted head grinned and grimaced at him. . . . With the stiff fingers of his bandaged hand he felt at his neck for the throb of his arteries. The morning was very cold, the steel blade felt like ice.

STORY THE NINTH

The Red Room

"I can assure you," said I, "that it will take a very tangible ghost to frighten me." And I stood up before the fire with my glass in my hand.

"It is your own choosing," said the man with the withered arm, and glanced at me askance.

"Eight-and-twenty years," said I, "I have lived, and never a ghost have I seen as yet."

The old woman sat staring hard into the fire, her pale eyes wide open. "Ay," she broke in; "and eight-and-twenty years you have lived and never seen the likes of this house, I reckon. There's a many things to see, when one's still but eight-and-twenty." She swayed her head slowly from side to side. "A many things to see and sorrow for."

I half suspected the old people were trying to enhance the spiritual terrors of their house by their droning insistence. I put down my empty glass on the table and looked about the room, and caught a glimpse of myself, abbreviated and broadened to an impossible sturdiness, in the queer old mirror at the end of the room. "Well," I said, "if I see anything to-night, I shall be so much the wiser. For I come to the business with an open mind."

"It's your own choosing," said the man with the withered arm once more.

I heard the sound of a stick and a shambling step on the flags in the passage outside, and the door creaked on its hinges as a second old man entered, more bent, more wrinkled, more aged even than the first. He supported himself by a single crutch, his eyes were covered by a shade, and his lower lip, half-averted, hung pale and pink from his decaying yellow teeth. He made straight for an arm-chair on the opposite side of the table, sat down clumsily, and began to cough. The man with the withered arm gave this new-comer a short glance of positive dislike; the old woman took no notice of his arrival, but remained with her eyes fixed steadily on the fire.

"I said—it's your own choosing," said the man with the withered arm, when the coughing had ceased for a while.

"It's my own choosing," I answered.

The man with the shade became aware of my presence for the first time, and threw his head back for a moment and sideways, to see me. I caught a momentary glimpse of his eyes, small and bright and inflamed. Then he began to cough and splutter again.

"Why don't you drink?" said the man with the withered arm, pushing the beer towards him. The man with the shade poured out a glassful with a shaky arm that splashed half as much again on the deal table. A monstrous shadow of him crouched upon the wall and mocked his action as he poured and drank. I must confess I had scarce expected these grotesque custodians. There is to my mind something inhuman in senility, something crouching and atavistic; the human qualities seem to drop from old people insensibly day by day. The three of them made me feel uncomfortable, with their gaunt silences, their bent carriage, their evident unfriendliness to me and to one another.

"If," said I, "you will show me to this haunted room of yours, I will make myself comfortable there."

The old man with the cough jerked his head back so suddenly that it startled me, and shot another glance of his red eyes at me from under the shade; but no one answered me. I waited a minute, glancing from one to the other.

"If," I said a little louder, "if you will show me to this haunted room of yours, I will relieve you from the task of entertaining me."

"There's a candle on the slab outside the door," said the

man with the withered arm, looking at my feet as he addressed me. "But if you go to the red room to-night"—

("This night of all nights!" said the old woman.)

"You go alone."

"Very well," I answered. "And which way do I go?"

"You go along the passage for a bit," said he, "until you come to a door, and through that is a spiral staircase, and half-way up that is a landing and another door covered with baize. Go through that and down the long corridor to the end, and the red room is on your left up the steps."

"Have I got that right?" I said, and repeated his directions. He corrected me in one particular.

"And are you really going?" said the man with the shade, looking at me again for the third time, with that queer, unnatural tilting of the face.

("This night of all nights!" said the old woman.)

"It is what I came for," I said, and moved towards the door. As I did so, the old man with the shade rose and staggered round the table, so as to be closer to the others and to the fire. At the door I turned and looked at them, and saw they were all close together, dark against the fire-light, staring at me over their shoulders, with an intent expression on their ancient faces.

"Good-night," I said, setting the door open.

"It's your own choosing," said the man with the withered arm.

I left the door wide open until the candle was well alight, and then I shut them in and walked down the chilly, echoing passage.

I must confess that the oddness of these three old pensioners in whose charge her ladyship had left the castle, and the deep-toned, old fashioned furniture of the house-keeper's room in which they foregathered, affected me in spite of my efforts to keep myself at a matter-of-fact phase. They seemed to belong to another age, an older age, an age when things spiritual were different from this of ours, less certain; an age when omens and witches were credible, and ghosts beyond denying. Their very existence was

P

spectral; the cut of their clothing, fashions born in dead brains. The ornaments and conveniences of the room about them were ghostly—the thoughts of vanished men, which still haunted rather than participated in the world of to-day. But with an effort I sent such thoughts to the right-about. The long, draughty subterranean passage was chilly and dusty, and my candle flared and made the shadows cower and quiver. The echoes rang up and down the spiral staircase, and a shadow came sweeping up after me, and one fled before me into the darkness overhead. I came to the landing and stopped there for a moment, listening to a rustling that I fancied I heard; then, satisfied of the absolute silence, I pushed open the baize-covered door and stood in the corridor.

The effect was scarcely what I expected, for the moonlight, coming in by the great window on the grand staircase, picked out everything in vivid black shadow or silvery illumination. Everything was in its place: the house might have been deserted on the yesterday instead of eighteen months ago. There were candles in the sockets of the sconces, and whatever dust had gathered on the carpets or upon the polished flooring was distributed so evenly as to be invisible in the moonlight. I was about to advance, and stopped abruptly. A bronze group stood upon the landing, hidden from me by the corner of the wall, but its shadow fell with marvellous distinctness upon the white panelling, and gave me the impression of someone crouching to waylay me. I stood rigid for half a minute perhaps. Then, with my hand in the pocket that held my revolver, I advanced, only to discover a Ganymede and Eagle glistening in the moonlight. That incident for a time restored my nerve, and a porcelain Chinaman on a buhl table, whose head rocked silently as I passed him, scarcely startled me.

The door to the red room and the steps up to it were in a shadowy corner. I moved my candle from side to side, in order to see clearly the nature of the recess in which I stood before opening the door. Here it was, thought I, that my predecessor was found, and the memory of that story gave me a sudden twinge of apprehension. I glanced over my shoulder at the Ganymede in the moonlight, and opened the door of the red room rather hastily, with my face half turned to the pallid silence of the landing.

I entered, closed the door behind me at once, turned the key I found in the lock within, and stood with the candle held aloft, surveying the scene of my vigil, the great red room of Lorraine Castle, in which the young duke had died. Or, rather, in which he had begun his dying, for he had opened the door and fallen headlong down the steps I had just ascended. That had been the end of his vigil, of his gallant attempt to conquer the ghostly tradition of the place; and never, I thought, had apoplexy better served the ends of superstition. And there were other and older stories that clung to the room, back to the half-credible beginning of it all, the tale of a timid wife and the tragic end that came to her husband's jest of frightening her. And looking around that large shadowy room, with its shadowy window bays, its recesses and alcoves, one could well understand the legends that had sprouted in its black corners, its germinating darkness. My candle was a little tongue of flame in its vastness, that failed to pierce the opposite end of the room, and left an ocean of mystery and suggestion beyond its island of light.

I resolved to make a systematic examination of the place at once, and dispel the fanciful suggestions of its obscurity before they obtained a hold upon me. After satisfying myself of the fastening of the door, I began to walk about the room, peering round each article of furniture, tucking up the valances of the bed, and opening its curtains wide. I pulled up the blinds and examined the fastenings of the several windows before closing the shutters, leant forward and looked up the blackness of the wide chimney, and tapped the dark oak panelling for any secret opening. There were two big mirrors in the room, each with a pair of sconces bearing candles, and on the mantelshelf, too, were more candles in china candlesticks. All these I lit one after the other. The fire was laid,—an unexpected consideration from the old housekeeper,—and I lit it, to keep down any disposition to shiver, and when it was burning well, I stood round with my back to it and regarded the room again. I had pulled up a chintz-covered arm-chair and a table, to form a kind of barricade before me, and on this lay my revolver ready to hand. My precise examination had done me good, but I still found the remoter darkness of the place, and its perfect stillness, too stimulating for the imagination. The echoing of the stir and crackling of the fire was no sort of comfort to me. The shadow in the alcove, at the end in particular, had that undefinable quality of a presence, that odd suggestion of a lurking living thing, that comes so easily in silence and solitude. At last, to reassure myself, I walked with a candle into it, and satisfied myself that there was nothing tangible there. I stood that candle upon the floor of the alcove, and left it in that position.

By this time I was in a state of considerable nervous tension, although to my reason there was no adequate cause for the condition. My mind, however, was perfectly clear. I postulated quite unreservedly that nothing supernatural could happen, and to pass the time I began to string some rhymes together, Ingoldsby fashion, of the original legend of the place. A few I spoke aloud, but the echoes were not pleasant. For the same reason I also abandoned, after a time, a conversation with myself upon the impossibility of ghosts and haunting. My mind reverted to the three old and distorted people downstairs, and I tried to keep it upon that topic. The sombre reds and blacks of the room troubled me; even with seven candles the place was merely dim. The one in the alcove flared in a draught, and the fire-flickering kept the shadows and penumbra perpetually shifting and stirring. Casting about for a remedy, I recalled the candles I had seen in the passage, and, with a slight effort, walked out into the moonlight, carrying a candle and leaving the door open, and presently returned with as many as ten. These I put in various knick-knacks of china with which the room was sparsely adorned, lit and placed where the shadows had lain deepest, some on the floor, some in the window recesses, until at last my seventeen candles were so arranged that not an inch of the room but had the direct light of at least one of them. It occurred to me that when the ghost came, I could warn him not to trip over them. The room was now quite brightly illuminated. There was something very cheery and reassuring in these little streaming flames, and snuffing them gave me an occupation, and afforded a reassuring sense of the passage of time.

Even with that, however, the brooding expectation of the vigil weighed heavily upon me. It was after midnight that the candle in the alcove suddenly went out, and the black shadow sprang back to its place. I did not see the candle go out; I simply turned and saw that the darkness was there, as one might start and see the unexpected presence of a stranger. "By Jove!" said I aloud; "that draught's a strong one!" and, taking the matches from the table, I walked across the room in a leisurely manner to relight the corner again. My first match would not strike, and as I succeeded with the second, something seemed to blink on the wall before me. I turned my head involuntarily, and saw that the two candles on the little table by the fireplace were extinguished. I rose at once to my feet.

"Odd!" I said. "Did I do that myself in a flash of absentmindedness?"

I walked back, relit one, and as I did so, I saw the candle in the right sconce of one of the mirrors wink and go right out, and almost immediately its companion followed it. There was no mistake about it. The flame vanished, as if the wicks had been suddenly nipped between a finger and a thumb, leaving the wick neither glowing nor smoking, but black. While I stood gaping, the candle at the foot of the bed went out, and the shadows seemed to take another step towards me.

"This won't do!" said I, and first one and then another candle on the mantelshelf followed.

"What's up?" I cried, with a queer high note getting into my voice somehow. At that the candle on the wardrobe went out, and the one I had relit in the alcove followed.

"Steady on!" I said. "These candles are wanted," speaking with a half-hysterical facetiousness, and scratching away at a match the while for the mantel candlesticks. My hands trembled so much that twice I missed the rough paper of the matchbox. As the mantel emerged from darkness again, two candles in the remoter end of the window were eclipsed. But with the same match I also relit the larger mirror candles, and those on the floor near the doorway, so that for the moment I seemed to gain on the extinctions. But then in a volley there vanished four lights at once in different corners of the room, and I struck another match in quivering haste, and stood hesitating whither to take it.

As I stood undecided, an invisible hand seemed to sweep out the two candles on the table. With a cry of terror, I dashed at the alcove, then into the corner, and then into the window, relighting three, as two more vanished by the fireplace; then, perceiving a better way, I dropped the matches on the iron-bound deed-box in the corner, and caught up the bedroom candlestick. With this I avoided the delay of striking matches; but for all that the steady process of extinction went on, and the shadows I feared and fought against returned, and crept in upon me, first a step gained on this side of me and then on that. It was like a ragged storm-cloud sweeping out the stars. Now and then one returned for a minute, and was lost again. I was now almost frantic with the horror of coming darkness, and my self-possession deserted me. I leaped panting and dishevelled from candle to candle, in a vain struggle against that remorseless advance.

I bruised myself on the thigh against the table, I sent a chair headlong, I stumbled and fell and whisked the cloth from the table in my fall. My candle rolled away from me, and I snatched another as I rose. Abruptly this was blown out, as I swung it off the table, by the wind of my sudden movement, and immediately the two remaining candles followed. But there was light still in the room, a red light that staved off the shadows from me. The fire! Of course, I could thrust my candle between the bars and relight it!

I turned to where the flames were dancing between the glowing coals, and splashing red reflections upon the furniture, made two steps towards the grate, and incontinently the flames dwindled and vanished, the glow vanished, the reflections rushed together and vanished, and as I thrust the candle between the bars, darkness closed upon me like the shutting of an eye, wrapped about me in a stifling embrace, sealed my vision, and crushed the last vestiges of reason from my brain. The candle fell from my hand. I flung out my arms in a vain effort to thrust that ponderous blackness away from me, and, lifting up my voice, screamed with all my might—once, twice, thrice. Then I think I must have staggered to my feet. I know I thought suddenly of the moonlit corridor, and, with my head bowed and my arms over my face, made a run for the door.

But I had forgotten the exact position of the door, and struck myself heavily against the corner of the bed. I staggered back, turned, and was either struck or struck myself against some other bulky furniture. I have a vague memory of battering myself thus, to and fro in the darkness, of a cramped struggle, and of my own wild crying as I darted to and fro, of a heavy blow at last upon my forehead, a horrible sensation of falling that lasted an age, of my last frantic effort to keep my footing, and then I remember no more.

I opened my eyes in daylight. My head was roughly bandaged, and the man with the withered arm was watching my face. I looked about me, trying to remember what had happened, and for a space I could not recollect. I turned to the corner, and saw the old woman, no longer abstracted, pouring out some drops of medicine from a little blue phial into a glass. "Where am I?" I asked. "I seem to remember you, and yet I cannot remember who you are."

They told me then, and I heard of the haunted Red Room as one who hears a tale. "We found you at dawn," said he, "and there was blood on your forehead and lips."

It was very slowly I recovered my memory of my

experience. "You believe now," said the old man, "that the room is haunted?" He spoke no longer as one who greets an intruder, but as one who grieves for a broken friend.

"Yes," said I; "the room is haunted."

"And you have seen it. And we, who have lived here all our lives, have never set eyes upon it. Because we have never dared. . . . Tell us, is it truly the old earl who"—

"No," said I; "it is not."

"I told you so," said the old lady, with the glass in her hand. "It is his poor young countess who was frightened"—

"It is not," I said. "There is neither ghost of earl nor ghost of countess in that room, there is no ghost there at all; but worse, far worse"—

"Well?" they said.

"The worst of all the things that haunt poor mortal man," said I; "and that is, in all its nakedness—*Fear!* Fear that will not have light nor sound, that will not bear with reason, that deafens and darkens and overwhelms. It followed me through the corridor, it fought against me in the room"—

I stopped abruptly. There was an interval of silence. My hand went up to my bandages.

Then the man with the shade sighed and spoke. "That is it," said he. "I knew that was it. A Power of Darkness. To put such a curse upon a woman! It lurks there always. You can feel it even in the daytime, even of a bright summer's day, in the hangings, in the curtains, keeping behind you however you face about. In the dusk it creeps along the corridor and follows you, so that you dare not turn. There is Fear in that room of hers—black Fear, and there will be—so long as this house of sin endures."

STORY THE TENTH

The Cone

THE night was hot and overcast, the sky red-rimmed with the lingering sunset of mid-summer. They sat at the open window, trying to fancy the air was fresher there. The trees and shrubs of the garden stood stiff and dark; beyond in the roadway a gas-lamp burnt, bright orange against the hazy blue of the evening. Farther were the three lights of the railway signal against the lowering sky. The man and woman spoke to one another in low tones.

"He does not suspect?" said the man, a little nervously.

"Not he," she said peevishly, as though that too irritated her. "He thinks of nothing but the works and the prices of fuel. He has no imagination, no poetry."

"None of these men of iron have," he said sententiously. "They have no hearts."

"*He* has not," she added. She turned her discontented face towards the window. The distant sound of a roaring and rushing drew nearer and grew in volume; the house quivered; one heard the metallic rattle of the tender. As the train passed, there was a glare of light above the cutting and a driving tumult of smoke; one, two, three, four, five, six, seven, eight black oblongs—eight trucks—passed across the dim grey of the embankment, and were suddenly extinguished one by one in the throat of the tunnel, which, with the last, seemed to swallow down train, smoke, and sound in one abrupt gulp.

"This country was all fresh and beautiful once," he said; "and now—it is Gehenna. Down that way—nothing but pot-banks and chimneys belching fire and dust into

457

the face of heaven. . . . But what does it matter? An
end comes, an end to all this cruelty. . . . *To-morrow.*"
He spoke the last word in a whisper.

"*To-morrow,*" she said, speaking in a whisper, too, and
still staring out of the window.

"Dear!" he said, putting his hand on hers.

She turned with a start, and their eyes searched one
another's. Hers softened to his gaze. "My dear one!"
she said, and then: "It seems so strange—that you should
have come into my life like this—to open"— She paused.

"To open?" he said.

"All this wonderful world"—she hesitated, and spoke
still more softly—"this world of *love* to me."

Then suddenly the door clicked and closed. They turned
their heads, and he started violently back. In the shadow
of the room stood a great shadowy figure—silent. They saw
the face dimly in the half-light, with unexpressive dark
patches under the penthouse brows. Every muscle in
Raut's body suddenly became tense. When could the
door have opened? What had he heard. Had he heard all?
What had he seen? A tumult of questions.

The new-comer's voice came at last, after a pause that
seemed interminable. "Well?" he said.

"I was afraid I had missed you, Horrocks," said the man
at the window, gripping the window-ledge with his hand.
His voice was unsteady.

The clumsy figure of Horrocks came forward out of the
shadow. He made no answer to Raut's remark. For a
moment he stood above them.

The woman's heart was cold within her. "I told Mr.
Raut it was just possible you might come back," she said,
in a voice that never quivered.

Horrocks, still silent, sat down abruptly in the chair by
her little work-table. His big hands were clenched; one
saw now the fire of his eyes under the shadow of his brows.
He was trying to get his breath. His eyes went from the
woman he had trusted to the friend he had trusted, and
then back to the woman.

By this time and for the moment all three half understood

one another. Yet none dared say a word to ease the pent-up things that choked them.

It was the husband's voice that broke the silence at last.

"You wanted to see me?" he said to Raut.

Raut started as he spoke. "I came to see you," he said, resolved to lie to the last.

"Yes," said Horrocks.

"You promised," said Raut, "to show me some fine effects of moonlight and smoke."

"I promised to show you some fine effects of moonlight and smoke," repeated Horrocks, in a colourless voice.

"And I thought I might catch you to-night before you went down to the works," proceeded Raut, "and come with you."

There was another pause. Did the man mean to take the thing coolly? Did he after all know? How long had he been in the room? Yet even at the moment when they heard the door, their attitudes . . . Horrocks glanced at the profile of the woman, shadowy pallid in the half-light. Then he glanced at Raut, and seemed to recover himself suddenly. "Of course," he said, "I promised to show you the works under their proper dramatic conditions. It's odd how I could have forgotten."

"If I am troubling you"—began Raut.

Horrocks started again. A new light had suddenly come into the sultry gloom of his eyes. "Not in the least," he said.

"Have you been telling Mr. Raut of all these contrasts of flame and shadow you think so splendid?" said the woman, turning now to her husband for the first time, her confidence creeping back again, her voice just one half-note too high. "That dreadful theory of yours that machinery is beautiful, and everything else in the world ugly. I thought he would not spare you Mr. Raut. It's his great theory, his one discovery in art."

"I am slow to make discoveries," said Horrocks grimly, damping her suddenly. "But what I discover . . ." He stopped.

"Well?" she said.

"Nothing;" and suddenly he rose to his feet.

"I promised to show you the works," he said to Raut,
and put his big, clumsy hand on his friend's shoulder.
"And you are ready to go?"

"Quite," said Raut, and stood up also.

There was another pause. Each of them peered through
the indistinctness of the dusk at the other two. Horrocks'
hand still rested on Raut's shoulder. Raut half fancied
still that the incident was trivial after all. But Mrs. Horrocks knew her husband better, knew that grim quiet
in his voice, and the confusion in her mind took a vague
shape of physical evil. "Very well," said Horrocks, and,
dropping his hand, turned towards the door.

"My hat?" Raut looked round in the half-light.

"That's my work-basket," said Mrs. Horrocks with a
gust of hysterical laughter. Their hands came together
on the back of the chair. "Here it is!" he said. She had
an impulse to warn him in an undertone, but she could
not frame a word. "Don't go!" and "Beware of him!"
struggled in her mind, and the swift moment passed.

"Got it?" said Horrocks, standing with the door half
open.

Raut stepped towards him. "Better say good-bye to Mrs.
Horrocks," said the ironmaster, even more grimly quiet in
his tone than before.

Raut started and turned. "Good-evening, Mrs. Horrocks,"
he said, and their hands touched.

Horrocks held the door open with a ceremonial politeness unusual in him towards men. Raut went out, and
then, after a wordless look at her, her husband followed.
She stood motionless while Raut's light footfall and her
husband's heavy tread, like bass and treble, passed down
the passage together. The front door slammed heavily.
She went to the window, moving slowly, and stood watching
—leaning forward. The two men appeared for a moment
at the gateway in the road, passed under the street lamp,
and were hidden by the black masses of the shrubbery. The
lamplight fell for a moment on their faces, showing only
unmeaning pale patches, telling nothing of what she still

feared, and doubted, and craved vainly to know. Then she sank down into a crouching attitude in the big armchair, her eyes wide open and staring out at the red lights from the furnaces that flickered in the sky. An hour after she was still there, her attitude scarcely changed.

The oppressive stillness of the evening weighed heavily upon Raut. They went side by side down the road in silence, and in silence turned into the cinder-made by-way that presently opened out the prospect of the valley.

A blue haze, half dust, half mist, touched the long valley with mystery. Beyond were Hanley and Etruria, grey and black masses, outlined thinly by the rare golden dots of the street-lamps, and here and there a gaslit window, or the yellow glare of some late-working factory or crowded public-house. Out of the masses, clear and slender against the evening sky, rose a multitude of tall chimneys, many of them reeking, a few smokeless during the season of "play." Here and there a pallid patch and ghostly stunted beehive shapes showed the position of a pot-bank, or a wheel, black and sharp against the hot lower sky, marked some colliery where they raise the iridescent coal of the place. Nearer at hand was the broad stretch of railway, and half invisible trains shunted—a steady puffing and rumbling, with every now and then a ringing concussion and a series of impacts, and a passage of intermittent puffs of white steam across the further view. And to the left, between the railway and the dark mass of the low hill beyond, dominating the whole view, colossal, inky-black, and crowned with smoke and fitful flames, stood the great cylinders of the Jeddah Company Blast Furnaces, the central edifices of the big ironworks of which Horrocks was the manager. They stood heavy and threatening, full of an incessant turmoil of flames and seething molten iron, and about the feet of them rattled the rolling-mills, and the steam hammer beat heavily and splashed the white iron sparks hither and thither. Even as they looked, a truckful of fuel was shot into one of the giants, and red flames gleamed out, and a confusion of smoke and black dust came boiling upwards towards the sky.

"Certainly you get some fine effects of colour with your furnaces," said Raut, breaking a silence that had become apprehensive.

Horrocks grunted. He stood with his hands in his pockets, frowning down at the dim steaming railway and the busy ironworks beyond, frowning as if he were thinking out some knotty problem.

Raut glanced at him and away again. "At present your moonlight effect is hardly ripe," he continued, looking upward; "the moon is still smothered by the vestiges of daylight."

Horrocks stared at him with the expression of a man who has suddenly awakened. "Vestiges of daylight? . . . Of course, of course." He too looked up at the moon, pale still in the midsummer sky. "Come along," he said suddenly, and, gripping Raut's arm in his hand, made a move towards the path that dropped from them to the railway.

Raut hung back. Their eyes met and saw a thousand things in a moment that their lips came near to say. Horrocks' hand tightened and then relaxed. He let go, and before Raut was aware of it, they were arm in arm, and walking, one unwillingly enough, down the path.

"You see the fine effect of the railway signals towards Burslem," said Horrocks, suddenly breaking into loquacity, striding fast and tightening the grip of his elbow the while. "Little green lights and red and white lights, all against the haze. You have an eye for effect, Raut. It's a fine effect. And look at those furnaces of mine, how they rise upon us as we come down the hill. That to the right is my pet—seventy feet of him. I packed him myself, and he's boiled away cheerfully with iron in his guts for five long years. I've a particular fancy for *him*. That line of red there—a lovely bit of warm orange you'd call it, Raut— that's the puddlers' furnaces, and there, in the hot light, three black figures—did you see the white splash of the steam-hammer then?—that's the rolling-mills. Come along! Clang, clatter, how it goes rattling across the floor! Sheet tin, Raut,—amazing stuff. Glass mirrors are not in it

when that stuff comes from the mill. And, squelch!—
there goes the hammer again. Come along!"

He had to stop talking to catch at his breath. His arm
twisted into Raut's with benumbing tightness. He had
come striding down the black path towards the railway as
though he was possessed. Raut had not spoken a word, had
simply hung back against Horrocks' pull with all his
strength.

"I say," he said now, laughing nervously, but with an
undertone of snarl in his voice, "why on earth are you
nipping my arm off, Horrocks, and dragging me along
like this?"

At length Horrocks released him. His manner changed
again. "Nipping your arm off?" he said. "Sorry. But it's
you taught me the trick of walking in that friendly way."

"You haven't learnt the refinements of it yet then," said
Raut, laughing artificially again. "By Jove! I'm black and
blue." Horrocks offered no apology. They stood now
near the bottom of the hill, close to the fence that bordered
the railway. The ironworks had grown larger and spread
out with their approach. They looked up to the blast
furnaces now instead of down; the further view of Etruria
and Hanley had dropped out of sight with their descent.
Before them, by the stile, rose a notice-board, bearing, still
dimly visible, the words, "BEWARE OF THE TRAINS," half
hidden by splashes of coaly mud.

"Fine effects," said Horrocks, waving his arm. "Here
comes a train. The puffs of smoke, the orange glare, the
round eye of light in front of it, the melodious rattle. Fine
effects! But these furnaces of mine used to be finer,
before we shoved cones in their throats, and saved the gas."

"How?" said Raut. "Cones?"

"Cones, my man, cones. I'll show you one nearer. The
flames used to flare out of the open throats, great—what
is it?—pillars of cloud by day, red and black smoke, and
pillars of fire by night. Now we run it off in pipes, and
burn it to heat the blast, and the top is shut by a cone.
You'll be interested in that cone."

"But every now and then," said Raut, "you get a burst of fire and smoke up there."

"The cone's not fixed, it's hung by a chain from a lever, and balanced by an equipoise. You shall see it nearer. Else, of course, there'd be no way of getting fuel into the thing. Every now and then the cone dips, and out comes the flare."

"I see," said Raut. He looked over his shoulder. "The moon gets brighter," he said.

"Come along," said Horrocks abruptly, gripping his shoulder again, and moving him suddenly towards the railway crossing. And then came one of those swift incidents, vivid, but so rapid that they leave one doubtful and reeling. Halfway across, Horrocks' hand suddenly clenched upon him like a vice, and swung him backward and through a half-turn, so that he looked up the line. And there a chain of lamp-lit carriage-windows telescoped swiftly as it came towards them, and the red and yellow lights of an engine grew larger and larger, rushing down upon them. As he grasped what this meant, he turned his face to Horrocks, and pushed with all his strength against the arm that held him back between the rails. The struggle did not last a moment. Just as certain as it was that Horrocks held him there, so certain was it that he had been violently lugged out of danger.

"Out of the way," said Horrocks, with a gasp, as the train came rattling by, and they stood panting by the gate into the ironworks.

"I did not see it coming," said Raut, still, even in spite of his own apprehensions, trying to keep up an appearance of ordinary intercourse.

Horrocks answered with a grunt. "The cone," he said, and then, as one who recovers himself, "I thought you did not hear."

"I didn't," said Raut.

"I wouldn't have had you run over then for the world," said Horrocks.

"For a moment I lost my nerve," said Raut.

Horrocks stood for a half a minute, then turned abruptly

towards the ironworks again. "See how fine these great mounds of mine, these clinker-heaps, look in the night! That truck yonder, up above there! Up it goes, and out-tilts the slag. See the palpitating red stuff go sliding down the slope. As we get nearer, the heap rises up and cuts the blast furnaces. See the quiver up above the big one. Not that way! This way, between the heaps. That goes to the puddling furnaces, but I want to show you the canal first." He came and took Raut by the elbow, and so they went along side by side. Raut answered Horrocks vaguely. What, he asked himself, had really happened on the line? Was he deluding himself with his own fancies, or had Horrocks actually held him back in the way of the train? Had he just been within an ace of being murdered?

Suppose this slouching, scowling monster *did* know anything? For a minute or two then Raut was really afraid for his life, but the mood passed as he reasoned with himself. After all, Horrocks might have heard nothing. At any rate, he pulled him out of the way in time. His odd manner might be due to the mere vague jealousy he had shown once before. He was talking now of the ash-heaps and the canal. "Eigh?" said Horrocks.

"What?" said Raut. "Rather! The haze in the moonlight. Fine!"

"Our canal," said Horrocks, stopping suddenly. "Our canal by moonlight and firelight is an immense effect. You've never seen it? Fancy that! You've spent too many of your evenings philandering up in Newcastle there. I tell you, for real florid effects— But you shall see. Boiling water . . ."

As they came out of the labyrinth of clinker-heaps and mounds of coal and ore, the noises of the rolling-mill sprang upon them suddenly, loud, near, and distinct. Three shadowy workmen went by and touched their caps to Horrocks. Their faces were vague in the darkness. Raut felt a futile impulse to address them, and before he could frame his words, they passed into the shadows. Horrocks pointed to the canal close before them now: a weird-looking place it seemed, in the blood-red reflections of the furnaces. The hot water that cooled the tuyères came into it, some fifty yards up—a tumultuous, almost boiling affluent, and the steam rose up from the water in silent white wisps and streaks, wrapping damply about them, an incessant succession of ghosts coming up from the black and red eddies, a white uprising that made the head swim. The shining black tower of the larger blast-furnace rose overhead out of the mist, and its tumultuous riot filled their ears. Raut kept away from the edge of the water, and watched Horrocks.

"Here it is red," said Horrocks, "blood-red vapour as red and hot as sin; but yonder there, where the moonlight falls on it, and it drives across the clinker-heaps, it is as white as death."

Raut turned his head for a moment, and then came back hastily to his watch on Horrocks. "Come along to the rolling-mills," said Horrocks. The threatening hold was not so evident that time, and Raut felt a little reassured. But all the same, what on earth did Horrocks mean about "white as death" and "red as sin"? Coincidence, perhaps?

They went and stood behind the puddlers for a little while, and then through the rolling-mills, where amidst an incessant din the deliberate steam-hammer beat the juice out of the succulent iron, and black, half-naked Titans rushed the plastic bars, like hot sealing-wax, between the wheels. "Come on," said Horrocks in Raut's ear, and they went and peeped through the little glass hole behind the tuyères, and saw the tumbled fire writhing in the pit of the blast-furnace. It left one eye blinded for a while. Then, with green and blue patches dancing across the dark, they went to the lift by which the trucks of ore and fuel and lime were raised to the top of the big cylinder.

And out upon the narrow rail that overhung the furnace, Raut's doubts came upon him again. Was it wise to be here? If Horrocks did know—everything! Do what he would, he could not resist a violent trembling. Right under foot was a sheer depth of seventy feet. It was a dangerous place. They pushed by a truck of fuel to get to the railing that crowned the place. The reek of the furnace, a sulphurous vapour streaked with pungent bitterness, seemed to make the distant hillside of Hanley quiver. The moon was riding out now from among a drift of clouds, half-way up the sky above the undulating wooded outlines of Newcastle. The steaming canal ran away from below them under an indistinct bridge, and vanished into the dim haze of the flat fields towards Burslem.

"That's the cone I've been telling you of," shouted Horrocks; "and, below that, sixty feet of fire and molten metal, with the air of the blast frothing through it like gas in soda-water."

Raut gripped the hand-rail tightly, and stared down at the cone. The heat was intense. The boiling of the iron and the tumult of the blast made a thunderous accompaniment to Horrocks' voice. But the thing had to be gone through now. Perhaps, after all . . .

"In the middle," bawled Horrocks, "temperature near a thousand degrees. If *you* were dropped into it . . . flash into flame like a pinch of gunpowder in a candle. Put your hand out and feel the heat of his breath. Why, even up here I've seen the rain-water boiling off the trucks. And that cone there. It's a damned sight too hot for roasting cakes. The top side of it's three hundred degrees."

"Three hundred degrees!" said Raut.

"Three hundred centigrade, mind!" said Horrocks. "It will boil the blood out of you in no time."

"Eigh?" said Raut, and turned.

"Boil the blood out of you in . . . No, you don't!"

"Let me go!" screamed Raut. "Let go my arm!"

With one hand he clutched at the hand-rail, then with both. For a moment the two men stood swaying. Then suddenly, with a violent jerk, Horrocks had twisted him from his hold. He clutched at Horrocks and missed, his foot went back into empty air; in mid-air he twisted himself, and then cheek and shoulder and knee struck the hot cone together.

He clutched the chain by which the cone hung, and the thing sank an infinitesimal amount as he struck it. A circle of glowing red appeared about him, and a tongue of flame, released from the chaos within, flickered up towards him. An intense pain assailed him at the knees, and he could smell the singeing of his hands. He raised himself to his feet, and tried to climb up the chain, and then something struck his head. Black and shining with the moonlight, the throat of the furnace rose about him.

Horrocks, he saw, stood above him by one of the trucks of fuel on the rail. The gesticulating figure was bright and white in the moonlight, and shouting, "Fizzle, you fool! Fizzle, you hunter of women! You hot-blooded hound! Boil! boil! boil!"

Suddenly he caught up a handful of coal out of the truck, and flung it deliberately, lump after lump, at Raut.

"Horrocks!" cried Raut. "Horrocks!"

He clung crying to the chain, pulling himself up from the burning of the cone. Each missile Horrocks flung hit him. His clothes charred and glowed, and as he struggled the cone dropped, and a rush of hot suffocating gas whooped out and burned round him in a swift breath of flame.

His human likeness departed from him. When the momentary red had passed, Horrocks saw a charred, blackened figure, its head streaked with blood, still clutching and fumbling with the chain, and writhing in agony—a cindery animal, an inhuman, monstrous creature that began a sobbing intermittent shriek.

Abruptly, at the sight, the ironmaster's anger passed. A deadly sickness came upon him. The heavy odour of burning flesh came drifting up to his nostrils. His sanity returned to him.

"God have mercy upon me!" he cried. "O God! what have I done?"

He knew the thing below him, save that it still moved and felt, was already a dead man—that the blood of the poor wretch must be boiling in his veins. An intense realisation of that agony came to his mind, and overcame every other feeling. For a moment he stood irresolute, and then, turning to the truck, he hastily tilted its contents upon the struggling thing that had once been a man. The mass fell with a thud, and went radiating over the cone. With the thud the shriek ended, and a boiling confusion of smoke, dust, and flame came rushing up towards him. As it passed, he saw the cone clear again.

Then he staggered back, and stood trembling, clinging to the rail with both hands. His lips moved, but no words came to them.

Down below was the sound of voices and running steps. The clangour of rolling in the shed ceased abruptly.

STORY THE ELEVENTH

The Purple Pileus

MR. COOMBES was sick of life. He walked away from his unhappy home, sick not only of his own existence, but of everybody else's, turned aside down Gaswork Lane to avoid the town, crossed the wooden bridge that goes over the canal to Starling's Cottages, and was presently alone in the damp pinewoods and out of sight and sound of human habitation. He would stand it no longer. He repeated aloud with blasphemies unusual to him that he would stand it no longer.

He was a pale-faced little man, with dark eyes and a fine and very black moustache. He had a very stiff, upright collar slightly frayed, that gave him an illusory double chin, and his overcoat (albeit shabby) was trimmed with astrachan. His gloves were a bright brown with black stripes over the knuckles, and split at the finger ends. His appearance, his wife had said once in the dear, dead days beyond recall,—before he married her, that is,—was military. But now she called him— It seems a dreadful thing to tell of between husband and wife, but she called him "a little grub." It wasn't the only thing she had called him, either.

The row had arisen about that beastly Jennie again. Jennie was his wife's friend, and by no invitation of Mr. Coombes she came in every blessed Sunday to dinner, and made a shindy all the afternoon. She was a big, noisy girl, with a taste for loud colours and a strident laugh; and this

470

Sunday she had outdone all her previous intrusions by bringing in a fellow with her, a chap as showy as herself. And Mr. Coombes, in a starchy, clean collar and his Sunday frock-coat, had sat dumb and wrathful at his own table, while his wife and her guests talked foolishly and undesirably, and laughed aloud. Well, he stood that, and after dinner (which, "as usual," was late), what must Miss Jennie do but go to the piano and play banjo tunes, for all the world as if it were a week-day! Flesh and blood could not endure such goings on. They would hear next door, they would hear in the road, it was a public announcement of their disrepute. He had to speak.

He had felt himself go pale, and a kind of rigour had affected his respiration as he delivered himself. He had been sitting on one of the chairs by the window—the new guest had taken possession of the arm-chair. He turned his head. "Sun Day!" he said over the collar, in the voice of one who warns. "Sun Day!" What people call a "nasty" tone it was.

Jennie had kept on playing, but his wife, who was looking through some music that was piled on the top of the piano, had stared at him. "What's wrong now?" she said; "can't people enjoy themselves?"

"I don't mind rational 'njoyment at all," said little Coombes, "but I ain't a-going to have week-day tunes playing on a Sunday in this house."

"What's wrong with my playing now?" said Jennie, stopping and twirling round on the music-stool with a monstrous rustle of flounces.

Coombes saw it was going to be a row, and opened too vigorously, as is common with your timid, nervous men all the world over. "Steady on with that music-stool!" said he; "it ain't made for 'eavy weights."

"Never you mind about weights," said Jennie, incensed. "What was you saying behind my back about my playing?"

"Surely you don't 'old with not having a bit of music on a Sunday, Mr. Coombes?" said the new guest, leaning back in the arm-chair, blowing a cloud of cigarette smoke and smiling in a kind of pitying way. And simultaneously his wife said something to Jennie about "Never mind 'im. You go on, Jinny."

"I do," said Mr. Coombes, addressing the new guest.

"May I arst why?" said the new guest, evidently enjoying both his cigarette and the prospect of an argument. He was, by the bye, a lank young man, very stylishly dressed in bright drab, with a white cravat and a pearl and silver pin. It had been better taste to come in a black coat, Mr. Coombes thought.

"Because," began Mr. Coombes, "it don't suit me. I'm a business man. I 'ave to study my connection. Rational 'njoyment"—

"His connection!" said Mrs. Coombes scornfully. "That's what he's always a-saying. We got to do this, and we got to do that"—

"If you don't mean to study my connection," said Mr. Coombes, "what did you marry me for?"

"I wonder," said Jennie, and turned back to the piano.

"I never saw such a man as you," said Mrs. Coombes. "You've altered all round since we were married. Before"—

Then Jennie began at the tum, tum, tum again.

"Look here!" said Mr. Coombes, driven at last to revolt, standing up and raising his voice. "I tell you I won't have that." The frock-coat heaved with his indignation.

"No vi'lence, now," said the long young man in drab, sitting up.

"Who the juice are you?" said Mr. Coombes fiercely.

Whereupon they all began talking at once. The new guest said he was Jennie's "intended," and meant to protect her, and Mr. Coombes said he was welcome to do so anywhere but in his (Mr. Coombes') house; and Mrs. Coombes said he ought to be ashamed of insulting his guests, and (as I have already mentioned) that he was getting a regular little grub; and the end was, that Mr. Coombes ordered his visitors out of the house, and they wouldn't go, and so he said he would go himself. With his face burning and tears of excitement in his eyes, he went into the passage, and as he struggled with his overcoat—his frock-coat sleeves got concertinaed up his arm—and gave a brush at his silk hat, Jennie began again at the piano, and strummed him insultingly out of the house. Tum, tum, tum. He slammed the shop door so that the house quivered. That, briefly, was the immediate making of his mood. You will perhaps begin to understand his disgust with existence.

As he walked along the muddy path under the firs,—it was late October, and the ditches and heaps of fir needles were gorgeous with clumps of fungi,—he recapitulated the melancholy history of his marriage. It was brief and commonplace enough. He now perceived with sufficient clearness that his wife had married him out of a natural curiosity and in order to escape from her worrying, laborious, and uncertain life in the workroom; and, like the majority of her class, she was far too stupid to realise that it was her duty to co-operate with him in his business. She was greedy of enjoyment, loquacious, and socially-minded, and evidently disappointed to find the restraints of poverty still hanging about her. His worries exasperated her, and the slightest attempt to control her proceedings resulted in a charge of "grumbling." Why couldn't he be nice—as he used to be? And Coombes was such a harmless little man, too, nourished mentally on *Self-Help,* and with a meagre ambition of self-denial and competition that was to end in a "sufficiency." Then Jennie came in as a female Mephistopheles, a gabbling chronicle of "fellers," and was always wanting his wife to go to theatres and "all that." And in addition were aunts of his wife, and cousins (male and female), to eat up capital, insult him personally, upset business arrangements, annoy good customers, and generally blight his life. It was not the first occasion by many that Mr. Coombes had fled his home in wrath and indignation and something like fear, vowing furiously and even aloud that he wouldn't stand it, and so frothing away his energy along the line of least resistance. But never before had he been quite so sick of life as on this particular Sunday afternoon. The Sunday dinner may have had its share in his despair—and the greyness of the sky. Perhaps, too, he was beginning to realise his unendurable frustration as a business man as the consequence of his marriage. Presently bankruptcy, and after that— Perhaps she might have reason to repent when it was too late. And destiny, as I have already intimated, had planted the path through the wood with evil-smelling fungi, thickly and variously planted it, not only on the right side but on the left.

A small shopman is in such a melancholy position if his wife turns out a disloyal partner. His capital is all tied up in his business, and to leave her, means to join the unemployed in some strange part of the earth. The luxuries of divorce are beyond him altogether. So that the good old tradition of marriage for better or worse holds inexorably for him, and things work up to tragic culminations. Bricklayers kick their wives to death, and dukes betray theirs; but it is among the small clerks and shopkeepers nowadays that it comes most often to a cutting of throats. Under the circumstances it is not so very remarkable—and you must take it as charitably as you can—that the mind of Mr. Coombes ran for a while on some such glorious close to his disappointed hopes, and that he thought of razors, pistols, bread-knives, and touching letters to the coroner denouncing his enemies by name, and praying piously for forgiveness. After a time his fierceness gave way to melancholia. He had been married in this very overcoat, in his first and only frock-coat that was buttoned up beneath it. He began to recall their courting along this very walk, his years of penurious saving to get capital, and the bright hopefulness of his marrying days. For it all to work out like this! Was there no sympathetic ruler anywhere in the world? He reverted to death as a topic.

He thought of the canal he had just crossed, and doubted whether he shouldn't stand with his head out, even in the middle, and it was while drowning was in his mind that the purple pileus caught his eye. He looked at it mechanically for a moment, and stopped and stooped towards it to pick it up, under the impression that it was some such small leather object as a purse. Then he saw that it was the purple top of a fungus, a peculiarly poisonous-looking purple: slimy, shiny, and emitting a sour odour.   He hesitated with his hand an inch or so from it, and the thought of poison crossed his mind. With that he picked the thing, and stood up again with it in his hand.

The odour was certainly strong—acrid, but by no means disgusting. He broke off a piece, and the fresh surface was a creamy white, that changed like magic in the space of ten seconds to a yellowsh-green colour. It was even an inviting-looking change. He broke off two other pieces to see it repeated. They were wonderful things these fungi, thought Mr. Coombes, and all of them the deadliest poisons, as his father had often told him. Deadly poisons!

There is no time like the present for a rash resolve. Why not here and now? thought Mr. Coombes. He tasted a little piece, a very little piece indeed—a mere crumb. It was so pungent that he almost spat it out again, then merely hot and full-flavoured. A kind of German mustard with a touch of horse-radish and—well, mushroom. He swallowed it in the excitement of the moment. Did he like it or did he not? His mind was curiously careless. He would try another bit. It really wasn't bad—it was good. He forgot his troubles in the interest of the immediate moment. Playing with death it was. He took another bite, and then deliberately finished a mouthful. A curious tingling sensation began in his finger-tips and toes. His pulse began to move faster. The blood in his ears sounded like a mill-race. "Try bi' more," said Mr. Coombes. He turned and looked about him, and found his feet unsteady. He saw and struggled towards a little patch of purple a dozen yards away. "Jol' goo' stuff," said Mr. Coombes. "E—lomore ye'." He pitched forward and fell on his face, his hands outstretched towards the cluster of pilei. But he did not eat any more of them. He forgot forthwith.

He rolled over and sat up with a look of astonishment on his face. His carefully brushed silk hat had rolled away towards the ditch. He pressed his hand to his brow. Something had happened, but he could not rightly determine what it was. Anyhow, he was no longer dull—he felt bright, cheerful. And his throat was afire. He laughed in the sudden gaiety of his heart. Had he been dull? He did not know; but at anyrate he would be dull no longer. He got up and stood unsteadily, regarding the universe with an agreeable smile. He began to remember. He could not remember very well, because of a steam roundabout that was beginning in his head. And he knew he had been disagreeable at home, just because they wanted to be happy. They were quite right; life should be as gay as possible. He would go home and make it up, and reassure them. And why not take some of this delightful toadstool with him, for them to eat? A hatful, no less. Some of those red ones with white spots as well, and a few yellow. He had been a dull dog, an enemy to merriment; he would make up for it. It would be gay to turn his coat-sleeves inside out, and stick some yellow gorse into his waistcoat pockets. Then home —singing—for a jolly evening.

After the departure of Mr. Coombes, Jennie discontinued playing, and turned round on the music-stool again. "What a fuss about nothing," said Jennie.

"You see, Mr. Clarence, what I've got to put up with," said Mrs. Coombes.

"He is a bit hasty," said Mr. Clarence judicially.

"He ain't got the slightest sense of our position," said Mrs. Coombes; "that's what I complain of. He cares for nothing but his old shop; and if I have a bit of company, or buy anything to keep myself decent, or get any little thing I want out of the housekeeping money, there's disagreeables. 'Economy,' he says; 'struggle for life,' and all that. He lies awake of nights about it, worrying how he can screw me out of a shilling. He wanted us to eat Dorset butter once. If once I was to give in to him—there!"

"Of course," said Jennie.

"If a man values a woman," said Mr. Clarence, lounging back in the arm-chair, "he must be prepared to make sacrifices for her. For my own part," said Mr. Clarence, with his eye on Jennie, "I shouldn't think of marrying till I was in a position to do the thing in style. It's downright selfishness. A man ought to go through the rough-and-tumble by himself, and not drag her"——

"I don't agree altogether with that," said Jennie. "I don't see why a man shouldn't have a woman's help, provided he doesn't treat her meanly, you know. It's meanness"——

"You wouldn't believe," said Mrs. Coombes. "But I was a fool to 'ave 'im. I might 'ave known. If it 'adn't been for my father, we shouldn't have had not a carriage to our wedding."

"Lord! he didn't stick out at that?" said Mr. Clarence, quite shocked.

"Said he wanted the money for his stock, or some such rubbish. Why, he wouldn't have a woman in to help me once a week if it wasn't for my standing out plucky. And the fusses he makes about money—comes to me, well, pretty near crying, with sheets of paper and figgers. 'If only we can tide over this year,' he says, 'the business is bound to go.' 'If only we can tide over this year.' I says; 'then it'll be, if only we can tide over next year. I know you,' I says. 'And you don't catch me screwing myself lean and ugly. Why didn't you marry a slavey?' I says, 'if you wanted one —instead of a respectable girl,' I says."

So Mrs. Coombes. But we will not follow this unedifying conversation further. Suffice it that Mr. Coombes was very satisfactorily disposed of, and they had a snug little time round the fire. Then Mrs. Coombes went to get the tea, and Jennie sat coquettishly on the arm of Mr. Clarence's chair until the tea-things clattered outside. "What was that I heard?" asked Mrs. Coombes playfully, as she entered, and there was badinage about kissing. They were just sitting down to the little circular table when the first intimation of Mr. Coombes' return was heard.

This was a fumbling at the latch of the front door.

"'Ere's my lord," said Mrs. Coombes. "Went out like a lion and comes back like a lamb, I'll lay."

Something fell over in the shop: a chair, it sounded like. Then there was a sound as of some complicated step exercise in the passage. Then the door opened and Coombes appeared. But it was Coombes transfigured. The immaculate collar had been torn carelessly from his throat. His carefully-brushed silk hat, half-full of a crush of fungi, was under one arm; his coat was inside out, and his waistcoat adorned with bunches of yellow-blossomed furze. These little eccentricities of Sunday costume, however, were quite overshadowed by the change in his face; it was livid white, his eyes were unnaturally large and bright, and his pale blue lips were drawn back in a cheerless grin. "Merry!" he said. He had stopped dancing to open the door. "Rational 'njoyment. Dance." He made three fantastic steps into the room, and stood bowing.

"Jim!" shrieked Mrs. Coombes, and Mr. Clarence sat petrified, with a dropping lower jaw.

"Tea," said Mr. Coombes. "Jol' thing, tea. Tose-stools, too. Brosher."

"He's drunk," said Jennie in a weak voice. Never before had she seen this intense pallor in a drunken man, or such shining, dilated eyes.

Mr. Coombes held out a handful of scarlet agaric to Mr. Clarence. "Jo' stuff," said he; "ta' some."

At that moment he was genial. Then at the sight of their startled faces he changed, with the swift transition of insanity, into overbearing fury. And it seemed as if he had suddenly recalled the quarrel of his departure. In such a huge voice as Mrs. Coombes had never heard before, he shouted, "My house. I'm master 'ere. Eat what I give yer!" He bawled this, as it seemed, without an effort, without a violent gesture, standing there as motionless as one who whispers, holding out a handful of fungus.

Clarence approved himself a coward. He could not meet the mad fury in Coombes' eyes; he rose to his feet, pushing back his chair, and turned, stooping. At that Coombes rushed at him. Jennie saw her opportunity, and, with the ghost of a shriek, made for the door. Mrs. Coombes followed her. Clarence tried to dodge. Over went the tea-table with a smash as Coombes clutched him by the collar and tried to thrust the fungus into his mouth. Clarence was content to leave his collar behind him, and shot out into the passage with red patches of fly agaric still adherent to his face. "Shut 'im in!" cried Mrs. Coombes, and would have closed the door, but her supports deserted her; Jennie saw the shop door open, and vanished thereby, locking it behind her, while Clarence went on hastily into the kitchen. Mr. Coombes came heavily against the door, and Mrs. Coombes, finding the key was inside, fled upstairs and locked herself in the spare bedroom.

So the new convert to *joie de vivre* emerged upon the passage, his decorations a little scattered, but that respectable hatful of fungi still under his arm. He hesitated at the three ways, and decided on the kitchen. Whereupon Clarence, who was fumbling with the key, gave up the attempt to imprison his host, and fled into the scullery, only to be captured before he could open the door into the yard. Mr. Clarence is singularly reticent of the details of what occurred. It seems that Mr. Coombes' transitory irritation had vanished again, and he was once more a genial play-fellow. And as there were knives and meat choppers about, Clarence very generously resolved to humour him and so avoid anything tragic. It is beyond dispute that Mr. Coombes played with Mr. Clarence to his heart's content; they could not have been more playful and familiar if they had known each other for years. He insisted gaily on Clarence trying the fungi, and after a friendly tussle, was smitten with remorse at the mess he was making of his guest's face. It also appears that Clarence was dragged under the sink and his face scrubbed with the blacking brush,—he being still resolved to humour the lunatic at any cost,—and that finally, in a somewhat dishevelled, chipped, and discoloured condition, he was assisted to his coat and shown out by the back door, the shopway being barred by Jennie. Mr. Coombes' wandering thoughts then turned to Jennie. Jennie had been unable to unfasten the shop door, but she shot the bolts against Mr. Coombes' latch-key, and remained in possession of the shop for the rest of the evening.

It would appear that Mr. Coombes then returned to the kitchen, still in pursuit of gaiety, and, albeit a strict Good Templar, drank (or spilt down the front of the first and only frock-coat) no less than five bottles of the stout Mrs.

Coombes insisted upon having for her health's sake. He made cheerful noises by breaking off the necks of the bottles with several of his wife's wedding-present dinner-plates, and during the earlier part of this great drunk he sang divers merry ballads. He cut his finger rather badly with one of the bottles,—the only bloodshed in this story,—and what with that, and the systematic convulsion of his inexperienced physiology by the liquorish brand of Mrs. Coombes' stout, it may be the evil of the fungus poison was somehow allayed. But we prefer to draw a veil over the concluding incidents of this Sunday afternoon. They ended in the coal cellar, in a deep and healing sleep.

An interval of five years elapsed. Again it was a Sunday afternoon in October, and again Mr. Coombes walked through the pinewood beyond the canal. He was still the same dark-eyed, black-moustached little man that he was at the outset of the story, but his double chin was now scarcely so illusory as it had been. His overcoat was new, with a velvet lapel, and a stylish collar with turn-down corners, free of any coarse starchiness, had replaced the original all-round article. His hat was glossy, his gloves newish—though one finger had split and been carefully mended. And a casual observer would have noticed about him a certain rectitude of bearing, a certain erectness of head that marks the man who thinks well of himself. He was a master now, with three assistants. Beside him walked a larger sunburnt parody of himself, his brother Tom, just back from Australia. They were recapitulating their early struggles, and Mr. Coombes had just been making a financial statement.

"It's a very nice little business, Jim," said brother Tom. "In these days of competition you're jolly lucky to have worked it up so. And you're jolly lucky, too, to have a wife who's willing to help like yours does."

"Between ourselves," said Mr. Coombes, "it wasn't always so. It wasn't always like this. To begin with, the missus was a bit giddy. Girls are funny creatures."

"Dear me!"

"Yes. You'd hardly think it, but she was downright extravagant, and always having slaps at me. I was a bit too easy and loving, and all that, and she thought the whole blessed show was run for her. Turned the 'ouse into a regular caravansery, always having her relations and girls from business in, and their chaps. Comic songs a' Sunday, it was getting to, and driving trade away. And she was making eyes at the chaps, too! I tell you, Tom, the place wasn't my own."

"Shouldn't 'a' thought it."

"It was so. Well—I reasoned with her. I said, 'I ain't a duke, to keep a wife like a pet animal. I married you for 'elp and company,' I said. 'You got to 'elp and pull the business through.' She wouldn't 'ear of it. 'Very well,' I says; 'I'm a mild man till I'm roused,' I says, 'and it's getting to that.' But she wouldn't 'ear of no warnings."

"Well?"

"It's the way with women. She didn't think I 'ad it in me to be roused. Women of her sort (between ourselves, Tom) don't respect a man until they're a bit afraid of him. So I just broke out to show her. In comes a girl named Jennie, that used to work with her, and her chap. We 'ad a bit of a row, and I came out 'ere—it was just such another day as this—and I thought it all out. Then I went back and pitched into them."

"You did?"

"I did. I was mad, I can tell you. I wasn't going to 'it 'er, if I could 'elp it, so I went back and licked into this chap, just to show 'er what I could do. 'E was a big chap, too. Well, I chucked him, and smashed things about, and gave 'er a scaring, and she ran up and locked 'erself into the spare room."

"Well?"

"That's all. I says to 'er the next morning, 'Now you know,' I says, 'what I'm like when I'm roused.' And I didn't 'ave to say anything more."

"And you've been happy ever after, eh?"

"So to speak. There's nothing like putting your foot down with them. If it 'adn't been for that afternoon I

Q

should 'a' been tramping the roads now, and she'd 'a' been grumbling at me, and all her family grumbling for bringing her to poverty—I know their little ways. But we're all right now. And it's a very decent little business, as you say."

They proceeded on their way meditatively. "Women are funny creatures," said brother Tom.

"They want a firm hand," says Coombes.

"What a lot of these funguses there are about here!" remarked brother Tom presently. "I can't see what use they are in the world."

Mr. Coombes looked. "I dessay they're sent for some wise purpose," said Mr. Coombes.

And that was as much thanks as the purple pileus ever got for maddening this absurd little man to the pitch of decisive action, and so altering the whole course of his life.

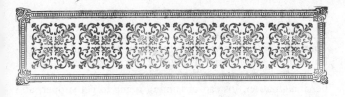

STORY THE TWELFTH

The Jilting of Jane

As I sit writing in my study, I can hear our Jane bumping her way downstairs with a brush and dustpan. She used in the old days to sing hymn tunes, or the British national song for the time being, to these instruments, but latterly she has been silent and even careful over her work. Time was when I prayed with fervour for such silence, and my wife with sighs for such care, but now they have come we are not so glad as we might have anticipated we should be. Indeed, I would rejoice secretly, though it may be unmanly weakness to admit it, even to hear Jane sing "Daisy," or by the fracture of any plate but one of Euphemia's best green ones, to learn that the period of brooding has come to an end.

Yet how we longed to hear the last of Jane's young man before we heard the last of him! Jane was always very free with her conversation to my wife, and discoursed admirably in the kitchen on a variety of topics—so well, indeed, that I sometimes left my study door open—our house is a small one—to partake of it. But after William came, it was always William, nothing but William; William this and William that; and when we thought William was worked out and exhausted altogether, then William all over again. The engagement lasted altogether three years; yet how she got introduced to William, and so became thus saturated with him, was always a secret. For my part, I believe it was at the street corner where the Rev. Barnabas Baux used

483

to hold an open-air service after evensong on Sundays. Young Cupids were wont to flit like moths round the paraffin flare of that centre of High Church hymn-singing. I fancy she stood singing hymns there, out of memory and her imagination, instead of coming home to get supper, and William came up beside her and said, "Hello!" "Hello yourself!" she said; and, etiquette being satisfied, they proceeded to converse.

As Euphemia has a reprehensible way of letting her servants talk to her, she soon heard of him. "He is *such* a respectable young man, ma'am," said Jane, "you don't know." Ignoring the slur cast on her acquaintance, my wife inquired further about this William.

"He is second porter at Maynard's, the draper's," said Jane, "and gets eighteen shillings—nearly a pound—a week, m'm; and when the head porter leaves he will be head porter. His relatives are quite superior people, m'm. Not labouring people at all. His father was a greengrosher, m'm, and had a chumor, and he was bankrup' twice. And one of his sisters is in a Home for the Dying. It will be a very good match for me, m'm," said Jane, "me being an orphan girl."

"Then you are engaged to him?" asked my wife.

"Not engaged, ma'am; but he is saving money to buy a ring—hammyfist."

"Well, Jane, when you are properly engaged to him you may ask him round here on Sunday afternoons, and have tea with him in the kitchen." For my Euphemia has a motherly conception of her duty towards her maid-servants. And presently the amethystine ring was being worn about the house, even with ostentation, and Jane developed a new way of bringing in the joint, so that this gage was evident. The elder Miss Maitland was aggrieved by it, and told my wife that servants ought not to wear rings. But my wife looked it up in *Enquire Within* and *Mrs. Motherly's Book of Household Management,* and found no prohibition. So Jane remained with this happiness added to her love.

The treasure of Jane's heart appeared to me to be what respectable people call a very deserving young man.

"William, ma'am," said Jane one day suddenly, with ill-concealed complacency, as she counted out the beer bottles, "William, ma'am, is a teetotaller. Yes, m'm; and he don't smoke. Smoking, ma'am," said Jane, as one who reads the heart, "*do* make such a dust about. Beside the waste of money. *And* the smell. However, I suppose it's necessary to some."

Possibly it dawned on Jane that she was reflecting a little severely upon Euphemia's comparative ill-fortune, and she added kindly, "I'm sure the master is a hangel when his pipe's alight. Compared to other times."

William was at first a rather shabby young man of the ready-made black coat school of costume. He had watery grey eyes, and a complexion appropriate to the brother of one in a Home for the Dying. Euphemia did not fancy him very much, even at the beginning. His eminent respectability was vouched for by an alpaca umbrella, from which he never allowed himself to be parted.

"He goes to chapel," said Jane. "His papa, ma'am"——

"His *what*, Jane?"

"His papa, ma'am, was Church; but Mr. Maynard is a Plymouth Brother, and William thinks it Policy, ma'am, to go there too. Mr. Maynard comes and talks to him quite friendly, when they ain't busy, about using up all the ends of string, and about his soul. He takes a lot of notice, do Maynard, of William, and the way he saves string and his soul, ma'am."

Presently we heard that the head porter at Maynard's had left, and that William was head porter at twenty-three shillings a week. "He is really kind of over the man who drives the van," said Jane, "and him married with three children." And she promised in the pride of her heart to make interest for us with William to favour us so that we might get our parcels of drapery from Maynard's with exceptional promptitude.

After this promotion a rapidly increasing prosperity came upon Jane's young man. One day, we learned that Mr. Maynard had given William a book. "Smiles' Elp Yourself, it's called," said Jane; "but it ain't comic. It tells you how to get on in the world, and some what William read to me was *lovely*, ma'am."

Euphemia told me of this laughing, and then she became suddenly grave. "Do you know, dear," she said, "Jane said one thing I did not like. She had been quiet for a minute, and then she suddenly remarked, 'William is a lot above me, ma'am, ain't he?'"

"I don't see anything in that," I said, though later my eyes were to be opened.

One Sunday afternoon about that time I was sitting at my writing-desk—possibly I was reading a good book—when a something went by the window. I heard a startled exclamation behind me, and saw Euphemia with her hands clasped together and her eyes dilated. "George," she said in an awestricken whisper, "did you see?"

Then we both spoke to one another at the same moment, slowly and solemnly: *"A silk hat! Yellow gloves! A new umbrella!"*

"It may be my fancy, dear," said Euphemia; "but his tie was very like yours. I believe Jane keeps him in ties. She told me a little while ago in a way that implied volumes about the rest of your costume, 'The master *do* wear pretty ties, ma'am.' And he echoes all your novelties."

The young couple passed our window again on their way to their customary walk. They were arm in arm. Jane looked exquisitely proud, happy, and uncomfortable, with new white cotton gloves, and William, in the silk hat, singularly genteel!

That was the culmination of Jane's happiness when she returned, "Ma. Maynard has been talking to William, ma'am," she said, "and he is to serve customers, just like the young shop gentlemen, during the next sale. And if he gets on, he is to be made an assistant, ma'am, at the first opportunity. He has got to be as gentlemanly as he can, ma'am; and if he ain't, ma'am, he says it won't be for want of trying. Mr. Maynard has took a great fancy to him."

"He *is* getting on, Jane," said my wife.

"Yes, ma'am," said Jane thoughtfully, "he *is* getting on." And she sighed.

That next Sunday, as I drank my tea, I interrogated my wife. "How is this Sunday different from all other Sundays, little woman? What has happened? Have you altered the curtains, or rearranged the furniture, or where is the indefinable difference of it? Are you wearing your hair in a new way without warning me? I clearly perceive a change in my environment, and I cannot for the life of me say what it is."

Then my wife answered in her most tragic voice, "George," she said, "that—that William has not come near the place to-day! And Jane is crying her heart out upstairs."

There followed a period of silence. Jane, as I have said, stopped singing about the house, and began to care for our brittle possessions, which struck my wife as being a very sad sign indeed. The next Sunday, and the next, Jane asked to go out, "to walk with William," and my wife, who never attempts to extort confidences, gave her permission, and asked no questions. On each occasion Jane came back looking flushed and very determined. At last one day she became communicative.

"William is being led away," she remarked abruptly, with a catching of the breath, apropos of tablecloths. "Yes, ma'am. She is a milliner, and she can play on the piano."

"I thought," said my wife, "that you went out with him on Sunday."

"Not out with him, m'm—after him. I walked along by the side of them, and told her he was engaged to me."

"Dear me, Jane, did you? What did they do?"

"Took no more notice of me than if I was dirt. So I told her she should suffer for it."

"It could not have been a very agreeable walk, Jane."

"Not for no parties, ma'am."

"I wish," said Jane, "I could play the piano, ma'am. But anyhow, I don't mean to let *her* get him away from me. She's older than him, and her hair ain't gold to the roots, ma'am."

It was on the August Bank Holiday that the crisis came. We do not clearly know the details of the fray, but only

such fragments as poor Jane let fall. She came home dusty,
excited, and with her heart hot within her.

The milliner's mother, the milliner, and William had
made a party to the Art Museum at South Kensington, I
think. Anyhow, Jane had calmly but firmly accosted them
somewhere in the streets, and asserted her right to what, in
spite of the consensus of literature, she held to be her
inalienable property. She did, I think, go so far as to lay
hands on him. They dealt with her in a crushingly superior
way. They "called a cab." There was a "scene," William
being pulled away into the four-wheeler by his future wife
and mother-in-law from the reluctant hands of our discarded Jane. There were threats of giving her "in charge."

"My poor Jane!" said my wife, mincing veal as though
she was mincing William. "It's a shame of them. I would
think no more of him. He is not worthy of you."

"No, m'm," said Jane. "He *is* weak."

"But it's that woman has done it," said Jane. She was
never known to bring herself to pronounce "that woman's"
name or to admit her girlishness. "I can't think what minds
some women must have—to try and get a girl's young man
away from her. But there, it only hurts to talk about it,"
said Jane.

Thereafter our house rested from William. But there was
something in the manner of Jane's scrubbing the front
doorstep or sweeping out the rooms, a certain viciousness,
that persuaded me that the story had not yet ended.

"Please, m'm, may I go and see a wedding to-morrow?"
said Jane one day.

My wife knew by instinct whose wedding. "Do you
think it wise, Jane?" she said.

"I would like to see the last of him," said Jane.

"My dear," said my wife, fluttering into my room about
twenty minutes after Jane had started, "Jane has been to
the boot-hole and taken all the left-off boots and shoes, and
gone off to the wedding with them in a bag. Surely she
cannot mean"——

"Jane," I said, "is developing character. Let us hope for
the best."

Jane came back with a pale, hard face. All the boots seemed to be still in her bag, at which my wife heaved a premature sigh of relief. We heard her go upstairs and replace the boots with considerable emphasis.

"Quite a crowd at the wedding, ma'am," she said presently, in a purely conversational style, sitting in our little kitchen, and scrubbing the potatoes; "and such a lovely day for them." She proceeded to numerous other details, clearly avoiding some cardinal incident.

"It was all extremely respectable and nice, ma'am, but *her* father didn't wear a black coat, and looked quite out of place, ma'am. Mr. Piddingquirk"——

"*Who?*"

"Mr. Piddingquirk—William that *was*, ma'am—had white gloves, and a coat like a clergyman, and a lovely chrysanthemum. He looked so nice, ma'am. And there was red carpet down, just like for gentlefolks. And they say he gave the clerk four shillings, ma'am. It was a real kerridge they had—not a fly. When they came out of church there was rice-throwing, and her two little sisters dropping dead flowers. And someone threw a slipper, and then I threw a boot"——

"Threw a *boot*, Jane!"

"Yes, ma'am. Aimed at *her*. But it hit *him*. Yes, ma'am, hard. Gev him a black eye, I should think. I only threw that one. I hadn't the heart to try again. All the little boys cheered when it hit him."

After an interval—"I am sorry the boot hit *him*."

Another pause. The potatoes were being scrubbed violently. "He always *was* a bit above me, you know, ma'am. And he was led away."

The potatoes were more than finished. Jane rose sharply, with a sigh, and rapped the basin down on the table.

"I don't care," she said. "I don't care a rap. He will find out his mistake yet. It serves me right. I was stuck up about him. I ought not to have looked so high. And I am glad things are as things are."

My wife was in the kitchen, seeing to the cookery. After the confession of the boot-throwing, she must have watched

poor Jane fuming with a certain dismay in those brown
eyes of hers.   But I imagine they softened again very
quickly, and then Jane's must have met them.

"Oh, ma'am," said Jane, with an astonishing change of
note, "think of all that *might* have been! Oh, ma'am, I
could have been so happy! I ought to have known, but I
didn't know . . . You're very kind to let me talk to you,
ma'am . . . for it's hard on me ma'am . . . it's har-r-r-r-d"——

And I gather that Euphemia so far forgot herself as to let
Jane sob out some of the fulness of her heart on a sympathetic shoulder. My Euphemia, thank Heaven, has never
properly grasped the importance of "keeping up her position." And since that fit of weeping, much of the accent of
bitterness has gone out of Jane's scrubbing and brush-work.

Indeed, something passed the other day with the butcher-
boy—but that scarcely belongs to this story. However, Jane
is young still, and time and change are at work with her.
We all have our sorrows, but I do not believe very much in
the existence of sorrows that never heal.

STORY THE
THIRTEENTH

In the Modern Vein: an Unsympathetic Love Story

OF course the cultivated reader has heard of Aubrey Vair. He has published on three several occasions volumes of delicate verses,—some, indeed, border on indelicacy,—and his column, "Of Things Literary" in the *Climax*, is well known. His Byronic visage and an interview have appeared in the *Perfect Lady*. It was Aubrey Vair, I believe, who demonstrated that the humour of Dickens was worse than his sentiment, and who detected "a subtle bourgeois flavour" in Shakespeare. However, it is not generally known that Aubrey Vair has had erotic experiences as well as erotic inspirations. He adopted Goethe some little time since as his literary prototype, and that may have had something to do with his temporary lapse from sexual integrity.

For it is one of the commonest things that undermine literary men, giving us landslips and picturesque effects along the otherwise even cliff of their respectable life, ranking next to avarice, and certainly above drink, this instability called genius, or, more fully, the consciousness of genius, such as Aubrey Vair possessed. Since Shelley set the fashion, your man of gifts has been assured that his duty to himself and his duty to his wife are incompatible, and his renunciation of the Philistine has been marked by such infidelity as his means and courage warranted. Most virtue is lack of imagination. At any rate, a minor genius without his affections twisted into an inextricable muddle, and who did not occasionally shed sonnets over his troubles, I have never met.

Even Aubrey Vair did this, weeping the sonnets overnight into his blotting-book, and pretending to write literary *causerie* when his wife came down in her bath slippers to see what kept him up. She did not understand him, of course. He did this even before the other woman appeared, so ingrained is conjugal treachery in the talented mind. Indeed, he wrote more sonnets before the other woman came than after that event, because thereafter he spent much of his leisure in cutting down the old productions, retrimming them, and generally altering this readymade clothing of his passion to suit her particular height and complexion.

Aubrey Vair lived in a little red villa with a lawn at the back and a view of the Downs behind Reigate. He lived upon discreet investment eked out by literary work. His wife was handsome, sweet, and gentle, and—such is the tender humility of good married women—she found her life's happiness in seeing that little Aubrey Vair had well-cooked variety for dinner, and that their house was the neatest and brightest of all the houses they entered. Aubrey Vair enjoyed the dinners, and was proud of the house, yet nevertheless he mourned because his genius dwindled. Moreover, he grew plump, and corpulence threatened him.

We learn in suffering what we teach in song, and Aubrey Vair knew certainly that his soul could give no creditable crops unless his affections were harrowed. And how to harrow them was the trouble, for Reigate is a moral neighbourhood.

So Aubrey Vair's romantic longings blew loose for a time, much as a seedling creeper might, planted in the midst of a flower-bed. But at last, in the fulness of time, the other woman came to the embrace of Aubrey Vair's yearning heart-tendrils, and his romantic episode proceeded as is here faithfully written down.

The other woman was really a girl, and Aubrey Vair met his first at a tennis party at Redhill. Aubrey Vair did not play tennis after the accident to Miss Morton's eye, and because latterly it made him pant and get warmer and moister than even a poet should be; and this young lady had only recently arrived in England, and could not play. So they gravitated into the two vacant basket chairs beside Mrs. Bayne's deaf aunt, in front of the hollyhocks, and were presently talking at their ease together.

The other woman's name was unpropitious,—Miss Smith, —but you would never have suspected it from her face and costume. Her parentage was promising, she was an orphan, her mother was a Hindoo, and her father an Indian civil servant; and Aubrey Vair—himself a happy mixture of Kelt and Teuton, as, indeed, all literary men have to be nowadays—naturally believed in the literary consequences of a mixture of races. She was dressed in white. She had finely moulded pale features, great depth of expression, and a cloud of delicately *frisé* black hair over her dark eyes, and shy, that contrasted admirably with the stereotyped frankness of your common Reigate girl.

"This is a splendid lawn—the best in Redhill," said Aubrey Vair in the course of the conversation; "and I like it all the better because the daisies are spared." He indicated the daisies with a graceful sweep of his rather elegant hand.

"They are sweet little flowers," said the lady in white, "and I have always associated them with England, chiefly, perhaps, through a picture I saw 'over there' when I was very little, of children making daisy chains. I promised myself that pleasure when I came home. But, alas! I feel now rather too large for such delights."

"I do not see why we should not be able to enjoy these simple pleasures as we grow older—why our growth should have in it so much forgetting. For my own part"——

"Has your wife got Jane's recipe for stuffing trout?" asked Mrs. Bayne's deaf aunt abruptly.

"I really don't know," said Aubrey Vair.

"That's all right," said Mrs. Bayne's deaf aunt. "It ought to please even you."

"Anything will please me," said Aubrey Vair; "I care very little"——

"Oh, it's a lovely dish," said Mrs. Bayne's deaf aunt, and relapsed into contemplation.

"I was saying," said Aubrey Vair, "that I think I still find my keenest pleasures in childish pastimes. I have a little nephew that I see a great deal of, and when we fly kites together, I am sure it would be hard to tell which of us is the happier. By the bye, you should get at your daisy chains in that way. Beguile some little girl."

"But I did. I took that Morton mite for a walk in the meadows, and timidly broached the subject. And she reproached me for suggesting 'frivolous pursuits.' It was a horrible disappointment."

"The governess here," said Aubrey Vair, "is robbing that child of its youth in a terrible way. What will a life be that has no childhood at the beginning?"

"Some human beings are never young," he continued, "and they never grow up. They lead absolutely colourless lives. They are—they are etiolated. They never love, and never feel the loss of it. They are—for the moment I can think of no better image—they are human flower-pots, in which no soul has been planted. But a human soul properly growing must begin in a fresh childishness."

"Yes," said the dark lady thoughtfully, "a careless childhood, running wild almost. That should be the beginning."

"Then we pass through the wonder and diffidence of youth."

"To strength and action," said the dark lady. Her dreamy eyes were fixed on the Downs, and her fingers tightened on her knees as she spoke. "Ah, it is a grand thing to live—as a man does—self-reliant and free."

"And so at last," said Aubrey Vair, "come to the culmination and crown of life." He paused and glanced hastily at her. Then he dropped his voice almost to a whisper—"And the culmination of life is love."

Their eyes met for a moment, but she looked away at once. Aubrey Vair felt a peculiar thrill and a catching in his breath, but his emotions were too complex for analysis. He had a certain sense of surprise, also, at the way his conversation had developed.

Mrs. Bayne's deaf aunt suddenly dug him in the chest with her ear-trumpet, and someone at tennis bawled, "Love all!"

"Did I tell you Jane's girls have had scarlet fever?" asked Mrs. Bayne's deaf aunt.

"No," said Aubrey Vair.

"Yes; and they are peeling now," said Mrs. Bayne's deaf aunt, shutting her lips tightly, and nodding in a slow, significant manner at both of them.

There was a pause. All three seemed lost in thought, too deep for words.

"Love," began Aubrey Vair presently, in a severely philosophical tone, leaning back in his chair, holding his hands like a praying saint's in front of him, and staring at the toe of his shoe,—"love is, I believe, the one true and real thing in life. It rises above reason, interest, or explanation. Yet I never read of an age when it was so much forgotten as it is now. Never was love expected to run so much in appointed channels, never was it so despised, checked, ordered, and obstructed. Policemen say, 'This way, Eros!' As a result, we relieve our emotional possibilities in the hunt for gold and notoriety. And after all, with the best fortune in these, we only hold up the gilded images of our success, and are weary slaves, with unsatisfied hearts, in the pageant of life."

Aubrey Vair sighed, and there was a pause. The girl looked at him out of the mysterious darkness of her eyes. She had read many books, but Aubrey Vair was her first literary man, and she took this kind of thing for genius—as girls have done before.

"We are," continued Aubrey Vair, conscious of a favourable impression,—"we are like fireworks, mere dead, inert things until the appointed spark comes; and then—if it is not damp—the dormant soul blazes forth in all its warmth and beauty. That is living. I sometimes think, do you know, that we should be happier if we could die soon after that golden time, like the Ephemerides. There is a decay sets in."

"Eigh?" said Mrs. Bayne's deaf aunt startlingly. "I didn't hear you."

"I was on the point of remarking," shouted Aubrey Vair, wheeling the array of his thoughts,—"I was on the point of remarking that few people in Redhill could match Mrs. Morton's fine broad green."

"Others have noticed it," Mrs. Bayne's deaf aunt shouted back. "It is since she has had in her new false teeth."

This interruption dislocated the conversation a little. However—

"I must thank you, Mr. Vair," said the dark girl, when they parted that afternoon, "for having given me very much to think about."

And from her manner, Aubrey Vair perceived clearly he had not wasted his time.

It would require a subtler pen than mine to tell how from that day a passion for Miss Smith grew like Jonah's gourd in the heart of Aubrey Vair. He became pensive, and in the prolonged absence of Miss Smith, irritable. Mrs. Aubrey Vair felt the change in him, and put it down to a vitriolic Saturday Reviewer. Indisputably the *Saturday* does at times go a little far. He re-read *Elective Affinities,* and lent it to Miss Smith. Incredible as it may appear to members of the Areopagus Club, where we know Aubrey Vair, he did also beyond all question inspire a sort of passion in that sombre-eyed, rather clever, and really very beautiful girl.

He talked to her a lot about love and destiny, and all that bric-à-brac of the minor poet. And they talked together about his genius. He elaborately, though discreetly, sought her society, and presented and read to her the milder of his unpublished sonnets. We consider his Byronic features pasty, but the feminine mind has its own laws. I suppose, also where a girl is not a fool, a literary man has an enormous advantage over anyone but a preacher, in the show he can make of his heart's wares.

At last a day in that summer came when he met her alone, possibly by chance, in a quiet lane towards Horley.

There were ample hedges on either side, rich with honey-suckle, vetch, and mullein.

They conversed intimately of his poetic ambitions, and then he read her those verses of his subsequently published in *Hobson's Magazine*: "Tenderly ever, since I have met thee." He had written these the day before; and though I think the sentiment is uncommonly trite, there is a redeeming note of sincerity about the lines not conspicuous in all Aubrey Vair's poetry.

He read rather well, and a swell of genuine emotion crept into his voice as he read, with one white hand thrown out to point the rhythm of the lines. "Ever, my sweet, for thee," he concluded, looking up into her face.

Before he looked up, he had been thinking chiefly of his poem and its effect. Straightway he forgot it. Her arms hung limply before her, and her hands were clasped to-gether. Her eyes were very tender.

"Your verses go to the heart," she said softly.

Her mobile features were capable of wonderful shades of expression. He suddenly forgot his wife and his position as a minor poet as he looked at her. It is possible that his classical features may themselves have undergone a certain transfiguration. For one brief moment—and it was always to linger in his memory—destiny lifted him out of his vain little self to a nobler level of simplicity. The copy of "Tenderly ever" fluttered from his hand. Considerations vanished. Only one thing seemed of importance.

"I love you," he said abruptly.

An expression of fear came into her eyes. The grip of her hands upon one another tightened convulsively. She became very pale.

Then she moved her lips as if to speak, bringing her face slightly nearer to his. There was nothing in the world at that moment for either of them but one another. They were both trembling exceedingly. In a whisper she said, "You love me?"

Aubrey Vair stood quivering and speechless, looking into her eyes. He had never seen such a light as he saw there before. He was in a wild tumult of emotion. He was dreadfully scared at what he had done. He could not say another word. He nodded.

"And this has come to me?" she said presently, in the same awe-stricken whisper, and then, "Oh, my love, my love!"

And thereupon Aubrey Vair had her clasped to himself, her cheek upon his shoulder and his lips to hers.

Thus it was that Aubrey Vair came by the cardinal memory of his life. To this day it recurs in his works.

A little boy clambering in the hedge some way down the lane saw this group with surprise, and then with scorn and contempt. Reckoning nothing of his destiny, he turned away feeling that he at least could never come to the unspeakable unmanliness of hugging girls. Unhappily for Reigate scandal, his shame for his sex was altogether too deep for words.

An hour after, Aubrey Vair returned home in a hushed mood. There were muffins after his own heart for his tea—Mrs. Aubrey Vair had had hers. And there were chrysanthemums, chiefly white ones,—flowers he loved,—set out in the china bowl he was wont to praise. And his wife came behind him to kiss him as he sat eating.

"De lill Jummuns," she remarked, kissing him under the ear.

Then it came into the mind of Aubrey Vair with startling clearness, while his ear was being kissed, and with his mouth full of muffin, that life is a singularly complex thing.

The summer passed at last into the harvest-time, and the leaves began falling. It was evening, the warm sunset light still touched the Downs, but up the valley a blue haze was creeping. One or two lamps in Reigate were already alight.

About half-way up the slanting road that scales the Downs, there is a wooden seat where one may obtain a fine view of the red villas scattered below, and of the succession of blue hills beyond. Here the girl with the shadowy face was sitting.

She had a book on her knees, but it lay neglected. She was leaning forward, her chin resting upon her hand. She was looking across the valley into the darkening sky, with troubled eyes.

Aubrey Vair appeared through the hazel-bushes, and sat down beside her. He held half a dozen dead leaves in his hand.

She did not alter her attitude. "Well?" she said.

"Is it to be flight?" he asked.

Aubrey Vair was rather pale. He had been having bad nights latterly, with dreams of the Continental Express, Mrs. Aubrey Vair possibly even in pursuit,—he always fancied her making the tragedy ridiculous by tearfully bringing additional pairs of socks, and any such trifles he had forgotten, with her,—all Reigate and Redhill in commotion. He had never eloped before, and he had visions of difficulties with hotel proprietors. Mrs. Aubrey Vair might telegraph ahead. Even he had had a prophetic vision of a headline in a halfpenny evening newspaper: "Young Lady abducts a Minor Poet." So there was a quaver in his voice as he asked, "Is it to be flight?"

"As you will," she answered, still not looking at him.

"I want you to consider particularly how this will affect you. A man," said Aubrey Vair, slowly, and staring hard at the leaves in his hand, "even gains a certain éclat in these affairs. But to a woman it is ruin—social, moral."

"This is not love," said the girl in white.

"Ah, my dearest! Think of yourself."

"Stupid!" she said, under her breath.

"You spoke?"

"Nothing."

"But cannot we go on, meeting one another, loving one another, without any great scandal or misery? Could we not"——

"That," interrupted Miss Smith, "would be unspeakably horrible."

"This is a dreadful conversation to me. Life is so intricate, such a web of subtle strands binds us this way and that. I cannot tell what is right. You must consider"——

"A man would break such strands."

"There is no manliness," said Aubrey Vair, with a sudden glow of moral exaltation, "in doing wrong. My love"——

"We could at least die together, dearest," she said.

"Good Lord!" said Aubrey Vair. "I mean—consider my wife."

"You have not considered her hitherto."

"There is a flavour—of cowardice, of desertion, about suicide," said Aubrey Vair. "Frankly, I have the English prejudice, and do not like any kind of running away."

Miss Smith smiled very faintly. "I see clearly now what I did not see. My love and yours are very different things."

"Possibly it is a sexual difference," said Aubrey Vair; and then, feeling the remark inadequate, he relapsed into silence.

They sat for some time without a word. The two lights in Reigate below multiplied to a score of bright points, and, above, one star had become visible. She began laughing, an almost noiseless, hysterical laugh that jarred unaccountably upon Aubrey Vair.

Presently she stood up. "They will wonder where I am," she said. "I think I must be going."

He followed her to the road. "Then this is the end?" he said, with a curious mixture of relief and poignant regret.

"Yes, this is the end," she answered, and turned away.

There straightway dropped into the soul of Aubrey Vair a sense of infinite loss. It was an altogether new sensation. She was perhaps twenty yards away, when he groaned aloud with the weight of it, and suddenly began running after her with his arms extended.

"Annie," he cried,—"Annie! I have been talking *rot*. Annie, now I know I love you! I cannot spare you. This must not be. I did not understand."

The weight was horrible.

"Oh, stop, Annie!" he cried, with a breaking voice, and there were tears on his face.

She turned upon him suddenly, and his arms fell by his side. His expression changed at the sight of her pale face.

"You do not understand," she said. "I have said good-bye."

She looked at him; he was evidently greatly distressed, a little out of breath, and he had just stopped blubbering. His contemptible quality reached the pathetic. She came up close to him, and, taking his damp Byronic visage between her hands, she kissed him again and again. "Good-bye, little man that I loved," she said; "and good-bye to this folly of love."

Then, with something that may have been a laugh or a sob,—she herself, when she came to write it all in her novel, did not know which,—she turned and hurried away again, and went out of the path that Aubrey Vair must pursue, at the cross-roads.

Aubrey Vair stood, where she had kissed him, with a mind as inactive as his body, until her white dress had disappeared. Then he gave an involuntary sigh, a large exhaustive expiration, and so awoke himself, and began walking, pensively dragging his feet through the dead leaves, home. Emotions are terrible things.

"Do you like the potatoes, dear?" asked Mrs. Aubrey Vair at dinner. "I cooked them myself."

Aubrey Vair descended slowly from cloudy, impalpable meditations to the level of fried potatoes. "These potatoes" —he remarked, after a pause during which he was struggling with recollection. "Yes. These potatoes have exactly the tints of the dead leaves of the hazel."

"What a fanciful poet it is!" said Mrs Aubrey Vair. "Taste them. They are very nice potatoes indeed."

STORY THE
FOURTEENTH

A Catastrophe

The little shop was not paying. The realisation came insensibly. Winslow was not the man for definite addition and subtraction and sudden discovery. He became aware of the truth in his mind gradually, as though it had always been there. A lot of facts had converged and led him there. There was that line of cretonnes—four half-pieces—untouched, save for half a yard sold to cover a stool. There were those shirtings at 4¾d.—Bandersnatch, in the Broadway, was selling them at 2¾d.—under cost, in fact. (Surely Bandersnatch might let a man live!) Those servants' caps a selling line, needed replenishing, and that brought back the memory of Winslow's sole wholesale dealers, Helter Skelter, & Grab. Why! how about their account?

Winslow stood with a big green box open on the counter before him when he thought of it. His pale grey eyes grew a little rounder; his pale, straggling moustache twitched. He had been drifting along, day after day. He went round to the ramshackle cash-desk in the corner—it was Winslow's weakness to sell his goods over the counter, give his customers a duplicate bill, and then dodge into the desk to receive the money, as though he doubted his own honesty. His lank forefinger, with the prominent joints, ran down the bright little calendar ("Clack's Cottons last for All Time"). "One—two—three; three weeks an' a day!" said Winslow

502

staring. "March! Only three weeks and a day. It *can't* be."

"Tea, dear," said Mrs. Winslow, opening the door with the glass window and the white blind that communicated with the parlour.

"One minute," said Winslow, and began unlocking the desk.

An irritable old gentleman, very hot and red about the face, and in a heavy fur-lined coat, came in noisily. Mrs. Winslow vanished.

"Ugh!" said the old gentleman. "Pocket-handkerchief."

"Yes, sir," said Winslow. "About what price"——

"Ugh!" said the old gentleman. "Poggit-handkerchief, quig!"

Winslow began to feel flustered. He produced two boxes.

"These sir"——began Winslow.

"Sheed tin!" said the old gentleman, clutching the stiffness of the linen. "Wad to blow my nose—not haggit about."

"A cotton one, p'raps, sir?" said Winslow.

"How much?" said the old gentleman over the handkerchief.

"Sevenpence, sir. There's nothing more I can show you? No ties, braces——?"

"Damn!" said the old gentleman, fumbling in his ticket-pocket, and finally producing half a crown. Winslow looked round for his metallic duplicate-book which he kept in various fixtures, according to circumstances, and then he caught the old gentleman's eye. He went straight to the desk at once and got the change, with an entire disregard of the routine of the shop.

Winslow was always more or less excited by a customer. But the open desk reminded him of his trouble. It did not come back to him all at once. He heard a finger-nail softly tapping on the glass, and, looking up, saw Minnie's eyes over the blind. It seemed like retreat opening. He shut and locked the desk, and went into the back room to tea.

But he was preoccupied. Three weeks and a day! He

took unusually large bites of his bread and butter, and stared hard at the little pot of jam. He answered Minnie's conversational advances distractedly. The shadow of Helter, Skelter, & Grab lay upon the tea-table. He was struggling with this new idea of failure, the tangible realisation that was taking shape and substance, condensing, as it were, out of the misty uneasiness of many days. At present it was simply one concrete fact; there were thirty-nine pounds left in the bank, and that day three weeks Messrs. Helter, Skelter, & Grab, those enterprising outfitters of young men, would demand their eighty pounds.

After tea there was a customer or so—small purchases: some muslin and buckram, dress-protectors, tape, and a pair of Lisle hose. Then, knowing that Black Care was lurking in the dusky corners of the shop, he lit the three lamps early and set to, refolding his cotton prints, the most vigorous and least meditative proceeding of which he could think. He could see Minnie's shadow in the other room as she moved about the table. She was busy turning an old dress. He had a walk after supper, looked in at the Y.M.C.A., but found no one to talk to, and finally went to bed. Minnie was already there. And there, too, waiting for him, nudging him gently, until about midnight he was hopelessly awake, sat Black Care.

He had had one or two nights lately in that company, but this was much worse. First came Messrs. Helter, Skelter, and Grab, and their demand for eighty pounds—an enormous sum when your original capital was only a hundred and seventy. They camped, as it were, before him, sat down and beleaguered him. He clutched feebly at the circumambient darkness for expedients. Suppose he had a sale, sold things for almost anything? He tried to imagine a sale miraculously successful in some unexpected manner, and mildly profitable, in spite of reductions below cost. Then Bandersnatch Limited, 101, 102, 103, 105, 106, 107 Broadway, joined the siege, a long caterpillar of frontage, a battery of shop fronts, wherein things were sold at a farthing above cost. How could he fight such an establishment? Besides, what had he to sell? He began to review his resources. What taking line was there to bait the sale? Then straightway came those pieces of cretonne, yellow and black, with a bluish-green flower; those discredited skirtings, prints without buoyancy, skirmishing haberdashery, some despairful four-button gloves by an inferior maker—a hopeless crew. And that was his force against Bandersnatch, Helter, Skelter, & Grab, and the pitiless world behind them. Whatever had made him think a mortal would buy such things? Why had he bought this and neglected that? He suddenly realised the intensity of his hatred for Helter, Skelter, & Grab's salesman. Then he drove towards an agony of self-reproach. He had spent too much on that cashdesk. What real need was there of a desk? He saw his vanity of that desk in a lurid glow of self-discovery. And the lamps? Five pounds! Then suddenly, with what was almost physical pain, he remembered the rent.

He groaned and turned over. And there, dim in the darkness, was the hummock of Mrs. Winslow's shoulder. That set him off in another direction. He became acutely sensible of Minnie's want of feeling. Here he was, worried to death about business, and she sleeping like a little child. He regretted having married, with that infinite bitterness that only comes to the human heart in the small hours of the morning. That hummock of white seemed absolutely without helpfulness, a burden, a responsibility. What fools men were to marry! Minnie's inert repose irritated him so much that he was almost provoked to wake her up and tell her that they were "Ruined." She would have to go back to her uncle; her uncle had always been against him: and as for his own future, Winslow was exceedingly uncertain. A shop assistant who has once set up for himself finds the utmost difficulty in getting into a situation again. He began to figure himself "crib-hunting" once more, going from this wholesale house to that, writing innumerable letters. How he hated writing letters! "Sir,—Referring to your advertisement in the *Christian World*." He beheld an infinite vista of discomfort and disappointment, ending—in a gulf.

He dressed, yawning, and went down to open the shop.

He felt tired before the day began. As he carried the shutters in, he kept asking himself what good he was doing. The end was inevitable, whether he bothered or not. The clear daylight smote into the place, and showed how old and rough and splintered was the floor, how shabby the second-hand counter, how hopeless the whole enterprise. He had been dreaming these past six months of a bright shop, of a happy couple, of a modest but comely profit flowing in. He had suddenly awakened from his dream. The braid that bound his decent black coat—it was a trifle loose—caught against the catch of the shop door, and was torn away. This suddenly turned his wretchedness to wrath. He stood quivering for a moment, then, with a spiteful clutch, tore the braid looser, and went in to Minnie.

"Here," he said, with infinite reproach; "look here! You might look after a chap a bit."

"I didn't see it was torn," said Minnie.

"You never do," said Winslow, with gross injustice, "until things are too late."

Minnie looked suddenly at his face. "I'll sew it now, Sid, if you like."

"Let's have breakfast first," said Winslow, "and do things at their proper time."

He was preoccupied at breakfast, and Minnie watched him anxiously. His only remark was to declare his egg a bad one. It wasn't; it was flavoury,—being one of those at fifteen a shilling,—but quite nice. He pushed it away from him, and then, having eaten a slice of bread and butter, admitted himself in the wrong by resuming the egg.

"Sid," said Minnie, as he stood up to go into the shop again, "you're not well."

"I'm *well* enough." He looked at her as though he hated her.

"Then there's something else the matter. You aren't angry with me, Sid, are you, about that braid? Do tell me what's the matter. You were just like this at tea yesterday, and at supper-time. It wasn't the braid then."

"And I'm likely to be."

She looked interrogation. "Oh, what *is* the matter?" she said.

It was too good a chance to miss, and he brought the evil news out with dramatic force. "Matter?" he said. "I done my best, and here we are. That's the matter! If I can't pay Helter, Skelter, & Grab eighty pounds, this day three weeks"—— Pause. "We shall be sold up! Sold up! That's the matter, Min! SOLD UP!"

"Oh, Sid!" began Minnie.

He slammed the door. For the moment he felt relieved of at least half his misery. He began dusting boxes that did not require dusting, and then reblocked a cretonne already faultlessly blocked. He was in a state of grim wretchedness; a martyr under the harrow of fate. At any-rate, it should not be said he failed for want of industry. And how he had planned and contrived and worked! All to this end! He felt horrible doubts. Providence and Bandersnatch—surely they were incompatible! Perhaps he was being "tried"? That sent him off upon a new tack, a very comforting one. The martyr pose, the gold-in-the-furnace attitude, lasted all the morning.

At dinner—"potato pie"—he looked up suddenly, and saw Minnie's face regarding him. Pale she looked, and a little red about the eyes. Something caught him suddenly with a queer effect upon his throat. All his thoughts seemed to wheel round into quite a new direction.

He pushed back his plate and stared at her blankly. Then he got up, went round the table to her—she staring at him. He dropped on his knees beside her without a word. "Oh, Minnie!" he said, and suddenly she knew it was peace, and put her arms about him, as he began to sob and weep.

He cried like a little boy, slobbering on her shoulder that he was a knave to have married her and brought her to this, that he hadn't the wits to be trusted with a penny, that it was all his fault; that he *"had* hoped *so"*—ending in a howl. And she, crying gently herself, patting his shoulders, said "S*sh!*" softly to his noisy weeping, and so soothed the out-break. Then suddenly the crazy bell upon the shop door began, and Winslow had to jump to his feet, and be a man again.

After that scene they "talked it over" at tea, at supper, in bed, at every possible interval in between, solemnly—quite inconclusively—with set faces and eyes for the most part staring in front of them—and yet with a certain mutual comfort. "What to do I don't know," was Winslow's main proposition. Minnie tried to take a cheerful view of service —with a probable baby. But she found she needed all her courage. And her uncle would help her again, perhaps, just at the critical time. It didn't do for folks to be too proud. Besides, "something might happen," a favourite formula with her.

One hopeful line was to anticipate a sudden afflux of customers. "Perhaps," said Minnie, "you might get together fifty. They know you well enough to trust you a bit." They debated that point. Once the possibility of Helter, Skelter, & Grab giving credit was admitted, it was pleasant to begin sweating the acceptable minimum. For some half-hour over tea the second day after Winslow's discoveries they were quite cheerful again, laughing even at their terrific fears. Even twenty pounds to go on with might be considered enough. Then in some mysterious way the pleasant prospect of Messrs. Helter, Skelter, & Grab tempering the wind to the shorn retailer vanished—vanished absolutely, and Winslow found himself again in the pit of despair.

He began looking about at the furniture, and wondering idly what it would fetch. The chiffonier was good, anyhow, and there were Minnie's old plates that her mother used to have. Then he began to think of desperate expedients for putting off the evil day. He had heard somewhere of Bills of Sale—there was to his ears something comfortingly substantial in the phrase. Then, why not "Go to the Money-Lenders"?

One cheering thing happened on Thursday afternoon a little girl came in with a pattern of "print," and he was able to match it. He had not been able to match anything out of his meagre stock before. He went in and told

Minnie. The incident is mentioned lest the reader should imagine it was uniform despair with him.

The next morning, and the next, after the discovery, Winslow opened shop late. When one has been awake most of the night, and has no hope, what *is* the good of getting up punctually? But as he went into the dark shop on Friday he saw something lying on the floor, something lit by the bright light that came under the ill-fitting door—a black oblong. He stooped and picked up an envelope with a deep mourning edge. It was addressed to his wife. Clearly a death in her family—perhaps her uncle. He knew the man too well to have expectations. And they would have to get mourning and go to the funeral. The brutal cruelty of people dying! He saw it all in a flash—he always visualised his thoughts. Black trousers to get, black crape, black gloves—none in stock—the railway fares, the shop closed for the day.

"I'm afraid there's bad news, Minnie," he said.

She was kneeling before the fireplace, blowing the fire. She had her housemaid's gloves on and the old country sun-bonnet she wore of a morning, to keep the dust out of her hair. She turned, saw the envelope, gave a gasp, and pressed two bloodless lips together.

"I'm afraid it's uncle," she said, holding the letter and staring with eyes wide open into Winslow's face. *"It's a strange hand!"*

"The postmark's Hull," said Winslow.

"The postmark's Hull."

Minnie opened the letter slowly, drew it out, hesitated, turned it over, saw the signature. "It's Mr. Speight!"

"What does he say?" said Winslow.

Minnie began to read. *"Oh!"* she screamed. She dropped the letter, collapsed into a crouching heap, her hands covering her eyes. Winslow snatched at it. "A most terrible accident has occurred," he read; "Melchior's chimney fell down yesterday evening right on the top of your uncle's house, and every living soul was killed—your uncle, your cousin Mary, Will and Ned, and the girl—every one of them, and smashed—you would hardly know them. I'm writing to you to break the news before you see it in the papers"—— The letter fluttered from Winslow's fingers. He put out his hand against the mantel to steady himself.

All of them dead! Then he saw, as in a vision, a row of seven cottages, each let at seven shillings a week, a timber yard, two villas, and the ruins—still marketable—of the avuncular residence. He tried to feel a sense of loss and could not. They were sure to have been left to Minnie's aunt. All dead! $7 \times 7 \times 52 \div 20$ began insensibly to work itself out in his mind, but discipline was ever weak in his mental arithmetic; figures kept moving from one line to another, like children playing at Widdy, Widdy Way. Was it two hundred pounds about—or one hundred pounds? Presently he picked up the letter again, and finished reading it. "You being the next of kin," said Mr. Speight.

"How *awful!*" said Minnie in a horror-struck whisper, and looking up at last. Winslow stared back at her, shaking his head solemnly. There were a thousand things running through his mind, but none that, even to his dull sense, seemed appropriate as a remark. "It was the Lord's will," he said at last.

"It seems so very, very terrible," said Minnie; "auntie, dear auntie—Ted—poor, dear uncle"——

"It was the Lord's will, Minnie," said Winslow, with infinite feeling. A long silence.

"Yes," said Minnie, very slowly, staring thoughtfully at the crackling black paper in the grate. The fire had gone out. "Yes, perhaps it was the Lord's will."

They looked gravely at one another. Each would have been terribly shocked at any mention of the property by the other. She turned to the dark fireplace and began tearing up an old newspaper slowly. Whatever our losses may be, the world's work still waits for us. Winslow gave a deep sigh and walked in a hushed manner towards the front door. As he opened it, a flood of sunlight came streaming into the dark shadows of the closed shop. Bandersnatch, Helter, Skelter, & Grab, had vanished out of his mind like the mists before the rising sun.

Presently he was carrying in the shutters, and in the

briskest way, the fire in the kitchen was crackling exhilarat-
ingly, with a little saucepan walloping above it, for Minnie
was boiling two eggs,—one for herself this morning, as well
as one for him,—and Minnie herself was audible, laying
breakfast with the greatest *éclat*. The blow was a sudden
and terrible one—but it behoves us to face such things
bravely in this sad, unaccountable world. It was quite
midday before either of them mentioned the cottages.

STORY THE FIFTEENTH

The Lost Inheritance

"My uncle," said the man with the glass eye, "was what you might call a hemi-semi-demi millionaire. He was worth about a hundred and twenty thousand. Quite. And he left me all his money."

I glanced at the shiny sleeve of his coat, and my eye travelled up to the frayed collar.

"Every penny," said the man with the glass eye, and I caught the active pupil looking at me with a touch of offence.

"I've never had any windfalls like that," I said, trying to speak enviously and propitiate him.

"Even a legacy isn't always a blessing," he remarked with a sigh, and with an air of philosophical resignation he put the red nose and the wiry moustache into his tankard for a space.

"Perhaps not," I said.

"He was an author, you see, and he wrote a lot of books."

"Indeed!"

"That was the trouble of it all." He stared at me with the available eye to see if I grasped his statement, then averted his face a little and produced a toothpick.

"You see," he said, smacking his lips after a pause, "it was like this. He was my uncle—my maternal uncle. And he had—what shall I call it?—a weakness for writing edifying literature. Weakness is hardly the word—downright mania is nearer the mark. He'd been librarian in a Polytechnic, and as soon as the money came to him he began to indulge his ambition. It's a simply extraordinary and incomprehensible thing to me. Here was a man of thirty-seven suddenly dropped into a perfect pile of gold, and he didn't go—not a day's bust on it. One would think a chap would go and get himself dressed a bit decent—say a couple of dozen pair of trousers at a West End tailor's; but he never did. You'd hardly believe it, but when he died he hadn't even a gold watch. It seems wrong for people like that to have money. All he did was just to take a house, and order in pretty nearly five tons of books and ink and paper, and set to writing edifying literature as hard as ever he could write. I *can't* understand it! But he did. The money came to him, curiously enough, through a maternal uncle of *his,* unexpected like, when he was seven-and-thirty. My mother, it happened, was his only relation in the wide, wide world, except some second cousins of his. And I was her only son. You follow all that? The second cousins had one only son, too, but they brought him to see the old man too soon. He was rather a spoilt youngster, was this son of theirs, and directly he set eyes on my uncle, he began bawling out as hard as he could. 'Take 'im away—er,' he says, 'take 'im away,' and so did for himself entirely. It was pretty straight sailing, you'd think, for me, eh? And my mother, being a sensible, careful woman, settled the business in her own mind long before he did.

"He was a curious little chap, was my uncle, as I remember him. I don't wonder at the kid being scared. Hair just like these Japanese dolls they sell, black and straight and stiff all round the brim and none in the middle, and below, a whitish kind of face and rather large dark grey eyes moving about behind his spectacles. He used to attach a great deal of importance to dress, and always wore a flapping overcoat and a big-brimmed felt hat of a most extraordinary size. He looked a rummy little beggar, I can tell you. Indoors it was, as a rule, a dirty red flannel dressing-gown and a black skullcap he had. That black skull-cap made him look like the portraits of all kinds of celebrated people. He was always moving about from house to house, was my uncle, with his

R

chair which had belonged to Savage Landor, and his two writing-tables, one of Carlyle's and the other of Shelley's, so the dealer told him, and the completest portable reference library in England, he said he had—and he lugged the whole caravan, now to a house at Down, near Darwin's old place, then to Reigate, near Meredith, then off to Haslemere, then back to Chelsea for a bit, and then up to Hampstead. He knew there was something wrong with his stuff, but he never knew there was anything wrong with his brains. It was always the air, or the water, or the altitude, or some tommy-rot like that. 'So much depends on environment,' he used to say, and stare at you hard, as if he half suspected you were hiding a grin at him somewhere under your face. 'So much depends on environment to a sensitive mind like mine.'

"What was his name? You wouldn't know it if I told you. He wrote nothing that anyone has ever read—nothing. No one *could* read it. He wanted to be a great teacher, he said, and he didn't know what he wanted to teach any more than a child. So he just blethered at large about Truth and Righteousness, and the Spirit of History, and all that. Book after book he wrote and published at his own expense. He wasn't quite right in his head, you know, really; and to hear him go on at the critics—not because they slated him, mind you—he liked that—but because they didn't take any notice of him at all. 'What do the nations want?' he would ask, holding out his brown old claw. 'Why, teaching—guidance! They are scattered upon the hills like sheep without a shepherd. There is War and Rumours of War, the unlaid Spirit of Discord abroad in the land, Nihilism, Vivisection, Vaccination, Drunkenness, Penury, Want, Socialistic Error, Selfish Capital! Do you see the clouds, Ted?'—My name, you know—'Do you see the clouds lowering over the land? and behind it all—the Mongol waits!' He was always very great on Mongols, and the Spectre of Socialism, and such-like things.

"Then out would come his finger at me, and with his eyes all afire and his skull-cap askew, he would whisper: 'And here am I. What do I want? Nations to teach.

Nations! I say it with all modesty, Ted, I *could*. I would guide them; nay! but I *will* guide them to a safe haven, to the land of Righteousness, flowing with milk and honey.'

"That's how he used to go on. Ramble, rave about the nations, and righteousness, and that kind of thing. Kind of mincemeat of Bible and blethers. From fourteen up to three-and-twenty, when I might have been improving my mind, my mother used to wash me and brush my hair (at least in the earlier years of it), with a nice parting down the middle, and take me, once or twice a week, to hear this old lunatic jabber about things he had read of in the morning papers, trying to do it as much like Carlyle as he could, and I used to sit according to instructions, and look intelligent and nice, and pretend to be taking it all in. Afterwards I used to go of my own free will, out of a regard for the legacy. I was the only person that used to go and see him. He wrote, I believe, to every man who made the slightest stir in the world, sending him a copy or so of his books, and inviting him to come and talk about the nations to him; but half of them didn't answer, and none ever came. And when the girl let you in—she was an artful bit of goods, that girl—there were heaps of letters on the hall-seat waiting to go off, addressed to Prince Bismarck, the President of the United States, and such-like people. And one went up the staircase and along the cobwebby passage,—the housekeeper drank like fury, and his passages were always cobwebby,—and found him at last, with books turned down all over the room, and heaps of torn paper on the floor, and telegrams and newspapers littered about, and empty coffee-cups and half-eaten bits of toast on the desk and the mantel. You'd see his back humped up, and his hair would be sticking out quite straight between the collar of that dressing-gown thing and the edge of the skull-cap.

"'A moment!' he would say. 'A moment!' over his shoulder. 'The *mot juste*, you know, Ted, *le mot juste*. Righteous thought righteously expressed—Aah!—concatenation. And now, Ted,' he'd say, spinning round in his study chair, 'how's Young England?' That was his silly name for me.

"Well, that was my uncle, and that was how he talked—to me, at any rate. With others about he seemed a bit shy. And he not only talked to me, but he gave me his books, books of six hundred pages or so, with cock-eyed headings, 'The Shrieking Sisterhood,' 'The Behemoth of Bigotry,' 'Crucibles and Cullenders,' and so on. All very strong, and none of them original. The very last time but one that I saw him he gave me a book. He was feeling ill even then, and his hand shook and he was despondent. I noticed it because I was naturally on the look-out for those little symptoms. 'My last book, Ted,' he said. 'My last book, my boy; my last word to the deaf and hardened nations;' and I'm hanged if a tear didn't go rolling down his yellow old cheek. He was regular crying because it was so nearly over, and he hadn't only written about fifty-three books of rubbish. 'I've sometimes thought, Ted'—he said, and stopped.

" 'Perhaps I've been a bit hasty and angry with this stiff-necked generation. A little more sweetness, perhaps, and a little less blinding light. I've sometimes thought—I might have swayed them. But I've done my best, Ted.'

"And then, with a burst, for the first and last time in his life he owned himself a failure. It showed he was really ill. He seemed to think for a minute, and then he spoke quietly and low, as sane and sober as I am now. 'I've been a fool, Ted,' he said. 'I've been flapping nonsense all my life. Only He who readeth the heart knows whether this is anything more than vanity. Ted, I don't. But He knows, He knows, and if I have done foolishly and vainly, in my heart—in my heart'—

"Just like that he spoke, repeating himself, and he stopped quite short and handed the book to me, trembling. Then the old shine came back into his eye. I remember it all fairly well, because I repeated it and acted it to my old mother when I got home, to cheer her up a bit. 'Take this book and read it,' he said. 'It's my last word, my very last word. I've left all my property to you, Ted, and may you use it better than I have done.' And then he fell a-coughing.

"I remember that quite well even now, and how I went home cock-a-hoop, and how he was in bed the next time I called. The housekeeper was downstairs drunk, and I fooled about—as a young man will—with the girl in the passage before I went to him. He was sinking fast. But even then his vanity clung to him.

" 'Have you read it?' he whispered.

" 'Sat up all night reading it,' I said in his ear to cheer him. 'It's the last,' said I, and then, with a memory of some poetry or other in my head, 'but it's the bravest and best.'

"He smiled a little and tried to squeeze my hand as a woman might do, and left off squeezing in the middle, and lay still. 'The bravest and the best,' said I again, seeing it pleased him. But he didn't answer. I heard the girl giggle outside the door, for occasionally we'd had just a bit of innocent laughter, you know, at his ways. I looked at his face, and his eyes were closed, and it was just as if somebody had punched in his nose on either side. But he was still smiling. It's queer to think of—he lay dead, lay dead there, an utter failure, with the smile of success on his face.

"That was the end of my uncle. You can imagine me and my mother saw that he had a decent funeral. Then, of course, came the hunt for the will. We began decent and respectful at first, and before the day was out we were ripping chairs, and smashing bureau panels, and sounding walls. Every hour we expected those others to come in. We asked the housekeeper, and found she'd actually witnessed a will—on an ordinary half-sheet of note-paper it was written, and very short, she said—not a month ago. The other witness was the gardener, and he bore her out word for word. But I'm hanged if there was that or any other will to be found. The way my mother talked must have made him turn in his grave. At last a lawyer at Reigate sprang one on us that had been made years ago during some temporary quarrel with my mother. I'm blest if that wasn't the only will to be discovered anywhere, and it left every penny he possessed to that 'Take 'im away' youngster of his second cousin's—a chap who'd never had to stand his talking not for one afternoon of his life."

The man with the glass eye stopped.

"I thought you said"—I began.

"Half a minute," said the man with the glass eye. "I had to wait for the end of the story till this very morning, and I was a blessed sight more interested than you are. You just wait a bit too. They executed the will, and the other chap inherited, and directly he was one-and-twenty he began to blew it. How he did blew it to be sure! He bet, he drank, he got in the papers for this and that. I tell you, it makes me wriggle to think of the times he had. He blewed every ha'penny of it before he was thirty, and the last I heard of him was—Holloway. Three years ago.

"Well, I naturally fell on hard times, because, as you see, the only trade I knew was legacy-cadging. All my plans were waiting over to begin, so to speak, when the old chap died. I've had my ups and downs since then. Just now it's a period of depression. I tell you frankly, I'm on the look-out for help. I was hunting round my room to find something to raise a bit on for immediate necessities, and the sight of all those presentation volumes—no one will buy them, not to wrap butter in, even—well, they annoyed me. I'd promised him not to part with them, and I never kept a promise easier. I let out at them with my boot, and sent them shooting across the room. One lifted at the kick and spun through the air. And out of it flapped—You guess.

"It was the will. He'd given it me himself in that very last volume of all."

He folded his arms on the table, and looked sadly with the active eye at his empty tankard. He shook his head slowly, and said softly, "I'd never *opened* the book, much more cut a page!" Then he looked up, with a bitter laugh for my sympathy. "Fancy hiding it there! Eigh? Of all places."

He began to fish absently for a dead fly with his finger

"It just shows you the vanity of authors," he said, looking up at me. "It wasn't no trick of his. He'd meant perfectly fair. He'd really thought I was really going home to read that blessed book of his through. But it shows you, don't it?"—his eye went down to the tankard again,—"It shows you, too, how we poor human beings fail to understand one another."

But there was no misunderstanding the eloquent thirst of his eye. He accepted with ill-feigned surprise. He said, in the usual subtle formula, that he didn't mind if he did.

STORY THE
SIXTEENTH

The Sad Story of a Dramatic Critic.

I WAS—you shall hear immediately why I am not now—
Egbert Craddock Cummins. The name remains. I am still
(Heaven help me!) Dramatic Critic to the *Fiery Cross*.
What I shall be in a little while I do not know. I write
in great trouble and confusion of mind. I will do what
I can to make myself clear in the face of terrible difficulties.
You must bear with me a little. When a man is rapidly
losing his own identity, he naturally finds a difficulty in
expressing himself. I will make it perfectly plain in a
minute, when once I get my grip upon the story. Let me
see—where *am* I? I wish I knew. Ah, I have it! Dead
self! Egbert Craddock Cummins!

In the past I should have disliked writing anything
quite so full of "I" as this story must be. It is full of "I's"
before and behind, like the beast in Revelation—the one
with a head like a calf, I am afraid. But my tastes have
changed since I became a Dramatic Critic and studied the
masters—G.R.S., G.B.S., G.A.S., and the others. Everything has changed since then. At least the story is about
myself—so that there is some excuse for me. And it is really
not egotism, because, as I say, since those days my identity
has undergone an entire alteration.

That past! ... I was—in those days—rather a nice fellow,
rather shy—taste for grey in my clothes, weedy little moustache, face "interesting," slight stutter which I had caught

520

in early life from a schoolfellow. Engaged to a very nice girl, named Delia. Fairly new, she was—cigarettes—liked me because I was human and original. Considered I was like Lamb—on the strength of the stutter, I believe. Father, an eminent authority on postage stamps. She read a great deal in the British Museum. (A perfect pairing ground for literary people, that British Museum—you should read George Egerton and Justin Huntly M'Carthy and Gissing and the rest of them.) We loved in our intellectual way, and shared the brightest hopes. (All gone now.) And her father liked me because I seemed honestly eager to hear about stamps. She had no mother. Indeed, I had the happiest prospects a young man could have. I never went to theatres in those days. My Aunt Charlotte before she died had told me not to.

Then Barnaby, the editor of the *Fiery Cross,* made me— in spite of my spasmodic efforts to escape—Dramatic Critic. He is a fine, healthy man, Barnaby, with an enormous head of frizzy black hair and a convincing manner, and he caught me on the staircase going to see Wembly. He had been dining, and was more than usually buoyant. "Hullo, Cummins!" he said. "The very man I want!" He caught me by the shoulder or the collar or something, ran me up the little passage, and flung me over the waste-paper basket into the arm-chair in his office. "Pray be seated," he said, as he did so. Then he ran across the room and came back with some pink and yellow tickets and pushed them into my hand. "Opera Comique," he said, "Thursday; Friday, the Surrey; Saturday, the Frivolity. That's all, I think."

"But"—I began.

"Glad you're free," he said, snatching some proofs off the desk and beginning to read.

"I don't quite understand," I said.

"*Eigh?*" he said, at the top of his voice, as though he thought I had gone, and was startled at my remark.

"Do you want me to criticise these plays?"

"Do something with 'em. . . . Did you think it was a treat?"

"But I can't."

"Did you call me a fool?"

"Well, I've never been to a theatre in my life."

"Virgin soil."

"But I don't know anything about it, you know."

"That's just it. New view. No habits. No *clichés* in stock. Ours is a live paper, not a bag of tricks. None of your clockwork professional journalism in this office. And I can rely on your integrity"—

"But I've conscientious scruples"—

He caught me up suddenly and put me outside his door. "Go and talk to Wembly about that," he said. "He'll explain."

As I stood perplexed, he opened the door again, said, "I forgot this," thrust a fourth ticket into my hand (it was for that night—in twenty minutes' time) and slammed the door upon me. His expression was quite calm, but I caught his eye.

I hate arguments. I decided that I would take his hint and become (to my own destruction) a Dramatic Critic. I walked slowly down the passage to Wembly. That Barnaby has a remarkably persuasive way. He has made few suggestions during our very pleasant intercourse of four years that he has not ultimately won me round to adopting. It may be, of course, that I am of a yielding disposition; certainly I am too apt to take my colour from my circumstances. It is, indeed, to my unfortunate susceptibility to vivid impressions that all my misfortunes are due. I have already alluded to the slight stammer I had acquired from a schoolfellow in my youth. However, this is a digression. ... I went home in a cab to dress.

I will not trouble the reader with my thoughts about the first-night audience, strange assembly as it is, those I reserve for my Memoirs, nor the humiliating story of how I got lost during the *entr'acte* in a lot of red plush passages, and saw the third act from the gallery. The only point upon which I wish to lay stress was the remarkable effect of the acting upon me. You must remember I had lived a quiet and retired life, and had never been to the theatre before, and that I am extremely sensitive to vivid impressions. At the risk of repetition I must insist upon these points.

The first effect was a profound amazement, not untinctured by alarm. The phenomenal unnaturalness of acting is a thing discounted in the minds of most people by early visits to the theatre. They get used to the fantastic gestures, the flamboyant emotions, the weird mouthings, melodious snortings, agonising yelps, lip-gnawings, glaring horrors, and other emotional symbolism of the stage. It becomes at last a mere deaf-and-dumb language to them, which they read intelligently *pari passu* with the hearing of the dialogue. But all this was new to me. The thing was called a modern comedy, the people were supposed to be English and were dressed like fashionable Americans of the current epoch, and I fell into the natural error of supposing that the actors were trying to represent human beings. I looked round on my first-night audience with a kind of wonder, discovered—as all new Dramatic Critics do—that it rested with me to reform the Drama, and, after a supper choked with emotion, went off to the office to write a column, piebald with "new paragraphs" (as all my stuff is—it fills out so) and purple with indignation. Barnaby was delighted.

But I could not sleep that night. I dreamt of actors—actors glaring, actors smiting their chests, actors flinging out a handful of extended fingers, actors smiling bitterly, laughing despairingly, falling hopelessly, dying idiotically. I got up at eleven with a slight headache, read my notice in the *Fiery Cross*, breakfasted, and went back to my room to shave. (It's my habit to do so.) Then an odd thing happened. I could not find my razor. Suddenly it occurred to me that I had not unpacked it the day before.

"Ah!" said I, in front of the looking-glass. Then "Hullo!"

Quite involuntarily, when I had thought of my portmanteau, I had flung up the left arm (fingers fully extended) and clutched at my diaphragm with my right hand. I am an acutely self-conscious man at all times. The gesture struck me as absolutely novel for me. I repeated it, for my own satisfaction. "Odd!" Then (rather puzzled) I turned to my portmanteau.

After shaving, my mind reverted to the acting I had seen, and I entertained myself before the cheval glass with some imitations of Jafferay's more exaggerated gestures. "Really, one might think it a disease," I said—"Stage-Walkitis!" (There's many a truth spoken in jest.) Then, if I remember rightly, I went off to see Wembly, and afterwards lunched at the British Museum with Delia. We actually spoke about our prospects, in the light of my new appointment.

But that appointment was the beginning of my downfall. From that day I necessarily became a persistent theatre-goer, and almost insensibly I began to change. The next thing I noticed after the gesture about the razor, was to catch myself bowing ineffably when I met Delia, and stooping in an old-fashioned, courtly way over her hand. Directly I caught myself, I straightened myself up and became very uncomfortable. I remember she looked at me curiously. Then, in the office, I found myself doing "nervous business," fingers on teeth, when Barnaby asked me a question I could not very well answer. Then, in some trifling difference with Delia, I clasped my hand to my brow. And I pranced through my social transactions at times singularly like an actor! I tried not to—no one could be more keenly alive to the arrant absurdity of the histrionic bearing. And I did!

It began to dawn on me what it all meant. The acting, I saw, was too much for my delicately-strung nervous system. I have always, I know, been too amenable to the suggestions of my circumstances. Night after night of concentrated attention to the conventional attitudes and intonation of the English stage was gradually affecting my speech and carriage. I was giving way to the infection of sympathetic imitation. Night after night my plastic nervous system took the print of some new amazing gesture, some new emotional exaggeration—and retained it. A kind of theatrical veneer threatened to plate over and obliterate my private individuality altogether. I saw myself in a kind of vision. Sitting by myself one night, my new self seemed to me to glide, posing and gesticulating, across the room. He clutched his throat, he opened his fingers, he opened his legs in walking like a high-class marionette. He went from attitude to attitude. He might have been clockwork. Directly after this I made an ineffectual attempt to resign my theatrical work. But Barnaby persisted in talking about the Polywhiddle Divorce all the time I was with him, and I could get no opportunity of saying what I wished.

And then Delia's manner began to change towards me. The ease of our intercourse vanished. I felt she was learning to dislike me. I grinned, and capered, and scowled, and posed at her in a thousand ways, and knew—with what a voiceless agony!—that I did it all the time. I tried to resign again, and Barnaby talked about "X" and "Z" and "Y" in the *New Review*, and gave me a strong cigar to smoke, and so routed me. And then I walked up the Assyrian Gallery in the manner of Irving to meet Delia, and so precipitated the crisis.

"Ah!—*Dear!*" I said, with more sprightliness and emotion in my voice than had ever been in all my life before I became (to my own undoing) a Dramatic Critic.

She held out her hand rather coldly, scrutinising my face as she did so. I prepared, with a new-won grace, to walk by her side.

"Egbert," she said, standing still, and thought. Then she looked at me.

I said nothing. I felt what was coming. I tried to be the old Egbert Craddock Cummins of shambling gait and stammering sincerity, whom she loved, but I felt even as I did so that I was a new thing, a thing of surging emotions and mysterious fixity--like no human being that ever lived, except upon the stage. "Egbert," she said, "you are not yourself."

"Ah!" Involuntarily I clutched my diaphragm and averted my head (as is the way with them).

"There!" she said.

"*What do you mean?*" I said, whispering in vocal italics --you know how they do it—turning on her, perplexity on face, right hand down, left on brow. I knew quite well what she meant. I knew quite well the dramatic unreality of my behaviour. But I struggled against it in vain. "What do you mean?" I said, and, in a kind of hoarse whisper, "I don't understand!"

She really looked as though she disliked me. "What do you keep on posing for?" she said. "I don't like it. You didn't use to."

"Didn't use to!" I said slowly, repeating this twice. I glared up and down the gallery, with short, sharp glances. "We are alone," I said swiftly. *"Listen!"* I poked my forefinger towards her—and glared at her.

"I am under a curse."

I saw her hands tighten upon her sunshade. "You are under some bad influence or other," said Delia. "You should give it up. I never knew anyone change as you have done."

"Delia!" I said, lapsing into the pathetic. "Pity me. Augh! Delia! *Pit*—y me!"

She eyed me critically. *"Why* you keep playing the fool like this I don't know," she said. "Anyhow, I really cannot go about with a man who behaves as you do. You made us both ridiculous on Wednesday. Frankly, I dislike you, as you are now. I met you here to tell you so—as it's about the only place where we can be sure of being alone together"—

"Delia!" said I, with intensity, knuckles of clenched hands white. "You don't mean"—

"I do," said Delia. "A woman's lot is sad enough at the best of times. But with you"—

I clapped my hand on my brow.

"So, good-bye," said Delia, without emotion.

"Oh, Delia!" I said. "Not *this?*"

"Good-bye, Mr. Cummins," she said.

By a violent effort I controlled myself and touched her hand. I tried to say some word of explanation to her. She looked into my working face and winced. "I *must* do it," she said hopelessly. Then she turned from me and began walking rapidly down the gallery.

Heavens! How the human agony cried within me! I

loved Delia. But nothing found expression—I was already too deeply crusted with my acquired self.

"Good-baye!" I said at last, watching her retreating figure. How I hated myself for doing it! After she had vanished, I repeated in a dreamy way, "Good-baye!" looking hopelessly round me. Then, with a kind of heart-broken cry, I shook my clenched fists in the air, staggered to the pedestal of a winged figure, buried my face in my arms, and made my shoulders heave. Something within me said "Ass!" as I did so. (I had the greatest difficulty in persuading the Museum policeman, who was attracted by my cry of agony, that I was not intoxicated, but merely suffering from a transient indisposition.)

But even this great sorrow has not availed to save me from my fate. I see it, everyone sees it; I grow more "theatrical" every day. And no one could be more painfully aware of the pungent silliness of theatrical ways. The quiet, nervous, but pleasing, E. C. Cummins vanishes. I cannot save him. I am driven like a dead leaf before the winds of March. My tailor even enters into the spirit of my disorder. He has a peculiar sense of what is fitting. I tried to get a dull grey suit from him this spring, and he foisted a brilliant blue upon me, and I see he has put braid down the sides of my new dress trousers. My hairdresser insists upon giving me a "wave."

I am beginning to associate with actors. I detest them, but it is only in their company that I can feel I am not glaringly conspicuous. Their talk infects me. I notice a growing tendency to dramatic brevity, to dashes and pauses in my style, to a punctuation of bows and attitudes. Barnaby has remarked it too. I offended Wembly by calling him "Dear Boy" yesterday. I dread the end, but I cannot escape from it.

The fact is, I am being obliterated. Living a grey, retired life all my youth, I came to the theatre a delicate sketch of a man, a thing of tints and faint lines. Their gorgeous colouring has effaced me altogether. People forget how much mode of expression, method of movement, are a matter of contagion. I have heard of stage-struck people before, and thought it a figure of speech. I spoke of it jestingly, as a disease. It is no jest. It *is* a disease. And I have got it bad! Deep down within me I protest against the wrong done to my personality—unavailingly. For three hours or more a week I have to go and concentrate my attention on some fresh play, and the suggestions of the drama strengthen their awful hold upon me. My manners grow so flamboyant, my passions so professional, that I doubt, as I said at the outset, whether it is really myself that behaves in such a manner. I feel merely the core of this dramatic casing, that grows thicker and presses upon me —me and mine. I feel like King John's abbot in his cope of lead.

I doubt, indeed, whether I should not abandon the struggle altogether—leave this sad world of ordinary life for which I am so ill-fitted, abandon the name of Cummins for some professional pseudonym, complete my self-efface-ment, and—a thing of tricks and tatters, of posing and pretence—go upon the stage. It seems my only resort—"to hold the mirror up to Nature." For in the ordinary life, I will confess, no one now seems to regard me as both sane and sober. Only upon the stage, I feel convinced, will people take me seriously. That will be the end of it. I *know* that will be the end of it. And yet . . . I will frankly confess . . . all that marks off your actor from your common man . . . I *detest*. I am still largely of my Aunt Charlotte's opinion, that playacting is unworthy of a pure-minded man's attention, much more participation. Even now I would resign my dramatic criticism and try a rest. Only I can't get hold of Barnaby. Letters of resignation he never notices. He says it is against the etiquette of journalism to write to your Editor. And when I go to see him, he gives me another big cigar and some strong whisky and soda, and then some-thing always turns up to prevent my explanation.

STORY THE
SEVENTEENTH

A Slip under the Microscope

OUTSIDE the laboratory windows was a watery-grey fog, and within a close warmth and the yellow light of the green-shaded gas lamps that stood two to each table down its narrow length. On each table stood a couple of glass jars containing the mangled vestiges of the crayfish, mussels, frogs, and guineapigs upon which the students had been working, and down the side of the room, facing the windows, were shelves bearing bleached dissections in spirits, surmounted by a row of beautifully executed anatomical drawings in whitewood frames and overhanging a row of cubical lockers. All the doors of the laboratory were panelled with black board, and on these were the half-erased diagrams of the previous day's work. The laboratory was empty, save for the demonstrator, who sat near the preparation-room door, and silent, save for a low, continuous murmur, and the clicking of the rocker microtome at which he was working. But scattered about the room were traces of numerous students: hand-bags, polished boxes of instruments, in one place a large drawing covered by newspapers, and in another a prettily bound copy of *News from Nowhere,* a book oddly at variance with its surroundings. These things had been put down hastily as the students had arrived and hurried at once to secure their seats in the adjacent lecture theatre. Deadened by the closed door, the measured accents of the professor sounded as a featureless muttering.

Presently, faint through the closed windows came the

sound of the Oratory clock striking the hour of eleven. The
clicking of the microtome ceased, and the demonstrator
looked at his watch, rose, thrust his hands into his pockets,
and walked slowly down the laboratory towards the lecture
theatre door. He stood listening for a moment, and then
his eye fell on the little volume by William Morris. He
picked it up, glanced at the title, smiled, opened it, looked
at the name on the fly-leaf, ran the leaves through with his
hand, and put it down. Almost immediately the even
murmur of the lecturer ceased, there was a sudden burst of
pencils rattling on the desks in the lecture theatre, a stirring, a scraping of feet, and a number of voices speaking
together. Then a firm footfall approached the door, which
began to open, and stood ajar as some indistinctly heard
question arrested the new-comer.

The demonstrator turned, walked slowly back past the
microtome, and left the laboratory by the preparation-room
door. As he did so, first one, and then several students carrying notebooks entered the laboratory from the lecture
theatre, and distributed themselves among the little tables, or
stood in a group about the doorway. They were an exceptionally heterogeneous assembly, for while Oxford and Cambridge still recoil from the blushing prospect of mixed
classes, the College of Science anticipated America in the
matter years ago—mixed socially too, for the prestige of the
College is high, and its scholarships, free of any age limit,
dredge deeper even than do those of the Scotch universities.
The class numbered one-and-twenty, but some remained in
the theatre questioning the professor, copying the black
board diagrams before they were washed off, or examining
the special specimens he had produced to illustrate the day's
teaching. Of the nine who had come into the laboratory
three were girls, one of whom, a little fair woman wearing
spectacles and dressed in greyish-green, was peering out of
the window at the fog, while the other two, both wholesome
looking, plain-faced schoolgirls, unrolled and put on the
brown holland aprons they wore while dissecting. Of the
men, two went down the laboratory to their places, one a
pallid, dark-bearded man, who had once been a tailor; the

other a pleasant-featured, ruddy young man of twenty, dressed in a well-fitting brown suit; young Wedderburn, the son of Wedderburn the eye specialist.   The others formed a little knot near the theatre door.   One of these, a dwarfed, spectacled figure with a hunch back, sat on a bent wood stool; two others, one a short, dark youngster and the other a flaxen-haired, reddish-complexioned young man, stood leaning side by side against the slate sink, while the fourth stood facing them, and maintained the larger share of the conversation.

The last person was named Hill.   He was a sturdily built young fellow, of the same age as Wedderburn; he had a white face, dark grey eyes, hair of an indeterminate colour, and prominent, irregular features.   He talked rather louder than was needful, and thrust his hands deeply into his pockets.   His collar was frayed and blue with the starch of a careless laundress, his clothes were evidently ready-made, and there was a patch on the side of his boot near the toe.   And as he talked or listened to the others, he glanced now and again towards the lecture theatre door.   They were discussing the depressing peroration of the lecture they had just heard, the last lecture it was in the introductory course in zoology.   "From ovum to ovum is the goal of the higher vertebrata," the lecturer had said in his melancholy tones, and so had neatly rounded off the sketch of comparative anatomy he had been developing.   The spectacled hunch-back had repeated it with noisy appreciation, had tossed it towards the fair-haired student with an evident provocation, and had started one of those vague, rambling discussions on generalities so unaccountably dear to the student mind all the world over.

"That is our goal, perhaps—I admit it, as far as science goes," said the fair-haired student, rising to the challenge. "But there are things above science."

"Science," said Hill confidently, "is systematic knowledge. Ideas that don't come into the system—must anyhow—be loose ideas."   He was not quite sure whether that was a clever saying or a fatuity until his hearers took it seriously.

"The thing I cannot understand," said the hunchback, a
large, "is whether Hill is a materialist or not."

"There is one thing above matter," said Hill promptly
feeling he made a better point this time, aware, too, o
someone in the doorway behind him, and raising his voice a
trifle for her benefit, "and that is, the delusion that there i
something above matter."

"So we have your gospel at last," said the fair student
"It's all a delusion, is it? All our aspirations to lead some
thing more than dogs' lives, all our work for anything be
yond ourselves. But see how inconsistent you are. You
socialism, for instance. Why do you trouble about the
interests of the race? Why do you concern yourself abou
the beggar in the gutter? Why are you bothering yoursel
to lend that book"—he indicated William Morris by a
movement of the head—"to everyone in the lab?"

"Girl," said the hunchback indistinctly, and glanced
guiltily over his shoulder.

The girl in brown, with the brown eyes, had come into
the laboratory, and stood on the other side of the table
behind him, with her rolled-up apron in one hand, looking
over her shoulder, listening to the discussion. She did no
notice the hunchback, because she was glancing from Hil
to his interlocutor. Hill's consciousness of her presence
betrayed itself to her only in his studious ignoring of the
fact; but she understood that, and it pleased her. "I se
no reason," said he, "why a man should live like a brute
because he knows of nothing beyond matter, and does no
expect to exist a hundred years hence."

"Why shouldn't he?" said the fair-haired student.

"Why *should* he?" said Hill.

"What inducement has he?"

"That's the way with all you religious people. It's all a
business of inducements. Cannot a man seek afte
righteousness for righteousness' sake?"

There was a pause. The fair man answered, with a kin
of vocal padding, "But—you see—inducement—when I said
inducement," to gain time. And then the hunchback came
to his rescue and inserted a question. He was a terrible

erson in the debating society with his questions, and they
nvariably took one form—a demand for a definition.
What's your definition of righteousness?" said the hunch-
ack at this stage.

Hill experienced a sudden loss of complacency at this
uestion, but even as it was asked, relief came in the person
f Brooks, the laboratory attendant, who entered by the
reparation-room door, carrying a number of freshly killed
uineapigs by their hind legs. "This is the last batch of
aterial this session," said the youngster who had not pre-
iously spoken. Brooks advanced up the laboratory, smack-
g down a couple of guineapigs at each table. The rest of
e class, scenting the prey from afar, came crowding in by
e lecture theatre door, and the discussion perished abruptly
s the students who were not already in their places hurried
 them to secure the choice of a specimen. There was a
oise of keys rattling on split rings as lockers were opened
d dissecting instruments taken out. Hill was already
anding by his table, and his box of scalpels was sticking
t of his pocket. The girl in brown came a step towards
im, and leaning over his table said softly, "Did you see
at I returned your book, Mr. Hill?"

During the whole scene she and the book had been vividly
resent in his consciousness; but he made a clumsy pretence
 looking at the book and seeing it for the first time. "Oh
es," he said, taking it up. "I see. Did you like it?"

"I want to ask you some questions about it—some time."

"Certainly," said Hill. "I shall be glad." He stopped
wkwardly. "You liked it?" he said.

"It's a wonderful book. Only some things I don't under-
and."

Then suddenly the laboratory was hushed by a curious
raying noise. It was the demonstrator. He was at the
ackboard ready to begin the day's instruction, and it was
is custom to demand silence by a sound midway between
e "Er" of common intercourse and the blast of a trumpet.
he girl in brown slipped back to her place: it was im-
ediately in front of Hill's, and Hill, forgetting her forth-
ith, took a notebook out of the drawer of his table, turned

over its leaves hastily, drew a stumpy pencil from hi
pocket, and prepared to make a copious note of the comin
demonstration. For demonstrations and lectures are th
sacred text of the College students. Books, saving only th
Professor's own, you may—it is even expedient to—ignore.

Hill was the son of a Landport cobbler, and had bee
hooked by a chance blue paper the authorities had throw
out to the Landport Technical College. He kept himself i
London on his allowance of a guinea a week, and foun
that, with proper care, this also covered his clothing a
lowance, an occasional waterproof collar, that is; and in
and needles and cotton and such-like necessaries for a ma
about town. This was his first year and his first session, bu
the brown old man in Landport had already got himse
detested in many public-houses by boasting of his son, "th
Professor." Hill was a vigorous youngster, with a seren
contempt for the clergy of all denominations, and a fir
ambition to reconstruct the world. He regarded his schola
ship as a brilliant opportunity. He had begun to read
seven, and had read steadily whatever came in his way, goo
or bad, since then. His worldly experience had been limite
to the island of Portsea, and acquired chiefly in the whol
sale boot factory in which he had worked by day, aft
passing the seventh standard of the Board school. He ha
a considerable gift of speech, as the College Debating S
ciety, which met amidst the crushing machines and mir
models in the metallurgical theatre downstairs, already reco
nised—recognised by a violent battering of desks whenev
he rose. And he was just at that fine emotional age whe
life opens at the end of a narrow pass like a broad valle
at one's feet, full of the promise of wonderful discoveries an
tremendous achievements. And his own limitations, sav
that he knew that he knew neither Latin nor French, wer
all unknown to him.

At first his interest had been divided pretty equal
between his biological work at the College and social an
theological theorising, an employment which he took
deadly earnest. Of a night, when the big museum libra
was not open, he would sit on the bed of his room i

Chelsea with his coat and a muffler on, and write out the lecture notes and revise his dissection memoranda until Thorpe called him out by a whistle—the landlady objected to open the door to attic visitors—and then the two would go prowling about the shadowy, shiny, gas-lit streets, talking, very much in the fashion of the sample just given, of the God Idea and Righteousness and Carlyle and the Reorganisation of Society. And in the midst of it all, Hill, arguing not only for Thorpe but for the casual passer-by, would lose the thread of his argument glancing at some pretty painted face that looked meaningly at him as he passed. Science and Righteousness! But once or twice lately there had been signs that a third interest was creeping into his life, and he had found his attention wandering from the fate of the mesoblastic somites or the probable meaning of the blastopore, to the thought of the girl with the brown eyes who sat at the table before him.

She was a paying student; she descended inconceivable social altitudes to speak to him. At the thought of the education she must have had, and the accomplishments she must possess, the soul of Hill became abject within him. She had spoken to him first over a difficulty about the alisphenoid of a rabbit's skull, and he had found that, in biology at least, he had no reason for self-abasement. And from that, after the manner of young people starting from any starting-point, they got to generalities, and while Hill attacked her upon the question of socialism, some instinct told him to spare her a direct assault upon her religion—she was gathering resolution to undertake what she told herself was his æsthetic education. She was a year or two older than he, though the thought never occurred to him. The loan of *News from Nowhere* was the beginning of a series of cross loans. Upon some absurd first principle of his, Hill had never "wasted time" upon poetry, and it seemed an appalling deficiency to her. One day in the lunch hour, when she chanced upon him alone in the little museum where the skeletons were arranged, shamefully eating the bun that constituted his mid-day meal, she retreated, and returned to lend him, with a slightly furtive air, a volume of

Browning. He stood sideways towards her and took the book rather clumsily, because he was holding the bun in the other hand. And in the retrospect his voice lacked the cheerful clearness he could have wished.

That occurred after the examination in comparative anatomy, on the day before the College turned out its students and was carefully locked up by the officials for the Christmas holidays. The excitement of cramming for the first trial of strength had for a little while dominated Hill to the exclusion of his other interests. In the forecasts of the result in which every one indulged he was surprised to find that no one regarded him as a possible competitor for the Harvey Commemoration Medal, of which this and the two subsequent examinations disposed. It was about this time that Wedderburn, who so far had lived inconspicuously on the uttermost margin of Hill's perceptions, began to take on the appearance of an obstacle. By a mutual agreement, the nocturnal prowlings with Thorpe ceased for the three weeks before the examination, and his landlady pointed out that she really could not supply so much lamp oil at the price. He walked to and fro from the College with little slips of mnemonics in his hand, lists of crayfish appendages, rabbits' skull-bones, and vertebrate nerves, for example, and became a positive nuisance to foot passengers in the opposite direction.

But, by a natural reaction, Poetry and the girl with the brown eyes ruled the Christmas holiday. The pending results of the examination became such a secondary consideration that Hill marvelled at his father's excitement. Even had he wished it, there was no comparative anatomy to read in Landport, and he was too poor to buy books, but the stock of poets in the library was extensive, and Hill's attack was magnificently sustained. He saturated himself with the fluent numbers of Longfellow and Tennyson, and fortified himself with Shakespeare; found a kindred soul in Pope and a master in Shelley, and heard and fled the siren voices of Eliza Cook and Mrs. Hemans. But he read no more Browning, because he hoped for the loan of other volumes from Miss Haysman when he returned to London.

He walked from his lodgings to the College with that volume of Browning in his shiny black bag, and his mind teeming with the finest general propositions about poetry. Indeed, he framed first this little speech and then that with which to grace the return. The morning was an exceptionally pleasant one for London; there was a clear hard frost and undeniable blue in the sky, a thin haze softened every outline, and warm shafts of sunlight struck between the house blocks and turned the sunny side of the street to amber and gold. In the hall of the College he pulled off his glove and signed his name with fingers so stiff with cold that the characteristic dash under the signature he cultivated became a quivering line. He imagined Miss Haysman about him everywhere. He turned at the staircase, and there, below, he saw a crowd struggling at the foot of the notice-board. This, possibly, was the biology list. He forgot Browning and Miss Haysman for the moment, and joined the scrimmage. And at last, with his cheek flattened against the sleeve of the man on the step above him, he read the list—

CLASS I

H. J. Somers Wedderburn
William Hill

and thereafter followed a second class that is outside our present sympathies. It was characteristic that he did not trouble to look for Thorpe on the physics list, but backed out of the struggle at once, and in a curious emotional state between pride over common second-class humanity and acute disappointment at Wedderburn's success, went on his way upstairs. At the top, as he was hanging up his coat in the passage, the zoological demonstrator, a young man from Oxford, who secretly regarded him as a blatant "mugger" of the very worst type, offered his heartiest congratulations.

At the laboratory door Hill stopped for a second to get his breath, and then entered. He looked straight up the laboratory and saw all five girl students grouped in their places, and Wedderburn, the once retiring Wedderburn, leaning rather gracefully against the window, playing with the blind tassel and talking apparently to the five of them. Now, Hill could talk bravely enough and even overbearingly to one girl, but this business of standing at ease and appreciating, fencing, and returning quick remarks round a group was, he knew, altogether beyond him. Coming up the staircase his feelings for Wedderburn had been generous, a certain admiration perhaps, a willingness to shake his hand conspicuously and heartily as one who had fought but the first round. But before Christmas Wedderburn had never gone up to that end of the room to talk. In a flash Hill's mist of vague excitement condensed abruptly to a vivid dislike of Wedderburn. Possibly his expression changed. As he came up to his place, Wedderburn nodded carelessly to him, and the others glanced round. Miss Haysman looked at him and away again, the faintest touch of her eyes. "I can't agree with you, Mr. Wedderburn," she said.

"I must congratulate you on your first class, Mr. Hill," said the spectacled girl in green, turning round and beaming at him.

"It's nothing," said Hill, staring at Wedderburn and Miss Haysman talking together, and eager to hear what they talked about.

"We poor folks in the second class don't think so," said the girl in spectacles.

What was it Wedderburn was saying? Something about William Morris? Hill did not answer the girl in spectacles, and the smile died out of his face. He could not hear, and failed to see how he could "cut in." Confound Wedderburn! He sat down, opened his bag, hesitated whether to return the volume of Browning forthwith, in the sight of all, and instead drew out his new notebooks for the short course in elementary botany that was now beginning, and which would terminate in February. As he did so, a fat heavy man with a white face and pale grey eyes—Bindon, the professor of botany, who came up from Kew for January and February —came in by the lecture theatre door, and passed, rubbing his hands together and smiling, in silent affability down the laboratory.

In the subsequent six weeks Hill experienced some very rapid and curiously complex emotional developments.   For the most part he had Wedderburn in focus—a fact that Miss Haysman never suspected. She told Hill (for in the comparative privacy of the museum she talked a good deal to him of socialism and Browning and general propositions) that she had met Wedderburn at the house of some people she knew, and "he's inherited his cleverness; for his father, you know, is the great eye specialist."

"*My* father is a cobbler," said Hill, quite irrelevantly, and perceived the want of dignity even as he said it. But the gleam of jealousy did not offend her. She conceived herself the fundamental source of it. He suffered bitterly from a sense of Wedderburn's unfairness, and a realisation of his own handicap. Here was this Wedderburn had picked up a prominent man for a father, and instead of his losing so many marks on the score of that advantage, it was counted to him for righteousness! And while Hill had to introduce himself and talk to Miss Haysman clumsily over mangled guineapigs in the laboratory, this Wedderburn, in some backstairs way, had access to her social altitudes, and could converse in a polished argot that Hill understood perhaps, but felt incapable of speaking. Not, of course, that he wanted to. Then it seemed to Hill that for Wedderburn to come there day after day with cuffs unfrayed, neatly tailored, precisely barbered, quiety perfect, was in itself an ill-bred, sneering sort of proceeding. Moreover, it was a stealthy thing for Wedderburn to behave insignificantly for a space, to mock modesty, to lead Hill to fancy that he himself was beyond dispute the man of the year, and then suddenly to dart in front of him, and incontinently to swell up in this fashion. In addition to these things, Wedderburn displayed an increasing disposition to join in any conversational grouping that included Miss Haysman; and would venture, and indeed seek occasion, to pass opinions derogatory to socialism and atheism. He goaded Hill to in-

civilities by neat, shallow, and exceedingly effective per-
sonalities about the socialist leaders, until Hill hated Ber-
nard Shaw's graceful egotisms, William Morris's limited
editions and luxurious wall-papers, and Walter Crane's
charmingly absurd ideal working men, about as much as he
hated Wedderburn. The dissertations in the laboratory,
that had been his glory in the previous term, became a
danger, degenerated into inglorious tussles with Wedderburn, and Hill kept to them only out of an obscure perception that his honour was involved. In the debating society
Hill knew quite clearly that, to a thunderous accompaniment of banged desks, he could have pulverised Wedderburn.   Only Wedderburn never attended the debating
society to be pulverised, because—nauseous affectation!—he
"dined late."

You must not imagine that these things presented themselves in quite such a crude form to Hill's perception. Hill
was a born generaliser. Wedderburn to him was not so
much an individual obstacle as a type, the salient angle of
a class. The economic theories that, after infinite ferment,
had shaped themselves in Hill's mind, became abruptly
concrete at the contact. The world became full of easy-
mannered,   graceful,   gracefully-dressed,   conversationally
dexterous, finally shallow Wedderburn's, Bishops Wedderburn, Wedderburn M.P.'s, Professors Wedderburn, Wedderburn landlords, all with finger-bowl shibboleths and
epigrammatic cities of refuge from a sturdy debater. And
everyone ill-clothed or ill-dressed, from the cobbler to the
cab-runner, was, to Hill's imagination, a man and a brother,
a fellow-sufferer. So that he became, as it were, a champion
of the fallen and oppressed, albeit to outward seeming only
a self-assertive, ill-mannered young man, and an unsuccessful champion at that. Again and again a skirmish over the
afternoon tea that the girl students had inaugurated left Hill
with flushed cheeks and a tattered temper, and the debating
society noticed a new quality of sarcastic bitterness in his
speeches.

You will understand now how it was necessary, if only
in the interests of humanity, that Hill should demolish

Wedderburn in the forthcoming examination and outshine him in the eyes of Miss Haysman; and you will perceive, too, how Miss Haysman fell into some common feminine misconceptions. The Hill-Wedderburn quarrel, for in his unostentatious way Wedderburn reciprocated Hill's ill-veiled rivalry, became a tribute to her indefinable charm; she was the Queen of Beauty in a tournament of scalpels and stumpy pencils.   To her confidential friend's secret annoyance, it even troubled her conscience, for she was a good girl, and painfully aware, through Ruskin and contemporary fiction, how entirely men's activities are determined by women's attitudes. And if Hill never by any chance mentioned the topic of love to her, she only credited him with the finer modesty for that omission.

So the time came on for the second examination, and Hill's increasing pallor confirmed the general rumour that he was working hard. In the aërated bread shop near South Kensington Station you would see him, breaking his bun and sipping his milk with his eyes intent upon a paper of closely written notes. In his bedroom there were propositions about buds and stems round his looking-glass, a diagram to catch his eye, if soap should chance to spare it, above his washing basin. He missed several meetings of the debating society, but he found the chance encounters with Miss Haysman in the spacious ways of the adjacent art museum, or in the little museum at the top of the College, or in the College corridors, more frequent and very restful. In particular, they used to meet in a little gallery full of wrought-iron chests and gates near the art library, and there Hill used to talk, under the gentle stimulus of her flattering attention, of Browning and his personal ambitions. A characteristic she found remarkable in him was his freedom from avarice. He contemplated quite calmly the prospect of living all his life on an income below a hundred pounds a year. But he was determined to be famous, to make, recognisably in his own proper person, the world a better place to live in. He took Bradlaugh and John Burns for his leaders and models, poor, even impecunious, great men. But Miss Haysman thought that such lives were deficient on the æsthetic side, by which, though she did not know it, she meant good wall-paper and upholstery, pretty books, tasteful clothes, concerts, and meals nicely cooked and respectfully served.

At last came the day of the second examination, and the professor of botany, a fussy, conscientious man, rearranged all the tables in a long narrow laboratory to prevent copying, and put his demonstrator on a chair on a table (where he felt, he said, like a Hindoo god), to see all the cheating, and stuck a notice outside the door, "Door closed," for no earthly reason that any human being could discover. And all the morning from ten till one the quill of Wedderburn shrieked defiance at Hill's, and the quills of the others chased their leaders in a tireless pack, and so also it was in the afternoon. Wedderburn was a little quieter than usual, and Hill's face was hot all day, and his overcoat bulged with textbooks and notebooks against the last moment's revision. And the next day, in the morning and in the afternoon, was the practical examination, when sections had to be cut and slides identified. In the morning Hill was depressed because he knew he had cut a thick section, and in the afternoon came the mysterious slip.

It was just the kind of thing that the botanical professor was always doing. Like the income tax, it offered a premium to the cheat. It was a preparation under the microscope, a little glass slip, held in its place on the stage of the instrument by light steel clips, and the inscription set forth that the slip was not to be moved. Each student was to go in turn to it, sketch it, write in his book of answers what he considered it to be, and return to his place. Now, to move such a slip is a thing one can do by a chance movement of the finger, and in a fraction of a second. The professor's reason for decreeing that the slip should not be moved depended on the fact that the object he wanted identified was characteristic of a certain tree stem. In the position in which it was placed it was a difficult thing to recognise, but once the slip was moved so as to bring other parts of the preparation into view, its nature was obvious enough.

Hill came to this, flushed from a contest with staining re-
agents, sat down on the little stool before the microscope,
turned the mirror to get the best light, and then, out of
sheer habit, shifted the slip. At once he remembered the
prohibition, and, with an almost continuous motion of his
hands, moved it back, and sat paralysed with astonishment
at his action.

Then, slowly, he turned his head. The professor was out
of the room; the demonstrator sat aloft on his impromptu ros-
trum, reading the *Q. Jour. Mi. Sci.*; the rest of the examinees
were busy, and with their backs to him. Should he own up
to the accident now? He knew quite clearly what the thing
was. It was a lenticel, a characteristic preparation from the
elder-tree. His eyes roved over his intent fellow-students
and Wedderburn suddenly glanced over his shoulder at him
with a queer expression in his eyes. The mental excitement
that had kept Hill at an abnormal pitch of vigour these two
days gave way to a curious nervous tension. His book of
answers was beside him. He did not write down what the
thing was, but with one eye at the microscope he began
making a hasty sketch of it. His mind was full of this grotesque puzzle in ethics that had suddenly been sprung upon
him. Should he identify it? or should he leave this question unanswered? In that case Wedderburn would probably
come out first in the second result. How could he tell now
whether he might not have identified the thing without
shifting it? It was possible that Wedderburn had failed to
recognise it, of course. Suppose Wedderburn too had
shifted the slide? He looked up at the clock. There were
fifteen minutes in which to make up his mind. He gathered
up his book of answers and the coloured pencils he used in
illustrating his replies and walked back to his seat.

He read through his manuscript, and then sat thinking
and gnawing his knuckle. It would look queer now if he
owned up. He *must* beat Wedderburn. He forgot the
examples of those starry gentlemen, John Burns and Bradlaugh. Besides, he reflected, the glimpse of the rest of the
slip he had had was after all quite accidental, forced upon
him by chance, a kind of providential revelation rather than

an unfair advantage. It was not nearly so dishonest to avail himself of that as it was of Broome, who believed in the efficacy of prayer, to pray daily for a first-class. "Five minutes more," said the demonstrator, folding up his paper and becoming observant. Hill watched the clock hands until two minutes remained; then he opened the book of answers, and, with hot ears and an affectation of ease, gave his drawing of the lenticel its name.

When the second pass list appeared, the previous positions of Wedderburn and Hill were reversed, and the spectacled girl in green, who knew the demonstrator in private life (where he was practically human), said that in the result of the two examinations taken together Hill had the advantage of a mark—167 to 166 out of a possible 200. Everyone admired Hill in a way, though the suspicion of "mugging" clung to him. But Hill was to find congratulations and Miss Haysman's enhanced opinion of him, and even the decided decline in the crest of Wedderburn, tainted by an unhappy memory. He felt a remarkable access of energy at first, and the note of a democracy marching to triumph returned to his debating society speeches; he worked at his comparative anatomy with tremendous zeal and effect, and he went on with his æsthetic education. But through it all, a vivid little picture was continually coming before his mind's eye—of a sneakish person manipulating a slide.

No human being had witnessed the act, and he was cocksure that no higher power existed to see it; but for all that it worried him. Memories are not dead things, but alive; they dwindle in disuse, but they harden and develop in all sorts of queer ways if they are being continually fretted. Curiously enough, though at the time he perceived clearly that the shifting was accidental, as the days wore on his memory became confused about it, until at last he was not sure—although he assured himself that he *was* sure—whether the movement had been absolutely involuntary. Then it is possible that Hill's dietary was conducive to morbid conscientiousness; a breakfast frequently eaten in a hurry, a midday bun, and, at such hours after five as chanced to be convenient, such meat as his means determined, usually in a chop-house in a back street off the Brompton Road. Occasionally he treated himself to threepenny or ninepenny classics, and they usually represented a suppression of potatoes or chops. It is indisputable that outbreaks of self-abasement and emotional revival have a distinct relation to periods of scarcity. But apart from this influence on the feelings, there was in Hill a distinct aversion to falsity that the blasphemous Landport cobbler had inculcated by strap and tongue from his earliest years. Of one fact about professed atheists I am convinced; they may be—they usually are—fools, void of subtlety, revilers of holy institutions, brutal speakers, and mischievous knaves, but they lie with difficulty. If it were not so, if they had the faintest grasp of the idea of compromise, they would simply be liberal churchmen. And, moreover, this memory poisoned his regard for Miss Haysman. For she now so evidently preferred him to Wedderburn that he felt sure he cared for her, and began reciprocating her attentions by timid marks of personal regard; at one time he even bought a bunch of violets, carried it about in his pocket, and produced it with a stumbling explanation, withered and dead, in the gallery of old iron. It poisoned, too, the denunciation of capitalist dishonesty that had been one of his life's pleasures. And, lastly, it poisoned his triumph in Wedderburn. Previously he had been Wedderburn's superior in his own eyes, and had raged simply at a want of recognition. Now he began to fret at the darker suspicion of positive inferiority. He fancied he found justifications for his position in Browning, but they vanished on analysis. At last—moved, curiously enough, by exactly the same motive forces that had resulted in his dishonesty—he went to Professor Bindon, and made a clean breast of the whole affair. As Hill was a paid student, Professor Bindon did not ask him to sit down, and he stood before the professor's desk as he made his confession.

"It's a curious story," said Professor Bindon, slowly realising how the thing reflected on himself, and then letting his anger rise,—"A most remarkable story. I can't understand your doing it, and I can't understand this avowal. You're a type of student—Cambridge men would never dream—I suppose I ought to have thought—Why *did* you cheat?"

"I didn't cheat," said Hill.

"But you have just been telling me you did."

"I thought I explained——"

"Either you cheated or you did not cheat."——

"I said my motion was involuntary."

"I am not a metaphysician, I am a servant of science—of fact. You were told not to move the slip. You did move the slip. If that is not cheating——"

"If I was a cheat," said Hill, with the note of hysterics in his voice, "should I come here and tell you?"

"Your repentance, of course, does you credit," said Professor Bindon, "but it does not alter the original facts."

"No, sir," said Hill, giving in in utter self-abasement.

"Even now you cause an enormous amount of trouble. The examination list will have to be revised."

"I suppose so, sir."

"Suppose so? Of course it must be revised. And I don't see how I can conscientiously pass you."

"Not pass me?" said Hill. "Fail me?"

"It's the rule in all examinations. Or where should we be? What else did you expect? You don't want to shirk the consequences of your own acts?"

"I thought, perhaps"—said Hill. And then, "Fail me? I thought, as I told you, you would simply deduct the marks given for that slip."

"Impossible!" said Bindon. "Besides, it would still leave you above Wedderburn. Deduct only the marks—Preposterous! The Departmental Regulations distinctly say——"

"But it's my own admission, sir."

"The Regulations say nothing whatever of the manner in which the matter comes to light. They simply provide——"

"It will ruin me. If I fail this examination, they won't renew my scholarship."

"You should have thought of that before."

"But, sir, consider all my circumstances——"

"I cannot consider anything. Professors in this College

are machines. The Regulations will not even let us recommend our students for appointments. I am a machine, and you have worked me. I have to do——"

"It's very hard, sir."

"Possibly it is."

"If I am to be failed this examination, I might as well go home at once."

"That is as you think proper." Bindon's voice softened a little; he perceived he had been unjust, and, provided he did not contradict himself, he was disposed to amelioration. "As a private person," he said, "I think this confession of yours goes far to mitigate your offence. But you have set the machinery in motion, and now it must take its course. I—I am really sorry you gave way."

A wave of emotion prevented Hill from answering. Suddenly, very vividly, he saw the heavily-lined face of the old Landport cobbler, his father. "Good God! What a fool I have been!" he said hotly and abruptly.

"I hope," said Bindon, "that it will be a lesson to you."

But, curiously enough, they were not thinking of quite the same indiscretion.

There was a pause.

"I would like a day to think, sir, and then I will let you know—about going home, I mean," said Hill, moving towards the door.

The next day Hill's place was vacant. The spectacled girl in green was, as usual, first with the news. Wedderburn and Miss Haysman were talking of a performance of *The Meistersingers* when she came up to them.

"Have you heard?" she said.

"Heard what?"

"There was cheating in the examination."

"Cheating!" said Wedderburn, with his face suddenly hot. "How?"

"That slide"——

"Moved? Never!"

"It was. That slide that we weren't to move"——

"Nonsense!" said Wedderburn. "Why! How could they find out? Who do they say——?"

"It was Mr. Hill."

"*Hill!*"

"Mr. Hill!"

"Not—surely not the immaculate Hill?" said Wedderburn recovering.

"I don't believe it," said Miss Haysman. "How do you know?"

"I *didn't*," said the girl in spectacles. "But I know it now for a fact. Mr. Hill went and confessed to Professor Bindon himself."

"By Jove!" said Wedderburn. "Hill of all people. But I am always inclined to distrust these philanthropists-on-principle"——

"Are you quite sure?" said Miss Haysman, with a catch in her breath.

"Quite. It's dreadful, isn't it? But, you know, what can you expect? His father is a cobbler."

Then Miss Haysman astonished the girl in spectacles.

"I don't care. I will not believe it," she said, flushing darkly under her warm-tinted skin. "I will not believe it until he has told me so himself—face to face. I would scarcely believe it then," and abruptly she turned her back on the girl in spectacles, and walked to her own place.

"It's true, all the same," said the girl in spectacles, peering and smiling at Wedderburn.

But Wedderburn did not answer her. She was indeed one of those people who seem destined to make unanswered remarks.

THE RECONCILIATION

THE
RECONCILIATION

TEMPLE had scarcely been with Findlay five minutes before he felt his old resentments, and the memory of that unforgettable wrong growing vivid again. But with the infatuation of his good resolution still upon him, he maintained the air of sham reconciliation that Findlay had welcomed so eagerly. They talked of this and that, carefully avoiding the matter of the separation. Temple at first spoke chiefly of his travels. He stood between the cabinet of minerals and the fireplace, his whisky on the mantel-board, while Findlay sat with his chair pushed back from his writing-desk, on which were scattered the dozen little skulls of hedgehogs and shrew mice upon which he had been working.

Temple's eye fell upon them, and abruptly brought his mind round from the topic of West Africa. "And you," said Temple. "While I have been wandering I suppose you have been going on steadily."

"Drumming along," said Findlay.

"To the Royal Society and fame and all the things we used to dream about—how long is it?"

"Five years—since our student days."

Temple glanced round the room, and his eye rested for a moment on a round greyish-drab object that lay in a corner near the door. "The same fat books and folios, only more of them, the same smell of old bones, and a dissection—is it the same one?—in the window. Fame is *your* mistress?"

"Fame," said Findlay. "But it's hardly fame. The herd outside say, 'Eminence in comparative anatomy.'"

"Eminence in comparative anatomy. No marrying—no avarice."

"None," said Findlay, glancing askance at him.

"I suppose it's the happiest way of living. But it wouldn't be the thing for me. Excitement—but, I say!"—his eye had fallen again on that fungoid shape of drabbish-grey—"there's a limit to scientific inhumanity. You really mustn't keep your door open with a human brainpan."

He went across the room as he spoke and picked the thing up. "Brainpan!" said Findlay. "Oh, *that!* Man alive, that's not a brainpan. Where's your science?"

"No. I see it's not," said Temple, carrying the object in his hand as he came back to his former position and scrutinising it curiously. "But what the devil is it?"

"Don't you know?" said Findlay.

The thing was about thrice the size of a man's hand, like a rough watch-pocket of thick bone.

Findlay laughed almost naturally. "You have a bad memory—it's a whale's ear-bone."

"Of course," said Temple, his appearance of interest vanishing. "The *bulla* of a whale. I've forgotten a lot of these things."

He half turned, and put the thing on the top of the cabinet beside Findlay's dumb-bells.

"If you are serious in your music-hall proposal," he said, reverting to a jovial suggestion of Findlay's, "I am at your service. I'm afraid—I may find myself a little old for that sort of thing—I haven't tried one for ages."

"But we are meeting to commemorate youth," said Findlay.

"And bury our early manhood," said Temple. "Well, well—yes, let us go to the music-hall, by all means, if you desire it. It is trivial—and appropriate. We want no tragic issues."

When the men returned to Findlay's study the little clock in the dimness on the mantel-shelf was pointing to half-past one. After the departure the little brown room, with its books and bones, was undisturbed, save for the two visits Findlay's attentive servant paid, to see to the fire and to pull down the blinds and draw the curtains. The ticking of the clock was the only sound in the quiet. Now and then the fire flickered and stirred, sending blood-red reflections chasing the shadows across the ceiling, and bringing into ghostly transitory prominence some grotesque grouping of animals' bones or skulls upon the shelves. At last the stillness was broken by the unlatching and slamming of the heavy street door and the sound of unsteady footsteps approaching along the passage. Then the door opened, and the two men came into the warm firelight.

Temple came in first, his brown face flushed with drink, his coat unbuttoned, his hands deep in his trousers' pockets. His Christmas resolution had long since dissolved in alcohol. He was a little puzzled to find himself in Findlay's company. And his fuddled brain insisted upon inopportune reminiscence. He walked straight to the fire and stood before it, an exaggerated black figure, staring down into the red glow. "After all," he said, "we are fools to quarrel—fools to quarrel about a little thing like that. Damned fools!"

Findlay went to the writing-table and felt about for the matches with quivering hands.

"It wasn't my doing," he said.

"It wasn't your doing," said Temple. "Nothing ever was your doing. You are always in the right—Findlay the all-right."

Findlay's attention was concentrated upon the lamp. His hand was unsteady, and he had some difficulty in turning up the wicks; one got jammed down and the other flared furiously. When at last it was lit and turned up, he came up to Temple. "Take your coat off, old man, and have some more whisky," he said. "That was a ripping little girl in the skirt dance."

"Fools to quarrel," said Temple slowly, and then woke up to Findlay's words. "Heigh?"

"Take off your coat and sit down," said Findlay, moving

up the little metal table and producing cigars and a siphon and whisky. "That lamp gives an infernally bad light, but it is all I have. Something wrong with the oil. Did you notice the dodge of that stone-smashing trick?"

Temple remained erect and gloomy, staring into the fire. "Fools to quarrel," he said. Findlay was now half drunk, and his finesse began to leave him. Temple had been drinking heavily, and was now in a curious rambling stage. And Findlay's one idea now was to close this curious reunion.

"There's no woman worth a man's friendship," said Temple abruptly.

He sat down in an easy chair, poured out and drank a dose of whisky and lithia. The idea of friendship took possession of him, and he became reminiscent of student days and student adventures. For some time it was, "Do you remember" this, and "Do you remember" that. And Findlay grew cheerful again.

"They were glorious times," said Findlay, pouring whisky into Temple's glass.

Then Temple startled him by abruptly reverting to that bitter quarrel. "No woman in the world," he said. "Curse them!"

He began to laugh stupidly. "After all," he said, "in the end."

"Oh, damn!" said Findlay.

"All very well for you to swear," said Temple, "but you forget about me. 'Tain't your place to swear. If only you'd left things alone——"

"I thought the password was forget," said Findlay.

Temple stared into the fire for a space. "Forget," he said, and then with a curious return to a clarity of speech, "Findlay, I'm getting drunk."

"Nonsense, man, take some more."

Temple rose out of his chair with a look of one awakening. "There's no reason why I should get drunk, because——"

"Drink," said Findlay, "and forget it."

"Faugh! I want to stick my head in water. I want to

think. What the deuce am I doing here, with *you* of all people?"

"Nonsense! *Talk* and forget it, if you won't drink. Do you remember old Jason and the boxing-gloves? I wonder whether you could put up your fives now."

Temple stood with his back to the fire, his brain spinning with drink, and the old hatred of Findlay came back in flood. He sought in his mind for some offensive thing to say, and his face grew dark. Findlay saw that a crisis was upon him and he cursed under his breath. His air of conviviality, his pose of hearty comforter, grew more and more difficult. But what else was there to do?

"Old Jason—full of science and as slow as an elephant!— but he made boxers of us. Do you remember our little set-to—at that place in Gower Street?"

To show his innocent liveliness, his freedom from preoccupation, Findlay pushed his chair aside, and stepped out into the middle of the room. There he began to pose in imitation of Jason, and to give a colourable travesty of the old prize-fighter's instructions. He picked up his boxing-gloves from the shelf in the recess, and slipped them on. Temple, lowering there on the brink of an explosion, was almost too much for his nerves. He felt his display of high spirits was a mistake, but he must go through with it now.

"Don't stand glooming there, man. You're in just that state when the world looks black as ink. Drink yourself merry again. There's no woman in the world worth a man's friendship—that's agreed upon. Come and have a bout with these gloves of mine—four-ounce gloves. There's nothing sets the blood and spirits stirring like that."

"All right," said Temple, quite mechanically. And then, waking up to what he was doing, "Where are the other gloves?"

"Over there in the corner. On the top of the mineral cabinet. By Jove! Temple, this is like old times!"

Temple, quivering strangely, went to the corner. He meant to thrash Findlay, and knew that in spite of his lighter weight he would do it. Yet it seemed puerile and inadequate to the pitch of absurdity, for the wrong Findlay had done him was great. And putting his hand on something pale in the shadow, he touched the *bulla* of the whale. The temptation was like a lightning flash. He slipped one glove on his left hand, and thrust the fingers of his right into the cavity of the *bulla*. It took all his fingers, and covered his knuckles and the back of his hand. And it was so oddly like a thumbless boxing glove! Just the very shape of the padded part. His spirits rose abruptly at the sudden prospect of a savage joke—how savage it would be, he did not know. Meanwhile Findlay, with a nervous alacrity, moved the lamp into the corner behind the arm-chair, and thrust his writing-desk into the window bay.

"Come on," said Findlay, behind him, and abruptly he turned.

Findlay looked straight into his eyes, on guard, his hands half open. He did not see the strange substitute for a glove that covered Temple's right hand. Both men were gone so far towards drunkenness that their power of observation was obscured. For a moment they stood squaring at one another, the host smiling, and his guest smiling also, but with his teeth set; two dark figures swaying in the firelight and the dim lamp-light. Then Findlay struck at his opponent's face with his left hand. As he did so Temple ducked slightly to the left, and struck savagely over Findlay's shoulder at his temple with the bone-covered fist. The blow was given with such tremendous force that it sent Findlay reeling sideways, half stunned, and overcome with astonishment. The thing struck his ear, and the side of his face went white at the blow. He struggled to keep his footing, and as he did so Temple's gloved right hand took him in the chest and sent him spinning to the foot of the cigar cabinet.

Findlay's eyes were wide open with astonishment. Temple was a lighter man by a stone or more than himself, and he did not understand how he had been felled. He was not stunned, although he was so dulled by the blow as not to notice the blood running down his cheek from his ear. He laughed insincerely, and, almost pulling the cigar cabinet over, scrambled to his feet, made as if he would speak, and put up his hand instinctively as Temple struck out at him again, a feint with the left hand. Findlay was an expert boxer, and, anticipating another right-hand blow over the ear, struck sharply at once with his own left hand in Temple's face, throwing his full weight into the blow, and dodging Temple's reply.

Temple's upper lip was cut against his teeth, and the taste of blood and the sight of it trickling down Findlay's cheek destroyed the last vestiges of restraint that drink had left him, stripped of all that education had ever done for him. There remained now only the savage man animal, the creature that thirsts for blood. With a half-bestial cry, he flung himself upon Findlay as he jumped back, and with a sudden sweep of his right arm cut down the defence, breaking Findlay's arm just above the wrist, and following with three rapid blows of the *bulla* upon the face. Findlay gave an inarticulate cry of astonishment, countered weakly once, and then went down like a felled ox. As he fell, Temple fell upon the top of him. There was a smash as the lamp went down.

The lamp was extinguished as it fell, and left the room red and black. Findlay struck heavily at Temple's ribs, and Temple, with his left elbow at Findlay's neck, swung up his right arm and struck down a sledge-hammer blow upon the face, and again, and yet again, until the body beneath his knees had ceased to writhe.

Then suddenly his frenzy left him at the voice of a woman shrieking so that it filled the room. He looked up and crouched motionless as he heard and saw the study door closing and heard the patter of feet retreating in panic. Then he looked down and saw the thing that had once been the face of Findlay. For an awful minute he remained kneeling agape.

Then he staggered to his feet and stood over Findlay's body in the glow of the dying fire, like a man awakening from a nightmare. Suddenly he perceived the *bulla* on his hand, covered with blood and hair, and began to understand what had happened. In a sudden horror he flung the diabolical thing from him. It struck the floor near the cigar cabinet, rolled for a yard or so on its edge, and came to rest in almost the position it had occupied when he had first set eyes on it. To Temple's excited imagination it seemed to be lying at exactly the same spot, the sole and sufficient cause of Findlay's death and his own.

MY FIRST AEROPLANE

MY FIRST
AEROPLANE

("ALAUDA MAGNA")

My First Aeroplane! What vivid memories of youth that
recalls!

Far back it was in the spring of 1912, that I acquired
"Alauda Magna," the great Lark, for so I christened her;
and I was then a slender young man of four-and-twenty,
with hair—beautiful blond hair—all over my adventurous
young head. I was a dashing young fellow enough, in spite
of the slight visual defect that obliged me to wear spectacles
on my prominent, aquiline, but by no means shapeless nose
—the typical flyer's nose. I was a good runner and swimmer,
a vegetarian as ever, an all-wooler, and an ardent advocate
of the extremist views in every direction about everything.
Precious little in the way of a movement got started that I
wasn't in. I owned two motor-bicycles, and an enlarged
photograph of me at that remote date, in leather skull-cap,
goggles, and gauntlets, still adorns my study fire-place. I
was also a great flyer of war-kites, and a voluntary scout-
master of high repute. From the first beginnings of the
boom in flying, therefore, I was naturally eager for the fray.

I chafed against the tears of my widowed mother for a
time, and at last told her I could endure it no longer. "If
I am not the first to fly in Mintonchester," I said, "I leave
Mintonchester. I'm your own son, mummy, and that's me!"

And it didn't take me a week to place my order when she
agreed.

561

I found one of the old price-lists the other day in a drawer, full of queer woodcuts of still queerer contrivances. What a time that was! An incredulous world had at last consented to believe that it could fly, and in addition to the motor-car people and the bicycle people, and so on, a hundred new, unheard-of firms were turning out aeroplanes of every size and pattern to meet the demand. Amazing prices they got for them, too—three hundred and fifty was cheap for the things! I find four hundred and fifty, five hundred, five hundred *guineas* in this list of mine; and many as capable of flight as oak trees! They were sold, too, without any sort of guarantee, and with the merest apology for instruction. Some of the early aeroplane companies paid nearly 200 per cent. on their ordinary shares in those early years.

How well I remember the dreams I had—and the doubts!

The dreams were all of wonder in the air. I saw myself rising gracefully from my mother's paddock, clearing the hedge at the end, circling up to get over the vicar's pear trees, and away between the church steeple and the rise of Withycombe, towards the market-place. Lord! how they would stare to see me! "Young Mr. Betts again!" they would say. "We *knew* he'd do it."

I would circle and perhaps wave a handkerchief, and then I meant to go over Lupton's gardens to the grounds of Sir Digby Foster. There a certain fair denizen might glance from the window. . . .

Ah, youth! Youth!

My doubts were all of the make I should adopt, the character of the engines I should choose. . . .

I remember my wild rush on my motor-bike to London to see the things and give my order, the day of muddy-traffic dodging as I went from one shop to another, my growing exasperation at hearing everywhere the same refrain, "Sold out! Can't undertake to deliver before the beginning of April."

Not me!

I got "Alauda Magna" at last at a little place in Blackfriars Road. She was an order thrown on the firm's hands at the eleventh hour by the death of the purchaser through another maker, and I ran my modest bank account into an overdraft to get her—to this day I won't confess the price I paid for her. Poor little Mumsy! Within a week she was in my mother's paddock, being put together after transport by a couple of not-too-intelligent mechanics.

The joy of it! And a sort of adventurous tremulousness. I'd had no lessons—all the qualified teachers were booked up at stupendous fees for months ahead; but it wasn't in my quality to stick at a thing like that! I couldn't have endured three days' delay. I assured my mother I had had lessons, for her peace of mind—it is a poor son who will not tell a lie to keep his parent happy.

I remember the exultant turmoil of walking round the thing as it grew into a credible shape, with the consciousness of half Mintonchester peering at me through the hedge, and only deterred by our new trespass-board and the disagreeable expression of Snape, our trusted gardener, who was partly mowing the grass and partly on sentry-go with his scythe, from swarming into the meadow. I lit a cigarette and watched the workmen sagely, and we engaged an elderly unemployed named Snorticombe to keep watch all night to save the thing from meddlers. In those days, you must understand, an aeroplane was a sign and a wonder.

"Alauda Magna" was a darling for her time, though nowadays I suppose she would be received with derisive laughter by every schoolboy in the land. She was a monoplane, and, roughly speaking, a Bleriot, and she had the dearest, neatest seven-cylinder forty horse-power G.K.C. engine, with its G.B.S. fly-wheel, that you can possibly imagine. I spent an hour or so tuning her up—she had a deafening purr, rather like a machine-gun in action—until the vicar sent round to say that he was writing a sermon upon "Peace" and was unable to concentrate his mind on that topic until I desisted. I took his objection in good part, and, after a culminating volley and one last lingering look, started for a stroll round the town.

In spite of every endeavour to be modest I could not but feel myself the cynosure of every eye. I had rather carelessly

forgotten to change my leggings and breeches I had bought
for the occasion, and I was also wearing my leather skull-
cap with ear-flaps carelessly adjusted, so that I could hear
what people were saying. I should think I had half the
population under fifteen at my heels before I was halfway
down the High Street.

"You going to fly, Mr. Betts?" says one cheeky youngster.

"Like a bird!" I said.

"Don't you fly till we comes out of school," says another.

It was a sort of Royal progress that evening for me. I
visited old Lupton, the horticulturist, and he could hardly
conceal what a great honour he thought it. He took me
over his new greenhouse—he had now got, he said, three
acres of surface under glass—and showed me all sorts of
clever dodges he was adopting in the way of intensive culture, and afterwards we went down to the end of his old
flower-garden and looked at his bees. When I came out my
retinue of kids was still waiting for me, reinforced. Then
I went round by Paramors and dropped into the Bull and
Horses, just as if there wasn't anything particular up, for a
lemon squash. Everybody was talking about my aeroplane.
They just shut up for a moment when I came in, and then
burst out with questions. It's odd nowadays to remember
all that excitement. I answered what they had to ask me
and refrained from putting on any side, and afterwards Miss
Flyteman and I went into the commercial-room and turned
over the pages of various illustrated journals and compared
the pictures with my machine in a quiet, unassuming sort of
way. Everybody encouraged me to go up—everybody.

I lay stress on that because, as I was soon to discover,
the tides and ebbs of popular favour are among the most
inexplicable and inconsistent things in the world.

I particularly remember old Cheeseman, the pork-butcher,
whose pigs I killed, saying over and over again, in a tone
of perfect satisfaction, "You won't 'ave any difficulty in
going *up*, you won't. There won't be any difficulty 'bout
going *up*." And winking and nodding to the other eminent
tradesmen there assembled.

I *hadn't* much difficulty in going up.   "Alauda Magna"

was a cheerful lifter, and the roar and spin of her engine had hardly begun behind me before she was off her wheels— snap, snap, they came up above the *ski* gliders—and swaying swiftly across the meadows towards the vicarage hedge. She had a sort of onward roll to her, rather like the movement of a corpulent but very buoyant woman.

I had just a glimpse of brave little mother, trying not to cry, and full of pride in me, on the veranda, with both the maids and old Snape beside her, and then I had to give all my attention to the steering-wheel if I didn't want to barge into the vicar's pear-trees.

I'd felt the faintest of tugs just as I came up, and fancied I heard a resounding whack on our new Trespassers will be Prosecuted board, and I saw the crowd of people in the lane running this way and that from my loud humming approach; but it was only after the flight was all over that I realised what that fool Snorticombe had been up to. It would seem he had thought the monster needed tethering—I won't attempt to explain the mysteries of his mind—and he had tied about a dozen yards of rope to the end of either wing and fixed them firmly to a couple of iron guy-posts that belonged properly to the Badminton net. Up they came at the tug of "Alauda," and now they were trailing and dancing and leaping along behind me, and taking the most vicious dives and lunges at everything that came within range of them. Poor old Templecom got it hottest in the lane, I'm told—a frightful whack on his bald head; and then we ripped up the vicar's cucumber frames, killed and scattered his parrot, smashed the upper pane of his study window, and just missed the housemaid as she stuck her head out of the upper bedroom window. I didn't, of course, know anything of this at the time—it was on a lower plane altogether from my proceedings. I was steering past his vicarage—a narrow miss—and trying to come round to clear the pear trees at the end of the garden—which I did with a graze—and the trailers behind me sent leaves and branches flying this way and that. I had reason to thank Heaven for my sturdy little G.K.C.'s.

Then I was fairly up for a time.

I found it much more confusing, than I had expected:
the engine made such an infernal whir-r-row for one thing,
and the steering tugged and struggled like a thing alive. But
I got her heading over the market-place all right. We
buzzed over Stunt's the greengrocer, and my trailers hopped
up his back premises and made a sanguinary mess of the tiles
on his roof, and sent an avalanche of broken chimney-pot
into the crowded street below. Then the thing dipped—I
suppose one of the guy-posts tried to anchor for a second in
Stunt's rafters— and I had the hardest job to clear the Bull
and Horses stables. I didn't, as a matter of fact, completely clear them. The ski-like alighting runners touched
the ridge for a moment and the left wing bent against the
top of the chimney-stack and floundered over it in an awkward, destructive manner.

I'm told that my trailers whirled about the crowded
market-place in the most diabolical fashion as I dipped and
recovered, but I'm inclined to think all this part of the
story has been greatly exaggerated. Nobody was killed,
and I couldn't have been half a minute from the time I
appeared over Stunt's to the time when I slid off the stable
roof and in among Lupton's glass. If people had taken
reasonable care of themselves instead of gaping at me, they
wouldn't have got hurt. I had enough to do without pointing out to people that they were likely to be hit by an iron
guy-post which had seen fit to follow me. If anyone ought
to have warned them it was that fool Snorticombe. Indeed,
what with the incalculable damage done to the left wing
and one of the cylinders getting out of rhythm and making
an ominous catch in the whirr, I was busy enough for anything on my own private personal account.

I suppose I am in a manner of speaking responsible for
knocking old Dudney off the station bus, but I don't see
that I can be held answerable for the subsequent evolutions
of the bus, which ended after a charge among the market
stalls in Cheeseman's shop-window, nor do I see that I am to
blame because an idle and ill-disciplined crowd chose to
stampede across a stock of carelessly distributed earthenware

and overturned a butter stall. I was a mere excuse for all this misbehaviour.

I didn't exactly fall into Lupton's glass, and I didn't exactly drive over it. I think ricochetting describes my passage across his premises as well as any single word can.

It was the queerest sensation, being carried along by this big, buoyant thing, which had, as it were, bolted with me, and feeling myself alternately lifted up and then dropped with a scrunch upon a fresh greenhouse-roof, in spite of all my efforts to get control. And the infinite relief when at last, at the fifth or sixth pounce, I rose—and kept on rising!

I seemed to forget everything disagreeable instantly. The doubt whether after all "Alauda Magna" was good for flying vanished. She was evidently very good. We whirred over the wall at the end, with my trailers still bumping behind, and beyond one of them hitting a cow, which died next day, I don't think I did the slightest damage to anything or anybody all across the breadth of Cheeseman's meadow. Then I began to rise, steadily but surely, and, getting the thing well in hand, came swooping round over his piggeries to give Mintonchester a second taste of my quality.

I meant to go up in a spiral until I was clear of all the trees and things and circle about the church spire. Hitherto I had been so concentrated on the plunges and tugs of the monster I was driving and so deafened by the uproar of my engine, that I had noticed little of the things that were going on below; but now I could make out a little lot of people, headed by Lupton with a garden fork, rushing obliquely across the corner of Cheeseman's meadow. It puzzled me for a second to imagine what they could think they were after.

Up I went, whirring and swaying, and presently got a glimpse down the High Street of the awful tangle everything had got into in the market-place. I didn't at the time connect that extraordinary smash-up with my transit.

It was the jar of my whack against the weathercock that really stopped my engines. I've never been able to make out quite how it was I hit the unfortunate vane; perhaps the twist I had given my left wing on Stunt's roof spoilt my steering; but, anyhow, I hit the gaudy thing and bent it, and for a lengthy couple of seconds I wasn't by any means sure whether I wasn't going to dive straight down into the market-place. I got her right by a supreme effort—I think the people I didn't smash might have squeezed out one drop of gratitude for that—drove pitching at the tree-tops of Withycombe, got round, and realised the engines were stopping. There wasn't any time to survey the country and arrange for a suitable landing place; there wasn't any chance of clearing the course. It wasn't my fault if a quarter of the population of Mintonchester was swarming out over Cheeseman's meadows. It was the only chance I had to land without a smash, and I took it. Down I came, a steep glide, doing the best I could for myself.

Perhaps I did bowl a few people over; but progress is progress.

And I had to kill his pigs. It was a case of either dropping among the pigs and breaking my rush, or going full tilt into the corrugated iron piggeries beyond. I might have been cut to ribbons. And pigs are born to die.

I stopped, and stood up stiffly upon the framework and looked behind me. It didn't take me a moment to realise that Mintonchester meant to take my poor efforts to give it an Aviation Day all to itself in a spirit of ferocious ingratitude.

The air was full of the squealing of the two pigs I had pinned under my machine and the bawling of the nearer spectators. Lupton occupied the middle distance with a garden fork, with the evident intention of jabbing it into my stomach. I am always pretty cool and quick-witted in an emergency. I dropped off poor "Alauda Magna" like a shot, dodged through the piggery, went up by Frobisher's orchard, nipped over the yard wall of Hinks's cottages, and was into the police-station by the back way before anyone could get within fifty feet of me.

"Halloa!" said Inspector Nenton; "smashed the thing?"

"No," I said; "but people seem to have got something the matter with them. I want to be locked in a cell." . . .

For a fortnight, do you know, I wasn't allowed to come

near my own machine. I went home from the police-station as soon as the first excitement had blown over a little, going round by Love Lane and the Chart so as not to arouse any febrile symptoms. I found mother frightfully indignant, you can be sure, at the way I had been treated. And there, as I say, was I, standing a sort of siege in the upstairs rooms, and sturdy little "Alauda Magna," away in Cheeseman's fields, being walked round and stared at by everybody in the world but me. Cheeseman's theory was that he had seized her. There came a gale one night, and the dear thing was blown clean over the hedge among Lupton's green-houses again, and then Lupton sent round a silly note to say that if we didn't remove her she would be sold to defray expenses, going off into a long tirade about damages and his solicitor. So mother posted off to Clamps', the fur-niture removers at Upnorton Corner, and they got hold of a timber-waggon, and popular feeling had allayed suffi-ciently before that arrived for me to go in person to superin-tend the removal. There she lay like a great moth above the *débris* of some cultural projects of Lupton's, scarcely damaged herself except for a hole or so and some bent rods and stays in the left wing and a smashed skid. But she was bespattered with pigs' blood and pretty dirty.

I went at once by instinct for the engines, and had them in perfect going order before the timber-waggon arrived.

A sort of popularity returned to me with that procession home. With the help of a swarm of men we got "Alauda Magna" poised on the waggon, and then I took my seat to see she balanced properly, and a miscellaneous team of seven horses started to tow her home. It was nearly one o'clock when we got to that, and all the children turned out to shout and jeer. We couldn't go by Pook's Lane and the vicarage, because the walls are too high and narrow, and so we headed across Cheeseman's meadows for Stokes' Waste and the Common, to get round by that *détour*.

I was silly, of course, to do what I did—I see that now—but sitting up there on my triumphal car with all the mul-titude about me excited me. I got a kind of glory on. I really only meant to let the propellers spin as a sort of hurrahing, but I was carried away. Whuz-z-z-z! It was like something blowing up, and behold! I was sailing and plunging away from my wain across the common for a second flight.

"Lord!" I said.

I fully meant to run up the air a little way, come about, and take her home to our paddock, but those early aeroplanes were very uncertain things.

After all, it wasn't such a very bad shot to land in the vicarage garden, and that practically is what I did. And I don't see that it was my fault that all the vicarage and a lot of friends should be having lunch on the lawn. They were doing that, of course, so as to be on the spot without having to rush out of the house when "Alauda Magna" came home again. Quiet exultation—that was their game. They wanted to gloat over every particular of my ignominious return. You can see that from the way they had arranged the table. I can't help it if Fate decided that my return wasn't to be so ignominious as all that, and swooped me down on the lot of them.

They were having their soup. They had calculated on me for the dessert, I suppose.

To this day I can't understand how it is I didn't kill the vicar. The forward edge of the left wing got him just under the chin and carried him back a dozen yards. He must have had neck vertebræ like steel; and even then I was amazed his head didn't come off. Perhaps he was holding on underneath; but I can't imagine where. If it hadn't been for the fascination of his staring face I think I could have avoided the veranda, but, as it was, that took me by surprise. That was a fair crumple up. The wood must have just rotted away under its green paint; but, anyhow, it and the climbing roses and the shingles above and everything snapped and came down like stage scenery, and I and the engines and the midde part drove clean through the French windows on to the drawing-room floor. It was jolly lucky for me, I think, that the French windows weren't shut. There's no unpleasanter way of getting hurt in the world than flying suddenly through thin window-glass;

and I think I ought to know. There was a frightful jaw-bation, but the vicar was out of action, that was one good thing. Those deep, sonorous sentences! But perhaps they would have calmed things. . . .

That was the end of "Alauda Magna," my first aeroplane. I never even troubled to take her away. I hadn't the heart to. . . .

And then the storm burst.

The idea seems to have been to make mother and me pay for everything that had ever tumbled down or got broken in Mintonchester since the beginning of things. Oh! and for any animal that had ever died a sudden death in the memory of the oldest inhabitant. The tariff ruled high, too. Cows were twenty-five to thirty pounds and upward; pigs about a pound each, with no reduction for killing a quantity; verandas—verandas were steady at forty-five guineas. Dinner services, too, were up, and so were tiling and all branches of the building trade. It seemed to certain persons in Mintonchester, I believe, that an era of unexampled prosperity had dawned upon the place—only limited, in fact, by the solvency of me and mother. The vicar tried the old "sold to defray expenses" racket, but I told him he might sell.

I pleaded defective machinery and the hand of God, did my best to shift the responsibility on to the firm in Black-friars Road, and, as an additional precaution, filed my petition in bankruptcy. I really hadn't any property in the world, thanks to mother's goodness, except my two motor-bicycles, which the brutes took, my photographic dark-room, and a lot of bound books on aeronautics and progress generally. Mother, of course, wasn't responsible. She hadn't lifted a wing.

Well, for all that, disagreeables piled up so heavily on me, what with being shouted after by a rag-tag and bobtail of schoolboys and golf caddies and hobbledehoys when I went out of doors, threatened with personal violence by stupid people like old Lupton, who wouldn't understand that a man can't pay what he hasn't got, pestered by the wives of various gentlemen who saw fit to become out-of-works on the strength of alleged injuries, and served with all sorts of silly summonses for all sorts of fancy offences, such as mischievous mischief and manslaughter and wilful damage and trespass, that I simply had to go away from Mintonchester to Italy, and leave poor little mother to manage them in her own solid, undemonstrative way. Which she did, I must admit, like a Brick.

They didn't get much out of her, anyhow, but she had to break up our little home at Mintonchester and join me at Arosa, in spite of her dislike of Italian cooking. She found me already a bit of a celebrity because I had made a record, so it seemed, by falling down three separate crevasses on three successive days. But that's another story altogether.

From start to finish I reckon that first aeroplane cost my mother over nine hundred pounds. If I hadn't put my foot down, and she had stuck to her original intention of paying all the damage, it would have cost her three thousand. . . . But it was worth it. It was worth it. I wish I could live it all over again; and many an old codger like me sits at home now and deplores those happy, vanished, adventurous times, when any lad of spirit was free to fly—and go anywhere—and smash anything—and discuss the question afterwards of just what the damages amounted to and what his legal liability might be.

LITTLE MOTHER UP THE MÖRDERBERG

LITTLE MOTHER UP THE MÖRDERBERG

I THINK I mentioned when I was telling how I sailed my first aeroplane that I made a kind of record at Arosa by falling down three separate crevasses on three successive days. That was before little mother followed me out there. When she came, I could see at a glance she was tired and jaded and worried, and so, instead of letting her fret about in the hotel and get into a wearing tangle of gossip, I packed her and two knapsacks up, and started off on a long, refreshing, easy-going walk northward, until a blister on her foot stranded us at the Magenruhe Hotel on the Sneejoch. She was for going on, blister or no blister—I never met pluck like mother's in all my life—but I said "No. This is a mountaineering inn, and it suits me down to the ground— or if you prefer it, up to the sky. You shall sit in the veranda by the telescope, and I'll prance about among the peaks for a bit."

"Don't have accidents," she said.

"Can't promise that, little mother," I said; "but I'll always remember I'm your only son."

So I pranced. . . .

I need hardly say that in a couple of days I was at loggerheads with all the mountaineers in that inn. They couldn't stand me. They didn't like my neck with its strong, fine Adam's apple—being mostly men with their heads *jammed* on—and they didn't like the way I bore myself and lifted my aviator's nose to the peaks. They didn't like my being a vegetarian and the way I evidently enjoyed it, and they didn't like the touch of colour, orange and green, in my rough serge suit. They were all of the dingy school—the sort of men I call gentlemanly owls—shy, correct-minded creatures, mostly from Oxford, and as solemn over their climbing as a cat frying eggs. Sage they were, great headnodders, and "I-wouldn't-venture-to-do-a-thing-like-that"-ers. They always did what the books and guides advised, and they classed themselves by their seasons; one was in his ninth season, and another in his tenth, and so on. I was a novice and had to sit with my mouth open for bits of humble-pie.

My style that! Rather!

I would sit in the smoking-room sucking away at a pipeful of hygienic herb tobacco—they said it smelt like burning garden rubbish—and waiting to put my spoke in and let a little light into their minds. They set aside their natural reticence altogether in their efforts to show how much they didn't like me.

"You chaps take these blessed mountains too seriously," I said. "They're larks, and you've got to lark with them."

They just slued their eyes round at me.

"I don't find the solemn joy in fussing you do. The old-style mountaineers went up with alpenstocks and ladders and light hearts. That's my idea of mountaineering."

"It isn't ours," said one red-boiled hero of the peaks, all blisters and peeling skin, and he said it with an air of crushing me.

"It's the right idea," I said serenely, and puffed at my herb tobacco.

"When you've had a bit of experience you'll know better," said another, an oldish young man with a small grey beard.

"Experience never taught *me* anything," I said.

"Apparently not," said someone, and left me one down and me to play. I kept perfectly tranquil.

"I mean to do the Mörderberg before I go down," I said quietly, and produced a sensation.

"When are you going down?"

"Week or so," I answered, unperturbed.

"It's not the climb a man ought to attempt in his first year," said the peeling gentleman.

"*You* particularly ought not to try it," said another.

"No guide will go with you."

"Foolhardy idea."

"Mere brag."

"Like to see him do it."

I just let them boil for a bit, and when they were back to the simmer I dropped in, pensively, with, "Very likely I'll take that little mother of mine. She's small, bless her, and she's as hard as nails."

But they saw they were being drawn by my ill-concealed smile; and this time they contented themselves with a few grunts and grunt-like remarks, and then broke up into little conversations in undertones that pointedly excluded me. It had the effect of hardening my purpose. I'm a stiff man when I'm put on my mettle, and I determined that the little mother *should* go up the Mörderberg, where half these solemn experts hadn't been, even if I had to be killed or orphaned in the attempt. So I spoke to her about it the next day. She was in a deck-chair on the veranda, wrapped up in rugs and looking at the peaks.

"Comfy?" I said.

"Very," she said.

"Getting rested?"

"It's so nice."

I strolled to the rail of the veranda. "See that peak there, mummy?"

She nodded happily, with eyes half shut.

"That's the Mörderberg. You and me have got to be up there the day after to-morrow."

Her eyes opened a bit. "Wouldn't it be rather a climb, dearest?" she said.

"I'll manage that all right," I said, and she smiled consentingly and closed her eyes.

"So long as you manage it," she said.

I went down the valley that afternoon to Daxdam to get gear and guides and porters, and I spent the next day in glacier and rock practice above the hotel. That didn't add

to my popularity. I made two little slips. One took me
down a crevasse—I've an extraordinary knack of going down
crevasses—and a party of three which was starting for the
Kinderspitz spent an hour and a half fishing me out; and
the other led to my dropping my ice-axe on a little string
of people going for the Humpi glacier. It didn't go within
thirty inches of anyone, but you might have thought from
the row they made that I had knocked out the collective
brains of the party. Quite frightful language they used, and
three ladies with them, too!

The next day there was something very like an organised
attempt to prevent our start. They brought out the landlord,
they remonstrated with mother, they did their best to
blacken the character of my two guides. The landlord's
brother had a first-class row with them.

"Two years ago," he said, "they lost their Herr!"

"No particular reason," I said, "why you shouldn't keep
yours on, is it?"

That settled him. He wasn't up to a polyglot pun, and
it stuck in his mind like a fishbone in the throat.

Then the peeling gentleman came along and tried to
overhaul our equipment. "Have you got this?" it was,
and "Have you got that?"

"Two things," I said, looking at his nose pretty hard,
"we haven't forgotten. One's blue veils and the other
vaseline."

I've still a bright little memory of the start, There was
the pass a couple of hundred feet or so below the hotel,
and the hotel—all name and windows—standing out in a
great, desolate, rocky place against lumpy masses of streaky
green rock, flecked here and there with patches of snow
and dark shelves of rhododendron, and rising perhaps a
thousand feet towards the western spur of the massif. Our
path ran before us, meandering among the boulders down to
stepping-stones over a rivulet, and then upward on the
other side of the stream towards the Magenruhe glacier,
where we had to go up the rocks to the left and then across
the icefall to shelves on the precipitous face on the west side.
It was dawn, the sun had still to rise, and everything looked

very cold and blue and vast about us. Everyone in the hotel had turned out to bear a hand in the row—some of the *deshabilles* were disgraceful—and now they stood in a silent group watching us recede. The last word I caught was, "They'll have to come back."

"We'll come back all right," I answered. "Never fear."

And so we went our way, cool and deliberate, over the stream and up and up towards the steep snowfields and icy shoulder of the Mörderberg. I remember that we went in absolute silence for a time, and then how suddenly the landscape gladdened with sunrise, and in an instant, as if speech had thawed, all our tongues were babbling.

I had one or two things in the baggage that I hadn't cared for the people at the inn to see, and I had made no effort to explain why I had five porters with the load of two and a half. But when we came to the icefall I showed my hand a little, and unslung a stout twine hammock for the mater. We put her in this with a rug round her, and sewed her in with a few stitches; then we roped off in line, with me last but one and a guide front and rear, and mummy in the middle carried by two of the porters. I stuck my alpenstock through two holes I had made in the shoulders of my jacket under my rucksack, T-shape to my body, so that when I went down a crevasse, as I did ever and again, I just stuck in its jaws and came up easy as the rope grew taut. And so, except for one or two bumps that made the mater chuckle, we got over without misadventure.

Then came the rock climb on the other side, requiring much judgment. We had to get from ledge to ledge as opportunity offered, and here the little mother was a perfect godsend. We unpacked her after we had slung her over the big fissure—I forget what you call it—that always comes between glacier and rock—and whenever we came to a bit of ledge within eight feet of the one we were working along, the two guides took her and slung her up, she being so light, and then she was able to give a foot for the next man to hold by and hoist himself. She said we were all pulling her leg, and that made her and me laugh so much that the whole party had to wait for us.

It was pretty tiring altogether doing that bit of the climb —two hours we had of it before we got to the loose masses of rock on the top of the arete. "It's worse going down," said the elder guide.

I looked back for the first time, and I confess it did make me feel a bit giddy. There was the glacier looking quite pretty, and with a black gash between itself and the rocks.

For a time it was pretty fair going up the rocky edge of the arete, and nothing happened of any importance, except that one of the porters took to grousing because he was hit on the shin by a stone I dislodged. "Fortunes of war," I said, but he didn't seem to see it, and when I just missed him with a second he broke out into a long, whining discourse in what I suppose he thought was German—I couldn't make head or tail of it.

"He says you might have killed him," said the little mother.

"They say," I quoted, "What say they? *Let* them say."

I was for stopping and filling him up with a feed, but the elder guide wouldn't have it. We had already lost time, he said, and the traverse round the other face of the mountain would be more and more subject to avalanches as the sun got up. So we went on. As we went round the corner to the other face I turned towards the hotel—it was the meanest little oblong spot by now—and made a derisive gesture or so for the benefit of anyone at the telescope.

We did get one rock avalanche that reduced the hindmost guide to audible prayer, but nothing hit us except a few bits of snow. The rest of the fall was a couple of yards and more out from us. We were on rock just then and overhung; before and afterwards we were edging along steps in an ice-slope cut by the foremost guide, and touched up by the porters. The avalanche was much more impressive before it came in sight, banging and thundering overhead, and it made a tremendous uproar in the blue deeps beneath, but in actual transit it seemed a mean show—mostly of stones smaller than I am.

"All right?" said the guide.

"Toned up," I answered.

"I suppose it *is* safe, dear?" asked the little mother.

"Safe as Trafalgar Square," I said. "Hop along, mummy-kins."

Which she did with remarkable agility.

The traverse took us on to old snow at last, and here we could rest for lunch—and pretty glad we were both of lunch and rest. But here the trouble with the guides and porters thickened. They were already a little ruffled about my animating way with loose rocks, and now they kicked up a tremendous shindy because instead of the customary brandy we had brought non-alcoholic ginger cordial. Would they even try it? Not a bit of it! It was a queer little dispute, high up in that rarefied air about food values and the advantages of making sandwiches with nuttar. They were an odd lot of men, invincibly set upon a vitiated and vitiating dietary. They wanted meat, they wanted alcohol, they wanted narcotics to smoke. You might have thought that men like these, living in almost direct contact with Nature, would have liked "Nature" foods, such as plasmon, protose, plobose, digestine, and so forth. Not them! They just craved for corruption. When I spoke of drinking pure water one of the porters spat in a marked, symbolic manner over the precipice. From that point onward discontent prevailed.

We started again about half-past eleven, after a vain attempt on the part of the head guide to induce us to turn back. We had now come to what is generally the most difficult part of the Mörderberg ascent, the edge that leads up to the snowfield below the crest. But here we came suddenly into a draught of warm air blowing from the south-west, and everything, the guide said, was unusual. Usually the edge is a sheet of ice over rock. To-day it was wet and soft, and one could kick steps in it and get one's toes into rock with the utmost ease.

"This is where Herr Tomlinson's party fell," said one of the porters, after we'd committed ourselves to the edge for ten minutes or so.

"Some people could fall out of a four-post bed," I said.

"It'll freeze hard again before we come back," said the

second guide, "and us with nothing but *verdammt* ginge
inside of us."

"You keep your rope taut," said I.

A friendly ledge came to the help of mother in the nic
of time, just as she was beginning to tire, and we sewed he
up all but the feet in her hammock again, and roped he
carefully. She bumped a bit, and at times she was jus
hanging over immensity and rotating slowly, with everybod
else holding like grim death.

"My dear," she said, the first time this happened, "is i
right for me to be doing this?"

"Quite right," I said, "but if you can get a foothol
presently again—it's rather better style."

"You're sure there's no danger, dear?"

"Not a scrap."

"And I don't fatigue you?"

"You're a stimulant."

"The view," she said, "is certainly becoming ver
beautiful."

But presently the view blotted itself out, and we were i
clouds and thin drift of almost thawing snowflakes.

We reached the upper snowfield about half-past one, an
the snow was extraordinarily soft. The elder guide wen
in up to his armpits.

"Frog it," I said, and spread myself out flat, in a sort o
swimming attitude. So we bored our way up to the cres
and along it. We went in little spurts and then stoppe
for breath, and we dragged the little mother after us i
her hammock-bag. Sometimes the snow was so good w
fairly skimmed the surface; sometimes it was so rotten w
plunged right into it and splashed about. I went too nea
the snow cornice once and it broke under me, but th
rope saved me, and we reached the summit about thre
o'clock without further misadventure. The summit wa
just bare rock with the usual cairn and pole. Nothing t
make a fuss about. The drift of snow and cloudwisp ha
passed, the sun was blazing hot overhead, and we seeme
to be surveying all Switzerland. The Magenruhe Hote
was at our toes, hidden, so to speak, by our chins. W

LITTLE MOTHER

squatted about the cairn, and the guides and porters were
reduced to ginger and vegetarian ham-sandwiches. I cut
and scratched an inscription, saying I had climbed on simple
food, and claiming a record.

Seen from the summit the snowfields on the north-east
side of the mountain looked extremely attractive, and I
asked the head guide why that way up wasn't used. He
said something in his peculiar German about precipices.

So far our ascent had been a fairly correct ascent in
rather slow time. It was in the descent that that strain in
me of almost unpremeditated originality had play. I wouldn't
have the rope returning across the upper snowfield, because
mother's feet and hands were cold, and I wanted her to jump
about a bit. And before I could do anything to prevent it she
had slipped, tried to get up by rolling over *down* the slope
instead of up, as she ought to have done, and was leading the
way, rolling over and over and over, down towards the guide's
blessed precipices above the lower snowfield.

I didn't lose an instant in flinging myself after her, axe
up, in glissading attitude. I'm not clear what I meant to
do, but I fancy the idea was to get in front of her and put
on the brake. I did not succeed, anyhow. In twenty
seconds I had slipped, and was sitting down and going
down out of my own control altogether.

Now, most great discoveries are the result of accident,
and I maintain that in that instant mother and I discovered
two distinct and novel ways of coming down a mountain.

It is necessary that there should be first a snow slope
above with a layer of softish, rotten snow on the top of ice,
then a precipice, with a snow-covered talus sloping steeply
at first and then less steeply, then more snow slopes and
precipices according to taste, ending in a snowfield or a
not-too-greatly-fissured glacier, or a reasonable, not-too-rocky
slope. Then it all becomes as easy as chuting the chutes.

Mother hit on the sideways method. She rolled. With
the snow in the adhesive state it had got into she had made
the jolliest little snowball of herself in half a minute, and
the nucleus of as clean and abundant a snow avalanche as
anyone could wish. There was plenty of snow going in

front of her, and that's the very essence of both our methods. You must fall on your snow, not your snow on you, or it smashes you. And you mustn't mix yourself up with loose stones.

I, on the other hand, went down feet first, and rather like a snow-plough; slower than she did, and if, perhaps, with less charm, with more dignity. Also I saw more. But it was certainly a tremendous rush. And I gave a sort of gulp when mummy bumped over the edge into the empty air and vanished.

It was like a toboggan ride gone mad down the slope until I took off from the edge of the precipice, and then it was like a dream.

I'd always thought falling must be horrible. It wasn't in the slightest degree. I might have hung with my clouds and lumps of snow about me for weeks, so great was my serenity. I had an impression then that I was as good as killed—and that it didn't matter. I wasn't afraid—that's nothing!—but I wasn't a bit uncomfortable. Whack! We'd hit something, and I expected to be flying to bits right and left. But we'd only got on to the snow-slope below, at so steep an angle that it was merely breaking the fall. Down we went again. I didn't see much of the view after that because the snow was all round and over my head, but I kept feet foremost and in a kind of sitting posture, and then I slowed and then I quickened again and bumped rather, and then harder, and bumped and then bumped again and came to rest. This time I was altogether buried in snow, and twisted sideways with a lot of heavy snow on my right shoulder.

I sat for a bit enjoying the stillness—and then I wondered what had become of mother, and set myself to get out of the snow about me. It wasn't so easy as you might think; the stuff was all in lumps and spaces like a gigantic sponge, and I lost my temper and struggled and swore a good deal, but at last I managed it. I crawled out and found myself on the edge of heaped masses of snow quite close to the upper part of the Magenruhe glacier. And far away, right up the glacier and near the other side, was a little thing like a black-beetle struggling in the heart of an immense split ball of snow.

I put my hands to my mouth and let out with my version of the yodel, and presently I saw her waving her hand.

It took me nearly twenty minutes to get to her. I knew my weakness, and I was very careful of every crevasse I came near. When I got up to her her face was anxious.

"What have you done with the guides?" she asked.

"They've got too much to carry," I said. "They're coming down another way. Did you like it?"

"Not very much, dear," she said; "but I dare say I shall get used to these things. Which way do we go now?"

I decided we'd find a snow-bridge across the *bergschrund* —that's the word I forgot just now—and so get on to the rocks on the east side of the glacier, and after that we had uneventful going right down to the hotel. . . .

Our return evoked such a strain of hostility and envy as I have never met before or since. First they tried to make out we'd never been to the top at all, but mother's little proud voice settled that sort of insult. And, besides, there was the evidence of the guides and porters following us down. When they asked about the guides, "They're following *your* methods," I said, "and I suppose they'll get back here to-morrow morning somewhere."

That didn't please them.

I claimed a record. They said my methods were illegitimate.

"If I see fit," I said, "to use an avalanche to get back by, what's that to you? You tell me me and mother can't do the confounded mountain anyhow, and when we do you want to invent a lot of rules to disqualify us. You'll say next one mustn't glissade. I've made a record, and you know I've made a record, and you're about as sour as you can be. The fact of it is, you chaps don't know your own silly business. Here's a good, quick way of coming down a mountain, and you ought to know about it——"

"The chance that both of you are not killed was one in a thousand."

"Nonsense! It's the proper way to come down for anyone who hasn't a hide-bound mind. You chaps ought to practise falling great heights in snow. It's perfectly easy and perfectly safe, if only you know how to set about it."

"Look here, young man," said the oldish young man with the little grey beard, "you don't seem to understand that you and that lady have been saved by a kind of miracle——"

"Theory!" I interrupted. "I'm surprised you fellows ever come to Switzerland. If I were your kind I'd just invent theoretical mountains and play for points. However, you're tired, little mummy. It's time you had some nice warm soup and tucked yourself up in bed. I shan't let you get up for six-and-thirty hours."

But it's queer how people detest a little originality.

THE STORY OF THE
LAST TRUMP

THE STORY OF
THE LAST TRUMP

§ 1

THE story of the Last Trump begins in Heaven and it ends
in all sorts of places round about the world. . . .

Heaven, you must know, is a kindly place, and the blessed
ones do not go on for ever singing Alleluia, whatever you
may have been told. For they too are finite creatures, and
must be fed with their eternity in little bits, as one feeds
a chick or a child. So that there are mornings and changes
and freshness, there is time to condition their lives. And
the children are still children, gravely eager about their playing and ready always for new things; just children they are,
but blessèd as you see them in the pictures beneath the
careless feet of the Lord God. And one of these blessèd
children routing about in an attic—for Heaven is, of course,
full of the most heavenly attics, seeing that it has children
—came upon a number of instruments stored away, and laid
its little chubby hands upon them. . . .

Now indeed I cannot tell what these instruments were,
for to do so would be to invade mysteries. . . . But one
I may tell of, and that was a great brazen trumpet which
the Lord God had made when He made the world—for the
Lord God finishes all His jobs—to blow when the time for
our Judgement came round. And He had made it and
left it; there it was, and everything was settled exactly as
child conceived one of those unaccountable passions of
childhood for its smoothness and brassiness, and he played

589

with it and tried to blow it, and trailed it about with him out of the attic into the gay and golden streets, and, after many fitful wanderings, to those celestial battlements of crystal of which you have doubtless read. And there the blessèd child fell to counting the stars, and forgot all about the Trumpet beside him until a flourish of his elbow sent it over. . . .

Down fell the trump, spinning as it fell, and for a day or so, which seemed but moments in heaven, the blessèd child watched its fall until it was a glittering little speck of brightness. . . .

When it looked a second time the trump was gone. . . .

I do not know what happened to that child when at last it was time for Judgement Day and that shining trumpet was missed. I know that Judgement Day is long overpassed, because of the wickedness of the world; I think perhaps it was in A.D. 1000 when the expected Day should have dawned that never came, but no other heavenly particulars do I know at all, because now my scene changes to the narrow ways of this Earth. . . .

And the Prologue in Heaven ends.

§ 2

And now the scene is a dingy little shop in Caledonian Market, where things of an incredible worthlessness lie in wait for such as seek after an impossible cheapness. In the window, as though it had always been there and never any-where else, lies a long, battered, discoloured trumpet of brass that no prospective purchaser has ever been able to sound. In it mice shelter, and dust and fluff have gathered after the fashion of this world. The keeper of the shop is a very old man, and he bought the shop long ago, but already this trumpet was there; he has no idea whence it came, nor its country of origin, nor anything about it. But once in a moment of enterprise that led to nothing he decided to call it an Ancient Ceremonial Shawm, though he ought to have known that whatever a shawm may be the last thing it was likely to be is a trumpet, seeing that they are always mentioned together. And above it hung concertinas and melodeons and cornets and tin whistles and mouth-organs and all that rubbish of musical instruments which delight the hearts of the poor. Until one day two blackened young men from the big motor works in the Pansophist Road stood outside the window and argued.

They argued about these instruments in stock and how you made these instruments sound, because they were fond of argument, and one asserted and the other denied that he could make every instrument in the place sound a note. And the argument rose high, and led to a bet.

"Supposing, of course, that the instrument is in order," said Hoskin, who was betting he could.

"That's understood," said Briggs.

And then they called as witnesses certain other young and black and greasy men in the same employment, and after much argument and discussion that lasted through the afternoon, they went in to the little old dealer about teatime, just as he was putting a blear-eyed, stinking paraffin-lamp, to throw an unfavourable light upon his always very unattractive window. And after great difficulty they arranged that for the sum of one shilling, paid in advance, Hoskin should have a try at every instrument in the shop that Briggs chose to indicate.

And the trial began.

The third instrument that was pitched upon by Briggs for the trial was the strange trumpet that lay at the bottom of the window, the trumpet that you, who have read the Introduction, know was the trumpet for the Last Trump. And Hoskin tried and tried again, and then, blowing desperately, hurt his ears. But he could get no sound from the trumpet. Then he examined the trumpet more carefully and discovered the mice and fluff and other things in it, and demanded that it should be cleaned; and the old dealer, nothing loth, knowing they were used to automobile-horns and such-like instruments, agreed to let them clean it on condition that they left it shiney. So the young men, after making a suitable deposit (which, as you shall hear, was presently confiscated), went off with the trumpet, proposing to clean it next day at the works and polish it with the peculiarly excellent brass polish employed upon the honk-honk horns of the firm. And this they did, and Hoskin tried again.

But he tried in vain. Whereupon there arose a great argument about the trumpet, whether it was in order or not, whether it was possible for any one to sound it. For if not, then clearly it was outside the condition of the bet.

Others among the young men tried it, including two who played wind instruments in a band and were musically knowing men. After their own failure they were strongly on the side of Hoskin and strongly against Briggs, and most of the other young men were of the same opinion.

"Not a bit of it," said Briggs, who was a man of resource. "I'll show you that it can be sounded."

And taking the instrument in his hand, he went towards a peculiarly powerful foot blow-pipe that stood at the far end of the toolshed. "Good old Briggs!" said one of the other young men, and opinion veered about.

Briggs removed the blow-pipe from its bellows and tube, and then adjusted the tube very carefully to the mouthpiece of the trumpet. Then with great deliberation he produced a piece of bees-waxed string from a number of other strange and filthy contents in his pocket and tied the tube to the mouthpiece. And then he began to work the treadle of the bellows.

"Good old Briggs!" said the one who had previously admired him.

And then something incomprehensible happened.

It was a flash. Whatever else it was, it was a flash. And a sound that seemed to coincide exactly with the flash.

Afterwards the young men agreed to it that the trumpet blew to bits. It blew to bits and vanished, and they were all flung upon their faces—not backward, be it noted, but on their faces—and Briggs was stunned and scared. The toolshed windows were broken and the various apparatus and cars around were much displaced, and no *traces of the trumpet were ever discovered.*

That last particular puzzled and perplexed poor Briggs very much. It puzzled and perplexed him the more because he had had an impression, so extraordinary, so incredible, that he was never able to describe it to any other living person. But his impression was this: that the flash that came with the sound came, not from the trumpet but to it, that it smote down to it and took it, and that its shape was in the exact likeness of a hand and arm of fire.

§ 3

And that was not all, that was not the only strange thing about the disappearance of that battered trumpet. There was something else, even more difficult to describe, an effect as though for one instant something opened. . . .

The young men who worked with Hoskin and Briggs had that clearness of mind which comes of dealing with machinery, and they all felt this indescribable something else, as if for an instant the world wasn't the world, but something lit and wonderful, larger——

This is what one of them said of it.

"I felt," he said, "just for a minute—as though I was blown to Kingdom Come."

"It is just how it took me," said another. " 'Lord,' I says, 'here's Judgement Day!' and then there I was sprawling among the files. . . ."

But none of the others felt that they could say anything more definite than that.

§ 4

Moreover, there was a storm. All over the world there was a storm that puzzled meteorology, a moment's gale that left the atmosphere in a state of wild swaygog, rains, tornadoes, depressions, irregularities for weeks. News came of it from all the quarters of the earth.

All over China, for example, that land of cherished graves,

there was a duststorm, dust leaped into the air. A kind of earthquake shook Europe—an earthquake that seemed to have at heart the peculiar interests of Mr. Algernon Ashton; everywhere it cracked mausoleums and shivered the pavements of cathedrals, swished the flower-beds of cemeteries, and tossed tombstones aside. A crematorium in Texas blew up. The sea was greatly agitated, and the beautiful harbour of Sydney, in Australia, was seen to be littered with sharks floating upside down in manifest distress. . . .

And all about the world a sound was heard like the sound of a trumpet instantly cut short.

§ 5

But this much is only the superficial dressing of the story. The reality is something different. It is this: that in an instant, and for an instant, the dead lived, and all that are alive in the world did for a moment see the Lord God and all His powers, His hosts of angels, and all His array looking down upon them. They saw Him as one sees by a flash of lightning in the darkness, and then instantly the world was opaque again, limited, petty, habitual. That is the tremendous reality of this story. Such glimpses have happened in individual cases before. The Lives of the saints abound in them. Such a glimpse it was that came to Devindranath Tagore upon the burning ghat at Benares. But this was not an individual but a world experience; the flash came to every one. Not always was it quite the same, and thereby the doubter found his denials, when presently a sort of discussion broke out in the obscurer Press. For this one testified that it seemed that "One stood very near to me," and another saw "all the hosts of heaven flame up towards the Throne."

And there were others who had a vision of brooding watchers, and others who imagined great sentinels before a veiled figure, and some one who felt nothing more divine than a sensation of happiness and freedom such as one gets from a sudden burst of sunshine in the spring. . . .

So that one is forced to believe that something more than wonderfully wonderful, something altogether strange, was seen, and that all these various things that people thought they saw were only interpretations drawn from their experiences and their imaginations. It was a light, it was beauty, it was high and solemn, it made this world seem a flimsy transparency.

Then it had vanished. . . .

And people were left with the question of what they had seen, and just how much it mattered.

§ 6

A little old lady sat by the fire in a small sitting-room in West Kensington. Her cat was in her lap, her spectacles were on her nose; she was reading the morning's paper, and beside her, on a little occasional table, was her tea and a buttered muffin. She had finished the crimes and she was reading about the Royal Family. When she had read all there was to read about the Royal Family, she put down the paper, deposited the cat on the hearthrug, and turned to her tea. She had poured out her first cup and she had just taken up a quadrant of muffin when the trump and the flash came. Through its instant duration she remained motionless with the quadrant of muffin poised halfway to her mouth. Then very slowly she put the morsel down.

"Now what was that?" she said.

She surveyed the cat, but the cat was quite calm. Then she looked very, very hard at her lamp. It was a patent safety lamp, and had always behaved very well. Then she stared at the window, but the curtains were drawn and everything was in order.

"One might think I was going to be ill," she said, and resumed her toast.

§ 7

Not far away from this old lady, not more than three-quarters of a mile at most, sat Mr. Parchester in his luxurious study, writing a perfectly beautiful, sustaining sermon about the Need of Faith in God. He was a handsome, earnest, modern preacher, he was rector of one of our big West End churches, and he had amassed a large, fashionable congregation. Every Sunday, and at convenient intervals during the week, he fought against Modern Materialism, Doubt, Levity, Selfish Individualism, Further Relaxation of the Divorce Laws, all the Evils of our Time—and anything else that was unpopular. He believed quite simply, he said, in all the old, simple, kindly things. He had the face of a saint, but he had rendered this generally acceptable by growing side whiskers. And nothing could tame the beauty of his voice.

He was an enormous asset in the spiritual life of the metropolis—to give it no harsher name—and his fluent periods had restored faith and courage to many a poor soul hovering on the brink of the dark river of thought. . . .

And just as beautiful Christian maidens played a wonderful part in the last days of Pompeii, in winning proud Roman hearts to a hated and despised faith, so Mr. Parchester's naturally graceful gestures, and his simple, melodious, trumpet voice won back scores of our half-pagan rich women to church attendance and the social work of which his church was the centre. . . .

And now by the light of an exquisitely shaded electric lamp he was writing this sermon of quiet, confident belief (with occasional hard smacks, perfect stingers in fact, at current unbelief and rival leaders of opinion) in the simple, divine faith of our fathers. . . .

When there came this truncated trump and this vision. . . .

§ 8

Of all the innumerable multitudes who for the infinitesimal fraction of a second had this glimpse of the Divinity, none were so blankly and profoundly astonished as Mr. Parchester. For—it may be because of his subtly spiritual nature—he *saw,* and seeing believed. He dropped his pen and let it roll across his manuscript, he sat stunned, every drop of blood fled from his face and his lips and his eyes dilated.

While he had just been writing and arguing about God, there *was* God!

The curtain had been snatched back for an instant; it had fallen again; but his mind had taken a photographic impression of everything that he had seen—the grave presences, the hierarchy, the effulgence, the vast concourse, the terrible, gentle eyes. He felt it, as though the vision still continued, behind the bookcases, behind the pictured wall and the curtained window: *even now there was judgement!*

For quite a long time he sat, incapable of more than apprehending this supreme realisation. His hands were held out limply upon the desk before him. And then very slowly his staring eyes came back to immediate things, and fell upon the scattered manuscript on which he had been engaged. He read an unfinished sentence and slowly recovered its intention. As he did so, a picture of his congregation came to him as he saw it from the pulpit during his evening sermon, as he had intended to see it on the Sunday evening that was at hand, with Lady Rupert in her sitting and Lady Blex in hers and Mrs. Munbridge, the rich and in her Jewish way very attractive Mrs. Munbridge, running them close in her adoration, and each with one or two friends they had brought to adore him, and behind them the Hexhams and the Wassinghams and behind them others and others and others, ranks and ranks of people, and the galleries on either side packed with worshippers of a less dominant class, and the great organ and his magnificent choir waiting to support him and supplement him, and the great altar to the left of him, and the beautiful new Lady Chapel, done by Roger Fry and Wyndham Lewis and all the latest people in Art, to the right. He thought of the listening multitude, seen through the haze of the thousand electric candles, and how he had planned the paragraphs of his discourse so that the notes of his beautiful voice should float slowly down, like golden leaves in autumn, into the smooth tarn of their silence word by word, phrase by phrase, until he came to——

"Now to God the Father, God the Son——"

And all the time he knew that Lady Blex would watch his face and Mrs. Munbridge, leaning those graceful shoulders of hers a little forward, would watch his face. . .

Many people would watch his face.

All sorts of people would come to Mr. Parchester's service at times. Once it was said Mr. Balfour had come. Just to hear him. After his sermons, the strangest people would come and make confessions in the beautifully furnished reception-room beyond the vestry. All sorts of people Once or twice he had asked people to come and listen to him; and one of them had been a very beautiful woman And often he had dreamt of the people who might come prominent people, influential people, remarkable people But never before had it occurred to Mr. Parchester that, a little hidden from the rest of the congregation, behind the thin veil of this material world, there was another auditorium. And that God also, God also, watched his face.

And watched him through and through.

Terror seized upon Mr. Parchester.

He stood up, as though Divinity had come into the room before him. He was trembling. He felt smitten and about to be smitten.

He perceived that it was hopeless to try and hide what he had written, what he had thought, the unclean egotism he had become.

"I did not know," he said at last.

The click of the door behind him warned him that he was not alone. He turned and saw Miss Skelton, his typist for it was her time to come for his manuscript and copy it out in the specially legible type he used. For a moment he stared at her strangely.

She looked at him with those deep, adoring eyes of hers

"Am I too soon, sir?" she asked in her slow, unhappy voice, and seemed prepared for a noiseless departure.

He did not answer immediately. Then he said: "Miss Skelton, the Judgement of God is close at hand!"

And seeing she stood perplexed, he said—

"Miss Skelton, how can you expect me to go on acting and mouthing this Tosh when the Sword of Truth hangs over us?"

Something in her face made him ask a question.

"Did *you* see anything?" he asked.

"I thought it was because I was rubbing my eyes."

"Then indeed there is a God! And he is watching us now. And all this about us, this sinful room, this foolish costume, this preposterous life of blasphemous pretension——!"

He stopped short, with a kind of horror on his face.

With a hopeless gesture he rushed by her. He appeared wild-eyed upon the landing before his man-servant, who was carrying a scuttle of coal upstairs.

"Brompton," he said, "what are you doing?"

"Coal, sir."

"Put it down, man!" he said. "Are you not an immortal soul? God is here! As close as my hand! Repent! Turn to Him! The Kingdom of Heaven is at hand!"

§ 9

Now if you are a policeman perplexed by a sudden and unaccountable collision between a taxicab and an electric standard, complicated by a blinding flash and a sound like an abbreviated trump from an automobile horn, you do not want to be bothered by a hatless clerical gentleman suddenly rushing out of a handsome private house and telling you that "the Kingdom of Heaven is at hand!" You are respectful to him because it is the duty of a policeman to be respectful to Gentlemen, but you say to him, "Sorry I can't attend to that now, sir. One thing at a time. I've got this little accident to see to." And if he persists in dancing

round the gathering crowd and coming at you again, you
say: "I'm afraid I must ask you just to get away from here,
sir. You aren't being a 'elp, sir." And if, on the other hand,
you are a well-trained clerical gentleman, who knows his
way about in the world, you do not go on pestering a policeman on duty after he has said that, even although you think
God is looking at you and Judgement is close at hand. You
turn away and go on, a little damped, looking for some one
else more likely to pay attention to your tremendous tidings.

And so it happened to the Reverend Mr. Parchester.

He experienced a curious little recession of confidence.
He went on past quite a number of people without saying
anything further, and the next person he accosted was a
flower woman sitting by her basket at the corner of
Chexington Square. She was unable to stop him at once
when he began to talk to her because she was tying up
a big bundle of white chrysanthemums and had an end of
string behind her teeth. And her daughter who stood
beside her was the sort of girl who wouldn't say "Boo!" to
a goose.

"Do you know, my good woman," said Mr. Parchester,
"that while we poor creatures of earth go about our poor
business here, while we sin and blunder and follow every
sort of base end, close to us, above us, around us, watching
us, judging us, are God and His holy angels? I have had
a vision, and I am not the only one. I have *seen*. We are
in the Kingdom of Heaven now and here, and Judgement
is all about us now! Have you seen nothing? No light?
No sound? No warning?"

By this time the old flower-seller had finished her bunch
of flowers and could speak. "I saw it," she said. "And
Mary—she saw it."

"Well?" said Mr. Parchester.

"But, Lord! It don't *mean* nothing!" said the old flower-
seller.

§ 10

At that a kind of chill fell upon Mr. Parchester. He went on across Chexington Square by his own inertia.

He was still about as sure that he had seen God as he had been in his study, but now he was no longer sure that the world would believe that he had. He felt perhaps that this idea of rushing out to tell people was precipitate and inadvisable. After all, a priest in the Church of England is only one unit in a great machine; and in a world-wide spiritual crisis it should be the task of that great machine to act as one resolute body. This isolated crying aloud in the street was unworthy of a consecrated priest. It was a dissenting kind of thing to do. A vulgar individualistic screaming. He thought suddenly that he would go and tell his Bishop, the great Bishop Wampach. He called a taxicab, and within half an hour he was in the presence of his commanding officer. It was an extraordinarily difficult and painful interview. . . .

You see, Mr. Parchester believed. The Bishop impressed him as being quite angrily resolved not to believe. And for the first time in his career Mr. Parchester realised just how much jealous hostility a beautiful, fluent, and popular preacher may arouse in the minds of the hierarchy. It wasn't, he felt, a conversation. It was like flinging oneself into the paddock of a bull that has long been anxious to gore one.

"Inevitably," said the Bishop, "this theatricalism, this star-turn business, with its extreme spiritual excitements, its exaggerated soul crisis and all the rest of it, leads to such a breakdown as afflicts you. Inevitably! You were at least wise to come to me. I can see you are only in the beginning of your trouble, that already in your mind fresh hallucinations are gathering to overwhelm you, voices, special charges and missions, strange revelations. . . . I wish I had the power to suspend you right away, to send you into retreat. . . ."

Mr. Parchester made a violent effort to control himself.

"But I tell you," he said, "that I saw God!" He added, as if to reassure himself: "More plainly, more certainly, than I see you."

"Of course," said the Bishop, "this is how strange new sects come into existence; this is how false prophets spring out of the bosom of the Church. Loose-minded, excitable men of your stamp——"

Mr. Parchester, to his own astonishment, burst into tears. "But I tell you," he wept, "He is here. I have seen. I know."

"Don't talk such nonsense!" said the Bishop. "There is no one here but you and I!"

Mr. Parchester expostulated. "But," he protested, "He is omnipotent."

The Bishop controlled an expression of impatience. "It is characteristic of your condition," he said, "that you are unable to distinguish between a matter of fact and a spiritual truth. . . . Now listen to me. If you value your sanity and public decency and the discipline of the Church, go right home from here and go to bed. Send for Broadhays, who will prescribe a safe sedative. And read something calming and graceful and purifying. For my own part, I should be disposed to recommend the 'Life of Saint Francis of Assisi.' . . ."

§ 11

Unhappily Mr. Parchester did not go home. He went out from the Bishop's residence stunned and amazed, and suddenly upon his desolation came the thought of Mrs. Munbridge. . . .

She would understand. . . .

He was shown up to her own little sitting-room. She had already gone up to her room to dress, but when she heard that he had called, and wanted very greatly to see her, she slipped on a loose, beautiful tea-gown négligé thing, and hurried to him. He tried to tell her everything, but she only kept saying, "There! there!" She was sure he wanted a cup of tea, he looked so pale and exhausted. She

rang to have the tea equipage brought back; she put the
dear saint in an arm-chair by the fire; she put cushions
about him, and ministered to him. And when she began
partially to comprehend what he had experienced, she suddenly realised that she too had experienced it. That vision
had been a brainwave between their two linked and sympathetic brains. And that thought glowed in her as she
brewed his tea with her own hands. He had been weeping!
How tenderly he felt all these things! He was more sensitive than a woman. What madness to have expected
understanding from a Bishop! But that was just like his
unworldliness. He was not fit to take care of himself. A
wave of tenderness carried her away. "Here is your tea!"
she said, bending over him, and fully conscious of her
fragrant warmth and sweetness, and suddenly, she could
never afterwards explain why she was so, she was moved
to kiss him on his brow. . . .

How indescribable is the comfort of a true-hearted
womanly friend! The safety of it! The consolation! . . .

About half-past seven that evening Mr. Parchester returned to his own home, and Brompton admitted him.
Brompton was relieved to find his employer looking quite
restored and ordinary again. "Brompton," said Mr. Parchester, "I will not have the usual dinner to-night. Just a
single mutton cutlet and one of those quarter-bottles of
Perrier Jouet on a tray in my study. I shall have to finish
my sermon to-night."

(And he had promised Mrs. Munbridge he would preach
that sermon specially for her.)

§ 12

And as it was with Mr. Parchester and Brompton and
Mrs. Munbridge, and the taxi-driver and the policeman and
the little old lady and the automobile mechanics and Mr.
Parchester's secretary and the Bishop, so it was with all the
rest of the world. If a thing is sufficiently strange and great
no one will perceive it. Men will go on in their own ways

though **one** rose from the dead to tell them that the Kingdom of Heaven was at hand, though the Kingdom itself
and all its glory became visible, blinding their eyes. They
and their ways are one. Men will go on in their ways as
rabbits will go on feeding in their hutches within a hundred
yards of a battery of artillery. For rabbits are rabbits, and
made to eat and breed, and men are human beings and
creatures of habit and custom and prejudice; and what has
made them, what will judge them, what will destroy them—
they may turn their eyes to it at times as the rabbits will
glance at the concussion of the guns, but it will never draw
them away from eating their lettuce and sniffing after their
does. . . .

THE GRISLY FOLK

THE GRISLY FOLK

"CAN these bones live?"

Could anything be more dead, more mute and inexpressive to the inexpert eye than the ochreous fragments of bone and the fractured lumps of flint that constitute the first traces of something human in the world? We see them in the museum cases, sorted out in accordance with principles we do not understand, labelled with strange names. Chellean, Mousterian, Solutrian and the like, taken mostly from the places Chelles, La Moustier, Solutre, and so forth where the first specimens were found. Most of us stare through the glass at them, wonder vaguely for a moment at that half-savage, half-animal past of our race, and pass on. "Primitive man," we say. "Flint implements. The mammoth used to chase him." Few of us realise yet how much the subtle indefatigable cross-examination of the scientific worker has been extracting from the evidence of these rusty and obstinate witnesses during the last few years.

One of the most startling results of this recent work is the gradual realisation that great quantities of these flint implements and some of the earlier fragments of bone that used to be ascribed to humanity are the vestiges of creatures, very manlike in many respects, but not, strictly speaking, belonging to the human species. Scientific men call these vanished races man (*Homo*), just as they call lions and tigers cats (*Felis*), but there are the soundest reasons for believing that these earlier so-called men were not of our blood, not our ancestors, but a strange and vanished animal, like us, akin to us, but different from us, as the mammoth was like, and akin to, and yet different from, the elephant. Flint and bone implements are found in deposits of very considerable antiquity; some in our museums may be a million years old or more, but the traces of really human creatures, mentally and anatomically like ourselves, do not go back much earlier than twenty or thirty thousand years ago. True men appeared in Europe then, and we do not know whence they came. These other tool-using, fire-making animals, the things that were like men and yet were not men, passed away before the faces of the true men.

Scientific authorities already distinguish four species of these pseudo-men, and it is probable that we shall learn from time to time of other species. One strange breed made the implements called Chellean. These are chiefly sole-shaped blades of stone found in deposits of perhaps 300,000 or 400,000 years ago. Chellean implements are to be seen in any great museum. They are huge implements, *four or five times as big as those made by any known race of true men,* and they are not ill made. Certainly some ceature with an intelligent brain made them. Big clumsy hands must have gripped and used these rocky chunks. But so far only one small fragment of a skeleton of this age has been found, a very massive chinless lower jawbone, with teeth rather *more* specialised than those of men to-day. We can only guess what strange foreshadowing of the human form once ate with that jaw, and struck at its enemies with those big but not unhandy flint blades. It may have been a tremendous fellow, probably much bigger in the body than a man. It may have been able to take bears by the scruff and the sabre-toothed lion by the throat. We do not know. We have just these great stone blades and that bit of a massive jaw and—the liberty to wonder.

Most fascinating riddle of all these riddles of the ages of ice and hardship, before the coming of the true men, is the riddle of the Mousterian men, because they were perhaps still living in the world when the true men came wandering into Europe. They lived much later than those unknown Chellean giants. They lived thirty or forty thousand years ago—a yesterday compared with the Chellean time. These Mousterians are also called Neandertalers. Until quite recently it was supposed that they were true men like ourselves. But now we begin to realise that they were different, so different that it is impossible that they can be very close relations of ours. They walked or shambled along with a peculiar slouch. they could not turn their heads up to the sky, and their teeth were very different from those of true men. One oddity about them is that in one or two points they were less like apes than we are. The dog tooth, the third tooth from the middle, which is so big in the gorilla, and which in man is pointed and still quite distinct from the other teeth, is not distinct at all in the Neandertaler. He had a very even row of teeth, and his cheek teeth also were very unlike ours, and less like the apes' than ours. He had more face and less brow than true men, but that is not because he had a lesser brain; his brain was as big as a modern man's but it was different, bigger behind and smaller in front, so that probably he thought and behaved differently from us. Perhaps he had a better memory and less reasoning power than real men, or perhaps he had more nervous energy and less intelligence. He had no chin, and the way his jawbones come together below make it very doubtful if he could have used any such sounds in speech as we employ. Probably he did not talk at all. He could not hold a pin between his finger and thumb. The more we learn about this beast-man the stranger he becomes to us and the less like the Australoid savage he was once supposed to be.

And as we realise the want of any close relationship between this ugly, strong, ungainly, manlike animal and mankind, the less likely it becomes that he had a naked skin and hair like ours and the more probable that he was different, and perhaps bristly or hairy in some queer inhuman fashion like the hairy elephant and the woolly rhinoceros who were his contemporaries. Like them he lived in a bleak land on the edge of the snows and glaciers that were even then receding northward. Hairy or grisly, with a big face like a mask, great brow ridges and no forehead, clutching an enormous flint, and running like a baboon with his head forward and not, like a man, with his head up, he must have been a fearsome creature for our forefathers to come upon.

U

Almost certainly they met, these grisly men and the true men. The true man must have come into the habitat of the Neandertaler, and the two must have met and fought. Some day we may come upon the evidences of this warfare.

Western Europe, which is the only part of the world that has yet been searched with any thoroughness for the remains of early men, was slowly growing warmer age by age; the glaciers that had once covered half the continent were receding, and wide stretches of summer pasture and thin woods of pine and birch were spreading slowly over the once icy land. South Europe then was like northern Labrador to-day. A few hardy beasts held out amidst the snows; the bears hibernated. With the spring grass and foliage came great herds of reindeer, wild horses, mammoth, elephant, and rhinoceros, drifting northward from the slopes of the great warm valley that is now filled up with water—the Mediterranean Sea. It was in those days before the ocean waters broke into the Mediterranean that the swallows and a multitude of other birds acquired the habit of coming north, a habit that nowadays impels them to brave the passage of the perilous seas that flow over and hide the lost secrets of the ancient Mediterranean valleys. The grisly men rejoiced at the return of life, came out of the caves in which they had lurked during the winter, and took their toll of the beasts.

These grisly men must have been almost solitary creatures.

The winter food was too scanty for communities. A male may have gone with a female or so; perhaps they parted in the winter and came together in the summer; when his sons grew big enough to annoy him, the grisly man killed them or drove them off. If he killed them he may have eaten them. If they escaped him they may have returned to kill him. The grisly folk may have had long unreasoning memories and very set purposes.

The true men came into Europe, we know not whence, out of the South. When they appeared in Europe their hands were as clever as ours; they could draw pictures we still admire, they could paint and carve; the implements they made were smaller than the Mousterian ones, far smaller than the Chellean, but better made and more various. They wore no clothes worth speaking of, but they painted themselves and probably they talked. And they came in little bands. They were already more social than the Neandertaler; they had laws and self-restraints; their minds had travelled a long way along that path of adaptation and self-suppression which has led to the intricate mind of man to-day with its concealed wishes, its confusions, and laughter and the fantasies and reveries and dreams. They were already held together, these men, and kept in order by the strange limitations of tabu.

They were still savages, very prone to violence and convulsive in their lusts and desires; but to the best of their poor ability they obeyed laws and customs already immemorably ancient, and they feared the penalties of wrongdoing. We can understand something of what was going on in their minds, those of us who can remember the fears, desires, fancies and superstitions of our childhood. Their moral struggles were ours—in cruder forms. They were our kind. But the grisly folk we cannot begin to understand. We cannot conceive in our different minds the strange ideas that chased one another through those queerly shaped brains. As well might we try to dream and feel as a gorilla dreams and feels.

We can understand how the true men drifted northward from the lost lands of the Mediterranean valley into the high Spanish valleys and the south and centre of France, and so on to what is now England—for there was no Channel then between England and France—and eastward to the Rhineland and over the broad wilderness which is now the North Sea, and the German plain. They would leave the snowy wilderness of the Alps, far higher then and covered with great glaciers, away on their right. These people drifted northward for the very good reason that their kind was multiplying and food diminishing. They would be oppressed by feuds and wars. They had no settled home; they were accustomed to drift with the seasons, every now and then some band would be pushed by hunger and fear a little farther northward into the unknown.

We can imagine the appearance of a little group of these wanderers, our ancestors, coming over some grassy crest into these northern lands. The time would be late spring or early summer, and they would probably be following up some grazing beasts, a reindeer herd or horses.

By a score of different means our anthropologists have been able to reconstruct the particulars of the appearance and habits of these early pilgrim fathers of mankind.

They would not be a very numerous band, because if they were there would be no reason why they should have been driven northward out of their former roving grounds. Two or three older men of thirty or so, eight or ten women and girls with a few young children, a few lads between fourteen and twenty, might make up the whole community. They would be a brownish brown-eyed people with wavy dark hair; the fairness of the European and the straight blue-black hair of the Chinaman had still to be evolved in the world. The older men would probably lead the band, the women and children would keep apart from the youths and men, fenced off by complex and definite tabus from any close companionship. The leaders would be tracking the herd they were following. Tracking was then the supreme accomplishment of mankind. By signs and traces that would be invisible to any modern civilised eye, they would be reading the story of the previous day's trek of the herd of sturdy little horses ahead of them. They would be so expert that they would go on from one faint sign to another with as little delay as a dog who follows a scent.

The horses they were following were only a little way ahead—so the trackers read the signs—they were numerous and nothing had alarmed them. They were grazing and moving only very slowly. There were no traces of wild dog or other enemies to stampede them. Some elephants were also going north, and twice our human tribe had crossed the spoor of woolly rhinoceros roaming westward.

The tribe travelled light. They were mainly naked, but all of them were painted with white and black and red and yellow ochre. At this distance of time it is difficult to see whether they were tattooed. Probably they were not. The babies and small children were carried by the women on their backs in slings or bags made of animal skins, and perhaps some or all of them wore mantles and loin bands of skin and had pouches and belts of leather. The men had stone-pointed spears, and carried sharpened flints in their hands.

There was no Old Man who was lord and master and father of this particular crowd. Weeks ago the Old Man had been charged and trampled to a jelly by a great bull in the swamp far away. Then two of the girls had been waylaid and carried off by the young men of another larger tribe. It was because of these losses that this remnant was now seeking new hunting grounds.

The landscape that spread before the eyes of this little band as they crested the hill was a bleaker, more desolate and altogether unkempt version of the landscape of western Europe to-day. About them was a grassy down athwart which a peewit flew with its melancholy cry. Before them stretched a great valley ridged with transverse purple hills over which the April cloud-shadows chased one another. Pinewoods and black heather showed where these hills became sandy, and the valleys were full of brown brushwood, and down their undrained troughs ran a bright green band of peaty swamps and long pools of weedy water. In the valley thickets many beasts lurked unseen, and where the winding streams had cut into the soil there were cliffs and caves. Far away along the northern slopes of the ridge that were now revealed, the wild ponies were to be seen grazing.

At a sign from the two leaders the little straggle of men-folk halted, and a woman who had been chattering in subdued tones to a little girl became silent. The brothers surveyed the wide prospect earnestly.

"Ugh!" said one abruptly and pointed.

"Ugh!" cried his brother.

The eyes of the whole tribe swung round to the pointing finger.

The group became one rigid stare.

Every soul of them stood still, astonishment had turned them into a tense group of statuettes.

Far away down the slope with his body in profile and his head turned towards them, frozen by an equal amazement, stood a hunched grey figure, bigger but shorter than a man. He had been creeping up behind a fold in the ground to peer at the ponies, and suddenly he had turned his eyes and seen the tribe. His head projected like a baboon's. In his hand he carried what seemed to the menfolk a great rock.

For a little while this animal scrutiny held discoverers and discovered motionless. Then some of the women and children began to stir and line out to see the strange creature better. "Man!" said an old crone of forty. "*Man!*" At the movement of the women the grisly man turned, ran clumsily for a score of yards or so towards a thicket of birch and budding thorn. Then he halted again for a moment to look at the newcomers, waved an arm strangely, and then dashed into cover.

The shadows of the thicket swallowed him up, and by hiding him seemed to make him enormous. It identified itself with him, and watched them with his eyes. Its tree stems became long silvery limbs, and a fallen trunk crouched and stared.

It was still early in the morning, and the leaders of the tribe had hoped to come up with the wild ponies as the day advanced and perhaps cut one off and drive it into difficulties among the bushes and swampy places below, and wound it and follow it up and kill it. Then they would have made a feast, and somewhere down in the valley they would have found water and dry bracken for litter and a fire before night. It had seemed a pleasant and hopeful morning to them until this moment. Now they were disconcerted. This grey figure was as if the sunny morning had suddenly made a horrible and inexplicable grimace.

The whole expedition stood gazing for a time, and then the two leaders exchanged a few words. Waugh, the elder, pointed. Click, his brother, nodded his head. They would go on, but instead of slanting down the slopes towards the thickets they would keep round the ridge.

"Come," said Waugh, and the little band began to move again. But now it marched in silence. When presently a little boy began a question his mother silenced him by a threat. Everybody kept glancing at the thickets below.

Presently a girl cried out sharply and pointed. All started and stopped short.

There was the grisly thing again. It was running across an open space, running almost on all fours, in joltering leaps. It was hunchbacked and very big and low, a grey hairy wolf-like monster. At times its long arms nearly touched the ground. It was nearer than it had been before. It vanished amidst the bushes again. It seemed to throw itself down among some red dead bracken. . . .

Waugh and Click took counsel.

A mile away was the head of the valley where the thickets had their beginning. Beyond stretched the woldy hills, bare of cover. The horses were grazing up towards the sun, and away to the north the backs of a herd of woolly rhinoceros were now visible on a crest—just the ridges of their backs showing like a string of black beads.

If the tribe struck across those grassy spaces, then the lurking prowler would have either to stay behind or come into the open. If he came into the open the dozen youths and men of the tribe would know how to deal with him.

So they struck across the grass. The little band worked round to the head of the valley, and there the men-folk stayed at the crest while the women and children pushed on ahead across the open.

For a time the watchers remained motionless, and then Waugh was moved to gestures of defiance. Click was not to be outdone. There were shouts at the hidden watcher, and then one lad, who was something of a clown, after certain grimaces and unpleasant gestures, obliged with an excellent imitation of the grey thing's lumbering run. At that, scare gave place to hilarity.

In those days laughter was a social embrace. Men could laugh, but there was no laughter in the grisly pre-man who watched and wondered in the shadow. He marvelled. The men rolled about and guffawed and slapped their thighs and one another. Tears ran down their faces.

Never a sign came from the thickets.

"Yahah," said the menfolk. "Yahah! Bzzzz. Yahah! Yah!"

They forgot altogether how frightened they had been.

And when Waugh thought the women and children had gone on a sufficient distance, he gave the word for the men to follow them.

In some such fashion it was that men, our ancestors, had their first glimpse of the pre-men of the wilderness of western Europe. . . .

The two breeds were soon to come to closer quarters.

The newcomers were pushing their way into the country of these grisly men. Presently came other glimpses of lurking semi-human shapes and grey forms that ran in the twilight. In the morning Click found long narrow footprints round the camp. . . .

Then one day one of the children, eating those little green thorn-buds that rustic English children speak of as bread and cheese, ventured too far from the others. There was a squeal and a scuffle and a thud, and something grey and hairy made off through the thickets carrying its victim, with Waugh and three of the younger men in hot pursuit. They chased the enemy into a dark gully, very much overgrown. This time it was not a solitary Neandertaler they had to deal with. Out of the bushes a big male came at them to cover the retreat of his mate, and hurled a rock that bowled over the youth it hit like a nine-pin, so that thereafter he limped always. But Waugh with his throwing spear got the grey monster in the shoulder, and he halted snarling.

No further sound came from the stolen child.

The female showed herself for a moment up the gully, snarling, bloodstained, and horrible, and the menfolk stood about afraid to continue their pursuit, and yet not caring to desist from it. One of them was already hobbling off with his hand to his knee.

How did that first fight go?

Perhaps it went against the men of our race. Perhaps the big Neandertaler male, his mane and beard bristling horribly, came down the gully with a thunderous roar, with a great rock in either hand. We do not know whether he threw these big discs of flint or whether he smote with them. Perhaps it was then that Waugh was killed in the act of running away. Perhaps it was bleak disaster then for the little tribe. Short of two of its members it presently made off over the hills as fast as it could go, keeping together for safety, and leaving the wounded youth far behind to limp along its tracks in lonely terror.

Let us suppose that he got back to the tribe at last—after nightmare hours.

Now that Waugh had gone, Click would become Old Man and he made the tribe camp that night and build their fire on the high ridges among the heather far away from the thickets in which the grisly folk might be lurking.

The grisly folk thought we knew not how about the menfolk, and the men thought about the grisly folk in such ways as we can understand; they imagined how their enemies might act in this fashion or that, and schemed to circumvent them. It may have been Click who had the first dim idea of getting at the gorge in which the Neandertalers had their lair, from above. For as we have said, the Neandertaler did not look up. Then the menfolk could roll a great rock upon him or pelt him with burning brands and set the dry bracken alight.

One likes to think of a victory for the human side. This Click we have conjured up had run in panic from the first onset of the grisly male, but as he brooded by the fire that night he heard again in imagination the cry of the lost girl, and he was filled with rage. In his sleep the grisly male came to him and Click fought in his dreams and started awake stiff with fury. There was a fascination for him in that gorge in which Waugh had been killed. He was compelled to go back and look again for the grisly beasts, to waylay them in their tracks, and watch them from an ambush. He perceived that the Neandertalers could not climb as easily as the menfolk could climb, nor hear so quickly, nor dodge with the same unexpectedness. These grisly men were to be dealt with as the bears were dealt with, the bears before whom you run and scatter, and then come at again from behind.

But one may doubt if the first human group to come into the grisly land was clever enough to solve the problems of the new warfare. Maybe they turned southward again to the gentler regions from which they had come, and were killed by or mingled with their own breathren again. Maybe they perished altogether in that new land of the grisly folk into which they had intruded. Yet the truth may be that they even held their own and increased. If they died there were others of their kind to follow them and achieve a better fate.

That was the beginning of a nightmare age for the little children of the human tribe. They knew they were watched.

Their steps were dogged. The legends of ogres and man-eating giants that haunt the childhood of the world may descend to us from those ancient days of fear. And for the Neandertalers it was the beginning of an incessant war that could end only in extermination.

The Neandertalers, albeit not so erect and tall as men, were the heavier, stronger creatures, but they were stupid, and they went alone or in twos and threes; the menfolk were swifter, quicker-witted, and more social—when they fought they fought in combination. They lined out and surrounded and pestered and pelted their antagonists from every side. They fought the men of that grisly race as dogs might fight a bear. They shouted to one another what each should do, and the Neandertaler had no speech; he did not understand. They moved too quickly for him and fought too cunningly.

Many and obstinate were the duels and battles these two sorts of men fought for this world in that bleak age of the windy steppes, thirty or forthy thousand years ago. The two races were intolerable to each other. They both wanted the caves and the banks by the rivers where the big flints were got. They fought over the dead mammoths that had been bogged in the marshes and over the reindeer stags that had been killed in the rutting season. When a human tribe found signs of the grisly folk near their cave and squatting place, they had perforce to track them down and kill them; their own safety and the safety of their little ones was only to be secured by that killing. The Neandertalers thought the little children of men fair game and pleasant eating.

How long the grisly folk lived on in that chill world of pines and silver birch between the steppes and the glaciers, after the true menfolk came, we do not know. For ages they may have held out, growing more cunning and dangerous as they became rare. The true men hunted them down by their spoor and by their tracks, and watched for the smoke of their fires, and made food scarce for them.

Great Paladins arose in that forgotten world, men who stood forth and smote the grey man-beast face to face and slew him. They made long spears of wood, hardened by fire at the tips; they raised shields of skin against his mighty blows. They struck at him with stones on cords, and slung them at him with slings. And it was not simply men who withstood the grisly beast but women. They stood over their children; they stood by their men against this eerie thing that was like and yet not like mankind. Unless the *savants* read all the signs awry, it was the women who were the makers of the larger tribes into which human families were already growing in those ancient times. It was the woman's subtle, love-guided wits which protected her sons from the fierce anger of the Old Man, and taught them to avoid his jealousy and wrath, and persuaded him to tolerate them and so have their help against the grisly enemy. It was woman, says Atkinson, in the beginning of things human, who taught the primary tabùs, that a son must go aside out of the way of his stepmother, and get himself a wife from another tribe, so as to keep the peace within the family. She came between the fratricides, and was the first peacemaker. Human societies in their beginnings were her work, done against the greater solitariness, the lonely fierceness of the adult male. Through her, men learnt the primary co-operation of sonship and brotherhood. The

grisly folk had not learnt even the rudest elements of co-
operation, and mankind had already spelt out the alphabet
of a unity that may some day comprehend the whole earth.
The menfolk kept together by the dozen and by the score.
By ones and twos and threes therefore the grisly folk were
beset and slain, until there were no more of them left in the
world.

Generation after generation, age after age, that long
struggle for existence went on between these men who were
not quite men and the men, our ancestors, who came out of
the south into western Europe. Thousands of fights and
hunts, sudden murders and headlong escapes there were
amidst the caves and thickets of that chill and windy world
between the last age of glaciers and our own warmer time.
Until at length the last poor grisly was brought to bay and
faced the spears of his pursuers in anger and despair.

What leapings of the heart were there not throughout
that long warfare! What moments of terror and triumph!
What acts of devotion and desperate wonders of courage!
And the strain of the victors was our strain; we are lineally
identical with those sun-brown painted beings who ran and
fought and helped one another, the blood in our veins
glowed in those fights and chilled in those fears of the forgotten past. For it was forgotten. Except perhaps for some
vague terrors in our dreaming life and for some lurking element of tradition in the legends and warnings of the nursery,
it has gone altogether out of the memory of our race. But
nothing is ever completely lost. Seventy or eighty years
ago a few curious *savants* began to suspect that there were
hidden memories in certain big chipped flints and scraps of
bone they found in ancient gravels. Much more recently
others have begun to find hints of remote strange experiences
in the dreams and odd kinks in modern minds. By degrees
these dry bones begin to live again.

This restoration of the past is one of the most astonishing
adventures of the human mind. As humanity follows the
gropings of scientific men among these ancient vestiges, it is
like a man who turns over the yellow pages of some long-
forgotten diary, some engagement book of his adolescence.

His dead youth lives again. Once more the old excitements stir him, the old happiness returns. But the old passions that once burnt, only warm him now, and the old fears and distresses signify nothing.

A day may come when these recovered memories may grow as vivid as if we in our own persons had been there and shared the thrill and the fear of those primordial days; a day may come when the great beasts of the past will leap to life again in our imaginations, when we shall walk again in vanished scenes, stretch painted limbs we thought were dust, and feel again the sunshine of a million years ago.

TALES OF SPACE AND TIME

STORY THE FIRST

The Crystal Egg

THERE was, until a year ago, a little and very grimy-looking shop near Seven Dials, over which, in weather-worn yellow lettering, the name of "C. Cave, Naturalist and Dealer in Antiquities," was inscribed. The contents of its window were curiously varied. They comprised some elephant tusks and an imperfect set of chessmen, beads and weapons, a box of eyes, two skulls of tigers and one human, several moth-eaten stuffed monkeys (one holding a lamp), an old-fashioned cabinet, a fly-blown ostrich egg or so, some fishing-tackle, and an extraordinarily dirty, empty glass fish-tank. There was also, at the moment the story begins, a mass of crystal, worked into the shape of an egg and brilliantly polished. And at that two people, who stood outside the window, were looking, one of them a tall, thin clergyman, the other a black-bearded young man of dusky complexion and unobtrusive costume. The dusky young man spoke with eager gesticulation, and seemed anxious for his companion to purchase the article.

While they were there, Mr. Cave came into his shop, his beard still wagging with the bread and butter of his tea. When he saw these men and the object of their regard, his countenance fell. He glanced guiltily over his shoulder, and softly shut the door. He was a little old man, with pale face and peculiar watery blue eyes; his hair was a dirty grey, and he wore a shabby blue frock-coat, an ancient silk hat, and carpet slippers very much down at heel. He remained watching the two men as they talked. The clergyman went deep into his trouser pocket, examined a handful

of money, and showed his teeth in an agreeable smile. Mr.
Cave seemed still more depressed when they came into the
shop.

The clergyman, without any ceremony, asked the price
of the crystal egg. Mr. Cave glanced nervously towards the
door leading into the parlour, and said five pounds. The
clergyman protested that the price was high, to his companion as well as to Mr. Cave—it was, indeed, very much
more than Mr. Cave had intended to ask, when he had
stocked the article—and an attempt at bargaining ensued.
Mr. Cave stepped to the shop-door, and held it open. "Five
pounds is my price," he said, as though he wished to save
himself the trouble of unprofitable discussion. As he did
so, the upper portion of a woman's face appeared above the
blind in the glass upper panel of the door leading into the
parlour, and stared curiously at the two customers. "Five
pounds is my price," said Mr. Cave, with a quiver in his
voice.

The swarthy young man had so far remained a spectator,
watching Cave keenly. Now he spoke. "Give him five
pounds," he said. The clergyman glanced at him to see if
he were in earnest, and, when he looked at Mr. Cave again,
he saw that the latter's face was white. "It's a lot of money,"
said the clergyman, and, diving into his pocket, began counting his resources. He had little more than thirty shillings,
and he appealed to his companion, with whom he seemed to
be on terms of considerable intimacy. This gave Mr. Cave
an opportunity of collecting his thoughts, and he began to
explain in an agitated manner that the crystal was not, as a
matter of fact, entirely free for sale. His two customers
were naturally surprised at this, and inquired why he had
not thought of that before he began to bargain. Mr. Cave
became confused, but he stuck to his story, that the crystal
was not in the market that afternoon, that a probable purchaser of it had already appeared. The two, treating this as
an attempt to raise the price still further, made as if they
would leave the shop. But at this point the parlour door
opened, and the owner of the dark fringe and the little eyes
appeared.

She was a coarse-featured, corpulent woman, younger and very much larger than Mr. Cave; she walked heavily, and her face was flushed. "That crystal *is* for sale," she said. "And five pounds is a good enough price for it. I can't think what you're about, Cave, not to take the gentleman's offer!"

Mr. Cave, greatly perturbed by the irruption, looked angrily at her over the rims of his spectacles, and, without excessive assurance, asserted his right to manage his business in his own way. An altercation began. The two customers watched the scene with interest and some amusement, occasionally assisting Mrs. Cave with suggestions. Mr. Cave, hard driven, persisted in a confused and impossible story of an enquiry for the crystal that morning, and his agitation became painful. But he stuck to his point with extraordinary persistence. It was the young Oriental who ended this curious controversy. He proposed that they should call again in the course of two days—so as to give the alleged enquirer a fair chance. "And then we must insist," said the clergyman. "Five pounds." Mrs. Cave took it on herself to apologise for her husband, explaining that he was sometimes "a little odd," and as the two customers left, the couple prepared for a free discussion of the incident in all its bearings.

Mrs. Cave talked to her husband with singular directness. The poor little man, quivering with emotion, muddled himself between his stories, maintaining on the one hand that he had another customer in view, and on the other asserting that the crystal was honestly worth ten guineas. "Why did you ask five pounds?" said his wife. "*Do* let me manage my business my own way!" said Mr. Cave.

Mr. Cave had living with him a step-daughter and a step-son, and at supper that night the transaction was re-discussed. None of them had a high opinion of Mr. Cave's business methods, and this action seemed a culminating folly.

"It's my opinion he's refused that crystal before," said the step-son, a loose-limbed lout of eighteen.

"But *Five Pounds!*" said the step-daughter, an argumentative young woman of six-and-twenty.

Mr. Cave's answers were wretched; he could only mumble
weak assertions that he knew his own business best. They
drove him from his half-eaten supper into the shop, to close
it for the night, his ears aflame and tears of vexation behind
his spectacles. "Why had he left the crystal in the window
so long? The folly of it!" That was the trouble closest in
his mind. For a time he could see no way of evading sale.

After supper his step-daughter and step-son smartened
themselves up and went out and his wife retired upstairs to
reflect upon the business aspects of the crystal, over a little
sugar and lemon and so forth in hot water. Mr. Cave went
into the shop, and stayed there until late, ostensibly to make
ornamental rockeries for gold-fish cases but really for a
private purpose that will be better explained later. The
next day Mrs. Cave found that the crystal had been removed
from the window, and was lying behind some second-hand
books on angling. She replaced it in a conspicuous position. But she did not argue further about it, as a nervous
headache disinclined her from debate. Mr. Cave was always
disinclined. The day passed disagreeably. Mr. Cave was,
if anything, more absent-minded than usual, and uncommonly irritable withal. In the afternoon, when his wife was
taking her customary sleep, he removed the crystal from the
window again.

The next day Mr. Cave had to deliver a consignment of
dog-fish at one of the hospital schools, where they were
needed for dissection. In his absence Mrs. Cave's mind reverted to the topic of the crystal, and the methods of expenditure suitable to a windfall of five pounds. She had already
devised some very agreeable expedients, among others a dress
of green silk for herself and a trip to Richmond, when a
jangling of the front door bell summoned her into the shop.
The customer was an examination coach who came to complain of the non-delivery of certain frogs asked for the previous day. Mrs. Cave did not approve of this particular
branch of Mr. Cave's business, and the gentleman, who had
called in a somewhat aggressive mood, retired after a brief

exchange of words—entirely civil so far as he was concerned. Mrs. Cave's eye then naturally turned to the window; for the sight of the crystal was an assurance of the five pounds and of her dreams. What was her surprise to find it gone!

She went to the place behind the locker on the counter, where she had discovered it the day before. It was not there; and she immediately began an eager search about the shop.

When Mr. Cave returned from his business with the dog-fish, about a quarter to two in the afternoon, he found the shop in some confusion, and his wife, extremely exasperated and on her knees behind the counter, routing among his taxidermic material. Her face came up hot and angry over the counter, as the jangling bell announced his return, and she forthwith accused him of "hiding it."

"Hid *what?*" asked Mr. Cave.

"The crystal!"

At that Mr. Cave, apparently much surprised, rushed to the window. "Isn't it here?" he said. "Great Heavens! what has become of it?"

Just then, Mr. Cave's step-son re-entered the shop from the inner room—he had come home a minute or so before Mr. Cave—and he was blaspheming freely. He was apprenticed to a second-hand furniture dealer down the road, but he had his meals at home, and he was naturally annoyed to find no dinner ready.

But, when he heard of the loss of the crystal, he forgot his meal, and his anger was diverted from his mother to his step-father. Their first idea, of course, was that he had hidden it. But Mr. Cave stoutly denied all knowledge of its fate—freely offering his bedabbled affidavit in the matter—and at last was worked up to the point of accusing, first, his wife and then his step-son of having taken it with a view to a private sale. So began an exceedingly acrimonious and emotional discussion, which ended for Mrs. Cave in a peculiar nervous condition midway between hysterics and amuck, and caused the step-son to be half-an-hour late at the furniture establishment in the afternoon. Mr. Cave took refuge from his wife's emotions in the shop.

In the evening the matter was resumed, with less passion and in a judicial spirit, under the presidency of the step-daughter. The supper passed unhappily and culminated in a painful scene. Mr. Cave gave way at last to extreme exasperation, and went out banging the front door violently. The rest of the family, having discussed him with the freedom his absence warranted, hunted the house from garret to cellar, hoping to light upon the crystal.

The next day the two customers called again. They were received by Mrs. Cave almost in tears. It transpired that no one *could* imagine all that she had stood from Cave at various times in her married pilgrimage. . . . She also gave a garbled account of the disappearance. The clergyman and the Oriental laughed silently at one another, and said it was very extraordinary. As Mrs. Cave seemed disposed to give them the complete history of his life they made to leave the shop. Thereupon Mrs. Cave, still clinging to hope, asked for the clergyman's address, so that, if she could get anything out of Cave, she might communicate it. The address was duly given, but apparently was afterwards mislaid. Mrs. Cave can remember nothing about it.

In the evening of that day, the Caves seem to have exhausted their emotions, and Mr. Cave, who had been out in the afternoon, supped in a gloomy isolation that contrasted pleasantly with the impassioned controversy of the previous days. For some time matters were very badly strained in the Cave household, but neither crystal nor customer reappeared.

Now, without mincing the matter, we must admit that Mr. Cave was a liar. He knew perfectly well where the crystal was. It was in the rooms of Mr. Jacoby Wace, Assistant Demonstrator at St. Catherine's Hospital, Westbourne Street. It stood on the sideboard partially covered by a black velvet cloth, and beside a decanter of American whisky. It is from Mr. Wace, indeed, that the particulars upon which this narrative is based were derived. Cave had taken off the thing to the hospital hidden in the dog-fish sack, and there had pressed the young investigator to keep it for him. Mr. Wace was a little dubious at first. His relationship to Cave was peculiar. He had a taste for singular characters, and he had more than once invited the old man to smoke and drink in his rooms, and to unfold his rather amusing views of life in general and of his wife in particular. Mr. Wace had encountered Mrs. Cave, too, on occasions when Mr. Cave was not at home to attend to him. He knew the constant interference to which Cave was subjected, and having weighed the story judicially, he decided to give the crystal a refuge. Mr. Cave promised to explain the reasons for his remarkable affection for the crystal more fully on a later occasion, but he spoke distinctly of seeing visions therein. He called on Mr. Wace the same evening.

He told a complicated story. The crystal he said had come into his possession with other oddments at the forced sale of another curiosity dealer's effects, and not knowing what its value might be, he had ticketed it at ten shillings. It had hung upon his hands at that price for some months, and he was thinking of "reducing the figure," when he made a singular discovery.

At that time his health was very bad—and it must be borne in mind that, throughout all this experience, his physical condition was one of ebb—and he was in considerable distress by reason of the negligence, the positive ill-treatment even, he received from his wife and step-children. His wife was vain, extravagant, unfeeling, and had a growing taste for private drinking; his step-daughter was mean and over-reaching; and his step-son had conceived a violent dislike for him, and lost no chance of showing it. The requirements of his business pressed heavily upon him, and Mr. Wace does not think that he was altogether free from occasional intemperance. He had begun life in a comfortable position, he was a man of fair education, and he suffered, for weeks at a stretch, from melancholia and insomnia. Afraid to disturb his family, he would slip quietly from his wife's side, when his thoughts became intolerable, and wander about the house. And about three o'clock one morning, late in August, chance directed him into the shop.

The dirty little place was impenetrably black except in one spot, where he perceived an unusual glow of light. Approaching this, he discovered it to be the crystal egg, which was standing on the corner of the counter towards the window. A thin ray smote through a crack in the shutters, impinged upon the object, and seemed as it were to fill its entire interior.

It occurred to Mr. Cave that this was not in accordance with the laws of optics as he had known them in his younger days. He could understand the rays being refracted by the crystal and coming to a focus in its interior, but this diffusion jarred with his physical conceptions. He approached the crystal nearly, peering into it and round it, with a transient revival of the scientific curiosity that in his youth had determined his choice of a calling. He was surprised to find the light not steady, but writhing within the substance of the egg, as though that object was a hollow sphere of some luminous vapour. In moving about to get different points of view, he suddenly found that he had come between it and the ray, and that the crystal none the less remained luminous. Greatly astonished, he lifted it out of the light ray and carried it to the darkest part of the shop. It remained bright for some four or five minutes, when it slowly faded and went out. He placed it in the thin streak of daylight, and its luminousness was almost immediately restored.

So far, at least, Mr. Wace was able to verify the remarkable story of Mr. Cave. He has himself repeatedly held this crystal in a ray of light (which had to be of a less diameter than one millimetre). And in a perfect darkness, such as could be produced by velvet wrapping, the crystal did undoubtedly appear very faintly phosphorescent. It would seem, however, that the luminousness was of some exceptional sort, and not equally visible to all eyes; for Mr. Harbinger—whose name will be familiar to the scientific reader in connection with the Pasteur Institute—was quite unable to see any light whatever. And Mr. Wace's own capacity for its appreciation was out of comparison inferior to that of Mr. Cave's. Even with Mr. Cave the power varied very considerably: his vision was most vivid during states of extreme weakness and fatigue.

Now from the outset this light in the crystal exercised an irresistible fascination upon Mr. Cave. And it says more for his loneliness of soul than a volume of pathetic writing could do, that he told no human being of his curious observations. He seems to have been living in such an atmosphere of petty spite that to admit the existence of a pleasure would have been to risk the loss of it. He found that as the dawn advanced, and the amount of diffused light increased, the crystal became to all appearance non-luminous. And for some time he was unable to see anything in it, except at night-time, in dark corners of the shop.

But the use of an old velvet cloth, which he used as a background for a collection of minerals, occurred to him, and by doubling this, and putting it over his head and hands, he was able to get a sight of the luminous movement within the crystal even in the day-time. He was very cautious lest he should be thus discovered by his wife, and he practised this occupation only in the afternoons, while she was asleep upstairs, and then circumspectly in a hollow under the counter. And one day, turning the crystal about in his hands, he saw something. It came and went like a flash, but it gave him the impression that the object had for a moment opened to him the view of a wide and spacious and strange country; and, turning it about, he did, just as the light faded, see the same vision again.

Now, it would be tedious and unnecessary to state all the phases of Mr. Cave's discovery from this point. Suffice that the effect was this: the crystal, being peered into at an angle of about 137 degrees from the direction of the illuminating ray, gave a clear and consistent picture of a wide and peculiar country-side. It was not dream-like at all; it produced a definite impression of reality, and the better the light the more real and solid it seemed. It was a moving picture: that is to say, certain objects moved in it, but slowly in an orderly manner like real things, and, according as the direction of the lighting and vision changed, the picture changed also. It must, indeed, have been like looking through an oval glass at a view, and turning the glass about to get at different aspects.

Mr. Cave's statements, Mr. Wace assures me, were extremely circumstantial, and entirely free from any of that emotional quality that taints hallucinatory impressions. But it must be remembered that all the efforts of Mr. Wace to see any similar clarity in the faint opalescence of the crystal were wholly unsuccessful, try as he would. The difference in intensity of the impressions received by the two men was very great, and it is quite conceivable that what was a view to Mr. Cave was a mere blurred nebulosity to Mr. Wace.

The view, as Mr. Cave described it, was invariably of an extensive plain, and he seemed always to be looking at it from a considerable height, as if from a tower or a mast. To the east and to the west the plain was bounded at a remote distance by vast reddish cliffs, which reminded him of those he had seen in some picture; but what the picture was Mr. Wace was unable to ascertain. These cliffs passed north and south—he could tell the points of the compass by the stars that were visible of a night—receding in an almost illimitable perspective and fading into the mists of the distance before they met. He was nearer the eastern set of cliffs, on the occasion of his first vision the sun was rising over them, and black against the sunlight and pale against their shadow appeared a multitude of soaring forms that Mr. Cave regarded as birds. A vast range of buildings spread below him; he seemed to be looking down upon them; and, as they approached the blurred and refracted edge of the picture, they became indistinct. There were also trees curious in shape, and in colouring, a deep mossy green and an exquisite grey, beside a wide and shining canal. And something great and brilliantly coloured flew across the picture. But the first time Mr. Cave saw these pictures he saw only in flashes, his hands shook, his head moved, the vision came and went, and grew foggy and indistinct. And at first he had the greatest difficulty in finding the picture again once the direction of it was lost.

His next clear vision, which came about a week after the first, the interval having yielded nothing but tan-

talising glimpses and some useful experience, showed him
the view down the length of the valley. The view was different, but he had a curious persuasion, which his subsequent observations abundantly confirmed, that he was regarding this strange world from exactly the same spot, although he was looking in a different direction. The long
façade of the great building, whose roof he had looked down
upon before, was now receding in perspective. He recognised the roof. In the front of the façade was a terrace of
massive proportions and extraordinary length, and down
the middle of the terrace, at certain intervals, stood huge
but very graceful masts, bearing small shiny objects which
reflected the setting sun. The import of these small objects
did not occur to Mr. Cave until some time after, as he was
describing the scene to Mr. Wace. The terrace overhung a
thicket of the most luxuriant and graceful vegetation, and
beyond this was a wide grassy lawn on which certain broad
creatures, in form like beetles but enormously larger, re-
posed. Beyond this again was a richly decorated causeway
of pinkish stone; and beyond that, and lined with dense *red*
weeds, and passing up the valley exactly parallel with the
distant cliffs, was a broad and mirror-like expanse of water.
The air seemed full of squadrons of great birds, manœuvring
in stately curves; and across the river was a multitude of
splendid buildings, richly coloured and glittering with
metallic tracery and facets, among a forest of moss-like and
lichenous trees. And suddenly something flapped repeatedly across the vision, like the fluttering of a jewelled
fan or the beating of a wing, and a face, or rather the upper
part of a face with very large eyes, came as it were close to
his own and as if on the other side of the crystal. Mr.
Cave was so startled and so impressed by the absolute reality
of these eyes, that he drew his head back from the crystal to
look behind it. He had become so absorbed in watching
that he was quite surprised to find himself in the cool darkness of his little shop, with its familiar odour of methyl,
mustiness, and decay. And, as he blinked about him, the
glowing crystal faded, and went out.

Such were the first general impressions of Mr. Cave. The

story is curiously direct and circumstantial. From the outset, when the valley first flashed momentarily on his senses, his imagination was strangely affected, and, as he began to appreciate the details of the scene he saw, his wonder rose to the point of a passion. He went about his business listless and distraught, thinking only of the time when he should be able to return to his watching. And then a few weeks after his first sight of the valley came the two customers, the stress and excitement of their offer, and the narrow escape of the crystal from sale, as I have already told.

Now while the thing was Mr. Cave's secret, it remained a mere wonder, a thing to creep to covertly and peep at, as a child might peep upon a forbidden garden. But Mr. Wace has, for a young scientific investigator, a particularly lucid and consecutive habit of mind. Directly the crystal and its story came to him, and he had satisfied himself, by seeing the phosphorescence with his own eyes, that there really was a certain evidence for Mr. Cave's statements, he proceeded to develop the matter systematically. Mr. Cave was only too eager to come and feast his eyes on this wonderland he saw, and he came every night from half-past eight until half-past ten, and sometimes, in Mr. Wace's absence, during the day. On Sunday afternoons, also, he came. From the outset Mr. Wace made copious notes, and it was due to his scientific method that the relation between the direction from which the initiating ray entered the crystal and the orientation of the picture was proved. And, by covering the crystal in a box perforated only with a small aperture to admit the exciting ray, and by substituting black holland for his buff blinds, he greatly improved the conditions of the observations; so that in a little while they were able to survey the valley in any direction they desired.

So having cleared the way, we may give a brief account of this visionary world within the crystal. The things were in all cases seen by Mr. Cave and the method of working was invariably for him to watch the crystal and report what he saw, while Mr. Wace (who as a science student had learnt the trick of writing in the dark) wrote a brief note of his report. When the crystal faded, it was put into its box in

the proper position and the electric light turned on. Mr.
Wace asked questions, and suggested observations to clear
up difficult points. Nothing, indeed, could have been less
visionary and more matter-of-fact.

The attention of Mr. Cave had been speedily directed to
the bird-like creatures he had seen so abundantly present in
each of his earlier visions. His first impression was soon
corrected, and he considered for a time that they might represent a diurnal species of bat. Then he thought, grotesquely
enough, that they might be cherubs. Their heads were
round, and curiously human, and it was the eyes of one of
them that had so startled him on his second observation.
They had broad, silvery wings, not feathered, but glistening almost as brilliantly as new-killed fish and with the
same subtle play of colour, and these wings were not built
on the plan of bird-wing, or bat, Mr. Wace learned, but
supported by curved ribs radiating from the body. (A sort
of butterfly wing with curved ribs seems best to express their
appearance.) The body was small, but fitted with two
bunches of prehensile organs, like long tentacles, immediately under the mouth. Incredible as it appeared to Mr.
Wace, the persuasion at last became irresistible, that it was
these creatures which owned the great quasi-human buildings and the magnificent garden that made the broad valley
so splendid. And Mr. Cave perceived that the buildings,
with other peculiarities, had no doors, but that the great
circular windows, which opened freely, gave the creatures
egress, and entrance. They would alight upon their tentacles, fold their wings to a smallness almost rod-like, and
hop into the interior. But among them was a multitude of
smaller-winged creatures, like great dragon-flies and moths
and flying beetles, and across the greensward brilliantly-
coloured gigantic ground-beetles crawled lazily to and fro.
Moreover, on the causeways and terraces, large-headed
creatures similar to the greater winged flies, but wingless,
were visible, hopping busily upon their hand-like tangle of
tentacles.

Allusion has already been made to the glittering objects
upon masts that stood upon the terrace of the nearer build-

ing. It dawned upon Mr. Cave, after regarding one of these masts very fixedly on one particularly vivid day, that the glittering object there was a crystal exactly like that into which he peered. And a still more careful scrutiny convinced him that each one in a vista of nearly twenty carried a similar object.

Occasionally one of the large flying creatures would flutter up to one, and, folding its wings and coiling a number of its tentacles about the mast, would regard the crystal fixedly for a space,—sometimes for as long as fifteen minutes. And a series of observations, made at the suggestion of Mr. Wace, convinced both watchers that, so far as this visionary world was concerned, the crystal into which they peered actually stood at the summit of the end-most mast on the terrace, and that on one occasion at least one of these inhabitants of this other world had looked into Mr. Cave's face while he was making these observations.

So much for the essential facts of this very singular story. Unless we dismiss it all as the ingenious fabrication of Mr. Wace ,we have to believe one of two things: either that Mr. Cave's crystal was in two worlds at once, and that, while it was carried about in one, it remained stationary in the other, which seems altogether absurd; or else that it had some peculiar relation of sympathy with another and exactly similar crystal in this other world, so that what was seen in the interior of the one in this world was, under suitable conditions, visible to an observer in the corresponding crystal in the other world; and *vice versa*. At present, indeed, we do not know of any way in which two crystals could so come *en rapport*, but nowadays we know enough to understand that the thing is not altogether impossible. This view of the crystals as *en rapport* was the supposition that occurred to Mr. Wace, and to me at least it seems extremely plausible. . . .

And where was this other world? On this, also, the alert intelligence of Mr. Wace speedily threw light. After sunset, the sky darkened rapidly—there was a very brief twilight interval indeed—and the stars shone out. They were recognisably the same as those we see, arranged in the same constellations. Mr. Cave recognised the Bear, the Pleiades, Aldebaran, and Sirius: so that the other world must be somewhere in the solar system, and, at the utmost, only a few hundreds of millions of miles from our own. Following up this clue, Mr. Wace learned that the midnight sky was a darker blue even than our midwinter sky, and that the sun seemed a little smaller. *And there were two small moons!* "like our moon but smaller, and quite differently marked," one of which moved so rapidly that its motion was clearly visible as one regarded it. These moons were never high in the sky, but vanished as they rose; that is, every time they revolved they were eclipsed because they were so near their primary planet. And all this answers quite completely, although Mr. Cave did not know it, to what must be the condition of things on Mars.

Indeed, it seems an exceedingly plausible conclusion that peering into this crystal Mr. Cave did actually see the planet Mars and its inhabitants. And, if that be the case, then the evening star that shone so brilliantly in the sky of that distant vision was neither more nor less than our own familiar earth.

For a time the Martians—if they were Martians—do not seem to have known of Mr. Cave's inspection. Once or twice one would come to peer, and go away very shortly to some other mast, as though the vision was unsatisfactory. During this time Mr. Cave was able to watch the proceedings of these winged people without being disturbed by their attentions, and, although his report is necessarily vague and fragmentary, it is nevertheless very suggestive. Imagine the impression of humanity a Martian observer would get who, after a difficult process of preparation and with considerable fatigue to the eyes, was able to peer at London from the steeple of St. Martin's Church for stretches, at longest, of four minutes at a time. Mr. Cave was unable to ascertain if the winged Martians were the same as the Martians who hopped about the causeways and terraces, and if the latter could put on wings at will. He several times saw certain clumsy bipeds, dimly suggestive of apes, white and partially translucent, feeding among certain of the lichenous trees, and once some of these fled before one of the hopping, round-headed Martians. The latter caught one in its tentacles, and then the picture faded suddenly and left Mr. Cave most tantalisingly in the dark. On another occasion a vast thing, that Mr. Cave thought at first was some gigantic insect, appeared advancing along the causeway beside the canal with extraordinary rapidity. As this drew nearer Mr. Cave perceived that it was a mechanism of shining metals and of extraordinary complexity. And then, when he looked again, it had passed out of sight.

After a time Mr. Cave aspired to attract the attention of the Martians, and the next time that the strange eyes of one of them appeared close to the crystal Mr. Cave cried out and sprang away, and they immediately turned on the light and began to gesticulate in a manner suggestive of signalling. But when at last Mr. Cave examined the crystal again the Martian had departed.

Thus far these observations had progressed in early November, and then Mr. Cave, feeling that the suspicions of his family about the crystal were allayed, began to take it to and fro with him in order that, as occasion arose in the daytime or night, he might comfort himself with what was fast becoming the most real thing in his existence.

In December Mr. Wace's work in connection with a forthcoming examination became heavy, the sittings were reluctantly suspended for a week, and for ten or eleven days— he is not quite sure which—he saw nothing of Cave. He then grew anxious to resume these investigations, and, the stress of his seasonal labours being abated, he went down to Seven Dials. At the corner he noticed a shutter before a bird fancier's window, and then another at a cobbler's. Mr. Cave's shop was closed.

He rapped and the door was opened by the step-son in black. He at once called Mrs. Cave, who was, Mr. Wace could not but observe, in cheap but ample widow's weeds of the most imposing pattern. Without any great surprise Mr. Wace learnt that Cave was dead and already buried. She was in tears, and her voice was a little thick. She had just returned from Highgate. Her mind seemed occupied with her own prospects and the honourable details of the obsequies, but Mr. Wace was at last able to learn the particulars of Cave's death. He had been found dead in his shop in the early morning, the day after his last visit to Mr. Wace, and the crystal had been clasped in his stone-cold hands. His face was smiling, said Mrs. Cave, and the velvet cloth from the minerals lay on the floor at his feet. He must have been dead five or six hours when he was found.

This came as a great shock to Wace, and he began to reproach himself bitterly for having neglected the plain symptoms of the old man's ill-health. But his chief thought was of the crystal. He approached that topic in a gingerly manner, because he knew Mrs. Cave's peculiarities. He was dumbfounded to learn that it was sold.

Mrs. Cave's first impulse, directly Cave's body had been taken upstairs, had been to write to the mad clergyman who had offered five pounds for the crystal, informing him of its recovery; but after a violent hunt in which her daughter joined her, they were convinced of the loss of his address. As they were without the means required to mourn and bury Cave in the elaborate style the dignity of an old Seven Dials inhabitant demands, they had appealed to a friendly fellow-tradesman in Great Portland Street. He had very kindly taken over a portion of the stock at a valuation. The valuation was his own and the crystal egg was included in one of the lots. Mr. Wace, after a few suitable consolatory observations, a little off-handedly proffered perhaps, hurried at once to Great Portland Street. But there he learned that the crystal egg had already been sold to a tall, dark man in grey. And there the material facts in this curious, and to me at least very suggestive story come abruptly to an end. The Great Portland Street dealer did not know who the tall dark man in grey was, nor had he observed him with sufficient attention to describe him minutely. He did not even know which way this person had gone after leaving the shop. For a time Mr. Wace remained in the shop, trying the dealer's patience with hopeless questions, venting his own exasperation. And at last, realising abruptly that the whole thing had passed out of his hands, had vanished like a vision of the night, he returned to his own rooms, a little astonished to find the notes he had made still tangible and visible upon his untidy table.

His annoyance and disappointment were naturally very great. He made a second call (equally ineffectual) upon the Great Portland Street dealer, and he resorted to advertisements in such periodicals as were likely to come into the hands of a bric-a-brac collector. He also wrote letters to *The Daily Chronicle* and *Nature*, but both those periodicals, suspecting a hoax, asked him to reconsider his action before they printed, and he was advised that such a strange story, unfortunately so bare of supporting evidence, might imperil his reputation as an investigator. Moreover, the calls of his proper work were urgent. So that after a month or so, save for an occasional reminder to certain dealers, he had reluctantly to abandon the quest for the crystal egg, and from that day to this it remains undiscovered. Occasionally however, he tells me, and I can quite believe him, he has bursts of zeal in which he abandons his more urgent occupation and resumes the search.

Whether or not it will remain lost for ever, with the material and origin of it, are things equally speculative at the present time. If the present purchaser is a collector, one would have expected the enquiries of Mr. Wace to have reached him through the dealers. He has been able to discover Mr. Cave's clergyman and "Oriental"—no other than the Rev. James Parker and the young Prince of Bosso-Kuni in Java. I am obliged to them for certain particulars. The object of the Prince was simply curiosity—and extravagance. He was so eager to buy, because Cave was so oddly reluctant to sell. It is just as possible that the buyer in the second instance was simply a casual purchaser and not a collector at all, and the crystal egg, for all I know, may at the present moment be within a mile of me, decorating a drawing-room or serving as a paper-weight—its remarkable functions all unknown. Indeed, it is partly with the idea of such a possibility that I have thrown this narrative into a form that will give it a chance of being read by the ordinary consumer of fiction.

My own ideas in the matter are practically identical with those of Mr. Wace. I believe the crystal on the mast in Mars and the crystal egg of Mr. Cave's to be in some physical, but at present quite inexplicable, way *en rapport,* and we both believe further that the terrestrial crystal must have been—possibly at some remote date—sent hither from that planet, in order to give the Martians a near view of our affairs. Possibly the fellows to the crystals in the other masts are also on our globe. No theory of hallucination suffices for the facts.

STORY THE SECOND

The Star

It was on the first day of the new year that the announcement was made, almost simultaneously from three observatories, that the motion of the planet Neptune, the outermost of all the plants that wheel about the sun, had become very erratic. Ogilvy had already called attention to a suspected retardation in its velocity in December. Such a piece of news was scarcely calculated to interest a world the greater portion of whose inhabitants were unaware of the existence of the planet Neptune, nor outside the astronomical profession did the subsequent discovery of a faint remote speck of light in the region of the perturbed planet cause any very great excitement. Scientific people, however, found the intelligence remarkable enough, even before it became known that the new body was rapidly growing larger and brighter, that its motion was quite different from the orderly progress of the planets, and that the deflection of Neptune and its satellite was becoming now of an unprecedented kind.

Few people without a training in science can realise the huge isolation of the solar system. The sun with its specks of planets, its dust of planetoids, and its impalpable comets, swims in a vacant immensity that almost defeats the imagination. Beyond the orbit of Neptune there is space, vacant so far as human observation has penetrated, without warmth or light or sound, blank emptiness, for twenty million times a million miles. That is the smallest estimate of the distance to be traversed before the very nearest of the stars is attained. And, saving a few comets more unsubstantial than the thinnest flame, no matter had ever to human knowledge crossed this gulf of space, until early in the twentieth century this strange wanderer appeared. A vast mass of matter it was, bulky, heavy, rushing without warning out of the black mystery of the sky into the radiance of the sun. By the second day it was clearly visible to any decent instrument, as a speck with a barely sensible diameter, in the constellation Leo near Regulus. In a little while an opera glass could attain it.

On the third day of the new year the newspaper readers of two hemispheres were made aware for the first time of the real importance of this unusual apparition in the heavens. "A Planetary Collision," one London paper headed the news, and proclaimed Duchaine's opinion that this strange new planet would probably collide with Neptune. The leader writers enlarged upon the topic. So that in most of the capitals of the world, on January 3rd, there was an expectation, however vague, of some imminent phenomenon in the sky; and as the night followed the sunset round the globe, thousands of men turned their eyes skyward to see— the old familiar stars just as they had always been.

Until it was dawn in London and Pollux setting and the stars overhead grown pale. The winter's dawn it was, a sickly filtering accumulation of daylight, and the light of gas and candles shone yellow in the windows to show where people were astir. But the yawning policeman saw the thing, the busy crowds in the markets stopped agape, workmen going to their work betimes, milkmen, the drivers of news-carts, dissipation going home jaded and pale, homeless wanderers, sentinels on their beats, and in the country, labourers trudging afield, poachers slinking home, all over the dusky quickening country it could be seen—and out at sea by seamen watching for the day—a great white star, come suddenly into the westward sky!

Brighter it was than any star in our skies; brighter than the evening star at its brightest. It still glowed out white and large, no mere twinkling spot of light, but a small round clear shining disc, an hour after the day had come. And where science has not reached, men stared and feared, telling one another of the wars and pestilences that are fore-shadowed by these fiery signs in the Heavens. Sturdy Boers, dusky Hottentots, Gold Coast negroes, Frenchmen, Spaniards, Portuguese, stood in the warmth of the sunrise watching the setting of this strange new star.

And in a hundred observatories there had been suppressed excitement, rising almost to shouting pitch, as the two remote bodies had rushed together, and a hurrying to and fro to gather photographic apparatus and spectroscope, and this appliance and that, to record this novel astonishing sight, the destruction of a world. For it was a world, a sister planet of our earth, far greater than our earth indeed, that had so suddenly flashed into flaming death. Neptune it was, had been struck, fairly and squarely, by the strange planet from outer space and the heat of the concussion had incontinently turned two solid globes into one vast mass of incandescence. Round the world that day, two hours before the dawn, went the pallid great white star, fading only as it sank westward and the sun mounted above it. Everywhere men marvelled at it, but of all those who saw it none could have marvelled more than those sailors, habitual watchers of the stars, who far away at sea had heard nothing of its advent and saw it now rise like a pigmy moon and climb zenithward and hang overhead and sink westward with the passing of the night.

And when next it rose over Europe everywhere were crowds of watchers on hilly slopes, on house-roofs, in open spaces, staring eastward for the rising of the great new star. It rose with a white glow in front of it, like the glare of a white fire, and those who had seen it come into existence the night before cried out at the sight of it. "It is larger," they cried. "It is brighter!" And, indeed the moon a quarter full and sinking in the west was in its apparent size beyond comparison, but scarcely in all its breadth had it as much brightness now as the little circle of the strange new star.

"It is brighter!" cried the people clustering in the streets. But in the dim observatories the watchers held their breath and peered at one another. "*It is nearer,*" they said. "*Nearer!*"

And voice after voice repeated, "It is nearer," and the clicking telegraph took that up, and it trembled along telephone wires, and in a thousand cities grimy compositors fingered the type. "It is nearer." Men writing in offices, struck with a strange realisation, flung down their pens; men talking in a thousand places suddenly came upon a grotesque possibility in those words, "It is nearer." It hurried along awakening streets, it was shouted down the frost-stilled ways of quiet villages, men who had read these things from the throbbing tape stood in yellow-lit doorways shouting the news to the passers-by. "It is nearer." Pretty women, flushed and glittering, heard the news told jestingly between the dances, and feigned an intelligent interest they did not feel. "Nearer! Indeed. How curious! How very, very clever people must be to find out things like that!"

Lonely tramps faring through the wintry night murmured those words to comfort themselves—looking skyward. "It has need to be nearer, for the night's as cold as charity. Don't seem much warmth from it if it *is* nearer, all the same."

"What is a new star to me?" cried the weeping woman kneeling beside her dead.

The schoolboy, rising early for his examination work, puzzled it out for himself—with the great white star, shining broad and bright through the frost flowers of his window. "Centrifugal, centripetal," he said, with his chin on his fist. "Stop a planet in its flight, rob it of its centrifugal force, what then? Centripetal has it, and down it falls into the sun! And this——!"

"Do *we* come in the way? I wonder——"

The light of that day went the way of its brethren, and with the later watches of the frosty darkness rose the strange star again. And it was now so bright that the waxing moon seemed but a pale yellow ghost of itself, hanging huge in the sunset. In a South African city a great man had married, and the streets were alight to welcome his return with his bride. "Even the skies have illuminated," said the flatterer. Under Capricorn, two negro lovers, daring the wild beasts and evil spirits, for love of one another, crouched together in a cane brake where the fire-flies hovered. "That is our star," they whispered, and felt strangely comforted by the sweet brilliance of its light.

The master mathematician sat in his private room and pushed the papers from him. His calculations were already finished. In a small white phial there still remained a little of the drug that had kept him awake and active for four long nights. Each day, serene, explicit, patient as ever, he had given his lecture to his students, and then had come back at once to this momentous calculation. His face was grave, a little drawn and hectic from his drugged activity. For some time he seemed lost in thought. Then he went to the window, and the blind went up with a click. Half way up the sky, over the clustering roofs, chimneys and steeples of the city, hung the star.

He looked at it as one might look into the eyes of a brave enemy. "You may kill me," he said after a silence. "But I can hold you—and all the universe for that matter—in the grip of this little brain. I would not change. Even now."

He looked at the little phial. "There will be no need of sleep again," he said. The next day at noon, punctual to the minute, he entered his lecture theatre, put his hat on the end of the tables as his habit was, and carefully selected a large piece of chalk. It was a joke among his students that he could not lecture without that piece of chalk to fumble in his fingers, and once he had been stricken to impotence by their hiding his supply. He came and looked under his grey eyebrows at the rising tiers of young fresh faces, and spoke with his accustomed studied commonness of phrasing. "Circumstances have arisen—circumstances beyond my control," he said and paused, "which will debar me from completing the course I had designed. It would seem, gentlemen, if I may put the thing clearly and briefly, that—Man has lived in vain."

The students glanced at one another. Had they heard aright? Mad? Raised eyebrows and grinning lips there were, but one or two faces remained intent upon his calm grey-fringed face. "It will be interesting," he was saying, "to devote this morning to an exposition, so far as I can make it clear to you, of the calculations that have led me to this conclusion. Let us assume——"

He turned towards the blackboard, meditating a diagram in the way that was usual to him. "What was that about 'lived in vain'?" whispered one student to another. "Listen," said the other, nodding towards the lecturer.

And presently they began to understand.

That night the star rose later, for its proper eastward motion had carried it some way across Leo towards Virgo, and its brightness was so great that the sky became a luminous blue as it rose, and every star was hidden in its turn, save only Jupiter near the zenith, Capella, Aldebaran, Sirius and the pointers of the Bear. It was very white and beautiful. In many parts of the world that night a pallid halo encircled it about. It was perceptibly larger; in the clear refractive sky of the tropics it seemed as if it were nearly a quarter the size of the moon. The frost was still on the ground in England, but the world was as brightly lit as if it were midsummer moonlight. One could see to read quite ordinary print by that cold clear light, and in the cities the lamps burnt yellow and wan.

And everywhere the world was awake that night, and throughout Christendom a sombre murmur hung in the keen air over the country side like the belling of bees in the heather, and this murmurous tumult grew to a clangour in the cities. It was the tolling of the bells in a million belfry towers and steeples, summoning the people to sleep no more, to sin no more, but to gather in their churches and pray. And overhead, growing larger and brighter as the earth rolled on its way and the night passed, rose the dazzling star.

And the streets and houses were alight in all the cities, the shipyards glared, and whatever roads led to high country were lit and crowded all night long. And in all the seas about the civilised lands, ships with throbbing engines, and ships with bellying sails, crowded with men and living creatures, were standing out to ocean and the north. For already the warning of the master mathematician had been telegraphed all over the world, and translated into a hundred tongues. The new planet and Neptune, locked in a fiery embrace, were whirling headlong, ever faster and faster towards the sun. Already every second this blazing mass flew a hundred miles, and every second its terrific velocity increased. As it flew now, indeed, it must pass a hundred million of miles wide of the earth and scarcely affect it. But near its destined path, as yet only slightly perturbed, spun the mighty planet Jupiter and his moons sweeping splendid round the sun. Every moment now the attraction between the fiery star and the greatest of the planets grew stronger. And the result of that attraction? Inevitably Jupiter would be deflected from his orbit into an elliptical path, and the burning star, swung by his attraction wide of its sunward rush, would "describe a curved path" and perhaps collide with, and certainly pass very close to, our earth. "Earthquakes, volcanic outbreaks, cyclones, sea waves, floods, and a steady rise in temperature to I know not what limit"—so prophesied the master mathematician.

And overhead, to carry out his words, lonely and cold and livid, blazed the star of the coming doom.

To many who stared at it that night until their eyes ached, it seemed that it was visibly approaching. And that night, too, the weather changed, and the frost that had gripped all Central Europe and France and England softened towards a thaw.

But you must not imagine because I have spoken of people praying through the night and people going aboard ships and people fleeing towards mountainous country that the whole world was already in a terror because of the star. As a matter of fact, use and wont still ruled the world, and save for the talk of idle moments and the splendour of the night, nine human beings out of ten were still busy at their common occupations. In all the cities and shops, save one here and there, opened and closed at their proper hours, the doctor and the undertaker plied their trades, the workers gathered in the factories, soldiers drilled, scholars studied, lovers sought one another, thieves lurked and fled, politicians planned their schemes. The presses of the newspapers roared through the nights, and many a priest of this church and that would not open his holy building to further what he considered a foolish panic. The newspapers insisted on the lesson of the year 1000—for then, too, people had anticipated the end. The star was no star—mere gas—a comet; and were it a star it could not possibly strike the earth. There was no precedent for such a thing. Common sense was sturdy everywhere, scornful, jesting, a little inclined to persecute the obdurate fearful. That night, at seven-fifteen by Greenwich time, the star would be at its nearest to Jupiter. Then the world would see the turn things would take. The master mathematician's grim warnings were treated by many as so much mere elaborate self-advertisement. Common sense at last, a little heated by argument, signified its unalterable convictions by going to bed. So, too, barbarism and savagery, already tired of the novelty, went about their mighty business, and save for a howling dog here and there, the beast world left the star unheeded.

And yet, when at last the watchers in the European States saw the star rise, an hour later it is true, but no larger than it had been the night before, there were still plenty awake to laugh at the master mathematician—to take the danger as if it had passed.

But hereafter the laughter ceased. The star grew—it grew with a terrible steadiness hour after hour, a little larger each hour, a little nearer the midnight zenith, and brighter and brighter, until it had turned night into a second day. Had it come straight to the earth instead of in a curved path, had it lost no velocity to Jupiter, it must have leapt the intervening gulf in a day, but as it was it took five days altogether to come by our planet. The next night it had become a third the size of the moon before it set to English eyes, and the thaw was assured. It rose over America near the size of the moon, but blinding white to look at, and *hot*; and a breath of hot wind blew now with its rising and gathering strength, and in Virginia, and Brazil, and down the St. Lawrence valley, it shone intermittently through a driving reek of thunder-clouds, flickering violet lightning, and hail unprecedented. In Manitoba was a thaw and devastating floods. And upon all the mountains of the earth the snow and ice began to melt that night, and all the rivers coming out of high country flowed thick and turbid, and soon—in their upper reaches—with swirling trees and the bodies of beasts and men. They rose steadily, steadily in the ghostly brilliance, and came trickling over their banks at last, behind the flying population of their valleys.

And along the coast of Argentina and up the South Atlantic the tides were higher than had ever been in the memory of man, and the storms drove the waters in many cases scores of miles inland, drowning whole cities. And so great grew the heat during the night that the rising of the sun was like the coming of a shadow. The earthquakes began and grew until all down America from the Arctic Circle to Cape Horn, hillsides were sliding, fissures were opening, and houses and walls crumbling to destruction. The whole side of Cotopaxi slipped out in one vast convulsion, and a tumult of lava poured out so high and broad and swift and liquid that in one day it reached the sea.

So the star, with the wan moon in its wake, marched across the Pacific, trailed the thunderstorms like the hem of a robe, and the growing tidal wave that toiled behind it, frothing and eager, poured over island and island and swept them clear of men. Until that wave came at last —in a blinding light and with the breath of a furnace, swift and terrible it came—a wall of water, fifty feet high, roaring hungrily, upon the long coasts of Asia, and swept inland across the plains of China. For a space the star, hotter now and larger and brighter than the sun in its strength, showed with pitiless brilliance the wide and populous country; towns and villages with their pagodas and trees, roads, wide cultivated fields, millions of sleepless people staring in helpless terror at the incandescent sky; and then, low and growing, came the murmur of the flood. And thus it was with millions of men that night—a flight no-whither, with limbs heavy with heat and breath fierce and scant, and the flood like a wall swift and white behind. And then death.

China was lit glowing white, but over Japan and Java and all the islands of Eastern Asia the great star was a ball of dull red fire because of the steam and smoke and ashes the volcanoes were spouting forth to salute its coming. Above was the lava, hot gases and ash, and below the seething floods, and the whole earth swayed and rumbled with the earthquake shocks. Soon the immemorial snows of Thibet and the Himalaya were melting and pouring down by ten million deepening converging channels upon the plains of Burmah and Hindostan. The tangled summits of the Indian jungles were aflame in a thousand places, and below the hurrying waters around the stems were dark objects that still struggled feebly and reflected the blood-red tongues of fire. And in a rudderless confusion a multitude of men and women fled down the broad river-ways to that one last hope of men—the open sea.

Larger grew the star, and larger, hotter, and brighter with a terrible swiftness now. The tropical ocean had lost its phosphorescence, and the whirling stream rose in ghostly wreaths from the black waves that plunged incessantly, speckled with storm-tossed ships.

And then came a wonder. It seemed to those who in Europe watched for the rising of the star that the world must have ceased its rotation. In a thousand open spaces of down and upland the people who had fled thither from the floods and the falling houses and sliding slopes of hill watched for that rising in vain. Hour followed hour through a terrible suspense, and the star rose not. Once again men set their eyes upon the old constellations they had counted lost to them forever. In England it was hot and clear overhead, though the ground quivered perpetually, but in the tropics, Sirius and Capella and Aldebaran showed through a veil of steam. And when at last the great star rose near ten hours late, the sun rose close upon it, and in the centre of its white heart was a disc of black.

Over Asia it was the star had begun to fall behind the movement of the sky, and then suddenly, as it hung over

India, its light had been veiled. All the plain of India
from the mouth of the Indus to the mouths of the Ganges
was a shallow waste of shining water that night, out of
which rose temples and palaces, mounds and hills, black
with people. Every minaret was a clustering mass of people,
who fell one by one into the turbid waters, as heat and
terror overcame them. The whole land seemed a-wailing,
and suddenly there swept a shadow across that furnace
of despair, and a breath of cold wind, and a gathering of
clouds, out of the cooling air. Men looking up, near
blinded, at the star, saw that a black disc was creeping
across the light. It was the moon, coming between the star
and the earth. And even as men cried to God at this respite,
out of the East with a strange inexplicable swiftness sprang
the sun. And then star, sun and moon rushed together
across the heavens.

So it was that presently, to the European watchers, star
and sun rose close upon each other, drove headlong for a
space and then slower, and at last came to rest, star and
sun merged into one glare of flame at the zenith of the
sky. The moon no longer eclipsed the star but was lost
to sight in the brilliance of the sky. And though those
who were still alive regarded it for the most part with that
dull stupidity that hunger, fatigue, heat and despair en-
gender, there were still men who could perceive the meaning
of these signs. Star and earth had been at their nearest,
had swung about one another, and the star had passed.
Already it was receding, swifter and swifter, in the last
stage of its headlong journey downward into the sun.

And then the clouds gathered, blotting out the vision of
the sky, the thunder and lightning wove a garment round
the world; all over the earth was such a downpour of rain
as men had never before seen, and where the volcanoes
flared red against the cloud canopy there descended tor-
rents of mud. Everywhere the waters were pouring off
the land, leaving mud-silted ruins, and the earth littered
like a storm-worn beach with all that had floated, and the
dead bodies of the men and brutes, its children. For days
the water streamed off the land, sweeping away soil and

trees and houses in the way, and piling huge dykes and scooping out Titanic gullies over the country side. Those were the days of darkness that followed the star and the heat. All through them, and for many weeks and months, the earthquakes continued.

But the star had passed, and men, hunger-driven and gathering courage only slowly, might creep back to their ruined cities, buried granaries, and sodden fields. Such few ships as had escaped the storms of that time came stunned and shattered and sounding their way cautiously through the new marks and shoals of once familiar ports. And as the storms subsided men perceived that everywhere the days were hotter than of yore, and the sun larger, and the moon, shrunk to a third of its former size, took now fourscore days between its new and new.

But of the new brotherhood that grew presently among men, of the saving of laws and books and machines, of the strange change that had come over Iceland and Greenland and the shores of Baffin's Bay, so that the sailors coming there presently found them green and gracious, and could scarce believe their eyes, this story does not tell. Nor of the movement of mankind now that the earth was hotter, northward and southward towards the poles of the earth. It concerns itself only with the coming and the passing of the Star.

The Martian astronomers—for there are astronomers on Mars, although they are very different beings from men—were naturally profoundly interested by these things. They saw them from their own standpoint of course. "Considering the mass and temperature of the missile that was flung through our solar system into the sun," one wrote, "it is astonishing what a little damage the earth, which it missed so narrowly, has sustained. All the familiar continental markings and the masses of the seas remain intact, and indeed the only difference seems to be a shrinkage of the white discoloration (supposed to be frozen water) round either pole." Which only shows how small the vastest of human catastrophes may seem, at a distance of a few million miles.

STORY THE THIRD

A Story of the Stone Age

1.—Ugh-lomi and Uya

This story is of a time beyond the memory of man, before the beginning of history, a time when one might have walked dryshod from France (as we call it now) to England, and when a broad and sluggish Thames flowed through its marshes to meet its father Rhine, flowing through a wide and level country that is under water in these latter days, and which we know by the name of the North Sea. In that remote age the valley which runs along the foot of the Downs did not exist, and the south of Surrey was a range of hills, fir-clad on the middle slopes, and snow-capped for the better part of the year. The cores of its summits still remain as Leith Hill, and Pitch Hill, and Hindhead. On the lower slopes of the range, below the grassy spaces where the wild horses grazed, were forests of yew and sweet-chestnut and elm, and the thickets and dark places hid the grizzly bear and the hyæna, and the grey apes clambered through the branches. And still lower amidst the woodland and marsh and open grass along the Wey did this little drama play itself out to the end that I have to tell. Fifty thousand years ago it was, fifty thousand years—if the reckoning of geologists is correct.

And in those days the spring-time was as joyful as it is now, and sent the blood coursing in just the same fashion. The afternoon sky was blue with piled white clouds sailing through it, and the southwest wind came like a soft

656

caress. The new-come swallows drove to and fro. The reaches of the river were spangled with white ranunculus, the marshy places were starred with lady's-smock and lit with marshmallow wherever the regiments of the sedges lowered their swords, and the northward moving hippopotami, shiny black monsters, sporting clumsily, came floundering and blundering through it all, rejoicing dimly and possessed with one clear idea, to splash the river muddy.

Up the river and well in sight of the hippopotami, a number of little buff-coloured animals dabbled in the water. There was no fear, no rivalry, and no enmity between them and the hippopotami. As the great bulks came crashing through the reeds and smashed the mirror of the water into silvery splashes, these little creatures shouted and gesticulated with glee. It was the surest sign of high spring. "Boloo!" they cried. "Baayah. Boloo!" They were the children of the men folk, the smoke of whose encampment rose from the knoll at the river's bend. Wild-eyed youngsters they were, with matted hair and little broad-nosed impish faces, covered (as some children are covered even nowadays) with a delicate down of hair. They were narrow in the loins and long in the arms. And their ears had no lobes, and had little pointed tips, a thing that still, in rare instances, survives. Stark-naked vivid little gipsies, as active as monkeys and as full of chatter, though a little wanting in words.

Their elders were hidden from the wallowing hippoptami by the crest of the knoll. The human squatting-place was a trampled area among the dead brown fronds of Royal Fern, through which the crosiers of this year's growth were unrolling to the light and warmth. The fire was a mouldering heap of char, light grey and black, replenished by the old women from time to time with brown leaves. Most of the men were asleep—they slept sitting with their foreheads on their knees. They had killed that morning a good quarry, enough for all, a deer that had been wounded by hunting dogs; so that there had been no quarrelling among them, and some of the women were still gnawing the bones that lay scattered about. Others were making a heap of leaves and sticks to feed Brother Fire when the darkness came again, that he might grow strong and tall therewith, and guard them against the beasts. And two were piling flints that they brought, an armful at a time, from the bend of the river where the children were at play.

None of these buff-skinned savages were clothed, but some wore about their hips rude girdles of adder-skin or crackling undressed hide, from which depended little bags, not made, but torn from the paws of beasts, and carrying the rudely-dressed flints that were men's chief weapons and tools. And one woman, the mate of Uya the Cunning Man, wore a wonderful necklace of perforated fossils— that others had worn before her. Beside some of the sleeping men lay the big antlers of the elk, with the tines chipped to sharp edges, and long sticks, hacked at the ends with flints into sharp points. There was little else save these things and the smouldering fire to mark these human beings off from the wild animals that ranged the country. But Uya the Cunning did not sleep, but sat with a bone in his hand and scraped busily thereon with a flint, a thing no animal would do. He was the oldest man in the tribe, beetle-browed, prognathous, lank-armed; he had a beard and his cheeks were hairy, and his chest and arms were black with thick hair. And by virtue both of his strength and cunning he was master of the tribe, and his share was always the most and the best.

Eudena had hidden herself among the alders, because she was afraid of Uya. She was still a girl, and her eyes were bright and her smile pleasant to see. He had given her a piece of the liver, a man's piece, and a wonderful treat for a girl to get; but as she took it the other woman with the necklace had looked at her, an evil glance, and Ugh-lomi had made a noise in his throat. At that, Uya had looked at him long and steadfastly, and Ugh-lomi's face had fallen. And then Uya had looked at her. She was frightened and she had stolen away, while the feeding was still going on, and Uya was busy with the marrow of a bone. Afterwards he had wandered about as if looking for her. And now she crouched among the alders, wonder ing mightily what Uya might be doing with the flint and the bone. And Ugh-lomi was not to be seen.

Presently a squirrel came leaping through the alders, and she lay so quiet the little man was within six feet of her before he saw her. Whereupon he dashed up a stem in a hurry and began to chatter and scold her. "What are you doing here," he asked, "away from the other men beasts?" "Peace," said Eudena, but he only chattered more, and then she began to break off the little black cones to throw at him. He dodged and defied her, and she grew excited and rose up to throw better, and then she saw Uya coming down the knoll. He had seen the movement of her pale arm amidst the thicket—he was very keen-eyed.

At that she forgot the squirrel and set off through the alders and reeds as fast as she could go. She did not care where she went so long as she escaped Uya. She splashed nearly knee-deep through a swampy place, and saw in front of her a slope of ferns—growing more slender and green as they passed up out of the light into the shade of the young chestnuts. She was soon amidst the trees—she was very fleet of foot, and she ran on and on until the forest was old and the vales great, and the vines about their stems where the light came were thick as young trees, and the ropes of ivy stout and tight. On she went, and she doubled and doubled again, and then at last lay down amidst some ferns in a hollow place near a thicket, and listened with her heart beating in her ears.

She heard footsteps presently rustling among the dead leaves, far off, and they died away and everything was still again, except the scandalising of the midges—for the evening was drawing on—and the incessant whisper of the leaves. She laughed silently to think the cunning Uya should go by her. She was not frightened. Sometimes, playing with the other girls and lads, she had fled into the wood, though never so far as this. It was pleasant to be hidden and alone.

She lay a long time there, glad of her escape, and then she sat up listening.

It was a rapid pattering growing louder and coming towards her, and in a little while she could hear grunting noises and the snapping of twigs. It was a drove of lean grisly wild swine. She turned about her, for a boar is an ill fellow to pass too closely, on account of the sideway slash of his tusks, and she made off slantingly through the trees. But the patter came nearer, they were not feeding as they wandered, but going fast—or else they would not overtake her—and she caught the limb of a tree, swung on to it, and ran up the stem with something of the agility of a monkey.

Down below the sharp bristling backs of the swine were already passing when she looked. And she knew the short, sharp grunts they made meant fear. What were they afraid of? A man? They were in a great hurry for just a man.

And then, so suddenly it made her grip on the branch tighten involuntarily, a fawn started in the brake and rushed after the swine. Something else went by, low and grey, with a long body; she did not know what it was, indeed she saw it only momentarily through the interstices of the young leaves; and then there came a pause.

She remained stiff and expectant, as rigid almost as though she was a part of the tree she clung to, peering down.

Then, far away among the trees, clear for a moment, then hidden, then visible knee-deep in ferns, then gone again, ran a man. She knew it was young Ugh-lomi by the fair colour of his hair, and there was red upon his face. Somehow his frantic flight and that scarlet mark made her feel sick. And then nearer, running heavily and breathing hard, came another man. At first she could not see, and then she saw, foreshortened and clear to her, Uya, running with great strides and his eyes staring. He was not going after Ugh-lomi. His face was white. It was Uya—*afraid!* He passed, and was still loud hearing, when something else, something large and with grizzled fur, swinging along with soft swift strides, came rushing in pursuit of him.

Eudena suddenly became rigid, ceased to breathe, her clutch convulsive, and her eyes starting.

She had never seen the thing before, she did not even see him clearly now, but she knew at once it was the Terror of the Woodshade. His name was a legend, the children would frighten one another, frighten even themselves with his name, and run screaming to the squatting-place. No man had ever killed any of his kind. Even the mighty mammoth feared his anger. It was the grizzly bear, the lord of the world as the world went then.

As he ran he made a continuous growling grumble. "Men in my very lair! Fighting and blood. At the very mouth of my lair. Men, men, men. Fighting and blood." For he was the lord of the wood and of the caves.

Long after he had passed she remained, a girl of stone, staring down through the branches. All her power of action had gone from her. She gripped by instinct with hands and knees and feet. It was some time before she could think, and then only one thing was clear in her mind, that the Terror was between her and the tribe—that it would be impossible to descend.

Presently when her fear was a little abated she clambered into a more comfortable position, where a great branch forked. The trees rose about her, so that she could see nothing of Brother Fire, who is black by day. Birds began to stir, and things that had gone into hiding for fear of her movements crept out. . . .

After a time the taller branches flamed out at the touch of the sunset. High overhead the rooks, who were wiser than men, went cawing home to their squatting-places among the elms. Looking down, things were clearer and darker. Eudena thought of going back to the squatting-place; she let herself down some way, and then the fear of the Terror of the Woodshade came again. While she hesitated a rabbit squealed dismally, and she dared not descend farther.

The shadows gathered, and the deeps of the forest began stirring. Eudena went up the tree again to be nearer the light. Down below the shadows came out of their hiding-places and walked abroad. Overhead the blue deepened.

A dreadful stillness came, and then the leaves began whispering.

Eudena shivered and thought of Brother Fire.

The shadows now were gathering in the trees, they sat on the branches and watched her. Branches and leaves were turned to ominous, quiet black shapes that would spring on her if she stirred. Then the white owl, flitting silently, came ghostly through the shades. Darker grew the world and darker, until the leaves and twigs against the sky were black, and the ground was hidden.

She remained there all night, an age-long vigil, straining her ears for the things that went on below in the darkness, and keeping motionless lest some stealthy beast should discover her. Man in those days was never alone in the dark, save for such rare accidents as this. Age after age he had learnt the lesson of its terror—a lesson we poor children of his have nowadays painfully to unlearn. Eudena, though in age a woman, was in heart like a little child. She kept as still, poor little animal, as a hare before it is started.

The stars gathered and watched her—her one grain of comfort. In one bright one she fancied there was something like Ugh-lomi. Then she fancied it *was* Ugh-lomi. And near him, red and duller, was Uya, and as the night passed Ugh-lomi fled before him up the sky.

She tried to see Brother Fire, who guarded the squatting-place from beasts, but he was not in sight. And far away she heard the mammoths trumpeting as they went down to the drinking-place, and once some huge bulk with heavy paces hurried along, making a noise like a calf, but what it was she could not see. But she thought from the voice it was Yaaa the rhinoceros, who stabs with his nose, goes always alone, and rages without cause.

At last the little stars began to hide, and then the larger ones. It was like all the animals vanishing before the Terror. The Sun was coming, lord of the sky, as the grizzly was lord of the forest. Eudena wondered what would happen if one star stayed behind. And then the sky paled to the dawn.

When the daylight came the fear of lurking things

assed, and she could descend. She was stiff, but not so
stiff as you would have been, dear young lady (by virtue
of your upbringing), and as she had not been trained to
at at least once in three hours, but instead had often fasted
three days, she did not feel uncomfortably hungry. She
crept down the tree very cautiously, and went her way
stealthily through the wood, and not a squirrel sprang or
deer started but the terror of the grizzly bear froze her
narrow.

Her desire was now to find her people again. Her dread
of Uya the Cunning was consumed by a greater dread of
loneliness. But she had lost her direction. She had run
needlessly overnight, and she could not tell whether the
squatting-place was sunward or where it lay. Ever and
again she stopped and listened, and at last, very far away,
she heard a measured chinking. It was so faint even in
the morning stillness that she could tell it must be far
away. But she knew the sound was that of a man sharpening a flint.

Presently the trees began to thin out, and then came a
regiment of nettles barring the way. She turned aside,
and then she came to a fallen tree that she knew, with a
noise of bees about it. And so presently she was in sight
of the knoll, very far off, and the river under it, and the
children and the hippopotami just as they had been yesterday, and the thin spire of smoke swaying in the morning
breeze. Far away by the river was the cluster of alders
where she had hidden. And at the sight of that the fear
of Uya returned, and she crept into a thicket of bracken,
out of which a rabbit scuttled, and lay awhile to watch the
squatting-place.

The men were mostly out of sight, saving Wau, the
flint-chopper; and at that she felt safer. They were away
hunting food, no doubt. Some of the women, too, were
down in the stream, stooping intent, seeking mussels, crayfish, and water-snails, and at the sight of their occupation
Eudena felt hungry. She rose, and ran through the fern,
designing to join them. As she went she heard a voice
among the bracken calling softly. She stopped. Then

suddenly she heard a rustle behind her, and turning, saw
Ugh-lomi rising out of the fern.   There were streaks
of brown blood and dirt on his face, and his eyes were
fierce, and the white stone of Uya, the white Fire Stone, that
none but Uya dared to touch, was in his hand.   In a stride
he was beside her, and gripped her arm.   He swung her
about, and thrust her before him towards the wood.
"Uya," he said, and waved his arms about.   She heard a
cry, looked back, and saw all the women standing up, and
two wading out of the stream.   Then came a nearer howling
and the old woman with the beard, who watched the fire
on the knoll, was waving her arms and Wau, the man who
had been chipping the flint, was getting to his feet.   The
little children too were hurrying and shouting.

"Come!" said Ugh-lomi, and dragged her by the arm.
She still did not understand.

"Uya has called the death word," said Ugh-lomi, and
she glanced back at the screaming curve of figures, and
understood.

Wau and all the women and children were coming towards them, a scattered array of buff shock-headed figures
howling, leaping, and crying.   Over the knoll two youths
hurried.   Down among the ferns to the right came a man
heading them off from the wood.   Ugh-lomi left her arm
and the two began running side by side, leaping the bracken
and stepping clear and wide.   Eudena, knowing her fleetness and the fleetness of Ugh-lomi, laughed aloud at the
unequal chase.   They were an exceptionally straight-limbed
couple for those days.

They soon cleared the open, and drew near the wood of
chestnut-trees again—neither afraid now because neither
was alone.   They slackened their pace, already not excessive.   And suddenly Eudena cried and swerved aside, pointing, and looking up through the tree-stems.   Ugh-lomi
saw the feet and legs of men running towards him.   Eudena
was already running off at a tangent.   And as he too turned
to follow her they heard the voice of Uya coming through
the trees, and roaring out his rage at them.

Then terror came in their hearts, not the terror that

numbs, but the terror that makes one silent and swift. They were cut off now on two sides. They were in a sort of corner of pursuit. On the right hand, and near by them, came the men swift and heavy, with bearded Uya, antler in hand, leading them; and on the left, scattered as one scatters corn, yellow dashes among the fern and grass, ran Wau and the women; and even the little children from the shallow had joined the chase. The two parties converged upon them. Off they went, with Eudena ahead.

They knew there was no mercy for them. There was no hunting so sweet to these ancient men as the hunting of men. Once the fierce passion of the chase was lit, the feeble beginnings of humanity in them were thrown to the winds. And Uya in the night had marked Ugh-lomi with the death word. Ugh-lomi was the day's quarry, the appointed feast.

They ran straight—it was their only chance—taking whatever ground came in the way—a spread of stinging nettles, an open glade, a clump of grass out of which a hyæna fled snarling. Then woods again, long stretches of shady leaf-mould and moss under the green trunks. Then a stiff slope, tree-clad, and long vistas of trees, a glade, a succulent green area of black mud, a wide open space again, and then a clump of lacerating brambles, with beast tracks through it. Behind them the chase trailed out and scattered, with Uya ever at their heels. Eudena kept the first place, running light and with her breath easy, for Ugh-lomi carried the Fire Stone in his hand.

It told on his pace—not at first, but after a time. His footsteps behind her suddenly grew remote. Glancing over her shoulder as they crossed another open space, Eudena saw that Ugh-lomi was many yards behind her, and Uya close upon him, with antler already raised in the air to strike him down. Wau and the others were but just emerging from the shadow of the woods.

Seeing Ugh-lomi in peril, Eudena ran sideways, looking back, threw up her arms and cried aloud, just as the antler flew. And young Ugh-lomi, expecting this and understanding her cry, ducked his head, so that the missile merely struck his scalp lightly, making but a trivial wound, and flew over him. He turned forthwith, the quartzite Fire Stone in both hands, and hurled it straight at Uya's body as he ran loose from the throw. Uya shouted, but could not dodge it. It took him under the ribs, heavy and flat, and he reeled and went down without a cry. Ugh-lomi caught up the antler—one tine of it was tipped with his own blood—and came running on again with a red trickle just coming out of his hair.

Uya rolled over twice, and lay a moment before he got up, and then he did not run fast. The colour of his face was changed. Wau overtook him, and then others, and he coughed and laboured in his breath. But he kept on.

At last the two fugitives gained the bank of the river, where the stream ran deep and narrow, and they still had fifty yards in hand of Wau, the foremost pursuer, the man who made the smiting stones. He carried one, a large flint, the shape of an oyster and double the size, chipped to a chisel edge, in either hand.

They sprang down the steep bank into the stream, rushed through the water, swam the deep current in two or three strokes, and came out wading again, dripping and refreshed, to clamber up the farther bank. It was undermined, and with willows growing thickly therefrom, so that it needed clambering. And while Eudena was still among the silvery branches and Ugh-lomi still in the water —for the antler had encumbered him—Wau came up against the sky on the opposite bank, and the smiting stone, thrown cunningly, took the side of Eudena's knee. She struggled to the top and fell.

They heard the pursuers shout to one another, and Ugh-lomi climbing to her and moving jerkily to mar Wau's aim, felt the second smiting stone graze his ear, and heard the water splash below him.

Then it was Ugh-lomi, the stripling, proved himself to have come to man's estate. For running on, he found Eudena fell behind, limping, and at that he turned, and crying savagely and with a face terrible with sudden wrath and trickling blood, ran swiftly past her back to the bank, whirling the antler round his head. And Eudena kept on, running stoutly still, though she must needs limp at every step, and the pain was already sharp.

So that Wau, rising over the edge and clutching the straight willow branches, saw Ugh-lomi towering over him, gigantic against the blue; saw his whole body swing round, and the grip of his hands upon the antler. The edge of the antler came sweeping through the air, and he saw no more. The water under the osiers whirled and eddied and went crimson six feet down the stream. Uya following stopped knee-high across the stream, and the man who was swimming turned about.

The other men who trailed after—they were none of them very mighty men (for Uya was more cunning than strong, brooking no sturdy rivals)—slackened momentarily at the sight of Ugh-lomi standing there above the willows, bloody and terrible, between them and the halting girl, with the huge antler waving in his hand. It seemed as though he had gone into the water a youth, and come out of it a man full grown.

He knew what there was behind him. A broad stretch of grass, and then a thicket, and in that Eudena could hide. That was clear in his mind, though his thinking powers were too feeble to see what should happen thereafter. Uya stood knee-deep, undecided and unarmed. His heavy mouth hung open, showing his canine teeth, and he panted heavily. His side was flushed and bruised under the hair. The other man beside him carried a sharpened stick. The rest of the hunters came up one by one to the top of the bank, hairy, long-armed men clutching flints and sticks. Two ran off along the bank down stream, and then clambered to the water, where Wau had come to the surface struggling weakly. Before they could reach him he went under again. Two others threatened Ugh-lomi from the bank.

He answered back, shouts, vague insults, gestures. Then Uya, who had been hesitating, roared with rage, and whirling his fists plunged into the water. His followers splashed after him.

Ugh-lomi glanced over his shoulder and found Eudena

already vanished into the thicket. He would perhaps hav
waited for Uya, but Uya preferred to spar in the wate
below him until the others were beside him. Human tac
tics in those days, in all serious fighting, were the tactic
of the pack. Prey that turned at bay they gathered aroun
and rushed. Ugh-lomi felt the rush coming, and hurling
the antler at Uya, turned about and fled.

When he halted to look back from the shadow of th
thicket, he found only three of his pursuers had followe
him across the river, and they were going back again. Uya
with a bleeding mouth, was on the farther side of th
stream again, but lower down, and holding his hand to hi
side. The others were in the river dragging something t
shore. For a time at least the chase was intermitted.

Ugh-lomi stood watching for a space, and snarled at th
sight of Uya. Then he turned and plunged into the thicket

In a minute, Eudena came hastening to join him, an
they went on hand in hand. He dimly perceived the pai
she suffered from the cut and bruised knee, and chose th
easier ways. But they went on all that day, mile afte
mile, through wood and thicket, until at last they came t
the chalk land, open grass with rare woods of beech, and
the birch growing near water, and they saw the Wealde
mountains nearer, and groups of horses grazing together
They went circumspectly, keeping always near thicket and
cover, for this was a strange region—even its ways were
strange. Steadily the ground rose, until the chestnu
forests spread wide and blue below them, and the Thame
marshes shone silvery, high and far. They saw no men
for in those days men were still only just come into thi
part of the world, and were moving but slowly along th
river-ways. Towards evening they came on the river again
but now it ran in a gorge, between high cliffs of whit
chalk that sometimes overhung it. Down the cliffs was
scrub of birches and there were many birds there. And
high up the cliff was a little shelf by a tree, whereon the
clambered to pass the night.

They had had scarcely any food; it was not the time o
year for berries, and they had no time to go aside to snare

or waylay. They tramped in a hungry weary silence, gnawing at twigs and leaves. But over the surface of the cliffs were a multitude of snails, and in a bush were the freshly laid eggs of a little bird, and then Ugh-lomi threw at and killed a squirrel in a beech-tree, so that at last they fed well. Ugh-lomi watched during the night, his chin on his knees; and he heard young foxes crying hard by, and the noise of mammoths down the gorge, and the hyænas yelling and laughing far away. It was chilly, but they dared not light a fire. Whenever he dozed, his spirit went abroad, and straightway met with the spirit of Uya, and they fought. And always Ugh-lomi was paralysed so that he could not smite nor run, and then he would awake suddenly. Eudena, too, dreamt evil things of Uya, so that they both awoke with the fear of him in their hearts, and by the light of the dawn they saw a woolly rhinoceros go blundering down the valley.

During the day they caressed one another and were glad of the sunshine, and Eudena's leg was so stiff she sat on the ledge all day. Ugh-lomi found great flints sticking out of the cliff face, greater than any he had seen, and he dragged some to the ledge and began chipping, so as to be armed against Uya when he came again. And at once he laughed heartily, and Eudena laughed, and they threw it about in derision. It had a hole in it. They stuck their fingers through it, it was very funny indeed. Then they peeped at one another through it. Afterwards, Ugh-lomi got himself a stick, and thrusting by chance at this foolish flint, the stick went in and stuck there. He had rammed it in too tightly to withdraw it. That was still stranger—scarcely funny, terrible almost, and for a time Ugh-lomi did not greatly care to touch the thing. It was as if the flint had bit and held with its teeth. But then he got familiar with the odd combination. He swung it about, and perceived that the stick with the heavy stone on the end struck a better blow than anything he knew. He went to and fro swinging it, and striking with it; but later he tired of it and threw it aside. In the afternoon he went up over the brow of the white cliff, and lay watch-

ing by a rabbit-warren until the rabbits came out to play
There were no men thereabouts, and the rabbits were heed
less. He threw a smiting stone he had made and got a kil

That night they made a fire from flint sparks and bracke
fronds, and talked and caressed by it. And in their slee
Uya's spirit came again, and suddenly, while Ugh-lom
was trying to fight vainly, the foolish flint on the stic
came into his hand, and he struck Uya with it, and behold
it killed him. But afterwards came other dreams of Uy
—for spirits take a lot of killing, and he had to be kille
again. Then after that the stone would not keep on th
stick. He awoke tired and rather gloomy, and was sulk
all the forenoon, in spite of Eudena's kindliness, and instea
of hunting he sat chipping a sharp edge to the singula
flint, and looking strangely at her. Then he bound th
perforated flint on to the stick with strips of rabbit skin
And afterwards he walked up and down the ledge, strikin
with it, and muttering to himself, and thinking of Uya
It felt very fine and heavy in the hand.

Several days, more than there was any counting in thos
days, five days, it may be, or six, did Ugh-lomi and Euden
stay on that shelf in the gorge of the river, and they los
all fear of men, and their fire burnt redly of a night. And
they were very merry together; there was food every day
sweet water, and no enemies. Eudena's knee was well i
a couple of days, for those ancient savages had quick-healin
flesh. Indeed, they were very happy.

On one of those days Ugh-lomi dropped a chunk of flin
over the cliff. He saw it fall, and go bounding across the rive
bank into the river, and after laughing and thinking i
over a little he tried another. This smashed a bush o
hazel in the most interesting way. They spent all th
morning dropping stones from the ledge, and in the after
noon they discovered this new and interesting pastime wa
also possible from the cliff-brow. The next day they had
forgotten this delight. Or at least, it seemed they had
forgotten.

But Uya came in dreams to spoil the paradise. Three
nights he came fighting Ugh-lomi. In the morning afte

these dreams Ugh-lomi would walk up and down, threatening him and swinging the axe, and at last came the night after Ugh-lomi brained the otter, and they had feasted. Uya went too far. Ugh-lomi awoke, scowling under his heavy brows, and he took his axe, and extending his hand towards Eudena he bade her wait for him upon the ledge. Then he clambered down the white declivity, glanced up once from the foot of it and flourished his axe, and without looking back again went striding along the river bank until the overhanging cliff at the bend hid him.

Two days and nights did Eudena sit alone by the fire on the ledge waiting, and in the night the beasts howled over the cliffs and down the valley, and on the cliff over against her the hunched hyænas prowled black against the sky. But no evil thing came near her save fear. Once, far away, she heard the roaring of a lion, following the horses as they came northward over the grass lands with the spring. All that time she waited—the waiting that is pain.

And the third day Ugh-lomi came back, up the river. The plumes of a raven were in his hair. The first axe was red-stained, and had long dark hairs upon it, and he carried the necklace that had marked the favourite of Uya in his hand. He walked in the soft places, giving no heed to his trail. Save a raw cut below his jaw there was not a wound upon him. "Uya!" cried Ugh-lomi exultant, and Eudena saw it was well. He put the necklace on Eudena, and they ate and drank together. And after eating he began to rehearse the whole story from the beginning, when Uya had cast his eyes on Eudena, and Uya and Ugh-lomi, fighting in the forest, had been chased by the bear, eking out his scanty words with abundant pantomime, springing to his feet and whirling the stone axe round when it came to the fighting. The last fight was a mighty one, stamping and shouting, and once a blow at the fire that sent a torrent of sparks up into the night. And Eudena sat red in the light of the fire, gloating on him, her face flushed and her eyes shining, and the necklace Uya had made

about her neck. It was a splendid time, and the stars that
look down on us looked down on her, our ancestor—who has
been dead now these fifty thousand years.

II.—The Cave Bear

In the days when Eudena and Ugh-lomi fled from the
people of Uya towards the fir-clad mountains of the Weald,
across the forests of sweet chestnut and the grass-clad
chalkland, and hid themselves at last in the gorge of the
river between the chalk cliffs, men were few and their
squatting-places far between. The nearest men to them
were those of the tribe, a full day's journey down the river,
and up the mountains there were none. Man was indeed
a newcomer to this part of the world in that ancient time,
coming slowly along the rivers, generation after generation,
from one squatting-place to another, from the south-westward. And the animals that held the land, the hippopota-
mus and rhinoceros of the river valleys, the horses of the
grass plains, the deer and swine of the woods, the grey apes
in the branches, the cattle of the uplands, feared him but
little—let alone the mammoths in the mountains and the
elephants that came through the land in the summer-time
out of the south. For why should they fear him, with but
the rough, chipped flints that he had not learnt to haft and
which he threw but ill, and the poor spear of sharpened
wood, as all the weapons he had against hoof and horn,
tooth and claw?

Andoo, the huge cave bear, who lived in the cave up
the gorge, had never even seen a man in all his wise and
respectable life, until midway through one night, as he was
prowling down the gorge along the cliff edge, he saw the
glare of Eudena's fire upon the ledge, and Eudena red
and shining, and Ugh-lomi, with a gigantic shadow mocking
him upon the white cliff, going to and fro, shaking his
mane of hair, and waving the axe of stone—the first axe of
stone—while he chanted of the killing of Uya. The cave
bear was far up the gorge, and he saw the thing slanting-

ways and far off. He was so surprised he stood quite still upon the edge, sniffing the novel odour of burning bracken, and wondering whether the dawn was coming up in the wrong place.

He was the lord of the rocks and caves, was the cave bear, as his slighter brother, the grizzly, was lord of the thick woods below, and as the dappled lion—the lion of those days was dappled—was lord of the thorn-thickets, reed-beds, and open plains. He was the greatest of all meat-eaters; he knew no fear, none preyed on him, and none gave him battle; only the rhinoceros was beyond his strength. Even the mammoth shunned his country. This invasion perplexed him. He noticed these new beasts were shaped like monkeys, and sparsely hairy like young pigs. "Monkey and young pig," said the cave bear. "It might not be so bad. But that red thing that jumps, and the black thing jumping with it yonder! Never in my life have I seen such things before!"

He came slowly along the brow of the cliff towards them, stopping thrice to sniff and peer, and the reek of the fire grew stronger. A couple of hyænas also were so intent upon the thing below that Andoo, coming soft and easy, was close upon them before they knew of him or he of them. They started guiltily and went lurching off. Coming round in a wheel, a hundred yards off, they began yelling and calling him names to revenge themselves for the start they had had. "Ya-ha!" they cried. "Who can't grub his own burrow? Who eats roots like a pig? . . . Ya-ha!" for even in those days the hyæna's manners were just as offensive as they are now.

"Who answers the hyæna?" growled Andoo, peering through the midnight dimness at them, and then going to look at the cliff edge.

There was Ugh-lomi still telling his story, and the fire getting low, and the scent of the burning hot and strong.

Andoo stood on the edge of the chalk cliff for some time, shifting his vast weight from foot to foot, and swaying his head to and fro, with his mouth open, his ears erect and twitching, and the nostrils of his big, black muzzle sniffing. He was very curious, was the cave bear, more curious than any of the bears that live now, and the flickering fire and the incomprehensible movements of the man, let alone the intrusion into his indisputable province, stirred him with a sense of strange new happenings. He had been after red deer fawn that night, for the cave bear was a miscellaneous hunter, but this quite turned him from that enterprise.

"Ya-ha!" yelled the hyænas behind. "Ya-ha-ha!"

Peering through the starlight, Andoo saw there were now three or four going to and fro against the grey hillside. "They will hang about me now all the night . . . until I kill," said Andoo. "Filth of the world!" And mainly to annoy them, he resolved to watch the red flicker in the gorge until the dawn came to drive the hyæna scum home. And after a time they vanished, and he heard their voices, like a party of Cockney beanfeasters, away in the beech-woods. Then they came slinking near again. Andoo yawned and went on along the cliff, and they followed. Then he stopped and went back.

It was a splendid night, beset with shining constellations, the same stars, but not the same constellations we know, for since those days all the stars have had time to move into new places. Far away across the open space beyond where the heavy-shouldered, lean-bodied hyænas blundered and howled, was a beech-wood, and the mountain slopes rose beyond, a dim mystery, until their snow-capped summits came out white and cold and clear, touched by the first rays of the yet unseen moon. It was a vast silence, save when the yell of the hyænas flung a vanishing discordance across its peace, or when from down the hills the trumpeting of the new-come elephants came faintly on the faint breeze. And below now, the red flicker had dwindled and was steady, and shone a deeper red, and Ugh-lomi had finished his story and was preparing to sleep, and Eudena sat and listened to the strange voices of unknown beasts, and watched the dark eastern sky growing deeply luminous at the advent of the moon. Down below, the river talked to itself, and things unseen went to and fro.

After a time the bear went away, but in an hour he was back again. Then, as if struck by a thought, he turned, and went up the gorge. . . .

The night passed, and Ugh-lomi slept on. The waning moon rose and lit the gaunt white cliff overhead with a light that was pale and vague. The gorge remained in a deeper shadow and seemed all the darker. Then by imperceptible degrees, the day came stealing in the wake of the moonlight. Eudena's eyes wandered to the cliff brow overhead once, and then again. Each time the line was sharp and clear against the sky, and yet she had a dim perception of something lurking there. The red of the fire grew deeper and deeper, grey scales spread upon it, its vertical column of smoke became more and more visible, and up and down the gorge things that had been unseen grew clear in a colourless illumination. She may have dozed.

Suddenly she started up from her squatting position, erect and alert, scrutinising the cliff up and down.

She made the faintest sound, and Ugh-lomi too, light-sleeping like an animal, was instantly awake. He caught up his axe and came noiselessly to her side.

The light was still dim, the world now all in black and dark grey, and one sickly star still lingered overhead. The ledge they were on was a little grassy space, six feet wide, perhaps, and twenty feet long, sloping outwardly, and with a handful of St. John's wort growing near the edge. Below it the soft, white rock fell away in a steep slope of nearly fifty feet to the thick bush of hazel that fringed the river. Down the river this slope increased, until some way off a thin grass held its own right up to the crest of the cliff. Overhead, forty or fifty feet of rock bulged into the great masses characteristic of chalk, but at the end of the ledge a gully, a precipitous groove of discoloured rock, slashed the face of the cliff, and gave a footing to a scrubby growth, by which Eudena and Ugh-lomi went up and down.

They stood as noiseless as startled deer, with every sense expectant. For a minute they heard nothing, and then came a faint rattling of dust down the gully, and the creaking of twigs.

Ugh-lomi gripped his axe, and went to the edge of the ledge, for the bulge of the chalk overhead had hidden the upper part of the gully. And forthwith, with a sudden contraction of the heart, he saw the cave bear half-way down from the brow, and making a gingerly backward step with his flat hind-foot. His hind-quarters were towards Ugh-lomi, and he clawed at the rocks and bushes so that he seemed flattened against the cliff. He looked none the less for that. From his shining snout to his stumpy tail he was a lion and a half, the length of two tall men. He looked over his shoulder, and his huge mouth was open with the exertion of holding up his great carcase, and his tongue lay out. . . .

He got his footing, and came down slowly, a yard nearer.

"Bear," said Ugh-lomi, looking round with his face white.

But Eudena, with terror in her eyes, was pointing down the cliff.

Ugh-lomi's mouth fell open. For down below, with her big fore-feet against the rock, stood another big brown-grey bulk—the she-bear. She was not so big as Andoo, but she was big enough for all that.

Then suddenly Ugh-lomi gave a cry, and catching up a handful of the litter of ferns that lay scattered on the ledge, he thrust it into the pallid ash of the fire. "Brother Fire!" he cried, "Brother Fire!" And Eudena, starting into activity, did likewise. "Brother Fire! Help, help! Brother Fire!"

Brother Fire was still red in his heart, but he turned to grey as they scattered him. "Brother Fire!" they screamed. But he whispered and passed, and there was nothing but ashes. Then Ugh-lomi danced with anger and struck the ashes with his first. But Eudena began to hammer the firestone against a flint. And the eyes of each were turning ever and again towards the gully by which Andoo was climbing down. Brother Fire!

Suddenly the huge furry hind-quarters of the bear came into view, beneath the bulge of the chalk that had hidden him. He was still clambering gingerly down the nearly vertical surface. His head was yet out of sight, but they could hear him talking to himself. "Pig and monkey," said the cave bear. "It ought to be good."

Eudena struck a spark and blew at it; it twinkled brighter and then—went out. At that she cast down flint and fire-stone and stared blankly. Then she sprang to her feet and scrambled a yard or so up the cliff above the ledge. How she hung on even for a moment I do not know, for the chalk was vertical and without grip for a monkey. In a couple of seconds she had slid back to the ledge again with bleeding hands.

Ugh-lomi was making frantic rushes about the ledge—now he would go to the edge, now to the gully. He did not know what to do, he could not think. The she-bear looked smaller than her mate—much. If they rushed down on her together, *one* might live. "Ugh?" said the cave bear, and Ugh-lomi turned again and saw his little eyes peering under the bulge of the chalk.

Eudena, cowering at the end of the ledge, began to scream like a gripped rabbit.

At that a sort of madness came upon Ugh-lomi. With a mighty cry, he caught up his axe and ran towards Andoo. The monster gave a grunt of surprise. In a moment Ugh-lomi was clinging to a bush right underneath the bear, and in another he was hanging to its back half buried in fur, with one fist clutched in the hair under its jaw. The bear was too astonished at this fantastic attack to do more than cling passive. And then the axe, the first of all axes, rang on its skull.

The bear's head twisted from side to side, and he began a petulant scolding growl. The axe bit within an inch of the left eye, and the hot blood blinded that side. At that the brute roared with surprise and anger, and his teeth gnashed six inches from Ugh-lomi's face. Then the axe, clubbed close, came down heavily on the corner of the jaw.

The next blow blinded the right side and called forth a roar, this time of pain. Eudena saw the huge, flat feet slipping and sliding, and suddenly the bear gave a clumsy leap sideways, as if for the ledge. Then everything vanished, and the hazels smashed, and a roar of pain and a tumult of shouts and growls came up from far below.

Eudena screamed and ran to the edge and peered over. For a moment, man and bears were a heap together, Ugh-lomi uppermost; and then he had sprung clear and was scaling the gully again, with the bears rolling and striking at one another among the hazels. But he had left his axe below, and three knob-ended streaks of carmine were shooting down his thigh. "Up!" he cried, and in a moment Eudena was leading the way to the top of the cliff.

In half a minute they were at the crest, their hearts pumping noisily, with Andoo and his wife far and safe below them. Andoo was sitting on his haunches, both paws at work, trying with quick exasperated movements to wipe the blindness out of his eyes, and the she-bear stood on all-fours a little way off, ruffled in appearance and growling angrily. Ugh-lomi flung himself flat on the grass, and lay panting and bleeding with his face on his arms.

For a second Eudena regarded the bears, then she came and sat beside him, looking at him. . . .

Presently she put forth her hand timidly and touched him, and made the guttural sound that was his name. He turned over and raised himself on his arm. His face was pale, like the face of one who is afraid. He looked at her steadfastly for a moment, and then suddenly he laughed. "Waugh!" he said exultantly.

"Waugh!" said she—a simple but expressive conversation.

Then Ugh-lomi came and knelt beside her, and on hands and knees peered over the brow and examined the gorge. His breath was steady now, and the blood on his leg had ceased to flow, though the scratches the she-bear had made were open and wide. He squatted up and sat staring at the footmarks of the great bear as they came to the gully— they were as wide as his head and twice as long. Then he jumped up and went along the cliff face until the ledge was visible. Here he sat down for some time thinking, while Eudena watched him. Presently she saw the bears had gone.

At last Ugh-lomi rose, as one whose mind is made up.

He returned towards the gully, Eudena keeping close by him, and together they clambered to the ledge. They took the firestone and a flint, and then Ugh-lomi went down to the foot of the cliff very cautiously, and found his axe. They returned to the cliff as quietly as they could, and set off at a brisk walk. The ledge was a home no longer, with such callers in the neighbourhood. Ugh-lomi carried the axe and Eudena the firestone. So simple was a Palæolithic removal.

They went up-stream, although it might lead to the very lair of the cave bear, because there was no other way to go. Down the stream was the tribe, and had not Ugh-lomi killed Uya and Wau? By the stream they had to keep—because of drinking.

So they marched through beech trees, with the gorge deepening until the river flowed, a frothing rapid, five hundred feet below them. Of all the changeful things in this world of change, the courses of rivers in deep valleys change least. It was the river Wey, the river we know to-day, and they marched over the very spots where nowadays stand little Guildford and Godalming—the first human beings to come into the land. Once a grey ape chattered and vanished, and all along the cliff edge, vast and even, ran the spoor of the great cave bear.

And then the spoor of the bear fell away from the cliff, showing, Ugh-lomi thought, that he came from some place to the left, and keeping to the cliff's edge, they presently came to an end. They found themselves looking down on a great semi-circular space caused by the collapse of the cliff. It had smashed right across the gorge, banking the up-stream water back in a pool which overflowed in a rapid. The slip had happened long ago. It was grassed over, but the face of the cliffs that stood about the semi-circle was still almost fresh-looking and white as on the day when the rock must have broken and slid down. Starkly exposed and black under the foot of these cliffs were the mouths of several caves. And as they stood there, looking at the space, and disinclined to skirt it, because they thought the bears' lair lay somewhere on the left in the direction they must needs take, they saw suddenly first one bear and then two coming up the grass slope to the right and going across the amphitheatre towards the caves. Andoo was first; he dropped a little on his fore-foot and his mien was despondent, and the she-bear came shuffling behind.

Eudena and Ugh-lomi stepped back from the cliff until they could just see the bears over the verge. Then Ugh-lomi stopped. Eudena pulled his arm, but he turned with a forbidding gesture, and her hand dropped. Ugh-lomi stood watching the bears, with his axe in his hand, until they had vanished into the cave. He growled softly, and shook the axe at the she-bear's receding quarters. Then to Eudena's terror, instead of creeping off with her, he lay flat down and crawled forward into such a position that he could just see the cave. It was bears—and he did it as calmly as if it had been rabbits he was watching!

He lay still, like a bared log, sun-dappled, in the shadow of the trees. He was thinking. And Eudena had learnt, even when a little girl, that when Ugh-lomi became still like that, jaw-bone on fist, novel things presently began to happen.

It was an hour before the thinking was over; it was noon when the two little savages had found their way to the cliff brow that overhung the bears' cave. And all the long afternoon they fought desperately with a great boulder of chalk; trundling it, with nothing but their unaided sturdy muscles, from the gully where it had hung like a loose tooth, towards the cliff top. It was full two yards about, it stood as high as Eudena's waist, it was obtuse-angled and toothed with flints. And when the sun set it was poised, three inches from the edge, above the cave of the great cave bear.

In the cave conversation languished during that afternoon. The she-bear snoozed sulkily in her corner—for she was fond of pig and monkey—and Andoo was busy licking the side of his paw and smearing his face to cool the smart and inflammation of his wounds. Afterwards he went and sat just within the mouth of the cave, blinking out at the afternoon sun with his uninjured eye, and thinking.

"I never was so startled in my life," he said at last. "They are the most extraordinary beasts. Attacking *me!*"

"I don't like them," said the she-bear, out of the darkness behind.

"A feebler sort of beast I *never* saw. I can't think what the world is coming to. Scraggy, weedy legs. . . . Wonder how they keep warm in winter?"

"Very likely they don't," said the she-bear.

"I suppose it's a sort of monkey gone wrong."

"It's a change," said the she-bear.

A pause.

"The advantage he had was merely accidental," said Andoo. "These things *will* happen at times."

"*I* can't understand why you let go," said the she-bear.

That matter had been discussed before, and settled. So Andoo, being a bear of experience, remained silent for a space. Then he resumed upon a different aspect of the matter. "He has a sort of claw—a long claw that he seemed to have first on one paw and then on the other. Just one claw. They're very odd things. The bright thing, too, they seemed to have—like that glare that comes in the sky in daytime—only it jumps about—it's really worth seeing. It's a thing with a root, too—like grass when it is windy."

"Does it bite?" asked the she-bear. "If it bites it can't be a plant."

"No——I don't know," said Andoo. "But it's curious, anyhow."

"I wonder if they *are* good eating?" said the she-bear.

"They look it," said Andoo, with appetite—for the cave bear, like the polar bear, was an incurable carnivore—no roots or honey for *him*.

The two bears fell into a meditation for a space. Then Andoo resumed his simple attentions to his eye. The sunlight up the green slope before the cave mouth grew warmer in tone and warmer, until it was a ruddy amber.

"Curious sort of thing—day," said the cave bear. "Lot too much of it, I think. Quite unsuitable for hunting. Dazzles me always. I can't smell nearly so well by day."

The she-bear did not answer, but there came a measured

crunching sound out of the darkness. She had turned up a bone. Andoo yawned. "Well," he said. He strolled to the cave mouth and stood with his head projecting, surveying the amphitheatre. He found he had to turn his head completely round to see objects on his right-hand side. No doubt that eye would be all right to-morrow.

He yawned again. There was a tap overhead, and a big mass of chalk flew out from the cliff face, dropped a yard in front of his nose, and starred into a dozen unequal fragments. It startled him extremely.

When he had recovered a little from his shock, he went and sniffed curiously at the representative pieces of the fallen projectile. They had a distinctive flavour, oddly reminiscent of the two drab animals of the ledge. He sat up and pawed the larger lump, and walked round it several times, trying to find a man about it somewhere. . . .

When night had come he went off down the river gorge to see if he could cut off either of the ledge's occupants. The ledge was empty, there were no signs of the red thing, but as he was rather hungry he did not loiter long that night, but pushed on to pick up a red deer fawn. He forgot about the drab animals. He found a fawn, but the doe was close by and made an ugly fight for her young. Andoo had to leave the fawn, but as her blood was up she stuck to the attack, and at last he got in a blow of his paw on her nose, and so got hold of her. More meat but less delicacy, and the she-bear, following, had her share. The next afternoon, curiously enough, the very fellow of the first white rock fell, and smashed precisely according to precedent.

The aim of the third, that fell the night after, however, was better. It hit Andoo's unspeculative skull with a crack that echoed up the cliff, and the white fragments went dancing to all the points of the compass. The she-bear, coming after him and sniffing curiously at him, found him lying in an odd sort of attitude, with his head wet and all out of shape. She was a young she-bear, and inexperienced, and having sniffed about him for some time and licked him a little, and so forth, she decided to leave him until the odd mood had passed, and went on her hunting alone.

She looked up the fawn of the red doe they had killed two nights ago, and found it. But it was lonely hunting without Andoo, and she returned caveward before dawn. The sky was grey and overcast, the trees up the gorge were black and unfamiliar, and into her ursine mind came a dim sense of strange and dreary happenings. She lifted up her voice and called Andoo by name. The sides of the gorge re-echoed her.

As she approached the caves she saw in the half light, and heard a couple of jackals scuttle off, and immediately after a hyæna howled and a dozen clumsy bulks went lumbering up the slope, and stopped and yelled derision. "Lord of the rocks and caves—ya-ha!" came down the wind. The dismal feeling in the she-bear's mind became suddenly acute. She shuffled across the amphitheatre.

"Ya-ha!" said the hyænas, retreating. "Ya-ha!"

The cave bear was not lying quite in the same attitude, because the hyænas had been busy, and in one place his ribs showed white. Dotted over the turf about him lay the smashed fragments of the three great lumps of chalk. And the air was full of the scent of death.

The she-bear stopped dead. Even now, that the great and wonderful Andoo was killed was beyond her believing. Then she heard far overhead a sound, a queer sound, a little like the shout of a hyæna but fuller and lower in pitch. She looked up, her little dawn-blinded eyes seeing little, her nostrils quivering. And there, on the cliff edge, far above her against the bright pink of dawn, were two little shaggy round dark things, the heads of Eudena and Ugh-lomi, as they shouted derision at her. But though she could not see them very distinctly she could hear, and dimly she began to apprehend. A novel feeling as of imminent strange evils came into her heart.

She began to examine the smashed fragments of chalk that lay about Andoo. For a space she stood still, looking about her and making a low continuous sound that was almost a moan. Then she went back incredulously to Andoo to make one last effort to rouse him.

III.—The First Horseman

In the days before Ugh-lomi there was little trouble
between the horses and men. They lived apart—the men
in the river swamps and thickets, the horses on the wide
grassy uplands between the chestnuts and the pines. Sometimes a pony would come straying into the clogging marshes
to make a flint-hacked meal, and sometimes the tribe would
find one, the kill of a lion, and drive off the jackals, and
feast heartily while the sun was high. These horses of the
old time were clumsy at the fetlock and dun-coloured, with
a rough tail and big head. They came every spring-time
north-westward into the country, after the swallows and
before the hippopotami, as the grass on the wide downland
stretches grew long. They came only in small bodies thus
far, each herd, a stallion and two or three mares and a
foal or so, having its own stretch of country, and they went
again when the chestnut-trees were yellow and the wolves
came down the Wealden mountains.

It was their custom to graze right out in the open, going
into cover only in the heat of the day. They avoided the
long stretches of thorn and beechwood, preferring an
isolated group of trees void of ambuscade, so that it was
hard to come upon them. They were never fighters; their
heels and teeth were for one another, but in the clear
country, once they were started, no living thing came near
them, though perhaps the elephant might have done so
had he felt the need. And in those days man seemed a
harmless thing enough. No whisper of prophetic intelligence
told the species of the terrible slavery that was to come, of
the whip and spur and bearing-rein, the clumsy load and the
slippery street, the insufficient food, and the knacker's yard,
that was to replace the wide grass-land and the freedom of
the earth.

Down in the Wey marshes Ugh-lomi and Eudena had
never seen the horses closely, but now they saw them every
day as the two of them raided out from their lair on the
ledge in the gorge, raiding together in search of food. They
had returned to the ledge after the killing of Andoo; for

of the she-bear they were not afraid. The she-bear had become afraid of them, and when she winded them she went aside. The two went together everywhere; for since they had left the tribe Eudena was not so much Ugh-lomi's woman as his mate; she learnt to hunt even—as much, that is, as any woman could. She was indeed a marvellous woman. He would lie for hours watching a beast, or planning catches in that shock head of his, and she would stay beside him, with her bright eyes upon him, offering no irritating suggestions—as still as any man. A wonderful woman!

At the top of the cliff was an open grassy lawn and then beechwoods, and going through the beechwoods one came to the edge of the rolling grassy expanse, and in sight of the horses. Here, on the edge of the wood and bracken, were the rabbit-burrows, and here among the fronds Eudena and Ugh-lomi would lie with their throwing-stones ready, until the little people came out to nibble and play in the sunset. And while Eudena would sit, a silent figure of watchfulness, regarding the burrows, Ugh-lomi's eyes were ever away across the greensward at those wonderful grazing strangers.

In a dim way he appreciated their grace and their supple nimbleness. As the sun declined in the evening-time, and the heat of the day passed, they would become active, would start chasing one another, neighing, dodging, shaking their manes, coming round in great curves, sometimes so close that the pounding of the turf sounded like hurried thunder. It looked so fine that Ugh-lomi wanted to join in badly. And sometimes one would roll over on the turf, kicking four hoofs heavenward, which seemed formidable and was certainly much less alluring.

Dim imaginings ran through Ugh-lomi's mind as he watched—by virtue of which two rabbits lived the longer. And sleeping, his brains were clearer and bolder—for that was the way in those days. He came near the horses, he dreamt, and fought, smiting-stone against hoof, but then the horses changed to men, or, at least, to men with horses' heads, and he awoke in a cold sweat of terror.

Yet the next day in the morning, as the horses were grazing, one of the mares whinnied, and they saw Ugh-lomi coming up the wind. They all stopped their eating and watched him. Ugh-lomi was not coming towards them, but strolling obliquely across the open, looking at anything in the world but horses. He had stuck three fern-fronds into the mat of his hair, giving him a remarkable appearance, and he walked very slowly. "What's up now?" said the Master Horse, who was capable, but inexperienced.

"It looks more like the first half of an animal than anything else in the world," he said. "Fore-legs and no hind."

"It's only one of those pink monkey things," said the Eldest Mare. "They're a sort of river monkey. They're quite common on the plains."

Ugh-lomi continued his oblique advance. The Eldest Mare was struck with the want of motive in his proceedings. "Fool!" said the Eldest Mare, in a quick conclusive way she had. She resumed her grazing. The Master Horse and the Second Mare followed suit.

"Look! he's nearer," said the Foal with a stripe.

One of the younger foals made uneasy movements. Ugh-lomi squatted down, and sat regarding the horses fixedly. In a little while he was satisfied that they meant neither flight nor hostilities. He began to consider his next procedure. He did not feel anxious to kill, but he had his axe with him, and the spirit of sport was upon him. How would one kill one of these creatures?—these great beautiful creatures!

Eudena, watching him with a fearful admiration from the cover of the bracken, saw him presently go on all fours, and so proceed again. But the horses preferred him a biped to a quadruped, and the Master Horse threw up his head and gave the word to move. Ugh-lomi thought they were off for good, but after a minute's gallop they came round in a wide curve, and stood winding him. Then, as a rise in the ground hid him, they tailed out, the Master Horse leading, and approached him spirally.

He was as ignorant of the possibilities of a horse as they were of his. And at this stage it would seem he funked.

He knew this kind of stalking would make red deer or buffalo charge, if it were persisted in. At any rate Eudena saw him jump up and come walking towards her with the fern plumes held in his hand.

She stood up, and he grinned to show that the whole thing was an immense lark, and that what he had done was just what he had planned to do from the very beginning. So that incident ended. But he was very thoughtful all that day.

The next day this foolish drab creature with the leonine mane, instead of going about the grazing or hunting he was made for, was prowling round the horses again. The Eldest Mare was all for silent contempt. "I suppose he wants to learn something from us," she said, and "*Let* him." The next day he was at it again. The Master Horse decided he meant absolutely nothing. But as a matter of fact, Ugh-lomi, the first of men to feel that curious spell of the horse that binds us even to this day, meant a great deal. He admired them unreservedly. There was a rudiment of the snob in him, I am afraid, and he wanted to be near these beautifully-curved animals. Then there were vague conceptions of a kill. If only they would let him come near them! But they drew the line, he found, at fifty yards. If he came nearer than that they moved off—with dignity. I suppose it was the way he had blinded Andoo that made him think of leaping on the back of one of them. But though Eudena after a time came out in the open too, and they did some unobtrusive stalking, things stopped there.

Then one memorable day a new idea came to Ugh-lomi. The horse looks down and level, but he does not look up. No animals look up—they have too much common-sense. It was only that fantastic creature, man, could waste his wits skyward. Ugh-lomi made no philosophical deductions, but he perceived the thing was so. So he spent a weary day in a beech that stood in the open, while Eudena stalked. Usually the horses went into the shade in the heat of the afternoon, but that day the sky was overcast, and they would not, in spite of Eudena's solicitude.

It was two days after that that Ugh-lomi had his desire.

The day was blazing hot, and the multiplying flies asserted themselves. The horses stopped grazing before mid-day, and came into the shadow below him, and stood in couples nose to tail, flapping.

The Master Horse, by virtue of his heels, came closest to the tree. And suddenly there was a rustle and a creak, a *thud*. . . . Then a sharp chipped flint bit him on the cheek. The Master Horse stumbled, came on one knee, rose to his feet, and was off like the wind. The air was full of the whirl of limbs, the prance of hoofs, and snorts of alarm. Ugh-lomi was pitched a foot in the air, came down again, up again, his stomach was hit violently, and then his knees got a grip of something between them. He found himself clutching with knees, feet, and hands, careering violently with extraordinary oscillation through the air —his axe gone heaven knows whither. "Hold tight," said Mother Instinct, and he did.

He was aware of a lot of coarse hair in his face, some of it between his teeth, and of green turf streaming past in front of his eyes. He saw the shoulder of the Master Horse, vast and sleek, with the muscles flowing swiftly under the skin. He perceived that his arms were round the neck, and that the violent jerkings he experienced had a sort of rhythm.

Then he was in the midst of a wild rush of tree-stems, and then there were fronds of bracken about, and then more open turf. Then a stream of pebbles rushing past, little pebbles flying sideways athwart the stream from the blow of the swift hoofs. Ugh-lomi began to feel frightfully sick and giddy, but he was not the stuff to leave go simply because he was uncomfortable.

He dared not leave his grip, but he tried to make himself more comfortable. He released his hug on the neck, gripping the mane instead. He slipped his knees forward, and pushing back, came into a sitting position where the quarters broadened. It was nervous work, but he managed it, and at last he was fairly seated astride, breathless indeed, and uncertain, but with that frightful pounding of his body at any rate relieved.

Slowly the fragments of Ugh-lomi's mind got into order again. The pace seemed to him terrific, but a kind of exultation was beginning to oust his first frantic terror. The air rushed by, sweet and wonderful, the rhythm of the hoofs changed and broke up and returned into itself again. They were on turf now, a wide glade—the beech-trees a hundred yards away on either side, and a succulent band of green starred with pink blossom and shot with silver water here and there, meandered down the middle. Far off was a glimpse of blue valley—far away. The exultation grew. It was man's first taste of pace.

Then came a wide space dappled with flying fallow deer scattering this way and that, and then a couple of jackals, mistaking Ugh-lomi for a lion, came hurrying after him. And when they saw it was not a lion they still came on out of curiosity. On galloped the horse, with his one idea of escape, and after him the jackals, with pricked ears and quickly-barked remarks. "Which kills which?" said the first jackal. "It's the horse being killed," said the second. They gave the howl of following, and the horse answered to it as a horse answers nowadays to the spur.

On they rushed, a little tornado through the quiet day, putting up startled birds, sending a dozen unexpected things darting to cover, raising a myriad of indignant dung-flies, smashing little blossoms, flowering complacently, back into their parental turf. Trees again, and then splash, splash across a torrent; then a hare shot out of a tuft of grass under the very hoofs of the Master Horse, and the jackals left them incontinently. So presently they broke into the open again, a wide expanse of turfy hillside—the very grassy downs that fall northward nowadays from the Epsom Stand.

The first hot bolt of the Master Horse was long since over. He was falling into a measured trot, and Ugh-lomi, albeit bruised exceedingly and quite uncertain of the future, was in a state of glorious enjoyment. And now came a new development. The pace broke again, the Master Horse came round on a short curve, and stopped dead. . . .

Ugh-lomi became alert. He wished he had a flint, but

the throwing flint he had carried in a thong about his waist was—like the axe—heaven knows where. The Master Horse turned his head, and Ugh-lomi became aware of an eye and teeth. He whipped his leg into a position of security, and hit at the cheek with his fist. Then the head went down somewhere out of existence apparently, and the back he was sitting on flew up into a dome. Ugh-lomi became a thing of instinct again—strictly prehensile; he held by knees and feet, and his head seemed sliding towards the turf. His fingers were twisted into the shock of mane, and the rough hair of the horse saved him. The gradient he was on lowered again, and then—"Whup!" said Ugh-lomi astonished, and the slant was the other way up. But Ugh-lomi was a thousand generations nearer the primordial than man: no monkey could have held on better. And the lion had been training the horse for countless generations against the tactics of rolling and rearing back. But he kicked like a master, and buck-jumped rather neatly. In five minutes Ugh-lomi lived a lifetime. If he came off the horse would kill him, he felt assured.

Then the Master Horse decided to stick to his old tactics again, and suddenly went off at a gallop. He headed down the slope, taking the steep places at a rush, swerving neither to the right nor to the left, and, as they rode down, the wide expanse of valley sank out of sight behind the approaching skirmishers of oak and hawthorn. They skirted a sudden hollow with the pool of a spring, rank weeds and silver bushes. The ground grew softer and the grass taller, and on the right-hand side and the left came scattered bushes of May—still splashed with belated blossom. Presently the bushes thickened until they lashed the passing rider, and little flashes and gouts of blood came out on horse and man. Then the way opened again.

And then came a wonderful adventure. A sudden squeal of unreasonable anger rose amidst the bushes, the squeal of some creature bitterly wronged. And crashing after them appeared a big, grey-blue shape. It was Yaaa, the big-horned rhinoceros, in one of those fits of fury of his, charging full tilt, after the manner of his kind. He had been startled at his feeding, and someone, it did not matter who, was to be ripped and trampled therefore. He was bearing down on them from the left, with his wicked little eye red, his great horn down and his tail like a jury-mast behind him. For a minute Ugh-lomi was minded to slip off and dodge, and then behold! the staccato of the hoofs grew swifter, and the rhinoceros and his stumpy hurrying little legs seemed to slide out at the back corner of Ugh-lomi's eye. In two minutes they were through the bushes of May, and out in the open, going fast. For a space he could hear the ponderous paces in pursuit receding behind him, and then it was just as if Yaaa had not lost his temper, as if Yaaa had never existed.

The pace never faltered, on they rode and on.

Ugh-lomi was now all exultation. To exult in those days was an insult. "Ya-ha! big nose!" he said, trying to crane back and see some remote speck of a pursuer. "Why don't you carry your smiting-stone in your fist?" he ended with a frantic whoop.

But that whoop was unfortunate, for coming close to the ear of the horse, and being quite unexpected, it startled the stallion extremely. He shied violently. Ugh-lomi suddenly found himself uncomfortable again. He was hanging on to the horse, he found, by one arm and one knee.

The rest of the ride was honourable but unpleasant. The view was chiefly of blue sky, and that was combined with the most unpleasant physical sensations. Finally, a bush of thorn lashed him and he let go.

He hit the ground with his cheek and shoulder, and then, after a complicated and extraordinarily rapid movement, hit it again with the end of his backbone. He saw splashes and sparks of light and colour. The ground seemed bouncing about just like the horse had done. Then he found he was sitting on turf, six yards beyond the bush. In front of him was a space of grass, growing greener and greener, and a number of human beings in the distance, and the horse was going round at a smart gallop quite a long way off to the right.

The human beings were on the opposite side of the river, some still in the water, but they were all running away as hard as they could go. The advent of a monster that took to pieces was not the sort of novelty they cared for. For quite a minute Ugh-lomi sat regarding them in a purely spectacular spirit. The bend of the river, the knoll among the reeds and the royal ferns, the thin streams of smoke going up to Heaven, were all perfectly familiar to him. It was the squatting-place of the Sons of Uya, of Uya from whom he had fled with Eudena, and whom he had waylaid in the chestnut woods and killed with the First Axe.

He rose to his feet, still dazed from his fall, and as he did so the scattering fugitives turned and regarded him. Some pointed to the receding horse and chattered. He walked slowly towards them, staring. He forgot the horse, he forgot his own bruises, in the growing interest of this encounter. There were fewer of them than there had been —he supposed the others must have hid—the heap of fern for the night fire was not so high. By the flint heaps should have sat Wau—but then he remembered he had killed Wau. Suddenly brought back to this familiar scene, the gorge and the bears and Eudena seemed things remote, things dreamt of.

He stopped at the bank and stood regarding the tribe. His mathematical abilities were of the slightest, but it was certain there were fewer. The men might be away, but there were fewer women and children. He gave the shout of home-coming. His quarrel had been with Uya and Wau —not with the others. "Children of Uya!" he cried. They answered with his name, a little fearfully because of the strange way he had come.

For a space they spoke together. Then an old woman lifted a shrill voice and answered him. "Our Lord is a Lion."

Ugh-lomi did not understand that saying. They answered him again several together, "Uya comes again. He comes as a Lion. Our Lord is a Lion. He comes at night. He slays whom he will. But none other may slay us, Ugh-lomi, none other may slay us."

Still Ugh-lomi did not understand.

"Our Lord is a Lion. He speaks no more to men."

Ugh-lomi stood regarding them. He had had dreams—he knew that though he had killed Uya, Uya still existed. And now they told him Uya was a Lion.

The shrivelled old woman, the mistress of the fire-minders, suddenly turned and spoke softly to those next to her. She was a very old woman indeed, she had been the first of Uya's wives, and he had let her live beyond the age to which it is seemly a woman should be permitted to live. She had been cunning from the first, cunning to please Uya and to get food. And now she was great in counsel. She spoke softly, and Ugh-lomi watched her shrivelled form across the river with a curious distaste. Then she called aloud, "Come over to us, Ugh-lomi."

A girl suddenly lifted up her voice. "Come over to us, Ugh-lomi," she said. And they all began crying, "Come over to us, Ugh-lomi."

It was strange how their manner changed after the old woman called.

He stood quite still watching them all. It was pleasant to be called, and the girl who had called first was a pretty one. But she made him think of Eudena.

"Come over to us, Ugh-lomi," they cried, and the voice of the shrivelled old woman rose above them all. At the sound of her voice his hesitation returned.

He stood on the river bank, Ugh-lomi—Ugh the Thinker —with his thoughts slowly taking shape. Presently one and then another paused to see what he would do. He was minded to go back, he was minded not to. Suddenly his fear or his caution got the upper hand. Without answering them he turned, and walked back towards the distant thorn-trees, the way he had come. Forthwith the whole tribe started crying to him again very eagerly. He hesitated and turned, then he went on, then he turned again, and then once again, regarding them with troubled eyes as they called. The last time he took two paces back, before his fear stopped him. They saw him stop once more, and suddenly shake his head and vanish among the hawthorn-trees.

Then all the women and children lifted up their voices together, and called to him in one last vain effort.

Far down the river the reeds were stirring in the breeze, where, convenient for his new sort of feeding, the old lion, who had taken to man-eating, had made his lair.

The old woman turned her face that way, and pointed to the hawthorn thickets. "Uya," she screamed, "there goes thine enemy! There goes thine enemy, Uya! Why do you devour us nightly? We have tried to snare him! There goes thine enemy, Uya!"

But the lion who preyed upon the tribe was taking his siesta. The cry went unheard. That day he had dined on one of the plumper girls, and his mood was a comfortable placidity. He really did not understand that he was Uya or that Ugh-lomi was his enemy.

So it was that Ugh-lomi rode the horse, and heard first of Uya the lion, who had taken the place of Uya the Master, and was eating up the tribe. And as he hurried back to the gorge his mind was no longer full of the horse, but of the thought that Uya was still alive, to slay or be slain. Over and over again he saw the shrunken band of women and children crying that Uya was a lion. Uya was a lion!

And presently, fearing the twilight might come upon him, Ugh-lomi began running.

IV.—Uya the Lion

The old lion was in luck. The tribe had a certain pride in their ruler, but that was all the satisfaction they got out of it. He came the very night that Ugh-lomi killed Uya the Cunning, and so it was they named him Uya. It was the old woman, the fire-minder, who first named him Uya. A shower had lowered the fires to a glow, and made the night dark. And as they conversed together, and peered at one another in the darkness, and wondered fearfully what

Uya would do to them in their dreams now that he was dead, they heard the mounting reverberations of the lion's roar close at hand. Then everything was still.

They held their breath, so that almost the only sounds were the patter of the rain and the hiss of the raindrops in the ashes. And then, after an interminable time, a crash, and a shriek of fear, and a growling. They sprang to their feet, shouting, screaming, running this way and that, but brands would not burn, and in a minute the victim was being dragged away through the ferns. It was Irk, the brother of Wau.

So the lion came.

The ferns were still wet from the rain the next night, and he came and took Click with the red hair. That sufficed for two nights. And then in the dark between the moons he came three nights, night after night, and that though they had good fires. He was an old lion with stumpy teeth, but very silent and very cool; he knew of fires before; these were not the first of mankind that had ministered to his old age. The third night he came between the outer fire and the inner, and he leapt the flint heap, and pulled down Irm the son of Irk, who had seemed like to be the leader. That was a dreadful night, because they lit great flares of fern and ran screaming, and the lion missed his hold of Irm. By the glare of the fire they saw Irm struggle up, and run a little way towards them, and then the lion in two bounds had him down again. That was the last of Irm.

So fear came, and all the delight of spring passed out of their lives. Already there were five gone out of the tribe, and four nights added three more to the number. Food-seeking became spiritless, none knew who might go next, and all day the women toiled, even the favourite women, gathering litter and sticks for the night fires. And the hunters hunted ill: in the warm spring-time hunger came again as though it was still winter. The tribe might have moved, had they had a leader, but they had no leader, and none knew where to go that the lion could not follow them. So the old lion waxed fat and thanked heaven for the kindly race of men. Two of the children and a youth died while the moon was still new, and then it was the shrivelled old fire-minder first bethought herself in a dream of Eudena and Ugh-lomi, and of the way Uya had been slain. She had lived in fear of Uya all her days, and now she lived in fear of the lion. That Ugh-lomi could kill Uya for good—Ugh-lomi whom she had seen born—was impossible. It was Uya still seeking his enemy!

And then came the strange return of Ugh-lomi, a wonderful animal seen galloping far across the river, that suddenly changed into two animals, a horse and a man. Following this portent, the vision of Ugh-lomi on the farther bank of the river. . . . Yes, it was all plain to her. Uya was punishing them, because they had not hunted down Ugh-lomi and Eudena.

The men came straggling back to the chances of the night while the sun was still golden in the sky. They were received with the story of Ugh-lomi. She went across the river with them and showed them his spoor hesitating on the farther bank. Siss the Tracker knew the feet for Ugh-lomi's. "Uya needs Ugh-lomi," cried the old woman, standing on the left of the bend, a gesticulating figure of flaring bronze in the sunset. Her cries were strange sounds, flitting to and fro on the borderland of speech, but this was the sense they carried: "The lion needs Eudena. He comes night after night seeking Eudena and Ugh-lomi. When he cannot find Eudena and Ugh-lomi, he grows angry and he kills. Hunt Eudena and Ugh-lomi, Eudena whom he pursued, and Ugh-lomi for whom he gave the death-word! Hunt Eudena and Ugh-lomi!"

She turned to the distant reed-bed, as sometimes she had turned to Uya in his life. "Is it not so, my lord?" she cried. And, as if in answer, the tall reeds bowed before a breath of wind.

Far into the twilight the sound of hacking was heard from the squatting-places. It was the men sharpening their ashen spears against the hunting of the morrow. And in the night, early before the moon rose, the lion came and took the girl of Siss the Tracker.

In the morning before the sun had risen, Siss the Tracker,

and the lad Wau-hau, who now chipped flints, and One Eye, and Bo, and the Snail-Eater, the two red-haired men, and Cat's-skin and Snake, all the men that were left alive of the Sons of Uya, taking their ash spears and their smiting-stones, and with throwing stones in the beast-paw bags, started forth upon the trail of Ugh-lomi through the haw-thorn thickets where Yaaa the Rhinoceros and his brothers were feeding, and up the bare downland towards the beech-woods.

That night the fires burnt high and fierce, as the waxing moon set, and the lion left the crouching women and children in peace.

And the next day, while the sun was still high, the hunters returned—all save One Eye, who lay dead with a smashed skull at the foot of the ledge. (When Ugh-lomi came back that evening from stalking the horses, he found the vultures already busy over him.) And with them the hunters brought Eudena, bruised and wounded, but alive. That had been the strange order of the shrivelled old woman, that she was to be brought alive—"She is no kill for us. She is for Uya the Lion." Her hands were tied with thongs, as though she had been a man, and she came weary and drooping—her hair over her eyes and matted with blood. They walked about her, and ever and again the Snail-Eater, whose name she had given, would laugh and strike her with his ashen spear. And after he had struck her with his spear, he would look over his shoulder like one who had done an over-bold deed. The others, too, looked over their shoulders ever and again, and all were in a hurry save Eudena. When the old woman saw them coming, she cried aloud with joy.

They made Eudena cross the river with her hands tied, although the current was strong, and when she slipped the old woman screamed, first with joy and then for fear she might be drowned. And when they had dragged Eudena to shore, she could not stand for a time, albeit they beat her sore. So they let her sit with her feet touching the water, and her eyes staring before her, and her face set, whatever they might do or say. All the tribe came down to the squat-ting-place, even curly little Haha, who as yet could scarcely toddle, and stood staring at Eudena and the old woman, as now we should stare at some strange wounded beast and its captor.

The old woman tore off the necklace of Uya that was about Eudena's neck, and put it on herself—she had been the first to wear it. Then she tore at Eudena's hair, and took a spear from Siss and beat her with all her might. And when she had vented the warmth of her heart on the girl she looked closely into her face. Eudena's eyes were closed and her features were set and she lay so still that for a moment the old woman feared she was dead. And then her nostrils quivered. At that the old woman slapped her face and laughed and gave the spear to Siss again, and went a little way off from her and began to talk and jeer at her after her manner.

The old woman had more words than any in the tribe. And her talk was a terrible thing to hear. Sometimes she screamed and moaned incoherently, and sometimes the shape of her guttural cries was the mere phantom of thoughts. But she conveyed to Eudena, nevertheless, much of the things that were yet to come, of the Lion and of the torment he would do her. "And Ugh-lomi! Ha, ha! Ugh-lomi is slain?"

And suddenly Eudena's eyes opened and she sat up again, and her look met the old woman's fair and level. "No," she said slowly, like one trying to remember, "I did not see my Ugh-lomi slain. I did not see my Ugh-lomi slain."

"Tell her," cried the old woman. "Tell her—he that killed him. Tell her how Ugh-lomi was slain."

She looked, and all the women and children there looked, from man to man.

None answered her. They stood shamefaced.

"Tell her," said the old woman. The men looked at one another.

Eudena's face suddenly lit.

"Tell her," she said. "Tell her, mighty men. Tell her the killing of Ugh-lomi."

The old woman rose and struck her sharply across her mouth.

"We could not find Ugh-lomi," said Siss the Tracker, slowly. "Who hunt two, kills none."

Then Eudena's heart leapt, but she kept her face hard. It was as well, for the old woman looked at her sharply, with murder in her eyes.

Then the old woman turned her tongue upon the men because they had feared to go on after Ugh-lomi. She dreaded no one now Uya was slain. She scolded them as one scolds children. And they scowled at her, and began to accuse one another. Until suddenly Siss the Tracker raised his voice and bade her hold her peace.

And so when the sun was setting they took Eudena and went—though their hearts sank within them—along the trail the old lion had made in the reeds. All the men went together. At one place was a group of alders, and here they hastily bound Eudena where the lion might find her when he came abroad in the twilight, and having done so they hurried back until they were near the squatting-place. Then they stopped. Siss stopped first and looked back again at the alders. They could see her head even from the squatting-place, a little black shock under the limb of the larger tree. That was as well.

All the women and children stood watching upon the crest of the mound. And the old woman stood and screamed for the lion to take her whom he sought, and counselled him on the torment he might do her.

Eudena was very weary now, stunned by beatings and fatigue and sorrow, and only the fear of the thing that was still to come upheld her. The sun was broad and blood-red between the stems of the distant chestnuts, and the west was all on fire; the evening breeze had died to a warm tranquillity. The air was full of midge swarms, the fish in the river hard by would leap at times, and now and again a cockchafer would drone through the air. Out of the corner of her eye Eudena could see a part of the squatting-knoll, and little figures standing and staring at her. And—a very little sound, but very clear,—she could hear the beating of the fire-stone. Dark and near to her and still was the reed-fringed thicket of the lair.

Presently the firestone ceased. She looked for the sun and found he had gone, and overhead and growing brighter was the waxing moon. She looked towards the thicket of the lair, seeking shapes in the reeds, and then suddenly she began to wriggle and wriggle, weeping and calling upon Ugh-lomi.

But Ugh-lomi was far away. When they saw her head moving with her struggles, they shouted together on the knoll, and she desisted and was still. And then came the bats, and the star that was like Ugh-lomi crept out of its blue hiding-place in the west. She called to it, but softly, because she feared the lion. And all through the coming of the twilight the thicket was still.

So the dark crept upon Eudena, and the moon grew bright, and the shadows of things that had fled up the hillside and vanished with the evening came back to them short and black. And the dark shapes in the thicket of reeds and alders where the lion lay, gathered, and a faint stir began there. But nothing came out therefrom all through the gathering of the darkness.

She looked at the squatting-place and saw the fires glowing smoky-red, and the men and women going to and fro. The other way, over the river, a white mist was rising. Then far away came the whimpering of young foxes and the yell of a hyæna.

There were long gaps of aching waiting. After a long time some animal splashed in the water, and seemed to cross the river at the ford beyond the lair, but what animal it was she could not see. From the distant drinking-pools she could hear the sound of splashing, and the noise of elephants—so still was the night.

The earth was now a colourless arrangement of white reflections and impenetrable shadows, under the blue sky. The silvery moon was already spotted with the filigree crests of the chestnut woods, and over the shadowy eastward hills the stars were multiplying. The knoll fires were bright red now, and black figures stood waiting against them. They were waiting for a scream. . . . Surely it would be soon.

The night suddenly seemed full of movement.   She

held her breath. Things were passing—one, two, three—
subtly sneaking shadows. . . . Jackals.

Then a long waiting again.

Then, asserting itself as real at once over all the sounds
her mind had imagined, came astir in the thicket, then a
vigorous movement. There was a snap. The reeds crashed
heavily, once, twice, thrice, and then everything was still
save a measured swishing. She heard a low tremulous
growl, and then everything was still again. The stillness
lengthened—would it never end? She held her breath; she
bit her lips to stop screaming. Then something scuttled
through the undergrowth. Her scream was involuntary.
She did not hear the answering yell from the mound.

Immediately the thicket woke up to vigorous movement
again. She saw the grass stems waving in the light of the
setting moon, the alders swaying. She struggled violently
—her last struggle. But nothing came towards her. A dozen
monsters seemed rushing about in that little place for a
couple of minutes, and then again came silence. The moon
sank behind the distant chestnuts and the night was dark.

Then an odd sound, a sobbing panting, that grew faster
and fainter. Yet another silence, and then dim sounds and
the grunting of some animal.

Everything was still again. Far away eastwards an elephant trumpeted, and from the woods came a snarling and
yelping that died away.

In the long interval the moon shone out again, between
the stems of the trees on the ridge, sending two great bars
of light and a bar of darkness across the reedy waste. Then
came a steady rustling, a splash, and the reeds swayed wider
and wider apart. And at last they broke open, cleft from
root to crest. . . . The end had come.

She looked to see the thing that had come out of the
reeds. For a moment it seemed certainly the great head and
jaw she expected, and then it dwindled and changed. It
was a dark low thing, that remained silent, but it was not
the lion. It became still—everything became still. She
peered. It was like some gigantic frog, two limbs and a

slanting body. Its head moved about searching the shadows. . . .

A rustle, and it moved clumsily, with a sort of hopping. And as it moved it gave a low groan.

The blood rushing through her veins was suddenly joy. "*Ugh-lomi!*" she whispered.

The thing stopped. "*Eudena,*" he answered softly with pain in his voice, and peering into the alders.

He moved again, and came out of the shadow beyond the reeds into the moonlight. All his body was covered with dark smears. She saw he was dragging his legs, and that he gripped his axe, the First Axe, in one hand. In another moment he had struggled into the position of all fours, and had staggered over to her. "The lion," he said in a strange mingling of exultation and anguish. "Wau!—I have slain a lion. With my own hand. Even as I slew the great bear." He moved to emphasise his words, and suddenly broke off with a faint cry. For a space he did not move.

"Let me free," whispered Eudena. . . .

He answered her no words but pulled himself up from his crawling attitude by means of the alder stem, and hacked at her thongs with the sharp edge of his axe. She heard him sob at each blow. He cut away the thongs about her chest and arms, and then his hand dropped. His chest struck against her shoulder and he slipped down beside her and lay still.

But the rest of her release was easy. Very hastily she freed herself. She made one step from the tree, and her head was spinning. Her last conscious movement was towards him. She reeled, and dropped. Her hand fell upon his thigh. It was soft and wet, and gave way under her pressure; he cried out at her touch, and writhed and lay still again.

Presently a dark dog-like shape came very softly through the reeds. Then stopped dead and stood sniffing, hesitated, and at last turned and slunk back into the shadows.

Long was the time they remained there motionless, with the light of the setting moon shining on their limbs. Very slowly, as slowly as the setting of the moon, did the shadow of the reeds towards the mound flow over them. Presently their legs were hidden, and Ugh-lomi was but a bust of silver. The shadow crept to his neck, crept over his face, and so at last the darkness of the night swallowed them up.

The shadow became full of instinctive stirrings. There was a patter of feet, and a faint snarling—the sound of a blow.

There was little sleep that night for the women and children at the squatting-place until they heard Eudena scream. But the men were weary and sat dozing. When Eudena screamed they felt assured of their safety, and hurried to get the nearest places to the fires. The old woman laughed at the scream, and laughed again because Si, the little friend of Eudena, whimpered. Directly the dawn came they were all alert and looking towards the alders. They could see that Eudena had been taken. They could not help feeling glad to think that Uya was appeased. But across the minds of the men the thought of Ugh-lomi fell like a shadow. They could understand revenge, for the world was old in revenge, but they did not think of rescue. Suddenly a hyæna fled out of the thicket, and came galloping across the reed space. His muzzle and paws were dark-stained. At that sight all the men shouted and clutched at throwing-stones and ran towards him, for no animal is so pitiful a coward as the hyæna by day. All men hated the hyæna because he preyed on children, and would come and bite when one was sleeping on the edge of the squatting-place. And Cat's-skin, throwing fair and straight, hit the brute shrewdly on the flank, whereat the whole tribe yelled with delight.

At the noise they made there came a flapping of wings from the lair of the lion, and three white-headed vultures rose slowly and circled and came to rest amidst the branches of an alder, overlooking the lair. "Our lord is abroad," said the old woman, pointing. "The vultures have their share of Eudena." For a space they remained there, and then first one and then another dropped back into the thicket.

Then over the eastern woods, and touching the whole

world of life and colour, poured, with the exaltation of a
trumpet blast, the light of the rising sun. At the sight of
him the children shouted together, and clapped their hands
and began to race off towards the water. Only little Si
lagged behind and looked wonderingly at the alders where
she had seen the head of Eudena overnight.

But Uya, the old lion, was not abroad, but at home, and
he lay very still, and a little on one side. He was not in his
lair, but a little way from it in a place of trampled grass.
Under one eye was a little wound, the feeble little bite of the
first axe. But all the ground beneath his chest was ruddy
brown with a vivid streak, and in his chest was a little hole
that had been made by Ugh-lomi's stabbing-spear. Along
his side and at his neck the vultures had marked their claims.
For so Ugh-lomi had slain him, lying stricken under his
paw and thrusting haphazard at his chest. He had driven
the spear in with all his strength and stabbed the giant to
the heart. So it was the reign of the lion, of the second incar-
nation of Uya the Master, came to an end.

From the knoll the bustle of preparation grew, the hacking of spears and throwing-stones. None spake the name
of Ugh-lomi for fear that it might bring him. The men
were going to keep together, close together, in the hunting
for a day or so. And their hunting was to be Ugh-lomi, lest
instead he should come a-hunting them.

But Ugh-lomi was lying very still and silent, outside the
lion's lair, and Eudena squatted beside him, with the ash
spear, all smeared with lion's blood, gripped in her hand.

V.—The Fight in the Lion's Thicket

Ugh-lomi lay still, his back against an alder, and his thigh
was a red mass terrible to see. No civilised man could have
lived who had been so sorely wounded, but Eudena got him
thorns to close his wounds, and squatted beside him day and
night, smiting the flies from him with a fan of reeds by day,
and in the night threatening the hyænas with the first axe
in her hand; and in a little while he began to heal. It was

high summer, and there was no rain. Little food they had
during the first two days his wounds were open. In the low
place where they hid were no roots nor little beasts, and the
stream, with its water-snails and fish, was in the open a hundred yards away. She could not go abroad by day for fear
of the tribe, her brothers and sisters, nor by night for fear of
the beasts, both on his account and hers. So they shared the
lion with the vultures. But there was a trickle of water near
by, and Eudena brought him plenty in her hands.

Where Ugh-lomi lay was well hidden from the tribe by
a thicket of alders, and all fenced about with bulrushes and
tall reeds. The dead lion he had killed lay near his old
lair on a place of trampled reeds fifty yards away, in sight
through the reed-stems, and the vultures fought each other
for the choicest pieces and kept the jackals off him. Very
soon a cloud of flies that looked like bees hung over him, and
Ugh-lomi could hear their humming. And when Ugh-lomi's
flesh was already healing—and it was not many days before
that began—only a few bones of the lion remained scattered
and shining white.

For the most part Ugh-lomi sat still during the day, looking before him at nothing; sometimes he would mutter of
the horses and bears and lions, and sometimes he would beat
the ground with the first axe and say the names of the tribe
—he seemed to have no fear of bringing the tribe—for hours
together. But chiefly he slept, dreaming little because of his
loss of blood and the slightness of his food. During the
short summer night both kept awake. All the while the
darkness lasted things moved about them, things they never
saw by day. For some nights the hyænas did not come, and
then one moonless night near a dozen came and fought for
what was left of the lion. The night was a tumult of
growling, and Ugh-lomi and Eudena could hear the bones
snap in their teeth. But they knew the hyæna dare not
attack any creature alive and awake, and so they were not
greatly afraid.

Of a daytime Eudena would go along the narrow path
the old lion had made in the reeds until she was beyond the
bend, and then she would creep into the thicket and watch

z

the tribe. She would lie close by the alders where they had bound her to offer her up to the lion, and thence she could see them on the knoll by the fire, small and clear, as she had seen them that night. But she told Ugh-lomi little of what she saw, because she feared to bring them by their names. For so they believed in those days, that naming called.

She saw the men prepare stabbing-spears and throwing-stones on the morning after Ugh-lomi had slain the lion, and go out to hunt him, leaving the women and children on the knoll. Little they knew how near he was as they tracked off in single file towards the hills, with Siss the Tracker leading them. And she watched the women and children, after the men had gone, gathering fern-fronds and twigs for the night fire, and the boys and girls running and playing together. But the very old woman made her feel afraid. Towards noon, when most of the others were down at the stream by the bend, she came and stood on the hither side of the knoll, a gnarled brown figure, and gesticulated so that Eudena could scarce believe she was not seen. Eudena lay like a hare in its form, with shining eyes fixed on the bent witch away there, and presently she dimly understood it was the lion the old woman was worshipping—the lion Ugh-lomi had slain.

And the next day the hunters came back weary, carrying a fawn, and Eudena watched the feast enviously. And then came a strange thing. She saw—distinctly she heard—the old woman shrieking and gesticulating and pointing towards her. She was afraid, and crept like a snake out of sight again. But presently curiosity overcame her and she was back at her spying-place, and as she peered her heart stopped, for there were all the men, with their weapons in their hands, walking together towards her from the knoll.

She dared not move lest her movement should be seen, but she pressed herself close to the ground. The sun was low and the golden light was in the faces of the men. She saw they carried a piece of rich red meat thrust through by an ashen stake. Presently they stopped. "Go on!" screamed the old woman. Cat's-skin grumbled, and they came on, searching the thicket with sun-dazzled eyes. "Here!" said

Siss. And they took the ashen stake with the meat upon it and thrust it into the ground. "Uya!" cried Siss, "behold thy portion. And Ugh-lomi we have slain. Of a truth we have slain Ugh-lomi. This day we slew Ugh-lomi, and to-morrow we will bring his body to you." And the others repeated the words.

They looked at each other and behind them, and partly turned and began going back. At first they walked half turned to the thicket, then, facing the mound, they walked faster, looking over their shoulders, then faster; soon they ran, it was a race at last, until they were near the knoll. Then Siss, who was hindmost, was first to slacken his pace.

The sunset passed and the twilight came, the fires glowed red against the hazy blue of the distant chestnut-trees, and the voices over the mound were merry. Eudena lay scarcely stirring, looking from the mound to the meat and then to the mound. She was hungry, but she was afraid. At last she crept back to Ugh-lomi.

He looked round at the little rustle of her approach. His face was in shadow. "Have you got me some food?" he said.

She said she could find nothing, but that she would seek further, and went back along the lion's path until she could see the mound again, but she could not bring herself to take the meat; she had the brute's instinct of a snare. She felt very miserable.

She crept back at last towards Ugh-lomi and heard him stirring and moaning. She turned back to the mound again; then she saw something in the darkness near the stake, and peering distinguished a jackal. In a flash she was brave and angry; she sprang up, cried out, and ran towards the offering. She stumbled and fell, and heard the growling of the jackal going off.

When she arose only the ashen stake lay on the ground, the meat was gone. So she went back, to fast through the night with Ugh-lomi; and Ugh-lomi was angry with her, because she had no food for him; but she told him nothing of the things she had seen.

Two days passed and they were near starving, when the tribe slew a horse. Then came the same ceremony, and a

haunch was left on the ashen stake; but this time Eudena
did not hesitate.

By acting and words she made Ugh-lomi understand, but
he ate most of the food before he understood; and then as
her meaning passed to him he grew merry with his food. "I
am Uya," he said; "I am the Lion. I am the Great Cave
Bear, I who was only Ugh-lomi. I am Wau the Cunning.
It is well that they should feed me, for presently I will kill
them all."

Then Eudena's heart was light, and she laughed with
him; and afterwards she ate what he had left of the horse-
flesh with gladness.

After that it was he had a dream, and the next day he
made Eudena bring him the lion's teeth and claws—so much
of them as she could find—and hack him a club of alder.
And he put the teeth and claws very cunningly into the
wood so that the points were outward. Very long it took
him, and he blunted two of the teeth hammering them in,
and was very angry and threw the thing away; but afterwards he dragged himself to where he had thrown it and
finished it—a club of a new sort set with teeth. That day
there was more meat for them both, an offering to the lion
from the tribe.

It was one day—more than a hand's fingers of days, more
than anyone had skill to count—after Ugh-lomi had made
the club, that Eudena while he was asleep was lying in the
thicket watching the squatting-place. There had been no
meat for three days. And the old woman came and worshipped after her manner. Now while she worshipped,
Eudena's little friend Si and another, the child of the first
girl Siss had loved, came over the knoll and stood regarding
her skinny figure, and presently they began to mock her.
Eudena found this entertaining, but suddenly the old
woman turned on them quickly and saw them. For a moment she stood and they stood motionless, and then with a
shriek of rage, she rushed towards them, and all three disappeared over the crest of the knoll.

Presently the children reappeared among the ferns beyond
the shoulder of the hill. Little Si ran first, for she was an

active girl, and the other child ran squealing with the old
woman close upon her. And over the knoll came Siss with a
bone in his hand, and Bo and Cat's-skin obsequiously behind
him, each holding a piece of food, and they laughed aloud
and shouted to see the old woman so angry. And with a
shriek the child was caught and the old woman set to work
slapping and the child screaming, and it was very good after-
dinner fun for them. Little Si ran on a little way and
stopped at last between fear and curiosity.

And suddenly came the mother of the child, with hair
streaming, panting, and with a stone in her hand, and the
old woman turned about like a wild cat. She was the equal
of any woman, was the chief of the fire-minders, in spite of
her years; but before she could do anything Siss shouted to
her and the clamour rose loud. Other shock heads came
into sight. It seemed the whole tribe was at home and feasting. But the old woman dared not go on wreaking herself
on the child Siss befriended.

Everyone made noises and called names—even little Si.
Abruptly the old woman let go of the child she had caught
and made a swift run at Si, for Si had no friends; and Si,
realising her danger when it was almost upon her, made off
headlong, with a faint cry of terror, not heeding whither she
ran, straight to the lair of the lion. She swerved aside into
the reeds presently, realising now whither she went.

But the old woman was a wonderful old woman, as active
as she was spiteful, and she caught Si by the streaming hair
within thirty yards of Eudena. All the tribe now was running down the knoll and shouting and laughing ready to see
the fun.

Then something stirred in Eudena; something that had
never stirred in her before; and, thinking all of little Si and
nothing of her fear, she sprang up from her ambush and
ran swiftly forward. The old woman did not see her, for
she was busy beating little Si's face with her hand, beating
with all her heart, and suddenly something hard and heavy
struck her cheek. She went reeling, and saw Eudena with
flaming eyes and cheeks between her and little Si. She
shrieked with astonishment and terror, and little Si, not un-

derstanding, set off towards the gaping tribe. They were quite close now, for the sight of Eudena had driven their fading fear of the lion out of their heads.

In a moment Eudena had turned from the cowering old woman and overtaken Si. "Si," she cried, "Si!" She caught the child up in her arms as it stopped, pressed the nail-lined face to hers, and turned about to run towards her lair, the lair of the old lion. The old woman stood waist-high in the reeds, and screamed foul things in inarticulate rage, but did not dare to intercept her; and at the bend of the path Eudena looked back and saw all the men of the tribe crying to one another and Siss coming at a trot along the lion's trail.

She ran straight along the narrow way through the reeds to the shady place where Ugh-lomi sat with his healing thigh, just awakened by the shouting and rubbing his eyes. She came to him, a woman, with little Si in her arms. Her heart throbbed in her throat. "Ugh-lomi!" she cried. "Ugh-lomi, the tribe comes!"

Ugh-lomi sat staring in stupid astonishment at her and Si.

She pointed with Si in one arm. She sought among her feeble store of words to explain. She could hear the men calling. Apparently they had stopped outside She put down Si and caught up the new club with the lion's teeth, and put it into Ugh-lomi's hand, and ran three yards and picked up the first axe.

"Ah!" said Ugh-lomi, waving the new club, and suddenly he perceived the occasion and, rolling over, began to struggle to his feet.

He stood but clumsily. He supported himself by one hand against the tree, and just touched the ground gingerly with the toe of his wounded leg. In the other hand he gripped the new club. He looked at his healing thigh; and suddenly the reeds began whispering, and ceased and whispered again, and coming cautiously along the track, bending down and holding his fire-hardened stabbing-stick of ash in his hand, appeared Siss. He stopped dead, and his eyes met Ugh-lomi's.

Ugh-lomi forgot he had a wounded leg. He stood firmly

on both feet. Something trickled. He glanced down and saw a little gout of blood had oozed out along the edge of the healing wound. He rubbed his hand there to give him the grip of his club, and fixed his eye again on Siss.

"Wau!" he cried, and sprang forward, and Siss, still stooping and watchful, drove his stabbing-stick up very quickly in an ugly thrust. It ripped Ugh-lomi's guarding arm and the club came down in a counter that Siss was never to understand. He fell, as an ox falls to the pole-axe, at Ugh-lomi's feet.

To Bo it seemed the strangest thing. He had a comforting sense of tall reeds on either side, and an impregnable rampart, Siss, between him and any danger. Snail-eater was close behind and there was no danger there. He was prepared to shove behind and send Siss to death or victory. That was his place as second man. He saw the butt of the spear Siss carried leap away from him, and suddenly a dull whack and the broad back fell away forward, and he looked Ugh-lomi in the face over his prostrate leader. It felt to Bo as if his heart had fallen down a well. He had a throwing-stone in one hand and an ashen stabbing-stick in the other. He did not live to the end of his momentary hesitation which to use.

Snail-eater was a readier man, and besides Bo did not fall forward as Siss had done, but gave at his knees and hips, crumpling up with the toothed club upon his head. The Snail-eater drove his spear forward swift and straight, and took Ugh-lomi in the muscle of the shoulder, and then he drove him hard with the smiting-stone in his other hand, shouting out as he did so. The new club swished ineffectually through the reeds. Eudena saw Ugh-lomi come staggering back from the narrow path into the open space, tripping over Siss and with a foot of ashen stake sticking out of him over his arm. And then the Snail-eater, whose name she had given, had his final injury from her, as his exultant face came out of the reeds after his spear. For she swung the first axe swift and high, and hit him fair and square on the temple; and down he went on Siss at prostrate Ugh-lomi's feet.

But before Ugh-lomi could get up, the two red-haired men were tumbling out of the reeds, spears and smiting-stones ready, and Snake hard behind them. One she struck on the neck, but not to fell him, and he blundered aside and spoilt his brother's blow at Ugh-lomi's head. In a moment Ugh-lomi dropped his club and had his assailant by the waist, and had pitched him sideways sprawling. He snatched at his club again and recovered it. The man Eudena had hit stabbed at her with his spear as he stumbled from her blow, and involuntarily she gave ground to avoid him. He hesitated between her and Ugh-lomi, half turned, gave a vague cry at finding Ugh-lomi so near, and in a moment Ugh-lomi had him by the throat, and the club had its third victim. As he went down Ugh-lomi shouted—no words, but an exultant cry.

The other red-haired man was six feet from her with his back to her, and a darker red streaking his head. He was struggling to his feet. She had an irrational impulse to stop his rising. She flung the axe at him, missed, saw his face in profile, and he had swerved beyond little Si, and was running through the reeds. She had a transitory vision of Snake standing in the throat of the path, half turned away from her, and then she saw his back. She saw the club whirling through the air, and the shock head of Ugh-lomi, with blood in the hair and blood upon the shoulder, vanishing below the reeds in pursuit. Then she heard Snake scream like a woman.

She ran past Si to where the handle of the axe stuck out of a clump of fern, and turning, found herself panting and alone with three motionless bodies. The air was full of shouts and screams. For a space she was sick and giddy, and then it came into her head that Ugh-lomi was being killed along the reed-path, and with an inarticulate cry she leapt over the body of Bo and hurried after him. Snake's feet lay across the path, and his head was among the reeds. She followed the path until it bent round and opened out by the alders, and then she saw all that was left of the tribe in the open, scattering like dead leaves before a gale, and going back over the knoll. Ugh-lomi was hard upon Cat's-skin.

But Cat's-skin was fleet of foot and got away, and so did young Wau-Hau when Ugh-lomi turned upon him, and Ugh-lomi pursued Wau-Hau far beyond the knoll before he desisted. He had the rage of battle on him now, and the wood thrust through his shoulder stung him like a spur. When she saw he was in no danger she stopped running and stood panting, watching the distant active figures run up and vanish one by one over the knoll. In a little time she was alone again. Everything had happened very swiftly. The smoke of Brother Fire rose straight and steady from the squatting-place, just as it had done ten minutes ago, when the old woman had stood yonder worshipping the lion.

And after a long time, as it seemed, Ugh-lomi reappeared over the knoll, and came back to Eudena, triumphant and breathing heavily. She stood, her hair about her eyes and hot-faced, with the blood-stained axe in her hand, at the place where the tribe had offered her as a sacrifice to the lion. "Wau!" cried Ugh-lomi at the sight of her, his face alight with the fellowship of battle, and he waved his new club, red now and hairy; and at the sight of his glowing face her tense pose relaxed somewhat, and she stood sobbing and rejoicing.

Ugh-lomi had a queer unaccountable pang at the sight of her tears; but he only shouted "Wau!" the louder and shook the axe east and west. He called manfully to her to follow him and turned back, striding, with the club swinging in his hand, towards the squatting-place, as if he had never left the tribe; and she ceased her weeping and followed quickly as a woman should.

So Ugh-lomi and Eudena came back to the squatting-place from which they had fled many days before from the face of Uya; and by the squatting-place lay a deer half eaten, just as there had been before Ugh-lomi was man or Eudena woman. So Ugh-lomi sat down to eat, and Eudena beside him like a man, and the rest of the tribe watched them from safe hiding-places. And after a time one of the elder girls came back timorously, carrying little Si in her arms, and Eudena called to them by name, and offered them food. But the elder girl was afraid and would not come, though

Si struggled to come to Eudena. Afterwards, when Ugh-lomi had eaten, he sat dozing, and at last he slept, and slowly the others came out of the hiding-places and drew near. And when Ugh-lomi woke, save that there were no men to be seen, it seemed as though he had never left the tribe.

Now, there is a thing strange but true: that all through this fight Ugh-lomi forgot that he was lame, and was not lame, and after he had rested behold! he was a lame man; and he remained a lame man to the end of his days.

Cat's-skin and the second red-haired man and Wau-Hau, who chipped flints cunningly, as his father had done before him, fled from the face of Ugh-lomi, and none knew where they hid. But two days after they came and squatted a good way off from the knoll among the bracken under the chestnuts and watched. Ugh-lomi's rage had gone, he moved to go against them and did not, and at sundown they went away. That day, too, they found the old woman among the ferns, where Ugh-lomi had blundered upon her when he had pursued Wau-Hau. She was dead and more ugly than ever, but whole. The jackals and vultures had tried her and left her;—she was ever a wonderful old woman.

The next day the three men came again and squatted nearer, and Wau-Hau had two rabbits to hold up, and the red-haired man a wood-pigeon, and Ugh-lomi stood before the women and mocked them.

The next day they sat again nearer—without stones or sticks, and with the same offerings, and Cat's-skin had a trout. It was rare men caught fish in those days, but Cat's-skin would stand silently in the water for hours and catch them with his hand. And the fourth day Ugh-lomi suffered these three to come to the squatting-place in peace, with the food they had with them. Ugh-lomi ate the trout. Thereafter for many moons Ugh-lomi was master and had his will in peace. And on the fullness of time he was killed and eaten even as Uya had been slain.

STORY THE FOURTH

A Story of the Days to Come

1.—The Cure for Love

THE excellent Mr. Morris was an Englishman, and he lived in the days of Queen Victoria the Good. He was a prosperous and very sensible man; he read *The Times* and went to church, and as he grew towards middle age an expression of quiet contented contempt for all who were not as himself settled on his face. He was one of those people who do everything that is right and proper and sensible with inevitable regularity. He always wore just the right and proper clothes, steering the narrow way between the smart and the shabby, always subscribed to the right charities, just the judicious compromise between ostentation and meanness, and never failed to have his hair cut to exactly the proper length.

Everything that it was right and proper for a man in his position to possess, he possessed; and everything that it was not right and proper for a man in his position to possess, he did not possess.

And among other right and proper possessions, this Mr. Morris had a wife and children. They were the right sort of wife, and the right sort and number of children, of course; nothing imaginative or highty-flighty about any of them, so far as Mr. Morris could see; they wore perfectly correct clothing, neither smart nor hygienic nor faddy in any way, but just sensible; and they lived in a nice sensible house in the later Victorian sham Queen Anne style of architecture,

715

with sham half-timbering of chocolate-painted plaster in the gables, Lincrusta Walton sham carved oakpanels, a terrace of terra cotta to imitate stone, and cathedral glass in the front door. His boys went to good solid schools, and were put to respectable professions; his girls, in spite of a fantastic protest or so, were all married to suitable, steady, oldish young men with good prospects. And when it was a fit and proper thing for him to do so, Mr. Morris died. His tomb was of marble, and, without any art nonsense or laudatory inscription, quietly imposing—such being the fashion of his time.

He underwent various changes according to the accepted custom in these cases, and long before this story begins his bones even had become dust, and were scattered to the four quarters of heaven. And his sons and his grandsons and his great-grandsons and his great-great-grandsons, they too were dust and ashes, and were scattered likewise. It was a thing he could not have imagined, that a day would come when even his great-great-grandsons would be scattered to the four winds of heaven. If any one had suggested it to him he would have resented it. He was one of those worthy people who take no interest in the future of mankind at all. He had grave doubts, indeed, if there was any future for mankind after he was dead.

It seemed quite impossible and quite uninteresting to imagine anything happening after he was dead. Yet the thing was so, and when even his great-great-grandson was dead and decayed and forgotten, when the sham half-timbered house had gone the way of all shams, and *The Times* was extinct, and the silk hat a ridiculous antiquity, and the modestly imposing stone that had been sacred to Mr. Morris had been burnt to make lime for mortar, and all that Mr. Morris had found real and important was sere and dead, the world was still going on, and people were still going about it, just as heedless and impatient of the Future, or, indeed, of anything but their own selves and property, as Mr. Morris had been.

And, strange to tell, and much as Mr. Morris would have been angered if any one had foreshadowed it to him, all over the world there were scattered a multitude of people, filled with the breath of life, in whose veins the blood of Mr. Morris flowed. Just as some day the life which is gathered now in the reader of this very story may also be scattered far and wide about this world, and mingled with a thousand alien strains, beyond all thought and tracing.

And among the descendants of this Mr. Morris was one almost as sensible and clear-headed as his ancestor. He had just the same stout, short frame as that ancient man of the nineteenth century, from whom his name of Morris—he spelt it Mwres—came; he had the same half-contemptuous expression of face. He was a prosperous person, too, as times went, and he disliked the "new-fangled," and bothers about the future and the lower classes, just as much as the ancestral Morris had done. He did not read *The Times*: indeed, he did not know there ever had been a *Times*—that institution had foundered somewhere in the intervening gulf of years; but the phonograph machine, that talked to him as he made his toilet of a morning, might have been the voice of a reincarnated Blowitz when it dealt with the world's affairs. This phonographic machine was the size and shape of a Dutch clock, and down the front of it were electric barometric indicators, and an electric clock and calendar, and automatic engagement reminders, and where the clock would have been was the mouth of a trumpet. When it had news the trumpet gobbled like a turkey, "Galloop, galloop," and then brayed out its message as, let us say, a trumpet might bray. It would tell Mwres in full, rich, throaty tones about the overnight accidents to the omnibus flying-machines that plied around the world, the latest arrivals at the fashionable resorts in Tibet, and of all the great monopolist company meetings of the day before, while he was dressing. If Mwres did not like hearing what it said, he had only to touch a stud, and it would choke a little and talk about something else.

Of course his toilet differed very much from that of his ancestor. It is doubtful which would have been the more shocked and pained to find himself in the clothing of the other. Mwres would certainly have sooner gone forth to the world stark naked than in the silk hat, frock coat, grey

trousers and watch-chain that had filled Mr. Morris with
sombre self-respect in the past. For Mwres there was no
shaving to do: a skilful operator had long ago removed every
hair-root from his face. His legs he encased in pleasant pink
and amber garments of an air-tight material, which with the
help of an ingenious little pump he distended so as to suggest enormous muscles. Above this he also wore pneumatic
garments beneath an amber silk tunic, so that he was clothed
in air and admirably protected against sudden extremes of
heat or cold. Over this he flung a scarlet cloak with its edge
fantastically curved. On his head, which had been skilfully
deprived of every scrap of hair, he adjusted a pleasant little
cap of bright scarlet, held on by suction and inflated with
hydrogen, and curiously like the comb of a cock. So his
toilet was complete; and, conscious of being soberly and be-
comingly attired, he was ready to face his fellow-beings with
a tranquil eye.

This Mwres—the civility of "Mr." had vanished ages ago
—was one of the officials under the Wind Vane and Waterfall Trust, the great company that owned every wind wheel
and waterfall in the world, and which pumped all the
water and supplied all the electric energy that people in
these latter days required. He lived in a vast hotel near
that part of London called Seventh Way, and had very large
and comfortable apartments on the seventeenth floor. Households and family life had long since disappeared with the
progressive refinement of manners; and indeed the steady
rise in rents and land values, the disappearance of domestic
servants, the elaboration of cookery, had rendered the separate domicile of Victorian times impossible, even had any
one desired such a savage seclusion. When his toilet was
completed he went towards one of the two doors of his apartment—there were doors at opposite ends, each marked with
a huge arrow pointing one one way and one the other—
touched a stud to open it, and emerged on a wide passage,
the centre of which bore chairs and was moving at a steady
pace to the left. On some of these chairs were seated
gaily-dressed men and women. He nodded to an acquaintance—it was not in those days etiquette to talk before break-

fast—and seated himself on one of these chairs, and in a few seconds he had been carried to the doors of a lift, by which he descended to the great and splendid hall in which his breakfast would be automatically served.

It was a very different meal from a Victorian breakfast. The rude masses of bread needing to be carved and smeared over with animal fat before they could be made palatable, the still recognisable fragments of recently killed animals, hideously charred and hacked, the eggs torn ruthlessly from beneath some protesting hen,—such things as these, though they constituted the ordinary fare of Victorian times, would have awakened only horror and disgust in the refined minds of the people of these latter days. Instead were pastes and cakes of agreeable and variegated design, without any suggestion in colour or form of the unfortunate animals from which their substance and juices were derived. They appeared on little dishes sliding out upon a rail from a little box at one side of the table. The surface of the table, to judge by touch and eye, would have appeared to a nineteenth-century person to be covered with fine white damask, but this was really an oxidised metallic surface, and could be cleaned instantly after a meal. There were hundreds of such little tables in the hall, and at most of them were other latter-day citizens singly or in groups. And as Mwres seated himself before his elegant repast, the invisible orchestra, which had been resting during an interval, resumed and filled the air with music.

But Mwres did not display any great interest either in his breakfast or the music; his eye wandered incessantly about the hall, as though he expected a belated guest. At last he rose eagerly and waved his hand, and simultaneously across the hall appeared a tall dark figure in a costume of yellow and olive green. As this person, walking amidst the tables with measured steps, drew near, the pallid earnestness of his face and the unusual intensity of his eyes became apparent. Mwres reseated himself and pointed to a chair beside him.

"I feared you would never come," he said. In spite of the intervening space of time, the English language was still

almost exactly the same as it had been in England under
Victoria the Good. The invention of the phonograph and
suchlike means of recording sound, and the gradual replace-
ment of books by such contrivances, had not only saved the
human eyesight from decay, but had also by the establishment of a sure standard arrested the process of change in
accent that had hitherto been so inevitable.

"I was delayed by an interesting case," said the man in
green and yellow. "A prominent politician—ahem!—suffering from overwork." He glanced at the breakfast and seated
himself. "I have been awake for forty hours."

"Eh dear!" said Mwres: "fancy that! You hypnotists have
your work to do."

The hypnotist helped himself to some attractive amber-
coloured jelly. "I happen to be a good deal in request,"
he said modestly.

"Heaven knows what we should do without you."

"Oh! we're not so indispensable as all that," said the
hypnotist, ruminating the flavour of the jelly. "The world
did very well without us for some thousands of years. Two
hundred years ago even—not one! In practice, that is.
Physicians by the thousand, of course—frightfully clumsy
brutes for the most part, and following one another like
sheep—but doctors of the mind, except a few empirical floun-
derers there were none."

He concentrated his mind on the jelly.

"But were people so sane——?" began Mwres.

The hypnotist shook his head. "It didn't matter then if
they were a bit silly or faddy. Life was so easy-going then.
No competition worth speaking of—no pressure. A human
being had to be very lopsided before anything happened.
Then, you know, they clapped 'em away in what they called
a lunatic asylum."

"I know," said Mwres. "In these confounded historical
romances that every one is listening to, they always rescue
a beautiful girl from an asylum or something of the sort. I
don't know if you attend to that rubbish."

"I must confess I do," said the hypnotist. "It carries one
out of oneself to hear of those quaint, adventurous, half-

civilised days of the nineteenth century, when men were stout and women simple. I like a good swaggering story before all things. Curious times they were, with their smutty railways and puffing old iron trains, their rum little houses and their horse vehicles. I suppose you don't read books?"

"Dear, no!" said Mwres. "I went to a modern school and we had none of that old-fashioned nonsense. Phonographs are good enough for me."

"Of course," said the hypnotist, "of course"; and surveyed the table for his next choice. "You know," he said, helping himself to a dark blue confection that promised well, "in those days our business was scarcely thought of. I daresay if any one had told them that in two hundred years' time a class of men would be entirely occupied in impressing things upon the memory, effacing unpleasant ideas, controlling and overcoming instinctive but undesirable impulses, and so forth, by means of hypnotism, they would have refused to believe the thing possible. Few people knew that an order made during a mesmeric trance, even an order to forget or an order to desire, could be given so as to be obeyed after the trance was over. Yet there were men alive then who could have told them the thing was as absolutely certain to come about as—well, the transit of Venus."

"They knew of hypnotism, then?"

"Oh, dear, yes! They used it—for painless dentistry and things like that! This blue stuff is confoundedly good: what is it?"

"Haven't the faintest idea," said Mwres, "but I admit it's very good. Take some more."

The hypnotist repeated his praises, and there was an appreciative pause.

"Speaking of these historical romances," said Mwres, with an attempt at an easy, off-hand manner, "brings me—ah—to the matter I—ah—had in mind when I asked you—when I expressed a wish to see you." He paused and took a deep breath.

The hypnotist turned an attentive eye upon him, and continued eating.

"The fact is," said Mwres, "I have a—in fact a—daughter. Well, you know I have given her—ah—every educational advantage. Lectures—not a solitary lecturer of ability in the world but she has had a telephone direct, dancing, deportment, conversation, philosophy, art criticism. . . ." He indicated catholic culture by a gesture of his hand. "I had intended her to marry a very good friend of mine—Bindon of the Lighting Commission—plain little man, you know, and a bit unpleasant in some of his ways, but an excellent fellow really—an excellent fellow."

"Yes," said the hypnotist, "go on. How old is she?"

"Eighteen."

"A dangerous age. Well?"

"Well: it seems that she has been indulging in these historical romances—excessively. Excessively. Even to the neglect of her philosophy. Filled her mind with unutterable nonsense about soldiers who fight—what is it?—Etruscans?"

"Egyptians."

"Egyptians—very probably. Hack about with swords and revolvers and things—blood-shed galore—horrible!—and about young men on torpedo catchers who blow up—Spaniards, I fancy—and all sorts of irregular adventurers. And she has got it into her head that she must marry for Love, and that poor little Bindon——"

"I've met similar cases," said the hypnotist. "Who is the other young man?"

Mwres maintained an appearance of resigned calm. "You may well ask," he said. "He is"—and his voice sank with shame—"a mere attendant upon the stage on which the flying-machines from Paris alight. He has—as they say in the romances—good looks. He is quite young and very eccentric. Affects the antique—he can read and write! So can she. And instead of communicating by telephone, like sensible people, they write and deliver—what is it?"

"Notes?"

"No—not notes. . . . Ah—poems."

The hypnotist raised his eyebrows. "How did she meet him?"

"Tripped coming down from the flying-machine from Paris—and fell into his arms. The mischief was done in a moment!"

"Yes?"

"Well—that's all. Things must be stopped. That is what I want to consult you about. What must be done? What *can* be done? Of course I'm not a hypnotist; my knowledge is limited. But you——?"

"Hypnotism is not magic," said the man in green, putting both arms on the table.

"Oh, precisely! But still——!"

"People cannot be hypnotised without their consent. If she is able to stand out against marrying Bindon, she will probably stand out against being hypnotised. But if once she can be hypnotised—even by somebody else—the thing is done."

"You can——?"

"Oh, certainly! Once we get her amenable, then we can suggest that she *must* marry Bindon—that that is her fate; or that the young man is repulsive, and that when she sees him she will be giddy and faint, or any little thing of that sort. Or if we can get her into a sufficiently profound trance we can suggest that she should forget him altogether——"

"Precisely."

"But the problem is to get her hypnotised. Of course no sort of proposal or suggestion must come from you—because no doubt she already distrusts you in the matter."

The hypnotist leant his head upon his arm and thought.

"It's hard a man cannot dispose of his own daughter," said Mwres irrelevantly.

"You must give me the name and address of the young lady," said the hypnotist, "and any information bearing upon the matter. And, by the bye, is there any money in the affair?"

Mwres hesitated.

"There's a sum—in fact, a considerable sum—invested in the Patent Road Company. From her mother. That's what makes the thing so exasperating."

"Exactly," said the hypnotist. And he proceeded to cross-examine Mwres on the entire affair.

It was a lengthy interview.

And meanwhile "Elizebeθ Mwres," as she spelt her name, or "Elizabeth Morris," as a nineteenth-century person would have put it, was sitting in a quiet waiting-place beneath the great stage upon which the flying-machine from Paris descended. And beside her sat her slender, handsome lover reading her the poem he had written that morning while on duty upon the stage. When he had finished they sat for a time in silence; and then, as if for their special entertainment, the great machine that had come flying through the air from America that morning rushed down out of the sky.

At first it was a little oblong, faint and blue amidst the distant fleecy clouds; and then it grew swiftly large and white, and larger and whiter, until they could see the separate tiers of sails, each hundreds of feet wide, and the lank body they supported, and at last even the swinging seats of the passengers in a dotted row. Although it was falling it seemed to them to be rushing up the sky, and over the roof-spaces of the city below its shadow leapt towards them. They heard the whistling rush of the air about it and its yelling siren, shrill and swelling, to warn those who were on its landing-stage of its arrival. And abruptly the note fell down a couple of octaves, and it had passed, and the sky was clear and void, and she could turn her sweet eyes again to Denton at her side.

Their silence ended; and Denton, speaking in a little language of broken English that was, they fancied, their private possession—though lovers have used such little languages since the world began—told her how they too would leap into the air one morning out of all the obstacles and difficulties about them, and fly to a sunlit city of delight he knew of in Japan, half-way about the world.

She loved the dream, but she feared the leap; and she put him off with "Some day, dearest one, some day," to all his pleading that it might be soon; and at last came a shrilling of whistles, and it was time for him to go back to his duties on the stage. They parted—as lovers have been wont to part for thousands of years. She walked down a passage to a lift, and so came to one of the streets of that latter-day London, all glazed in with glass from the weather, and with incessant moving platforms that went to all parts of the city. And by one of these she returned to her apartments in the Hotel for Women where she lived, the apartments that were in telephonic communication with all the best lecturers in the world. But the sunlight of the flying stage was in her heart, and the wisdom of all the best lecturers in the world seemed folly in that light.

She spent the middle part of the day in the gymnasium, and took her midday meal with two other girls and their common chaperone—for it was still the custom to have a chaperone in the case of motherless girls of the more prosperous classes. The chaperone had a visitor that day, a man in green and yellow, with a white face and vivid eyes, who talked amazingly. Among other things, he fell to praising a new historical romance that one of the great popular story-tellers of the day had just put forth. It was, of course, about the spacious times of Queen Victoria; and the author, among other pleasing novelties, made a little argument before each section of the story, in imitation of the chapter headings of the old-fashioned books: as for example, "How the Cabmen of Pimlico stopped the Victoria Omnibuses, and of the Great Fight in Palace Yard," and "How the Piccadilly Policeman was slain in the midst of his Duty." The man in green and yellow praised this innovation. "These pithy sentences," he said, "are admirable. They show at a glance those headlong, tumultuous times, when men and animals jostled in the filthy streets, and death might wait for one at every corner. Life was life then! How great the world must have seemed then! How marvellous! There were still parts of the world absolutely unexplored. Nowadays we have almost abolished wonder, we lead lives so trim and orderly that courage, endurance, faith, all the noble virtues seem fading from mankind."

And so on, taking the girls' thoughts with him, until the life they led, life in the vast and intricate London of the twenty-second century, a life interspersed with soaring excur-

sions to every part of the globe, seemed to them a monotonous misery compared with the dædal past.

At first Elizabeth did not join in the conversation, but
after a time the subject became so interesting that she made
a few shy interpolations. But he scarcely seemed to notice
her as he talked. He went on to describe a new method of
entertaining people. They were hypnotised, and then suggestions were made to them so skilfully that they seemed
to be living in ancient times again. They played out a little
romance in the past as vivid as reality, and when at last
they awakened they remembered all they had been through
as though it were a real thing.

"It is a thing we have sought to do for years and years,"
said the hypnotist. "It is practically an artificial dream.
And we know the way at last. Think of all it opens out to
us—the enrichment of our experience, the recovery of
adventure, the refuge it offers from this sordid, competitive
life in which we live! Think!"

"And you can do that!" said the chaperone eagerly.

"The thing is possible at last," the hypnotist said. "You
may order a dream as you wish."

The chaperone was the first to be hypnotised, and the
dream, she said, was wonderful, when she came to again.

The other two girls, encouraged by her enthusiasm, also
placed themselves in the hands of the hypnotist and had
plunges into the romantic past. No one suggested that
Elizabeth should try this novel entertainment; it was at her
own request at last that she was taken into that land of
dreams where there is neither any freedom of choice nor
will. . . .

And so the mischief was done.

One day, when Denton went down to that quiet seat beneath the flying stage, Elizabeth was not in her wonted
place. He was disappointed, and a little angry. The next
day she did not come, and the next also. He was afraid.
To hide his fear from himself, he set to work to write sonnets for her when she should come again. . . .

For three days he fought against his dread by such distraction, and then the truth was before him clear and cold,

and would not be denied. She might be ill, she might be dead; but he would not believe that he had been betrayed. There followed a week of misery. And then he knew she was the only thing on earth worth having, and that he must seek her, however hopeless the search, until she was found once more.

He had some small private means of his own, and so he threw over his appointment on the flying stage, and set himself to find this girl who had become at last all the world to him. He did not know where she lived, and little of her circumstances; for it had been part of the delight of her girlish romance that he should know nothing of her, nothing of the difference of their station. The ways of the city opened before him east and west, north and south. Even in Victorian days London was a maze, that little London with its poor four millions of people; but the London he explored, the London of the twenty-second century, was a London of thirty million souls. At first he was energetic and headlong, taking time neither to eat nor sleep. He sought for weeks and months, he went through every imaginable phase of fatigue and despair, over-excitement and anger. Long after hope was dead, by the sheer inertia of his desire he still went to and fro, peering into faces and looking this way and that, in the incessant ways and lifts and passages of that interminable hive of men.

At last chance was kind to him, and he saw her.

It was in a time of festivity. He was hungry; he had paid the inclusive fee and had gone into one of the gigantic dining-places of the city; he was pushing his way among the tables and scrutinising by mere force of habit every group he passed.

He stood still, robbed of all power of motion, his eyes wide, his lips apart. Elizabeth sat scarcely twenty yards away from him, looking straight at him. Her eyes were as hard to him, as hard and expressionless and void of recognition, as the eyes of a statue.

She looked at him for a moment, and then her gaze passed beyond him.

Had he had only her eyes to judge by he might have

doubted if it was indeed Elizabeth, but he knew her by the gesture of her hand, by the grace of a wanton little curl that floated over her ear as she moved her head. Something was said to her, and she turned smiling tolerantly to the man beside her, a little man in foolish raiment knobbed and spiked like some odd reptile with pneumatic horns—the Bindon of her father's choice.

For a moment Denton stood white and wild-eyed; then came a terrible faintness, and he sat before one of the little tables. He sat down with his back to her, and for a time he did not dare to look at her again. When at last he did, she and Bindon and two other people were standing up to go. The others were her father and her chaperone.

He sat as if incapable of action until the four figures were remote and small, and then he rose up possessed with the one idea of pursuit. For a space he feared he had lost them, and then he came upon Elizabeth and her chaperone again in one of the streets of moving platforms that intersected the city. Bindon and Mwres had disappeared.

He could not control himself to patience. He felt he must speak to her forthwith, or die. He pushed forward to where they were seated, and sat down beside them. His white face was convulsed with half-hysterical excitement.

He laid his hand on her wrist. "Elizabeth?" he said.

She turned in unfeigned astonishment. Nothing but the fear of a strange man showed in her face.

"Elizabeth," he cried, and his voice was strange to him: "dearest—you *know* me?"

Elizabeth's face showed nothing but alarm and perplexity. She drew herself away from him. The chaperone, a little grey-headed woman with mobile features, leant forward to intervene. Her resolute bright eyes examined Denton. "*What* do you say?" she asked.

"This young lady," said Denton,—"she knows me."

"Do you know him, dear?"

"No," said Elizabeth in a strange voice, and with a hand to her forehead, speaking almost as one who repeats a lesson. "No, I do not know him. I *know*—I do not know him."

"But—but ... Not know me! It is I—Denton. Denton! To whom you used to talk. Don't you remember the flying stages? The little seat in the open air? The verses—"

"No," cried Elizabeth,—"no. I do not know him. I do not know him. There is something ... But I don't know. All I know is that I do not know him." Her face was a face of infinite distress.

The sharp eyes of the chaperone flitted to and fro from the girl to the man. "You see?" she said, with the faint shadow of a smile. "She does not know you."

"I do not know you," said Elizabeth. "Of that I am sure."

"But, dear—the songs—the little verses—"

"She does not know you," said the chaperone. "You must not ... You have made a mistake. You must not go on talking to us after that. You must not annoy us on the public ways."

"But—" said Denton, and for a moment his miserably haggard face appealed against fate.

"You must not persist, young man," protested the chaperone.

"*Elizabeth!*" he cried.

Her face was the face of one who is tormented. "I do not know you," she cried, hand to brow. "Oh, I do not know you!"

For an instant Denton sat stunned. Then he stood up and groaned aloud.

He made a strange gesture of appeal towards the remote glass roof of the public way, then turned and went plunging recklessly from one moving platform to another, and vanished amidst the swarms of people going to and fro thereon. The chaperone's eyes followed him, and then she looked at the curious faces about her.

"Dear," asked Elizabeth, clasping her hand, and too deeply moved to heed observation, "who was that man? Who *was* that man?"

The chaperone raised her eyebrows. She spoke in a clear, audible voice. "Some half-witted creature. I have never set eyes on him before."

"Never?"

"Never, dear. Do not trouble your mind about a thing like this."

And soon after this the celebrated hypnotist who dressed in green and yellow had another client. The young man paced his consulting-room, pale and disordered. "I want to forget," he cried. "I *must* forget."

The hypnotist watched him with quiet eyes, studied his face and clothes and bearing. "To forget anything—pleasure or pain—is to be, by so much—*less*. However, you know your own concern. My fee is high."

"If only I can forget—"

"That's easy enough with you. You wish it. I've done much harder things. Quite recently. I hardly expected to do it: the thing was done against the will of the hypnotised person. A love affair too—like yours. A girl. So rest assured."

The young man came and sat beside the hypnotist. His manner was a forced calm. He looked into the hypnotist's eyes. "I will tell you. Of course you will want to know what it is. There was a girl. Her name was Elizabeth Mwres. Well . . ."

He stopped. He had seen the instant surprise on the hypnotist's face. In that instant he knew. He stood up. He seemed to dominate the seated figure by his side. He gripped the shoulder of green and gold. For a time he could not find words.

"*Give her me back!*" he said at last. "Give her me back!"

"What do you mean?" gasped the hypnotist.

"Give her me back."

"Give whom?"

"Elizabeth Mwres—the girl—"

The hypnotist tried to free himself; he rose to his feet. Denton's grip tightened.

"Let go!" cried the hypnotist, thrusting an arm against Denton's chest.

In a moment the two men were locked in a clumsy wrestle. Neither had the slightest training—for athleticism, except for exhibition and to afford opportunity for betting, had faded out of the earth—but Denton was not only the younger but the stronger of the two. They swayed across the room, and then the hypnotist had gone down under his antagonist. They fell together. . . .

Denton leaped to his feet, dismayed at his own fury; but the hypnotist lay still, and suddenly from a little white mark where his forehead had struck a stool shot a hurrying band of red. For a space Denton stood over him irresolute, trembling.

A fear of the consequences entered his gently nurtured mind. He turned towards the door. "No," he said aloud, and came back to the middle of the room. Overcoming the instinctive repugnance of one who had seen no act of violence in all his life before, he knelt down beside his antagonist and felt his heart. Then he peered at the wound. He rose quickly and looked about him. He began to see more of the situation.

When presently the hypnotist recovered his senses, his head ached severely, his back was against Denton's knees and Denton was sponging his face.

The hypnotist did not speak. But presently he indicated by a gesture that in his opinion he had been sponged enough. "Let me get up," he said.

"Not yet," said Denton.

"You have assaulted me, you scoundrel!"

"We are alone," said Denton, "and the door is secure."

There was an interval of thought.

"Unless I sponge," said Denton, "your forehead will develop a tremendous bruise."

"You can go on sponging," said the hypnotist sulkily.

There was another pause.

"We might be in the Stone Age," said the hypnotist. "Violence! Struggle!"

"In the Stone Age no man dared to come between man and woman," said Denton.

The hypnotist thought again.

"What are you going to do?" he asked.

"While you were insensible I found the girl's address on your tablets. I did not know it before. I telephoned. She will be here soon. Then—"

"She will bring her chaperone."

"That is all right."

"But what—? I don't see. What do you mean to do?"

"I looked about for a weapon also. It is an astonishing thing how few weapons there are nowadays. If you consider that in the Stone Age men owned scarcely anything *but* weapons. I hit at last upon this lamp. I have wrenched off the wires and things, and I hold it so." He extended it over the hypnotist's shoulders. "With that I can quite easily smash your skull. I *will*—unless you do as I tell you."

"Violence is no remedy," said the hypnotist, quoting from the "Modern Man's Book of Moral Maxims."

"It's an undesirable disease,'" said Denton.

"Well?"

"You will tell that chaperone you are going to order the girl to marry that knobby little brute with the red hair and ferrety eyes. I believe that's how things stand?"

"Yes—that's how things stand."

"And, pretending to do that, you will restore her memory of me."

"It's unprofessional."

"Look here! If I cannot have that girl I would rather die than not. I don't propose to respect your little fancies. If anything goes wrong you shall not live five minutes. This is a rude makeshift of a weapon, and it may quite conceivably be painful to kill you. But I will. It is unusual, I know, nowadays to do things like this—mainly because there is so little in life that is worth being violent about."

"The chaperone will see you directly she comes—"

"I shall stand in that recess. Behind you."

The hypnotist thought. "You are a determined young man," he said, "and only half civilised. I have tried to do my duty to my client, but in this affair you seem likely to get your own way. . . ."

"You mean to deal straightly."

"I'm not going to risk having my brains scattered in a petty affair like this."

"And afterwards?"

"There is nothing a hypnotist or doctor hates so much as a scandal. I at least am no savage. I am annoyed. . . . But in a day or so I shall bear no malice. . . ."

"Thank you. And now that we understand each other, there is no necessity to keep you sitting any longer on the floor."

II.—The Vacant Country

The world, they say, changed more between the year 1800 and the year 1900 than it had done in the previous five hundred years. That century, the nineteenth century, was the dawn of a new epoch in the history of mankind— the epoch of the great cities, the end of the old order of country life.

In the beginning of the nineteenth century the majority of mankind still lived upon the countryside, as their way of life had been for countless generations. All over the world they dwelt in little towns and villages then, and engaged either directly in agriculture, or in occupations that were of service to the agriculturist. They travelled rarely, and dwelt close to their work, because swift means of transit had not yet come. The few who travelled went either on foot, or in slow sailing-ships, or by means of jogging horses incapable of more than sixty miles a day. Think of it!—sixty miles a day. Here and there, in those sluggish times, a town grew a little larger than its neighbours, as a port or as a centre of government; but all the towns in the world with more than a hundred thousand inhabitants could be counted on a man's fingers. So it was in the beginning of the nineteenth century. By the end, the invention of railways, telegraphs, steamships, and complex agricultural machinery, had changed all these things: changed them beyond all hope of return. The vast shops, the varied pleasures, the countless conveniences of the larger towns were suddenly possible, and no sooner existed than they were brought into competition with the homely resources of the rural centres. Mankind were drawn to the cities by an overwhelming attraction. The demand for labour fell with the increase of machinery, the local markets were entirely superseded, and there was a rapid growth of the larger centres at the expense of the open country.

The flow of population townward was the constant preoccupation of Victorian writers. In Great Britain and New England, in India and China, the same thing was remarked everywhere a few swollen towns were visibly replacing the ancient order. That this was an inevitable result of improved means of travel and transport—that, given swift means of transit, these things must be—was realised by few; and the most puerile schemes were devised to overcome the mysterious magnetism of the urban centres, and keep the people on the land.

Yet the developments of the nineteenth century were only the dawning of the new order. The first great cities of the new time were horribly inconvenient, darkened by smoky fogs, insanitary and noisy; but the discovery of new methods of building, new methods of heating, changed all this. Between 1900 and 2000 the march of change was still more rapid; and between 2000 and 2100 the continually accelerated progress of human invention made the reign of Victoria the Good seem at last an almost incredible vision of idyllic tranquil days.

The introduction of railways was only the first step in that development of those means of locomotion which finally revolutionised human life. By the year 2000 railways and roads had vanished together. The railways, robbed of their rails, had become weedy ridges and ditches upon the face of the world; the old roads, strange barbaric tracks of flint and soil, hammered by hand or rolled by rough iron rollers, strewn with miscellaneous filth, and cut by iron hoofs and wheels into ruts and puddles often many inches deep, had been replaced by patent tracks made of a substance called Eadhamite. This Eadhamite—it was named after its patentee—ranks with the invention of printing and steam as one of the epoch-making discoveries of the world's history.

When Eadham discovered the substance, he probably thought of it as a mere cheap substitute for india-rubber; it cost a few shillings a ton. But you can never tell all an invention will do. It was the genius of a man named Warming that pointed to the possibility of using it, not only for the tires of wheels, but as a road substance, and who organised the enormous network of public ways that speedily covered the world.

These public ways were made with longitudinal divisions. On the outer on either side went foot cyclists and conveyances travelling at a less speed than twenty-five miles an hour; in the middle, motors capable of speed up to a hundred; and the inner, Warming (in the face of enormous ridicule) reserved for vehicles travelling at speeds of a hundred miles an hour and upward.

For ten years his inner ways were vacant. Before he died they were the most crowded of all, and vast light frameworks with wheels of twenty and thirty feet in diameter, hurled along them at paces that year after year rose steadily towards two hundred miles an hour. And by the time this revolution was accomplished, a parallel revolution had transformed the ever-growing cities. Before the development of practical science the fogs and filth of Victorian times vanished. Electric heating replaced fires (in 2013 the lighting of a fire that did not absolutely consume its own smoke was made an indictable nuisance), and all the city ways, all public squares and places, were covered in with a recently invented glass-like substance. The roofing of London became practically continuous. Certain short-sighted and foolish legislation against tall buildings was abolished, and London, from a squat expanse of petty houses —feebly archaic in design—rose steadily towards the sky. To the municipal responsibility for water, light, and drainage, was added another, and that was ventilation.

But to tell of all the changes in human convenience that these two hundred years brought about, to tell of the long foreseen invention of flying, to describe how life in households was steadily supplanted by life in interminable hotels, how at last even those who were still concerned in agricultural work came to live in the towns and to go to and fro to their work every day, to describe how at last in all England only four towns remained, each with many millions of people, and how there were left no inhabited houses in all the countryside: to tell all this would take us far from our story of Denton and his Elizabeth. They had been separated and reunited, and still they could not marry. For Denton—it was his only fault—had no money. Neither had Elizabeth until she was twenty-one, and as yet she was only eighteen. At twenty-one all the property of her mother would come to her, for that was the custom of the time. She did not know that it was possible to anticipate her fortune, and Denton was far too delicate a lover to suggest such a thing. So things stuck hopelessly between them. Elizabeth said that she was very unhappy, and that nobody understood her but Denton, and that when she was away from him she was wretched; and Denton said that his heart longed for her day and night. And they met as often as they could to enjoy the discussion of their sorrows.

They met one day at their little seat upon the flying stage. The precise site of this meeting was where in Victorian times the road from Wimbledon came out upon the common. They were, however, five hundred feet above that point. Their seat looked far over London. To convey the appearance of it all to a nineteenth-century reader would have been difficult. One would have had to tell him to think of the Crystal Palace, of the newly built "mammoth" hotels—as those little affairs were called—of the larger railway stations of his time, and to imagine such buildings enlarged to vast proportions and run together and continuous over the whole metropolitan area. If then he was told that this continuous roof-space bore a huge forest of rotating wind-wheels, he would have begun very dimly to appreciate what to these young people was the commonest sight in their lives.

To their eyes it had something of the quality of a prison, and they were talking, as they had talked a hundred times before, of how they might escape from it and be at last happy together: escape from it, that is, before the appointed three years were at an end. It was, they both agreed, not only impossible but almost wicked, to wait three years. "Before that," said Denton—and the notes of his voice told of a splendid chest—"*we might both be dead!*"

Their vigorous young hands had to grip at this, and then Elizabeth had a still more poignant thought that brought the tears from her wholesome eyes and down her healthy cheeks. "*One* of us," she said, "*one* of us might be—"

She choked; she could not say the word that is so terrible to the young and happy.

Yet to marry and be very poor in the cities of that time was—for any one who had lived pleasantly—a very dreadful thing. In the old agricultural days that had drawn to an end in the eighteenth century there had been a pretty proverb of love in a cottage; and indeed in those days the poor of the countryside had dwelt in flower-covered, diamond-windowed cottages of thatch and plaster, with the sweet air and earth about them, amidst tangled hedges and the song of birds, and with the ever-changing sky overhead. But all this had changed (the change was already beginning in the nineteenth century), and a new sort of life was opening for the poor—in the lower quarters of the city.

In the nineteenth century the lower quarters were still beneath the sky; they were areas of land on clay or other unsuitable soil, liable to floods or exposed to the smoke of more fortunate districts, insufficiently supplied with water, and as insanitary as the great fear of infectious diseases felt by the wealthier classes permitted. In the twenty-second century, however, the growth of the city storey above storey, and the coalescence of buildings, had led to a different arrangement. The prosperous people lived in a vast series of sumptuous hotels in the upper storeys and halls of the city fabric; the industrial population dwelt beneath in the tremendous ground-floor and basement, so to speak, of the place.

AA

In the refinement of life and manners these lower classes differed little from their ancestors, the East-enders of Queer Victoria's time; but they had developed a distinct dialect of their own. In these under ways they lived and died, rarely ascending to the surface except when work took them there. Since for most of them this was the sort of life to which they had been born, they found no great misery in such circumstances; but for people like Denton and Elizabeth, such a plunge would have seemed more terrible than death.

"And yet what else is there?" asked Elizabeth.

Denton professed not to know. Apart from his own feeling of delicacy, he was not sure how Elizabeth would like the idea of borrowing on the strength of her expectations.

The passage from London to Paris even, said Elizabeth, was beyond their means; and in Paris, as in any other city in the world, life would be just as costly and impossible as in London.

Well might Denton cry aloud: "If only we had lived in those days, dearest! If only we had lived in the past!" For to their eyes even nineteenth-century Whitechapel was seen through a mist of romance.

"Is there *nothing?*" cried Elizabeth, suddenly weeping. "Must we really wait for those three long years? Fancy *three* years—six-and-thirty months!" The human capacity for patience had not grown with the ages.

Then suddenly Denton was moved to speak of something that had already flickered across his mind. He had hit upon it at last. It seemed to him so wild a suggestion that he made it only half seriously. But to put a thing into words has ever a way of making it seem more real and possible than it seemed before. And so it was with him.

"Suppose," he said, "we went into the country?"

She looked at him to see if he was serious in proposing such an adventure.

"The country?"

"Yes—beyond there. Beyond the hills."

"How could we live?" she said. *"Where* could we live?"

"It is not impossible," he said. "People used to live in the country."

"But then there were houses."

"There are the ruins of villages and towns now. On the clay lands they are gone, of course. But they are still left on the grazing land, because it does not pay the Food Company to remove them. I know that—for certain. Besides, one sees them from the flying machines, you know. Well, we might shelter in some one of these, and repair it with our hands. Do you know, the thing is not so wild as it seems. Some of the men who go out every day to look after the crops and herds might be paid to bring us food. . . ."

She stood in front of him. "How strange it would be if one really could. . . ."

"Why not?"

"But no one dares."

"That is no reason."

"It would be—oh! it would be so romantic and strange. If only it were possible."

"Why not possible?"

"There are so many things. Think of all the things we have, things that we should miss."

"Should we miss them? After all, the life we lead is very unreal—very artificial." He began to expand his idea, and as he warmed to his exposition the fantastic quality of his first proposal faded away.

She thought. "But I have heard of prowlers—escaped criminals."

He nodded. He hesitated over his answer because he thought it sounded boyish. He blushed. "I could get some one I know to make me a sword."

She looked at him with enthusiasm growing in her eyes. She had heard of swords, had seen one in a museum; she thought of those ancient days when men wore them as a common thing. His suggestion seemed an impossible dream to her, and perhaps for that reason she was eager for more detail. And inventing for the most part as he went along, he told her how they might live in the country as the old-world people had done. With every detail her interest grew, for she was one of those girls for whom romance and adventure have a fascination.

His suggestion seemed, I say, an impossible dream to her on that day, but the next day they talked about it again, and it was strangely less impossible.

"At first we should take food," said Denton. "We could carry food for ten or twelve days." It was an age of compact artificial nourishment, and such a provision had none of the unwieldy suggestion it would have had in the nineteenth century.

"But—until our house," she asked—"until it was ready, where should we sleep?"

"It is summer."

"But . . . What do you mean?"

"There was a time when there were no houses in the world; when all mankind slept always in the open air."

"But for us! The emptiness! No walls—no ceiling!"

"Dear," he said, "in London you have many beautiful ceilings. Artists paint them and stud them with lights. But I have seen a ceiling more beautiful than any in London. . . ."

"But where?"

"It is the ceiling under which we two would be alone. . . ."

"You mean . . . ?"

"Dear," he said, "it is something the world has forgotten. It is Heaven and all the host of stars."

Each time they talked the thing seemed more possible and more desirable to them. In a week or so it was quite possible. Another week, and it was the inevitable thing they had to do. A great enthusiasm for the country seized hold of them and possessed them. The sordid tumult of the town, they said, overwhelmed them. They marvelled that this simple way out of their troubles had never come upon them before.

One morning near Midsummer-day, there was a new minor official upon the flying stage, and Denton's place was to know him no more.

Our two young people had secretly married, and were

going forth manfully out of the city in which they and
their ancestors before them had lived all their days. She
wore a new dress of white cut in an old-fashioned pattern,
and he had a bundle of provisions strapped athwart his
back, and in his hand he carried—rather shamefacedly it
is true, and under his purple cloak—an implement of
archaic form, a cross-hilted thing of tempered steel.

Imagine that going forth! In their days the sprawling
suburbs of Victorian times with their vile roads, petty
houses, foolish little gardens of shrub and geranium, and
all their futile, pretentious privacies, had disappeared: the
towering buildings of the new age, the mechanical ways,
the electric and water mains, all came to an end together,
like a wall, like a cliff, near four hundred feet in height,
abrupt and sheer. All about the city spread the carrot,
swede, and turnip fields of the Food Company, vegetables
that were the basis of a thousand varied foods, and weeds
and hedgerow tangles had been utterly extirpated. The
incessant expense of weeding that went on year after year
in the petty, wasteful and barbaric farming of the ancient
days, the Food Company had economised for ever more by
a campaign of extermination. Here and there, however,
neat rows of bramble standards and apple trees with white-
washed stems, intersected the fields, and at places groups
of gigantic teazles reared their favoured spikes. Here and
there huge agricultural machines hunched under waterproof covers. The mingled waters of the Wey and Mole and
Wandle ran in rectangular channels; and wherever a gentle
elevation of the ground permitted a fountain of deodorised
sewage distributed its benefits athwart the land and made
a rainbow of the sunlight.

By a great archway in that enormous city wall emerged
the Eadhamite road to Portsmouth, swarming in the morning
sunshine with an enormous traffic bearing the blue-clad
servants of the Food Company to their toil. A rushing
traffic, beside which they seemed two scarce-moving dots.
Along the outer tracks hummed and rattled the tardy
little old-fashioned motors of such as had duties within
twenty miles or so of the city; the inner ways were filled

742 TALES OF SPACE AND TIME

with vaster mechanisms—swift monocycles bearing a score of men, lank multicycles, quadricycles sagging with heavy loads, empty gigantic produce carts that would come back again filled before the sun was setting, all with throbbing engines and noiseless wheels and a perpetual wild melody of horns and gongs.

Along the very verge of the outermost way our young people went in silence, newly wed and oddly shy of one another's company. Many were the things shouted to them as they tramped along, for in 2100 a foot-passenger on an English road was almost as strange a sight as a motor car would have been in 1800. But they went on with stead-fast eyes into the country, paying no heed to such cries.

Before them in the south rose the Downs, blue at first, and as they came nearer changing to green, surmounted by the row of gigantic wind-wheels that supplemented the wind-wheels upon the roof-spaces of the city, and broken and restless with the long morning shadows of those whirling vanes. By midday they had come so near that they could see here and there little patches of pallid dots—the sheep the Meat Department of the Food Company owned. In another hour they had passed the clay and the root crops and the single fence that hedged them in, and the prohibition against trespass no longer held: the levelled roadway plunged into a cutting with all its traffic, and they could leave it and walk over the greensward and up the open hillside.

Never had these children of the latter days been together in such a lonely place.

They were both very hungry and footsore—for walking was a rare exercise—and presently they sat down on the weedless, close-cropped grass, and looked back for the first time at the city from which they had come, shining wide and splendid in the blue haze of the valley of the Thames. Elizabeth was a little afraid of the unenclosed sheep away up the slope—she had never been near big unrestrained animals before—but Denton reassured her. And overhead a white-winged bird circled in the blue.

They talked but little until they had eaten, and then

their tongues were loosened. He spoke of the happiness that was now certainly theirs, of the folly of not breaking sooner out of that magnificent prison of latter-day life, of the old romantic days that had passed from the world for ever. And then he became boastful. He took up the sword that lay on the ground beside him, and she took it from his hand and ran a tremulous finger along the blade.

"And you could," she said, "*you*—could raise this and strike a man?"

"Why not? If there were need."

"But," she said, "it seems so horrible. It would slash. . . . There would be"—her voice sank—"*blood*."

"In the old romances you have read often enough . . ."

"Oh, I know: in those—yes. But that is different. One knows it is not blood, but just a sort of red ink. . . And *you* —killing!"

She looked at him doubtfully, and then handed him back the sword.

After they had rested and eaten, they rose up and went on their way towards the hills. They passed quite close to a huge flock of sheep, who stared and bleated at their unaccustomed figures. She had never seen sheep before, and she shivered to think such gentle things must needs be slain for food. A sheep-dog barked from a distance, and then a shepherd appeared amidst the supports of the wind-wheels, and came down towards them.

When he drew near he called out asking whither they were going.

Denton hesitated, and told him briefly that they sought some ruined house among the Downs, in which they might live together. He tried to speak in an offhand manner, as though it was a usual thing to do. The man stared incredulously.

"Have you *done* anything?" he asked.

"Nothing," said Denton. "Only we don't want to live in a city any longer. Why should we live in cities?"

The shepherd stared more incredulously than ever. "You can't live here," he said.

"We mean to try."

The shepherd stared from one to the other. "You'll go back to-morrow," he said. "It looks pleasant enough in the sunlight. . . . Are you sure you've done nothing? We shepherds are not such *great* friends of the police."

Denton looked at him steadfastly. "No," he said. "But we are too poor to live in the city, and we can't bear the thought of wearing clothes of blue canvas and doing drudgery. We are going to live a simple life here, like the people of old."

The shepherd was a bearded man with a thoughtful face. He glanced at Elizabeth's fragile beauty.

"*They* had simple minds," he said.

"So have we," said Denton.

The shepherd smiled.

"If you go along here," he said, "along the crest beneath the wind-wheels, you will see a heap of mounds and ruins on your right-hand side. That was once a town called Epsom. There are no houses there, and the bricks have been used for a sheep pen. Go on, and another heap on the edge of the root-land is Leatherhead; and then the hill turns away along the border of a valley, and there are woods of beech. Keep along the crest. You will come to quite wild places. In some parts, in spite of all the weeding that is done, ferns and bluebells and other such useless plants are growing still. And through it all underneath the wind-wheels runs a straight lane paved with stones, a roadway of the Romans two thousand years old. Go to the right of that, down into the valley and follow it along by the banks of the river. You come presently to a street of houses, many with the roofs still sound upon them. There you may find shelter."

They thanked him.

"But it's a quiet place. There is no light after dark there, and I have heard tell of robbers. It is lonely. Nothing happens there. The phonographs of the story-tellers, the kinematograph entertainments, the news machines—none of them are to be found there. If you are hungry there is no food, if you are ill no doctor. . . ." He stopped.

"We shall try it," said Denton, moving to go on. Then

a thought struck him, and he made an agreement with the shepherd, and learnt where they might find him, to buy and bring them anything of which they stood in need, out of the city.

And in the evening they came to the deserted village, with its houses that seemed so small and odd to them: they found it golden in the glory of the sunset, and desolate and still. They went from one deserted house to another, marvelling at their quaint simplicity, and debating which they should choose. And at last, in a sunlit corner of a room that had lost its outer wall, they came upon a wild flower, a little flower of blue that the weeders of the Food Company had overlooked.

That house they decided upon; but they did not remain in it long that night, because they were resolved to feast upon nature. And moreover the houses became very gaunt and shadowy after the sunlight had faded out of the sky. So after they had rested a little time they went to the crest of the hill again to see with their own eyes the silence of heaven set with stars, about which the old poets had had so many things to tell. It was a wonderful sight, and Denton talked like the stars, and when they went down the hill at last the sky was pale with dawn. They slept but little, and in the morning when they woke a thrush was singing in a tree.

So these young people of the twenty-second century began their exile. That morning they were very busy exploring the resources of this new home in which they were going to live the simple life. They did not explore very fast or very far, because they went everywhere hand-in-hand; but they found the beginnings of some furniture. Beyond the village was a store of winter fodder for the sheep of the Food Company, and Denton dragged great armfuls to the house to make a bed; and in several of the houses were old fungus-eaten chairs and tables—rough, barbaric, clumsy furniture, it seemed to them, and made of wood. They repeated many of the things they had said on the previous day, and towards evening they found another flower, a harebell. In the late afternoon some Company shepherds went down the river valley riding on a big multicycle; but they hid from them, because their presence, Elizabeth said, seemed to spoil the romance of this old-world place altogether.

In this fashion they lived a week. For all that week the days were cloudless, and the nights, nights of starry glory, that were invaded each a little more by a crescent moon.

Yet something of the first splendour of their coming faded—faded imperceptibly day after day; Denton's eloquence became fitful, and lacked fresh topics of inspiration; the fatigue of their long march from London told in a certain stiffness of the limbs, and each suffered from a slight unaccountable cold. Moreover, Denton became aware of unoccupied time. In one place among the carelessly heaped lumber of the old times he found a rust-eaten spade, and with this he made a fitful attack on the razed and grass-grown garden—though he had nothing to plant or sow. He returned to Elizabeth with a sweat-streaming face, after half an hour of such work.

"There were giants in those days," he said, not understanding what wont and training will do. And their walk that day led them along the hills until they could see the city shimmering far away in the valley. "I wonder how things are going on there," he said.

And then came a change in the weather. "Come out and see the clouds," she cried; and behold! they were a sombre purple in the north and east, streaming up to ragged edges at the zenith. And as they went up the hill these hurrying streamers blotted out the sunset. Suddenly the wind set the beech-trees swaying and whispering, and Elizabeth shivered. And then far away the lightning flashed, flashed like a sword that is drawn suddenly, and the distant thunder marched about the sky, and even as they stood astonished, pattering upon them came the first headlong raindrops of the storm. In an instant the last streak of sunset was hidden by a falling curtain of hail, and the lightning flashed again, and the voice of the thunder roared louder, and all about them the world scowled dark and strange.

Seizing hands, these children of the city ran down the hill to their home, in infinite astonishment. And ere they reached it, Elizabeth was weeping with dismay, and the darkling ground about them was white and brittle and active with the pelting hail.

Then began a strange and terrible night for them. For the first time in their civilised lives they were in absolute darkness; they were wet and cold and shivering, all about them hissed the hail, and through the long neglected ceilings of the derelict home came noisy spouts of water and formed pools and rivulets on the creaking floors. As the gusts of the storm struck the worn-out building, it groaned and shuddered, and now a mass of plaster from the wall would slide and smash, and now some loosened tile would rattle down the roof and crash into the empty greenhouse below. Elizabeth shuddered, and was still; Denton wrapped his gay and flimsy city cloak about her, and so they crouched in the darkness. And ever the thunder broke louder and nearer, and ever more lurid flashed the lightning, jerking into a momentary gaunt clearness the steaming, dripping room in which they sheltered.

Never before had they been in the open air save when the sun was shining. All their time had been spent in the warm and airy ways and halls and rooms of the latter-day city. It was to them that night as if they were in some other world, some disordered chaos of stress and tumult, and almost beyond hoping that they should ever see the city ways again.

The storm seemed to last interminably, until at last they dozed between the thunderclaps, and then very swiftly it fell and ceased. And as the last patter of the rain died away they heard an unfamiliar sound.

"What is that?" cried Elizabeth.

It came again. It was the barking of dogs. It drove down the desert lane and passed; and through the window, whitening the wall before them and throwing upon it the shadow of the window-frame and of a tree in black silhouette, shone the light of the waxing moon.

Just as the pale dawn was drawing the things about them into sight, the fitful barking of dogs came near again, and stopped. They listened. After a pause they heard the quick pattering of feet seeking round the house, and short, half-smothered barks. Then again everything was still.

"Ssh!" whispered Elizabeth, and pointed to the door of their room.

Denton went half-way towards the door, and stood listening. He came back with a face of affected unconcern. "They must be the sheep-dogs of the Food Company," he said. "They will do us no harm."

He sat down again beside her. "What a night it has been!" he said, to hide how keenly he was listening.

"I don't like dogs," answered Elizabeth, after a long silence.

"Dogs never hurt any one," said Denton. "In the old days—in the nineteenth century—everybody had a dog."

"There was a romance I heard once. A dog killed a man."

"Not this sort of dog," said Denton confidently. "Some of those romances—are exaggerated."

Suddenly a half bark and a pattering up the staircase; the sound of panting. Denton sprang to his feet and drew the sword out of the damp straw upon which they had been lying. Then in the doorway appeared a gaunt sheep-dog, and halted there. Behind it stared another. For an instant man and brute faced each other, hesitating.

Then Denton, being ignorant of dogs, made a sharp step forward. "Go away," he said, with a clumsy motion of his sword.

The dog started and growled. Denton stopped sharply. "Good dog!" he said.

The growling jerked into a bark.

"Good dog!" said Denton. The second dog growled and barked. A third out of sight down the staircase took up the barking also. Outside others gave tongue—a large number it seemed to Denton.

"This is annoying," said Denton, without taking his eye off the brutes before him. "Of course the shepherds won't come out of the city for hours yet. Naturally these dogs don't quite make us out."

"I can't hear," shouted Elizabeth. She stood up and came to him.

Denton tried again, but the barking still drowned his voice. The sound had a curious effect upon his blood. Odd disused emotions began to stir; his face changed as he shouted. He tried again; the barking seemed to mock him, and one dog danced a pace forward, bristling. Suddenly he turned, and uttering certain words in the dialect of the underways, words incomprehensible to Elizabeth, he made for the dogs. There was a sudden cessation of the barking, a growl and a snapping. Elizabeth saw the snarling head of the foremost dog, its white teeth and retracted ears, and the flash of the thrust blade. The brute leapt into the air and was flung back.

Then Denton, with a shout, was driving the dogs before him. The sword flashed above his head with a sudden new freedom of gesture, and then he vanished down the staircase. She made six steps to follow him, and on the landing there was blood. She stopped, and hearing the tumult of dogs and Denton's shouts pass out of the house, ran to the window.

Nine wolfish sheep-dogs were scattering, one writhed before the porch; and Denton, tasting that strange delight of combat that slumbers still in the blood of even the most civilised man, was shouting and running across the garden space. And then she saw something that for a moment he did not see. The dogs circled round this way and that, and came again. They had him in the open.

In an instant she divined the situation. She would have called to him. For a moment she felt sick and helpless, and then, obeying a strange impulse, she gathered up her white skirt and ran downstairs. In the hall was the rusting spade. That was it! She seized it and ran out.

She came none too soon. One dog rolled before him, well-nigh slashed in half; but a second had him by the thigh, a third gripped his collar behind, and a fourth had the blade of the sword between his teeth, tasting its own blood. He parried the leap of a fifth with his left arm.

It might have been the first century instead of the twenty-second, so far as she was concerned. All the gentleness of her eighteen years of city life vanished before this primordial need. The spade smote hard and sure, and cleft a dog's skull. Another, crouching for a spring, yelped with dismay at this unexpected antagonist, and rushed aside. Two wasted precious moments on the binding of a feminine skirt.

The collar of Denton's cloak tore and parted as he staggered back; and that dog too felt the spade, and ceased to trouble him. He sheathed his sword in the brute at his thigh.

"To the wall!" cried Elizabeth; and in three seconds the fight was at an end, and our young people stood side by side, while a remnant of five dogs, with ears and tails of disaster, fled shamefully from the stricken field.

For a moment they stood panting and victorious, and then Elizabeth, dropping her spade, covered her face, and and sank to the ground in a paroxysm of weeping. Denton looked about him, thrust the point of his sword into the ground so that it was at hand, and stooped to comfort her.

At last their more tumultuous emotions subsided, and they could talk again. She leant upon the wall, and he sat upon it so that he could keep an eye open for any returning dogs. Two, at any rate, were up on the hillside and keeping up a vexatious barking.

She was tear-stained, but not very wretched now, because for half an hour he had been repeating that she was brave and had saved his life. But a new fear was growing in her mind.

"They are the dogs of the Food Company," she said. "There will be trouble."

"I am afraid so. Very likely they will prosecute us for trespass."

A pause.

"In the old times," he said, "this sort of thing happened day after day."

"Last night!" she said. "I could not live through another such night."

He looked at her. Her face was pale for want of sleep, and drawn and haggard. He came to a sudden resolution. "We must go back," he said.

She looked at the dead dogs, and shivered. "We cannot stay here," she said.

"We must go back," he repeated, glancing over his shoulder to see if the enemy kept their distance. "We have been happy for a time. . . . But the world is too civilised. Ours is the age of cities. More of this will kill us."

"But what are we to do? How can we live there?"

Denton hesitated. His heel kicked against the wall on which he sat. "It's a thing I haven't mentioned before," he said, and coughed; "but . . ."

"Yes?"

"You could raise money on your expectations," he said.

"Could I?" she said eagerly.

"Of course you could. What a child you are!"

She stood up, and her face was bright. "Why did you not tell me before?" she asked. "And all this time we have been here!"

He looked at her for a moment, and smiled. Then the smile vanished. "I thought it ought to come from you," he said. "I didn't like to ask for your money. And besides —at first I thought this would be rather fine."

There was a pause.

"It *has* been fine," he said; and glanced once more over his shoulder. "Until all this began."

"Yes," she said, "those first days. The first three days."

They looked for a space into one another's faces, and then Denton slid down from the wall and took her hand.

"To each generation," he said, "the life of its time. I see it all plainly now. In the city—that is the life to which we were born. To live in any other fashion . . . Coming here was a dream, and this—is the awakening."

"It was a pleasant dream," she said,—"in the beginning."

For a long space neither spoke.

"If we would reach the city before the shepherds come here, we must start," said Denton. "We must get our food out of the house and eat as we go."

Denton glanced about him again, and, giving the dead dogs a wide berth, they walked across the garden space and into the house together. They found the wallet with their food, and descended the blood-stained stairs again. In the hall Elizabeth stopped. "One minute," she said. "There is something here."

She led the way into the room in which that one little blue flower was blooming. She stooped to it, she touched it with her hand.

"I want it," she said; and then, "I cannot take it. . . ."

Impulsively she stooped and kissed its petals.

Then silently, side by side, they went across the empty garden-space into the old high road, and set their faces resolutely towards the distant city—towards the complex mechanical city of those latter days, the city that had swallowed up mankind.

III.—The Ways of the City

Prominent if not paramount among world-changing inventions in the history of man is that series of contrivances in locomotion that began with the railway and ended for a century or more with the motor and the patent road. That these contrivances, together with the device of limited liability joint stock companies and the supersession of agricultural labourers by skilled men with ingenious machinery, would necessarily concentrate mankind in cities of unparalleled magnitude and work an entire revolution in human life, became, after the event, a thing so obvious that it is a matter of astonishment it was not more clearly anticipated. Yet that any steps should be taken to anticipate the miseries such a revolution might entail does not appear even to have been suggested; and the idea that the moral prohibitions and sanctions, the privileges and concessions, the conception of property and responsibility, of comfort and beauty, that had rendered the mainly agricultural states of the past prosperous and happy, would fail in the rising torrent of novel opportunities and novel stimulations, never seems to have entered the nineteenth-century mind. That a citizen, kindly and fair in his ordinary life, could as a shareholder become almost murderously greedy; that commercial methods that were reasonable and honourable on the old-fashioned countryside, should on an enlarged scale be deadly and overwhelming; that ancient charity was modern pauperisation, and ancient employment modern sweating; that, in fact, a revision and enlargement of the duties and rights of man had become urgently necessary, were things it could not entertain, nourished as it was on an archaic system of education and profoundly retrospective and legal in all its habits of thought. It was known that the accumulation of men in cities involved unprecedented dangers of pestilence; there was an energetic development of sanitation; but that the diseases of gambling and usury, of luxury and tyranny should become endemic, and produce horrible consequences was beyond the scope of nineteenth-century thought. And so, as if it were some inorganic process, practically unhindered by the creative will of man, the growth of the swarming unhappy cities that mark the twenty-first century accomplished itself.

The new society was divided into three main classes. At the summit slumbered the property owner, enormously rich by accident rather than design, potent save for the will and aim, the last *avatar* of Hamlet in the world. Below was the enormous multitude of workers employed by the gigantic companies that monopolised control; and between these two the dwindling middle class, officials of innumerable sorts, foremen, managers, the medical, legal, artistic, and scholastic classes, and the minor rich, a middle class whose members led a life of insecure luxury and precarious speculation amidst the movements of the great managers.

Already the love story and the marrying of two persons of this middle class have been told: how they overcame the obstacles between them, and how they tried the simple old-fashioned way of living on the countryside and came back speedily enough into the city of London. Denton had no means, so Elizabeth borrowed money on the securities that her father Mwres held in trust for her until she was one-and-twenty.

The rate of interest she paid was of course high, because of the uncertainty of her security, and the arithmetic of lovers is often sketchy and optimistic. Yet they had very glorious times after that return. They determined they would not go to a Pleasure city nor waste their days rushing through the air from one part of the world to the other, for in spite of one disillusionment, their tastes were still old-fashioned. They furnished their little room with quaint old Victorian furniture, and found a shop on the forty-second floor in Seventh Way where printed books of the old sort were still to be bought. It was their pet affectation to read print instead of hearing phonographs. And when presently there came a sweet little girl, to unite them further if it were possible, Elizabeth would not send it to a *Creche*, as the custom was, but insisted on nursing it at home. The rent of their apartments was raised on account of this singular proceeding, but that they did not mind. It only meant borrowing a little more.

Presently Elizabeth was of age, and Denton had a business interview with her father that was not agreeable. An exceedingly disagreeable interview with their money-lender followed, from which he brought home a white face. On his return Elizabeth had to tell him of a new and marvellous intonation of "Goo" that their daughter had devised, but Denton was inattentive. In the midst, just as she was at the cream of her description, he interrupted. "How much money do you think we have left, now that everything is settled?"

She stared and stopped her appreciative swaying of the Goo genius that had accompanied her description.

"You don't mean . . . ?"

"Yes," he answered. "Ever so much. We have been wild. It's the interest. Or something. And the shares you had, slumped. Your father did not mind. Said it was not his business, after what had happened. He's going to marry again. . . . Well—we have scarcely a thousand left!"

"Only a thousand?"

"Only a thousand."

And Elizabeth sat down. For a moment she regarded him with a white face, then her eyes went about the quaint, old-fashioned room, with its middle Victorian furniture and genuine oleographs, and rested at last on the little lump of humanity within her arms.

Denton glanced at her and stood downcast. Then he swung round on his heel and walked up and down very rapidly.

"I must get something to do," he broke out presently. "I am an idle scoundrel. I ought to have thought of this before. I have been a selfish fool. I wanted to be with you all day . . ."

He stopped, looking at her white face. Suddenly he came and kissed her and the little face that nestled against her breast.

"It's all right, dear," he said, standing over her; "you won't be lonely now—now Dings is beginning to talk to you. And I can soon get something to do, you know. Soon . . . Easily . . . It's only a shock at first. But it will come all right. It's sure to come right. I will go out again as soon as I have rested, and find what can be done. For the present it's hard to think of anything . . ."

"It would be hard to leave these rooms," said Elizabeth; "but—"

"There won't be any need of that—trust me."

"They are expensive."

Denton waved that aside. He began talking of the work he could do. He was not very explicit what it would be; but he was quite sure that there was something to keep them comfortably in the happy middle class, whose way of life was the only one they knew.

"There are three-and-thirty million people in London," he said; "some of them *must* have need of me."

"Some *must.*"

"The trouble is . . . Well—Bindon, that brown little old

man your father wanted you to marry. He's an important person. . . . I can't go back to my flying-stage work, because he is now a Commissioner of the Flying Stage Clerks."

"I didn't know that," said Elizabeth.

"He was made that in the last few weeks . . . or things would be easy enough, for they liked me on the flying stage. But there's dozens of other things to be done—dozens. Don't you worry, dear. I'll rest a little while, and then we'll dine, and then I'll start on my rounds. I know lots of people—lots."

So they rested, and then they went to the public dining-room and dined, and then he started on his search for employment. But they soon realised that in the matter of one convenience the world was just as badly off as it had ever been, and that was a nice, secure, honourable, remunerative employment, leaving ample leisure for the private life, and demanding no special ability, no violent exertion nor risk, and no sacrifice of any sort for its attainment. He evolved a number of brilliant projects, and spent many days hurrying from one part of the enormous city to another in search of influential friends; and all his influential friends were glad to see him, and very sanguine until it came to definite proposals, and then they became guarded and vague. He would part with them coldly, and think over their behaviour, and get irritated on his way back, and stop at some telephone office and spend money on an animated but unprofitable quarrel. And as the days passed, he got so worried and irritated that even to seem kind and careless before Elizabeth cost him an effort—as she, being a loving woman, perceived very clearly.

After an extremely complex preface one day, she helped him out with a painful suggestion. He had expected her to weep and give way to despair when it came to selling all their joyfully bought early Victorian treasures, their quaint objects of art, their antimacassars, bead mats, repp curtains, veneered furniture, gold-framed steel engravings and pencil drawings, wax flowers under shades, stuffed birds, and all sorts of choice old things; but it was she who made the proposal. The sacrifice seemed to fill her with pleasure, and so did the idea of shifting to apartments ten or twelve floors lower in another hotel. "So long as Dings is with us, nothing matters," she said. "It's all experience." So he kissed her, said she was braver than when she fought the sheep-dogs, called her Boadicea, and abstained very carefully from reminding her that they would have to pay a considerably higher rent on account of the little voice with which Dings greeted the perpetual uproar of the city.

His idea had been to get Elizabeth out of the way when it came to selling the absurd furniture about which their affections were twined and tangled; but when it came to the sale it was Elizabeth who haggled with the dealer while Denton went about the running ways of the city, white and sick with sorrow and the fear of what was still to come. When they moved into their sparsely furnished pink-and-white apartments in a cheap hotel, there came an outbreak of furious energy on his part, and then nearly a week of lethargy during which he sulked at home. Through those days Elizabeth shone like a star, and at the end Denton's misery found a vent in tears. And then he went out into the city ways again, and—to his utter amazement—found some work to do.

His standard of employment had fallen steadily until at last it had reached the lowest level of independent workers. At first he had aspired to some high official position in the great Flying or Windvane or Water Companies, or to an appointment on one of the General Intelligence Orgonisations that had replaced newspapers, or to some professional partnership, but those were the dreams of the beginning. From that he had passed to speculation, and three hundred gold "lions" out of Elizabeth's thousand had vanished one evening in the share market. Now he was glad his good looks secured him a trial in the position of salesman to the Suzannah Hat Syndicate, a Syndicate dealing in ladies' caps, hair decorations, and hats—for though the city was completely covered in, ladies still wore extremely elaborate and beautiful hats at the theatres and places of public worship.

It would have been amusing if one could have confronted

a Regent Street shopkeeper of the nineteenth century with the development of his establishment in which Denton's duties lay. Nineteenth Way was still sometimes called Regent Street, but it was now a street of moving platforms and nearly eight hundred feet wide. The middle space was immovable and gave access by staircases descending into subterranean ways to the houses on either side. Right and left were an ascending series of continuous platforms each of which travelled about five miles an hour faster than the one internal to it, so that one could step from platform to platform until one reached the swiftest outer way and so go about the city. The establishment of the Suzannah Hat Syndicate projected a vast *façade* upon the outer way, sending out overhead at either end an overlapping series of huge white glass screens, on which gigantic animated pictures of the faces of well-known beautiful living women wearing novelties in hats were thrown. A dense crowd was always collected in the stationary central way watching a vast kinematograph which displayed the changing fashion. The whole front of the building was in perpetual chromatic change, and all down the *façade*—four hundred feet it measured—and all across the street of moving ways, laced and winked and glittered in a thousand varieties of colour and lettering the inscription—

SUZANNA! 'ETS! SUZANNA! 'ETS!

A broadside of gigantic phonographs drowned all conversation in the moving way and roared *"hats"* at the passer-by, while far down the street and up, other batteries counselled the public to "walk down for Suzannah," and queried, "Why *don't* you buy the girl a hat?"

For the benefit of those who chanced to be deaf—and deafness was not uncommon in the London of that age, inscriptions of all sizes were thrown from the roof above upon the moving platforms themselves, and on one's hand or on the bald head of the man before one, or on a lady's shoulders, or in a sudden jet of flame before one's feet, the moving finger wrote in unanticipated letters of fire "'ets r chip t'de," or simply "'ets." And spite of all these efforts so high was the pitch at which the city lived, so trained became one's eyes and ears to ignore all sorts of advertisement, that many a citizen had passed that place thousands of times and was still unaware of the existence of the Suzannah Hat Syndicate.

To enter the building one descended the staircase in the middle way and walked through a public passage in which pretty girls promenaded, girls who were willing to wear a ticketed hat for a small fee. The entrance chamber was a large hall in which wax heads fashionably adorned rotated gracefully upon pedestals, and from this one passed through a cash office to an interminable series of little rooms, each room with its salesman, its three or four hats and pins, its mirrors, its kinematographs, telephones and hat slides in communication with the central depôt, its comfortable lounge and tempting refreshments. A salesman in such an apartment did Denton now become. It was his business to attend to any of the incessant stream of ladies who chose to stop with him, to behave as winningly as possible, to offer refreshment, to converse on any topic the possible customer chose, and to guide the conversation dexterously but not insistently towards hats. He was to suggest trying on various types of hat and to show by his manner and bearing, but without any coarse flattery, the enhanced impression made by the hats he wished to sell. He had several mirrors, adapted by various subtleties of curvature and tint to different types of face and complexion, and much depended on the proper use of these.

Denton flung himself at these curious and not very congenial duties with a good will and energy that would have amazed him a year before; but all to no purpose. The Senior Manageress, who had selected him for appointment and conferred various small marks of favour upon him, suddenly changed in her manner, declared for no assignable cause that he was stupid, and dismissed him at the end of six weeks of salesmanship. So Denton had to resume his ineffectual search for employment.

This second search did not last very long. Their money

was at the ebb. To eke it out a little longer they resolved
to part with their darling Dings, and took that small person
to one of the public *creches* that abounded in the city.
That was the common use of the time. The industrial
emancipation of women, the correlated disorganisation of
the secluded "home," had rendered *creches* a necessity for
all but very rich and exceptionally-minded people. Therein
children encountered hygienic and educational advantages
impossible without such organisation. *Creches* were of all
classes and types of luxury, down to those of the Labour
Company, where children were taken on credit, to be re-
deemed in labour as they grew up.

But both Denton and Elizabeth being, as I have explained,
strange old-fashioned young people, full of nineteenth-
century ideas, hated these convenient *creches* exceedingly
and at last took their little daughter to one with extreme
reluctance. They were received by a motherly person in
a uniform who was very brisk and prompt in her manner
until Elizabeth wept at the mention of parting from her
child. The motherly person, after a brief astonishment
at this unusual emotion, changed suddenly into a creature
of hope and comfort, and so won Elizabeth's gratitude for
life. They were conducted into a vast room presided over
by several nurses and with hundreds of two-year-old girls
grouped about the toy-covered floor. This was the Two-
year-old Room. Two nurses came forward, and Elizabeth
watched their bearing towards Dings with jealous eyes.
They were kind—it was clear they felt kind, and yet . . .

Presently it was time to go. By that time Dings was
happily established in a corner, sitting on the floor with
her arms filled, and herself, indeed, for the most part
hidden by an unaccustomed wealth of toys. She seemed
careless of all human relationships as her parents receded.

They were forbidden to upset her by saying good-bye.

At the door Elizabeth glanced back for the last time, and
behold! Dings had dropped her new wealth and was standing
with a dubious face. Suddenly Elizabeth gasped, and the
motherly nurse pushed her forward and closed the door.

"You can come again soon, dear," she said, with unexpected tenderness in her eyes. For a moment Elizabeth stared at her with a blank face. "You can come again soon," repeated the nurse. Then with a swift transition Elizabeth was weeping in the nurse's arms. So it was that Denton's heart was won also.

And three weeks after our young people were absolutely penniless, and only one way lay open. They must go to the Labour Company. So soon as the rent was a week overdue their few remaining possessions were seized, and with scant courtesy they were shown the way out of the hotel. Elizabeth walked along the passage towards the staircase that ascended to the motionless middle way, too dulled by misery to think. Denton stopped behind to finish a stinging and unsatisfactory argument with the hotel porter, and then came hurrying after her, flushed and hot. He slackened his pace as he overtook her, and together they ascended to the middle way in silence. There they found two seats vacant and sat down.

"We need not go there—*yet?*" said Elizabeth.

"No—not till we are hungry," said Denton.

They said no more.

Elizabeth's eyes sought a resting-place and found none. To the right roared the eastward ways, to the left the ways in the opposite direction, swarming with people. Backwards and forwards along a cable overhead rushed a string of gesticulating men, dressed like clowns, each marked on back and chest with one gigantic letter, so that altogether they spelt out:

"Purkinje's Digestive Pills."

An anæmic little woman in horrible coarse blue canvas pointed a little girl to one of this string of hurrying advertisements.

"Look!" said the anæmic woman: "there's yer father."

"Which?" said the little girl.

"'Im wiv his nose coloured red," said the anæmic woman.

The little girl began to cry, and Elizabeth could have cried too.

"Ain't 'e kickin' 'is legs!—*just!*" said the anæmic woman in blue, trying to make things bright again. "Looky—*now!*"

On the *façade* to the right a huge intensely bright disc of weird colour span incessantly, and letters of fire that came and went spelt out—

"DOES THIS MAKE YOU GIDDY?"

Then a pause, followed by

"TAKE A PURKINJE'S DIGESTIVE PILL."

A vast and desolating braying began. "If you love Swagger Literature, put your telephone on to Bruggles, the Greatest Author of all Time. The Greatest Thinker of all Time. Teaches you Morals up to your Scalp! The very image of Socrates, except the back of his head, which is like Shakespeare. He has six toes, dresses in red, and never cleans his teeth. Hear HIM!"

Denton's voice became audible in a gap in the uproar. "I never ought to have married you," he was saying. "I have wasted your money, ruined you, brought you to misery. I am a scoundrel . . . Oh, this accursed world!"

She tried to speak, and for some moments could not. She grasped his hand. "No," she said at last.

A half-formed desire suddenly became determination. She stood up. "Will you come?"

He rose also. "We need not go there yet."

"Not that. But I want you to come to the flying stages—where we met. You know? The little seat."

He hesitated. "*Can* you?" he said, doubtfully.

"Must," she answered.

He hesitated still for a moment, then moved to obey her will.

And so it was they spent their last half-day of freedom out under the open air in the little seat under the flying stages where they had been wont to meet five short years ago. There she told him, what she could not tell him in the tumultuous public ways, that she did not repent even now of their marriage—that whatever discomfort and misery life still had for them, she was content with the things that had been. The weather was kind to them, the seat was sunlit and warm, and overhead the shining aeroplanes went and came.

At last towards sunsetting their time was at an end, and they made their vows to one another and clasped hands, and then rose up and went back into the ways of the city, a shabby-looking, heavy-hearted pair, tired and hungry. Soon they came to one of the pale blue signs that marked a Labour Company Bureau. For a space they stood in the middle way regarding this and at last descended, and entered the waiting-room.

The Labour Company had originally been a charitable organisation; its aim was to supply food, shelter, and work to all comers. This it was bound to do by the conditions of its incorporation, and it was also bound to supply food and shelter and medical attendance to all incapable of work who chose to demand its aid. In exchange these incapables paid labour notes, which they had to redeem upon recovery. They signed these labour notes with thumb-marks, which were photographed and indexed in such a way that this world-wide Labour Company could identify any one of its two or three hundred million clients at the cost of an hour's inquiry. The day's labour was defined as two spells in a treadmill used in generating electrical force, or its equivalent, and its due performance could be enforced by law. In practice the Labour Company found it advisable to add to its statutory obligations of food and shelter a few pence a day as an inducement to effort; and its enterprise had not only abolished pauperisation altogether, but supplied practically all but the very highest and most responsible labour throughout the world. Nearly a third of the population of the world were its serfs and debtors from the cradle to the grave.

In this practical, unsentimental way the problem of the unemployed had been most satisfactorily met and overcome. No one starved in the public ways, and no rags, no costume less sanitary and sufficient than the Labour Company's hygienic but inelegant blue canvas, pained the eye throughout the whole world. It was the constant theme of the phonographic newspapers how much the world had progressed since nineteenth-century days, when the bodies of those killed by the vehicular traffic or dead of starvation, were, they alleged, a common feature in all the busier streets.

Denton and Elizabeth sat apart in the waiting-room until their turn came. Most of the others collected there seemed limp and taciturn, but three or four young people gaudily dressed made up for the quietude of their companions. They were life clients of the Company, born in the Company's *creche* and destined to die in its hospital, and they had been out for a spree with some shillings or so of extra pay. They talked vociferously in a later development of the Cockney dialect, manifestly very proud of themselves.

Elizabeth's eyes went from these to the less assertive figures. One seemed exceptionally pitiful to her. It was a woman of perhaps forty-five, with gold-stained hair and a painted face, down which abundant tears had trickled; she had a pinched nose, hungry eyes, lean hands and shoulders, and her dusty worn-out finery told the story of her life. Another was a grey-bearded old man in the costume of a bishop of one of the high episcopal sects—for religion was now also a business, and had its ups and downs. And besides him a sickly, dissipated-looking boy of perhaps two-and-twenty glared at Fate.

Presently Elizabeth and then Denton interviewed the manageress—for the Company preferred women in this capacity—and found she possessed an energetic face, a contemptuous manner, and a particularly unpleasant voice. They were given various cheques, including one to certify that they need not have their heads cropped; and when they had given their thumb-marks, learnt the number corresponding thereunto, and exchanged their shabby middle-class clothes for duly numbered blue canvas suits, they repaired to the huge plain dining-room for their first meal under these new conditions. Afterwards they were to return to her for instructions about their work.

When they had made the exchange of their clothing Elizabeth did not seem able to look at Denton at first; but he looked at her, and saw with astonishment that even in blue canvas she was still beautiful. And then their soup and bread came sliding on its little rail down the long table towards them and stopped with a jerk, and he forgot the matter. For they had had no proper meal for three days.

After they had dined they rested for a time. Neither talked—there was nothing to say; and presently they got up and went back to the manageress to learn what they had to do.

The manageress referred to a tablet. "Y'r rooms won't be here; it'll be in the Highbury Ward, ninety-seventh way, number two thousand and seventeen. Better make a note of it on y'r card. *You,* nought nought nought, type seven, sixty-four, b.c.d., *gamma* forty-one, female; you 'ave to go to the Metal-beating Company, and try that for a day— fourpence bonus if ye're satisfactory; and *you,* nought seven one, type four, seven hundred and nine, g.f.b., *pi* five and ninety, male; you 'ave to go to the Photographic Company on Eighty-first way, and learn something or other—*I* don't know—thrippence. 'Ere's y'r cards. That's all. Next! *What?* Didn't catch it all? Lor! So suppose I must go over it all again. Why don't you listen? Keerless, unprovident people! One'd think these things didn't matter."

Their ways to their work lay together for a time. And now they found they could talk. Curiously enough, the worst of their depression seemed over now that they had actually donned the blue. Denton could talk with interest even of the work that lay before them. "Whatever it is," he said, "it can't be so hateful as that hat shop. And after we have paid for Dings, we shall still have a whole penny a day between us even now. Afterwards—we may improve, —get more money."

Elizabeth was less inclined to speech. "I wonder why work should seem so hateful," she said.

"It's odd," said Denton. "I suppose it wouldn't be if it were not the thought of being ordered about. . . . I hope we shall have decent managers."

Elizabeth did not answer. She was not thinking of that. She was tracing out some thoughts of her own.

"Of course," she said presently, "we have been using up work all our lives. It's only fair—"

She stopped. It was too intricate.

"We paid for it," said Denton, for at that time he had not troubled himself about these complicated things.

"We did nothing—and yet we paid for it. That's what I cannot understand."

"Perhaps we are paying," said Elizabeth presently—for her theology was old-fashioned and simple.

Presently it was time for them to part, and each went to the appointed work. Denton's was to mind a complicated hydraulic press that seemed almost an intelligent thing. This press worked by the sea-water that was destined finally to flush the city drains—for the world had long since abandoned the folly of pouring drinkable water into its sewers. This water was brought close to the eastward edge of the city by a huge canal, and then raised by an enormous battery of pumps into reservoirs at a level of four hundred feet above the sea, from which it spread by a billion arterial branches over the city. Thence it poured down, cleansing, sluicing, working machinery of all sorts, through an infinite variety of capillary channels into the great drains, the *cloacae maximae*, and so carried the sewage out to the agricultural areas that surrounded London on every side.

The press was employed in one of the processes of the photographic manufacture, but the nature of the process it did not concern Denton to understand. The most salient fact to his mind was that it had to be conducted in ruby light, and as a consequence the room in which he worked was lit by one coloured globe that poured a lurid and painful illumination about the room. In the darkest corner stood the press whose servant Denton had now become; it was a huge, dim, glittering thing with a projecting hood that had a remote resemblance to a bowed head, and, squatting like some metal Buddha in this weird light that ministered to its needs, it seemed to Denton in certain

moods almost as if this must needs be the obscure idol
to which humanity in some strange aberration had offered
up his life. His duties had a varied monotony. Such items
as the following will convey an idea of the service of the
press. The thing worked with a busy clicking so long as
things went well; but if the paste that came pouring through
a feeder from another room and which it was perpetually
compressing into thin plates, changed in quality the rhythm
of its click altered and Denton hastened to make certain
adjustments. The slightest delay involved a waste of paste
and the docking of one or more of his daily pence. If the
supply of paste waned—there were hand processes of a
peculiar sort involved in its preparation, and sometimes the
workers had convulsions which deranged their output—
Denton had to throw the press out of gear. In the painful
vigilance a multitude of such trivial attentions entailed,
painful because of the incessant effort its absence of natural
interest required, Denton had now to pass one-third of his
days. Save for an occasional visit from the manager, a
kindly but singularly foul-mouthed man, Denton passed his
working hours in solitude.

Elizabeth's work was of a more social sort. There was
a fashion for covering the private apartments of the very
wealthy with metal plates beautifully embossed with repeated
patterns. The taste of the time demanded, however, that
the repetition of the patterns should not be exact—not
mechanical, but "natural"—and it was found that the most
pleasing arrangement of pattern irregularity was obtained
by employing women of refinement and natural taste to
punch out the patterns with small dites. So many square
feet of plates was exacted from Elizabeth as a minimum, and
for whatever square feet she did in excess she received a
small payment. The room, like most rooms of women
workers, was under a manageress: men had been found
by the Labour Company not only less exacting but extremely
liable to excuse favoured ladies from a proper share of their
duties. The manageress was a not unkindly, taciturn person,
with the hardened remains of beauty of the brunette type;
and the other women workers, who of course hated her,

associated her name scandalously with one of the metal-work directors in order to explain her position.

Only two or three of Elizabeth's fellow-workers were born labour serfs; plain, morose girls, but most of them corresponded to what the nineteenth century would have called a "reduced" gentlewoman. But the ideal of what constituted a gentlewoman had altered: the faint, faded, negative virtue, the modulated voice and restrained gesture of the old-fashioned gentlewoman had vanished from the earth. Most of her companions showed in discoloured hair, ruined complexions, and the texture of their reminiscent conversations, the vanished glories of a conquering youth. All of these artistic workers were much older than Elizabeth, and two openly expressed their surprise that any one so young and pleasant should come to share their toil. But Elizabeth did not trouble them with her old-world moral conceptions.

They were permitted, and even encouraged to converse with each other, for the directors very properly judged that anything that conduced to variations of mood made for pleasing fluctuations in their patterning; and Elizabeth was almost forced to hear the stories of these lives with which her own interwove: garbled and distorted they were by vanity indeed and yet comprehensible enough. And soon she began to appreciate the small spites and cliques, the little misunderstandings and alliances that enmeshed about her. One woman was excessively garrulous and descriptive about a wonderful son of hers; another had cultivated a foolish coarseness of speech, that she seemed to regard as the wittiest expression of originality conceivable; a third mused for ever on dress, and whispered to Elizabeth how she saved her pence day after day, and would presently have a glorious day of freedom, wearing . . . and then followed hours of description; two others sat always together, and called one another pet names, until one day some little thing happened, and they sat apart, blind and deaf as it seemed to one another's being. And always from them all came an incessant tap, tap, tap, tap, and the manageress listened always to the rhythm to mark if one fell away.

Tap, tap, tap, tap: so their days passed, so their lives must pass. Elizabeth sat among them, kindly and quiet, gray-hearted, marvelling at Fate: tap, tap, tap; tap, tap, tap; tap, tap, tap.

So there came to Denton and Elizabeth a long succession of laborious days, that hardened their hands, wove strange threads of some new and sterner substance into the soft prettiness of their lives, and drew grave lines and shadows on their faces. The bright, convenient ways of the former life had receded to an inaccessible distance; slowly they learnt the lesson of the under-world—sombre and laborious, vast and pregnant. There were many little things happened: things that would be tedious and miserable to tell, things that were bitter and grievous to bear—indignities, tyrannies, such as must ever season the bread of the poor in cities; and one thing that was not little, but seemed like the utter blackening of life to them, which was that the child they had given life to sickened and died. But that story, that ancient, perpetually recurring story, has been told so often, has been told so beautifully, that there is no need to tell it over again here. There was the same sharp fear, the same long anxiety, the deferred inevitable blow, and the black silence. It has always been the same; it will always be the same. It is one of the things that must be.

And it was Elizabeth who was the first to speak, after an aching, dull interspace of days: not, indeed, of the foolish little name that was a name no longer, but of the darkness that brooded over her soul. They had come through the shrieking, tumultuous ways of the city together; the clamour of trade, of yelling competitive religions, of political appeal, had beat upon deaf ears; the glare of focused lights, of dancing letters, and fiery advertisements, had fallen upon the set, miserable faces unheeded. They took their dinner in the dining-hall at a place apart. "I want," said Elizabeth clumsily, "to go out to the flying stages—to that seat. Here, one can say nothing. . . ."

Denton looked at her. "It will be night," he said.

"I have asked,—it is a fine night." She stopped.

He perceived she could find no words to explain herself.

Suddenly he understood that she wished to see the stars once more, the stars they had watched together from the open downland in that wild honeymoon of theirs five years ago. Something caught at his throat. He looked away from her.

"There will be plenty of time to go," he said, in a matter-of-fact tone.

And at last they came out to their little seat on the flying stage, and sat there for a long time in silence. The little seat was in shadow, but the zenith was pale blue with the effulgence of the stage overhead, and all the city spread below them, squares and circles and patches of brilliance caught in a mesh-work of light. The little stars seemed very faint and small: near as they had been to the old-world watcher, they had become now infinitely remote. Yet one could see them in the darkened patches amidst the glare, and especially in the northward sky, the ancient constellations gliding steadfast and patient about the pole.

Long our two people sat in silence, and at last Elizabeth sighed.

"If I understood," she said, "if I could understand. When one is down there the city seems everything—the noise, the hurry, the voices—you must live, you must scramble. Here —it is nothing; a thing that passes. One can think in peace."

"Yes," said Denton. "How flimsy it all is! From here more than half of it is swallowed by the night. . . . It will pass."

"We shall pass first," said Elizabeth.

"I know," said Denton. "If life were not a moment, the whole of history would seem like the happening of a day. . . . Yes—we shall pass. And the city will pass, and all the things that are to come. Man and the Overman and wonders unspeakable. And yet . . ."

He paused, and then began afresh. "I know what you feel. At least I fancy. . . . Down there one thinks of one's work, one's little vexations and pleasures, one's eating and drinking and ease and pain. One lives, and one must die.

Down there and everyday—our sorrow seemed the end of life. . . .

"Up here it is different. For instance, down there it would seem impossible almost to go on living if one were horribly disfigured, horribly crippled, disgraced. Up here— under these stars—none of those things would matter. They don't matter. . . . They are a part of something. One seems just to touch that something—under the stars. . . ."

He stopped. The vague, impalpable things in his mind, cloudy emotions half shaped towards ideas, vanished before the rough grasp of words. "It is hard to express," he said lamely.

They sat through a long stillness.

"It is well to come here," he said at last. "We stop— our minds are very finite. After all we are just poor animals rising out of the brute, each with a mind, the poor beginning of a mind. We are so stupid. So much hurts. And yet. . . ."

"I know, I know—and some day we shall *see*.

"All this frightful stress, all this discord will resolve to harmony, and we shall know it. Nothing is but it makes for that. Nothing. All the failures—every little thing makes for that harmony. Everything is necessary to it, we shall find. We shall find. Nothing, not even the most dreadful thing, could be left out. Not even the most trivial. Every tap of your hammer on the brass, every moment of work, my idleness even. . . . Dear one! every movement of our poor little one. . . . All these things go on for ever. And the faint impalpable things. We, sitting here together.—Everything. . . .

"The passion that joined us, and what has come since. It is not passion now. More than anything else it is sorrow. Dear. . . ."

He could say no more, could follow his thoughts no further.

Elizabeth made no answer—she was very still; but presently her hand sought his and found it.

IV.—Underneath

Under the stars one may reach upward and touch resignation, whatever the evil thing may be, but in the heat and stress of the day's work we lapse again, come disgust and anger and intolerable moods. How little is all our magnanimity—an accident! a phase! The very Saints of old had first to flee the world. And Denton and Elizabeth could not flee their world, no longer were there open roads to unclaimed lands where men might live freely—however hardly—and keep their souls in peace. The city had swallowed up mankind.

For a time these two Labour Serfs were kept at their original occupations, she at her brass stamping and Denton at his press; and then came a move for him that brought with it fresh and still bitterer experiences of life in the underways of the great city. He was transferred to the care of a rather more elaborate press in the central factory of the London Tile Trust.

In this new situation he had to work in a long vaulted room with a number of other men, for the most part born Labour Serfs. He came to this intercourse reluctantly. His upbringing had been refined, and, until his ill fortune had brought him to that costume, he had never spoken in his life, except by way of command or some immediate necessity, to the white-faced wearers of the blue canvas. Now at last came contact; he had to work beside them, share their tools, eat with them. To both Elizabeth and himself this seemed a further degradation.

His taste would have seemed extreme to a man of the nineteenth century. But slowly and inevitably in the intervening years a gulf had opened between the wearers of the blue canvas and the classes above, a difference not simply of circumstances and habits of life, but of habits of thought—even of language. The underways had developed a dialect of their own: above, too, had arisen a dialect, a code of thought, a language of "culture," which aimed by a sedulous search after fresh distinction to widen perpetually the space between itself and "vulgarity." The bond of a common faith, moreover, no longer held the race together. The last years of the nineteenth century were distinguished by the rapid development among the prosperous idle of esoteric perversions of the popular religion: glosses and interpretations that reduced the broad teachings of the carpenter of Nazareth to the exquisite narrowness of their lives. And, spite of their inclination towards the ancient fashion of living, neither Elizabeth nor Denton had been sufficiently original to escape the suggestion of their surroundings. In matters of common behaviour they had followed the ways of their class, and so when they fell at last to be Labour Serfs it seemed to them almost as though they were falling among offensive inferior animals; they felt as a nineteenth-century duke and duchess might have felt who were forced to take rooms in the Jago.

Their natural impulse was to maintain a "distance." But Denton's first idea of a dignified isolation from his new surroundings was soon rudely dispelled. He had imagined that his fall to the position of a Labour Serf was the end of his lesson, that when their little daughter had died he had plumbed the deeps of life; but indeed these things were only the beginning. Life demands something more from us than acquiescence. And now in a roomful of machine minders he was to learn a wider lesson, to make the acquaintance of another factor in life, a factor as elemental as the loss of things dear to us, more elemental even than toil.

His quiet discouragement of conversation was an immediate cause of offence—was interpreted, rightly enough I fear, as disdain. His ignorance of the vulgar dialect, a thing upon which he had hitherto prided himself, suddenly took upon itself a new aspect. He failed to perceive at once that his reception of the coarse and stupid but genially intended remarks that greeted his appearance must have stung the makers of these advances like blows in their faces. "Don't understand," he said rather coldly, and at hazard, "No, thank you."

The man who had addressed him stared, scowled, and turned away.

A second, who also failed at Denton's unaccustomed ear, took the trouble to repeat his remark, and Denton discovered he was being offered the use of an oil can. He expressed polite thanks, and this second man embarked upon a penetrating conversation. Denton, he remarked, had been a swell, and he wanted to know how he had come to wear the blue. He clearly expected an interesting record of vice and extravagance. Had Denton ever been at a Pleasure City? Denton was speedily to discover how the existence of these wonderful places of delight permeated and defiled the thought and honour of these unwilling, hopeless workers of the underworld.

His aristocratic temperament resented these questions. He answered "No" curtly. The man persisted with a still more personal question, and this time it was Denton who turned away.

"Gorblimey!" said his interlocutor, much astonished.

It presently forced itself upon Denton's mind that this remarkable conversation was being repeated in indignant tones to more sympathetic hearers, and that it gave rise to astonishment and ironical laughter. They looked at Denton with manifestly enhanced interest. A curious perception of isolation dawned upon him. He tried to think of his press and its unfamiliar peculiarities. . . .

The machines kept everybody pretty busy during the first spell, and then came a recess. It was only an interval for refreshment, too brief for any one to go out to a Labour Company dining-room. Denton followed his fellow-workers into a short gallery, in which were a number of bins and refuse from the presses.

Each man produced a packet of food. Denton had no packet. The manager, a careless young man who held his position by influence, had omitted to warn Denton that it was necessary to apply for this provision. He stood apart, feeling hungry. The others drew together in a group and talked in undertones, glancing at him ever and again. He became uneasy. His appearance of disregard cost him an increasing effort. He tried to think of the levers of his new press.

Presently one, a man shorter but much broader and stouter than Denton, came forward to him. Denton turned to him as unconcernedly as possible. "Here!" said the delegate—as Denton judged him to be—extending a cube of bread in a not too clean hand. He had a swart, broad-nosed face, and his mouth hung down towards one corner.

Denton felt doubtful for the instant whether this was meant for civility or insult. His impulse was to decline. "No, thanks," he said; and, at the man's change of expression, "I'm not hungry."

There came a laugh from the group behind. "Told you so," said the man who had offered Denton the loan of an oil can. "He's top side, he is. You ain't good enough for 'im."

The swart face grew a shade darker.

"Here," said its owner, still extending the bread, and speaking in a lower tone; "you got to eat this. See?"

Denton looked into the threatening face before him, and odd little currents of energy seemed to be running through his limbs and body.

"I don't want it," he said, trying a pleasant smile that twitched and failed.

The thickset man advanced his face, and the bread became a physical threat in his hand. Denton's mind rushed together to the one problem of his antagonist's eyes.

"Eat it," said the swart man.

There came a pause, and then they both moved quickly. The cube of bread described a complicated path, a curve that would have ended in Denton's face; and then his fist hit the wrist of the hand that gripped it, and it flew upward, and out of the conflict—its part played.

He stepped back quickly, fists clenched and arms tense. The hot, dark countenance receded, became an alert hostility, watching its chance. Denton for one instant felt confident, and strangely buoyant and serene. His heart beat quickly. He felt his body alive, and glowing to the tips.

"Scrap, boys!" shouted some one, and then the dark figure

had leapt forward, ducked back and sideways, and come in
again. Denton struck out, and was hit. One of his eyes
seemed to him to be demolished, and he felt a soft lip under
his fist just before he was hit again—this time under the
chin. A huge fan of fiery needles shot open. He had a
momentary persuasion that his head was knocked to pieces,
and then something hit his head and back from behind,
and the fight became an uninteresting, an impersonal thing.

He was aware that time—seconds or minutes—had passed,
abstract, uneventful time. He was lying with his head
in a heap of ashes, and something wet and warm ran
swiftly into his neck. The first shock broke up into discrete
sensations. All his head throbbed; his eye and his chin
throbbed exceedingly, and the taste of blood was in his
mouth.

"He's all right," said a voice. "He's opening his eyes."

"Serve him——well right," said a second.

His mates were standing about him. He made an effort
and sat up. He put his hand to the back of his head, and
his hair was wet and full of cinders. A laugh greeted the
gesture. His eye was partially closed. He perceived what
had happened. His momentary anticipation of a final
victory had vanished.

"Looks surprised," said some one.

"'Ave any more?" said a wit; and then, imitating Denton's
refined accent: "No, thank you."

Denton perceived the swart man with a blood-stained
handkerchief before his face, and somewhat in the background.

"Where's that bit of bread he's got to eat?" said a little
ferret-faced creature; and sought with his foot in the ashes
of the adjacent bin.

Denton had a moment of internal debate. He knew the
code of honour requires a man to pursue a fight he has
begun to the bitter end; but this was his first taste of the
bitterness. He was resolved to rise again, but he felt
no passionate impulse. It occured to him—and the thought
was no very violent spur—that he was perhaps after all a
coward. For a moment his will was heavy, a lump of lead.

" 'Ere it is," said the little ferret-faced man, and stooped to pick up a cindery cube. He looked at Denton, then at the others.

Slowly, unwillingly, Denton stood up.

A dirty-faced albino extended a hand to the ferret-faced man.

"Gimme that toke," he said. He advanced threateningly, bread in hand, to Denton. "So you ain't 'ad your bellyful yet," he said. "Eh?"

Now it was coming. "No, I haven't," said Denton, with a catching of the breath, and resolved to try this brute behind the ear before he himself got stunned again. He knew he would be stunned again. He was astonished how ill he had judged himself beforehand. A few ridiculous lunges, and down he would go again. He watched the albino's eyes. The albino was grinning confidently, like a man who plans an agreeable trick. A sudden perception of impending indignities stung Denton.

"You leave 'im alone, Jim," said the swart man suddenly over the blood-stained rag. "He ain't done nothing to you."

The albino's grin vanished. He stopped. He looked from one to the other. It seemed to Denton that the swart man demanded the privilege of his destruction. The albino would have been better.

"You leave 'im alone," said the swart man. "See? 'E's 'ad 'is licks."

A clattering bell lifted up its voice and solved the situation. The albino hesitated. "Lucky for you," he said, adding a foul metaphor, and turned with the others towards the press-room again. "Wait for the end of the spell, mate," said the albino over his shoulder—an afterthought. The swart man waited for the albino to precede him. Denton realised that he had a reprieve.

The men passed towards an open door. Denton became aware of his duties, and hurried to join the tail of the queue. At the doorway of the vaulted gallery of presses a yellow-uniformed labour policeman stood ticking a card. He had ignored the swart man's hæmorrhage.

"Hurry up there!" he said to Denton.

"Hello!" he said, at the sight of his facial disarray. "Who's been hitting *you*?"

"That's my affair," said Denton.

"Not if it spiles your work, it ain't," said the man in yellow. "You mind that."

Denton made no answer. He was a rough—a labourer. He wore the blue canvas. The laws of assault and battery, he knew, were not for the likes of him. He went to his press.

He could feel the skin of his brow and chin and head lifting themselves to noble bruises, felt the throb and pain of each aspiring contusion. His nervous system slid down to lethargy; at each movement in his press adjustment he felt he lifted a weight. And as for his honour—that too throbbed and puffed. How did he stand? What precisely had happened in the last ten minutes? What would happen next? He knew that there was enormous matter for thought, and he could not think save in disordered snatches.

His mood was a sort of stagnant astonishment. All his conceptions were overthrown. He had regarded his security from physical violence as inherent, as one of the conditions of life. So, indeed, it had been while he wore his middle-class costume, had his middle-class property to serve for his defence. But who would interfere among Labour roughs fighting together? And indeed in those days no man would. In the Underworld there was no law between man and man; the law and machinery of the state had become for them something that held men down, fended them off from much desirable property and pleasure, and that was all. Violence, that ocean in which the brutes live for ever, and from which a thousand dykes and contrivances have won our hazardous civilised life, had flowed in again upon the sinking underways and submerged them. The fist ruled. Denton had come right down at last to the elemental—fist and trick and the stubborn heart and fellowship—even as it was in the beginning.

The rhythm of his machine changed, and his thoughts were interrupted.

Presently he could think again. Strange how quickly

things had happened! He bore these men who had thrashed him no very vivid ill-will. He was bruised and enlightened. He saw with absolute fairness now the reasonableness of his unpopularity. He had behaved like a fool. Disdain, seclusion, are the privilege of the strong. The fallen aristocrat still clinging to his pointless distinction is surely the most pitiful creature of pretence in all this clamant universe. Good heavens! what was there for him to despise in these men?

What a pity he had not appreciated all this better five hours ago!

What would happen at the end of the spell? He could not tell. He could not imagine. He could not imagine the thoughts of these men. He was sensible only of their hostility and utter want of sympathy. Vague possibilities of shame and violence chased one another across his mind. Could he devise some weapon? He recalled his assault upon the hypnotist, but there were no detachable lamps here. He could see nothing that he could catch up in his defence.

For a space he thought of a headlong bolt for the security of the public ways directly the spell was over. Apart from the trivial consideration of his self-respect, he perceived that this would be only a foolish postponement and aggravation of his trouble. He perceived the ferret-faced man and the albino talking together with their eyes towards him. Presently they were talking to the swart man, who stood with his broad back studiously towards Denton.

At last came the end of the second spell. The lender of oil cans stopped his press sharply and turned round, wiping his mouth with the back of his hand. His eyes had the quiet expectation of one who seats himself in a theatre.

Now was the crisis, and all the little nerves of Denton's being seemed leaping and dancing. He had decided to show fight if any fresh indignity was offered him. He stopped his press and turned. With an enormous affectation of ease he walked down the vault and entered the passage of the ash pits, only to discover he had left his jacket—which he had taken off because of the heat of the vault—beside his press. He walked back. He met the albino eye to eye.

He heard the ferret-faced man in expostulation. " 'E reely ought, eat it," said the ferret-faced man. " 'E did reely."

"No—you leave 'im alone," said the swart man.

Apparently nothing further was to happen to him that day. He passed out to the passage and staircase that led up to the moving platforms of the city.

He emerged on the livid brilliance and streaming movement of the public street. He became acutely aware of his disfigured face, and felt his swelling bruises with a limp, investigatory hand. He went up to the swiftest platform, and seated himself on a Labour Company bench.

He lapsed into a pensive torpor. The immediate dangers and stresses of his position he saw with a sort of static clearness. What would they do to-morrow? He could not tell. What would Elizabeth think of his brutalisation? He could not tell. He was exhausted. He was aroused presently by a hand upon his arm.

He looked up, and saw the swart man seated beside him. He started. Surely he was safe from violence in the public way!

The swart man's face retained no traces of his share in the fight; his expression was free from hostility—seemed almost deferential. " 'Scuse me," he said, with a total absence of truculence. Denton realised that no assault was intended. He stared, awaiting the next development.

It was evident the next sentence was premeditated. "Whad—I—was—going—to say—was this," said the swart man, and sought through a silence for further words.

"Whad—I—was—going—to say—was this," he repeated.

Finally he abandoned that gambit. "*You're* aw right," he cried, laying a grimy hand on Denton's grimy sleeve. "*You're* aw right. You're a ge'man. Sorry—very sorry. Wanted to tell you that."

Denton realised that there must exist motives beyond a mere impulse to abominable proceedings in the man. He meditated, and swallowed an unworthy pride.

"I did not mean to be offensive to you," he said, "in refusing that bit of bread."

"Meant it friendly," said the swart man, recalling the scene; "but—in front of that blarsted Whitey and his snigger—well—I 'ad to scrap."

"Yes," said Denton with sudden fervour: "I was a fool."

"Ah," said the swart man, with great satisfaction. "That's aw right. Shake!"

And Denton shook.

The moving platform was rushing by the establishment of a face moulder, and its lower front was a huge display of mirror, designed to stimulate the thirst for more symmetrical features. Denton caught the reflection of himself and his new friend, enormously twisted and broadened. His own face was puffed, one-sided, and blood-stained; a grin of idiotic and insincere amiability distorted its latitude. A wisp of hair occluded one eye. The trick of the mirror presented the swart man as a gross expansion of lip and nostril. They were linked by shaking hands. Then abruptly this vision passed—to return to memory in the anæmic meditations of a waking dawn.

As he shook, the swart man made some muddled remark, to the effect that he had always known he could get on with a gentleman if one came his way. He prolonged the shaking until Denton, under the influence of the mirror, withdrew his hand. The swart man became pensive, spat impressively on the platform, and resumed his theme.

"Whad I was going to say was this," he said; was gravelled, and shook his head at his foot.

Denton became curious. "Go on," he said, attentive.

The swart man took the plunge. He grasped Denton's arm, became intimate in his attitude. "'Scuse me," he said. "Fact is, you done know 'ow to scrap. Done know 'ow to. Why—you done know 'ow to begin. You'll get killed if you don't mind. 'Ouldin' your 'ands— There!"

He reinforced his statement by objurgation, watching the effect of each oath with a wary eye.

"F'r instance. You're tall. Long arms. You got a longer reach than any one in the brasted vault. Gobblimey, but I thought I'd got a Tough on. 'Stead of which . . . 'Scuse me. I wouldn't have *'it* you if I'd known. It's like fighting sacks. 'Tisn't right. Y'r arms seemed 'ung on 'ooks. Reg'lar 'ung on 'ooks. There!"

Denton stared, and then surprised and hurt his battered chin by a sudden laugh. Bitter tears came into his eyes.

"Go on," he said.

The swart man reverted to his formula. He was good enough to say he liked the look of Denton, thought he had stood up "amazing plucky. On'y pluck ain't no good ain't no brasted good—if you don't 'old your 'ands.

"Whad I was going to say was this," he said. "Lemme show you 'ow to scrap. Just lemme. You're ig'nant, you ain't no class; but you might be a very decent scrapper— very decent. Shown. That's what I meant to say."

Denton hesitated. "But——" he said, "I can't give you anything——"

"That's the ge'man all over," said the swart man. "Who arst you to?"

"But your time?"

"If you don't get learnt scrapping you'll get killed,—don't you make no bones of that."

Denton thought. "I don't know," he said.

He looked at the face beside him, and all its native coarseness shouted at him. He felt a quick revulsion from his transient friendliness. It seemed to him incredible that it should be necessary for him to be indebted to such a creature.

"The chaps are always scrapping," said the swart man. "Always. And, of course—if one gets waxy and 'its you vital . . ."

"By God!" cried Denton; "I wish one would."

"Of course, if you feel like that——"

"You don't understand."

"P'raps I don't," said the swart man; and lapsed into a fuming silence.

When he spoke again his voice was less friendly, and he prodded Denton by way of address. "Look see!" he said: "are you going to let me show you 'ow to scrap?"

"It's tremendously kind of you," said Denton; "but——"

There was a pause. The swart man rose and bent over Denton.

"Too much ge'man," he said—"eh? I got a red face. . . . By gosh! you are—you *are* a brasted fool!"

He turned away, and instantly Denton realised the truth of this remark.

The swart man descended with dignity to a cross way, and Denton, after a momentary impulse to pursuit, remained on the platform. For a time the things that had happened filled his mind. In one day his graceful system of resignation had been shattered beyond hope. Brute force, the final, the fundamental, had thrust its face through all his explanations and glosses and consolations and grinned enigmatically. Though he was hungry and tired, he did not go on directly to the Labour Hotel, where he would meet Elizabeth. He found he was beginning to think, he wanted very greatly to think; and so, wrapped in a monstrous cloud of meditation, he went the circuit of the city on his moving platform twice.   You figure him, tearing through the glaring, thunder-voiced city at a pace of fifty miles an hour, the city upon the planet that spins along its chartless path through space many thousands of miles an hour, funking most terribly, and trying to understand why the heart and will in him should suffer and keep alive.

When at last he came to Elizabeth, she was white and anxious. He might have noted she was in trouble, had it not been for his own preoccupation. He feared most that she would desire to know every detail of his indignities, that she would be sympathetic or indignant. He saw her eyebrows rise at the sight of him.

"I've had rough handling," he said, and gasped. "It's too fresh—too hot. I don't want to talk about it." He sat down with an unavoidable air of sullennness.

She stared at him in astonishment, and as she read something of the significant hieroglyphic of his battered face, her lips whitened. Her hand—it was thinner now than in the days of their prosperity, and her first finger was a little altered by the metal punching she did—clenched convulsively. "This horrible world!" she said, and said no more.

In these latter days they had become a very silent couple; they said scarcely a word to each other that night, but each followed a private train of thought. In the small hours, as Elizabeth lay awake, Denton started up beside her suddenly —he had been lying as still as a dead man.

"I cannot stand it!" cried Denton. "I *will* not stand it!"

She saw him dimly, sitting up; saw his arm lunge as if in a furious blow at the enshrouding night. Then for a space he was still.

"It is too much—it is more than one can bear!"

She could say nothing. To her, also, it seemed that this was as far as one could go. She waited through a long stillness. She could see that Denton sat with his arms about his knees, his chin almost touching them.

Then he laughed.

"No," he said at last, "I'm going to stand it. That's the peculiar thing. There isn't a grain of suicide in us—not a grain. I suppose all the people with a turn that way have gone. We're going through with it—to the end."

Elizabeth thought grayly, and realised that this also was true.

"We're going through with it. To think of all who have gone through with it: all the generations—endless—endless. Little beasts that snapped and snarled, snapping and snarling, snapping and snarling, generation after generation."

His monotone, ended abruptly, resumed after a vast interval.

"There were ninety thousand years of stone age. A Denton somewhere in all those years. Apostolic succession. The grace of going through. Let me see! Ninety—nine hundred—three nines, twenty-seven—*three thousand* generations of men!—men more or less. And each fought, and was bruised, and shamed, and somehow held his own—going through with it—passing it on. . . . And thousands more to come perhaps—thousands!

"Passing it on. I wonder if they will thank us."

His voice assumed an argumentative note. "If one could

find something definite. . . . . If one could say, 'This is why
—this is why it goes on. . . . ' "

He became still, and Elizabeth's eyes slowly separated him
from the darkness until at last she could see how he sat
with his head resting on his hand. A sense of the enormous
remoteness of their minds came to her; that dim suggestion
of another being seemed to her a figure of their mutual understanding.   What could he be thinking now?   What
might he not say next?   Another age seemed to elapse before
he sighed and whispered: "No. I don't understand it. No!"
Then a long interval, and he repeated this. But the second
time it had the tone almost of a solution.

She became aware that he was preparing to lie down.
She marked his movements, perceived with astonishment
how he adjusted his pillow with a careful regard to
comfort. He lay down with a sigh of contentment almost.
His passion had passed. He lay still, and presently his
breathing became regular and deep.

But Elizabeth remained with eyes wide open in the darkness, until the clamour of a bell and the sudden brilliance of
the electric light warned them that the Labour Company
had need of them for yet another day.

That day came a scuffle with the albino Whitey and the
little ferret-faced man. Blunt, the swart artist in scrapping,
having first let Denton grasp the bearing of his lesson, in-
tervened, not without a certain quality of patronage. "Drop
'is 'air, Whitey, and let the man be," said his gross voice
through a shower of indignities. "Can't you see 'e don't
know 'ow to scrap?"   And Denton, lying shamefully in
the dust, realised that he must accept that course of instruction after all.

He made his apology straight and clean. He scrambled
up and walked to Blunt. "I was a fool, and you are right,"
he said. "If it isn't too late . . ."

That night, after the second spell, Denton went with
Blunt to certain waste and slime-soaked vaults under the
Port of London, to learn the first beginnings of the high art
of scrapping as it had been perfected in the great world of
the underways: how to hit or kick a man so as to hurt him

excruciatingly or make him violently sick, how to hit c
kick "vital," how to use glass in one's garments as a clu
and to spread red ruin with various domestic implements
how to anticipate and demolish your adversary's intention
in other directions; all the pleasant devices, in fact, that ha
grown up among the disinherited of the great cities of th
twentieth and twenty-first centuries, were spread out by
gifted exponent for Denton's learning. Blunt's bashfulnes
fell from him as the instruction proceeded, and he develope
a certain expert dignity, a quality of fatherly consideratior
He treated Denton with the utmost consideration, onl
"flicking him up a bit" now and then, to keep the interes
hot, and roaring with laugher at a happy fluke of Denton
that covered his mouth with blood.

"I'm always keerless of my mouth," said Blunt, admittin
a weakness. "Always. It don't seem to matter, like, ju
getting bashed in the mouth—not if your chin's all righ
Tastin' blood does me good. Always. But I better not '
you again."

Denton went home, to fall asleep exhausted and wake i
the small hours with aching limbs and all his bruise
tingling. Was it worth while that he should go on living
He listened to Elizabeth's breathing, and remembering tha
he must have awaked her the previous night, he lay ver
still. He was sick with infinite disgust at the new condition
of his life. He hated it all, hated even the genial savag
who had protected him so generously. The monstrous frau
of civilisation glared stark before his eyes; he saw it as a va
lunatic growth, producing a deepening torrent of savager
below, and above ever more flimsy gentility and silly wast
fulness. He could see no redeeming reason, no touch c
honour, either in the life he had led or in this life to whic
he had fallen. Civilisation presented itself as some cata
trophic product as little concerned with men—save as victim
—as a cyclone or a planetary collision. He, and therefor
all mankind, seemed living utterly in vain. His min
sought some strange expedients of escape, if not for himsel
then at least for Elizabeth. But he meant them for hin
self. What if he hunted up Mwres and told him of thei

disaster? It came to him as an astonishing thing how utterly Mwres and Bindon had passed out of his range. Where were they? What were they doing? From that he passed to thoughts of utter dishonour. And finally, not arising in any way out of this mental tumult, but ending it as dawn ends the night, came the clear and obvious conclusion of the night before: the conviction that he had to go through with things; that, apart from any remoter view and quite sufficient for all his thought and energy, he had to stand up and fight among his fellows and quit himself like a man.

The second night's instruction was perhaps less dreadful than the first; and the third was even endurable, for Blunt dealt out some praise. The fourth day Denton chanced upon the fact that the ferret-faced man was a coward. There passed a fortnight of smouldering days and feverish instruction at night; Blunt, with many blasphemies, testified that never had he met so apt a pupil; and all night long Denton dreamt of kicks and counters and gouges and cunning tricks. For all that time no further outrages were attempted, for fear of Blunt; and then came the second crisis. Blunt did not come one day—afterwards he admitted his deliberate intention—and through the tedious morning Whitey awaited the interval between the spells with an ostentatious impatience. He knew nothing of the scrapping lessons, and he spent the time in telling Denton and the vault generally of certain disagreeable proceedings he had in mind.

Whitey was not popular, and the vault disgorged to see him haze the new man with only a languid interest. But matters changed when Whitey's attempt to open the proceedings by kicking Denton in the face was met by an excellently executed duck, catch and throw, and completed the flight of Whitey's foot in its orbit and brought Whitey's head into the ash-heap that had once received Denton's. Whitey arose a shade whiter, and now blasphemously bent upon vital injuries. There were indecisive passages, foiled enterprises that deepened Whitey's evidently growing perplexity; and then things developed into a grouping of Denton uppermost with Whitey's throat in his hand, his knee on Whitey's chest, and a tearful Whitey with a black face, protruding tongue and broken finger endeavouring to explain the misunderstanding by means of hoarse sounds. Moreover, it was evident that among the bystanders there had never been a more popular person than Denton.

Denton, with proper precaution, released his antagonist and stood up. His blood seemed changed to some sort of fluid fire, his limbs felt light and supernaturally strong. The idea that he was a martyr in the civilisation machine had vanished from his mind. He was a man in a world of men.

The little ferret-faced man was the first in the competition to pat him on the back. The lender of oil cans was a radiant sun of genial congratulation. . . . It seemed incredible to Denton that he had ever thought of despair.

Denton was convinced that not only had he to go through with things, but that he could. He sat on the canvas pallet expounding this new aspect to Elizabeth. One side of his face was bruised. She had not recently fought, she had not been patted on the back, there were no hot bruises upon her face, only a pallor and a new line or so about the mouth. She was taking the woman's share. She looked steadfastly at Denton in his new mood of prophecy. "I feel that there is something," he was saying, "something that goes on, a Being of Life in which we live and move and have our being, something that began fifty—a hundred million years ago, perhaps, that goes on—on: growing, spreading, to things beyond us—things that will justify us all. . . . That will explain and justify my fighting—these bruises, and all the pain of it. It's the chisel—yes, the chisel of the Maker. If only I could make you feel as I feel, if I could make you! You *will*, dear, I know you will."

"No," she said in a low voice. "No, I shall not."

"So I might have thought——"

She shook her head. "No," she said, "I have thought as well. What you say—doesn't convince me."

She looked at his face resolutely. "I hate it," she said, and caught at her breath. "You do not understand, you do not think. There was a time when you said things and I believed them. I am growing wiser. You are a man, you can fight, force your way. You do not mind bruises. You

can be coarse and ugly, and still a man.   Yes—it makes
you.   It makes you.   You are right.   Only a woman is not
like that.   We are different.   We have let ourselves get
civilised too soon.   This underworld is not for us."

She paused and began again.

"I hate it!   I hate this horrible canvas!   I hate it more
than—more than the worst that can happen.   It hurts my
fingers to touch it.   It is horrible to the skin.   And the
women I work with day after day!   I lie awake at nights
and think how I may be growing like them. . . ."

She stopped.   "I *am* growing like them," she cried passionately.

Denton stared at her distress.   "But——" he said and
stopped.

"You don't understand.   What have I?   What have I to
save me?   *You* can fight.   Fighting is man's work.   But
women—women are different. . . .   I have thought it all
out, I have done nothing but think night and day.   Look at
the colour of my face!   I cannot go on.   I cannot endure
this life. . . .   I cannot endure it."

She stopped.   She hesitated.

"You do not know all," she said abruptly, and for an instant her lips had a bitter smile.   "I have been asked to
leave you."

"Leave me!"

She made no answer save an affirmative movement of the
head.

Denton stood up sharply.   They stared at one another
through a long silence.

Suddenly she turned herself about, and flung face downward upon their canvas bed.   She did not sob, she made no
sound.   She lay still upon her face.   After a vast, distressful
void her shoulders heaved and she began to weep silently.

"Elizabeth!" he whispered—"Elizabeth!"

Very softly he sat down beside her, bent down, put his
arm across her in a doubtful caress, seeking vainly for some
clue to this intolerable situation.

"Elizabeth," he whispered in her ear.

She thrust him from her with her hand.   "I cannot bear a

child to be a slave!" and broke out into loud and bitter weeping.

Denton's face changed—became blank dismay. Presently he slipped from the bed and stood on his feet. All the complacency had vanished from his face, had given place to impotent rage. He began to rave and curse at the intolerable forces which pressed upon him, at all the accidents and hot desires and heedlessness that mock the life of man. His little voice rose in that little room, and he shook his fist, this animalcule of the earth, at all that environed him about, at the millions about him, at his past and future and all the insensate vastness of the overwhelming city.

V.—Bindon Intervenes

In Bindon's younger days he had dabbled in speculation and made three brilliant flukes. For the rest of his life he had the wisdom to let gambling alone, and the conceit to believe himself a very clever man. A certain desire for influence and reputation interested him in the business intrigues of the giant city in which his flukes were made. He became at last one of the most influential shareholders in the company that owned the London flying-stages to which the aeroplanes came from all parts of the world. This much for his public activities. In his private life he was a man of pleasure. And this is the story of his heart.

But before proceeding to such depths, one must devote a little time to the exterior of this person. Its physical basis was slender, and short, and dark; and the face, which was fine-featured and assisted by pigments, varied from an insecure self-complacency to an intelligent uneasiness. His face and head had been depilated, according to the cleanly and hygienic fashion of the time, so that the colour and contour of his hair varied with his costume. This he was constantly changing.

At times he would distend himself with pneumatic vestments in the rococo vein. From among the billowy developments of this style, and beneath a translucent and illuminated head-dress, his eye watched jealously for the respect

of the less fashionable world. At other times he emphasised
his elegant slenderness in close-fitting garments of black
satin. For effects of dignity he would assume broad pneumatic shoulders, from which hung a robe of carefully arranged folds of China silk, and a classical Bindon in pink
tights was also a transient phenomenon in the eternal
pageant of Destiny. In the days when he hoped to marry
Elizabeth, he sought to impress and charm her, and at the
same time to take off something of his burthen of forty
years, by wearing the last fancy of the contemporary buck,
a costume of elastic material with distensible warts and
horns, changing in colour as he walked, by an ingenious
arrangement of versatile chromatophores. And no doubt, if
Elizabeth's affection had not been already engaged by the
worthless Denton, and if her tastes had not had that odd
bias for old-fashioned ways, this extremely *chic* conception
would have ravished her. Bindon had consulted Elizabeth's
father before presenting himself in this garb—he was one
of those men who always invite criticism of their costume—
and Mwres had pronounced him all that the heart of woman
could desire. But the affair of the hypnotist proved that his
knowledge of the heart of woman was incomplete.

Bindon's idea of marrying had been formed some little
time before Mwres threw Elizabeth's budding womanhood
in his way. It was one of Bindon's most cherished secrets
that he had a considerable capacity for a pure and simple
life of a grossly sentimental type. The thought imparted a
sort of pathetic seriousness to the offensive and quite incon-
sequent and unmeaning excesses, which he was pleased to
regard as dashing wickedness, and which a number of good
people also were so unwise as to treat in that desirable manner. As a consequence of these excesses, and perhaps by
reason also of an inherited tendency to early decay, his liver
became seriously affected, and he suffered increasing inconvenience when travelling by aeroplane. It was during his
convalescence from a protracted bilious attack that it occurred to him that in spite of all the terrible fascinations of
Vice, if he found a beautiful, gentle, good young woman of
a not too violently intellectual type to devote her life to him,

he might yet be saved to Goodness, and even rear a spirited family in his likeness to solace his declining years. But like so many experienced men of the world, he doubted if there were any good women. Of such as he had heard tell he was outwardly sceptical and privately much afraid.

When the aspiring Mwres effected his introduction to Elizabeth, it seemed to him that his good fortune was complete. He fell in love with her at once. Of course, he had always been falling in love since he was sixteen, in accordance with the extremely varied recipes to be found in the accumulated literature of many centuries. But this was different. This was real love. It seemed to him to call forth all the lurking goodness in his nature. He felt that for her sake he could give up a way of life that had already produced the gravest lesions on his liver and nervous system. His imagination presented him with idyllic pictures of the life of the reformed rake. He would never be sentimental with her, or silly; but always a little cynical and bitter, as became the past. Yet he was sure she would have an intuition of his real greatness and goodness. And in due course he would confess things to her, pour his version of what he regarded as his wickedness—showing what a complex of Goethe, and Benvenuto Cellini, and Shelley, and all those other chaps he really was—into her shocked, very beautiful, and no doubt sympathetic ear. And preparatory to these things he wooed her with infinite subtlety and respect. And the reserve with which Elizabeth treated him seemed nothing more nor less than an exquisite modesty touched and enhanced by an equally exquisite lack of ideas.

Bindon knew nothing of her wandering affections, nor of the attempt made by Mwres to utilise hypnotism as a corrective to this digression of her heart; he conceived he was on the best of terms with Elizabeth, and had made her quite successfully various significant presents of jewellery and the more virtuous cosmetics, when her elopement with Denton threw the world out of gear for him. His first aspect of the matter was rage begotten of wounded vanity, and as Mwres was the most convenient person, he vented the first brunt of it upon him.

He went immediately and insulted the desolate father grossly, and then spent an active and determined day going to and fro about the city and interviewing people in a consistent and partly-successful attempt to ruin that matrimonial speculator. The effectual nature of these activities gave him a temporary exhilaration, and he went to the dining-place he had frequented in his wicked days in a devil-may-care frame of mind, and dined altogether too amply and cheerfully with two other golden youths in the early forties. He threw up the game; no woman was worth being good for, and he astonished even himself by the strain of witty cynicism he developed. One of the other desperate blades, warmed with wine, made a facetious allusion to his disappointment, but at the time this did not seem unpleasant.

The next morning found his liver and temper inflamed. He kicked his phonographic-news machine to pieces, dismissed his valet, and resolved that he would perpetrate a terrible revenge upon Elizabeth. Or Denton. Or somebody. But anyhow, it was to be a terrible revenge; and the friend who had made fun at him should no longer see him in the light of a foolish girl's victim. He knew something of the little property that was due to her, and that this would be the only support of the young couple until Mwres should relent. If Mwres did not relent, and if unpropitious things should happen to the affair in which Elizabeth's expectations lay, they would come upon evil times and be sufficiently amenable to temptation of a sinister sort. Bindon's imagination, abandoning its beautiful idealism altogether, expanded the idea of temptation of a sinister sort. He figured himself as the implacable, the intricate and powerful man of wealth pursuing this maiden who had scorned him. And suddenly her image came upon his mind vivid and dominant, and for the first time in his life Bindon realised something of the real power of passion.

His imagination stood aside like a respectful footman who has done his work in ushering in the emotion.

"My God!" cried Bindon: "I will have her! If I have to kill myself to get her! And that other fellow——!"

After an interview with his medical man and a penance

for his overnight excesses in the form of bitter drugs, a mitigated but absolutely resolute Bindon sought out Mwres. Mwres he found properly smashed, and impoverished and humble, in a mood of frantic self-preservation, ready to sell himself body and soul, much more any interest in a disobedient daughter, to recover his lost position in the world. In the reasonable discussion that followed, it was agreed that these misguided young people should be left to sink into distress, or possibly even assisted towards that improving discipline by Bindon's financial influence.

"And then?" said Mwres.

"They will come to the Labour Company," said Bindon. "They will wear the blue canvas."

"And then?"

"She will divorce him," he said, and sat for a moment intent upon that prospect. For in those days the austere limitations of divorce of Victorian times were extraordinarily relaxed, and a couple might separate on a hundred different scores.

Then suddenly Bindon astonished himself and Mwres by jumping to his feet. "She *shall* divorce him!" he cried. "I will have it so—I will work it so. By God! it shall be so. He shall be disgraced, so that she must. He shall be smashed and pulverised."

The idea of smashing and pulverising inflamed him further. He began a Jovian pacing up and down the little office. "I will have her," he cried. "I *will* have her! Heaven and Hell shall not save her from me!" His passion evaporated in its expression, and left him at the end simply histrionic. He struck an attitude and ignored with heroic determination a sharp twinge of pain about the diaphragm. And Mwres sat with his pneumatic cap deflated and himself very visibly impressed.

And so, with a fair persistence, Bindon set himself to the work of being Elizabeth's malignant providence, using with ingenious dexterity every particle of advantage wealth in those days gave a man over his fellow-creatures. A resort to the consolations of religion hindered these operations not at all. He would go and talk with an interesting, experienced and sympathetic Father of the Huysmanite sect of the Isis cult, about all the irrational little proceedings he was pleased to regard as his Heaven-dismaying wickedness, and the interesting, experienced and sympathetic Father representing Heaven dismayed, would with a pleasing affectation of horror, suggest simple and easy penances, and recommend a monastic foundation that was airy, cool, hygienic, and not vulgarised, for viscerally disordered penitent sinners of the refined and wealthy type. And after these excursions, Bindon would come back to London quite active and passionate again. He would machinate with really considerable energy, and repair to a certain gallery high above the street of moving ways, from which he could view the entrance to the barrack of the Labour Company in the ward which sheltered Denton and Elizabeth. And at last one day he saw Elizabeth go in, and thereby his passion was renewed.

So in the fullness of time the complicated devices of Bindon ripened, and he could go to Mwres and tell him that the young people were near despair.

"It's time for you," he said, "to let your parental affections have play. She's been in blue canvas some months, and they've been cooped together in one of those Labour dens, and the little girl is dead. She knows now what his manhood is worth to her, by way of protection, poor girl. She'll see things now in a clearer light. You go to her—I don't want to appear in this affair yet—and point out to her how necessary it is that she should get a divorce from him. . . ."

"She's obstinate," said Mwres doubtfully.

"Spirit!" said Bindon. "She's a wonderful girl—a wonderful girl!"

"She'll refuse."

"Of course she will. But leave it open to her. Leave it open to her. And some day—in that stuffy den, in that irksome, toilsome life they can't help it—*they'll have a quarrel*. And then——"

Mwres meditated over the matter, and did as he was told.

Then Bindon, as he had arranged with his spiritual adviser, went into retreat. The retreat of the Huysmanite sect was a beautiful place, with the sweetest air in London,

lit by natural sunlight, and with restful quadrangles of rea
grass open to the sky, where at the same times the peniten
man of pleasure might enjoy all the pleasures of loafing anc
all the satisfaction of distinguished austerity. And, sav<
for participation in the simple and wholesome dietary of th<
place and in certain magnificent chants, Bindon spent al
his time in meditation upon the theme of Elizabeth, anc
the extreme purification his soul had undergone since h<
first saw her, and whether he would be able to get a dispen
sation to marry her from the experienced and sympathetic
Father in spite of the approaching "sin" of her divorce; anc
then . . . Bindon would lean against a pillar of the quad
rangle and lapse into reveries on the superiority of virtuou:
love to any other form of indulgence. A curious feeling ir
his back and chest that was trying to attract his attention
a disposition to be hot or shiver, a general sense of ill-health
and cutaneous discomfort he did his best to ignore. Al
that of course belonged to the old life that he was shaking
off.

When he came out of retreat he went at once to Mwres
to ask for news of Elizabeth. Mwres was clearly under the
impression that he was an exemplary father, profoundly
touched about the heart by his child's unhappiness. "She
was pale," he said, greatly moved; "She was pale. When I
asked her to come away and leave him—and be happy—she
put her head down upon the table"—Mwres sniffed—"and
cried."

His agitation was so great that he could say no more.

"Ah!" said Bindon, respecting this manly grief. "Oh!"
said Bindon quite suddenly, with his hand to his side.

Mwres looked up sharply out of the pit of his sorrows,
startled. "What's the matter?" he asked, visibly concerned.

"A most violent pain. Excuse me! You were telling me
about Elizabeth."

And Mwres, after a decent solicitude for Bindon's pain,
proceeded with his report. It was even unexpectedly hopeful. Elizabeth, in her first emotion at discovering that her
father had not absolutely deserted her, had been frank with
him about her sorrows and disgusts.

"Yes," said Bindon, magnificently, "I shall have her yet."
And then that novel pain twitched him for the second
time.

For these lower pains the priest was comparatively ineffectual, inclining rather to regard the body and them as
mental illusions amenable to contemplation; so Bindon
took it to a man of a class he loathed, a medical man of
extraordinary repute and incivility. "We must go all over
you," said the medical man, and did so with the most disgusting frankness. "Did you ever bring any children into
the world?" asked this gross materialist among other impertinent questions.

"Not that I know of," said Bindon, too amazed to stand
upon his dignity.

"Ah!" said the medical man, and proceeded with his
punching and sounding. Medical science in those days
was just reaching the beginnings of precision. "You'd better
go right away," said the medical man, "and make the
Euthanasia. The sooner the better."

Bindon gasped. He had been trying not to understand
the technical explanations and anticipations in which the
medical man had indulged.

"I say!" he said. "But do you mean to say . . . Your
science . . ."

"Nothing," said the medical man. "A few opiates. The
thing is your own doing, you know, to a certain extent."

"I was sorely tempted in my youth."

"It's not that so much. But you come of a bad stock.
Even if you'd have taken precautions you'd have had bad
times to wind up with. The mistake was getting born.
The indiscretions of the parents.   And you've shirked
exercise, and so forth."

"I had no one to advise me."

"Medical men are always willing."

"I was a spirited young fellow."

"We won't argue; the mischief's done now. You've lived.
We can't start you again. You ought never to have started
at all. Frankly—the Euthanasia!"

Bindon hated him in silence for a space. Every word of

this brutal expert jarred upon his refinements. He was s
gross, so impermeable to all the subtler issues of being. Bu
it is no good picking a quarrel with a doctor. "My religiou
beliefs," he said. "I don't approve of suicide."

"You've been doing it all your life."

"Well, anyhow, I've come to take a serious view of lit
now."

"You're bound to, if you go on living. You'll hurt. Bu
for practical purposes it's late. However, if you mean to d
that—perhaps I'd better mix you a little something. You'
hurt a great deal. These little twinges . . ."

"Twinges!"

"Mere preliminary notices."

"How long can I go on? I mean, before I hurt—really."

"You'll get it hot soon. Perhaps three days."

Bindon tried to argue for an extension of time, and i
the midst of his pleading gasped, put his hand to his sid
Suddenly the extraordinary pathos of his life came to hi
clear and vivid. "It's hard," he said. "It's infernally har
I've been no man's enemy but my own. I've always treate
everybody quite fairly."

The medical man stared at him without any sympathy f
some seconds. He was reflecting how excellent it was th
there were no more Bindons to carry on that line of patho
He felt quite optimistic. Then he turned to his telephor
and ordered up a prescription from the Central Pharmacy.

He was interrupted by a voice behind him. "By God"
cried Bindon; "I'll have her yet."

The physician stared over his shoulder at Bindon's expre
sion, and then altered the prescription.

So soon as this painful interview was over, Bindon gav
way to rage. He settled that the medical man was not on.
an unsympathetic brute and wanting in the first beginnin;
of a gentleman, but also highly incompetent; and he wei
off to four other practitioners in succession, with a view i
the establishment of this intuition. But to guard again
surprises he kept that little prescription in his pocket. Wit
each he began by expressing his grave doubts of the fir
doctor's intelligence, honesty and professional knowledg

and then stated his symptoms, suppressing only a few more
material facts in each case. These were always subsequently elicited by the doctor. In spite of the welcome de-
preciation of another practitioner, none of these eminent spe-
cialists would give Bindon any hope of eluding the anguish
and helplessness that loomed now close upon him. To the
last of them he unburthened his mind of an accumulated
disgust with medical science. "After centuries and centuries," he exclaimed hotly; "and you can do nothing—except
admit your helplessness. I say, 'save me'—and what do you
do?"

"No doubt it's hard on you," said the doctor. "But you
should have taken precautions."

"How was I to know?"

"It wasn't our place to run after you," said the medical
man, picking a thread of cotton from his purple sleeve.
"Why should we save *you* in particular? You see—from one
point of view—people with imaginations and passions like
yours have to go—they have to go."

"Go?"

"Die out. It's an eddy."

He was a young man with a serene face. He smiled at
Bindon. "We get on with research, you know; we give
advice when people have the sense to ask for it. And we
bide our time."

"Bide your time?"

"We hardly know enough yet to take over the management, you know."

"The management?"

"You needn't be anxious. Science is young yet. It's got
to keep on growing for a few generations. We know
enough now to know we don't know enough yet. . . . But
the time is coming, all the same. *You* won't see the time.
But, between ourselves, you rich men and party bosses, with
your natural play of the passions and patriotism and religion
and so forth, have made rather a mess of things; haven't you?
These Underways! And all that sort of thing. Some of us
have a sort of fancy that in time we may know enough to

take over a little more than the ventilation and drains. Knowledge keeps on piling up, you know. It keeps on growing. And there's not the slightest hurry for a generation or so. Some day—some day, men will live in a different way." He looked at Bindon and meditated. "There'll be a lot of dying out before that day can come."

Bindon attempted to point out to this young man how silly and irrelevant such talk was to a sick man like himself, how impertinent and uncivil it was to him, an older man occuping a position in the official world of extraordinary power and influence. He insisted that a doctor was paid to cure people—he laid great stress on *"paid"*—and had no business to glance even for a moment at "those other questions." "But we do," said the young man, insisting upon facts, and Bindon lost his temper.

His indignation carried him home. That these incompetent imposters, who were unable to save the life of a really influential man like himself, should dream of some day robbing the legitimate property owners of social control, of inflicting one knew not what tyranny upon the world. Curse science! He fumed over the intolerable prospect for some time, and then the pain returned, and he recalled the made-up prescription of the first doctor, still happily in his pocket. He took a dose forthwith.

It calmed and soothed him greatly, and he could sit down in his most comfortable chair beside his library (of phonographic records), and think over the altered aspect of affairs. His indignation passed, his anger and his passion crumbled under the subtle attack of that prescription, pathos became his sole ruler. He stared about him, at his magnificent and voluptuously appointed apartment, at his statuary and discreetly veiled pictures, and all the evidences of a cultivated and elegant wickedness; he touched a stud and the sad pipings of Tristan's shepherd filled the air. His eye wandered from one object to another. They were costly and gross and florid—but they were his. They presented in concrete form his ideals, his conceptions of beauty and desire, his idea of all that is precious in life. And now—he must leave it all like a common man. He was, he felt, a slender and delicate flame, burning out. So must all life flame up and pass, he thought. His eyes filled with tears.

Then it came into his head that he was alone. Nobody cared for him, nobody needed him! at any moment he might begin to hurt vividly. He might even howl. Nobody would mind. According to all the doctors he would have excellent reason for howling in a day or so. It recalled what his spiritual adviser had said of the decline of faith and fidelity, the degeneration of the age. He beheld himself as a pathetic proof of this; he, the subtle, able, important, voluptuous, cynical, complex Bindon, possibly howling, and not one faithful simple creature in all the world to howl in sympathy. Not one faithful simple soul was there—no shepherd to pipe to him! Had all such faithful simple creatures vanished from this harsh and urgent earth? He wondered whether the horrid vulgar crowd that perpetually went about the city could possibly know what he thought of them. If they did he felt sure *some* would try to earn a better opinion. Surely the world went from bad to worse. It was becoming impossible for Bindons. Perhaps some day . . . He was quite sure that the one thing he had needed in life was sympathy. For a time he regretted that he left no sonnets—no enigmatical pictures or something of that sort behind him to carry on his being until at last the sympathetic mind should come. . . .

It seemed incredible to him that this that came was extinction. Yet his sympathetic spiritual guide was in this matter annoyingly figurative and vague. Curse science! It had undermined all faith—all hope. To go out, to vanish from theatre and street, from office and dining-place, from the dear eyes of womankind. And not to be missed! On the whole to leave the world happier!

He reflected that he had never worn his heart upon his sleeve. Had he after all been *too* unsympathetic? Few people could suspect how subtly profound he really was beneath the mask of that cynical gaiety of his. They would not understand the loss they had suffered. Elizabeth, for example, had not suspected. . . .

He had reserved that. His thoughts having come to Elizabeth gravitated about her for some time. How *little* Elizabeth understood him!

That thought became intolerable. Before all other things he must set that right. He realised that there was still something for him to do in life, his struggle against Elizabeth was even yet not over. He could never overcome her now, as he had hoped and prayed. But he might still impress her!

From that idea he expanded. He might impress her profoundly—he might impress her so that she should for evermore regret her treatment of him. The thing that she must realise before everything else was his magnanimity. His magnanimity! Yes! he had loved her with amazing greatness of heart. He had not seen it so clearly before—but of course he was going to leave her all his property. He saw it instantly, as a thing determined and inevitable. She would think how good he was, how spaciously generous; surrounded by all that makes life tolerable from his hand, she would recall with infinite regret her scorn and coldness. And when she sought expression for that regret, she would find that occasion gone forever, she should be met by a locked door, by a disdainful stillness, by a white dead face. He closed his eyes and remained for a space imagining himself that white dead face.

From that he passed to other aspects of the matter, but his determination was assured. He meditated elaborately before he took action, for the drug he had taken inclined him to a lethargic and dignified melancholy. In certain respects he modified details. If he left all his property to Elizabeth it would include the voluptuously appointed room he occupied, and for many reasons he did not care to leave that to her. On the other hand, it had to be left to some one. In his clogged condition this worried him extremely.

In the end he decided to leave it to the sympathetic exponent of the fashionable religious cult whose conversation had been so pleasing in the past. "*He* will understand," said Bindon with a sentimental sigh. "He knows what Evil means—he understands something of the Stupendous Fascination of the Sphinx of Sin. Yes—he will understand."

By that phrase it was that Bindon was pleased to dignify certain unhealthy and undignified departures from sane conduct to which a misguided vanity and an ill-controlled curiosity had led him. He sat for a space thinking how very Hellenic and Italian and Neronic, and all those things, he had been. Even now—might one not try a sonnet? A penetrating voice to echo down the ages, sensuous, sinister, and sad. For a space he forgot Elizabeth. In the course of half an hour he spoilt three phonographic coils, got a headache, took a second dose to calm himself, and reverted to magnanimity and his former design.

At last he faced the unpalatable problem of Denton. It needed all his newborn magnanimity before he could swallow the thought of Denton; but at last this greatly misunderstood man, assisted by his sedative and the near approach of death, effected even that. If he was at all exclusive about Denton, if he should display the slightest distrust, if he attempted any specific exclusion of that young man, she might—*misunderstand*. Yes—she should have her Denton still. His magnanimity must go even to that. He tried to think only of Elizabeth in the matter.

He rose with a sigh, and limped across to the telephonic apparatus that communicated with his solicitor. In ten minutes a will duly attested and with its proper thumb-mark signature lay in the solicitor's office three miles away. And then for a space Bindon sat very still.

Suddenly he started out of a vague reverie and pressed an investigatory hand to his side.

Then he jumped eagerly to his feet and rushed to the telephone. The Euthanasia Company had rarely been called by a client in a greater hurry.

So it came at last that Denton and his Elizabeth, against all hope, returned unseparated from the labour servitude to which they had fallen. Elizabeth came out from her cramped subterranean den of metal-beaters and all the sordid circumstances of blue canvas, as one comes out of a nightmare. Back towards the sunlight their fortune took them; once the bequest was known to them, the bare thought of another day's hammering became intolerable.

They went up long lifts and stairs to levels that they had not seen since the days of their disaster. At first she was full of this sensation of escape; even to think of the underways was intolerable; only after many months could she begin to recall with sympathy the faded women who were still below there, murmuring scandals and reminiscences and folly, and tapping away their lives.

Her choice of the apartments they presently took expressed the vehemence of her release. They were rooms upon the very verge of the city; they had a roof space and a balcony upon the city wall, wide open to the sun and wind, the country and the sky.

And in that balcony comes the last scene in this story. It was a summer sunsetting, and the hills of Surrey were very blue and clear. Denton leant upon the balcony regarding them, and Elizabeth sat by his side. Very wide and spacious was the view, for their balcony hung five hundred feet above the ancient level of the ground. The oblongs of the Food Company, broken here and there by the ruins—grotesque little holes and sheds—of the ancient suburbs, and intersected by shining streams of sewage, passed at last into a remote diapering at the foot of the distant hills. There once had been the squatting-place of the children of Uya. On those further slopes gaunt machines of unknown import worked slackly at the end of their spell, and the hill crest was set with stagnant wind vanes. Along the great south road the Labour Company's field workers in huge wheeled mechanical vehicles were hurrying back to their meals, their last spell finished. And through the air a dozen little private aëropiles sailed down towards the city. Familiar scene as it was to the eyes of Denton and Elizabeth, it would have filled the minds of their ancestors with incredulous amazement. Denton's thoughts fluttered towards the future in a vain attempt at what that scene might be in another two hundred years, and, recoiling, turned towards the past.

He shared something of the growing knowledge of the time; he could picture the quaint smoke-grimed Victorian city with its narrow little roads of beaten earth, its wide common-land, ill-organised, ill-built suburbs, and irregular enclosures; the old countryside of the Stuart times, with its little villages and its petty London; the England of the monasteries, the far older England of the Roman dominion, and then before that a wild country with here and there the huts of some warring tribe. These huts must have come and gone and come again through a space of years that made the Roman camp and villa seem but yesterday; and before those years, before even the huts, there had been men in the valley. Even then—so recent had it all been when one judged it by the standards of geological time—this valley had been here; and those hills yonder, higher, perhaps, and snow-tipped, had still been yonder hills, and the Thames had flowed down from the Cotswolds to the sea. But the men had been but the shapes of men, creatures of darkness and ignorance, victims of beasts and floods, storms and pestilence and incessant hunger. They had held a precarious foothold amidst bears and lions and all the monstrous violence of the past. Already some at least of these enemies were overcome. . . .

For a time Denton pursued the thoughts of this spacious vision, trying in obedience to his instinct to find his place and proportion in the scheme.

"It has been chance," he said, "it has been luck. We have come through. It happens we have come through. Not by any strength of our own. . . .

"And yet . . . No. I don't know."

He was silent for a long time before he spoke again.

"After all—there is a long time yet. There have scarcely been men for twenty thousand years—and there has been life for twenty millions. And what are generations? What are generations? It is enormous, and we are so little. Yet we know—we feel. We are not dumb atoms, we are part of it—part of it—to the limits of our strength and will. Even to die is part of it. Whether we die or live, we are in the making. . . .

"As time goes on—*perhaps*—men will be wiser. . . . Wiser. . . .

"Will they ever understand?"

He became silent again. Elizabeth said nothing to these

things, but she regarded his dreaming face with infinite affection. Her mind was not very active that evening.   A great contentment possessed her. After a time she laid a gentle hand on his beside her. He fondled it softly, still looking out upon the spacious gold-woven view. So they sat as the sun went down.   Until presently Elizabeth shivered.

Denton recalled himself abruptly from these spacious issues of his leisure, and went in to fetch her a shawl.

STORY THE FIFTH

The Man who could work Miracles

A Pantoum in Prose

It is doubtful whether the gift was innate. For my own part, I think it came to him suddenly. Indeed, until he was thirty he was a sceptic, and did not believe in miraculous powers. And here, since it is the most convenient place, I must mention that he was a little man, and had eyes of a hot brown, very erect red hair, a moustache with ends that he twisted up, and freckles. His name was George McWhirter Fotheringay—not the sort of name by any means to lead to any expectation of miracles—and he was clerk at Gomshott's. He was greatly addicted to assertive argument. It was while he was asserting the impossibility of miracles that he had his first intimation of his extraordinary powers. This particular argument was being held in the bar of the Long Dragon, and Toddy Beamish was conducting the opposition by a monotonous but effective "So *you* say," that drove Mr. Fotheringay to the very limit of his patience.

There were present, besides these two, a very dusty cyclist, landlord Cox, and Miss Maybridge, the perfectly respectable and rather portly barmaid of the Dragon. Miss Maybridge was standing with her back to Mr. Fotheringay, washing glasses; the others were watching him, more or less amused by the present ineffectiveness of the assertive method. Goaded by the Torres Vedras tactics of Mr. Beamish, Mr. Fotheringay determined to make an unusual rhetorical effort. "Looky here, Mr. Beamish," said Mr. Fotheringay.

"Let us clearly understand what a miracle is. It's something contrariwise to the course of nature done by power of Will, something what couldn't happen without being specially willed."

"So *you* say," said Mr. Beamish, repulsing him.

Mr. Fotheringay appealed to the cyclist, who had hitherto been a silent auditor, and received his assent—given with a hesitating cough and a glance at Mr. Beamish. The landlord would express no opinion, and Mr. Fotheringay, returning to Mr. Beamish, received the unexpected concession of a qualified assent to his definition of a miracle.

"For instance," said Mr. Fotheringay, greatly encouraged. "Here would be a miracle. That lamp, in the natural course of nature, couldn't burn like that upsy-down, could it, Beamish?"

"*You* say it couldn't," said Beamish.

"And you?" said Fotheringay. "You don't mean to say—eh?"

"No," said Beamish reluctantly. "No, it couldn't."

"Very well," said Mr. Fotheringay. "Then here comes someone, as it might be me, along here, and stands as it might be here, and says to that lamp, as I might do, collecting all my will—'Turn upsy-down without breaking, and go on burning steady,' and——Hullo!"

It was enough to make anyone say "Hullo!" The impossible, the incredible, was visible to them all. The lamp hung inverted in the air, burning quietly with its flame pointing down. It was as solid, as indisputable as ever a lamp was, the prosaic common lamp of the Long Dragon bar.

Mr. Fotheringay stood with an extended forefinger and the knitted brows of one anticipating a catastrophic smash. The cyclist, who was sitting next the lamp, ducked and jumped across the bar. Everybody jumped, more or less. Miss Maybridge turned and screamed. For nearly three seconds the lamp remained still. A faint cry of mental distress came from Mr. Fotheringay. "I can't keep it up," he said, "any longer." He staggered back, and the inverted lamp suddenly flared, fell against the corner of the bar, bounced aside, smashed upon the floor, and went out.

It was lucky it had a metal receiver, or the whole place would have been in a blaze. Mr. Cox was the first to speak, and his remark, shorn of needless excrescences, was to the effect that Fotheringay was a fool. Fotheringay was beyond disputing even so fundamental a proposition as that! He was astonished beyond measure at the thing that had occurred. The subsequent conversation threw absolutely no light on the matter so far as Fotheringay was concerned; the general opinion not only followed Mr. Cox very closely but very vehemently. Everyone accused Fotheringay of a silly trick, and presented him to himself as a foolish destroyer of comfort and security. His mind was in a tornado of perplexity, he was himself inclined to agree with them, and he made a remarkably ineffectual opposition to the proposal of his departure.

He went home flushed and heated, coat-collar crumpled, eyes smarting and ears red. He watched each of the ten street lamps nervously as he passed it. It was only when he found himself alone in his little bedroom in Church Row that he was able to grapple seriously with his memories of the occurrence, and ask, "What on earth happened?"

He had removed his coat and boots, and was sitting on the bed with his hands in his pockets repeating the text of his defence for the seventeenth time, "I didn't want the confounded thing to upset," when it occurred to him that at the precise moment he had said the commanding words he had inadvertently willed the thing he said, and that when he had seen the lamp in the air he had felt that it depended on him to maintain it there without being clear how this was to be done. He had not a particularly complex mind, or he might have stuck for a time at that "inadvertently willed," embracing, as it does, the abstrusest problems of voluntary action; but as it was, the idea came to him with a quite acceptable haziness. And from that, following, as I must admit, no clear logical path, he came to the test of experiment.

He pointed resolutely to his candle and collected his

mind, though he felt he did a foolish thing. "Be raised up," he said. But in a second that feeling vanished. The candle was raised, hung in the air one giddy moment, and as Mr. Fotheringay gasped, fell with a smash on his toilet-table, leaving him in darkness save for the expiring glow of its wick.

For a time Mr. Fotheringay sat in the darkness, perfectly still. "It did happen, after all," he said. "And 'ow I'm to explain it I *don't* know." He sighed heavily, and began feeling in his pockets for a match. He could find none, and he rose and groped about the toilet-table. "I wish I had a match," he said. He resorted to his coat, and there were none there, and then it dawned upon him that miracles were possible even with matches. He extended a hand and scowled at it in the dark. "Let there be a match in that hand," he said. He felt some light object fall across his palm, and his fingers closed upon a match.

After several ineffectual attempts to light this, he discovered it was a safety-match. He threw it down, and then it occurred to him that he might have willed it lit. He did, and perceived it burning in the midst of his toilet-table mat. He caught it up hastily, and it went out. His perception of possibilities enlarged, and he felt for and replaced the candle in its candlestick. "Here! *you* be lit," said Mr. Fotheringay, and forthwith the candle was flaring, and he saw a little black hole in the toilet-cover, with a wisp of smoke rising from it. For a time he stared from this to the little flame and back, and then looked up and met his own gaze in the looking glass. By this help he communed with himself in silence for a time.

"How about miracles now?" said Mr. Fotheringay at last, addressing his reflection.

The subsequent meditations of Mr. Fotheringay were of a severe but confused description. So far as he could see, it was a case of pure willing with him. The nature of his first experiences disinclined him for any further experiments except of the most cautious type. But he lifted a sheet of paper, and turned a glass of water pink and then green, and he created a snail, which he miraculously annihilated, and got himself a miraculous new tooth-brush. Somewhen in the small hours he had reached the fact that his will-power must be of a particularly rare and pungent quality, a fact of which he had certainly had inklings before, but no certain assurance. The scare and perplexity of his first discovery was now qualified by pride in this evidence of singularity and by vague intimations of advantage. He became aware that the church clock was striking one, and as it did not occur to him that his daily duties at Gomshott's might be miraculously dispensed with, he resumed undressing, in order to get to bed without further delay. As he struggled to get his shirt over his head, he was struck with a brilliant idea. "Let me be in bed," he said, and found himself so. "Undressed," he stipulated; and, finding the sheets cold, added hastily, "and in my nightshirt—no, in a nice soft woollen nightshirt. Ah!" he said with immense enjoyment. "And now let me be comfortably asleep. . . ."

He awoke at his usual hour and was pensive all through breakfast-time, wondering whether his overnight experience might not be a particularly vivid dream. At length his mind turned again to cautious experiments. For instance, he had three eggs for breakfast; two his landlady had supplied, good, but shoppy, and one was a delicious fresh goose-egg, laid, cooked, and served by his extraordinary will. He hurried off to Gomshott's in a state of profound but carefully concealed excitement, and only remembered the shell of the third egg when his landlady spoke of it that night. All day he could do no work because of this astonishingly new self-knowledge, but this caused him no inconvenience, because he made up for it miraculously in his last ten minutes.

As the day wore on his state of mind passed from wonder to elation, albeit the circumstances of his dismissal from the Long Dragon were still disagreeable to recall, and a garbled account of the matter that had reached his colleagues led to some badinage. It was evident he must be careful how he lifted frangible articles, but in other ways his gift promised more and more as he turned it over in his mind. He intended among other things to increase his personal property by unostentatious acts of creation. He called into existence

a pair of very splendid diamond studs, and hastily annihilated them again as young Gomshott came across the
counting-house to his desk. He was afraid young Gomshott might wonder how he had come by them. He saw
quite clearly the gift required caution and watchfulness in
its exercise, but so far as he could judge the difficulties
attending its mastery would be no greater than those he had
already faced in the study of cycling. It was that analogy,
perhaps, quite as much as the feeling that he would be
unwelcome in the Long Dragon, that drove him out after
supper into the lane beyond the gas-works, to rehearse a few
miracles in private.

There was possibly a certain want of originality in his
attempts, for apart from his will-power Mr. Fotheringay was
not a very exceptional man. The miracle of Moses' rod
came to his mind, but the night was dark and unfavourable
to the proper control of large miraculous snakes. Then he
recollected the story of "Tannhäuser" that he had read on
the back of the Philharmonic programme. That seemed to
him singularly attractive and harmless. He stuck his walking-stick—a very nice Poona-Penang lawyer—into the turf
that edged the footpath, and commanded the dry wood to
blossom. The air was immediately full of the scent of roses,
and by means of a match he saw for himself that this beautiful miracle was indeed accomplished. His satisfaction was
ended by advancing footsteps. Afraid of a premature discovery of his powers, he addressed the blossoming stick
hastily: "Go back." What he meant was "Change back;"
but of course he was confused. The stick receded at a considerable velocity, and incontinently came a cry of anger and
a bad word fom the approaching person. "Who are you
throwing brambles at, you fool?" cried a voice. "That got
me on the shin."

"I'm sorry, old chap," said Mr. Fotheringay, and then
realising the awkward nature of the explanation, caught
nervously at his moustache. He saw Winch, one of the
three Immering constables, advancing.

"What d'yer mean by it?" asked the constable. "Hullo!

It's you, is it? The gent that broke the lamp at the Long Dragon!"

"I don't mean anything by it," said Mr. Fotheringay. "Nothing at all."

"What d'yer do it for then?"

"Oh, bother!" said Mr. Fotheringay.

"Bother indeed! D'yer know that stick hurt? What d'yer do it for, eh?"

For the moment Mr. Fotheringay could not think what he had done it for. His silence seemed to irritate Mr. Winch. "You've been assaulting the police, young man, this time. That's what *you* done."

"Look here, Mr. Winch," said Mr. Fotheringay, annoyed and confused, "I'm very sorry. The fact is——"

"Well?"

He could think of no way but the truth. "I was working a miracle." He tried to speak in an off-hand way, but try as he would he couldn't.

"Working a——! 'Ere, don't you talk rot. Working a miracle, indeed! Miracle! Well, that's downright funny! Why, you's the chap that don't believe in miracles. . . . Fact is, this is another of your silly conjuring tricks—that's what this is. Now, I tell you——"

But Mr. Fotheringay never heard what Mr. Winch was going to tell him. He realised he had given himself away, flung his valuable secret to all the winds of heaven. A violent gust of irritation swept him to action. He turned on the constable swiftly and fiercely. "Here," he said, "I've had enough of this, I have! I'll show you a silly conjuring trick, I will! Go to Hades! Go, now!"

He was alone!

Mr. Fotheringay performed no more miracles that night, nor did he trouble to see what had become of his flowering stick. He returned to the town, scared and very quiet, and went to his bedroom. "Lord!" he said, "it's a powerful gift —an extremely powerful gift. I didn't hardly mean as much as that. Not really. . . . I wonder what Hades is like!"

He sat on the bed taking off his boots. Struck by a happy thought he transferred the constable to San Francisco, and without any more interference with normal causation went soberly to bed. In the night he dreamt of the anger of Winch.

The next day Mr. Fotheringay heard two interesting items of news. Someone had planted a most beautiful climbing rose against the elder Mr. Gumshott's private house in the Lullaborough Road, and the river as far as Rawling's Mill was to be dragged for Constable Winch.

Mr. Fotheringay was abstracted and thoughtful all that day, and performed no miracles except certain provisions for Winch, and the miracle of completing his day's work with punctual perfection in spite of all the bee-swarm of thoughts that hummed through his mind. And the extraordinary abstraction and meekness of his manner was remarked by several people, and made a matter for jesting. For the most part he was thinking of Winch.

On Sunday evening he went to chapel, and oddly enough, Mr. Maydig, who took a certain interest in occult matters, preached about "things that are not lawful." Mr. Fotheringay was not a regular chapel goer, but the system of assertive scepticism, to which I have already alluded, was now very much shaken. The tenor of the sermon threw an entirely new light on these novel gifts, and he suddenly decided to consult Mr. Maydig immediately after the service. So soon as that was determined, he found himself wondering why he had not done so before.

Mr. Maydig, a lean, excitable man with quite remarkably long wrists and neck, was gratified at a request for a private conversation from a young man whose carelessness in religious matters was a subject for general remark in the town. After a few necessary delays, he conducted him to the study of the Manse, which was contiguous to the chapel, seated him comfortably, and, standing in front of a cheerful fire—his legs threw a Rhodian arch of shadow on the opposite wall—requested Mr. Fotheringay to state his business.

At first Mr. Fotheringay was a little abashed, and found some difficulty in opening the matter. "You will scarcely believe me, Mr. Maydig, I am afraid"—— and so forth for some time. He tried a question at last, and asked Mr. Maydig his opinion of miracles.

Mr. Maydig was still saying "Well" in an extremely judicial tone, when Mr. Fotheringay interrupted again: "You don't believe, I suppose, that some common sort of person—like myself, for instance—as it might be sitting here now, might have some sort of twist inside him that made him able to do things by his will."

"It's possible," said Mr. Maydig. "Something of the sort, perhaps, is possible."

"If I might make free with something here, I think I might show you by a sort of experiment," said Mr. Fotheringay. "Now, take that tobacco-jar on the table, for instance. What I want to know is whether what I am going to do with it is a miracle or not. Just half a minute, Mr. Maydig, please."

He knitted his brows, pointed to the tobacco-jar and said: "Be a bowl of vi'lets."

The tobacco-jar did as it was ordered.

Mr. Maydig started violently at the change, and stood looking from the thaumaturgist to the bowl of flowers. He said nothing. Presently he ventured to lean over the table and smell the violets; they were fresh-picked and very fine ones. Then he stared at Mr. Fotheringay again.

"How did you do that?" he asked.

Mr. Fotheringay pulled his moustache. "Just told it—and there you are. Is that a miracle, or is it black art, or what is it? And what do you think's the matter with me? That's what I want to ask."

"It's a most extraordinary occurrence."

"And this day last week I knew no more that I could do things like that than you did. It came quite sudden. It's something odd about my will, I suppose, and that's as far as I can see."

"Is *that*—the only thing. Could you do others things besides that?"

"Lord, yes!" said Mr. Fotheringay. "Just anything." He thought, and suddenly recalled a conjuring entertainment he had seen. "Here!" He pointed. "Change into a bowl of fish—no, not that—change into a glass bowl full of water with goldfish swimming in it. That's better! You see that, Mr. Maydig?"

"It's astonishing. It's incredible. You are either a most extraordinary . . . But no——"

"I could change it into anything," said Mr. Fotheringay. "Just anything. Here! be a pigeon, will you?"

In another moment a blue pigeon was fluttering round the room, and making Mr. Maydig duck every time it came near him. "Stop there, will you," said Mr. Fotheringay; and the pigeon hung motionless in the air. "I could change it back to a bowl of flowers," he said, and after replacing the pigeon on the table worked that miracle. "I expect you will want your pipe in a bit," he said, and restored the tobacco-jar.

Mr. Maydig had followed all these later changes in a sort of ejaculatory silence. He stared at Mr. Fotheringay and, in a very gingerly manner, picked up the tobacco-jar, examined it, replaced it on the table. "*Well!*" was the only expression of his feelings.

"Now, after that it's easier to explain what I came about," said Mr. Fotheringay; and proceeded to a lengthy and involved narrative of his strange experiences, beginning with the affair of the lamp in the Long Dragon and complicated by persistent allusions to Winch. As he went on, the transient pride Mr. Maydig's consternation had caused passed away; he became the very ordinary Mr. Fotheringay of everyday intercourse again. Mr. Maydig listened intently, the tobacco-jar in his hand, and his bearing changed also with the course of the narrative. Presently, while Mr. Fotheringay was dealing with the miracle of the third egg, the minister interrupted with a fluttering extended hand—

"It is possible," he said. "It is credible. It is amazing, of course, but it reconciles a number of difficulties. The power to work miracles is a gift—a peculiar quality like genius or second sight—hitherto it has come very rarely and to exceptional people. But in this case . . . I have always wondered at the miracles of Mahomet, and at Yogi's miracles, and the miracles of Madame Blavatsky. But, of course!

Yes, it is simply a gift! It carries out so beautifully the arguments of that great thinker"—Mr. Maydig's voice sank—"his Grace the Duke of Argyll. Here we plumb some profounder law—deeper than the ordinary laws of nature. Yes—yes. Go on. Go on!"

Mr. Fotheringay proceeded to tell of his misadventure with Winch, and Mr. Maydig, no longer overawed or scared, began to jerk his limbs about and interject astonishment. "It's this what troubled me most," proceeded Mr. Fotheringay; "it's this I'm most mijitly in want of advice for; of course he's at San Francisco—wherever San Francisco may be—but of course it's awkward for both of us, as you'll see, Mr. Maydig. I don't see how he can understand what has happened, and I dare say he's scared and exasperated something tremendous, and trying to get at me. I dare say he keeps on starting off to come here. I send him back, by a miracle, every few hours, when I think of it. And of course, that's a thing he won't be able to understand, and it's bound to annoy him; and, of course, if he takes a ticket every time it will cost him a lot of money. I done the best I could for him, but of course it's difficult for him to put himself in my place. I thought afterwards that his clothes might have got scorched, you know—if Hades is all it's supposed to be—before I shifted him. In that case I suppose they'd have locked him up in San Francisco. Of course I willed him a new suit of clothes on him directly I thought of it. But, you see, I'm already in a deuce of a tangle——"

Mr. Maydig looked serious. "I see you are in a tangle. Yes, it's a difficult position. How you are to end it . . ." He became diffuse and inconclusive.

"However, we'll leave Winch for a little and discuss the larger question. I don't think this is a case of the black art or anything of the sort. I don't think there is any taint of criminality about it at all, Mr. Fotheringay—none whatever, unless you are suppressing material facts. No, it's miracles —pure miracles—miracles, if I may say so, of the very highest class."

He began to pace the hearthrug and gesticulate, while Mr. Fotheringay sat with his arm on the table and his head on his arm, looking worried. "I don't see how I'm to manage about Winch," he said.

"A gift of working miracles—apparently a very powerful gift," said Mr. Maydig, "will find a way about Winch—never fear. My dear Sir, you are a most important man—a man of the most astonishing possibilities. As evidence, for example! And in other ways, the things you may do . . ."

"Yes, I've thought of a thing or two," said Mr. Fotheringay. "But—some of the things came a bit twisty. You saw that fish at first? Wrong sort of bowl and wrong sort of fish. And I thought I'd ask someone."

"A proper course," said Mr. Maydig, "a very proper course—altogether the proper course." He stopped and looked at Mr. Fotheringay. "It's practically an unlimited gift. Let us test your powers, for instance. If they really are . . . If they really are all they seem to be."

And so, incredible as it may seem, in the study of the little house behind the Congregational Chapel, on the evening of Sunday, Nov. 10, 1896, Mr. Fotheringay, egged on and inspired by Mr. Maydig, began to work miracles. The reader's attention is specially and definitely called to the date. He will object, probably has already objected, that certain points in this story are improbable, that if any things of the sort already described had indeed occurred, they would have been in all the papers a year ago. The details immediately following he will find particularly hard to accept, because among other things they involve the conclusion that he or she, the reader in question, must have been killed in a violent and unprecedented manner more than a year ago. Now a miracle is nothing if not improbable, and as a matter of fact the reader *was* killed in a violent and un-precedented manner a year ago. In the subsequent course of this story that will become perfectly clear and credible, as every right-minded and reasonable reader will admit. But this is not the place for the end of the story, being but little beyond the hither side of the middle. And at first the mir-acles worked by Mr. Fotheringay were timid little miracles —little things with the cups and parlour fitments, as feeble as the miracles of Theosophists, and, feeble as they were, they were received with awe by his collaborator. He would have preferred to settle the Winch business out of hand, but Mr. Maydig would not let him. But after they had worked a dozen of these domestic trivialities, their sense of power grew, their imagination began to show signs of stimulation, and their ambition enlarged. Their first larger enterprise was due to hunger and the negligence of Mrs. Minchin, Mr. Maydig's housekeeper. The meal to which the minister conducted Mr. Fotheringay was certainly ill-laid and uninviting as refreshment for two industrious miracle-workers; but they were seated, and Mr. Maydig was descanting in sorrow rather than in anger upon his housekeeper's shortcomings, before it occurred to Mr. Fotheringay that an opportunity lay before him. "Don't you think, Mr. Maydig," he said, "if it isn't a liberty, I——"

"My dear Mr. Fotheringay! Of course! No—I didn't think."

Mr Fotheringay waved his hand. "What shall we have?" he said, in a large, inclusive spirit, and, at Mr. Maydig's order, revised the supper very thoroughly. "As for me," he said, eyeing Mr. Maydig's selection, "I am always particularly fond of a tankard of stout and a nice Welsh rarebit, and I'll order that. I ain't much given to Burgundy," and forthwith stout and Welsh rarebit promptly appeared at his command. They sat long at their supper, talking like equals, as Mr. Fotheringay presently perceived with a glow of surprise and gratification, of all the miracles they would presently do. "And, by the bye, Mr. Maydig," said Mr. Fotheringay, "I might perhaps be able to help you—in a domestic way."

"Don't quite follow," said Mr. Maydig, pouring out a glass of miraculous old Burgundy.

Mr. Fotheringay helped himself to a second Welsh rarebit out of vacancy, and took a mouthful. "I was thinking," he said, "I might be able (*chum, chum*) to work (*chum, chum*) a miracle with Mrs. Minchin (*chum, chum*)—make her a better woman."

Mr. Maydig put down the glass and looked doubtful. "She's—— She strongly objects to interference, you know,

Mr. Fotheringay. And—as a matter of fact—it's well past eleven and she's probably in bed and asleep. Do you think, on the whole——"

Mr. Fotheringay considered these objections. "I don't see that it shouldn't be done in her sleep."

For a time Mr. Maydig opposed the idea, and then he yielded. Mr. Fotheringay issued his orders, and a little less at their ease, perhaps, the two gentlemen proceeded with their repast. Mr. Maydig was enlarging on the changes he might expect in his housekeeper next day, with an optimism that seemed even to Mr. Fotheringay's supper senses a little forced and hectic, when a series of confused noises from upstairs began. Their eyes exchanged interrogations, and Mr. Maydig left the room hastily. Mr. Fotheringay heard him calling up to his housekeeper and then his footsteps going softly up to her.

In a minute or so the minister returned, his step light, his face radiant. "Wonderful!" he said, "and touching! Most touching!"

He began pacing the hearthrug. "A repentance—a most touching repentance—through the crack of the door. Poor woman! A most wonderful change! She had got up. She must have got up at once. She had got up out of her sleep to smash a private bottle of brandy in her box. And to confess it too! . . . But this gives us—it opens—a most amazing vista of possibilities. If we can work this miraculous change in *her* . . ."

"The thing's unlimited seemingly," said Mr. Fotheringay. "And about Mr. Winch——"

"Altogether unlimited." And from the hearthrug Mr. Maydig, waving the Winch difficulty aside, unfolded a series of wonderful proposals—proposals he invented as he went along.

Now what those proposals were does not concern the essentials of this story. Suffice it that they were designed in a spirit of infinite benevolence, the sort of benevolence that used to be called post-prandial. Suffice it, too, that the problem of Winch remained unsolved. Nor is it necessary to describe how far that series got to its fulfilment. There were astonishing changes. The small hours found Mr. Maydig and Mr. Fotheringay careering across the chilly market-square under the still moon, in a sort of ecstasy of thaumaturgy, Mr. Maydig all flap and gesture, Mr. Fotheringay short and bristling, and no longer abashed at his greatness. They had reformed every drunkard in the Parliamentary division, changed all the beer and alcohol to water (Mr. Maydig had overruled Mr. Fotheringay on this point), they had, further, greatly improved the railway communication of the place, drained Flinder's swamp, improved the soil of One Tree Hill, and cured the Vicar's wart. And they were going to see what could be done with the injured pier at South Bridge. "The place," gasped Mr. Maydig, "won't be the same place to-morrow. How surprised and thankful everyone will be!" And just at that moment the church clock struck three.

"I say," said Mr. Fotheringay, "that's three o'clock! I must be getting back. I've got to be at business by eight. And besides, Mrs. Wimms——"

"We're only beginning," said Mr. Maydig, full of the sweetness of unlimited power. "We're only beginning. Think of all the good we're doing. When people wake——"

"But——," said Mr. Fotheringay.

Mr. Maydig gripped his arm suddenly. His eyes were bright and wild. "My dear chap," he said, "there's no hurry. Look"—he pointed to the moon at the zenith—"Joshua!"

"Joshua?" said Mr. Fotheringay.

"Joshua," said Mr. Maydig. "Why not? Stop it."

Mr. Fotheringay looked at the moon.

"That's a bit tall," he said after a pause.

"Why not?" said Mr Maydig. "Of course it doesn't stop. You stop the rotation of the earth, you know. Time stops. It isn't as if we were doing harm."

"H'm!" said Mr. Fotheringay. "Well." He sighed. "I'll try. Here——"

He buttoned up his jacket and addressed himself to the habitable globe, with as good an assumption of confidence as lay in his power. "Jest stop rotating, will you," said Mr. Fotheringay.

Incontinently he was flying head over heels through the air at the rate of dozens of miles a minute. In spite of the innumerable circles he was describing per second, he thought; for thought is wonderful—sometimes as sluggish as flowing pitch, sometimes as instantaneous as light. He thought in a second and willed. "Let me come down safe and sound. Whatever else happens, let me down safe and sound."

He willed it only just in time, for his clothes, heated by his rapid flight through the air, were already beginning to singe. He came down with a forcible, but by no means injurious bump in what appeared to be a mound of fresh-turned earth. A large mass of metal and masonry, extraordinarily like the clock-tower in the midde of the market-square, hit the earth near him, ricochetted over him, and flew into stonework, bricks, and masonry, like a bursting bomb. A hurtling cow hit one of the larger blocks and smashed like an egg. There was a crash that made all the most violent crashes of his past life seem like the sound of falling dust, and this was followed by a descending series of lesser crashes. A vast wind roared throughout earth and heaven, so that he could scarcely lift his head to look. For a while he was too breathless and astonished even to see where he was or what had happened. And his first movement was to feel his head and reassure himself that his streaming hair was still his own.

"Lord!" gasped Mr. Fotheringay, scarce able to speak for the gale. "I've had a squeak! What's gone wrong? Storms and thunder. And only a minute ago a fine night. It's Maydig set me on to this sort of thing. *What* a wind! If I go on fooling in this way I'm bound to have a thundering accident! . . .

"Where's Maydig?

"What a confounded mess everything's in!"

He looked about him so far as his flapping jacket would permit. The appearance of things was really extremely strange. "The sky's all right anyhow," said Mr. Fotheringay.

"And that's about all that is all right. And even there it looks like a terrific gale coming up. But there's the moon overhead. Just as it was just now. Bright as midday. But as for the rest—— Where's the village? Where's—where's anything? And what on earth set this wind a-blowing? *I* didn't order no wind."

Mr. Fotheringay struggled to get to his feet in vain, and after one failure, remained on all fours, holding on. He surveyed the moonlit world to leeward, with the tails of his jacket streaming over his head. "There's something seriously wrong," said Mr. Fotheringay. "And what it is—goodness knows."

Far and wide nothing was visible in the white glare through the haze of dust that drove before a screaming gale but tumbled masses of earth and heaps of inchoate ruins, no trees, no houses, no familiar shapes, only a wilderness of disorder vanishing at last into the darkness beneath the whirling columns and streamers, the lightnings and thunderings of a swiftly rising storm. Near him in the livid glare was something that might once have been an elm-tree, a smashed mass of splinters, shivered from boughs to base, and further a twisted mass of iron girders—only too evidently the viaduct—rose out of the piled confusion.

You see, when Mr. Fotheringay had arrested the rotation of the solid globe, he had made no stipulation concerning the trifling movables upon its surface. And the earth spins so fast that the surface at its equator is travelling at rather more than a thousand miles an hour, and in the latitudes at more than half that pace. So that the village, and Mr. Maydig, and Mr. Fotheringay, and everybody and everything had been jerked violently forward at about nine miles per second—that is to say, much more violently than if they had been fired out of a cannon. And every human being, every living creature, every house, and every tree— all the world as we know it—had been so jerked and smashed and utterly destroyed. That was all.

These things Mr. Fotheringay did not, of course, fully appreciate. But he perceived that his miracle had miscarried, and with that a great disgust of miracles came upon him. He was in darkness now, for the clouds had swept together and blotted out his momentary glimpse of the moon, and the air was full of fitful struggling tortured wraiths of hail. A great roaring of wind and waters filled earth and sky, and, peering under his hand through the dust and sleet to windward, he saw by the play of the lightnings a vast wall of water pouring towards him.

"Maydig!" screamed Mr. Fotheringay's feeble voice, amid the elemental uproar. "Here!—Maydig!"

"Stop!" cried Mr. Fotheringay to the advancing water. "Oh, for goodness' sake, stop"

"Just a moment," said Mr. Fotheringay to the lightnings and thunder. "Stop jest a moment while I collect my thoughts. . . . And now what shall I do?" he said. "What *shall* I do? Lord! I wish Maydig was about."

"I know," said Mr. Fotheringay. "And for goodness' sake let's have it right *this* time."

He remained on all fours, leaning against the wind, very intent to have everything right.

"Ah!" he said. "Let nothing what I'm going to order happen until I say 'Off!' . . . Lord! I wish I'd thought of that before"

He shifted his little voice against the whirlwind, shouting louder and louder in the vain desire to hear himself speak. "Now then!—here goes! Mind about that what I said just now. In the first place, when all I've got to say is done, let me lose my miraculous power, let my will become just like anybody else's will, and all these dangerous miracles be stopped. I don't like them. I'd rather I didn't work 'em. Ever so much. That's the first thing. And the second is—let me be back just before the miracles begin; let everything be just as it was before that blessed lamp turned up. It's a big job, but it's the last. Have you got it? No more miracles; everything as it was—me back in the Long Dragon just before I drank my half-pint. That's it! Yes."

He dug his fingers into the mould, closed his eyes, and said "Off!"

Everything became perfectly still. He perceived that he was standing erect.

"So *you* say," said a voice.

He opened his eyes. He was in the bar of the Long Dragon, arguing about miracles with Toddy Beamish. He had a vague sense of some great thing forgotten that instantaneously passed. You see, except for the loss of his miraculous powers, everything was back as it had been; his mind and memory therefore were now just as they had been at the time when this story began. So that he knew absolutely nothing of all that is told here, knows nothing of all that is told here to this day. And among other things, of course, he still did not believe in miracles.

"I tell you that miracles, properly speaking, can't possibly happen," he said, "whatever you like to hold. And I'm prepared to prove it up to the hilt."

"That's what *you* think," said Toddy Beamish, and "Prove it if you can."

"Looky here, Mr. Beamish," said Mr. Fotheringay. "Let us clearly understand what a miracle is. It's something contrariwise to the course of nature done by power of Will. . . ."

TWELVE STORIES AND A DREAM

STORY THE FIRST

Filmer

IN truth the mastery of flying was the work of thousands of men—this man a suggestion and that an experiment, until at last only one vigorous intellectual effort was needed to finish the work. But the inexorable injustice of the popular mind has decided that of all these thousands, one man, and that a man who never flew, should be chosen as the discoverer, just as it has chosen to honour Watt as the discoverer of steam and Stephenson of the steam-engine. And surely of all honoured names none is so grotesquely and tragically honoured as poor Filmer's, the timid, intellectual creature who solved the problem over which the world had hung perplexed and a little fearful for so many generations, the man who pressed the button that has changed peace and warfare and well-nigh every condition of human life and happiness. Never has that recurring wonder of the littleness of the scientific man in the face of the greatness of his science found such an amazing exemplification. Much concerning Filmer is, and must remain, profoundly obscure—Filmers attract no Boswells—but the essential facts and the concluding scene are clear enough, and there are letters, and notes, and casual allusions to piece the whole together. And this is the story one makes, putting this thing with that, of Filmer's life and death.

The first authentic trace of Filmer on the page of history is a document in which he applies for admission as a paid student in physics to the Government laboratories at South Kensington, and therein he describes himself as the son of a "military bootmaker" ("cobbler" in the vulgar tongue)

of Dover, and lists his various examination proofs of a high
proficiency in chemistry and mathematics. With a certain
want of dignity he seeks to enhance these attainments by
a profession of poverty and disadvantages, and he writes of
the laboratory as the "gaol" of his ambitions, a slip which
reinforces his claim to have devoted himself exclusively
to the exact sciences. The document is endorsed in a
manner that shows Filmer was admitted to this coveted
opportunity; but until quite recently no traces of his success
in the Government institution could be found.

It has now, however, been shown that in spite of his
professed zeal for research, Filmer, before he had held this
scholarship a year, was tempted by the possibility of a
small increase in his immediate income, to abandon it
in order to become one of the nine-pence-an-hour com-
puters employed by a well-known Professor in his vicarious
conduct of those extensive researches of his in solar physics
—researches which are still a matter of perplexity to
astronomers. Afterwards, for the space of seven years,
save for the pass lists of the London University, in which
he is seen to climb slowly to a double first-class B.Sc., in
mathematics and chemistry, there is no evidence of how
Filmer passed his life. No one knows how or where he
lived, though it seems highly probable that he continued
to support himself by teaching while he prosecuted the
studies necessary for this distinction. And then, oddly
enough, one finds him mentioned in the correspondence of
Arthur Hicks, the poet.

"You remember Filmer," Hicks writes to his friend Vance;
"well *he* hasn't altered a bit, the same hostile mumble and
the nasty chin—how *can* a man contrive to be always three
days from shaving?—and a sort of furtive air of being
engaged in sneaking in front of one; even his coat and
that frayed collar of his show no further signs of the passing
of years. He was writing in the library and I sat down
beside him in the name of God's charity, whereupon he
deliberately insulted me by covering up his memoranda.
It seems he has some brilliant research on hand that he
suspects me of all people—with a Bodley Booklet a-printing!

—of stealing. He has taken remarkable honours at the University—he went through them with a sort of hasty slobber, as though he feared I might interrupt him before he had told me all—and he spoke of taking his D.Sc. as one might speak of taking a cab. And he asked what I was doing—with a sort of comparative accent, and his arm was spread nervously, positively a protecting arm, over the paper that hid the precious idea—his one hopeful idea.

"'Poetry,' he said, 'poetry. And what do you profess to teach in it, Hicks?'

"The thing's a provincial professorling in the very act of budding, and I thank the Lord devoutly that but for the precious gift of indolence I also might have gone this way to D.Sc. and destruction . . ."

A curious little vignette that I am inclined to think caught Filmer in or near the very birth of his discovery.

Hicks was wrong in anticipating a provincial professorship for Filmer. Our next glimpse of him is lecturing on "rubber and rubber substitutes," to the Society of Arts —he had become manager to a great plastic-substance manufactory—and at that time, it is now known, he was a member of the Aeronautical Society, albeit he contributed nothing to the discussions of that body, preferring no doubt to mature his great conception without external assistance. And within two years of that paper before the Society of Arts he was hastily taking out a number of patents and proclaiming in various undignified ways the completion of the divergent inquiries which made his flying machine possible. The first definite statement to that effect appeared in a half-penny evening paper through the agency of a man who lodged in the same house with Filmer. His final haste after his long laborious secret patience seems to have been due to a needless panic, Bootle, the notorious American scientific quack, having made an announcement that Filmer interpreted wrongly as an anticipation of his idea.

Now what precisely was Filmer's idea? Really a very simple one. Before his time the pursuit of aeronautics had taken two divergent lines, and had developed on the one hand balloons—large apparatus lighter than air, easy in ascent, and comparatively safe in descent, but floating helplessly before any breeze that took them; and on the other, flying machines that flew only in theory—vast flat structures heavier than air, propelled and kept up by heavy engines and for the most part smashing at the first descent. But, neglecting the fact that the inevitable final collapse rendered them impossible, the weight of the flying machines gave them this theoretical advantage, that they could go through the air against a wind, a necessary condition if aerial navigation was to have any practical value. It is Filmer's particular merit that he perceived the way in which the contrasted and hitherto incompatible merits of balloon and heavy flying machine might be combined in one apparatus, which should be at choice either heavier or lighter than air. He took hints from the contractile bladders of fish and the pneumatic cavities of birds. He devised an arrangement of contractile and absolutely closed balloons which when expanded could lift the actual flying apparatus with ease, and when retracted by the complicated "musculature" he wove about them, were withdrawn almost completely into the frame; and he built the large framework which these balloons sustained, of hollow, rigid tubes, the air in which, by an ingenious contrivance, was automatically pumped out as the apparatus fell, and which then remained exhausted so long as the aeronaut desired. There were no wings or propellors to his machine, such as there had been to all previous aeroplanes, and the only engine required was the compact and powerful little appliance needed to contract the balloons. He perceived that such an apparatus as he had devised might rise with frame exhausted and balloons expanded to a considerable height, might then contract its balloons and let the air into its frame, and by an adjustment of its weights slide down the air in any desired direction. As it fell it would accumulate velocity and at the same time lose weight, and the momentum accumulated by its down-rush could be utilised by means of a shifting of its weights to drive it up in the air again as the balloons expanded. This con-

ception, which is still the structural conception of all successful flying machines, needed, however, a vast amount
of toil upon its details before it could actually be realised,
and such toil Filmer—as he was accustomed to tell the
numerous interviewers who crowded upon him in the
heyday of his fame—"ungrudgingly and unsparingly gave."
His particular difficulty was the elastic lining of the contractile balloon. He found he needed a new substance,
and in the discovery and manufacture of that new substance he had, as he never failed to impress upon the
interviewers, "performed a far more arduous work than
even in the actual achievement of my seemingly greater
discovery."

But it must not be imagined that these interviews followed
hard upon Filmer's proclamation of his invention. An
interval of nearly five years elapsed during which he timidly
remained at his rubber factory—he seems to have been
entirely dependent on his small income from this source—
making misdirected attempts to assure a quite indifferent
public that he really *had* invented what he had invented.
He occupied the greater part of his leisure in the composition of letters to the scientific and daily press, and so forth,
stating precisely the net result of his contrivances, and
demanding financial aid. That alone would have sufficed
for the suppression of his letters. He spent such holidays
as he could arrange in unsatisfactory interviews with the
door-keepers of leading London papers—he was singularly
not adapted for inspiring hall-porters with confidence—and
he positively attempted to induce the War Office to take
up his work with him. There remains a confidential letter
from Major-General Volleyfire to the Earl of Frogs. "The
man's a crank and a bounder to boot," says the Major-
General in his bluff, sensible, army way, and so left it
open for the Japanese to secure, as they subsequently did,
the priority in this side of warfare—a priority they still to
our great discomfort retain.

And then by a stroke of luck the membrane Filmer had
invented for his contractile balloon was discovered to be
useful for the valves of a new oil-engine, and he obtained

the means for making a trial model of his invention. He threw up his rubber factory appointment, desisted from all further writing, and, with a certain secrecy that seems to have been an inseparable characteristic of all his proceedings, set to work upon the apparatus. He seems to have directed the making of its parts and collected most of it in a room in Shoreditch, but its final putting together was done at Dymchurch, in Kent. He did not make the affair large enough to carry a man, but he made an extremely ingenious use of what were then called the Marconi rays to control its flight. The first flight of this first practicable flying machine took place over some fields near Burford Bridge, near Hythe, in Kent, and Filmer followed and controlled its flight upon a specially constructed motor tricycle.

The flight was, considering all things, an amazing success. The apparatus was brought in a cart from Dymchurch to Burford Bridge, ascended there to a height of nearly three hundred feet, swooped thence very nearly back to Dymchurch, came about in its sweep, rose again, circled, and finally sank uninjured in a field behind the Burford Bridge Inn. At its descent a curious thing happened. Filmer got off his tricycle, scrambled over the intervening dyke, advanced perhaps twenty yards towards his triumph, threw out his arms in a strange gesticulation, and fell down in a dead faint. Everyone could then recall the ghastliness of his features and all the evidences of extreme excitement they had observed throughout the trial, things they might otherwise have forgotten. Afterwards in the inn he had an unaccountable gust of hysterical weeping.

Altogether there were not twenty witnesses of this affair, and those for the most part uneducated men. The New Romney doctor saw the ascent but not the descent, his horse being frightened by the electrical apparatus on Filmer's tricycle and giving him a nasty spill. Two members of the Kent constabulary watched the affair from a cart in an unofficial spirit, and a grocer calling round the Marsh for orders and two lady cyclists seem almost to complete the list of educated people. There were two reporters present, one representing a Folkestone paper and

the other being a fourth-class interviewer and "symposium"
journalist, whose expenses down, Filmer, anxious as ever
for adequate advertisement—and now quite realising the
way in which adequate advertisement may be obtained—
had paid. The latter was one of those writers who can
throw a convincing air of unreality over the most credible
events, and his half-facetious account of the affair appeared
in the magazine page of a popular journal. But, happily
for Filmer, this person's colloquial methods were more convincing. He went to offer some further screed upon the
subject to Banghurst, the proprietor of the *New Paper*, and
one of the ablest and most unscrupulous men in London
journalism, and Banghurst instantly seized upon the situation. The interviewer vanishes from the narrative, no
doubt very doubtfully remunerated, and Banghurst, Banghurst himself, double chin, grey twill suit, abdomen, voice,
gestures and all, appears at Dymchurch, following his large,
unrivalled journalistic nose. He had seen the whole thing
at a glance, just what it was and what it might be.

At his touch, as it were, Filmer's long-pent investigations
exploded into fame. He instantly and most magnificently
was a Boom. One turns over the files of the journals of
the year 1907 with a quite incredulous recognition of how
swift and flaming the boom of those days could be. The
July papers know nothing of flying, see nothing in flying,
state by a most effective silence that men never would,
could, or should fly. In August flying and Filmer and
flying and parachutes and aerial tactics and the Japanese
Government and Filmer and again flying, shouldered the
war in Yunnan and the gold mines of Upper Greenland
off the leading page. And Banghurst had given ten
thousand pounds, and, further, Banghurst was giving five
thousand pounds, and Banghurst had devoted his well-
known, magnificent (but hitherto sterile) private laboratories
and several acres of land near his private residence on the
Surrey hills to the strenuous and violent completion—
Banghurst fashion—of the life-size practicable flying
machine. Meanwhile, in the sight of privileged multitudes
in the walled-garden of the Banghurst town residence in

Fulham, Filmer was exhibited at weekly garden parties putting the working model through its paces. At enormous initial cost, but with a final profit, the *New Paper* presented its readers with a beautiful photographic souvenir of the first of these occasions.

Here again the correspondence of Arthur Hicks and his friend Vance comes to our aid.

"I saw Filmer in his glory," he writes, with just the touch of envy natural to his position as a poet *passé*. "The man is brushed and shaved, dressed in the fashion of a Royal-Institution-Afternoon Lecturer, the very newest shape in frock-coats and long patent shoes, and altogether in a state of extraordinary streakiness between an owlish great man and a scared abashed self-conscious bounder cruelly exposed. He hasn't a touch of colour in the skin of his face, his head juts forward, and those queer little dark amber eyes of his watch furtively round him for his fame. His clothes fit perfectly and yet sit upon him as though he had bought them ready-made. He speaks in a mumble still, but he says, you perceive indistinctly, enormous self-assertive things, he backs into the rear of groups by instinct if Banghurst drops the line for a minute, and when he walks across Banghurst's lawn one perceives him a little out of breath and going jerky, and that his weak white hands are clenched. His is a state of tension—horrible tension. And he is the Greatest Discoverer of This or Any Age—the Greatest Discoverer of This or Any Age! What strikes one so forcibly about him is that he didn't somehow quite expect it ever, at any rate, not at all like this. Banghurst is about everywhere, the energetic M.C. of his great little catch, and I swear he will have everyone down on his lawn there before he has finished with the engine; he had bagged the prime minister yesterday, and he, bless his heart! didn't look particularly outsize, on the very first occasion. Conceive it! Filmer! Our obscure unwashed Filmer, the Glory of British science! Duchesses crowd upon him, beautiful, bold peeresses say in their beautiful, clear loud voices—have you noticed how penetrating the great lady is becoming nowadays?—'Oh, Mr. Filmer, how *did* you do it?'

"Common men on the edge of things are too remote for the answer. One imagines something in the way of that interview, 'toil ungrudgingly and unsparingly given, Madam, and, perhaps—I don't know—but perhaps a little special aptitude.'"

So far Hicks, and the photographic supplement to the *New Paper* is in sufficient harmony with the description. In one picture the machine swings down towards the river, and the tower of Fulham church appears below it through a gap in the elms, and in another, Filmer sits at his guiding batteries, and the great and beautiful of the earth stand around him, with Banghurst massed modestly but resolutely in the rear. The grouping is oddly apposite. Occluding much of Banghurst, and looking with a pensive, speculative expression at Filmer, stands the Lady Mary Elkinghorn, still beautiful, in spite of the breath of scandal and her eight-and-thirty years, the only person whose face does not admit a perception of the camera that was in the act of snapping them all.

So much for the exterior facts of the story, but, after all, they are very exterior facts. About the real interest of the business one is necessarily very much in the dark. How was Filmer feeling at the time? How much was a certain unpleasant anticipation present inside that very new and fashionable frock-coat? He was in the halfpenny, penny, sixpenny, and more expensive papers alike, and acknowledged by the whole world as "the Greatest Discoverer of This or Any Age." He had invented a practicable flying machine, and every day down among the Surrey hills the life-sized model was getting ready. And when it was ready, it followed as a clear inevitable consequence of his having invented and made it—everybody in the world, indeed, seemed to take it for granted; there wasn't a gap anywhere in that serried front of anticipation—that he would proudly and cheerfully get aboard it, ascend with it, and fly.

But we know now pretty clearly that simple pride and cheerfulness in such an act were singularly out of harmony with Filmer's private constitution. It occurred to no one at the time, but there the fact is. We can guess with some confidence now that it must have been drifting about in his mind a great deal during the day, and, from a little note to his physician complaining of persistent insomnia, we have the soundest reason for supposing it dominated his nights—the idea that it would be after all, in spite of his theoretical security, an abominally sickening, uncomfortable, and dangerous thing for him to flap about in nothingness a thousand feet or so in the air. It must have dawned upon him quite early in the period of being the Greatest Discoverer of This or Any Age, the vision of doing this and that with an extensive void below. Perhaps somewhere in his youth he had looked down a great height or fallen down in some excessively uncomfortable way; perhaps some habit of sleeping on the wrong side had resulted in that disagreeable falling nightmare one knows, and given him his horror; of the strength of that horror there remains now not a particle of doubt.

Apparently he had never weighed this duty of flying in his earlier days of research; the machine had been his end, but now things were opening out beyond his end, and particularly this giddy whirl up above there. He was a Discoverer and he had Discovered. But he was not a Flying Man, and it was only now that he was beginning to perceive clearly that he was expected to fly. Yet, however much the thing was present in his mind he gave no expression to it until the very end, and meanwhile he went to and fro from Banghurst's magnificent laboratories, and was interviewed and lionised, and wore good clothes, and ate good food, and lived in an elegant flat, enjoying a very abundant feast of such good, coarse, wholesome Fame and Success as a man, starved for all his years as he had been starved, might be reasonably expected to enjoy.

After a time, the weekly gatherings in Fulham ceased. The model had failed one day just for a moment to respond to Filmer's guidance, or he had been distracted by the compliments of an archbishop. At any rate, it suddenly dug its nose into the air just a little too steeply as the archbishop was sailing through a Latin quotation for all the world like an archbishop in a book, and it came down in the Fulham Road within three yards of a 'bus horse. It stood for a second perhaps, astonishing and in its attitude astonished, then it crumpled, shivered into pieces, and the 'bus horse was incidentally killed.

Filmer lost the end of the archiepiscopal compliment. He stood up and stared as his invention swooped out of sight and reach of him. His long, white hands still gripped his useless apparatus. The archbishop followed his skyward stare with an apprehension unbecoming in an archbishop.

Then came the crash and the shouts and uproar from the road to relieve Filmer's tension. "My God!" he whispered, and sat down.

Everyone else almost was staring to see where the machine had vanished, or rushing into the house.

The making of the big machine progressed all the more rapidly for this. Over its making presided Filmer, always a little slow and very careful in his manner, always with a growing preoccupation in his mind. His care over the strength and soundness of the apparatus was prodigious. The slightest doubt, and he delayed everything until the doubtful part could be replaced. Wilkinson, his senior assistant, fumed at some of these delays, which, he insisted, were for the most part unnecessary. Banghurst magnified the patient certitude of Filmer in the *New Paper*, and reviled it bitterly to his wife, and MacAndrew, the second assistant, approved Filmer's wisdom. "We're not wanting a *fiasco*, man," said MacAndrew. "He's perfectly well advised."

And whenever an opportunity arose Filmer would expound to Wilkinson and MacAndrew just exactly how every part of the flying machine was to be controlled and worked, so that in effect they would be just as capable, and even more capable, when at last the time came, of guiding it through the skies.

Now I should imagine that if Filmer had seen fit at this stage to define just what he was feeling, and to take a

definite line in the matter of his ascent, he might have
escaped that painful ordeal quite easily. If he had had it
clearly in his mind he could have done endless things.
He would surely have found no difficulty with a specialist
to demonstrate a weak heart, or something gastric or pul-
monary, to stand in his way—that is the line I am astonished
he did not take—or he might, had he been man enough,
have declared simply and finally that he did not intend
to do the thing. But the fact is, though the dread was
hugely present in his mind, the thing was by no means
sharp and clear. I fancy that all through this period he
kept telling himself that when the occasion came he would
find himself equal to it. He was like a man just gripped
by a great illness, who says he feels a little out of sorts,
and expects to be better presently. Meanwhile he delayed
the completion of the machine, and let the assumption
that he was going to fly it take root and flourish exceedingly
about him. He even accepted anticipatory compliments on
his courage. And, barring this secret squeamishness, there
can be no doubt he found all the praise and distinction
and fuss he got a delightful and even intoxicating draught.

The Lady Mary Elkinghorn made things a little more
complicated for him.

How *that* began was a subject of inexhaustible speculation to Hicks. Probably in the beginning she was just a
little "nice" to him with that impartial partiality of hers,
and it may be that to her eyes, standing out conspicuously
as he did ruling his monster in the upper air, he had a
distinction that Hicks was not disposed to find. And
somehow they must have had a moment of sufficient isolation, and the great Discoverer a moment of sufficient courage
for something just a little personal to be mumbled or
blurted. However it began, there is no doubt that it did
begin, and presently became quite perceptible to a world
accustomed to find in the proceedings of the Lady Mary
Elkinghorn a matter of entertainment. It complicated
things, because the state of love in such a virgin mind as
Filmer's would brace his resolution, if not sufficiently, at
any rate considerably, towards facing a danger he feared,

and hampered him in such attempts at evasion as would
otherwise be natural and congenial.

It remains a matter for speculation just how the Lady
Mary felt for Filmer and just what she thought of him.
At thirty-eight one may have gathered much wisdom and
still be not altogether wise, and the imagination still functions actively enough in creating glamours and effecting
the impossible. He came before her eyes as a very central man,
and that always counts, and he had powers, unique powers
as it seemed, at any rate in the air. The performance with
the model had just a touch of the quality of a potent incantation, and women have ever displayed an unreasonable
disposition to imagine that when a man has powers he
must necessarily have Power. Given so much, and what
was not good in Filmer's manner and appearance became an
added merit. He was modest, he hated display, but given
an occasion where *true* qualities are needed, then—then one
would see!

The late Mrs. Bampton thought it wise to convey to Lady
Mary her opinion that Filmer, all things considered, was
rather a "grub." "He's certainly not a sort of man I have
have ever met before," said the Lady Mary, with a quite
unruffled serenity. And Mrs. Bampton, after a swift,
imperceptible glance at that serenity, decided that so far as
saying anything to Lady Mary went, she had done as much
as could be expected of her. But she said a great deal to
other people.

And at last, without any undue haste or unseemliness,
the day dawned, the great day, when Banghurst had promised
his public—the world in fact—that flying should be finally
attained and overcome. Filmer saw it dawn, watched even
in the darkness before it dawned, watched its stars fade and
the grey and pearly pinks give place at last to the clear
blue sky of a sunny, cloudless day. He watched it from
the window of his bedroom in the new-built wing of Banghurst's Tudor house. And as the stars were overwhelmed
and the shapes and substances of things grew into being
out of the amorphous dark, he must have seen more and
more distinctly the festive preparations beyond the beech

clumps near the green pavilion in the outer park, the three stands for the privileged spectators, the raw, new fencing of the enclosure, the sheds and workshops, the Venetian masts and fluttering flags that Banghurst had considered essential, black and limp in the breezeless dawn, and amidst all these things a great shape covered with tarpauling. A strange and terrible portent for humanity was that shape, a beginning that must surely spread and widen and change and dominate all the affairs of men, but to Filmer it is very doubtful whether it appeared in anything but a narrow and personal light. Several people heard him pacing in the small hours—for the vast place was packed with guests by a proprietor editor who, before all things, understood compression. And about five o'clock, if not before, Filmer left his room and wandered out of the sleeping house into the park, alive by that time with sunlight and birds and squirrels and the fallow deer. MacAndrew, who was also an early riser, met him near the machine, and they went and had a look at it together.

It is doubtful if Filmer took any breakfast, in spite of the urgency of Banghurst. So soon as the guests began to be about in some number he seems to have retreated to his room. Thence about ten he went into the shrubbery, very probably because he had seen the Lady Mary Elkinghorn there. She was walking up and down, engaged in conversation with her old school friend, Mrs. Brewis-Craven, and although Filmer had never met the latter lady before, he joined them and walked beside them for some time. There were several silences in spite of Lady Mary's brilliance. The situation was a difficult one, and Mrs. Brewis-Craven did not master its difficulty. "He struck me," she said afterwards with a luminous self-contradiction, "as a very unhappy person who had something to say, and wanted before all things to be helped to say it. But how was one to help him when one didn't know what it was?"

At half-past eleven the enclosures for the public in the outer park were crammed, there was an intermittent stream of equipages along the belt which circles the outer park, and the house party was dotted over the lawn and shrubbery and the corner of the inner park, in a series of brilliantly attired knots, all making for the flying machine. Filmer walked in a group of three with Banghurst, who was supremely and conspicuously happy, and Sir Theodore Hickle, the president of the Aeronautical Society. Mrs. Banghurst was close behind with the Lady Mary Elkinghorn, Georgina Hickle, and the Dean of Stays. Banghurst was large and copious in speech, and such interstices as he left were filled in by Hickle with complimentary remarks to Filmer. And Filmer walked between them saying not a word except by way of unavoidable reply. Behind, Mrs. Banghurst listened to the admirably suitable and shapely conversation of the Dean with that fluttered attention to the ampler clergy ten years of social ascent and ascendency had not cured in her; and the Lady Mary watched, no doubt with an entire confidence in the world's disillusionment, the drooping shoulders of the sort of man she had never met before.

There was some cheering as the central party came into view of the enclosures, but it was not very unanimous nor invigorating cheering. They were within fifty yards of the apparatus when Filmer took a hasty glance over his shoulder to measure the distance of the ladies behind them, and decided to make the first remark he had initiated since the house had been left. His voice was just a little hoarse, and he cut in on Banghurst in mid-sentence on Progress.

"I say, Banghurst," he said, and stopped.

"Yes," said Banghurst.

"I wish——" He moistened his lips. "I'm not feeling well." Banghurst stopped dead. "Eh?" he shouted.

"A queer feeling." Filmer made to move on, but Banghurst was immovable. "I don't know. I may be better in a minute. If not—perhaps. . . . MacAndrew——"

"You're not feeling *well?*" said Banghurst, and stared at his white face.

"My dear!" he said, as Mrs. Banghurst came up with them, "Filmer says he isn't feeling *well.*"

"A little queer," exclaimed Filmer, avoiding the Lady Mary's eyes. "It may pass off——"

There was a pause.

It came to Filmer that he was the most isolated person in the world.

"In any case," said Banghurst, "the ascent must be made. Perhaps if you were to sit down somewhere for a moment——"

"It's the crowd, I think," said Filmer.

There was a second pause. Banghurst's eye rested in scrutiny on Filmer, and then swept the sample of public in the enclosure.

"It's unfortunate," said Sir Theodore Hickle; "but still—I suppose—— Your assistants—— Of course, if you feel out of condition and disinclined——"

"I don't think Mr. Filmer would permit *that* for a moment," said the Lady Mary.

"But if Mr. Filmer's nerve is run—— It might even be dangerous for him to attempt——" Hickle coughed.

"It's just because it's dangerous," began the Lady Mary, and felt she had made her point of view and Filmer's plain enough.

Conflicting motives struggled for Filmer.

"I feel I ought to go up," he said, regarding the ground. He looked up and met the Lady Mary's eyes. "I want to go up," he said, and smiled whitely at her. He turned towards Banghurst. "If I could just sit down somewhere for a moment out of the crowd and sun——"

Banghurst, at least, was beginning to understand the case. "Come into my little room in the green pavilion," he said. "It's quite cool there." He took Filmer by the arm.

Filmer turned his face to the Lady Mary Elkinghorn again. "I shall be all right in five minutes," he said. "I'm tremendously sorry——"

The Lady Mary Elkinghorn smiled at him. "I couldn't think," he said to Hickle, and obeyed the compulsion of Banghurst's pull.

The rest remained watching the two recede.

"He's so fragile," said the Lady Mary.

"He's certainly a highly nervous type," said the Dean, whose weakness it was to regard the whole world, except married clergymen with enormous families, as "neurotic."

"Of course," said Hickle, "it isn't absolutely necessary for him to go up because he has invented——"

"How *could* he avoid it?" asked the Lady Mary, with the faintest shadow of scorn.

"It's certainly most unfortunate if he's going to be ill now," said Mrs. Banghurst a little severely.

"He's not going to be ill," said the Lady Mary, and certainly she had met Filmer's eye.

"*You'll* be all right," said Banghurst, as they went towards the pavilion. "All you want is a nip of brandy. It ought to be you, you know. You'll be—you'd get it rough, you know, if you let another man——"

"Oh, I want to go," said Filmer. "I shall be all right. As a matter of fact I'm almost inclined *now*—— No! I think I'll have that nip of brandy first."

Banghurst took him into the little room and routed out an empty decanter. He departed in search of a supply. He was gone perhaps five minutes.

The history of those five minutes cannot be written. At intervals Filmer's face could be seen by the people on the easternmost of the stands erected for spectators, against the window pane peering out, and then it would recede and fade. Banghurst vanished shouting behind the grand stand, and presently the butler appeared going pavilionward with a tray.

The apartment in which Filmer came to his last solution was a pleasant little room very simply furnished with green furniture and an old bureau—for Banghurst was simple in all his private ways. It was hung with little engravings after Morland and it had a shelf of books. But as it happened, Banghurst had left a rook rifle he sometimes played with on the top of the desk, and on the corner of the mantel-shelf was a tin with three or four cartridges remaining in it. As Filmer went up and down that room wrestling with his intolerable dilemma he went first towards the neat little rifle athwart the blotting-pad and then towards the neat little red label

"·22 LONG."

The thing must have jumped into his mind in a moment.

Nobody seems to have connected the report with him, though the gun, being fired in a confined space, must have sounded loud, and there were several people in the billiard-room, separated from him only by a lath-and-plaster partition. But directly Banghurst's butler opened the door and smelt the sour smell of the smoke, he knew, he says, what had happened. For the servants at least of Banghurst's household had guessed something of what was going on in Filmer's mind.

All through that trying afternoon Banghurst behaved as he held a man should behave in the presence of hopeless disaster, and his guests for the most part succeeded in not insisting upon the fact—though to conceal their perception of it altogether was impossible—that Banghurst had been pretty elaborately and completely swindled by the deceased. The public in the enclosure, Hicks told me, dispersed "like a party that has been ducking a welsher," and there wasn't a soul in the train to London, it seems, who hadn't known all along that flying was a quite impossible thing for man. "But he might have tried it," said many, "after carrying the thing so far."

In the evening, when he was comparatively alone, Banghurst broke down and went on like a man of clay. I have been told he wept, which must have made an imposing scene, and he certainly said Filmer had ruined his life, and offered and sold the whole apparatus to MacAndrew for half-a-crown. "I've been thinking——" said MacAndrew at the conclusion of the bargain, and stopped.

The next morning the name of Filmer was, for the first time, less conspicuous in the *New Paper* than in any other daily paper in the world. The rest of the world's instructors, with varying emphasis, according to their dignity and the degree of competition between themselves and the *New Paper*, proclaimed the "Entire Failure of the New Flying Machine," and "Suicide of the Impostor." But in the district of North Surrey the reception of the news was tempered by a perception of unusual aerial phenomena.

Overnight Wilkinson and MacAndrew had fallen into

violent argument on the exact motives of their principal's rash act.

"The man was certainly a poor, cowardly body, but so far as his science went he was *no* impostor," said Mac-Andrew, "and I'm prepared to give that proposition a very practical demonstration, Mr. Wilkinson, so soon as we've got the place a little more to ourselves. For I've no faith in all this publicity for experimental trials."

And to that end, while all the world was reading of the certain failure of the new flying machine, MacAndrew was soaring and curvetting with great amplitude and dignity over the Epsom and Wimbledon divisions; and Banghurst, restored once more to hope and energy, and regardless of public security and the Board of Trade, was pursuing his gyrations and trying to attract his attention, on a motor car and in his pyjamas—he had caught sight of the ascent when pulling up the blind of his bedroom window—equipped, among other things, with a film camera that was subsequently discovered to be jammed.

And Filmer was lying on the billiard table in the green pavilion with a sheet about his body.

STORY THE SECOND

The Magic Shop

I HAD seen the Magic Shop from afar several times; I had passed it once or twice, a shop window of alluring little objects, magic balls, magic hens, wonderful cones, ventriloquist dolls, the material of the basket trick, packs of cards that *looked* all right, and all that sort of thing, but never had I thought of going in until one day, almost without warning, Gip hauled me by my finger right up to the window, and so conducted himself that there was nothing for it but to take him in. I had not thought the place was there, to tell the truth—a modest-sized frontage in Regent Street, between the picture shop and the place where the chicks run about just out of patent incubators—but there it was sure enough. I had fancied it was down nearer the Circus, or round the corner in Oxford Street, or even in Holborn; always over the way and a little inaccessible it had been, with something of the mirage in its position; but here it was now quite indisputably, and the fat end of Gip's pointing finger made a noise upon the glass.

"If I was rich," said Gip, dabbing a finger at the Disappearing Egg, "I'd buy myself that. And that"—which was The Crying Baby, Very Human—"and that," which was a mystery, and called, so a neat card asserted, "Buy One and Astonish Your Friends."

"Anything," said Gip, "will disappear under one of those cones. I have read about it in a book.

"And there, dadda, is the Vanishing Halfpenny—only they've put it this way up so's we can't see how it's done."

Gip, dear boy, inherits his mother's breeding, and he did

not propose to enter the shop or worry in any way; only, you know, quite unconsciously he lugged my finger door-ward, and he made his interest clear.

"That," he said, and pointed to the Magic Bottle.

"If you had that?" I said; at which promising inquiry he looked up with a sudden radiance.

"I could show it to Jessie," he said, thoughtful as ever of others.

"It's less than a hundred days to your birthday, Gibbles," I said, and laid my hand on the door-handle.

Gip made no answer, but his grip tightened on my finger, and so we came into the shop.

It was no common shop this; it was a magic shop, and all the prancing precedence Gip would have taken in the matter of mere toys was wanting. He left the burthen of the conversation to me.

It was a little, narrow shop, not very well lit, and the door-bell pinged again with a plaintive note as we closed it behind us. For a moment or so we were alone and could glance about us. There was a tiger in *papier-maché* on the glass case that covered the low counter—a grave, kind-eyed tiger that waggled his head in a methodical manner; there were several crystal spheres, a china hand holding magic cards, a stock of magic fish-bowls in various sizes, and an immodest magic hat that shamelessly displayed its springs. On the floor were magic mirrors; one to draw you out long and thin, one to swell your head and vanish your legs, and one to make you short and fat like a draught; and while we were laughing at these the shopman, as I suppose, came in.

At any rate, there he was behind the counter—a curious, sallow, dark man, with one ear larger than the other and a chin like the toe-cap of a boot.

"What can we have the pleasure?" he said, spreading his long, magic fingers on the glass case; and so with a start we were aware of him.

"I want," I said, "to buy my little boy a few simple tricks."

"Legerdemain?" he asked. "Mechanical? Domestic?"

"Anything amusing?" said I.

"Um!" said the shopman, and scratched his head for a moment as if thinking. Then, quite distinctly, he drew from his head a glass ball. "Something in this way?" he said, and held it out.

The action was unexpected. I had seen the trick done at entertainments endless times before—it's part of the common stock of conjurers—but I had not expected it here. "That's good," I said, with a laugh.

"Isn't it?" said the shopman.

Gip stretched out his disengaged hand to take this object and found merely a blank palm.

"It's in your pocket," said the shopman, and there it was!

"How much will that be?" I asked.

"We make no charge for glass balls," said the shopman, politely. "We get them"—he picked one out of his elbow as he spoke—"free." He produced another from the back of his neck, and laid it beside its predecessor on the counter. Gip regarded his glass ball sagely, then directed a look of inquiry at the two on the counter, and finally brought his round-eyed scrutiny to the shopman, who smiled. "You may have those too," said the shopman, "and if you *don't* mind, one from my mouth— *So!*"

Gip counselled me mutely for a moment, and then in a profound silence put away the four balls, resumed my reassuring finger, and nerved himself for the next event.

"We get all our smaller tricks in that way," the shopman remarked.

I laughed in the manner of one who subscribes to a jest. "Instead of going to the wholesale shop," I said. "Of course, it's cheaper."

"In a way," the shopman said. "Though we pay in the end. But not so heavily—as people suppose. . . . Our larger tricks, and our daily provisions and all the other things we want, we get out of that hat. . . . And you know, sir, if you'll excuse my saying it, there *isn't* a wholesale shop, not for Genuine Magic goods, sir. I don't know if you noticed our inscription—the Genuine Magic shop." He drew a business-card from his cheek and handed it to me.

"Genuine," he said, with his finger on the word, and added, "There is absolutely no deception, sir."

He seemed to be carrying out the joke pretty thoroughly, I thought.

He turned to Gip with a smile of remarkable affability. "You, you know, are the Right Sort of Boy."

I was surprised at his knowing that, because, in the interests of discipline, we keep it rather a secret even at home; but Gip received it in unflinching silence, keeping a steadfast eye on him.

"It's only the Right Sort of Boy gets through that doorway."

And as if by way of illustration, there came a rattling at the door, and a squeaking little voice could be faintly heard. "N yar! I *warn* a' go in there dadda, I WARN 'a go in there. Nya-a-a-ah!" and then the accents of a down-trodden parent, urging consolations and propitiations. "It's locked, Edward," he said.

"But it isn't," said I.

"It is, sir," said the shopman, "always—for that sort of child," and as he spoke we had a glimpse of the other youngster, a small, white face, pallid from sweet-eating and over-sapid food, and distorted by evil passions, a ruthless little egotist, pawing at the enchanted pane. "It's no good, sir," said the shopman, as I moved, with my natural helpfulness, doorward, and presently the spoilt child was carried off howling.

"How do you manage that?" I said, breathing more freely.

"Magic!" said the shopman, with a careless wave of the hand, and behold! sparks of coloured fire flew out of his fingers and vanished into the shadows of the shop.

"You were saying," he said, addressing himself to Gip, "before you came in, that you would like one of our 'Buy One and Astonish your Friends' boxes?"

Gip, after a gallant effort, said "Yes."

"It's in your pocket."

And leaning over the counter—he really had an extraordinarily long body—this amazing person produced the article in the customary conjurer's manner. "Paper," he

said, and took a sheet out of the empty hat with the springs;
"string," and behold his mouth was a string-box, from
which he drew an unending thread, which when he had
tied his parcel he bit off—and, it seemed to me, swallowed
the ball of string. And then he lit a candle at the nose
of one of the ventriloquist's dummies, stuck one of his
fingers (which had become sealing-wax red) into the flame,
and so sealed the parcel. "Then there was the Disappearing
Egg," he remarked, and produced one from within my coat-
breast and packed it, and also The Crying Baby, Very
Human. I handed each parcel to Gip as it was ready, and
he clasped them to his chest.

He said very little, but his eyes were eloquent; the clutch
of his arms was eloquent. He was the playground of unspeakable emotions. These, you know, were *real* Magics.

Then, with a start, I discovered something moving about
in my hat—something soft and jumpy. I whipped it off, and
a ruffled pigeon—no doubt a confederate—dropped out and
ran on the counter, and went, I fancy, into a cardboard
box behind the *papier-maché* tiger.

"Tut, tut!" said the shopman, dexterously relieving me of
my headdress; "careless bird, and—as I live—nesting!"

He shook my hat, and shook out into his extended hand
two or three eggs, a large marble, a watch, about half-a-dozen
of the inevitable glass balls, and the crumpled, crinkled
paper, more and more and more, talking all the time of
the way in which people neglect to brush their hats *inside*
as well as out, politely, of course, but with a certain personal application. "All sorts of things accumulate, sir. . . .
Not *you*, of course, in particular. . . . Nearly every customer. . . . Astonishing what they carry about with them.
. . ." The crumpled paper rose and billowed on the counter
more and more and more, until he was nearly hidden from
us, until he was altogether hidden, and still his voice went
on and on. "We none of us know what the fair semblance
of a human being may conceal, sir. Are we all then no
better than brushed exteriors, whited sepulchres——"

His voice stopped—exactly like when you hit a neigh-
bour's gramophone with a well-aimed brick, the same instant

silence, and the rustle of the paper stopped, and everything was still. . . .

"Have you done with my hat?" I said, after an interval.

There was no answer.

I stared at Gip, and Gip stared at me; and there were our distortions in the magic mirrors, looking very rum, and grave, and quiet. . . .

"I think we'll go now," I said. "Will you tell me how much all this comes to? . . .

"I say," I said, on a rather louder note, "I want the bill; and my hat, please."

It might have been a sniff from behind the paper pile. . . .

"Let's look behind the counter, Gip," I said. "He's making fun of us."

I led Gip round the head-wagging tiger, and what do you think there was behind the counter? No one at all! Only my hat on the floor, and a common conjurer's lop-eared white rabbit lost in meditation, and looking as stupid and crumpled as only a conjurer's rabbit can do. I resumed my hat, and the rabbit lolloped a lollop or so out of my way.

"Dadda!" said Gip, in a guilty whisper.

"What is it, Gip?" said I.

"I *do* like this shop, dadda."

"So should I," I said to myself, "if the counter wouldn't suddenly extend itself to shut one off from the door." But I didn't call Gip's attention to that. "Pussy!" he said, with a hand out to the rabbit as it came lolloping past us; "Pussy, do Gip a magic!" and his eyes followed it as it squeezed through a door I had certainly not remarked a moment before. Then this door opened wider, and the man with one ear larger than the other appeared again. He was smiling still, but his eye met mine with something between amusement and defiance. "You'd like to see our showroom, sir," he said, with an innocent suavity. Gip tugged my finger forward. I glanced at the counter and met the shopman's eye again. I was beginning to think the magic just a little too genuine. "We haven't *very* much time,"

I said. But somehow we were inside the showroom before I could finish that.

"All goods of the same quality," said the shopman, rubbing his flexible hands together, "and that is the Best. Nothing in the place that isn't genuine Magic, and warranted thoroughly rum. Excuse me, sir!"

I felt him pull at something that clung to my coat-sleeve, and then I saw he held a little, wriggling red demon by the tail—the little creature bit and fought and tried to get at his hand—and in a moment he tossed it carelessly behind a counter. No doubt the thing was only an image of twisted indiarubber, but for the moment——! And his gesture was exactly that of a man who handles some petty biting bit of vermin. I glanced at Gip, but Gip was looking at a magic rocking-horse. I was glad he hadn't seen the thing. "I say," I said, in an undertone, and indicating Gip and the red demon with my eyes, "you haven't many things like *that* about, have you?"

"None of ours! Probably brought it with you," said the shopman—also in an undertone, and with a more dazzling smile than ever. "Astonishing what people *will* carry about with them unawares!" And then to Gip, "Do you see anything you fancy here?"

There were many things that Gip fancied there.

He turned to this astonishing tradesman with mingled confidence and respect. "Is that a Magic Sword?" he said.

"A Magic Toy Sword. It neither, bends, breaks, nor cuts the fingers. It renders the bearer invincible in battle against anyone under eighteen. Half-a-crown to seven and sixpence, according to size. These panoplies on cards are for juvenile knights-errant and very useful—shield of safety, sandals of swiftness, helmet of invisibility."

"Oh, dadda!" gasped Gip.

I tried to find out what they cost, but the shopman did not heed me. He had got Gip now; he had got him away from my finger; he had embarked upon the exposition of all his confounded stock, and nothing was going to stop him. Presently I saw with a qualm of distrust and something very like jealousy that Gip had hold of this person's finger as usually he has hold of mine. No doubt the fellow was interesting, I thought, and had an interestingly faked lot of stuff, really *good* faked stuff, still——

I wandered after them, saying very little, but keeping an eye on this prestidigital fellow. After all, Gip was enjoying it. And no doubt when the time came to go we should be able to go quite easily.

It was a long, rambling place, that showroom, a gallery broken up by stands and stalls and pillars, with archways leading off to other departments, in which the queerest-looking assistants loafed and stared at one, and with perplexing mirrors and curtains. So perplexing, indeed, were these that I was presently unable to make out the door by which we had come.

The shopman showed Gip magic trains that ran without steam or clockwork, just as you set the signals, and then some very, very valuable boxes of soldiers that all came alive directly you took off the lid and said—— I myself haven't a very quick ear and it was a tongue-twisting sound, but Gip—he has his mother's ear—got it in no time. "Bravo!" said the shopman, putting the men back into the box unceremoniously and handing it to Gip. "Now," said the shopman, and in a moment Gip had made them all alive again.

"You'll take that box?" asked the shopman.

"We'll take that box," said I, "unless you charge its full value. In which case it would need a Trust Magnate——"

"Dear heart! *No!*" and the shopman swept the little men back again, shut the lid, waved the box in the air, and there it was, in brown paper, tied up and—*with Gip's full name and address on the paper!*

The shopman laughed at my amazement.

"This is the genuine magic," he said. "The real thing."

"It's almost too genuine for my taste," I said again.

After that he fell to showing Gip tricks, odd tricks, and still odder the way they were done. He explained them, he turned them inside out, and there was the dear little chap nodding his busy bit of a head in the sagest manner.

I did not attend as well as I might. "Hey, presto!" said

the Magic Shopman, and then would come the clear, small "Hey, presto!" of the boy. But I was distracted by other things. It was being borne in upon me just how tremendously rum this place was; it was, so to speak, inundated by a sense of rumness. There was something vaguely rum about the fixtures even, about the ceiling, about the floor, about the casually distributed chairs. I had a queer feeling that whenever I wasn't looking at them straight they went askew, and moved about, and played a noiseless puss-in-the-corner behind my back. And the cornice had a serpentine design with masks—masks altogether too expressive for proper plaster.

Then abruptly my attention was caught by one of the odd-looking assistants. He was some way off and evidently unaware of my presence—I saw a sort of three-quarter length of him over a pile of toys and through an arch—and, you know, he was leaning against a pillar in an idle sort of way doing the most horrid things with his features! The particularly horrid thing he did was with his nose. He did it just as though he was idle and wanted to amuse himself. First of all it was a short, blobby nose, and then suddenly he shot it out like a telescope, and then out it flew and became thinner and thinner until it was like a long, red, flexible whip. Like a thing in a nightmare it was! He flourished it about and flung it forth as a fly-fisher flings his line.

My instant thought was that Gip mustn't see him. I turned about, and there was Gip quite preoccupied with the shopman, and thinking no evil. They were whispering together and looking at me. Gip was standing on a stool, and the shopman was holding a sort of big drum in his hand.

"Hide and seek, dadda!" cried Gip. "You're He!"

And before I could do anything to prevent it, the shopman had clapped the big drum over him.

I saw what was up directly. "Take that off," I cried, "this instant! You'll frighten the boy. Take it off!"

The shopman with the unequal ears did so without a word, and held the big cylinder towards me to show its emptiness. And the stool was vacant! In that instant my boy had utterly disappeared! . . .

You know, perhaps, that sinister something that comes like a hand out of the unseen and grips your heart about. You know it takes your common self away and leaves you tense and deliberate, neither slow nor hasty, neither angry nor afraid. So it was with me.

I came up to this grinning shopman and kicked his stool aside.

"Stop this folly!" I said. "Where is my boy?"

"You see," he said, still displaying the drum's interior, "there is no deception——"

I put out my hand to grip him, and he eluded me by a dexterous movement. I snatched again, and he turned from me and pushed open a door to escape. "Stop!" I said, and he laughed, receding. I leapt after him—into utter darkness.

Thud!

"Lor' bless my 'eart! I didn't see you coming, sir!"

I was in Regent Street, and I had collided with a decent-looking working man; and a yard away, perhaps, and looking extremely perplexed with himself, was Gip. There was some sort of apology, and then Gip had turned and come to me with a bright little smile, as though for a moment he had missed me.

And he was carrying four parcels in his arm!

He secured immediate possession of my finger.

For the second I was rather at a loss. I stared round to see the door of the magic shop, and, behold, it was not there! There was no door, no shop, nothing, only the common pilaster between the shop where they sell pictures and the window with the chicks! . . .

I did the only thing possible in that mental tumult; I walked straight to the kerbstone and held up my umbrella for a cab.

"'Ansoms," said Gip, in a note of culminating exultation.

I helped him in, recalled my address with an effort, and got in also. Something unusual proclaimed itself in my tail-coat pocket, and I felt and discovered a glass ball. With a petulant expression I flung it into the street.

Gip said nothing.

For a space neither of us spoke.

"Dadda!" said Gip, at last, "that *was* a proper shop!"

I came round with that to the problem of just how the whole thing had seemed to him. He looked completely undamaged—so far, good; he was neither scared nor unhinged, he was simply tremendously satisfied with the afternoon's entertainment, and there in his arms were the four parcels.

Confound it! what could be in them?

"Um!" I said. "Little boys can't go to shops like that every day."

He received this with his usual stoicism, and for a moment I was sorry I was his father and not his mother, and so couldn't suddenly there, *coram publico*, in our hansom, kiss him. After all, I thought, the thing wasn't so very bad.

But it was only when we opened the parcels that I really began to be reassured. Three of them contained boxes of soldiers, quite ordinary lead soldiers, but of so good a quality as to make Gip altogether forget that originally these parcels had been Magic Tricks of the only genuine sort, and the fourth contained a kitten, a little living white kitten, in excellent health and appetite and temper.

I saw this unpacking with a sort of provisional relief. I hung about in the nursery for quite an unconscionable time. . . .

That happened six months ago. And now I am beginning to believe it is all right. The kitten had only the magic natural to all kittens, and the soldiers seem as steady a company as any colonel could desire. And Gip——?

The intelligent parent will understand that I have to go cautiously with Gip.

But I went so far as this one day. I said, "How would you like your soldiers to come alive, Gip, and march about by themselves?"

"Mine do," said Gip. "I just have to say a word I know before I open the lid."

"Then they march about alone?"

"Oh, *quite*, dadda. I shouldn't like them if they didn't do that."

I displayed no unbecoming surprise, and since then I have taken occasion to drop in upon him once or twice, unannounced, when the soldiers were about, but so far I have never discovered them performing in anything like a magical manner. . . .

It's so difficult to tell.

There's also a question of finance. I have an incurable habit of paying bills. I have been up and down Regent Street several times, looking for that shop. I am inclined to think, indeed, that in that matter honour is satisfied, and that, since Gip's name and address are known to them, I may very well leave it to these people, whoever they may be, to send in their bill in their own time.

STORY THE THIRD

The Valley of Spiders

TOWARDS midday the three pursuers came abruptly round a bend in the torrent bed upon the sight of a very broad and spacious valley. The difficult and winding trench of pebbles along which they had tracked the fugitives for so long, expanded to a broad slope, and with a common impulse the three men left the trail, and rode to a low eminence set with olive-dun trees, and there halted, the two others, as became them, a little behind the man with the silver-studded bridle.

For a space they scanned the great expanse below them with eager eyes. It spread remoter and remoter, with only a few clusters of sere thorn bushes here and there, and the dim suggestions of some now waterless ravine to break its desolation of yellow grass. Its purple distances melted at last into the bluish slopes of the further hills—hills it might be of a greener kind—and above them invisibly supported, and seeming indeed to hang in the blue, were the snow-clad summits of mountains—that grew larger and bolder to the north-westward as the sides of the valley grew together. And westward the valley opened until a distant darkness under the sky told where the forests began. But the three men looked neither east nor west, but only steadfastly across the valley.

The gaunt man with the scarred lip was the first to speak. "Nowhere," he said, with a sigh of disappointment in his voice. "But after all, they had a full day's start."

"They don't know we are after them," said the little man on the white horse.

"*She* would know," said the leader bitterly, as if speaking to himself.

"Even then they can't go fast. They've got no beast but the mule, and all to-day the girl's foot has been bleeding——"

The man with the silver bridle flashed a quick intensity of rage on him. "Do you think I haven't seen that?" he snarled.

"It helps, anyhow," whispered the little man to himself.

The gaunt man with the scarred lip stared impassively. "They can't be over the valley," he said. "If we ride hard——"

He glanced at the white horse and paused.

"Curse all white horses!" said the man with the silver bridle, and turned to scan the beast his curse included.

The little man looked down between the melancholy ears of his steed.

"I did my best," he said.

The two others stared again across the valley for a space. The gaunt man passed the back of his hand across the scarred lip.

"Come up!" said the man who owned the silver bridle, suddenly. The little man started and jerked his rein, and the horse hoofs of the three made a multitudinous faint pattering upon the withered grass as they turned back towards the trail. . . .

They rode cautiously down the long slope before them, and so came through a waste of prickly twisted bushes and strange dry shapes of horny branches that grew amongst the rocks, into the level below. And there the trail grew faint, for the soil was scanty, and the only herbage was this scorched dead straw that lay upon the ground. Still, by hard scanning, by leaning beside the horse's necks and pausing ever and again, even these white men could contrive to follow after their prey.

There were trodden places, bent and broken blades of the coarse grass, and ever and again the sufficient intimation of a footmark. And once the leader saw a brown smear of blood where the half-caste girl may have trod. And at that under his breath he cursed her for a fool.

The gaunt man checked his leader's tracking, and the little man on the white horse rode behind, a man lost in a dream. They rode one after another, the man with the silver bridle led the way, and they spoke never a word. After a time it came to the little man on the white horse that the world was very still. He started out of his dream. Besides the minute noises of their horses and equipment, the whole great valley kept the brooding quiet of a painted scene.

Before him went his master and his fellow, each intently leaning forward to the left, each impassively moving with the paces of his horse; their shadows went before them— still, noiseless, tapering attendants; and nearer a crouched cool shape was his own. He looked about him. What was it had gone? Then he remembered the reverberation from the banks of the gorge and the perpetual accompaniment of shifting, jostling pebbles. And, moreover——? There was no breeze. That was it! What a vast, still place it was, a monotonous afternoon slumber. And the sky open and blank, except for a sombre veil of haze that had gathered in the upper valley.

He straightened his back, fretted with his bridle, puckered his lips to whistle, and simply sighed. He turned in his saddle for a time, and stared at the throat of the mountain gorge out of which they had come. Blank! Blank slopes on either side, with never a sign of a decent beast or tree— much less a man. What a land it was! What a wilderness! He dropped again into his former pose.

It filled him with a momentary pleasure to see a wry stick of purple black flash out into the form of a snake, and vanish amidst the brown. After all, the infernal valley *was* alive. And then, to rejoice him still more, came a breath across his face, a whisper that came and went, the faintest inclination of a stiff black-antlered bush upon a crest, the first intimations of a possible breeze. Idly he wetted his finger, and held it up.

He pulled up sharply to avoid a collision with the gaunt man, who had stopped at fault upon the trail. Just at that guilty moment he caught his master's eye looking towards him.

For a time he forced an interest in the tracking. Then, as they rode on again, he studied his master's shadow and hat and shoulder appearing and disappearing behind the gaunt man's nearer contours. They had ridden four days out of the very limits of the world into this desolate place, short of water, with nothing but a strip of dried meat under their saddles, over rocks and mountains, where surely none but these fugitives had ever been before—for *that!*

And all this was for a girl, a mere wilful child! And the man had whole cityfuls of people to do his basest bidding—girls, women! Why in the name of passionate folly *this* one in particular? asked the little man, and scowled at the world, and licked his parched lips with a blackened tongue. It was the way of the master, and that was all he knew. Just because she sought to evade him. . . .

His eye caught a whole row of high plumed canes bending in unison, and then the tails of silk that hung before his neck flapped and fell. The breeze was growing stronger. Somehow it took the stiff stillness out of things—and that was well.

"Hullo!" said the gaunt man.

All three stopped abruptly.

"What?" asked the master. "What?"

"Over there," said the gaunt man, pointing up the valley. "What?"

"Something coming towards us."

And as he spoke a yellow animal crested a rise and came bearing down upon them. It was a big wild dog, coming before the wind, tongue out, at a steady pace, and running with such an intensity of purpose that he did not seem to see the horsemen he approached. He ran with his nose up, following, it was plain, neither scent nor quarry. As he drew nearer the little man felt for his sword. "He's mad," said the gaunt rider.

"Shout!" said the little man and shouted.

The dog came on. Then when the little man's blade was already out, it swerved aside and went panting by them and past. The eyes of the little man followed its flight. "There was no foam," he said. For a space the man with the silver-studded bridle stared up the valley. "Oh, come on!" he cried at last. "What does it matter?" and jerked his horse into movement again.

The little man left the insoluble mystery of a dog that fled from nothing but the wind, and lapsed into profound musings on human character. "Come on!" he whispered to himself. "Why should it be given to one man to say 'Come on!' with that stupendous violence of effect. Always, all his life, the man with the silver bridle has been saying that. If *I* said it——!" thought the little man. But people marvelled when the master was disobeyed even in the wildest things. This half-caste girl seemed to him, seemed to everyone, mad—blasphemous almost. The little man, by way of comparison, reflected on the gaunt rider with the scarred lip, as stalwart as his master, as brave and, indeed, perhaps braver, and yet for him there was obedience, nothing but to give obedience duly and stoutly. . . .

Certain sensations of the hands and knees called the little man back to more immediate things. He became aware of something. He rode up beside his gaunt fellow. "Do you notice the horses?" he said in an undertone.

The gaunt face looked interrogation.

"They don't like this wind," said the little man, and dropped behind as the man with the silver bridle turned upon him.

"It's all right," said the gaunt-faced man.

They rode on again for a space in silence. The foremost two rode downcast upon the trail, the hindmost man watched the haze that crept down the vastness of the valley, nearer and nearer, and noted how the wind grew in strength moment by moment. Far away on the left he saw a line of dark bulks—wild hog perhaps, galloping down the valley, but of that he said nothing, nor did he remark again upon the uneasiness of the horses.

And then he saw first one and then a second great white ball, a great shining white ball like a gigantic head of thistledown, that drove before the wind athwart the path. These balls soared high in the air, and dropped and rose again and caught for a moment, and hurried on and passed, but at the sight of them the restlessness of the horses increased.

Then presently he saw that more of these drifting globes —and then soon very many more—were hurrying towards him down the valley.

They became aware of a squealing. Athwart the path a huge boar rushed, turning his head but for one instant to glance at them, and then hurling on down the valley again. And at that, all three stopped and sat in their saddles, staring into the thickening haze that was coming upon them.

"If it were not for this thistledown——" began the leader.

But now a big globe came drifting past within a score of yards of them. It was really not an even sphere at all, but a vast, soft, ragged, filmy thing, a sheet gathered by the corners, an ærial jelly-fish, as it were, but rolling over and over as it advanced, and trailing long, cobwebby threads and streamers that floated in its wake.

"It isn't thistledown," said the little man.

"I don't like the stuff," said the gaunt man.

And they looked at one another.

"Curse it!" cried the leader. "The air's full of it up there. If it keeps on at this pace long, it will stop us altogether."

An instinctive feeling, such as lines out a herd of deer at the approach of some ambiguous thing, prompted them to turn their horses to the wind, ride forward for a few paces, and stare at that advancing multitude of floating masses. They came on before the wind with a sort of smooth swiftness, rising and falling noiselessly, sinking to earth, rebounding high, soaring—all with a perfect unanimity, with a still, deliberate assurance.

Right and left of the horsemen the pioneers of this strange army passed. At one that rolled along the ground, breaking shapelessly and trailing out reluctantly into long grappling ribbons and bands, all three horses began to shy and dance.

EE

The master was seized with a sudden, unreasonable impatience. He cursed the drifting globes roundly. "Get on!" he cried; "get on! What do these things matter? How *can* they matter? Back to the trail!" He fell to swearing at his horse and sawed the bit across its mouth.

He shouted aloud with rage. "I will follow that trail, I tell you," he cried. "Where is the trail?"

He gripped the bridle of his prancing horse and searched amidst the grass. A long and clinging thread fell across his face, a grey streamer dropped about his bridle arm, some big, active thing with many legs ran down the back of his head. He looked up to discover one of those grey masses anchored as it were above him by these things and flapping out ends as a sail flaps when a boat comes about—but noiselessly.

He had an impression of many eyes, of a dense crew of squat bodies, of long, many-jointed limbs hauling at their mooring ropes to bring the thing down upon him. For a space he stared up, reining in his prancing horse with the instinct born of years of horsemanship. Then the flat of a sword smote his back, and a blade flashed overhead and cut the drifting balloon of spider-web free, and the whole mass lifted softly and drove clear and away.

"Spiders!" cried the voice of the gaunt man. "The things are full of big spiders! Look, my lord!"

The man with the silver bridle still followed the mass that drove away.

"Look, my lord!"

The master found himself staring down at a red smashed thing on the ground that, in spite of partial obliteration, could still wriggle unavailing legs. Then when the gaunt man pointed to another mass that bore down upon them, he drew his sword hastily. Up the valley now it was like a fog bank torn to rags. He tried to grasp the situation.

"Ride for it!" the little man was shouting. "Ride for it down the valley."

What happened then was like the confusion of a battle. The man with the silver bridle saw the little man go past him slashing furiously at imaginary cobwebs, saw him

cannon into the horse of the gaunt man and hurl it and
its rider to earth. His own horse went a dozen paces
before he could rein it in. Then he looked up to avoid
imaginary dangers, and then back again to see a horse
rolling on the ground, the gaunt man standing and slashing
over it at a rent and fluttering mass of grey that streamed
and wrapped about them both. And thick and fast as
thistledown on waste land on a windy day in July, the
cobweb masses were coming on.

The little man had dismounted, but he dared not release
his horse. He was endeavouring to lug the struggling
brute back with the strength of one arm, while with the
other he slashed aimlessly. The tentacles of a second grey
mass had entangled themselves with the struggle, and this
second grey mass came to its moorings, and slowly sank.

The master set his teeth, gripped his bridle, lowered his
head, and spurred his horse forward. The horse on the
ground rolled over, there was blood and moving shapes
upon the flanks, and the gaunt man suddenly leaving it,
ran forward towards his master, perhaps ten paces. His
legs were swathed and encumbered with grey; he made
ineffectual movements with his sword. Grey streamers
waved from him; there was a thin veil of grey across his
face. With his left hand he beat at something on his
body, and suddenly he stumbled and fell. He struggled
to rise, and fell again, and suddenly, horribly, began to
howl, "Oh—ohoo, ohooh!"

The master could see the great spiders upon him, and
others upon the ground.

As he strove to force his horse nearer to this gesticulating,
screaming grey object that struggled up and down, there
came a clatter of hoofs, and the little man, in act of mounting, swordless, balanced on his belly athwart the white
horse, and clutching its mane, whirled past. And again a
clinging thread of grey gossamer swept across the master's
face. All about him, and over him, it seemed this drifting,
noiseless cobweb circled and drew nearer him. . . .

To the day of his death he never knew just how the event
of that moment happened. Did he, indeed, turn his horse,

or did it really of its own accord stampede after its fellow? Suffice it that in another second he was galloping full tilt down the valley with his sword whirling furiously overhead. And all about him on the quickening breeze, the spiders' air-ships, their air bundles and air sheets, seemed to him to hurry in a conscious pursuit.

Clatter, clatter, thud, thud—the man with the silver bridle rode, heedless of his direction, with his fearful face looking up now right, now left, and his sword arm ready to slash. And a few hundred yards ahead of him, with a tail of torn cobweb trailing behind him, rode the little man on the white horse, still but imperfectly in the saddle. The reeds bent before them, the wind blew fresh and strong, over his shoulder the master could see the webs hurrying to overtake. . . .

He was so intent to escape the spiders' webs that only as his horse gathered together for a leap did he realise the ravine ahead. And then he realised it only to misunderstand and interfere. He was leaning forward on his horse's neck and sat up and back all too late.

But if in his excitement he had failed to leap, at any rate he had not forgotten how to fall. He was horseman again in mid-air. He came off clear with a mere bruise upon his shoulder, and his horse rolled, kicking spasmodic legs, and lay still. But the master's sword drove its point into the hard soil, and snapped clean across, as though Chance refused him any longer as her Knight, and the splintered end missed his face by an inch or so.

He was on his feet in a moment, breathlessly scanning the onrushing spider-webs. For a moment he was minded to run, and then thought of the ravine, and turned back. He ran aside once to dodge one drifting terror, and then he was swiftly clambering down the precipitous sides, and out of the touch of the gale.

There under the lee of the dry torrent's steeper banks he might crouch, and watch these strange, grey masses pass and pass in safety till the wind fell, and it became possible to escape. And there for a long time he crouched, watching the strange, grey, ragged masses trail their streamers across his narrowed sky.

Once a stray spider fell into the ravine close beside him— a full foot it measured from leg to leg, and its body was half a man's hand—and after he had watched its monstrous alacrity of search and escape for a little while, and tempted it to bite his broken sword, he lifted up his iron heeled boot and smashed it into a pulp. He swore as he did so, and for a time sought up and down for another.

Then presently, when he was surer these spider swarms could not drop into the ravine, he found a place where he could sit down, and sat and fell into deep thought and began after his manner to gnaw his knuckles and bite his nails. And from this he was moved by the coming of the man with the white horse.

He heard him long before he saw him, as a clattering of hoofs, stumbling footsteps, and a reassuring voice. Then the little man appeared, a rueful figure, still with a tail of white cobweb trailing behind him. They approached each other without speaking, without a salutation. The little man was fatigued and shamed to the pitch of hopeless bitterness, and came to a stop at last, face to face with his seated master. The latter winced a little under his dependant's eye. "Well?" he said at last, with no pretence of authority.

"You left him?"

"My horse bolted."

"I know. So did mine."

He laughed at his master mirthlessly.

"I say my horse bolted," said the man who once had a silver-studded bridle.

"Cowards both," said the little man.

The other gnawed his knuckle through some meditative moments, with his eye on his inferior.

"Don't call me a coward," he said at length.

"You are a coward like myself."

"A coward possibly. There is a limit beyond which every man must fear. That I have learnt at last. But not like yourself. That is where the difference comes in."

"I never could have dreamt you would have left him. He saved your life two minutes before. . . . Why are you our lord?"

The master gnawed his knuckles again, and his countenance was dark.

"No man calls me a coward," he said. "No. . . . A broken sword is better than none. . . . One spavined white horse cannot be expected to carry two men a four days' journey. I hate white horses, but this time it cannot be helped. You begin to understand me? . . . I perceive that you are minded, on the strength of what you have seen and fancy, to taint my reputation. It is men of your sort who unmake kings. Besides which—I never liked you."

"My lord!" said the little man.

"No," said the master. "*No!*"

He stood up sharply as the little man moved. For a minute perhaps they faced one another. Overhead the spiders' balls went driving. There was a quick movement among the pebbles; a running of feet, a cry of despair, a gasp and a blow. . . .

Towards nightfall the wind fell. The sun set in a calm serenity, and the man who had once possessed the silver bridle came at last very cautiously and by an easy slope out of the ravine again; but now he led the white horse that once belonged to the little man. He would have gone back to his horse to get his silver-mounted bridle again, but he feared night and a quickening breeze might still find him in the valley, and besides he disliked greatly to think he might discover his horse all swathed in cobwebs and perhaps unpleasantly eaten.

And as he thought of those cobwebs and of all the dangers he had been through, and the manner in which he had been preserved that day, his hand sought a little reliquary that hung about his neck, and he clasped it for a moment with heartfelt gratitude. As he did so his eyes went across the valley.

"I was hot with passion," he said, "and now she has met her reward. They also, no doubt——"

And behold! Far away out of the wooded slopes across the valley, but in the clearness of the sunset distinct and unmistakable, he saw a little spire of smoke.

At that his expression of serene resignation changed to an amazed anger. Smoke? He turned the head of the white horse about, and hesitated. And as he did so a little rustle of air went through the grass about him. Far away upon some reeds swayed a tattered sheet of grey. He looked at the cobwebs; he looked at the smoke.

"Perhaps, after all, it is not them," he said at last.

But he knew better.

After he had stared at the smoke for some time, he mounted the white horse.

As he rode, he picked his way amidst stranded masses of web. For some reason there were many dead spiders on the ground, and those that lived feasted guiltily on their fellows. At the sound of his horse's hoofs they fled.

Their time had passed. From the ground, without either a wind to carry them or a winding sheet ready, these things, for all their poison, could do him no evil.

He flicked with his belt at those he fancied came too near. Once, where a number ran together over a bare place, he was minded to dismount and trample them with his boots, but this impulse he overcame. Ever and again he turned in his saddle, and looked back at the smoke.

"Spiders," he muttered over and over again. "Spiders! Well, well. . . . The next time I must spin a web."

STORY THE FOURTH

The Truth about Pyecraft

HE sits not a dozen yards away. If I glance over my shoulder I can see him. And if I catch his eye—and usually I catch his eye—it meets me with an expression——

It is mainly an imploring look—and yet with suspicion in it.

Confound his suspicion! If I wanted to tell on him I should have told long ago. I don't tell and I don't tell, and he ought to feel at his ease. As if anything so gross and fat as he could feel at ease! Who would believe me if I did tell?

Poor old Pyecraft! Great, uneasy jelly of substance! The fattest clubman in London.

He sits at one of the little club tables in the huge bay by the fire, stuffing. What is he stuffing? I glance judiciously and catch him biting at the round of hot buttered tea-cake, with his eyes on me. Confound him!—with his eyes on me!

That settles it, Pyecraft! Since you *will* be abject, since you *will* behave as though I was not a man of honour, here, right under your embedded eyes, I write the thing down—the plain truth about Pyecraft. The man I helped, the man I shielded, and who has requited me by making my club unendurable, absolutely unendurable, with his liquid appeal, with the perpetual "don't tell" of his looks.

And, besides, why does he keep on eternally eating?

Well, here goes for the truth, the whole truth, and nothing but the truth!

Pyecraft—— I made the acquaintance of Pyecraft in this

872

very smoking-room. I was a young, nervous new member, and he saw it. I was sitting all alone, wishing I knew more of the members, and suddenly he came, a great rolling front of chins and abdomina, towards me, and grunted and sat down in a chair close by me, and wheezed for a space, and scraped for a space with a match and lit a cigar, and then addressed me. I forget what he said—something about the matches not lighting properly, and afterwards as he talked he kept stopping the waiters one by one as they went by, and telling them about the matches in that thin, fluty voice he has. But, anyhow, it was in some such way we began our talking.

He talked about various things and came round to games. And thence to my figure and complexion. "You ought to be a good cricketer," he said. I suppose I am slender, slender to what some people would call lean, and I suppose I am rather dark, still—— I am not ashamed of having a Hindu great-grandmother, but, for all that, I don't want casual strangers to see through me at a glance to *her*. So that I was set against Pyecraft from the beginning.

But he only talked about me in order to get to himself.

"I expect," he said, "you take no more exercise than I do, and probably you eat no less." (Like all excessively obese people he fancied he ate nothing.) "Yet"—and he smiled an oblique smile—"we differ."

And then he began to talk about his fatness and his fatness; all he did for his fatness and all he was going to do for his fatness; what people had advised him to do for his fatness and what he had heard of people doing for fatness similar to his. "*A priori*," he said, "one would think a question of nutrition could be answered by dietary and a question of assimilation by drugs." It was stifling. It was dumpling talk. It made me feel swelled to hear him.

One stands that sort of thing once in a way at a club, but a time came when I fancied I was standing too much. He took to me altogether too conspicuously. I could never go into the smoking-room but he would come wallowing towards me, and sometimes he came and gormandised round and about me while I had my lunch. He seemed at times almost to be clinging to me. He was a bore, but not so fearful a bore as to be limited to me; and from the first there was something in his manner—almost as though he knew, almost as though he penetrated to the fact that I *might*—that there was a remote, exceptional chance in me that no one else presented.

"I'd give anything to get it down," he would say—"any-thing," and peer at me over his vast cheeks and pant.

Poor old Pyecraft! He has just gonged, no doubt to order another buttered teacake!

He came to the actual thing one day. "Our Pharma-copœia," he said, "our Western Pharmacopœia, is anything but the last word of medical science. In the East, I've been told——"

He stopped and stared at me. It was like being at an aquarium.

I was quite suddenly angry with him. "Look here," I said, "who told you about my great-grandmother's recipes?"

"Well," he fenced.

"Every time we've met for a week," I said—"and we've met pretty often—you've given me a broad hint or so about that little secret of mine."

"Well," he said, "now the cat's out of the bag, I'll admit, yes, it is so. I had it——"

"From Pattison?"

"Indirectly," he said, which I believe was lying, "yes."

"Pattison," I said, "took that stuff at his own risk."

He pursed his mouth and bowed.

"My great-grandmother's recipes," I said, "are queer things to handle. My father was near making me promise——"

"He didn't?"

"No. But he warned me. He himself used one—once."

"Ah! . . . But do you think——? Suppose—suppose there did happen to be one——"

"The things are curious documents," I said. "Even the smell of 'em. . . . No!"

But after going so far Pyecraft was resolved I should go farther. I was always a little afraid if I tried his patience too much he would fall on me suddenly and smother me. I own I was weak. But I was also annoyed with Pyecraft. I had got to that state of feeling for him that disposed me to say, "Well, *take* the risk!" The little affair of Pattison to which I have alluded was a different matter altogether. What it was doesn't concern us now, but I knew, anyhow, that the particular recipe I used then was safe. The rest I didn't know so much about, and, on the whole, I was inclined to doubt their safety pretty completely.

Yet even if Pyecraft got poisoned——

I must confess the poisoning of Pyecraft struck me as an immense undertaking.

That evening I took that queer, odd-scented sandalwood box out of my safe and turned the rustling skins over. The gentleman who wrote the recipes for my great-grandmother evidently had a weakness for skins of a miscellaneous origin, and his handwriting was cramped to the last degree. Some of the things are quite unreadable to me—though my family, with its Indian Civil Service associations, has kept up a knowledge of Hindustani from generation to generation—and none are absolutely plain sailing. But I found the one that I knew was there soon enough, and sat on the floor by my safe for some time looking at it.

"Look here," said I to Pyecraft next day, and snatched the slip away from his eager grasp.

"So far as I can make it out, this is a recipe for Loss of Weight. ("Ah!" said Pyecraft.) I'm not absolutely sure, but I think it's that. And if you take my advice you'll leave it alone. Because, you know—I blacken my blood in your interest, Pyecraft—my ancestors on that side were, so far as I can gather, a jolly queer lot. See?"

"Let me try it," said Pyecraft.

I leant back in my chair. My imagination made one mighty effort and fell flat within me. "What in Heaven's name, Pyecraft," I asked, "do you think you'll look like when you get thin?"

He was impervious to reason. I made him promise never to say a word to me about his disgusting fatness again whatever happened—never, and then I handed him that little piece of skin.

"It's nasty stuff," I said.

"No matter," he said, and took it.

He goggled at it. "But—but——" he said.

He had just discovered that it wasn't English.

"To the best of my ability," I said, "I will do you a translation."

I did my best. After that we didn't speak for a fortnight. Whenever he approached me I frowned and motioned him away, and he respected our compact, but at the end of the fortnight he was as fat as ever. And then he got a word in.

"I must speak," he said. "It isn't fair. There's something wrong. It's done me no good. You're not doing your great-grandmother justice."

"Where's the recipe?"

He produced it gingerly from his pocket-book.

I ran my eye over the items. "Was the egg addled?" I asked.

"No. Ought it to have been?"

"That," I said, "goes without saying in all my poor dear great-grandmother's recipes. When condition or quality is not specified you must get the worst. She was drastic or nothing. . . . And there's one or two possible alternatives to some of these other things. You got *fresh* rattlesnake venom?"

"I got rattlesnake from Jamrach's. It cost—it cost——"

"That's your affair, anyhow. This last item——"

"I know a man who——"

"Yes. H'm. Well, I'll write the alternatives down. So far as I know the language, the spelling of this recipe is particularly atrocious. By-the-bye, dog here probably means pariah dog."

For a month after that I saw Pyecraft constantly at the club and as fat and anxious as ever. He kept our treaty, but at times he broke the spirit of it by shaking his head despondently. Then one day in the cloak-room he said, "Your great-grandmother——"

"Not a word against her," I said: and he held his peace.

I could have fancied he had desisted, and I saw him one day talking to three new members about his fatness as though he was in search of other recipes. And then, quite unexpectedly his telegram came.

"Mr. Formalyn!" bawled a page-boy under my nose and I took the telegram and opened it at once.

"For Heaven's sake come.—Pyecraft."

"H'm," said I, and to tell the truth I was so pleased at the rehabilitation of my great-grandmother's reputation this evidently promised that I made a most excellent lunch.

I got Pyecraft's address from the hall porter. Pyecraft inhabited the upper half of a house in Bloomsbury, and I went there as soon as I had done my coffee and Trappistine. I did not wait to finish my cigar.

"Mr. Pyecraft?" said I, at the front door.

They believed he was ill; he hadn't been out for two days.

"He expects me," said I, and they sent me up.

I rang the bell at the lattice-door upon the landing.

"He shouldn't have tried it, anyhow," I said to myself. "A man who eats like a pig ought to look like a pig."

An obviously worthy woman, with an anxious face and a carelessly placed cap, came and surveyed me through the lattice.

I gave my name and she opened his door for me in a dubious fashion.

"Well?" said I, as we stood together inside Pyecraft's piece of the landing.

"'E said you was to come in if you came," she said, and regarded me, making no motion to show me anywhere. And then, confidentially, "'E's locked in, sir."

"Locked in?"

"Locked himself in yesterday morning and 'asn't let anyone in since, sir. And ever and again *swearing*. Oh, my!"

I stared at the door she indicated by her glances. "In there?" I said.

"Yes, sir."

"What's up?"

She shook her head sadly. " 'E keeps on calling for vittles, sir. *'Eavy* vittles 'e wants. I get 'im what I can. Pork 'e's 'ad, sooit puddin', sossiges, noo bread. Everythink like that. Left outside, if you please, and me go away. 'E's eatin' sir, somethink *awful*."

There came a piping bawl from inside the door: "That Formalyn?"

"'That you, Pyecraft?" I shouted, and went and banged the door.

"Tell her to go away."

I did.

Then I could hear a curious pattering upon the door, almost like someone feeling for the handle in the dark, and Pyecraft's familiar grunts.

"It's all right," I said, "she's gone."

But for a long time the door didn't open.

I heard the key turn. Then Pyecraft's voice said, "Come in."

I turned the handle and opened the door. Naturally I expected to see Pyecraft.

Well, you know, he wasn't there!

I never had such a shock in my life. There was his sitting-room in a state of untidy disorder, plates and dishes among the books and writing things, and several chairs overturned, but Pyecraft—

"It's all right, o' man; shut the door," he said, and then I discovered him.

There he was right up close to the cornice in the corner by the door, as though someone had glued him to the ceiling. His face was anxious and angry. He panted and gesticulated. "Shut the door," he said. "If that woman gets hold of it—"

I shut the door, and went and stood away from him and stared.

"If anything gives way and you tumble down," I said, "you'll break your neck, Pyecraft."

"I wish I could," he wheezed.

"A man of your age and weight getting up to kiddish gymnastics——"

"Don't," he said, and looked agonized. "Your damned great-grandmother——"

"Be careful," I warned him.

"I'll tell you," he said, and gesticulated.

"How the deuce," said I, "are you holding on up there?"

And then abruptly I realized that he was not holding on at all, that he was floating up there—just as a gas-filled bladder might have floated in the same position. He began a struggle to thrust himself away from the ceiling and to clamber down the wall to me. It's that prescription," he panted, as he did so. "Your great-gran——"

"*No!*" I cried.

He took hold of a framed engraving rather carelessly as he spoke and it gave way, and he flew back to the ceiling again, while the picture smashed on to the sofa. Bump he went against the ceiling, and I knew then why he was all over white on the more salient curves and angles of his person. He tried again more carefully, coming down by way of the mantel.

It was really a most extraordinary spectacle, that great, fat, apoplectic-looking man upside down and trying to get from the ceiling to the floor. "That prescription," he said. "Too successful."

"How?"

"Loss of weight—almost complete."

And then, of course, I understood.

"By Jove, Pyecraft," said I, "what you wanted was a cure for fatness! But you always called it weight. You would call it weight."

Somehow I was extremely delighted. I quite liked Pyecraft for the time. "Let me help you!" I said, and took his hand and pulled him down. He kicked about, trying to get foothold somewhere. It was very like holding a flag on a windy day.

"That table," he said, pointing, "is solid mahogany and very heavy. If you can put me under that——"

I did, and there he wallowed about like a captive balloon, while I stood on his hearthrug and talked to him.

I lit a cigar. "Tell me," I said, "what happened?"

"I took it," he said.

"How did it taste?"

"Oh, *beastly!*"

I should fancy they all did. Whether one regards the ingredients or the probable compound or the possible results, almost all my great-grandmother's remedies appear to me at least to be extraordinarily uninviting. For my own part——

"I took a little sip first."

"Yes?"

"And as I felt lighter and better after an hour, I decided to take the draught."

"My dear Pyecraft!"

"I held my nose," he explained. "And then I kept on getting lighter and lighter—and helpless, you know."

He gave way suddenly to a burst of passion. "What the goodness am I to *do?*" he said.

"There's one thing pretty evident," I said, "that you mustn't do. If you go out of doors you'll go up and up." I waved an arm upward. "They'd have to send Santos-Dumont after you to bring you down again."

"I suppose it will wear off?"

I shook my head. "I don't think you can count on that," I said.

And then there was another burst of passion, and he kicked out at adjacent chairs and banged the floor. He behaved just as I should have expected a great, fat, self-indulgent man to behave under trying circumstances—that is to say, very badly. He spoke of me and of my great-grandmother with an utter want of discretion.

"I never asked you to take the stuff," I said.

And generously disregarding the insults he was putting upon me, I sat down in his armchair and began to talk to him in a sober, friendly fashion.

I pointed out to him that this was a trouble he had

brought upon himself, and that it had almost an air of poetical justice. He had eaten too much. This he disputed, and for a time we argued the point.

He became noisy and violent, so I desisted from this aspect of his lesson. "And then," said I, "you committed the sin of euphuism. You called it, not Fat, which is just and inglorious, but Weight. You——"

He interrupted to say that he recognised all that. What was he to *do?*

I suggested he should adapt himself to his new conditions. So we came to the really sensible part of the business. I suggested that it would not be difficult for him to learn to walk about on the ceiling with his hands——

"I can't sleep," he said.

But that was no great difficulty. It was quite possible, I pointed out, to make a shake-up under a wire mattress, fasten the under things on with tapes, and have a blanket, sheet, and coverlid to button at the side. He would have to confide in his housekeeper, I said; and after some squabbling he agreed to that. (Afterwards it was quite delightful to see the beautifully matter-of-fact way with which the good lady took all these amazing inversions.) He could have a library ladder in his room, and all his meals could be laid on the top of his bookcase. We also hit on an ingenious device by which he could get to the floor whenever he wanted, which was simply to put the *British Encyclopædia* (tenth edition) on the top of his open shelves. He just pulled out a couple of volumes and held on, and down he came. And we agreed there must be iron staples along the skirting, so that he could cling to those whenever he wanted to get about the room on the lower level.

As we got on with the thing I found myself almost keenly interested. It was I who called in the housekeeper and broke matters to her, and it was I chiefly who fixed up the inverted bed. In fact, I spent two whole days at his flat. I am a handy, interfering sort of man with a screwdriver, and I made all sorts of ingenious adaptations for him—ran a wire to bring his bells within reach, turned all his electric lights up instead of down, and so on. The whole affair was extremely curious and interesting to me, and it was delightful to think of Pyecraft like some great, fat blow-fly, crawling about on his ceiling and clambering round the lintel of his doors from one room to another, and never, never, never coming to the club any more. . . .

Then, you know, my fatal ingenuity got the better of me. I was sitting by his fire drinking his whisky, and he was up in his favourite corner by the cornice, tacking a Turkey carpet to the ceiling, when the idea struck me. "By Jove, Pyecraft!" I said, "all this is totally unnecessary."

And before I could calculate the complete consequences of my notion I blurted it out. "Lead underclothing," said I, and the mischief was done.

Pyecraft received the thing almost in tears. "To be right ways up again——" he said.

I gave him the whole secret before I saw where it would take me. "Buy sheet lead," I said, "stamp it into discs. Sew 'em all over your underclothes until you have enough. Have lead-soled boots, carry a bag of solid lead, and the thing is done! Instead of being a prisoner here you may go abroad again, Pyecraft! you may travel——"

A still happier idea came to me. "You need never fear a shipwreck. All you need do is just slip off some or all of your clothes, take the necessary amount of luggage in your hand, and float up in the air——"

In his emotion he dropped the tack-hammer within an ace of my head. "By Jove!" he said, "I shall be able to come back to the club again."

The thing pulled me up short. "By Jove!" I said, faintly. "Yes. Of course—you will."

He did. He does. There he sits behind me now stuffing —as I live!—a third go of buttered teacake. And no one in the whole world knows—except his housekeeper and me— that he weighs practically nothing; that he is a mere boring mass of assimilatory matters, mere clouds in clothing, *niente*, *nefas*, and most inconsiderable of men. There he sits watching until I have done this writing. Then, if he can, he will waylay me. He will come billowing up to me. . . .

He will tell me over again all about it, how it feels, how it doesn't feel, how he sometimes hopes it is passing off a little. And always somewhere in that fat, abundant discourse he will say, "The secret's keeping, eh? If anyone knew of it—I should be so ashamed. . . . Makes a fellow look such a fool, you know. Crawling about on a ceiling and all that. . . ."

And now to elude Pyecraft, occupying, as he does, an admirable strategic position between me and the door.

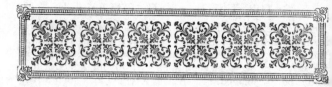

STORY THE FIFTH

Mr. Skelmersdale in Fairyland

"There's a man in that shop," said the Doctor, "who has been in Fairyland."

"Nonsense!" I said, and stared back at the shop. It was the usual village shop, post-office, telegraph wire on its brow, zinc pans and brushes outside, boots, shirtings, and potted meats in the window. "Tell me about it," I said, after a pause.

"I don't know," said the Doctor. "He's an ordinary sort of lout—Skelmersdale is his name. But everybody about here believes it like Bible truth."

I reverted presently to the topic.

"I know nothing about it," said the Doctor, "and I don't *want* to know. I attended him for a broken finger—Married and Single cricket match—and that's when I struck the nonsense. That's all. But it shows you the sort of stuff I have to deal with, anyhow, eh? Nice to get modern sanitary ideas into a people like this!"

"Very," I said in a mildly sympathetic tone, and he went on to tell me about that business of the Bonham drain. Things of that kind, I observe, are apt to weigh on the minds of Medical Officers of Health. I was as sympathetic as I knew how, and when he called the Bonham people "asses," I said they were " thundering asses," but even that did not allay him.

Afterwards, later in the summer, an urgent desire to seclude myself, while finishing my chapter on Spiritual Pathology—it was really, I believe, stiffer to write than it is

884

to read—took me to Bignor. I lodged at a farmhouse, and presently found myself outside that little general shop again, in search of tobacco. "Skelmersdale," said I to myself at the sight of it, and went in.

I was served by a short, but shapely, young man, with a fairy downy complexion, good, small teeth, blue eyes, and a languid manner. I scrutinised him curiously. Except for a touch of melancholy in his expression, he was nothing out of the common. He was in the shirt-sleeves and tucked-up apron of his trade, and a pencil was thrust behind his inoffensive ear. Athwart his black waistcoat was a gold chain, from which dangled a bent guinea.

"Nothing more to-day, sir?" he inquired. He leant forward over my bill as he spoke.

"Are you Mr. Skelmersdale?" said I.

"I am, sir," he said, without looking up.

"Is it true that you have been in Fairyland?"

He looked up at me for a moment with wrinkled brows, with an aggrieved, exasperated face. "O sHUT it!" he said, and, after a moment of hostility, eye to eye, he went on adding up my bill. "Four, six and a half," he said, after a pause. "Thank you, sir."

So, unpropitiously, my acquaintance with Mr. Skelmersdale began.

Well, I got from that to confidence—through a series of toilsome efforts. I picked him up again in the Village Room, where of a night I went to play billiards after my supper, and mitigate the extreme seclusion from my kind that was so helpful to work during the day. I contrived to play with him and afterwards to talk with him. I found the one subject to avoid was Fairyland. On everything else he was open and amiable in a commonplace sort of way, but on that he had been worried—it was a manifest taboo. Only once in the room did I hear the slightest allusion to his experience in his presence, and that was by a cross-grained farm hand who was losing to him. Skelmersdale had run a break into double figures, which, by the Bignor standards, was uncommonly good play. "Steady on!" said his adversary. "None of your fairy flukes!"

Skelmersdale stared at him for a moment, cue in hand, then flung it down and walked out of the room.

"Why can't you leave 'im alone?" said a respectable elder who had been enjoying the game, and in the general murmur of disapproval, the grin of satisfied wit faded from the ploughboy's face.

I scented my opportunity. "What's this joke," said I, "about Fairyland?"

"'Taint no joke about Fairyland, not to young Skelmersdale," said the respectable elder, drinking.

A little man with rosy cheeks was more communicative. "They *do* say, sir," he said, "that they took him into Aldington Knoll an' kep' him there a matter of three weeks."

And with that the gathering was well under weigh. Once one sheep had started, others were ready enough to follow, and in a little time I had at least the exterior aspect of the Skelmersdale affair. Formerly, before he came to Bignor, he had been in that very similar little shop at Aldington Corner, and there whatever it was did happen had taken place. The story was clear that he had stayed out late one night on the Knoll and vanished for three weeks from the sight of men, and had returned with "his cuffs as clean as when he started," and his pockets full of dust and ashes. He returned in a state of moody wretchedness that only slowly passed away, and for many days he would give no account of where it was he had been. The girl he was engaged to at Clapton Hill tried to get it out of him, and threw him over partly because he refused, and partly because, as she said, he fairly gave her the "'ump." And then when, some time after, he let out to someone carelessly that he had been in Fairyland and wanted to go back, and when the thing spread and the simple badinage of the countryside came into play, he threw up his situation abruptly, and came to Bignor to get out of the fuss. But as to what had happened in Fairyland none of these people knew. There the gathering in the Village Room went to pieces like a pack at fault. One said this, and another said that.

Their air in dealing with this marvel was ostensibly critical and sceptical, but I could see a considerable amount of

belief showing through their guarded qualifications. I took
a line of intelligent interest, tinged with a reasonable doubt
of the whole story.

"If Fairyland's inside Aldington Knoll," I said, "why don't
you dig it out?"

"That's what I says," said the young ploughboy.

"There's a-many have tried to dig on Aldington Knoll,"
said the respectable elder, solemnly, "one time and another.
But there's none as goes about to-day to tell what they got
by digging."

The unanimity of vague belief that surrounded me was
rather impressive; I felt there must surely be *something* at
the root of so much conviction, and the already pretty keen
curiosity I felt about the real facts of the case was distinctly whetted. If these real facts were to be got from anyone, they were to be got from Skelmersdale himself; and I
set myself, therefore, still more assiduously to efface the first
bad impression I had made and win his confidence to the
pitch of voluntary speech. In that endeavour I had a social
advantage. Being a person of affability and no apparent
employment, and wearing tweeds and knickerbockers, I was
naturally classed as an artist in Bignor, and in the remarkable
code of social precedence prevalent in Bignor an artist ranks
considerably higher than a grocer's assistant. Skelmersdale,
like too many of his class, is something of a snob; he had
told me to "shut it" only under sudden, excessive provocation, and with, I am certain, a subsequent repentance; he
was, I knew, quite glad to be seen walking about the village
with me. In due course, he accepted the proposal of a pipe
and whisky in my rooms readily enough, and there, scenting
by some happy instinct that there was trouble of the heart in
this, and knowing that confidences beget confidences, I
plied him with much of interest and suggestion from my
real and fictitious past. And it was after the third whisky
of the third visit of that sort, if I remember rightly, *àpropos*
of some artless expansion of a little affair that had touched
and left me in my teens, that he did at last, of his own free
will and motion, break the ice. "It was like that with me,"
he said, "over there at Aldington. It's just that that's so

rum. First I didn't care a bit and it was all her, and afterwards, when it was too late, it was, in a manner of speaking, all me."

I forbore to jump upon this allusion, and so he presently threw out another, and in a little while he was making it as plain as daylight that the one thing he wanted to talk about now was this Fairyland adventure he had sat tight upon for so long. You see, I'd done the trick with him, and from being just another half-incredulous, would-be facetious stranger, I had, by all my wealth of shameless self-exposure, become the possible confidant. He had been bitten by the desire to show that he, too, had lived and felt many things, and the fever was upon him.

He was certainly confoundedly allusive at first, and my eagerness to clear him up with a few precise questions was only equalled and controlled by my anxiety not to get to this sort of thing too soon. But in another meeting or so the basis of confidence was complete; and from first to last I think I got most of the items and aspects—indeed, I got quite a number of times over almost everything that Mr. Skelmersdale, with his very limited powers of narration, will ever be able to tell. And so I come to the story of his adventure, and I piece it all together again. Whether it really happened, whether he imagined it or dreamt it, or fell upon it in some strange hallucinatory trance, I do not profess to say. But that he invented it I will not for one moment entertain. The man simply and honestly believes the thing happened as he says it happened; he is transparently incapable of any lie so elaborate and sustained, and in the belief of the simple, yet often keenly penetrating, rustic minds about him I find a very strong confirmation of his sincerity. He believes—and nobody can produce any positive fact to falsify his belief. As for me, with this much of endorsement, I transmit his story—I am a little old now to justify or explain.

He says he went to sleep on Aldington Knoll about ten o'clock one night—it was quite possibly Midsummer night, though he has never thought of the date, and he cannot be sure within a week or so—and it was a fine night and wind-

less, with a rising moon. I have been at the pains to visit
this Knoll thrice since his story grew up under my persuasions, and once I went there in the twilight summer
moonrise on what was, perhaps, a similar night to that of
his adventure. Jupiter was great and splendid above the
moon, and in the north and north-west the sky was green
and vividly bright over the sunken sun. The Knoll stands
out bare and bleak under the sky, but surrounded at a little
distance by dark thickets, and as I went up towards it there
was a mighty starting and scampering of ghostly or quite
invisible rabbits. Just over the crown of the Knoll, but
nowhere else, was a multitudinous thin trumpeting of
midges. The Knoll is, I believe, an artificial mound, the
tumulus of some great prehistoric chieftain, and surely no
man ever chose a more spacious prospect for a sepulchre.
Eastward one sees along the hills to Hythe, and thence
across the Channel to where, thirty miles and more, perhaps,
away, the great white lights by Gris Nez and Boulogne wink
and pass and shine. Westward lies the whole tumbled
valley of the Weald, visible as far as Hindhead and Leith
Hill, and the valley of the Stour opens the Downs in the
north to interminable hills beyond Wye. All Romney Marsh
lies southward at one's feet, Dymchurch and Romney and
Lydd, Hastings and its hill are in the middle distance, and
the hills multiply vaguely far beyond where Eastbourne
rolls up to Beachy Head.

And out upon all this it was that Skelmersdale wandered,
being troubled in his earlier love affair, and as he says, "not
caring *where* he went." And there he sat down to think it
over, and so, sulking and grieving, was overtaken by sleep.
And so he fell into the fairies' power.

The quarrel that had upset him was some trivial matter
enough between himself and the girl at Clapton Hill to
whom he was engaged. She was a farmer's daughter, said
Skelmersdale, and "very respectable," and no doubt an
excellent match for him; but both girl and lover were very
young and with just that mutual jealousy, that intolerantly
keen edge of criticism, that irrational hunger for a beautiful
perfection, that life and wisdom do presently and most

mercifully dull. What the precise matter of quarrel was I have no idea. She may have said she liked men in gaiters when he hadn't any gaiters on, or he may have said he liked her better in a different sort of hat, but however it began, it got by a series of clumsy stages to bitterness and tears. She no doubt got tearful and smeary, and he grew dusty and drooping, and she parted with invidious comparisons, grave doubts whether she ever had *really* cared for him, and a clear certainty she would never care again. And with this sort of thing upon his mind he came out upon Aldington Knoll grieving, and presently, after a long interval, perhaps, quite inexplicably, fell asleep.

He woke to find himself on a softer turf than ever he had slept on before, and under the shade of very dark trees that completely hid the sky. Always, indeed, in Fairyland the sky is hidden, it seems. Except for one night when the fairies were dancing, Mr. Skelmersdale, during all his time with them, never saw a star. And of that night I am in doubt whether he was in Fairyland proper or out where the rings and rushes are, in those low meadows near the railway line at Smeeth.

But it was light under these trees for all that, and on the leaves and amidst the turf shone a multitude of glow-worms, very bright and fine. Mr. Skelmersdale's first impression was that he was *small*, and the next that quite a number of people still smaller were standing all about him. For some reason, he says, he was neither surprised nor frightened, but sat up quite deliberately and rubbed the sleep out of his eyes. And there all about him stood the smiling elves who had caught him sleeping under their privileges and had brought him into Fairyland.

What these elves were like I have failed to gather, so vague and imperfect is his vocabulary, and so unobservant of all minor detail does he seem to have been. They were clothed in something very light and beautiful, that was neither wool, nor silk, nor leaves, nor the petals of flowers. They stood all about him as he sat and waked, and down the glade towards him, down a glow-worm avenue and fronted by a star, came at once that Fairy Lady who is the

chief personage of his memory and tale. Of her I gathered
more. She was clothed in filmy green, and about her little
waist was a broad silver girdle. Her hair waved back from
her forehead on either side; there were curls not too way-
ward and yet astray, and on her brow was a little tiara, set
with a single star. Her sleeves were some sort of open
sleeves that gave little glimpses of her arms; her throat, I
think, was a little displayed, because he speaks of the beauty
of her neck and chin. There was a necklace of coral about
her white throat, and in her breast a coral-coloured flower.
She had the soft lines of a little child in her chin and cheeks
and throat. And her eyes, I gather, were of a kindled
brown, very soft and straight and sweet under her level
brows. You see by these particulars how greatly this lady
must have loomed in Mr. Skelmersdale's picture. Certain
things he tried to express and could not express; "the way
she moved," he said several times; and I fancy a sort of
demure joyousness radiated from this Lady.

And it was in the company of this delightful person, as
the guest and chosen companion of this delightful person,
that Mr. Skelmersdale set out to be taken into the intimacies
of Fairyland. She welcomed him gladly and a little warmly
—I suspect a pressure of his hand in both of hers and a lit
face to his. After all, ten years ago young Skelmersdale
may have been a very comely youth. And once she took his
arm, and once, I think, she led him by the hand adown
the glade that the glow-worms lit.

Just how things chanced and happened there is no telling
from Mr. Skelmersdale's disarticulated skeleton of description. He gives little unsatisfactory glimpses of strange
corners and doings, of places where there were many fairies
together, of "toadstool things that shone pink," of fairy food,
of which he could only say "you should have tasted it!" and
of fairy music, "like a little musical box," that came out of
nodding flowers. There was a great open place where
fairies rode and raced on "things," but what Mr. Skelmersdale meant by "these here things they rode," there is no
telling. Larvæ, perhaps, or crickets, or the little beetles
that elude us so abundantly. There was a place where

water splashed and gigantic king-cups grew, and there in the hotter times the fairies bathed together. There were games being played and dancing and much elvish love-making, too, I think, among the moss branch thickets. There can be no doubt that the Fairy Lady made love to Mr. Skelmersdale, and no doubt either that this young man set himself to resist her. A time came, indeed, when she sat on a bank beside him, in a quiet secluded place "all smelling of vi'lets," and talked to him of love.

"When her voice went low and she whispered," said Mr. Skelmersdale, "and laid 'er 'and on my 'and, you know, and came close with a soft, warm friendly way she 'ad, it was as much as I could do to keep my 'ead."

It seems he kept his head to a certain limited unfortunate extent. He saw "'ow the wind was blowing," he says, and so, sitting there in a place all smelling of violets, with the touch of this lovely Fairy Lady about him, Mr. Skelmersdale broke it to her gently—that he was engaged!

She had told him she loved him dearly, that he was a sweet human lad for her, and whatever he would ask of her he should have—even his heart's desire.

And Mr. Skelmersdale, who, I fancy, tried hard to avoid looking at her little lips as they just dropped apart and came together, led up to the more intimate question by saying he would like enough capital to start a little shop. He'd just like to feel, he said, he had money enough to do that. I imagine a little surprise in those brown eyes he talked about, but she seemed sympathetic for all that, and she asked him many questions about the little shop, "laughing like" all the time. So he got to the complete statement of his affianced position, and told her all about Millie.

"All?" said I.

"Everything," said Mr. Skelmersdale, "just who she was, and where she lived, and everything about her. I sort of felt I 'ad to all the time, I did."

"'Whatever you want you shall have,' said the Fairy Lady. 'That's as good as done. You *shall* feel you have the money just as you wish. And now, you know—*you must kiss me.*'"

And Mr. Skelmersdale pretended not to hear the latter part of her remark, and said she was very kind. That he really didn't deserve she should be so kind. And——"

The Fairy Lady suddenly came quite close to him and whispered "Kiss me!"

"And," said Mr. Skelmersdale, "like a fool, I did."

There are kisses and kisses, I am told, and this must have been quite the other sort from Millie's resonant signals of regard. There was something magic in that kiss; assuredly it marked a turning point. At any rate, this is one of the passages that he thought sufficiently important to describe most at length. I have tried to get it right, I have tried to disentangle it from the hints and gestures through which it came to me, but I have no doubt that it was all different from my telling and far finer and sweeter, in the soft filtered light and the subtly stirring silences of the fairy glades. The Fairy Lady asked him more about Millie, and was she very lovely, and so on—a great many times. As to Millie's loveliness, I conceive him answering that she was "all right." And then, or on some such occasion, the Fairy Lady told him she had fallen in love with him as he slept in the moonlight, and so he had been brought into Fairyland, and she had thought, not knowing of Millie, that perhaps he might chance to love her. "But now you know you can't," she said, "so you must stop with me just a little while, and then you must go back to Millie." She told him that, and you know Skelmersdale was already in love with her, but the pure inertia of his mind kept him in the way he was going. I imagine him sitting in a sort of stupefaction amidst all these glowing beautiful things, answering about his Millie and the little shop he projected and the need of a horse and cart. . . . And that absurd state of affairs must have gone on for days and days. I see this little lady, hovering about him and trying to amuse him, too dainty to understand his complexity and too tender to let him go. And he, you know, hypnotised as it were by his earthly position, went his way with her hither and thither, blind to everything in Fairyland but this wonderful intimacy that had come to him. It is hard, it is impossible, to give in print the effect of her radiant sweetness shining through the jungle of poor Skelmersdale's rough and broken sentences. To me, at least, she shone clear amidst the muddle of his story like a glow-worm in a tangle of weeds.

There must have been many days of things while all this was happening—and once, I say, they danced under the moonlight in the fairy rings that stud the meadows near Smeeth—but at last it all came to an end. She led him into a great cavernous place, lit by "a red nightlight sort of thing," where there were coffers piled on coffers, and cups and golden boxes, and a great heap of what certainly seemed to all Mr. Skelmersdale's senses—coined gold. There were little gnomes amidst this wealth, who saluted her at her coming, and stood aside. And suddenly she turned on him there with brightly shining eyes.

"And now," she said, "you have been kind to stay with me so long, and it is time I let you go. You must go back to your Millie. You must go back to your Millie, and here—just as I promised you—they will give you gold."

"She choked like," said Mr. Skelmersdale. "At that, I had a sort of feeling——" (he touched his breastbone) "as though I was fainting here. I felt pale, you know, and shivering, and even then—I a'dn't a thing to say."

He paused. "Yes," I said.

The scene was beyond his describing. But I know that she kissed him good-bye.

"And you said nothing?"

"Nothing," he said. "I stood like a stuffed calf. She just looked back once, you know, and stood smiling like and crying—I could see the shine of her eyes—and then she was gone, and there was all these little fellows bustling about me, stuffing my 'ands and my pockets and the back of my collar and everywhere with gold."

And then it was, when the Fairy Lady had vanished, that Mr. Skelmersdale really understood and knew. He suddenly began plucking out the gold they were thrusting upon him, and shouting out at them to prevent their giving him more. "'I don't *want* yer gold,' I said. 'I 'aven't done yet.

I'm not going. I want to speak to that Fairy Lady again.' I started off to go after her and they held me back. Yes, stuck their little 'ands against my middle and shoved me back. They kept giving me more and more gold until it was running all down my trouser legs and dropping out of my 'ands. 'I don't *want* yer gold,' I says to them, 'I want just to speak to the Fairy Lady again.'"

"And did you?"

"It came to a tussle."

"Before you saw her?"

"I didn't see her. When I got out from them she wasn't anywhere to be seen."

So he ran in search of her out of this red-lit cave, down a long grotto, seeking her, and thence he came out in a great and desolate place athwart which a swarm of will-o'-the-wisps were flying to and fro. And about him elves were dancing in derision, and the little gnomes came out of the cave after him, carrying gold in handfuls and casting it after him, shouting, "Fairy love and fairy gold! Fairy love and fairy gold!"

And when he heard these words, came a great fear that it was all over, and he lifted up his voice and called to her by her name, and suddenly set himself to run down the slope from the mouth of the cavern, through a place of thorns and briers, calling after her very loudly and often. The elves danced about him unheeded, pinching him and pricking him, and the will-o'-the-wisps circled round him and dashed into his face, and the gnomes pursued him shouting and pelting him with fairy gold. As he ran with all this strange rout about him and distracting him, suddenly he was knee-deep in a swamp, and suddenly he was amidst thick twisted roots, and he caught his foot in one and stumbled and fell. . . .

He fell and he rolled over, and in that instant he found himself sprawling upon Aldington Knoll, all lonely under the stars.

He sat up sharply at once, he says, and found he was very stiff and cold, and his clothes were damp with dew. The first pallor of dawn and a chilly wind were coming up together.   He could have believed the whole thing a strangely vivid dream until he thrust his hands into his side pocket and found it stuffed with ashes.   Then he knew for certain it was fairy gold they had given him.   He could feel all their pinches and pricks still, though there was never a bruise upon him.   And in that manner, and so suddenly, Mr. Skelmersdale came out of Fairyland back into this world of men.   Even then he fancied the thing was but the matter of a night until he returned to the shop at Aldington Corner and discovered amidst their astonishment that he had been away three weeks.

"Lor! the trouble I 'ad!" said Mr. Skelmersdale.

"How?"

"Explaining.   I suppose you've never had anything like that to explain."

"Never," I said, and he expatiated for a time on the behaviour of this person and that.   One name he avoided for a space.

"And Millie?" said I at last.

"I didn't seem to care a bit for seeing Millie," he said.

"I expect she seemed changed?"

"Everyone was changed.   Changed for good.   Everyone seemed big, you know, and coarse.   And their voices seemed loud.   Why the sun, when it rose in the morning, fair hit me in the eye!"

"And Millie?"

"I didn't want to see Millie."

"And when you did?"

"I came up against her Sunday, coming out of church. 'Where you been?' she said, and I saw there was a row.   I didn't care if there was.   I seemed to forget about her even while she was there a-talking to me.   She was just nothing. I couldn't make out whatever I 'ad seen in 'er ever, or what there could 'ave been.   Sometimes when she wasn't about, I did get back a little, but never when she was there.   Then it was always the other came up and blotted her out. . . . Any'ow, it didn't break her heart."

"Married?" I asked.

"Married 'er cousin," said Mr. Skelmersdale, and reflected on the pattern of the tablecloth for a space.

When he spoke again it was clear that his former sweetheart had clean vanished from his mind, and that the talk had brought back the Fairy Lady triumphant in his heart. He talked of her—soon he was letting out the oddest things, queer love secrets it would be treachery to repeat. I think, indeed, that was the queerest thing in the whole affair, to hear that neat little grocer man after his story was done, with a glass of whisky beside him and a cigar between his fingers, witnessing, with sorrow still, though now, indeed, with a time blunted anguish, of the inappeasable hunger of the heart that presently came upon him. "I couldn't eat," he said, "I couldn't sleep. I made mistakes in orders and got mixed with change. There she was day and night, drawing me and drawing me. Oh, I wanted her. Lord! how I wanted her! I was up there, most evenings I was up there on the Knoll, often even when it rained. I used to walk over the Knoll and round it and round it, calling for them to let me in. Shouting. Near blubbering I was at times. Daft I was and miserable. I kept on saying it was all a mistake. And every Sunday afternoon I went up there, wet and fine, though I knew as well as you do it wasn't no good by day. And I've tried to go to sleep there."

He stopped sharply and decided to drink some whisky.

"I've tried to go to sleep there," he said, and I could swear his lips trembled. "I've tried to go to sleep there, often and often. And, you know, I couldn't, sir—never. I've thought if I could go to sleep there, there might be something. . . . But I've sat up there and laid up there, and I couldn't—not for thinking and longing. It's the longing. . . . I've tried——"

He blew, drank up the rest of his whisky spasmodically, stood up suddenly and buttoned his jacket, staring closely and critically at the cheap oleographs beside the mantel meanwhile. The little black notebook in which he recorded the orders of his daily round projected stiffly from his breast pocket. When all the buttons were quite done, he patted his chest and turned on me suddenly. "Well," he said, "I must be going."

There was something in his eyes and manner that was too difficult for him to express in words. "One gets talking," he said at last at the door, and smiled wanly, and so vanished from my eyes. And that is the tale of Mr. Skelmersdale in Fairyland just as he told it to me.

STORY THE SIXTH

The Inexperienced Ghost

THE scene amidst which Clayton told his last story comes back very vividly to my mind. There he sat, for the greater part of the time, in the corner of the authentic settle by the spacious open fire, and Sanderson sat beside him smoking the Broseley clay that bore his name. There was Evans, and that marvel among actors, Wish, who is also a modest man. We had all come down to the Mermaid Club that Saturday morning, except Clayton, who had slept there overnight—which indeed gave him the opening of his story. We had golfed until golfing was invisible; we had dined, and we were in that mood of tranquil kindliness when men will suffer a story. When Clayton began to tell one, we naturally supposed he was lying. It may be that indeed he was lying—of that the reader will speedily be able to judge as well as I. He began, it is true, with an air of matter-of-fact anecdote, but that we thought was only the incurable artifice of the man.

"I say!" he remarked, after a long consideration of the upward rain of sparks from the log that Sanderson had thumped, "you know I was alone here last night?"

"Except for the domestics," said Wish.

"Who sleep in the other wing," said Clayton. "Yes. Well——" He pulled at his cigar for some little time as though he still hesitated about his confidence. Then he said, quite quietly, "I caught a ghost!"

"Caught a ghost, did you?" said Sanderson. "Where is it?"

And Evans, who admires Clayton immensely and has been

four weeks in America, shouted, "*Caught* a ghost, did you, Clayton? I'm glad of it! Tell us all about it right now."

Clayton said he would in a minute, and asked him to shut the door.

He looked apologetically at me. "There's no eavesdropping of course, but we don't want to upset our very excellent service with any rumours of ghosts in the place. There's too much shadow and oak panelling to trifle with that. And this, you know, wasn't a regular ghost. I don't think it will come again—ever."

"You mean to say you didn't keep it?" said Sanderson.

"I hadn't the heart to," said Clayton.

And Sanderson said he was surprised.

We laughed, and Clayton looked aggrieved. "I know," he said, with the flicker of a smile, "but the fact is it really *was* a ghost, and I'm as sure of it as I am that I am talking to you now. I'm not joking. I mean what I say."

Sanderson drew deeply at his pipe, with one reddish eye on Clayton, and then emitted a thin jet of smoke more eloquent than many words.

Clayton ignored the comment. "It is the strangest thing that has ever happened in my life. You know I never believed in ghosts or anything of the sort, before, ever; and then, you know, I bag one in a corner; and the whole business is in my hands."

He meditated still more profoundly and produced and began to pierce a second cigar with a curious little stabber he affected.

"You talked to it?" asked Wish.

"For the space, probably, of an hour."

"Chatty?" I said, joining the party of the sceptics.

"The poor devil was in trouble," said Clayton, bowed over his cigar-end and with the very faintest note of reproof.

"Sobbing?" someone asked.

Clayton heaved a realistic sigh at the memory. "Good Lord!" he said; "yes." And then, "Poor fellow! yes."

"Where did you strike it?" asked Evans, in his best American accent.

"I never realised," said Clayton, ignoring him, "the poor

sort of thing a ghost might be," and he hung us up again for a time, while he sought for matches in his pocket and lit and warmed to his cigar.

"I took an advantage," he reflected at last.

We were none of us in a hurry. "A character," he said, "remains just the same character for all that it's been disembodied. That's a thing we too often forget. People with a certain strength or fixity of purpose may have ghosts of a certain strength and fixity of purpose—most haunting ghosts, you know, must be as one-idea'd as monomaniacs and as obstinate as mules to come back again and again. This poor creature wasn't." He suddenly looked up rather queerly, and his eye went round the room. "I say it," he said, "in all kindliness, but that is the plain truth of the case. Even at the first glance he struck me as weak."

He punctuated with the help of his cigar.

"I came upon him, you know, in the long passage. His back was towards me and I saw him first. Right off I knew him for a ghost. He was transparent and whitish; clean through his chest I could see the glimmer of the little window at the end. And not only his physique but his attitude struck me as being weak. He looked, you know, as though he didn't know in the slightest whatever he meant to do. One hand was on the panelling and the other fluttered to his mouth. Like—*so!*"

"What sort of physique?" said Sanderson.

"Lean. You know that sort of young man's neck that has two great flutings down the back, here and here—so! And a little, meanish head with scrubby hair and rather bad ears. Shoulders bad, narrower than the hips; turndown collar, ready-made short jacket, trousers baggy and a little frayed at the heels. That's how he took me. I came very quietly up the staircase. I did not carry a light, you know—the candles are on the landing table and there is that lamp—and I was in my list slippers, and I saw him as I came up. I stopped dead at that—taking him in. I wasn't a bit afraid. I think that in most of these affairs one is never nearly so afraid or excited as one imagines one would be. I was surprised and interested. I thought, 'Good Lord! Here's a ghost at last!

And I haven't believed for a moment in ghosts during the last five-and-twenty years.'"

"Um," said Wish.

"I suppose I wasn't on the landing a moment before he found out I was there. He turned on me sharply, and I saw the face of an immature young man, a weak nose, a scrubby little moustache, a feeble chin. So for an instant we stood —he looking over his shoulder at me—and regarded one another. Then he seemed to remember his high calling. He turned round, drew himself up, projected his face, raised his arms, spread his hands in approved ghost fashion—came towards me. As he did so his little jaw dropped, and he emitted a faint, drawn-out 'Boo." No, it wasn't—not a bit dreadful. I'd dined. I'd had a bottle of champagne, and being all alone, perhaps two or three—perhaps even four or five—whiskies, so I was as solid as rocks and no more frightened than if I'd been assailed by a frog. 'Boo!' I said. 'Nonsense. You don't belong to *this* place. What are you doing here?'

"I could see him wince. 'Boo—oo,' he said.

"'Boo—be hanged! Are you a member?' I said; and just to show I didn't care a pin for him I stepped through a corner of him and made to light my candle. 'Are you a member?' I repeated, looking at him sideways.

"He moved a little so as to stand clear of me, and his bearing became crestfallen. 'No,' he said, in answer to the persistent interrogation of my eye; 'I'm not a member—I'm a ghost.'

"'Well, that doesn't give you the run of the Mermaid Club. Is there anyone you want to see, or anything of that sort?' And doing it as steadily as possible for fear that he should mistake the carelessness of whisky for the distraction of fear, I got my candle alight. I turned on him, holding it. 'What are you doing here?' I said.

"He had dropped his hands and stopped his booing, and there he stood, abashed and awkward, the ghost of a weak, silly, aimless young man. 'I'm haunting,' he said.

"'You haven't any business to,' I said in a quiet voice.

"'I'm a ghost,' he said, as if in defence.

"'That may be, but you haven't any business to haunt here. This is a respectable private club; people often stop here with nursemaids and children, and, going about in the careless way you do, some poor little mite could easily come upon you and be scared out of her wits. I suppose you didn't think of that?'

"'No, sir,' he said, 'I didn't.'

"'You should have done. You haven't any claim on the place, have you? Weren't murdered here, or anything of that sort?'

"'None, sir; but I thought as it was old and oak-panelled——'

"'That's *no* excuse,' I regarded him firmly. 'Your coming here is a mistake,' I said, in a tone of friendly superiority. I feigned to see if I had my matches, and then looked up at him frankly. 'If I were you I wouldn't wait for cock-crow —I'd vanish right away.'

"He looked embarrassed. 'The fact is, sir——' he began.

"'I'd vanish,' I said, driving it home.

"'The fact is, sir, that—somehow—I can't.'

"'You *can't?*'

"'No, sir. There's something I've forgotten. I've been hanging about here since midnight last night, hiding in the cupboards of the empty bedrooms and things like that. I'm flurried. I've never come haunting before, and it seems to put me out.'

"'Put you out?'

"'Yes, sir. I've tried to do it several times, and it doesn't come off. There's some little thing has slipped me, and I can't get back.'

"That, you know, rather bowled me over. He looked at me in such an abject way that for the life of me I couldn't keep up quite the high hectoring vein I had adopted. 'That's queer,' I said, and as I spoke I fancied I heard someone moving about down below. 'Come into my room and tell me more about it,' I said. 'I didn't of course, understand this,' and I tried to take him by the arm. But, of course, you might as well have tried to take hold of a puff of smoke! I had forgotten my number, I think; anyhow, I remember going into several bedrooms—it was lucky I was the only soul in that wing—until I saw my traps. 'Here we are,' I said, and sat down in the armchair; 'sit down and tell me all about it. It seems to me you have got yourself into a jolly awkward position, old chap.'

"Well, he said he wouldn't sit down; he'd prefer to flit up and down the room if it was all the same to me. And so he did, and in a little while we were deep in a long and serious talk. And presently, you know, something of those whiskies and sodas evaporated out of me, and I began to realise just a little what a thundering rum and weird business it was that I was in. There he was, semi-transparent—the proper conventional phantom, and noiseless except for his ghost of a voice—flitting to and fro in that nice, clean, chintz-hung old bedroom. You could see the gleam of the copper candlesticks through him, and the lights on the brass fender, and the corners of the framed engravings on the wall, and there he was telling me all about this wretched little life of his that had recently ended on earth. He hadn't a particularly honest face, you know, but being transparent, of course, he couldn't avoid telling the truth."

"Eh?" said Wish, suddenly sitting up in his chair.

"What?" said Clayton.

"Being transparent—couldn't avoid telling the truth—I don't see it," said Wish.

"*I* don't see it," said Clayton, with inimitable assurance. "But it *is* so, I can assure you nevertheless. I don't believe he got once a nail's breadth off the Bible truth. He told me how he had been killed—he went down into a London basement with a candle to look for a leakage of gas—and described himself as a senior English master in a London private school when that release occurred."

"Poor wretch!" said I.

"That's what I thought, and the more he talked the more I thought it. There he was, purposeless in life and purposeless out of it. He talked of his father and mother and his schoolmaster, and all who had ever been anything to him in the world, meanly. He had been too sensitive, too nervous; none of them had ever valued him properly or understood him, he said. He had never had a real friend in the world, I think; he had never had a success. He had shirked games and failed examinations. 'It's like that with some people,' he said; 'whenever I got into the examination-room or anywhere everything seemed to go.' Engaged to be married of course to another over-sensitive person, I suppose—when the indiscretion with the gas escape ended his affairs. 'And where are you now?' I asked. 'Not in——?'

"He wasn't clear on that point at all. The impression he gave me was of a sort of vague, intermediate state, a special reserve for souls too non-existent for anything so positive as either sin or virtue. I don't know. He was much too egotistical and unobservant to give me any clear idea of the kind of place, kind of country, there is on the Other Side of Things. Wherever he was, he seems to have fallen in with a set of kindred spirits: ghosts of weak Cockney young men, who were on a footing of Christian names, and among these there was certainly a lot of talk about 'going haunting' and things like that. Yes—going haunting! They seemed to think 'haunting' a tremendous adventure, and most of them funked it all the time. And so primed, you know, he had come."

"But really!" said Wish to the fire.

"These are the impressions he gave me, anyhow," said Clayton, modestly. "I may, of course, have been in a rather uncritical state, but that was the sort of background he gave to himself. He kept flitting up and down, with his thin voice going—talking, talking about his wretched self, and never a word of clear, firm statement from first to last. He was thinner and sillier and more pointless than if he had been real and alive. Only then, you know, he would not have been in my bedroom here—if he *had* been alive. I should have kicked him out."

"Of course," said Evans, "there *are* poor mortals like that."

"And there's just as much chance of their having ghosts as the rest of us," I admitted.

"What gave a sort of point to him, you know, was the fact that he did seem within limits to have found himself out. The mess he had made of haunting had depressed him terribly. He had been told it would be a 'lark'; he had come expecting it to be a 'lark,' and here it was, nothing but another failure added to his record! He proclaimed himself an utter out-and-out failure. He said, and I can quite believe it, that he had never tried to do anything all his life that he hadn't made a perfect mess of—and through all the wastes of eternity he never would. If he had had sympathy, perhaps— He paused at that, and stood regarding me. He remarked that, strange as it might seem to me, nobody, not anyone, ever, had given him the amount of sympathy I was doing now. I could see what he wanted straight away, and I determined to head him off at once. I may be a brute, you know, but being the Only Real Friend, the recipient of the confidences of one of these egotistical weaklings, ghost or body, is beyond my physical endurance. I got up briskly. 'Don't you brood on these things too much,' I said. 'The thing you've got to do is to get out of this—get out of this sharp. You pull yourself together and *try*.' 'I can't,' he said. 'You try,' I said, and try he did."

"Try!" said Sanderson. "*How?*"

"Passes," said Clayton.

"Passes?"

"Complicated series of gestures and passes with the hands. That's how he had come in and that's how he had to get out again. Lord! what a business I had!"

"But how could *any* series of passes——" I began.

"My dear man," said Clayton, turning on me and putting a great emphasis on certain words, "you want *everything* clear. I don't know *how*. All I know is that you *do*—that *he* did, anyhow, at least. After a fearful time, you know, he got his passes right and suddenly disappeared."

"Did you," said Sanderson slowly, "observe the passes?"

"Yes," said Clayton, and seemed to think. "It was tremendously queer," he said. "There we were, I and this thin vague ghost, in that silent room, in this silent, empty inn, in this silent little Friday-night town. Not a sound except our voices and a faint panting he made when he swung. There was the bedroom candle, and one candle on the dressing-table alight, that was all—sometimes one or other would flare up into a tall, lean, astonished flame for a space. And queer things happened. 'I can't,' he said; 'I shall never——!' And suddenly he sat down on a little chair at the foot of the bed and began to sob and sob. Lord! what a harrowing, whimpering thing he seemed!

" 'You pull yourself together,' I said, and tried to pat him on the back, and . . . my confounded hand went through him! By that time, you know, I wasn't nearly so—massive as I had been on the landing. I got the queerness of it full. I remember snatching back my hand out of him, as it were, with a little thrill, and walking over to the dressing-table. 'You pull yourself together,' I said to him, 'and try.' And in order to encourage and help him I began to try as well."

"What!" said Sanderson, "the passes?"

"Yes, the passes."

"But——" I said, moved by an idea that eluded me for a space.

"This is interesting," said Sanderson, with his finger in his pipe-bowl. "You mean to say this ghost of yours gave way——"

"Did his level best to give away the whole confounded barrier? *Yes.*"

"He didn't," said Wish; "he couldn't. Or you'd have gone there too."

"That's precisely it," I said, finding my elusive idea put into words for me.

"That *is* precisely it," said Clayton, with thoughtful eyes upon the fire.

For just a little while there was silence.

"And at last he did it?" said Sanderson.

"At last he did it. I had to keep him up to it hard, but he did it at last—rather suddenly. He despaired, we had a scene, and then he got up abruptly and asked me to go through the whole performance, slowly, so that he might see. 'I believe,' he said, 'if I could *see* I should spot what was wrong at once.' And he did. 'I know,' he said. 'What do you know?' said I. 'I know,' he repeated. Then he said, peevishly, 'I *can't* do it, if you look at me—I really *can't*; it's been that, partly, all along. I'm such a nervous fellow that you put me out.' Well, we had a bit of an argument. Naturally I wanted to see; but he was as obstinate as a mule, and suddenly I had come over as tired as a dog—he tired me out. 'All right,' I said, 'I won't look at you,' and turned towards the mirror, on the wardrobe, by the bed.

"He started off very fast. I tried to follow him by looking in the looking-glass, to see just what it was had hung. Round went his arms and his hands, so, and so, and so, and then with a rush came to the last gesture of all—you stand erect and open out your arms—and so, don't you know, he stood. And then he didn't! He didn't! He wasn't! I wheeled round from the looking-glass to him. There was nothing! I was alone, with the flaring candles and a staggering mind. What had happened? Had anything happened? Had I been dreaming? . . . And then, with an absurd note of finality about it, the clock upon the landing discovered the moment was ripe for striking *one*. So!—Ping! And I was as grave and sober as a judge, with all my champagne and whisky gone into the vast serene. Feeling queer, you know—confoundedly *queer*! Queer! Good Lord!"

He regarded his cigar-ash for a moment. "That's all that happened," he said.

"And then you went to bed?" asked Evans.

"What else was there to do?"

I looked Wish in the eye. We wanted to scoff, and there was something, something perhaps in Clayton's voice and manner, that hampered our desire.

"And about these passes?" said Sanderson.

"I believe I could do them now."

"Oh!" said Sanderson, and produced a pen-knife and set himself to grub the dottel out of the bowl of his clay.

"Why don't you do them now?" said Sanderson, shutting his pen-knife with a click.

"That's what I'm going to do," said Clayton.

"They won't work," said Evans.

"If they do——" I suggested.

"You know, I'd rather you didn't," said Wish, stretching out his legs.

"Why?" asked Evans.

"I'd rather he didn't," said Wish.

"But he hasn't got 'em right," said Sanderson, plugging too much tobacco into his pipe.

"All the same, I'd rather he didn't," said Wish.

We argued with Wish. He said that for Clayton to go through those gestures was like mocking a serious matter. "But you don't believe——?" I said. Wish glanced at Clayton, who was staring into the fire, weighing something in his mind. "I do—more than half, anyhow, I do," said Wish.

"Clayton," said I, "you're too good a liar for us. Most of it was all right. But that disappearance . . . happened to be convincing. Tell us, it's a tale of cock and bull."

He stood up without heeding me, took the middle of the hearthrug, and faced me. For a moment he regarded his feet thoughtfully, and then for all the rest of the time his eyes were on the opposite wall, with an intent expression. He raised his two hands slowly to the level of his eyes and so began. . . .

Now, Sanderson is a Freemason, a member of the lodge of the Four Kings, which devotes itself so ably to the study and elucidation of all the mysteries of Masonry past and present, and among the students of this lodge Sanderson is by no means the least. He followed Clayton's motions with a singular interest in his reddish eye. "That's not bad," he said, when it was done. "You really do, you know, put things together, Clayton, in a most amazing fashion. But there's one little detail out."

"I know," said Clayton. "I believe I could tell you which."

"Well?"

"This," said Clayton, and did a queer little twist and writhing and thrust of the hands.

"Yes."

"That, you know, was what *he* couldn't get right," said Clayton. "But how do *you*——?"

"Most of this business, and particularly how you invented it, I don't understand at all," said Sanderson, "but just that phase—I do." He reflected. "These happen to be a series of gestures—connected with a certain branch of esoteric

Masonry— Probably you know. Or else—— *How?*" He reflected still further. "I do not see I can do any harm in telling you just the proper twist. After all, if you know, you know; if you don't, you don't."

"I know nothing," said Clayton, "except what the poor devil let out last night."

"Well, anyhow," said Sanderson, and placed his church-warden very carefully upon the shelf over the fireplace. Then very rapidly he gesticulated with his hands.

"So?" said Clayton, repeating.

"So," said Sanderson, and took his pipe in hand again.

"Ah, *now*," said Clayton, "I can do the whole thing— right."

He stood up before the waning fire and smiled at us all. But I think there was just a little hesitation in his smile. "If I begin——" he said.

"I wouldn't begin," said Wish.

"It's all right!" said Evans. "Matter is indestructible. You don't think any jiggery-pokery of this sort is going to snatch Clayton into the world of shades. Not it! You may try, Clayton, so far as I'm concerned, until your arms drop off at the wrists."

"I don't believe that," said Wish, and stood up and put his arm on Clayton's shoulder. "You've made me half believe in that story somehow, and I don't want to see the thing done."

"Goodness!" said I, "here's Wish frightened!"

"I am," said Wish, with real or admirably feigned intensity. "I believe that if he goes through these motions right he'll *go*."

"He'll not do anything of the sort," I cried. "There's only one way out of this world for men, and Clayton is thirty years from that. Besides . . . And such a ghost! Do you think——?"

Wish interrupted me by moving. He walked out from among our chairs and stopped beside the table and stood there. "Clayton," he said, "you're a fool."

Clayton, with a humorous light in his eyes, smiled back at him. "Wish," he said, "is right and all you others are wrong. I shall go. I shall get to the end of these passes, and as the last swish whistles through the air, Presto!—this hearthrug will be vacant, the room will be blank amazement, and a respectably dressed gentleman of fifteen stone will plump into the world of shades. I'm certain. So will you be. I decline to argue further. Let the thing be tried."

"*No*," said Wish, and made a step and ceased, and Clayton raised his hands once more to repeat the spirit's passing.

By that time, you know, we were all in a state of tension—largely because of the behaviour of Wish. We sat all of us with our eyes on Clayton—I, at least, with a sort of tight, stiff feeling about me as though from the back of my skull to the middle of my thighs my body had been changed to steel. And there, with a gravity that was imperturbably serene, Clayton bowed and swayed and waved his hands and arms before us. As he drew towards the end one piled up, one tingled in one's teeth. The last gesture, I have said, was to swing the arms out wide open, with the face held up. And when at last he swung out to this closing gesture I ceased even to breathe. It was ridiculous, of course, but you know that ghost-story feeling. It was after dinner, in a queer, old shadowy house. Would he, after all——?

There he stood for one stupendous moment, with his arms open and his upturned face, assured and bright, in the glare of the hanging lamp. We hung through that moment as if it were an age, and then came from all of us something that was half a sigh of infinite relief and half a reassuring "*No!*" For visibly—he wasn't going. It was all nonsense. He had told an idle story, and carried it almost to conviction, that was all! . . . And then in that moment the face of Clayton changed.

It changed. It changed as a lit house changes when its lights are suddenly extinguished. His eyes were suddenly eyes that were fixed, his smile was frozen on his lips, and he stood there still. He stood there, very gently swaying.

That moment, too, was an age. And then, you know, chairs were scraping, things were falling, and we were all moving. His knees seemed to give, and he fell forward, and Evans rose and caught him in his arms. . . .

It stunned us all. For a minute I suppose no one said a coherent thing. We believed it, yet could not believe it. . . . I came out of a muddled stupefaction to find myself kneeling beside him, and his vest and shirt were torn open, and Sanderson's hand lay on his heart. . . .

Well—the simple fact before us could very well wait our convenience; there was no hurry for us to comprehend. It lay there for an hour; it lies athwart my memory, black and amazing still, to this day. Clayton had, indeed, passed into the world that lies so near to and so far from our own, and he had gone thither by the only road that mortal man may take. But whether he did indeed pass there by that poor ghost's incantation, or whether he was stricken suddenly by apoplexy in the midst of an idle tale—as the coroner's jury would have us believe—is no matter for my judging; it is just one of those inexplicable riddles that must remain unsolved until the final solution of all things shall come. All I certainly know is that, in the very moment, in the very instant, of concluding those passes, he changed, and staggered, and fell down before us—dead!

STORY THE SEVENTH

Jimmy Goggles the God

"It isn't everyone who's been a god," said the sunburnt man. "But it's happened to me. Among other things."

I intimated my sense of his condescension.

"It don't leave much for ambition, does it?" said the sunburnt man.

"I was one of those men who were saved from the *Ocean Pioneer*. Gummy! how time flies! It's twenty years ago. I doubt if you'll remember anything of the *Ocean Pioneer*?"

The name was familiar, and I tried to recall when and where I had read it. The *Ocean Pioneer*? "Something about gold dust," I said vaguely, "but the precise——"

"That's it," he said. "In a beastly little channel she hadn't no business in—dodging pirates. It was before they'd put the kybosh on that business. And there'd been volcanoes or something and all the rocks was wrong. There's places about by Soona where you fair have to follow the rocks about to see where they're going next. Down she went in twenty fathoms before you could have dealt for whist, with fifty thousand pounds worth of gold aboard, it was said, in one form or another."

"Survivors?"

"Three."

"I remember the case now," I said. "There was something about salvage——"

But at the word salvage the sunburnt man exploded into language so extraordinarily horrible that I stopped aghast. He came down to more ordinary swearing, and pulled himself up abruptly. "Excuse me," he said, "but—salvage!"

He leant over towards me. "I was in that job," he said.
"Tried to make myself a rich man, and got made a god
instead. I've got my feelings——

"It ain't all jam being a god," said the sunburnt man, and
for some time conversed by means of such pithy but unpro-
gressive axioms. At last he took up his tale again.

"There was me," said the sunburnt man, "and a seaman
name Jacobs, and Always, the mate of the *Ocean Pioneer*.
And him it was that set the whole thing going. I remember
him now, when we was in the jolly boat, suggesting it all to
our minds just by one sentence. He was a wonderful hand
at suggesting things. 'There was forty thousand pounds,'
he said, 'on that ship, and it's for me to say just where she
went down.' It didn't need much brains to tumble to that.
And he was the leader from the first to the last. He got
hold of the Sanderses and their brig; they were brothers, and
the brig was the *Pride of Banya*, and he it was bought the
diving dress—a second-hand one with a compressed air apparatus instead of pumping. He'd have done the diving too,
if it hadn't made him sick going down. And the salvage
people were mucking about with a chart he'd cooked up, as
solemn as could be, at Starr Race, a hundred and twenty
miles away.

"I can tell you we was a happy lot aboard that brig, jokes
and drink and bright hopes all the time. It all seemed so
neat and clean and straightforward, and what rough chaps
call a 'cert.' And we used to speculate how the other blessed
lot, the proper salvagers, who'd started two days before us,
were getting on, until our sides fairly ached. We all messed
together in the Sanderses' cabin—it was a curious crew, all
officers and no men—and there stood the diving-dress waiting its turn. Young Sanders was a humorous sort of chap,
and there certainly was something funny in the confounded
thing's great fat head and its stare, and he made us see it too.
'Jimmy Goggles,' he used to call it, and talk to it like a
Christian. Asked him if he was married, and how Mrs.
Goggles was, and all the little Goggleses. Fit to make you
split. And every blessed day all of us used to drink the
health of Jimmy Goggles in rum, and unscrew his eye and

pour a glass of rum in him, until, instead of that nasty mackintosheriness, he smelt as nice in his inside as a cask of rum. It was jolly times we had in those days, I can tell you —little suspecting, poor chaps! what was a-coming.

"We weren't going to throw away our chances by any blessed hurry, you know, and we spent a whole day sounding our way towards where the *Ocean Pioneer* had gone down, right between two chunks of ropy grey rock—lava rocks that rose nearly out of the water. We had to lay off about half a mile to get a safe anchorage, and there was a thundering row who should stop on board. And there she lay just as she had gone down, so that you could see the top of the masts that was still standing perfectly distinctly. The row ended in all coming in the boat. I went down in the diving-dress on Friday morning directly it was light.

"What a surprise it was! I can see it all now quite distinctly. It was a queer-looking place, and the light was just coming. People over here think every blessed place in the tropics is a flat shore and palm trees and surf, bless 'em! This place, for instance, wasn't a bit that way. Not common rocks they were, undermined by waves; but great curved banks like ironwork cinder heaps, with green slime below, and thorny shrubs and things just waving upon them here and there, and the water glassy calm and clear, and showing you a kind of dirty grey-black shine, with huge flaring red-brown weeds spreading motionless, and crawling and darting things going through it. And far away beyond the ditches and pools and the heaps was a forest on the mountain flank, growing again after the fires and cinder showers of the last eruption. And the other way forest, too, and a kind of broken—what is it?—amby-theatre of black and rusty cinders rising out of it all, and the sea in a kind of bay in the middle.

"The dawn, I say, was just coming, and there wasn't much colour about things, and not a human being but ourselves anywhere in sight up or down the channel. Except the *Pride of Banya*, lying out beyond a lump of rocks towards the line of the sea.

"Not a human being in sight," he repeated, and paused.

"I don't know where they came from, not a bit. And we were feeling so safe that we were all alone that poor young Sanders was a-singing. I was in Jimmy Goggles, all except the helmet. 'Easy,' says Always, 'there's her mast.' And after I'd had just one squint over the gunwhale, I caught up the bogey and almost tipped out as old Sanders brought the boat round. When the windows were screwed and everything was all right, I shut the valve from the air belt in order to help my sinking, and jumped overboard, feet foremost—for we hadn't a ladder. I left the boat pitching, and all of them staring down into the water after me, as my head sank down into the weeds and blackness that lay about the mast. I suppose nobody, not the most cautious chap in the world, would have bothered about a look-out at such a desolate place. It stunk of solitude.

"Of course you must understand that I was a greenhorn at diving. None of us were divers. We'd had to muck about with the thing to get the way of it, and this was the first time I'd been deep. It feels damnable. Your ears hurt beastly. I don't know if you've ever hurt yourself yawning or sneezing, but it takes you like that, only ten times worse. And a pain over the eyebrows here—splitting—and a feeling like influenza in the head. And it isn't all heaven in your lungs and things. And going down feels like the beginning of a lift, only it keeps on. And you can't turn your head to see what's above you, and you can't get a fair squint at what's happening to your feet without bending down something painful. And being deep it was dark, let alone the blackness of the ashes and mud that formed the bottom. It was like going down out of the dawn back into the night, so to speak.

"The mast came up like a ghost out of the black, and then a lot of fishes, and then a lot of flapping red seaweed, and then whack I came with a kind of dull bang on the deck of the *Ocean Pioneer*, and the fishes that had been feeding on the dead rose about me like a swarm of flies from road stuff in summer time. I turned on the compressed air again— for the suit was a bit thick and mackintoshery after all, in spite of the rum—and stood recovering myself. It struck coolish down there, and that helped take off the stuffiness a bit.

"When I began to feel easier, I started looking about me. It was an extraordinary sight. Even the light was extraordinary, a kind of reddy coloured twilight, on account of the streamers of seaweed that floated up on either side of the ship. And far overhead just a moony, deep green blue. The deck of the ship, except for a slight list to starboard, was level, and lay all dark and long between the weeds, clear except where the masts had snapped when she rolled, and vanishing into black night towards the forecastle. There weren't any dead on the decks, most were in the weeds alongside, I suppose; but afterwards I found two skeletons lying in the passengers' cabins, where death had come to them. It was curious to stand on that deck and recognise it all, bit by bit; a place against the rail where I'd been fond of smoking by starlight, and the corner where an old chap from Sydney used to flirt with a widow we had aboard. A comfortable couple they'd been, only a month ago, and now you couldn't have got a meal for a baby crab off either of them.

"I've always had a bit of philosophical turn, and I dare say I spent the best part of five minutes in such thoughts before I went below to find where the blessed dust was stored. It was slow work hunting, feeling it was for the most part, pitchy dark, with confusing blue gleams down the companion. And there were things moving about, a dab at my glass once, and once a pinch at my leg. Crabs I expect. I kicked a lot of loose stuff that puzzled me, and stooped and picked up something all knobs and spikes. What do you think? Backbone! But I never had any particular feeling for bones. We had talked the affair over pretty thoroughly, and Always knew just where the stuff was stowed. I found it that trip. I lifted a box one end an inch or more."

He broke off in his story. "I've lifted it," he said, "as near as that! Forty thousand pounds worth of pure gold! Gold! I shouted inside my helmet as a kind of cheer and hurt my ears. I was getting confounded stuffy and tired by this time—I must have been down twenty-five minutes or more—and I thought this was good enough. I went up the companion again, and as my eyes came up flush with the deck, a thundering great crab gave a kind of hysterical jump and went scutting off sideways. Quite a start it gave me. I stood up clear on deck and shut the valve behind the helmet to let the air accumulate to carry me up again—I noticed a kind of whacking from above, as though they were hitting the water with an oar, but I didn't look up. I fancied they were signalling me to come up.

"And then something shot down by me—something heavy, and stood a-quiver in the planks. I looked, and there was a long knife I'd seen young Sanders handling. Thinks I, he's dropped it, and I was still calling him this kind of fool and that—for it might have hurt me serious—when I began to lift and drive up towards the daylight. Just about the level of the top spars of the *Ocean Pioneer*, whack! I came against something sinking down, and a boot knocked in front of my helmet. Then something else, struggling frightful. It was a big weight atop of me, whatever it was, and moving and twisting about. I'd have thought it a big octopus, or some such thing, if it hadn't been for the boot. But octopuses don't wear boots. It was all in a moment, of course. I felt myself sinking down again, and I threw my arms about to keep steady, and the whole lot rolled free of me and shot down as I went up——"

He paused.

"I saw young Sanders's face, over a naked black shoulder, and a spear driven clean through his neck, and out of his mouth and neck what looked like spirts of pink smoke in the water. And down they went clutching one another, and turning over, and both too far gone to leave go. And in another second my helmet came a whack, fit to split, against the niggers' canoe. It was niggers! Two canoes full.

"It was lively times, I tell you! Overboard came Always with three spears in him. There was the legs of three or four black chaps kicking about me in the water. I couldn't see much, but I saw the game was up at a glance, gave my valve a tremendous twist, and went bubbling down again after poor Always, in as awful a state of scare and astonishment as you can well imagine. I passed young Sanders and the nigger going up again and struggling still a bit, and in another moment I was standing in the dim again on the deck of the *Ocean Pioneer*.

" 'Gummy,' thinks I, 'here's a fix! Niggers?' At first I couldn't see anything for it but Stifle below or Stabs above. I didn't properly understand how much air there was to last me out, but I didn't feel like standing very much more of it down below. I was hot and frightfully heady quite apart from the blue funk I was in. We'd never reckoned with these beastly natives, filthy Papuan beasts. It wasn't any good coming up where I was, but I had to do something. On the spur of the moment, I clambered over the side of the brig and landed among the weeds, and set off through the darkness as fast as I could. I just stopped once and knelt, and twisted back my head in the helmet and had a look up. It was a most extraordinary bright green-blue above, and the two canoes and the boat floating there very small and distant like a kind of twisted H. And it made me feel sick to squint up at it, and think what the pitching and swaying of the three meant.

"It was just about the most horrible ten minutes I ever had, blundering about in that darkness—pressure something awful, like being buried in sand, pain across the chest, sick with funk, and breathing nothing as it seemed but the smell of rum and mackintosh. Gummy! After a bit, I found myself going up a steepish sort of slope. I had another squint to see if anything was visible of the canoes and boats, and then kept on. I stopped with my head a foot from the surface, and tried to see where I was going, but, of course, nothing was to be seen but the reflection of the bottom. Then out I dashed like knocking my head through a mirror. Directly I got my eyes out of the water, I saw I'd come up a kind of beach near the forest. I had a look round, but the natives and the brig were both hidden by a big hummucky heap of twisted lava. The born fool in me suggested a run for the woods. I didn't take the helmet off, but I eased open one of the windows, and, after a bit of a pant, went on out of the water. You'd hardly imagine how clean and light the air tasted.

"Of course, with four inches of lead in your boot soles, and your head in a copper knob the size of a football, and been thirty-five minutes under water, you don't break any records running. I ran like a ploughboy going to work. And halfway to the trees I saw a dozen niggers or more, coming out in a gaping, astonished sort of way to meet me.

"I just stopped dead, and cursed myself for all the fools out of London. I had about as much chance of cutting back to the water as a turned turtle. I just screwed up my window again to leave my hands free, and waited for them. There wasn't anything else for me to do.

"But they didn't come on very much. I began to suspect why. 'Jimmy Goggles,' I says, 'it's your beauty does it.' I was inclined to be a little light-headed, I think, with all these dangers about and the change in the pressure of the blessed air. 'Who're ye staring at?' I said, as if the savages could hear me. 'What d'ye take me for? I'm hanged if I don't give you something to stare at,' I said, and with that I screwed up the escape valve and turned on the compressed air from the belt, until I was swelled out like a blown frog. Regular imposing it must have been. I'm blessed if they'd come on a step; and presently one and then another went down on their hands and knees. They didn't know what to make of me, and they was doing the extra polite, which was very wise and reasonable of them. I had half a mind to edge back seaward and cut and run, but it seemed too hopeless. A step back and they'd have been after me. And out of sheer desperation I began to march towards them up the beach, with slow, heavy steps, and waving my blown-out arms about, in a dignified manner. And inside of me I was singing as small as a tomtit.

"But there's nothing like a striking appearance to help a man over a difficulty—I've found that before and since.

People like ourselves, who're up to diving dresses by the time we're seven, can scarcely imagine the effect of one on a simple-minded savage. One or two of these niggers cut and run, the others started in a great hurry trying to knock their brains out on the ground. And on I went as slow and solemn and silly-looking and artful as a jobbing plumber. It was evident they took me for something immense.

"Then up jumped one and began pointing, making extraordinary gestures to me as he did so, and all the others began sharing their attention between me and something out at sea. 'What's the matter now,' I said. I turned slowly on account of my dignity, and there I saw, coming round a point, the poor old *Pride of Banya* towed by a couple of canoes. The sight fairly made me sick. But they evidently expected some recognition, so I waved my arms in a striking sort of non-committal manner. And then I turned and stalked on towards the trees again. At that time I was praying like mad, I remember, over and over again: 'Lord help me through with it! Lord help me through with it!' It's only fools who know nothing of dangers can afford to laugh at praying.

"But these niggers weren't going to let me walk through and away like that. They started a kind of bowing dance about me, and sort of pressed me to take a pathway that lay through the trees. It was clear to me they didn't take me for a British citizen, whatever else they thought of me, and for my own part I was never less anxious to own up to the old country.

"You'd hardly believe it, perhaps, unless you're familiar with savages, but these poor misguided, ignorant creatures took me straight to their kind of joss place to present me to the blessed old black stone there. By this time I was beginning to sort of realise the depth of their ignorance, and directly I set eyes on this deity I took my cue. I started a baritone howl, 'wow-wow,' very long on one note, and began waving my arms about a lot, and then very slowly and ceremoniously turned their image over on its side and sat down on it. I wanted to sit down badly, for diving dresses ain't much wear in the tropics. Or, to put it different like, they're a sight too much. It took away their breath, I could see, my sitting on their joss, but in less time than a minute they made up their minds and were hard at work worshipping me. And I can tell you I felt a bit relieved to see things turning out so well, in spite of the weight on my shoulders and feet.

"But what made me anxious was that the chaps in the canoes might think when they came back. If they'd seen me in the boat before I went down, and without the helmet on—for they might have been spying and hiding since over night—they would very likely take a different view from the others. I was in a deuce of a stew about that for hours, as it seemed, until the shindy of the arrival began.

"But they took it down—the whole blessed village took it down. At the cost of sitting up stiff and stern, as much like those sitting Egyptian images one sees as I could manage, for pretty nearly twelve hours, I should guess at least, on end, I got over it. You'd hardly think what it meant in that heat and stink. I don't think any of them dreamt of the man inside. I was just a wonderful leathery great joss that had come up with luck out of the water. But the fatigue! the heat! the beastly closeness! the mackintosheriness and the rum! and the fuss! They lit a stinking fire on a kind of lava slab there was before me, and brought in a lot of gory muck—the worst parts of what they were feasting on outside, the Beasts—and burnt it all in my honour. I was getting a bit hungry, but I understand now how gods manage to do without eating, what with the smell of burnt offerings about them. And they brought in a lot of the stuff they'd got off the brig and, among other stuff, what I was a bit relieved to see, the kind of pneumatic pump that was used for the compressed air affair, and then a lot of chaps and girls came in and danced about me something disgraceful. It's extraordinary the different ways different people have of showing respect. If I'd had a hatchet handy I'd have gone for the lot of them—they made me feel that wild. All this time I sat as stiff as company, not knowing anything better to do. And at last, when nightfall came, and the wattle joss-house place got a bit too shadowy for their taste—all these here savages are afraid of the dark you know—and I started a sort of 'Moo' noise, they built big bonfires outside and left me alone in peace in the darkness of my hut, free to unscrew my windows a bit and think things over, and feel just as bad as I liked. And, Lord! I was sick.

"I was weak and hungry, and my mind kept on behaving like a beetle on a pin, tremendous activity and nothing done at the end of it. Come round just where it was before. There was sorrowing for the other chaps, beastly drunkards certainly, but not deserving such a fate, and young Sanders with the spear through his neck wouldn't go out of my mind. There was the treasure down there in the *Ocean Pioneer*, and how one might get it and hide it somewhere safer, and get away and come back for it. And there was the puzzle where to get anything to eat. I tell you I was fair rambling. I was afraid to ask by signs for food, for fear of behaving too human, and so there I sat and hungered until very near the dawn. Then the village got a bit quiet, and I couldn't stand it any longer, and I went out and got some stuff like artichokes in a bowl and some sour milk. What was left of these I put away among the other offerings, just to give them a hint of my tastes. And in the morning they came to worship, and found me sitting up stiff and respectable on their previous god, just as they'd left me overnight. I'd got my back against the central pillar of the hut, and, practically, I was asleep. And that's how I became a god among the heathen—a false god no doubt, and blasphemous, but one can't always pick and choose.

"Now, I don't want to crack myself up as a god beyond my merits, but I must confess that while I was god to these people they was extraordinary successful. I don't say there's anything in it, mind you. They won a battle with another tribe—I got a lot of offerings I didn't want through it— they had wonderful fishing, and their crop of pourra was exceptionally fine. And they counted the capture of the brig among the benefits I brought 'em. I must say I don't think that was a poor record for a perfectly new hand. And, though perhaps you'd scarcely credit it, I was the tribal god of those beastly savages for pretty nearly four months. . . .

"What else could I do, man? But I didn't wear that diving-dress all the time. I made 'em rig me up a sort of holy of holies, and a deuce of a time I had too, making them understand what it was I wanted them to do. That indeed was the great difficulty—making them understand my wishes. I couldn't let myself down by talking their lingo badly—even if I'd been able to speak at all—and I couldn't go flapping a lot of gestures at them. So I drew pictures in sand and sat down beside them and hooted like one o'clock. Sometimes they did the things I wanted all right, and sometimes they did them all wrong. They was always very willing, certainly. All the while I was puzzling how I was to get the confounded business settled. Every night before the dawn I used to march out in full rig and go off to a place where I could see the channel in which the *Ocean Pioneer* lay sunk, and once even, one moonlight night, I tried to walk out to her, but the weeds and rocks and dark clean beat me. I didn't get back till full day, and then I found all those silly niggers out on the beach praying their sea-god to return to them. I was that vexed and tired, messing and tumbling about, and coming up and going down again, I could have punched their silly heads all round when they started rejoicing. I'm hanged if I like so much ceremony.

"And then came the missionary. That missionary! It was in the afternoon, and I was sitting in state in my outer temple place, sitting on that old black stone of theirs when he came. I heard a row outside and jabbering, and then his voice speaking to an interpreter. 'They worship stocks and stones,' he said, and I knew what was up, in a flash. I had one of my windows out for comfort, and I sang out straight away on the spur of the moment. 'Stocks and stones!' I says. 'You come inside,' I said, 'and I'll punch your blooming head.' There was a kind of silence and more jabbering, and in he came, Bible in hand, after the manner of them—a little sandy chap in specks and a pith helmet. I flatter myself that me sitting there in the shadows, with my copper head and my big goggles, struck him a bit of a heap at first. 'Well,' I says, 'how's the trade in calico?' for I don't hold with missionaries.

"I had a lark with that missionary. He was a raw hand, and quite outclassed with a man like me. He gasped out who was I, and I told him to read the inscription at my feet if he wanted to know. Down he goes to read, and his interpreter, being of course as superstitious as any of them, took it as an act of worship and plumped down like a shot. All my people gave a howl of triumph, and there wasn't any more business to be done in my village after that journey, not by the likes of him.

"But, of course, I was a fool to choke him off like that. If I'd had any sense I should have told him straight away of the treasure and taken him into Co. I've no doubt he'd have come into Co. A child, with a few hours to think it over, could have seen the connection between my diving dress and the loss of the *Ocean Pioneer*. A week after he left I went out one morning and saw the *Motherhood*, the salver's ship from Starr Race, towing up the channel and sounding. The whole blessed game was up, and all my trouble thrown away. Gummy? How wild I felt! And guying it in that stinking silly dress! Four months!"

The sunburnt man's story degenerated again. "Think of it," he said, when he emerged to linguistic purity once more. "Forty thousand pounds' worth of gold."

"Did the little missionary come back?" I asked.

"Oh yes? Bless him! And he pledged his reputation there was a man inside the god, and started out to see as much with tremendous ceremony. But there wasn't—he got sold again. I always did hate scenes and explanations, and long before he came I was out of it all—going home to Banya along the coast, hiding in bushes by day, and thieving food from the villages by night. Only weapon, a spear. No clothes, no money. Nothing. My face was my fortune, as the saying is. And just a squeak of eight thousand pounds of gold—fifth share. But the natives cut up rusty, thank goodness, because they thought it was him had driven their luck away."

STORY THE EIGHTH

The New Accelerator

CERTAINLY, if ever a man found a guinea when he was looking for a pin it is my good friend Professor Gibberne. I have heard before of investigators overshooting the mark, but never quite to the extent that he has done. He has really, this time at any rate, without any touch of exaggeration in the phrase, found something to revolutionise human life. And that when he was simply seeking an all-round nervous stimulant to bring languid people up to the stresses of these pushful days. I have tasted the stuff now several times, and I cannot do better than describe the effect the thing had on me. That there are astonishing experiences in store for all in search of new sensations will become apparent enough.

Professor Gibberne, as many people know, is my neighbour in Folkestone. Unless my memory plays me a trick, his portrait at various ages has already appeared in *The Strand Magazine*—I think late in 1899; but I am unable to look it up because I have lent that volume to someone who has never sent it back. The reader may, perhaps, recall the high forehead and the singularly long black eyebrows that give such a Mephistophelian touch to his face. He occupies one of those pleasant detached houses in the mixed style that make the western end of the Upper Sandgate Road so interesting. His is the one with the Flemish gables and the Moorish portico, and it is in the room with the mullioned bay window that he works when he is down here, and in which of an evening we have so often smoked and talked together. He is a mighty jester, but, besides, he

927

likes to talk to me about his work; he is one of those men who find a help and stimulus in talking, and so I have been able to follow the conception of the New Accelerator right up from a very early stage. Of course, the greater portion of his experimental work is not done in Folkestone, but in Gower Street, in the fine new laboratory next to the hospital that he has been the first to use.

As everyone knows, or at least as all intelligent people know, the special department in which Gibberne has gained so great and deserved a reputation among physiologists is the action of drugs upon the nervous system. Upon soporifics, sedatives, and anæsthetics he is, I am told, unequalled. He is also a chemist of considerable eminence, and I suppose in the subtle and complex jungle of riddles that centres about the ganglion cell and the axis fibre there are little cleared places of his making, glades of illumination, that, until he sees fit to publish his results, are inaccessible to every other living man. And in the last few years he has been particularly assiduous upon this question of nervous stimulants, and already, before the discovery of the New Accelerator, very successful with them. Medical science has to thank him for at least three distinct and absolutely safe invigorators of unrivalled value to practising men. In cases of exhaustion the preparation known as Gibberne's B Syrup has, I suppose, saved more lives already than any lifeboat round the coast.

"But none of these things begin to satisfy me yet," he told me nearly a year ago. "Either they increase the central energy without affecting the nerves or they simply increase the available energy by lowering the nervous conductivity; and all of them are unequal and local in their operation. One wakes up the heart and viscera and leaves the brain stupefied, one gets at the brain champagne fashion and does nothing good for the solar plexus, and what I want—and what, if it's an earthly possibility, I mean to have—is a stimulant that stimulates all round, that wakes you up for a time from the crown of your head to the tip of your great toe, and makes you go two—or even three to everybody else's one. Eh? That's the thing I'm after."

"It would tire a man," I said.

"Not a doubt of it. And you'd eat double or treble—and all that. But just think what the thing would mean. Imagine yourself with a little phial like this"—he held up a bottle of green glass and marked his points with it—"and in this precious phial is the power to think twice as fast, move twice as quickly, do twice as much work in a given time as you could otherwise do."

"But is such a thing possible?"

"I believe so. If it isn't, I've wasted my time for a year. These various preparations of the hypophosphites, for example, seem to show that something of the sort. . . . Even if it was only one and a half times as fast it would do."

"It *would* do," I said.

"If you were a statesman in a corner, for example, time rushing up against you, something urgent to be done, eh?"

"He could dose his private secretary," I said.

"And gain—double time. And think if *you*, for example, wanted to finish a book."

"Usually," I said, "I wish I'd never begun 'em."

"Or a doctor, driven to death, wants to sit down and think out a case. Or a barrister—or a man cramming for an examination."

"Worth a guinea a drop," said I, "and more—to men like that."

"And in a duel again," said Gibberne, "where it all depends on your quickness in pulling the trigger."

"Or in fencing," I echoed.

"You see," said Gibberne, "if I get it as an all-round thing it will really do you no harm at all—except perhaps to an infinitesimal degree it brings you nearer old age. You will just have lived twice to other people's once——"

"I suppose," I meditated, "in a duel—it would be fair?"

"That's a question for the seconds," said Gibberne.

I harked back further. "And you really think such a thing *is* possible?" I said.

"As possible," said Gibberne, and glanced at something

that went throbbing by the window, "as a motor-bus. As a matter of fact——"

He paused and smiled at me deeply, and tapped slowly on the edge of his desk with the green phial. "I think I know the stuff. . . . Already I've got something coming." The nervous smile upon his face betrayed the gravity of his revelation. He rarely talked of his actual experimental work unless things were very near the end. "And it may be, it may be—I shouldn't be surprised—it may even do the thing at a greater rate than twice."

"It will be rather a big thing," I hazarded.

"It will be, I think, rather a big thing."

But I don't think he quite knew what a big thing it was to be, for all that.

I remember we had several subsequent talks about the stuff. "The New Accelerator" he called it, and his tone about it grew more confident on each occasion. Sometimes he talked nervously of unexpected physiological results its use might have, and then he would get a bit unhappy; at others he was frankly mercenary, and we debated long and anxiously how the preparation might be turned to commercial account. "It's a good thing," said Gibberne, " a tremendous thing. I know I'm giving the world something, and I think it only reasonable we should expect the world to pay. The dignity of science is all very well, but I think somehow I must have the monopoly of the stuff for, say, ten years. I don't see why *all* the fun in life should go to the dealers in ham."

My own interest in the coming drug certainly did not wane in the time. I have always had a queer twist towards metaphysics in my mind. I have always been given to paradoxes about space and time, and it seemed to me that Gibberne was really preparing no less than the absolute acceleration of life. Suppose a man repeatedly dosed with such a preparation: he would live an active and record life indeed, but he would be an adult at eleven, middle-aged at twenty-five, and by thirty well on the road to senile decay. It seemed to me that so far Gibberne was only going to do for anyone who took this drug exactly what Nature has done for the Jews and Orientals, who are men in their teens and aged by fifty, and quicker in thought and act than we are all the time. The marvel of drugs has always been great to my mind; you can madden a man, calm a man, make him incredibly strong and alert or a helpless log, quicken this passion and allay that, all by means of drugs, and here was a new miracle to be added to this strange armoury of phials the doctors use! But Gibberne was far too eager upon his technical points to enter very keenly into my aspect of the question.

It was the 7th or 8th of August when he told me the distillation that would decide his failure or success for a time was going forward as we talked, and it was on the 10th that he told me the thing was done and the New Accelerator a tangible reality in the world. I met him as I was going up the Sandgate Hill towards Folkestone—I think I was going to get my hair cut; and he came hurrying down to meet me—I suppose he was coming to my house to tell me at once of his success. I remember that his eyes were unusually bright and his face flushed, and I noted even then the swift alacrity of his step.

"It's done," he cried, and gripped my hand, speaking very fast; "it's more than done. Come up to my house and see."

"Really?"

"Really!" he shouted. "Incredibly! Come up and see."

"And it does—twice?"

"It does more, much more. It scares me. Come up and see the stuff. Taste it! Try it! It's the most amazing stuff on earth. He gripped my arm and, walking at such a pace that he forced me into a trot, went shouting with me up the hill. A whole charabancful of people turned and stared at us in unison after the manner of people in charabancs. It was one of those hot, clear days that Folkestone sees so much of, every colour incredibly bright and every outline hard. There was a breeze, of course, but not so much breeze as sufficed under these conditions to keep me cool and dry. I panted for mercy.

"I'm not walking fast, am I?" cried Gibberne, and slackened his pace to a quick march.

"You've been taking some of this stuff," I puffed.

"No," he said. "At the utmost a drop of water that stood in a beaker from which I had washed out the last traces of the stuff. I took some last night, you know. But that is ancient history, now."

"And it goes twice?" I said, nearing his doorway in a grateful perspiration.

"It goes a thousand times, many thousand times!" cried Gibberne, with a dramatic gesture, flinging open his Early English carved oak gate.

"Phew!" said I, and followed him to the door.

"I don't know how many times it goes," he said, with his latch-key in his hand.

"And you——"

"It throws all sorts of light on nervous physiology, it kicks the theory of vision into a perfectly new shape! . . . Heaven knows how many thousand times. We'll try all that after—— The thing is to try the stuff now."

"Try the stuff?" I said, as we went along the passage.

"Rather," said Gibberne, turning on me in his study. "There it is in that little green phial there! Unless you happen to be afraid?"

I am a careful man by nature, and only theoretically adventurous. I *was* afraid. But on the other hand there is pride.

"Well," I haggled. "You say you've tried it?"

"I've tried it," he said, "and I don't look hurt by it, do I? I don't even look livery and I *feel*——"

I sat down. "Give me the potion," I said. "If the worst comes to the worst it will save having my hair cut, and that I think is one of the most hateful duties of a civilised man. How do you take the mixture?"

"With water," said Gibberne, whacking down a carafe.

He stood up in front of his desk and regarded me in his easy chair; his manner was suddenly reflected by a touch of the Harley Street specialist. "It's rum stuff, you know," he said.

I made a gesture with my hand.

"I must warn you in the first place as soon as you've got it down to shut your eyes, and open them very cautiously in a minute or so's time. One still sees. The sense of vision is a question of length of vibration, and not of multitude of impacts; but there's a kind of shock to the retina, a nasty giddy confusion just at the time if the eyes are open. Keep 'em shut."

"Shut," I said. "Good!"

"And the next thing is, keep still. Don't begin to whack about. You may fetch something a nasty rap if you do. Remember you will be going several thousand times faster than you ever did before, heart, lungs, muscles, brain—everything—and you will hit hard without knowing it. You won't know it, you know. You'll feel just as you do now. Only everything in the world will seem to be going ever so many thousand times slower than it ever went before. That's what makes it so deuced queer."

"Lor'," I said. "And you mean——"

"You'll see," said he, and took up a measure. He glanced at the material on his desk. "Glasses," he said, "water. All here. Mustn't take too much for the first attempt."

The little phial glucked out its precious contents. "Don't forget what I told you," he said, turning the contents of the measure into a glass in the manner of an Italian waiter measuring whisky. "Sit with the eyes tightly shut and in absolute stillness for two minutes," he said. "Then you will hear me speak."

He added an inch or so of water to the dose in each glass.

"By-the-bye," he said, "don't put your glass down. Keep it in your hand and rest your hand on your knee. Yes—so. And now——"

He raised his glass.

"The New Accelerator," I said.

"The New Accelerator," he answered, and we touched glasses and drank, and instantly I closed my eyes.

You know that blank non-existence into which one drops when one has taken "gas." For an indefinite interval it was like that. Then I heard Gibberne telling me to wake up, and I stirred and opened my eyes. There he stood as he had been standing, glass still in hand. It was empty, that was all the difference.

"Well?" said I.

"Nothing out of the way?"

"Nothing. A slight feeling of exhilaration, perhaps. Nothing more."

"Sounds?"

"Things are still," I said. "By Jove! yes! They *are* still. Except the sort of faint pat, patter, like rain falling on different things. What is it?"

"Analysed sounds," I think he said, but I am not sure. He glanced at the window. "Have you ever seen a curtain before a window fixed in that way before?"

I followed his eyes, and there was the end of the curtain, frozen, as it were, corner high, in the act of flapping briskly in the breeze.

"No," said I; "that's odd."

"And here," he said, and opened the hand that held the glass. Naturally I winced, expecting the glass to smash. But so far from smashing it did not even seem to stir; it hung in mid-air—motionless. "Roughly speaking," said Gibberne, "an object in these latitudes falls 16 feet in the first second. This glass is falling 16 feet in a second now. Only, you see, it hasn't been falling yet for the hundredth part of a second. That gives you some idea of the pace of my Accelerator." And he waved his hand round and round, over and over the slowly sinking glass. Finally he took it by the bottom, pulled it down and placed it very carefully on the table. "Eh?" he said to me, and laughed.

"That seems all right," I said, and began very gingerly to raise myself from my chair. I felt perfectly well, very light and comfortable, and quite confident in my mind. I was going fast all over. My heart, for example, was beating a thousand times a second, but that caused me no discomfort at all. I looked out of the window. An immovable cyclist, head down and with a frozen puff of dust behind his driving-wheel, scorched to overtake a galloping charabanc that did not stir. I gaped in amazement at this incredible spectacle. "Gibberne," I cried, "how long will this confounded stuff last?"

"Heaven knows!" he answered. "Last time I took it I went to bed and slept it off. I tell you, I was frightened. It must have lasted some minutes, I think—it seemed like hours. But after a bit it slows down rather suddenly, I believe."

I was proud to observe that I did not feel frightened—I suppose because there were two of us. "Why shouldn't we go out?" I asked.

"Why not?"

"They'll see us."

"Not they. Goodness, no! Why, we shall be going a thousand times faster than the quickest conjuring trick that was ever done. Come along! Which way shall we go? Window, or door?"

And out by the window we went.

Assuredly of all the strange experiences that I have ever had, or imagined, or read of other people having or imagining, that little raid I made with Gibberne on the Folkestone Leas, under the influence of the New Accelerator, was the strangest and maddest of all. We went out by his gate into the road, and there we made a minute examination of the statuesque passing traffic. The tops of the wheels and some of the legs of the horses of this charabanc, the end of the whip-lash and the lower jaw of the conductor— who was just beginning to yawn—were perceptibly in motion, but all the rest of the lumbering conveyance seemed still. And quite noiseless except for a faint rattling that came from one man's throat! And as parts of this frozen edifice there were a driver, you know, and a conductor, and eleven people! The effect as we walked about the thing began by being madly queer and ended by being—disagreeable. There they were, people like ourselves and yet not like ourselves, frozen in careless attitudes, caught in mid-gesture. A girl and a man smiled at one another, a leering smile that threatened to last for ever-more; a woman in a floppy capelline rested her arm on the rail and stared at Gibberne's house with the unwinking stare of eternity; a man stroked his moustache like a figure of wax, and another stretched a tiresome stiff hand with extended fingers towards his loosened hat. We stared at them, we laughed at them, we made faces at them, and then a sort of disgust of them came upon us, and we turned away and walked round in front of the cyclist towards the Leas.

"Goodness!" cried Gibberne, suddenly; "look there!"

He pointed, and there at the tip of his finger and sliding down the air with wings flapping slowly and at the speed of an exceptionally languid snail—was a bee.

And so we came out upon the Leas. There the thing seemed madder than ever. The band was playing in the upper stand, though all the sound it made for us was a low-pitched, wheezy rattle, a sort of prolonged last sigh that passed at times into a sound like the slow muffled ticking of some monstrous clock. Frozen people stood erect; strange, silent, self-conscious-looking dummies hung unstably in mid-stride, promenading upon the grass. I passed close to a poodle dog suspended in the act of leaping, and watched the slow movement of his legs as he sank to earth. "Lord, look *here!*" cried Gibberne, and we halted for a moment before a magnificent person in white faint-striped flannels, white shoes, and a Panama hat, who turned back to wink at two gaily dressed ladies he had passed. A wink, studied with such leisurely deliberation as we could afford, is an unattractive thing. It loses any quality of alert gaiety, and one remarks that the winking eye does not completely close, that under its drooping lid appears the lower edge of an eyeball and a line of white. "Heaven give me memory," said I, "and I will never wink again."

"Or smile," said Gibberne, with his eye on the lady's answering teeth.

"It's infernally hot, somehow," said I. "Let's go slower."

"Oh, come along!" said Gibberne.

We picked our way among the bath-chairs in the path. Many of the people sitting in the chairs seemed almost natural in their passive poses, but the contorted scarlet of the bandsmen was not a restful thing to see. A purple-faced gentleman was frozen in the midst of a violent struggle to refold his newspaper against the wind; there were many evidences that all these people in their sluggish way were exposed to a considerable breeze, a breeze that had no existence so far as our sensations went. We came out and walked a little way from the crowd, and turned and regarded it. To see all that multitude changed to a picture, smitten rigid, as it were, into the semblance of realistic wax, was impossibly wonderful. It was absurd, of course; but it filled me with an irrational, an exultant sense of superior advantage. Consider the wonder of it! All that I had said and thought and done since the stuff had begun to work in my veins had happened, so far as those people, so far as the world in general went, in the twinkling of an eye. "The New Accelerator——" I began, but Gibberne interrupted me.

"There's that infernal old woman!" he said.

"What old woman?"

"Lives next door to me," said Gibberne. "Has a lapdog that yaps. Gods! The temptation is strong!"

There is something very boyish and impulsive about Gibberne at times. Before I could expostulate with him he had dashed forward, snatched the unfortunate animal out of visible existence, and was running violently with it towards the cliff of the Leas. It was most extraordinary. The little brute, you know, didn't bark or wriggle or make the slightest sign of vitality. It kept quite stiffly in an attitude of somnolent repose, and Gibberne held it by the neck. It was like running about with a dog of wood. "Gibberne," I cried, "put it down!" Then I said something else. "If you run like that, Gibberne," I cried, "you'll set your clothes on fire. Your linen trousers are going brown as it is!"

He clapped his hand on his thigh and stood hesitating on the verge. "Gibberne," I cried, coming up, "put it down.

This heat is too much! It's our running so! Two or three miles a second! Friction of the air!"

"What?" he said, glancing at the dog.

"Friction of the air," I shouted. "Friction of the air. Going too fast. Like meteorites and things. Too hot. And, Gibberne! Gibberne! I'm all over pricking and a sort of perspiration. You can see people stirring slightly. I believe the stuff's working off! Put that dog down."

"Eh?" he said.

"It's working off," I repeated. "We're too hot and the stuff's working off! I'm wet through."

He stared at me. Then at the band, the wheezy rattle of whose performance was certainly going faster. Then with a tremendous sweep of the arm he hurled the dog away from him and it went spinning upward, still inanimate and hung at last over the grouped parasols of a knot of chattering people. Gibberne was gripping my elbow. "By Jove!" he cried. "I believe it is! A sort of hot pricking and—yes. That man's moving his pocket-handkerchief! Perceptibly. We must get out of this sharp."

But we could not get out of it sharply enough. Luckily perhaps! For we might have run, and if we had run we should, I believe, have burst into flames. Almost certainly we should have burst into flames! You know we had neither of us thought of that. . . . But before we could even begin to run the action of the drug had ceased. It was the business of a minute fraction of a second. The effect of the New Accelerator passed like the drawing of a curtain, vanished in the movement of a hand. I heard Gibberne's voice in infinite alarm. "Sit down," he said, and flop, down upon the turf at the edge of the Leas I sat—scorching as I sat. There is a patch of burnt grass there still where I sat down. The whole stagnation seemed to wake up as I did so, the disarticulated vibration of the band rushed together into a blast of music, the promenaders put their feet down and walked their ways, the papers and flags began flapping, smiles passed into words, the winker finished his wink and went on his way complacently, and all the seated people moved and spoke.

The whole world had come alive again, was going as fast as we were, or rather we were going no faster than the rest of the world. It was like slowing down as one comes into a railway station. Everything seemed to spin round for a second or two, I had the most transient feeling of nausea, and that was all. And the little dog which had seemed to hang for a moment when the force of Gibberne's arm was expended fell with a swift acceleration clean through a lady's parasol!

That was the saving of us. Unless it was for one corpulent old gentleman in a bath-chair, who certainly did start at the sight of us and afterwards regarded us at intervals with a darkly suspicious eye, and finally, I believe, said something to his nurse about us, I doubt if a solitary person remarked our sudden appearance among them. Plop! We must have appeared abruptly. We ceased to smoulder almost at once, though the turf beneath me was uncomfortably hot. The attention of everyone—including even the Amusements' Association band, which on this occasion, for the only time in its history, got out of tune—was arrested by the amazing fact, and the still more amazing yapping and uproar caused by the fact, that a respectable, over-fed lapdog sleeping quietly to the east of the bandstand should suddenly fall through the parasol of a lady on the west—in a slightly singed condition due to the extreme velocity of its movements through the air. In these absurd days, too, when we are all trying to be as psychic and silly and superstitious as possible! People got up and trod on other people, chairs were overturned, the Leas policeman ran. How the matter settled itself I do not know—we were much too anxious to disentangle ourselves from the affair and get out of range of the eye of the old gentleman in the bath-chair to make minute inquiries. As soon as we were sufficiently cool and sufficiently recovered from our giddiness and nausea and confusion of mind to do so we stood up and, skirting the crowd, directed our steps back along the road below the Metropole towards Gibberne's house. But amidst the din I heard very distinctly the gentleman who had been sitting beside the lady of the ruptured sunshade using quite unjustifiable threats and language to one of those chair-attendants who have "Inspector" written on their caps. "If you didn't throw the dog," he said, "who *did*?"

The sudden return of movement and familiar noises, and our natural anxiety about ourselves (our clothes were still dreadfully hot, and the fronts of the thighs of Gibberne's white trousers were scorched a drabbish brown), prevented the minute observations I should have liked to make on all these things. Indeed, I really made no observations of any scientific value on that return. The bee, of course, had gone. I looked for that cyclist, but he was already out of sight as we came into the Upper Sandgate Road or hidden from us by traffic; the charabanc, however, with its people now all alive and stirring, was clattering along at a spanking pace almost abreast of the nearer church.

We noted, however, that the window-sill on which we had stepped in getting out of the house was slightly singed, and that the impressions of our feet on the gravel of the path were unusually deep.

So it was I had my first experience of the New Accelerator. Practically we had been running about and saying and doing all sorts of things in the space of a second or so of time. We had lived half an hour while the band had played, perhaps, two bars. But the effect it had upon us was that the whole world had stopped for our convenient inspection. Considering all things, and particularly considering our rashness in venturing out of the house, the experience might certainly have been much more disagreeable than it was. It showed, no doubt, that Gibberne has still much to learn before his preparation is a manageable convenience, but its practicability it certainly demonstrated beyond all cavil.

Since that adventure he has been steadily bringing its use under control, and I have several times, and without the slightest bad result, taken measured doses under his direction; though I must confess I have not yet ventured abroad again while under its influence. I may mention, for example, that this story has been written at one sitting and without interruption, except for the nibbling of some chocolate, by its means. I began at 6.25, and my watch is now very nearly at the minute past the half-hour. The convenience of securing a long, uninterrupted spell of work in the midst of a day full of engagements cannot be exaggerated. Gibberne is now working at the quantitative handling of his preparation, with especial reference to its distinctive effects upon different types of constitution. He then hopes to find a Retarder with which to dilute its present rather excessive potency. The Retarder will, of course, have the reverse effect to the Accelerator; used alone it should enable the patient to spread a few seconds over many hours of ordinary time, and so to maintain an apathetic inaction, a glacierlike absence of alacrity, amidst the most animated or irritating surroundings. The two things together must necessarily work an entire revolution in civilised existence. It is the beginning of our escape from that Time Garment of which Carlyle speaks. While this Accelerator will enable us to concentrate ourselves with tremendous impact upon any moment or occasion that demands our utmost sense and vigour, the Retarder will enable us to pass in passive tranquillity through infinite hardship and tedium. Perhaps I am a little optimistic about the Retarder, whch has indeed still to be discovered, but about the Accelerator there is no possible sort of doubt whatever. Its appearance upon the market in a convenient, controllable, and assimilable form is a matter of the next few months. It will be obtainable of all chemists and druggists, in small green bottles, at a high but, considering its extraordinary qualities, by no means excessive price. Gibberne's Nervous Accelerator it will be called, and he hopes to be able to supply it in three strengths: one in 200, one in 900, and one in 2000, distinguished by yellow, pink, and white labels respectively.

No doubt its use renders a great number of very extraordinary things possible; for, of course, the most remarkable and, possibly, even criminal proceedings may be effected with impunity by thus dodging, as it were, into the interstices of time. Like all potent preparations it will be liable to abuse. We have, however, discussed this aspect of the question very thoroughly, and we have decided that this is purely a matter of medical jurisprudence and altogether outside our province. We shall manufacture and sell the Accelerator, and, as for the consequences—we shall see.

STORY THE NINTH

Mr. Ledbetter's Vacation

My friend, Mr. Ledbetter, is a round-faced little man, whose natural mildness of eye is gigantically exaggerated when you catch the beam through his glasses, and whose deep, deliberate voice irritates irritable people. A certain elaborate clearness of enunciation has come with him to his present vicarage from his scholastic days, an elaborate clearness of enunciation and a certain nervous determination to be firm and correct upon all issues, important and unimportant alike. He is a sacerdotalist and a chess player, and suspected by many of the secret practice of the higher mathematics—creditable rather than interesting things. His conversation is copious and given much to needless detail. By many, indeed, his intercourse is condemned, to put it plainly, as "boring," and such have even done me the compliment to wonder why I countenance him. But, on the other hand, there is a large faction who marvel at his countenancing such a dishevelled, discreditable acquaintance as myself. Few appear to regard our friendship with equanimity. But that is because they do not know of the link that binds us, of my amiable connection *via* Jamaica with Mr. Ledbetter's past.

About that past he displays an anxious modesty. "I do not know *what* I should do if it became known," he says; and repeats, impressively, "I do not know *what* I should do." As a matter of fact, I doubt if he would do anything except get very red about the ears. But that will appear later; nor will I tell here of our first encounter, since, as a general rule—though I am prone to break it—the end of a story

944   TWELVE STORIES AND A DREAM

should come after, rather than before, the beginning. And the beginning of the story goes a long way back; indeed, it is now nearly twenty years since Fate, by a series of complicated and startling manœuvres, brought Mr. Ledbetter, so to speak, into my hands.

In those days I was living in Jamaica, and Mr. Ledbetter was a schoolmaster in England. He was in orders, and already recognisably the same man that he is to-day: the same rotundity of visage, the same or similar glasses, and the same faint shadow of surprise in his resting expression. He was, of course, dishevelled when I saw him, and his collar less of a collar than a wet bandage, and that may have helped to bridge the natural gulf between us—but of that, as I say, later.

The business began at Hithergate-on-Sea, and simultaneously with Mr. Ledbetter's summer vacation. Thither he came for a greatly needed rest, with a bright brown portmanteau marked "F. W. L.," a new white and black straw hat, and two pairs of white flannel trousers. He was naturally exhilarated at his release from school—for he was not very fond of the boys he taught. After dinner he fell into a discussion with a talkative person established in the boarding-house to which, acting on the advice of his aunt, he had resorted. This talkative person was the only other man in the house. Their discussion concerned the melancholy disappearance of wonder and adventure in these latter days, this prevalence of globe-trotting, the abolition of distance by steam and electricity, the vulgarity of advertisement, the degradation of men by civilisation, and many such things. Particularly was the talkative person eloquent on the decay of human courage through security, a security Mr. Ledbetter rather thoughtlessly joined him in deploring. Mr. Ledbetter, in the first delight of emancipation from "duty," and being anxious, perhaps, to establish a reputation for manly conviviality, partook, rather more freely than was advisable, of the excellent whisky the talkative person produced. But he did not become intoxicated, he insists.

He was simply eloquent beyond his sober wont, and with

the finer edge gone from his judgment. And after that long talk of the brave old days that were past for ever, he went out into moonlit Hithergate alone and up the cliff road where the villas cluster together.

He had bewailed, and now as he walked up the silent road he still bewailed, the fate that had called him to such an uneventful life as a pedagogue's. What a prosaic existence he led, so stagnant, so colourless! Secure, methodical, year in year out, what call was there for bravery? He thought enviously of those roving, mediæval days, so near and so remote, of quests and spies and *condottieri* and many a risky blade-drawing business. And suddenly came a doubt, a strange doubt, springing out of some chance thought of tortures, and destructive altogether of the position he had assumed that evening.

Was he—Mr. Ledbetter—really, after all, so brave as he assumed? Would he really be so pleased to have railways, policemen, and security vanish suddenly from the earth?

The talkative man had spoken enviously of crime. "The burglar," he said, "is the only true adventurer left on earth. Think of his singlehanded fight—against the whole civilised world!" And Mr. Ledbetter had echoed his envy. "They *do* have some fun out of life," Mr. Ledbetter had said. "And about the only people who do. Just think how it must feel to wire a lawn!" And he had laughed wickedly. Now, in this franker intimacy of self-communion he found himself instituting a comparison between his own brand of courage and that of the habitual criminal. He tried to meet these insidious questionings with blank assertion. "I could do all that," said Mr. Ledbetter. "I long to do all that. Only I do not give way to my criminal impulses. My moral courage restrains me." But he doubted, even while he told himself these things.

Mr. Ledbetter passed a large villa standing by itself. Conveniently situated above a quiet, practicable balcony was a window, gaping black, wide open. At the time he scarcely marked it, but the picture of it came with him, wove into his thoughts. He figured himself climbing up that balcony, crouching—plunging into that dark, mysterious interior. "Bah! You would not dare," said the Spirit of Doubt. "My duty to my fellow-men forbids," said Mr. Ledbetter's self-respect.

It was nearly eleven, and the little seaside town was already very still. The whole world slumbered under the moonlight. Only one warm oblong of window-blind far down the road spoke of waking life. He turned and came back slowly towards the villa of the open window. He stood for a time outside the gate, a battlefield of motives. "Let us put things to the test," said Doubt. "For the satisfaction of these intolerable doubts, show that you dare go into that house. Commit a burglary in blank. That, at any rate, is no crime." Very softly he opened and shut the gate and slipped into the shadow of the shrubbery. "This is foolish," said Mr. Ledbetter's caution. "I expected that," said Doubt. His heart was beating fast, but he was certainly not afraid. He was *not* afraid. He remained in that shadow for some considerable time.

The ascent of the balcony, it was evident, would have to be done in a rush, for it was all in clear moonlight, and visible from the gate in the avenue. A trellis thinly set with young, ambitious climbing roses made the ascent ridiculously easy. There, in that black shadow by the stone vase of flowers, one might crouch and take a closer view of this gaping breach in the domestic defences, the open window. For a while Mr. Ledbetter was as still as the night, and then that insidious whisky tipped the balance. He dashed forward. He went up the trellis with quick, convulsive movements, swung his legs over the parapet of the balcony, and dropped panting in the shadow even as he had designed. He was trembling violently, short of breath, and his heart pumped noisily, but his mood was exultation. He could have shouted to find he was so little afraid.

A happy line that he had learnt from Wills's "Mephistopheles" came into his mind as he crouched there. "I feel like a cat on the tiles," he whispered to himself. It was far better than he had expected—this adventurous exhilaration. He was sorry for all poor men to whom burglary was unknown. Nothing happened. He was quite safe. And he was acting in the bravest manner!

And now for the window, to make the burglary complete! Must he dare do that? Its position above the front door defined it as a landing or passage, and there were no looking-glasses or any bedroom signs about it, or any other window on the first floor, to suggest the possibility of a sleeper within. For a time he listened under the ledge, then raised his eyes above the sill and peered in. Close at hand, on a pedestal, and a little startling at first, was a nearly life-size gesticulating bronze. He ducked, and after some time he peered again. Beyond was a broad landing, faintly gleaming; a flimsy fabric of bead curtain, very black and sharp, against a further window, a broad staircase, plunging into a gulf of darkness below; and another ascending to the second floor. He glanced behind him, but the stillness of the night was unbroken. "Crime," he whispered, "crime," and scrambled softly and swiftly over the sill into the house. His feet fell noiselessly on a mat of skin. He was a burglar indeed!

He crouched for a time, all ears and peering eyes. Outside was a scampering and rustling, and for a moment he repented of his enterprise. A short "miaow," a spitting, and a rush into silence, spoke reassuringly of cats. His courage grew. He stood up. Everyone was abed, it seemed. So easy is it to commit a burglary, if one is so minded. He was glad he had put it to the test. He determined to take some petty trophy, just to prove his freedom from any abject fear of the law, and depart the way he had come.

He peered about him, and suddenly the critical spirit arose again. Burglars did far more than such mere elementary entrance as this: they went into rooms, they forced safes. Well—he was not afraid. He could not force safes, because that would be a stupid want of consideration for his hosts. But he would go into rooms—he would go upstairs. More: he told himself that he was perfectly secure; an empty house could not be more reassuringly still. He had to clench his hands, nevertheless, and summon all his resolution before he began very softly to ascend the dim staircase, pausing for several seconds between each step. Above was a square landing with one open and several closed doors; and all the house was still. For a moment he stood wondering what would happen if some sleeper woke suddenly and emerged. The open door showed a moonlit bedroom, the coverlet white and undisturbed. Into this room he crept in three interminable minutes, and took a piece of soap for his plunder—his trophy. He turned to descend even more softly than he had ascended. It was as easy as—— Hist! . . .

Footsteps! On the gravel outside the house—and then the noise of a latchkey, the yawn and bang of a door, and the spitting of a match in the hall below. Mr. Ledbetter stood petrified by the sudden discovery of the folly upon which he had come. "How on earth am I to get out of this?" said Mr. Ledbetter.

The hall grew bright with a candle flame, some heavy object bumped against the umbrella-stand, and feet were ascending the staircase. In a flash Mr. Ledbetter realised that his retreat was closed. He stood for a moment, a pitiful figure of penitent confusion. "My goodness! What a *fool* I have been!" he whispered, and then darted swiftly across the shadowy landing into the empty bedroom from which he had just come. He stood listening—quivering. The footsteps reached the first-floor landing.

Horrible thought! This was possibly the late-comer's room! Not a moment was to be lost! Mr. Ledbetter stooped beside the bed, thanked Heaven for a valance, and crawled within its protection not ten seconds too soon. He became motionless on hands and knees. The advancing candle-light appeared through the thinner stitches of the fabric, the shadows ran wildly about, and became rigid as the candle was put down.

"Lord, what a day!" said the newcomer, blowing noisily, and it seemed he deposited some heavy burthen on what Mr. Ledbetter, judging by the feet, decided to be a writing-table. The unseen then went to the door and locked it, examined the fastenings of the windows carefully and pulled down the blinds, and returning sat down upon the bed with startling ponderosity.

"*What* a day!" he said. "Good Lord!" and blew again, and Mr. Ledbetter inclined to believe that the person was mopping his face. His boots were good stout boots; the shadows of his legs upon the valance suggested a formidable stoutness of aspect. After a time he removed some upper garments—a coat and waistcoat, Mr. Ledbetter inferred—and casting them over the rail of the bed remained breathing less noisily, and as it seemed cooling from a considerable temperature. At intervals he muttered to himself, and once he laughed softly. And Mr. Ledbetter muttered to himself, but he did not laugh. "Of all the foolish things," said Mr. Ledbetter. "What on earth am I to do now?"

His outlook was necessarily limited. The minute apertures between the stitches of the fabric of the valance admitted a certain amount of light, but permitted no peeping. The shadows upon this curtain, save for those sharply defined legs, were enigmatical, and intermingled confusingly with the florid patterning of the chintz. Beneath the edge of the valance a strip of carpet was visible, and, by cautiously depressing his eye, Mr. Ledbetter found that this strip broadened until the whole area of the floor came into view. The carpet was a luxurious one, the room spacious, and, to judge by the castors and so forth of the furniture, well equipped.

What he should do he found it difficult to imagine. To wait until this person had gone to bed, and then, when he seemed to be sleeping, to creep to the door, unlock it, and bolt headlong for that balcony seemed the only possible thing to do. Would it be possible to jump from the balcony? The danger of it! When he thought of the chances against him, Mr. Ledbetter despaired. He was within an ace of thrusting forth his head beside the gentleman's legs, coughing if necessary to attract his attention, and then, smiling, apologising and explaining his unfortunate intrusion by a few well-chosen sentences. But he found these sentences hard to choose. "No doubt, sir, my appearance is peculiar," or, "I trust, sir, you will pardon my somewhat ambiguous appearance from beneath you," was about as much as he could get.

Grave possibilities forced themselves on his attention. Suppose they did not believe him, what would they do to him? Would his unblemished high character count for nothing? Technically he was a burglar, beyond dispute. Following out this train of thought, he was composing a lucid apology for "this technical crime I have committed," to be delivered before sentence in the dock, when the stout gentleman got up and began walking about the room. He locked and unlocked drawers, and Mr. Ledbetter had a transient hope that he might be undressing. But, no! He seated himself at the writing-table, and began to write and then tear up documents. Presently the smell of burning cream-laid paper mingled with the odour of cigars in Mr. Ledbetter's nostrils.

"The position I had assumed," said Mr. Ledbetter when he told me of these things, "was in many respects an ill-advised one. A transverse bar beneath the bed depressed my head unduly, and threw a disproportionate share of my weight upon my hands. After a time, I experienced what is called, I believe, a crick in the neck. The pressure of my hands on the coarsely-stitched carpet speedily became painful. My knees, too, were painful, my trousers being drawn tightly over them. At that time I wore rather higher collars than I do now—two and a half inches, in fact—and I discovered what I had not remarked before, that the edge of the one I wore was frayed slightly under the chin. But much worse than these things was an itching of my face, which I could only relieve by violent grimacing—I tried to raise my head, but the rustle of the sleeve alarmed me. After a time I had to desist from this relief also, because happily in time—I discovered that my facial contortions were shifting my glasses down my nose. Their fall would, of course, have exposed me, and as it was they came to rest in an oblique position of by no means stable equilibrium. In addition I had a slight cold, and an intermittent desire to sneeze or sniff caused me inconvenience. In fact, quite apart from the extreme anxiety of my position, my physical discomfort became in a short time very considerable indeed. But I had to stay there motionless, nevertheless."

After an interminable time, there began a chinking sound. This deepened into a rhythm: chink, chink, chink—twenty-five chinks—a rap on the writing-table, and a grunt from the owner of the stout legs. It dawned upon Mr. Ledbetter that this chinking was the chinking of gold. He became incredulously curious as it went on. His curiosity grew. Already, if that was the case, this extraordinary man must have counted some hundreds of pounds. At last Mr. Ledbetter could resist it no longer, and he began very cautiously to fold his arms and lower his head to the level of the floor, in the hope of peeping under the valance. He moved his feet, and one made a slight scraping on the floor. Suddenly the chinking ceased. Mr. Ledbetter became rigid. After a while the chinking was resumed. Then it ceased again, and everything was still, except Mr. Ledbetter's heart—that organ seemed to him to be beating like a drum.

The stillness continued. Mr. Ledbetter's head was now on the floor, and he could see the stout legs as far as the shins. They were quite still. The feet were resting on the toes and drawn back, as it seemed, under the chair of the owner. Everything was quite still, everything continued still. A wild hope came to Mr. Ledbetter that the unknown was in a fit or suddenly dead, with his head upon the writing-table. . . .

The stillness continued. What had happened? The desire to peep became irresistible. Very cautiously Mr. Ledbetter shifted his hand forward, projected a pioneer finger, and began to lift the valance immediately next his eye. Nothing broke the stillness. He saw now the stranger's knees, saw the back of the writing-table, and then—he was staring at the barrel of a heavy revolver pointed over the writing-table at his head.

"Come out of that, you scoundrel!" said the voice of the stout gentleman in a tone of quiet concentration. "Come out. This side, and now. None of your hanky-panky—come right out now."

Mr. Ledbetter came right out, a little reluctantly perhaps, but without any hanky-panky, and at once, even as he was told.

"Kneel," said the stout gentleman. "And hold up your hands."

The valance dropped again behind Mr. Ledbetter, and he rose from all fours and held up his hands. "Dressed like a parson," said the stout gentleman. "I'm blest if he isn't! A little chap, too! You *scoundrel!* What the deuce possessed you to come here to-night? What the deuce possessed you to get under my bed?"

He did not appear to require an answer, but proceeded at once to several very objectionable remarks upon Mr. Ledbetter's personal appearance. He was not a very big man, but he looked strong to Mr. Ledbetter: he was as stout as his legs had promised, he had rather delicately-chiselled small features distributed over a considerable area of whitish face, and quite a number of chins. And the note of his voice had a sort of whispering undertone.

"What the deuce, I say, possessed you to get under my bed?"

Mr. Ledbetter, by an effort, smiled a wan, propitiatory smile. He coughed. "I can quite understand——" he said.

"Why! What on earth . . .? It's *soap!* No!—you scoundrel! Don't you move that hand."

"It's soap," said Mr. Ledbetter. "From your washstand. No doubt if——"

"Don't talk," said the stout man. "I see it's soap. Of all incredible things."

"If I might explain——"

"Don't explain. It's sure to be a lie, and there's no time for explanations. What was I going to ask you? Ah! Have you any mates?"

"In a few minutes, if you——"

"Have you any mates? Curse you. If you start any soapy palaver I'll shoot. Have you any mates?"

"No," said Mr. Ledbetter.

"I suppose it's a lie," said the stout man. "But you'll

pay for it if it is. Why the deuce didn't you floor me when I came upstairs? You won't get a chance to now, anyhow. Fancy getting under the bed! I reckon it's a fair cop, anyhow, so far as you are concerned."

"I don't see how I could prove an *alibi*," remarked Mr. Ledbetter, trying to show by his conversation that he was an educated man. There was a pause. Mr. Ledbetter perceived that on a chair beside his captor was a large black bag on a heap of crumpled papers, and that there were torn and burnt papers on the table. And in front of these, and arranged methodically along the edge, were rows and rows of little yellow rouleaux—a hundred times more gold than Mr. Ledbetter had seen in all his life before. The light of two candles, in silver candlesticks, fell upon these. The pause continued. "It is rather fatiguing holding up my hands like this," said Mr. Ledbetter, with a deprecatory smile.

"That's all right," said the fat man. "But what to do with you I don't exactly know."

"I know my position is ambiguous."

"Lord!" said the fat man, "ambiguous! And goes about with his own soap, and wears a thundering great clerical collar! You *are* a blooming burglar, you are—if ever there was one!"

"To be strictly accurate," said Mr. Ledbetter, and suddenly his glasses slipped off and clattered against his vest buttons.

The fat man changed countenance, a flash of savage resolution crossed his face, and something in the revolver clicked. He put his other hand to the weapon. And then he looked at Mr. Ledbetter, and his eye went down to the dropped *pince-nez*.

"Full-cock now, anyhow," said the fat man, after a pause, and his breath seemed to catch. "But I'll tell you, you've never been so near death before. Lord! *I'm* almost glad. If it hadn't been that the revolver wasn't cocked, you'd be lying dead there now."

Mr. Ledbetter said nothing, but he felt that the room was swaying.

"A miss is as good as a mile. It's lucky for both of us it wasn't. Lord!" He blew noisily. "There's no need for you to go pale-green for a little thing like that."

"I can assure you, sir——" said Mr. Ledbetter, with an effort.

"There's only one thing to do. If I call in the police, I'm bust—a little game I've got on is bust. That won't do. If I tie you up and leave you—again, the thing may be out to-morrow. To-morrow's Sunday, and Monday's Bank Holiday—I've counted on three clear days. Shooting you's murder—and hanging; and besides, it will bust the whole blooming kernooze. I'm hanged if I can think what to do—I'm hanged if I can."

"Will you permit me——"

"You gas as much as if you were a real parson, I'm blessed if you don't. Of all the burglars you are the—— Well! No—I won't permit you. There isn't time. If you start off jawing again, I'll shoot right in your stomach. See? But I know now—I know now! What we're going to do first, my man, is an examination for concealed arms—an examination for concealed arms. And look here! When I tell you to do a thing, don't start off at a gabble—do it brisk."

And with many elaborate precautions, and always pointing the pistol at Mr. Ledbetter's head, the stout man stood him up and searched him for weapons. "Why, you *are* a burglar!" he said. "You're a perfect amateur. You haven't even a pistol-pocket in the back of your breeches. No, you don't! Shut up, now."

So soon as the issue was decided, the stout man made Mr. Ledbetter take off his coat and roll up his shirt-sleeves, and, with the revolver at one ear, proceed with the packing his appearance had interrupted. From the stout man's point of view that was evidently the only possible arrangement, for if he had packed, he would have had to put down the revolver. So that even the gold on the table was handled by Mr. Ledbetter. This noctural packing was peculiar. The stout man's idea was evidently to dis-tribute the weight of the gold as unostentatiously as possible through his luggage. It was by no means an inconsiderable weight. There was, Mr. Ledbetter says, altogether nearly £18,000 in gold in the black bag and on the table. There were also many little rolls of £5 bank-notes. Each rouleau of £25 was wrapped by Mr. Ledbetter in paper. These rouleaux were then put neatly in cigar-boxes and distributed between a travelling trunk, a Gladstone bag, and a hat-box. About £600 went in a tobacco tin in a dressing-bag. £10 in gold and a number of £5 notes the stout man pocketed. Occasionally he objurgated Mr. Ledbetter's clumsiness, and urged him to hurry, and several times he appealed to Mr. Ledbetter's watch for information.

Mr. Ledbetter strapped the trunk and bag, and returned the stout man the keys. It was then ten minutes to twelve, and until the stroke of midnight the stout man made him sit on the Gladstone bag, while he sat at a reasonably safe distance on the trunk and held the revolver handy and waited. He appeared to be now in a less aggressive mood, and having watched Mr. Ledbetter for some time, he offered a few remarks.

"From your accent I judge you are a man of some education," he said, lighting a cigar. "No—*don't* begin that explanation of yours. I know it will be long-winded from your face, and I am much too old a liar to be interested in other men's lying. You are, I say, a person of education. You do well to dress as a curate. Even among educated people you might pass as a curate."

"I *am* a curate," said Mr. Ledbetter, "or, at least——"

"You are trying to be. I know. But you didn't ought to burgle. You are not the man to burgle. You are, if I may say it—the thing will have been pointed out to you before—a coward."

"Do you know," said Mr. Ledbetter, trying to get a final opening, "it was that very question——"

The stout man waved him into silence.

"You waste your education in burglary. You should do one of two things. Either you should forge or you should embezzle. For my own part, I embezzle. Yes; I embezzle. What do you think a man could be doing with all this gold but for that? Ah! Listen! Midnight! . . . Ten. Eleven. Twelve. There is something very impressive to me in that slow beating of the hours. Time—space; what mysteries they are! What mysteries. . . . It's time for us to be moving. Stand up!"

And then kindly, but firmly, he induced Mr. Ledbetter to sling the dressing bag over his back by a string across his chest, to shoulder the trunk, and, over-ruling a gasping protest, to take the Gladstone bag in his disengaged hand. So encumbered, Mr. Ledbetter struggled perilously downstairs. The stout gentleman followed with an overcoat, the hat-box, and the revolver, making derogatory remarks about Mr. Ledbetter's strength, and assisting him at the turnings of the stairs.

"The back door," he directed, and Mr. Ledbetter staggered through a conservatory, leaving a wake of smashed flowerpots behind him. "Never mind the crockery," said the stout man; "it's good for trade. We wait here until a quarter past. You can put those things down. You have!"

Mr. Ledbetter collapsed panting on the trunk. "Last night," he gasped, "I was asleep in my little room, and I no more dreamt——"

"There's no need for you to incriminate yourself," said the stout gentleman, looking at the lock of the revolver. He began to hum. Mr. Ledbetter made to speak, and thought better of it.

There presently came the sound of a bell, and Mr. Ledbetter was taken to the back door and instructed to open it. A fair-haired man in yachting costume entered. At the sight of Mr. Ledbetter he started violently and clapped his hand behind him. Then he saw the stout man. "Bingham!" he cried, "who's this?"

"Only a little philanthropic do of mine—burglar I'm trying to reform. Caught him under my bed just now. He's all right. He's a frightful ass. He'll be useful to carry some of our things."

The newcomer seemed inclined to resent Mr. Ledbetter's presence at first, but the stout man reassured him.

"He's quite alone. There's not a gang in the world would own him. No——! Don't start talking, for goodness' sake."

They went out into the darkness of the garden, with the trunk still bowing Mr. Ledbetter's shoulders. The man in yachting costume walked in front with the Gladstone bag and a pistol; then came Mr. Ledbetter like Atlas; Mr. Bingham followed with the hat-box, coat, and revolver as before. The house was one of those that have their gardens right up to the cliff. At the cliff was a steep wooden stairway, descending to a bathing tent dimly visible on the beach. Below was a boat pulled up, and a silent little man with a black face stood beside it. "A few moments' explanation," said Mr. Ledbetter; "I can assure you——" Somebody kicked him, and he said no more.

They made him wade to the boat carrying the trunk, they pulled him aboard by the shoulders and hair, they called him no better name than "scoundrel" and "burglar" all that night. But they spoke in undertones so that the general public was happily unaware of his ignominy. They hauled him aboard a yacht manned by strange, unsympathetic Orientals, and partly they thrust him and partly he fell down a gangway into a noisome, dark place, where he was to remain many days—how many he does not know, because he lost count among other things when he was sea-sick. They fed him on biscuits and incomprehensible words; they gave him water to drink mixed with unwished-for rum. And there were cockroaches where they put him —night and day there were cockroaches, and in the night-time there were rats. The Orientals emptied his pockets and took his watch—but Mr. Bingham being appealed to, took that himself. And five or six times the five Lascars— if they were Lascars—and the Chinaman and the negro who constituted the crew, fished him out and took him aft to Bingham and his friend to play cribbage and euchre and three-handed whist, and to listen to their stories and boastings in an interested manner.

Then these principals would talk to him as men talk to those who have lived a life of crime. Explanations they would never permit, though they made it abundantly clear to him that he was the rummiest burglar they had ever set eyes on. They said as much again and again. The fair man was of a taciturn disposition and irascible at play; but Mr. Bingham, now that the evident anxiety of his departure from England was assuaged, displayed a vein of genial philosophy. He enlarged upon the mystery of space and time, and quoted Kant and Hegel—or, at least, he said he did. Several times Mr. Ledbetter got as far as: "My position under your bed, you know——" but then he always had to cut, or pass the whisky, or do some such intervening thing. After his third failure, the fair man got quite to look for this opening, and whenever Mr. Ledbetter began after that, he would roar with laughter and hit him violently on the back. "Same old start, same old story; good old burglar!" the fair-haired man would say.

So Mr. Ledbetter suffered for many days, twenty perhaps; and one evening he was taken, together with some tinned provisions, over the side and put ashore on a rocky little island with a spring. Mr. Bingham came in the boat with him, giving him good advice all the way, and waving his last attempts at an explanation aside.

"I am really *not* a burglar," said Mr. Ledbetter.

"You never will be," said Mr. Bingham. "You'll never make a burglar. I'm glad you are beginning to see it. In choosing a profession a man must study his temperament. If you don't, sooner or later you will fail. Compare myself, for example. All my life I have been in banks—I have got on in banks. I have even been a bank manager. But was I happy? No. Why wasn't I happy? Because it did not suit my temperament. I am too adventurous—too versatile. Practically I have thrown it over. I do not suppose I shall ever manage a bank again. They would be glad to get me, no doubt, but I have learnt the lesson of my temperament—at last. . . . No! I shall never manage a bank again.

"Now, your temperament unfits you for crime—just as

mine unfits me for respectability. I know you better than I did, and now I do not even recommend forgery. Go back to respectable courses, my man. *Your* lay is the philanthropic lay—that is your lay. With that voice—the Association for the Promotion of Snivelling among the Young—something in that line. You think it over.

"The island we are approaching has no name apparently—at least, there is none on the chart. You might think out a name for it while you are there——while you are thinking about all these things. It has quite drinkable water, I understand. It is one of the Grenadines—one of the Windward Islands. Yonder, dim and blue, are others of the Grenadines. There are quantities of Grenadines, but the majority are out of sight. I have often wondered what these islands are for—now, you see, I am wiser. This one at least is for you. Sooner or later some simple native will come along and take you off. Say what you like about us then—abuse us, if you like—we shan't care a solitary Grenadine! And here—here is half a sovereign's worth of silver. Do not waste that in foolish dissipation when you return to civilisation. Properly used, it may give you a fresh start in life. And do not—— Don't beach her, you beggars, he can wade!—— Do not waste the precious solitude before you in foolish thoughts. Properly used, it may be a turning-point in your career. Waste neither money nor time. You will die rich. I'm sorry, but I must ask you to carry your tucker to land in your arms. No; it's not deep. Curse that explanation of yours! There's not time. No, no, no! I won't listen. Overboard you go!"

And the falling night found Mr. Ledbetter—the Mr. Ledbetter who had complained that adventure was dead—sitting beside his cans of food, his chin resting upon his drawn-up knees, staring through his glasses in dismal mildness over the shining, vacant sea.

He was picked up in the course of three days by a negro fisherman and taken to St. Vincent's, and from St. Vincent's he got, by the expenditure of his last coins, to Kingston, in Jamaica. And there he might have foundered. Even nowadays he is not a man of affairs, and then he was a singularly helpless person. He had not the remotest idea what he ought to do. The only thing he seems to have done was to visit all the ministers of religion he could find in the place to borrow a passage home. But he was much too dirty and incoherent—and his story far too incredible for them. I met him quite by chance. It was close upon sunset, and I was walking out after my siesta on the road to Dunn's Battery, when I met him—I was rather bored, and with a whole evening on my hands—luckily for him. He was trudging dismally towards the town. His woebegone face and the quasi-clerical cut of his dust-stained, filthy costume caught my humour. Our eyes met. He hesitated. "Sir," he said with a catching of the breath, "could you spare a few minutes for what I fear will seem an incredible story?"

"Incredible!" I said.

"Quite," he answered eagerly. "No one will believe it, alter it though I may. Yet I can assure you, sir——"

He stopped hopelessly. The man's tone tickled me. He seemed an odd character. "I am," he said, "one of the most unfortunate beings alive."

"Among other things, you haven't dined?" I said, struck with an idea.

"I have not," he said solemnly, "for many days."

"You'll tell it better after that," I said; and without more ado led the way to a low place I knew, where such a costume as his was unlikely to give offence. And there—with certain omissions which he subsequently supplied, I got his story. At first I was incredulous, but as the wine warmed him, and the faint suggestion of cringing which his misfortunes had added to his manner disappeared, I began to believe. At last, I was so far convinced of his sincerity that I got him a bed for the night, and next day verified the banker's reference he gave me through my Jamaica banker. And that done, I took him shopping for underwear and suchlike equipments of a gentleman at large. Presently came the verified reference. His astonishing story was true. I will not amplify our subsequent proceedings. He started for England in three days' time.

"I do not know how I can possibly thank you enough," began the letter he wrote me from England, "for all your kindness to a total stranger," and proceeded for some time in a similar strain. "Had it not been for your generous assistance, I could certainly never have returned in time for the resumption of my scholastic duties, and my few minutes of reckless folly would, perhaps, have proved my ruin. As it is, I am entangled in a tissue of lies and evasions, of the most complicated sort, to account for my sunburnt appearance and my whereabouts. I have rather carelessly told two or three different stories, not realising the trouble this would mean for me in the end. The truth I dare not tell. I have consulted a number of law books in the British Museum, and there is not the slightest doubt that I have connived at and abetted and aided a felony. That scoundrel Bingham was the Hithergate bank manager, I find, and guilty of the most flagrant embezzlement. Please, please burn this letter when read—I trust you implicitly. The worst of it is, neither my aunt nor her friend who kept the boarding-house at which I was staying seem altogether to believe a guarded statement I have made them—practically of what actually happened. They suspect me of some discreditable adventure, but what sort of discreditable adventure they suspect me of, I do not know. My aunt says she would forgive me if I told her everything. I have—I have told her *more* than everything, and still she is not satisfied. It would never do to let them know the truth of the case, of course, and so I represent myself as having been waylaid and gagged upon the beach. My aunt wants to know *why* they waylaid and gagged me, why they took me away in their yacht. I do not know. Can you suggest any reason? I can think of nothing. If, when you wrote, you could write on two sheets so that I could show her one, and on that one if you could show clearly that I really *was* in Jamaica this summer, and had come there by being removed from a ship, it would be of great service to me. It would certainly add to the load of my obligation to you—a load

HH

that I fear I can never fully repay. Although if gratitude.
. . ." And so forth. At the end he repeated his request for
me to burn the letter.

So the remarkable story of Mr. Ledbetter's Vacation ends.
That breach with his aunt was not of long duration. The
old lady had forgiven him before she died.

STORY THE TENTH

The Stolen Body

MR. BESSEL was the senior partner in the firm of Bessel,
Hart, and Brown, of St. Paul's Churchyard, and for many
years he was well known among those interested in
psychical research as a liberal-minded and conscientious
investigator. He was an unmarried man, and instead of
living in the suburbs, after the fashion of his class, he
occupied rooms in the Albany, near Piccadilly. He was
particularly interested in the questions of thought transference and of apparitions of the living, and in November,
1896, he commenced a series of experiments in conjunction
with Mr. Vincey, of Staple Inn, in order to test the alleged
possibility of projecting an apparition of oneself by force
of will through space.

Their experiments were conducted in the following
manner: At a pre-arranged hour Mr. Bessel shut himself
in one of his rooms in the Albany and Mr. Vincey in his
sitting-room in Staple Inn, and each then fixed his mind
as resolutely as possible on the other. Mr. Bessel had
acquired the art of self-hypnotism, and, so far as he could,
he attempted first to hypnotise himself and then to project
himself as a "phantom of the living" across the intervening
space of nearly two miles into Mr. Vincey's apartment.
On several evenings this was tried without any satisfactory
result, but on the fifth or sixth occasion Mr. Vincey did
actually see or imagine he saw an apparition of Mr. Bessel
standing in his room. He states that the appearance,
although brief, was very vivid and real. He noticed that
Mr. Bessel's face was white and his expression anxious, and,

963

moreover, that his hair was disordered. For a moment Mr. Vincey, in spite of his state of expectation, was too surprised to speak or move, and in that moment it seemed to him as though the figure glanced over its shoulder and incontinently vanished.

It had been arranged that an attempt should be made to photograph any phantasm seen, but Mr. Vincey had not the instant presence of mind to snap the camera that lay ready on the table beside him, and when he did so he was too late. Greatly elated, however, even by this partial success, he made a note of the exact time, and at once took a cab to the Albany to inform Mr. Bessel of this result.

He was surprised to find Mr. Bessel's outer door standing open to the night, and the inner apartments lit and in an extraordinary disorder. An empty champagne magnum lay smashed upon the floor; its neck had been broken off against the inkpot on the bureau and lay beside it. An octagonal occasional table, which carried a bronze statuette and a number of choice books, had been rudely overturned, and down the primrose paper of the wall inky fingers had been drawn, as it seemed for the mere pleasure of defilement. One of the delicate chintz curtains had been violently torn from its rings and thrust upon the fire, so that the smell of its smouldering filled the room. Indeed the whole place was disarranged in the strangest fashion. For a few minutes Mr. Vincey, who had entered sure of finding Mr. Bessel in his easy chair awaiting him, could scarcely believe his eyes, and stood staring helplessly at these unanticipated things.

Then, full of a vague sense of calamity, he sought the porter at the entrance lodge. "Where is Mr. Bessel?" he asked. "Do you know that all the furniture is broken in Mr. Bessel's room?" The porter said nothing, but, obeying his gestures, came at once to Mr. Bessel's apartment to see the state of affairs. "This settles it," he said, surveying the lunatic confusion. "I didn't know of this. Mr. Bessel's gone off. He's mad!"

He then proceeded to tell Mr. Vincey that about half an hour previously, that is to say, at about the time of

Mr. Bessel's apparition in Mr. Vincey's rooms, the missing gentleman had rushed out of the gates of the Albany into Vigo Street, hatless and with disordered hair, and had vanished into the direction of Bond Street. "And as he went past me," said the porter, "he laughed—a sort of gasping laugh, with his mouth open and his eyes glaring— I tell you, sir, he fair scared me!—like this."

According to his imitation it was anything but a pleasant laugh. "He waved his hand, with all his fingers crooked and clawing—like that. And he said, in a sort of fierce whisper, 'Life.' Just that one word, 'Life!'"

"Dear me," said Mr. Vincey. "Tut, tut," and "Dear me!" He could think of nothing else to say. He was naturally very much surprised. He turned from the room to the porter and from the porter to the room in the gravest perplexity. Beyond his suggestion that probably Mr. Bessel would come back presently and explain what had happened, their conversation was unable to proceed. "It might be a sudden tooth-ache," said the porter, "a very sudden and violent tooth-ache, jumping on him suddenly-like and driving him wild. I've broken things myself before now in such a case . . ." He thought. "If it was, why should he say 'life' to me as he went past?"

Mr. Vincey did not know. Mr. Bessel did not return, and at last Mr. Vincey, having done some more helpless staring, and having addressed a note of brief inquiry and left it in a conspicuous position on the bureau, returned in a very perplexed frame of mind to his own premises in Staple Inn. This affair had given him a shock. He was at a loss to account for Mr. Bessel's conduct on any sane hypothesis. He tried to read, but he could not do so; he went for a short walk, and was so preoccupied that he narrowly escaped a cab at the top of Chancery Lane; and at last—a full hour before his usual time—he went to bed. For a considerable time he could not sleep because of his memory of the silent confusion of Mr. Bessel's apartment, and when at length he did attain an uneasy slumber it was at once disturbed by a very vivid and distressing dream of Mr. Bessel.

He saw Mr. Bessel gesticulating wildly, and with his face white and contorted. And, inexplicably mingled with his appearance, suggested perhaps by his gestures, was an intense fear, an urgency to act. He even believes that he heard the voice of his fellow experimenter calling distressfully to him, though at the time he considered this to be an illusion. The vivid impression remained though Mr. Vincey awoke. For a space he lay awake and trembling in the darkness, possessed with that vague, unaccountable terror of unknown possibilities that comes out of dreams upon even the bravest men. But at last he roused himself, and turned over and went to sleep again, only for the dream to return with enhanced vividness.

He awoke with such a strong conviction that Mr. Bessel was in overwhelming distress and need of help that sleep was no longer possible. He was persuaded that his friend had rushed out to some dire calamity. For a time he lay reasoning vainly against this belief, but at last he gave way to it. He arose, against all reason, lit his gas and dressed, and set out through the deserted streets—deserted, save for a noiseless policeman or so and the early news carts —towards Vigo Street to inquire if Mr. Bessel had returned.

But he never got there. As he was going down Long Acre some unaccountable impulse turned him aside out of that street towards Covent Garden, which was just waking to its nocturnal activities. He saw the market in front of him —a queer effect of glowing yellow lights and busy black figures. He became aware of a shouting, and perceived a figure turn the corner by the hotel and run swiftly towards him. He knew at once that it was Mr. Bessel. But it was Mr. Bessel transfigured. He was hatless and dishevelled, his collar was torn open, he grasped a bone-handled walking-cane near the ferrule end, and his mouth was pulled awry. And he ran, with agile strides, very rapidly. Their encounter was the affair of an instant. "Bessel!" cried Vincey.

The running man gave no sign of recognition either of Mr. Vincey or of his own name. Instead, he cut at his friend savagely with the stick, hitting him in the face within an inch of the eye. Mr. Vincey, stunned and astonished, staggered back, lost his footing, and fell heavily on the pavement. It seemed to him that Mr. Bessel leapt over him as he fell. When he looked again Mr. Bessel had vanished, and a policeman and a number of garden porters and salesmen were rushing past towards Long Acre in hot pursuit.

With the assistance of several passers-by—for the whole street was speedily alive with running people—Mr. Vincey struggled to his feet. He at once became the centre of a crowd greedy to see his injury. A multitude of voices competed to reassure him of his safety, and then to tell him of the behaviour of the madman, as they regarded Mr. Bessel. He had suddenly appeared in the middle of the market screaming "Life! Life!" striking left and right with a blood-stained walking-stick, and dancing and shouting with laughter at each successful blow. A lad and two women had broken heads, and he had smashed a man's wrist; a little child had been knocked insensible, and for a time he had driven everyone before him, so furious and resolute had his behaviour been. Then he made a raid upon a coffee stall, hurled its paraffin flare through the window of the post office, and fled laughing, after stunning the foremost of the two policemen who had the pluck to charge him.

Mr. Vincey's first impulse was naturally to join in the pursuit of his friend, in order if possible to save him from the violence of the indignant people. But his action was slow, the blow had half stunned him, and while this was still no more than a resolution came the news, shouted through the crowd, that Mr. Bessel had eluded his pursuers. At first Mr. Vincey could scarcely credit this, but the universality of the report, and presently the dignified return of two futile policemen, convinced him. After some aimless inquiries he returned towards Staple Inn, padding a handkerchief to a now very painful nose.

He was angry and astonished and perplexed. It appeared to him indisputable that Mr. Bessel must have gone violently mad in the midst of his experiment in thought transference, but why that should make him appear with a sad white face in Mr. Vincey's dreams seemed a problem beyond solution.

He racked his brains in vain to explain this. It seemed to him at last that not simply Mr. Bessel, but the order of things must be insane. But he could think of nothing to do. He shut himself carefully into his room, lit his fire—it was a gas fire with asbestos bricks—and, fearing fresh dreams if he went to bed, remained bathing his injured face, or holding up books in a vain attempt to read, until dawn. Throughout that vigil he had a curious persuasion that Mr. Bessel was endeavouring to speak to him, but he would not let himself attend to any such belief.

About dawn, his physical fatigue asserted itself, and he went to bed and slept at last in spite of dreaming. He rose late, unrested and anxious and in considerable facial pain. The morning papers had no news of Mr. Bessel's aberration—it had come too late for them. Mr. Vincey's perplexities, to which the fever of his bruise added fresh irritation, became at last intolerable, and, after a fruitless visit to the Albany, he went down to St. Paul's Church-yard to Mr. Hart, Mr. Bessel's partner, and so far as Mr. Vincey knew, his nearest friend.

He was surprised to learn that Mr. Hart, although he knew nothing of the outbreak, had also been disturbed by a vision, the very vision that Mr. Vincey had seen—Mr. Bessel, white and dishevelled, pleading earnestly by his gestures for help. That was his impression of the import of his signs. "I was just going to look him up in the Albany when you arrived," said Mr. Hart. "I was so sure of some-thing being wrong with him."

As the outcome of their consultation the two gentlemen decided to inquire at Scotland Yard for news of their missing friend. "He is bound to be laid by the heels," said Mr. Hart. "He can't go on at that pace for long." But the police authorities had not laid Mr. Bessel by the heels. They confirmed Mr. Vincey's overnight experiences and added fresh circumstances, some of an even graver character than those he knew—a list of smashed glass along the upper half of Tottenham Court Road, an attack upon a policeman in Hampstead Road, and an atrocious assault upon a woman. All these outrages were committed between half-past twelve and a quarter to two in the morning, and between those hours—and, indeed, from the very moment of Mr. Bessel's first rush from his rooms at half-past nine in the evening—they could trace the deepening violence of his fantastic career. For the last hour, at least from before one, that is, until a quarter to two, he had run amuck through London, eluding with amazing agility every effort to stop or capture him.

But after a quarter to two he had vanished. Up to that hour witnesses were multitudinous. Dozens of people had seen him, fled from him or pursued him, and then things suddenly came to an end. At a quarter to two he had been seen running down the Euston Road towards Baker Street, flourishing a can of burning colza oil and jerking splashes of flame therefrom at the windows of the houses he passed. But none of the policemen on Euston Road beyond the Waxwork Exhibition, nor any of those in the side streets down which he must have passed had he left the Euston Road, had seen anything of him. Abruptly he disappeared. Nothing of his subsequent doings came to light in spite of the keenest inquiry.

Here was a fresh astonishment for Mr. Vincey. He had found considerable comfort in Mr. Hart's conviction: "He is bound to be laid by the heels before long," and in that assurance he had been able to suspend his mental perplexities. But any fresh development seemed destined to add new impossibilities to a pile already heaped beyond the powers of his acceptance. He found himself doubting whether his memory might not have played him some grotesque trick, debating whether any of these things could possibly have happened; and in the afternoon he hunted up Mr. Hart again to share the intolerable weight on his mind. He found Mr. Hart engaged with a well-known private detective, but as that gentleman accomplished nothing in this case, we need not enlarge upon his proceedings.

All that day Mr. Bessel's whereabouts eluded an unceasingly active inquiry, and all that night. And all that day there was a persuasion in the back of Mr. Vincey's mind that Mr. Bessel sought his attention, and all through the night Mr. Bessel with a tear-stained face of anguish pursued him through his dreams. And whenever he saw Mr. Bessel in his dreams he also saw a number of other faces, vague but malignant, that seemed to be pursuing Mr. Bessel.

It was on the following day, Sunday, that Mr. Vincey recalled certain remarkable stories of Mrs. Bullock, the medium, who was then attracting attention for the first time in London. He determined to consult her. She was staying at the house of that well-known inquirer, Dr. Wilson Paget, and Mr. Vincey, although he had never met that gentleman before, repaired to him forthwith with the intention of invoking her help. But scarcely had he mentioned the name of Bessel when Doctor Paget interrupted him. "Last night—just at the end," he said, "we had a communication."

He left the room, and returned with a slate on which were certain words written in a handwriting, shaky indeed, but indisputably the handwriting of Mr. Bessel!

"How did you get this?" said Mr. Vincey. "Do you mean?——"

"We got it last night," said Doctor Paget. With numerous interruptions from Mr. Vincey, he proceeded to explain how the writing had been obtained. It appears that in her *séances*, Mrs. Bullock passes into a condition of trance, her eyes rolling up in a strange way under her eyelids, and her body becoming rigid. She then begins to talk very rapidly, usually in voices other than her own. At the same time one or both of her hands may become active, and if slates and pencils are provided they will then write messages simultaneously with and quite independently of the flow of words from her mouth. By many she is considered an even more remarkable medium than the celebrated Mrs. Piper. It was one of these messages, the one written by her left hand, that Mr. Vincey now had before him. It consisted of eight words written disconnectedly "George Bessel . . . trial excav^n . . . Baker Street . . . help . . . starvation." Curiously enough, neither Doctor Paget nor the two other inquirers who were present had heard of the disappearance of Mr. Bessel—the news of it appeared only in the evening papers of Saturday—and they had put the message aside with many others of a vague and enigmatical sort that Mrs. Bullock has from time to time delivered.

When Doctor Paget heard Mr. Vincey's story, he gave himself at once with great energy to the pursuit of this clue to the discovery of Mr. Bessel. It would serve no useful purpose here to describe the inquiries of Mr. Vincey and himself; suffice it that the clue was a genuine one, and that Mr. Bessel was actually discovered by its aid.

He was found at the bottom of a detached shaft which had been sunk and abandoned at the commencement of the work for the new electric railway near Baker Street Station. His arm and leg and two ribs were broken. The shaft is protected by a hoarding nearly 20 feet high, and over this, incredible as it seems, Mr. Bessel, a stout, middle-aged gentleman, must have scrambled in order to fall down the shaft. He was saturated in colza oil, and the smashed tin lay beside him, but luckily the flame had been extinguished by his fall. And his madness had passed from him altogether. But he was, of course, terribly enfeebled, and at the sight of his rescuers he gave way to hysterical weeping.

In view of the deplorable state of his flat, he was taken to the house of Dr. Hatton in Upper Baker Street. Here he was subjected to a sedative treatment, and anything that might recall the violent crisis through which he had passed was carefully avoided. But on the second day he volunteered a statement.

Since that occasion Mr. Bessel has several times repeated this statement—to myself among other people—varying the details as the narrator of real experiences always does, but never by any chance contradicting himself in any particular. And the statement he makes is in substance as follows.

In order to understand it clearly it is necessary to go back to his experiments with Mr. Vincey before his remarkable attack. Mr. Bessel's first attempts at self-projection, in his experiments with Mr. Vincey, were, as the reader will remember, unsuccessful. But through all of them he was concentrating all his power and will upon getting out of the body—"willing it with all my might," he says. At last, almost against expectation, came success. And Mr. Bessel asserts that he, being alive, did actually, by an effort of will, leave his body and pass into some place or state outside this world.

The release was, he asserts, instantaneous. "At one moment I was seated in my chair, with my eyes tightly shut, my hands gripping the arms of the chair, doing all I could to concentrate my mind on Vincey, and then I perceived myself outside my body—saw my body near me, but certainly not containing me, with the hands relaxing and the head drooping forward on the breast."

Nothing shakes him in his assurance of that release. He describes in a quiet, matter-of-fact way the new sensation he experienced. He felt he had become impalpable—so much he had expected, but he had not expected to find himself enormously large. So, however, it would seem he became. "I was a great cloud—if I may express it that way—anchored to my body. It appeared to me, at first, as if I had discovered a greater self of which the conscious being in my brain was only a little part. I saw the Albany and Piccadilly and Regent Street and all the rooms and places in the houses, very minute and very bright and distinct, spread out below me like a little city seen from a balloon. Every now and then vague shapes like drifting wreaths of smoke made the vision a little indistinct, but at first I paid little heed to them. The thing that astonished me most, and which astonishes me still, is that I saw quite distinctly the insides of the houses as well as the streets, saw little people dining and talking in the private houses, men and women dining, playing billiards, and drinking in restaurants and hotels, and several places of entertainment crammed with people. It was like watching the affairs of a glass hive."

Such were Mr. Bessel's exact words as I took them down when he told me the story. Quite forgetful of Mr. Vincey, he remained for a space observing these things. Impelled by curiosity, he says, he stooped down, and with the shadowy arm he found himself possessed of attempted to touch a man walking along Vigo Street. But he could not do so, though his finger seemed to pass through the man. Something prevented his doing this, but what it was he finds it hard to describe. He compares the obstacle to a sheet of glass.

"I felt as a kitten may feel," he said, "when it goes for the first time to pat its reflection in a mirror." Again and again, on the occasion when I heard him tell this story, Mr. Bessel returned to that comparison of the sheet of glass. Yet it was not altogether a precise comparison, because, as the reader will speedily see, there were interruptions of this generally impermeable resistance, means of getting through the barrier to the material world again. But, naturally, there is a very great difficulty in expressing these unprecedented impressions in the language of everyday experience.

A thing that impressed him instantly, and which weighed upon him throughout all this experience, was the stillness of this place—he was in a world without sound.

At first Mr. Bessel's mental state was an unemotional wonder. His thought chiefly concerned itself with where he might be. He was out of the body—out of his material body, at any rate—but that was not all. He believes, and I for one believe also, that he was somewhere out of space, as we understand it, altogether. By a strenuous effort of will he had passed out of his body into a world beyond this world, a world undreamt of, yet lying so close to it and so strangely situated with regard to it that all things on this earth are clearly visible both from without and from within in this other world about us. For a long time, as it seemed to him, this realisation occupied his mind to the exclusion of all other matters, and then he recalled the engagement with Mr. Vincey, to which this astonishing experience was, after all, but a prelude.

He turned his mind to locomotion in this new body in which he found himself. For a time he was unable to shift himself from his attachment to his earthly carcass. For a time this new strange cloud body of his simply swayed, contracted, expanded, coiled, and writhed with his efforts to free himself, and then quite suddenly the link that bound him snapped. For a moment everything was hidden by what appeared to be whirling spheres of dark vapour, and then through a momentary gap he saw his drooping body collapse limply, saw his lifeless head drop sideways, and found he was driving along like a huge cloud in a strange place of shadowy clouds that had the luminous intricacy of London spread like a model below.

But now he was aware that the fluctuating vapour about him was something more than vapour, and the temerarious excitement of his first essay was shot with fear. For he perceived, at first indistinctly, and then suddenly very clearly, that he was surrounded by *faces!* that each roll and coil of the seeming cloud-stuff was a face. And such faces! Faces of thin shadow, faces of gaseous tenuity. Faces like those faces that glare with intolerable strangeness upon the sleeper in the evil hours of his dreams. Evil, greedy eyes that were full of a covetous curiosity, faces with knit brows and snarling, smiling lips; their vague hands clutched at Mr. Bessel as he passed, and the rest of their bodies was but an elusive streak of trailing darkness. Never a word they said, never a sound from the mouths that seemed to gibber. All about him they pressed in that dreamy silence, passing freely through the dim mistiness that was his body, gathering ever more numerously about him. And the shadowy Mr. Bessel, now suddenly fear-stricken, drove through the silent, active multitude of eyes and clutching hands.

So inhuman were these faces, so malignant their staring eyes, and shadowy, clawing gestures, that it did not occur to Mr. Bessel to attempt intercourse with these drifting creatures. Idiot phantoms, they seemed, children of vain desire, beings unborn and forbidden the boon of being, whose only expressions and gestures told of the envy and craving for life that was their one link with existence.

It says much for his resolution that, amidst the swarming cloud of these noiseless spirits of evil, he could still think of Mr. Vincey. He made a violent effort of will and found himself, he knew not how, stooping towards Staple Inn, saw Vincey sitting attentive and alert in his armchair by the fire.

And clustering also about him, as they clustered ever about all that lives and breathes, was another multitude of these vain voiceless shadows, longing, desiring, seeking some loophole into life.

For a space Mr. Bessel sought ineffectually to attract his friend's attention. He tried to get in front of his eyes, to move the objects in his room, to touch him. But Mr. Vincey remained unaffected, ignorant of the being that was so close to his own. The strange something that Mr. Bessel has compared to a sheet of glass separated them impermeably.

And at last Mr. Bessel did a desperate thing. I have told how that in some strange way he could see not only the outside of a man as we see him, but within. He extended his shadowy hand and thrust his vague black fingers, as it seemed, through the heedless brain.

Then, suddenly, Mr. Vincey started like a man who recalls his attention from wandering thoughts, and it seemed to Mr. Bessel that a little dark-red body situated in the middle of Mr. Vincey's brain swelled and glowed as he did so. Since that experience he has been shown anatomical figures of the brain, and he knows now that this is that useless structure, as doctors call it, the pineal eye. For, strange as it will seem to many, we have, deep in our brains —where it cannot possibly see any earthly light—an eye! At the time this, with the rest of the internal anatomy of the brain, was quite new to him. At the sight of its changed appearance, however, he thrust forth his finger, and, rather fearful still of the consequences, touched this little spot. And instantly Mr. Vincey started, and Mr. Bessel knew that he was seen.

And at that instant it came to Mr. Bessel that evil had happened to his body, and behold! a great wind blew through all that world of shadows and tore him away. So strong was this persuasion that he thought no more of Mr. Vincey, but turned about forthwith, and all the countless faces drove back with him like leaves before a gale. But he returned too late. In an instant he saw the body that he had left inert and collapsed—lying, indeed, like the body of a man just dead—had arisen, had arisen by virtue of some strength and will beyond his own. It stood with staring eyes, stretching its limbs in dubious fashion.

For a moment he watched it in wild dismay, and then he stooped towards it. But the pane of glass had closed against him again, and he was foiled. He beat himself passionately against this, and all about him the spirits of evil grinned and pointed and mocked. He gave way to furious anger. He compares himself to a bird that has fluttered heedlessly into a room and is beating at the window-pane that holds it back from freedom.

And behold! the little body that had once been his was now dancing with delight. He saw it shouting, though he could not hear its shouts; he saw the violence of its movements grow. He watched it fling his cherished furniture about in the mad delight of existence, rend his books apart, smash bottles, drink heedlessly from the jagged fragments, leap and smite in a passionate acceptance of living. He watched these actions in paralyzed astonishment. Then once more he hurled himself against the impassable barrier, and then, with all that crew of mocking ghosts about him, hurried back in dire confusion to Vincey to tell him of the outrage that had come upon him.

But the brain of Vincey was now closed against apparitions, and the disembodied Mr. Bessel pursued him in vain as he hurried out into Holborn to call a cab. Foiled and terror-stricken, Mr. Bessel swept back again, to find his desecrated body whooping in a glorious frenzy down the Burlington Arcade. . . .

And now the attentive reader begins to understand Mr. Bessel's interpretation of the first part of this strange story. The being whose frantic rush through London had inflicted so much injury and disaster had indeed Mr. Bessel's body, but it was not Mr. Bessel. It was an evil spirit out of that strange world beyond existence, into which Mr. Bessel had so rashly ventured. For twenty hours it held possession of him, and for all those twenty hours the dispossessed spirit-body of Mr. Bessel was going to and fro in that unheard-of middle world of shadows seeking help in vain.

He spent many hours beating at the minds of Mr. Vincey

and of his friend Mr. Hart. Each, as we know, he roused by his efforts. But the language that might convey his situation to these helpers across the gulf he did not know; his feeble fingers groped vainly and powerlessly in their brains. Once, indeed, as we have already told, he was able to turn Mr. Vincey aside from his path so that he encountered the stolen body in its career, but he could not make him understand the thing that had happened: he was unable to draw any help from that encounter. . . .

All through those hours the persuasion was overwhelming in Mr. Bessel's mind that presently his body would be killed by its furious tenant, and he would have to remain in this shadow-land for evermore. So that those long hours were a growing agony of fear. And ever as he hurried to and fro in his ineffectual excitement innumerable spirits of that world about him mobbed him and confused his mind. And ever an envious applauding multitude poured after their successful fellow as he went upon his glorious career.

For that, it would seem, must be the life of these bodiless things of this world that is the shadow of our world. Ever they watch, coveting a way into a mortal body, in order that they may descend, as furies and frenzies, as violent lusts and mad, strange impulses, rejoicing in the body they have won. For Mr. Bessel was not the only human soul in that place. Witness the fact that he met first one, and afterwards several shadows of men, men like himself, it seemed, who had lost their bodies even it may be as he had lost his, and wandered, despairingly, in that lost world that is neither life nor death. They could not speak because that world is silent, yet he knew them for men because of their dim human bodies, and because of the sadness of their faces.

But how they had come into that world he could not tell, nor where the bodies they had lost might be, whether they still raved about the earth, or whether they were closed for ever in death against return. That they were the spirits of the dead neither he nor I believe. But Doctor Wilson Paget thinks they are the rational souls of men who are lost in madness on the earth.

At last Mr. Bessel chanced upon a place where a little crowd of such disembodied silent creatures was gathered, and thrusting through them he saw below a brightly-lit room, and four or five quiet gentlemen and a woman, a stoutish woman dressed in black bombazine and sitting awkwardly in a chair with her head thrown back. He knew her from her portraits to be Mrs. Bullock, the medium. And he perceived that tracts and structures in her brain glowed and stirred as he had seen the pineal eye in the brain of Mr. Vincey glow. The light was very fitful; sometimes it was a broad illumination, and sometimes merely a faint twilight spot, and it shifted slowly about her brain. She kept on talking and writing with one hand. And Mr. Bessel saw that the crowding shadows of men about him, and a great multitude of the shadow spirits of that shadow land, were all striving and thrusting to touch the lighted regions of her brain. As one gained her brain or another was thrust away, her voice and the writing of her hand changed. So that what she said was disorderly and confused for the most part; now a fragment of one soul's message, and now a fragment of another's, and now she babbled the insane fancies of the spirits of vain desire. Then Mr. Bessel understood that she spoke for the spirit that had touch of her, and he began to struggle very furiously towards her. But he was on the outside of the crowd and at that time he could not reach her, and at last, growing anxious, he went away to find what had happened meanwhile to his body.

For a long time he went to and fro seeking it in vain and fearing that it must have been killed, and then he found it at the bottom of the shaft in Baker Street, writhing furiously and cursing with pain. Its leg and an arm and two ribs had been broken by its fall. Moreover, the evil spirit was angry because his time had been so short and because of the pain—making violent movements and casting his body about.

And at that Mr. Bessel returned with redoubled earnestness to the room where the *séance* was going on, and so soon as he had thrust himself within sight of the place he saw one of the men who stood about the medium looking at his watch as if he meant that the *séance* should presently end. At that a great number of the shadows who had been striving turned away with gestures of despair. But the thought that the *séance* was almost over only made Mr. Bessel the more earnest, and he struggled so stoutly with his will against the others that presently he gained the woman's brain. It chanced that just at that moment it glowed very brighly, and in that instant she wrote the message that Doctor Wilson Paget preserved. And then the other shadows and the cloud of evil spirits about him had thrust Mr. Bessel away from her, and for all the rest of the *séance* he could regain her no more.

So he went back and watched through the long hours at the bottom of the shaft where the evil spirit lay in the stolen body it had maimed, writhing and cursing, and weeping and groaning, and learning the lesson of pain. And towards dawn the thing he had waited for happened, the brain glowed brightly and the evil spirit came out, and Mr. Bessel entered the body he had feared he should never enter again. As he did so, the silence—the brooding silence—ended; he heard the tumult of traffic and the voices of people overhead, and that strange world that is the shadow of our world —the dark and silent shadows of ineffectual desire and the shadows of lost men—vanished clean away.

He lay there for the space of about three hours before he was found. And in spite of the pain and suffering of his wounds, and of the dim damp place in which he lay; in spite of the tears—wrung from him by his physical distress —his heart was full of gladness to know that he was nevertheless back once more in the kindly world of men.

STORY THE
ELEVENTH

Mr. Brisher's Treasure

"You can't be *too* careful *who* you marry," said Mr. Brisher, and pulled thoughtfully with a fat-wristed hand at the lank moustache that hides his want of chin.

"That's why——" I ventured.

"Yes," said Mr. Brisher, with a solemn light in his bleary, blue-grey eyes, moving his head expressively and breathing alcohol intimately at me. "There's lots as 'ave 'ad a try at me—many as I could name in *this* town—but none 'ave done it—none."

I surveyed the flushed countenance, the equatorial expansion, the masterly carelessness of his attire, and heaved a sigh to think that by reason of the unworthiness of women he must needs be the last of his race.

"I was a smart young chap when I was younger," said Mr. Brisher. "I 'ad my work cut out. But I was very careful—very. And I got through . . ."

He leant over the taproom table and thought visibly on the subject of my trustworthiness. I was relieved at last by his confidence.

"I was engaged once," he said at last, with a reminiscent eye on the shuv-a'penny board.

"So near as that?"

He looked at me. "So near as that. Fact is——" He looked about him, brought his face close to mine, lowered his voice, and fenced off an unsympathetic world with a grimy hand. "If she ain't dead or married to someone else

or anything—I'm engaged still. Now." He confirmed this
statement with nods and facial contortions. "*Still,*" he said,
ending the pantomime, and broke into a reckless smile at my
surprise. "*Me!*

"Run away," he explained further, with coruscating eyebrows. "Come 'ome.

"That ain't all.

"You'd hardly believe it," he said, "but I found a treasure.
Found a regular treasure."

I fancied this was irony, and did not, perhaps, greet it
with proper surprise. "Yes," he said, "I found a treasure. And
come 'ome. I tell you I could surprise you with things that
has happened to me." And for some time he was content
to repeat that he had found a treasure—and left it.

I made no vulgar clamour for a story, but I became attentive to Mr. Brisher's bodily needs, and presently I led
him back to the deserted lady.

"She was a nice girl," he said—a little sadly, I thought.
"*And* respectable."

He raised his eyebrows and tightened his mouth to
express extreme respectability—beyond the likes of us elderly
men.

"It was a long way from 'ere. Essex, in fact. Near Colchester. It was when I was up in London—in the buildin'
trade. I was a smart young chap then, I can tell you.
Slim. 'Ad best clo'es 's good as anybody. 'At—*silk* 'at, mind
you." Mr. Brisher's hand shot above his head towards the
infinite to indicate a silk hat of the highest. "Umbrella—
nice umbrella with a 'orn 'andle. Savin's. Very careful
I was. . . ."

He was pensive for a little while, thinking, as we must
all come to think sooner or later, of the vanished brightness
of youth. But he refrained, as one may do in taprooms, from
the obvious moral.

"I got to know 'er through a chap what was engaged to
'er sister. She was stopping in London for a bit with a
naunt that 'ad a 'am an' beef shop. This aunt was very
particular—they was all very particular people, all 'er people
was—and wounldn't let 'er sister go out with this feller except

'er other sister, *my* girl that is, went with them. So 'e brought me into it, sort of to ease the crowding. We used to go walks in Battersea Park of a Sunday afternoon. Me in my topper, and 'im in 'is; and the girls—well—stylish. There wasn't many in Battersea Park 'ad the larf of us. She wasn't what you'd call pretty, but a nicer girl I never met. *I* liked 'er from the start, and, well—though I say it who shouldn't—she liked me. You know 'ow it is, I dessay?"

I pretended I did.

"And when this chap married 'er sister—'im and me was great friends—what must 'e do but arst me down to Colchester, close by where She lived. Naturally I was introjuced to 'er people, and well, very soon, her and me was engaged."

He repeated "engaged."

"She lived at 'ome with 'er father and mother, quite the lady, in a very nice little 'ouse with a garden—and remarkable respectable people they was. Rich you might call 'em a'most. They owned their own 'ouse—got it out of the Building Society, and cheap because the chap who had it before was a burglar and in prison—and they 'ad a bit of free'old land, and some cottages and money 'nvested—all nice and tight: they was what you'd call snug and warm. I tell you, I was On. Furniture too. Why! They 'ad a pianner. Jane—'er name was Jane—used to play it Sundays, and very nice she played too. There wasn't 'ardly a 'im toon in the book she *couldn't* play. . . .

"Many's the evenin' we've met and sung 'ims there, me and 'er and the family.

"'Er father was quite a leadin' man in chapel. You should ha' seen him Sundays, interruptin' the minister and givin' out 'ims. He had gold spectacles, I remember, and used to look over 'em at you while he sang hearty—he was always great on singing 'earty to the Lord—and when *he* got out o' toon 'arf the people went after 'im—always. 'E was that sort of man. And to walk be'ind 'im in 'is nice black clo'es—'is 'at was a brimmer—made one regular proud to be engaged to such a father-in-law. And when the summer came I went down there and stopped a fortnight.

"Now, you know there was a sort of Itch," said Mr. Brisher. "We wanted to marry, me and Jane did, and get things settled. But 'E said I 'ad to get a proper position first. Consequently there was a Itch. Consequently, when I went down there, I was anxious to show that I was a good useful sort of chap like. Show I could do pretty nearly everything like. See?"

I made a sympathetic noise.

"And down at the bottom of their garden was a bit of wild part like. So I says to 'im, 'Why don't you 'ave a rockery 'ere?' I says. 'It 'ud look nice.'

" 'Too much expense,' he says.

" 'Not a penny,' says I. 'I'm a dab at rockeries. Lemme make you one.' You see, I'd 'elped my brother make a rockery in the beer garden be'ind 'is tap, so I knew 'ow to do it to rights. 'Lemme make you one,' I says. 'It's 'oli-days, but I'm that sort of chap, I 'ate doing nothing,' I says. 'I'll make you one to rights.' And the long and the short of it was, he said I might.

"And that's 'ow I come on the treasure."

"What treasure?" I asked.

"Why!" said Mr. Brisher, "the treasure I'm telling you about, what's the reason why I never married."

"What!—a treasure—dug up?"

"Yes—buried wealth—treasure trove. Come out of the ground. What I kept on saying—regular treasure. . . ." He looked at me with unusual disrespect.

"It wasn't more than a foot deep, not the top of it," he said. "I'd 'ardly got thirsty like, before I come on the corner."

"Go on," I said. "I didn't understand."

"Why! Directly I 'it the box I knew it was treasure. A sort of instinct told me. Something seemed to shout inside of me—'Now's your chance—lie low.' It's lucky I knew the laws of treasure trove or I'd 'ave been shoutin' there and then. I dare say you know——?"

"Crown bags it," I said, "all but one per cent. Go on. It's a shame. What did you do?"

"Uncovered the top of the box. There wasn't anybody

in the garden or about like. Jane was 'elping 'er mother do the 'ouse. I *was* excited—I tell you. I tried the lock and then gave a whack at the hinges. Open it came. Silver coins—full! Shining. It made me tremble to see 'em. And jest then—I'm blessed if the dustman didn't come round the back of the 'ouse. It pretty nearly gave me 'eart disease to think what a fool I was to 'ave that money showing. And directly after I 'eard the chap next door—'e was 'olidaying too—I 'eard him watering 'is beans. If only 'e'd looked over the fence!"

"What did you do?"

"Kicked the lid on again and covered it up like a shot, and went on digging about a yard away from it—like mad. And my face, so to speak, was laughing on its own account till I had it hid. I tell you I was regular scared like at my luck. I jest thought that it 'ad to be kep' close and that was all. 'Treasure,' I kep' whisperin' to myself, 'Treasure' and ''undreds of pounds, 'undreds, 'undreds of pounds.' Whispering to myself like, and digging like blazes. It seemed to me the box was regular sticking out and showing, like your legs do under the sheets in bed, and I went and put all the earth I'd got out of my 'ole for the rockery slap on top of it. I *was* in a sweat. And in the midst of it all out toddles 'er father. He didn't say anything to me, jest stood behind me and stared, but Jane tole me afterwards when he went indoors, 'e says, 'That there jackanapes of yours, Jane'—he always called me a jackanapes some'ow—'knows 'ow to put 'is back into it after all.' Seemed quite impressed by it, 'e did."

"How long was the box?" I asked, suddenly.

"'Ow long?" said Mr. Brisher.

"Yes—in length?"

"Oh! 'bout so—by so." Mr. Brisher indicated a moderate-sized trunk.

"*Full?*" said I.

"Full up of silver coins—arf-crowns, I believe."

"Why!" I cried, "that would mean—hundreds of pounds."

"Thousands," said Mr. Brisher, in a sort of sad calm. "I calc'lated it out."

"But how did they get there?"

"All I know is what I found. What I thought at the time was this. The chap who'd owned the 'ouse before 'er father 'd been a regular slap-up burglar. What you'd call a 'igh-class criminal. Used to drive 'is trap—like Peace did." Mr. Brisher meditated on the difficulties of narration and embarked on a complicated parenthesis. "I don't know if I told you it'd been a burglar's 'ouse before it was my girl's father's, and I knew 'e'd robbed a mail train once, I did know that. It seemed to me——"

"That's very likely," I said. "But what did you do?"

"Sweated," said Mr. Brisher. "Regular run orf me. All that morning," said Mr. Brisher, "I was at it, pretending to make that rockery and wondering what I should do. I'd 'ave told 'er father p'r'aps, only I was doubtful of 'is honesty— I was afraid he might rob me of it like, and give it up to the authorities—and besides, considering I was marrying into the family, I thought it would be nicer like if it came through me. Put me on a better footing, so to speak. Well, I 'ad three days before me left of my 'olidays, so there wasn't no hurry, so I covered it up and went on digging, and tried to puzzle out 'ow I was to make sure of it. Only I couldn't.

"I thought," said Mr. Brisher, "*and* I thought. Once I got regular doubtful whether I'd seen it or not, and went down to it and 'ad it uncovered again, just as her ma came out to 'ang up a bit of washin' she'd done. Jumps again! Afterwards I was just thinking I'd 'ave another go at it, when Jane comes to tell me dinner was ready. 'You'll want it,' she said, 'seeing all the 'ole you've dug.'

"I was in a regular daze all dinner, wondering whether that chap next door wasn't over the fence and filling 'is pockets. But in the afternoon I got easier in my mind— it seemed to me it must 'ave been there so long it was pretty sure to stop a bit longer—and I tried to get up a bit of a discussion to dror out the old man and see what '*e* thought of treasure trove."

Mr. Brisher paused, and affected amusement at the memory.

"The old man was a scorcher," he said; "a regular scorcher."

"What!" said I; "did he——?"

"It was like this," explained Mr. Brisher, laying a friendly hand on my arm and breathing into my face to calm me. "Just to dror 'im out, I told a story of a chap I said I knew—pretendin', you know—who'd found a sovring in a novercoat 'e'd borrowed. I said 'e stuck to it, but I said I wasn't sure whether that was right or not. And then the old man began. Lor! 'e *did* let me 'ave it!" Mr. Brisher affected an insincere amusement. " 'E was, well—what you might call a rare 'and at snacks. Said that was the sort of friend 'e'd naturally expect me to 'ave. Said 'e'd naturally expect that from the friend of a out-of-work loafer who took up with daughters who didn't belong to 'im. There! I couldn't tell you *'arf* 'e said. 'E went on most outrageous. I stood up to 'im about it, just to dror 'im out. 'Wouldn't you stick to a arf-sov,' not if you found it in the street?' I says. 'Certainly not,' 'e says; 'certainly I wouldn't.' 'What! not if you found it as a sort of treasure?' 'Young man,' 'e says, 'there's 'i'er 'thority than mine—Render unto Cæsar—what is it? Yes. Well, he fetched up that. A rare 'and at 'itting you over the 'ed with the Bible, was the old man. And so he went on. 'E got to such Snacks about me at last I couldn't stand it. I'd promised Jane not to answer 'im back, but it got a bit *too* thick. I— I give it 'im . . ."

Mr. Brisher, by means of enigmatical facework, tried to make me think he had had the best of that argument, but I knew better.

"I went out in a 'uff at last. But not before I was pretty sure I 'ad to lift that treasure by myself. The only thing that kep' me up was thinking 'ow I'd take it out of 'im when I 'ad the cash. . . ."

There was a lengthy pause.

"Now, you'd 'ardly believe it, but all them three days I never 'ad a chance at the blessed treasure, never got out not even a 'arf-crown. There was always a Somethink—always.

" 'Stonishing thing it isn't thought of more," said Mr. Brisher. "Finding treasure's no great shakes. It's gettin' it. I don't suppose I slep' a wink any of those nights, thinking where I was to take it, what I was to do with it, 'ow I was to explain it. It made me regular ill. And days I was that dull, it made Jane regular 'uffy. 'You ain't the same chap you was in London,' she says, several times. I tried to lay it on 'er father and 'is Snacks, but bless you, she knew better. What must she 'ave but that I'd got another girl on my mind! Said I wasn't True. Well, we had a bit of a row. But I was that set on the Treasure, I didn't seem to mind a bit Anything she said.

"Well, at last I got a sort of plan. I was always a bit good at planning, though carrying out isn't so much in my line. I thought it all out and settled on a plan. First, I was going to take all my pockets full of these 'ere 'arf-crowns—see?—and afterwards—as I shall tell.

"Well, I got to that state I couldn't think of getting at the Treasure again in the daytime, so I waited until the night before I had to go, and then, when everything was still, up I gets and slips down to the back door, meaning to get my pockets full. What must I do in the scullery but fall over a pail? Up gets 'er father with a gun—'e was a light sleeper was 'er father, and very suspicious—and there was me: 'ad to explain I'd come down to the pump for a drink because my water-bottle was bad. 'E didn't let me off a Snack or two over that bit, you lay a bob."

"And you mean to say——" I began.

"Wait a bit," said Mr. Brisher. "I say, I'd made my plan. That put the kybosh on one bit, but it didn't 'urt the general scheme not a bit. I went and I finished that rockery next day, as though there wasn't a Snack in the world; cemented over the stones, I did, dabbed it green and everythink. I put a dab of green just to show where the box was. They all came and looked at it, and said 'ow nice it was—even 'e was a bit softer like to see it, and all he said was, 'It's a pity you can't always work like that, then you might get something definite to do,' he says.

" 'Yes,' I says—I couldn't 'elp it—'I put a lot in that

rockery,' I says, like that. See? 'I put a lot in that rockery'—meaning——"

"I see," said I—for Mr. Brisher is apt to over-elaborate his jokes.

"'E didn't," said Mr. Brisher. "Not then, anyhow.

"Ar-ever—after all that was over, off I set for London. . . . Orf I set for London. . . ."

Pause.

"On'y I wasn't going to no London," said Mr. Brisher, with sudden animation, and thrusting his face into mine. "no fear! What do *you* think?

"I didn't go no further than Colchester—not a yard.

"I'd left the spade just where I could find it. I'd got everything planned and right. I 'ired a little trap in Colchester, and pretended I wanted to go to Ipswich and stop the night, and come back next day, and the chap I 'ired it from made me leave two sovrings on it right away, and off I set.

"I didn't go to no Ipswich neither.

"Midnight the 'orse and trap was 'itched by the little road that ran by the cottage where 'e lived—not sixty yards off, it wasn't—and I was at it like a good 'un. It was jest the night for such games—overcast—but a trifle too 'ot, and all round the sky there was summer lightning and presently a thunderstorm. Down it came. First big drops in a sort of fizzle, then 'ail. I kep' on. I whacked at it—I didn't dream the old man would 'ear. I didn't even trouble to go quiet with the spade, and the thunder and lightning and 'ail seemed to excite me like. I shouldn't wonder if I was singing. I got so 'ard at it I clean forgot the thunder and the 'orse and trap. I precious soon got the box showing, and started to lift it. . . ."

"Heavy?" I said.

"I couldn't no more lift it than fly. I *was* sick. I'd never thought of that! I got regular wild—I tell you, I cursed. I got sort of outrageous. I didn't think of dividing it like for the minute, and even then I couldn't 'ave took money about loose in a trap. I hoisted one end sort of wild like, and over the whole show went with a tremenjous

noise. Perfeck smash of silver. And then right on the
heels of that, Flash! Lightning like the day! and there was
the back door open and the old man coming down the
garden with 'is blooming old gun. He wasn't not a 'undred
yards away!

"I tell you I was that upset—I didn't think what I was
doing. I never stopped—not even to fill my pockets. I went
over the fence like a shot, and ran like one o'clock for the
trap, cussing and swearing as I went. I *was* in a state. . . .

"And will you believe me, when I got to the place where
I'd left the 'orse and trap, they'd gone. Orf! When I
saw that I 'adn't a cuss left for it. I jest danced on the
grass, and when I'd danced enough I started off to London.
. . . I was done."

Mr. Brisher was pensive for an interval. "I was done,"
he repeated, very bitterly.

"Well?" I said.

"That's all," said Mr. Brisher.

"You didn't go back?"

"No fear. I'd 'ad enough of *that* blooming treasure,
any'ow for a bit. Besides, I didn't know what was done
to chaps who tried to collar a treasure trove. I started off
for London there and then. . . ."

"And you never went back?"

"Never."

"But about Jane? Did you write?"

"Three times, fishing like. And no answer. We'd
parted in a bit of a 'uff on account of 'er being jealous. So
that I couldn't make out for certain what it meant.

"I didn't know what to do. I didn't even know whether
the old man knew it was me. I sort of kep' an eye open
on papers to see when he'd give up that treasure to the
Crown, as I hadn't a doubt 'e would considering 'ow respectable he'd always been."

"And did he?"

Mr. Brisher pursed his mouth and moved his head
slowly from side to side. "Not '*im*," he said.

"Jane was a nice girl," he said, "a thorough nice girl
mind you, *if* jealous, and there's no knowing I mightn't

'ave gone back to 'er after a bit. I thought if he didn't give up the treasure I might 'ave a sort of 'old on 'im. . . . Well, one day I looks as usual under Colchester—and there I saw 'is name. What for d'yer think?"

I could not guess.

Mr. Brisher's voice sank to a whisper, and once more he spoke behind his hand. His manner was suddenly suffused with a positive joy. "Issuing counterfeit coins," he said. "Counterfeit coins!"

"You don't mean to say——?"

"Yes—It. Bad. Quite a long case they made of it. But they got 'im, though he dodged tremenjous. Traced 'is 'aving passed, oh!—nearly a dozen bad 'arf-crowns."

"And you didn't——?"

"No fear. And it didn't do 'im much good to say it was treasure trove."

STORY THE
TWELFTH

Miss Winchelsea's Heart

MISS WINCHELSEA was going to Rome. The matter had
filled her mind for a month or more, and had overflowed
so abundantly into her conversation that quite a number
of people who were not going to Rome, and who were
not likely to go to Rome, had made it a personal grievance
against her. Some indeed had attempted quite unavailingly
to convince her that Rome was not nearly such a desirable
place as it was reported to be, and others had gone so far
as to suggest behind her back that she was dreadfully "stuck
up" about "that Rome of hers." And little Lily Hardhurst
had told her friend Mr. Binns that so far as she was concerned Miss Winchelsea might "go to her old Rome and
stop there; *she* (Miss Lily Hardhurst) wouldn't grieve." And
the way in which Miss Winchelsea put herself upon terms
of personal tenderness with Horace and Benvenuto Cellini
and Raphael and Shelley and Keats—if she had been
Shelley's widow she could not have professed a keener
interest in his grave—was a matter of universal astonishment. Her dress was a triumph of tactful discretion, sensible but not too "touristy"—Miss Winchelsea had a great
dread of being "touristy"—and her Baedeker was carried
in a cover of grey to hide its glaring red. She made a prim
and pleasant little figure on the Charing Cross platform,
in spite of her swelling pride, when at last the great day
dawned and she could start for Rome. The day was bright,
the Channel passage would be pleasant, and all the omens

promised well. There was the gayest scene of adventure in this unprecedented departure.

She was going with two friends who had been fellow-students with her at the training college, nice honest girls both, though not so good at history and literature as Miss Winchelsea. They both looked up to her immensely, though physically they had to look down, and she anticipated some pleasant times to be spent in "stirring them up" to her own pitch of æsthetic and historical enthusiasm. They had secured seats already, and welcomed her effusively at the carriage door. In the instant criticism of the encounter she noted that Fanny had a slightly "touristy" leather strap, and that Helen had succumbed to a serge jacket with side pockets, into which her hands were thrust. But they were much too happy with themselves and the expedition for their friend to attempt any hint at the moment about these things. As soon as the first ecstasies were over—Fanny's enthusiasm was a little noisy and crude, and consisted mainly in emphatic repetitions of "Just *fancy!* we're going to Rome, my dear!—Rome!"—they gave their attention to their fellow-travellers. Helen was anxious to secure a compartment to themselves, and, in order to discourage intruders, got out and planted herself firmly on the step. Miss Winchelsea peeped out over her shoulder, and made sly little remarks about the accumulating people on the platform, at which Fanny laughed gleefully.

They were travelling with one of Mr. Thomas Gunn's parties—fourteen days in Rome for fourteen pounds. They did not belong to the personally conducted party of course—Miss Winchelsea had seen to that—but they travelled with it because of the convenience of that arrangement. The people were the oddest mixture, and wonderfully amusing. There was a vociferous red-faced polyglot personal conductor in a pepper and salt suit, very long in the arms and legs and very active. He shouted proclamations. When he wanted to speak to people he stretched out an arm and held them until his purpose was accomplished. One hand was full of papers, tickets, counterfoils of tourists. The people of the personally conducted party were, it seemed, of two sorts; people the conductor wanted and could not find, and people he did not want and who followed him in a steadily growing tail up and down the platform. These people seemed, indeed, to think that their one chance of reaching Rome lay in keeping close to him. Three little old ladies were particularly energetic in his pursuit, and at last maddened him to the pitch of clapping them into a carriage and daring them to emerge again. For the rest of the time, one, two, or three of their heads protruded from the window wailing enquiries about "a little wicker-work box" whenever he drew near. There was a very stout man with a very stout wife in shiny black; there was a little old man like an aged ostler.

"What *can* such people want in Rome?" asked Miss Winchelsea. "What can it mean to them?" There was a tall curate in a very small straw hat, and a short curate encumbered by a long camera stand. The contrast amused Fanny very much. Once they heard someone calling for "Snooks." "I always thought that name was invented by novelists," said Miss Winchelsea. "Fancy! Snooks. I wonder which *is* Mr. Snooks." Finally they picked out a stout and resolute little man in a large check suit. "If he isn't Snooks, he ought to be," said Miss Winchelsea.

Presently the conductor discovered Helen's attempt at a corner in carriages. "Room for five," he bawled with a parallel translation on his fingers. A party of four together —mother, father, and two daughters—blundered in, all greatly excited. "It's all right, Ma—you let *me*," said one of the daughters, hitting her mother's bonnet with a handbag she struggled to put in the rack. Miss Winchelsea detested people who banged about and called their mother "Ma." A young man travelling alone followed. He was not at all "touristy" in his costume, Miss Winchelsea observed; his Gladstone bag was of good pleasant leather with labels reminiscent of Luxembourg and Ostend, and his boots, though brown, were not vulgar. He carried an overcoat on his arm. Before these people had properly settled in their places, came an inspection of tickets and a slamming of doors, and behold! they were gliding out of Charing Cross station on their way to Rome.

"Fancy!" cried Fanny, "we are going to Rome, my dear! Rome! I don't seem to believe it, even now."

Miss Winchelsea suppressed Fanny's emotions with a little smile, and the lady who was called "Ma" explained to people in general why they had "cut it so close" at the station. The two daughters called her "Ma" several times, toned her down in a tactless effective way, and drove her at last to the muttered inventory of a basket of travelling requisites. Presently she looked up. "Lor!" she said, "I didn't bring *them!*" Both the daughters said "Oh, Ma!" but what "them" was did not appear. Presently Fanny produced Hare's *Walks in Rome,* a sort of mitigated guide-book very popular among Roman visitors; and the father of the two daughters began to examine his books of tickets minutely, apparently in a search after English words. When he had looked at the tickets for a long time right way up, he turned them upside down. Then he produced a fountain pen and dated them with considerable care. The young man having completed an unostentatious survey of his fellow travellers produced a book and fell to reading. When Helen and Fanny were looking out of the window at Chislehurst—the place interested Fanny because the poor dear Empress of the French used to live there—Miss Winchelsea took the opportunity to observe the book the young man held. It was not a guide-book but a thin volume of poetry—*bound.* She glanced at his face—it seemed a refined pleasant face to her hasty glance. He wore a gilt *pince-nez.* "Do you think she lives there now?" said Fanny, and Miss Winchelsea's inspection came to an end.

For the rest of the journey Miss Winchelsea talked little, and what she said was as pleasant and as stamped with refinement as she could make it. Her voice was always low and clear and pleasant, and she took care that on this occasion it was particularly low and clear and pleasant. As they came under the white cliffs the young man put his book of poetry away, and when at last the train stopped beside the boat, he displayed a graceful alacrity with the impedimenta of Miss Winchelsea and her friends. Miss Winchelsea "hated nonsense," but she was pleased to see the young man perceived at once that they were ladies, and helped them without any violent geniality; and how nicely he showed that his civilities were to be no excuse for further intrusions. None of her party had been out of England before, and they were all excited and nervous at the Channel passage. They stood in a little group in a good place near the middle of the boat—the young man had taken Miss Winchelsea's hold-all there and had told her it was a good place—and they watched the white shores of Albion recede and quoted Shakespeare and made quiet fun of their fellow travellers in the English way.

They were particularly amused at the precautions the bigger-sized people had taken against the waves—cut lemons and flasks prevailed, one lady lay full length in a deck chair with a handkerchief over her face, and a very broad resolute man in a bright brown "touristy" suit walked all the way from England to France along the deck, with his legs as widely apart as Providence permitted. These were all excellent precautions, and nobody was ill. The personally conducted party pursued the conductor about the deck with enquiries, in a manner that suggested to Helen's mind the rather vulgar image of hens with a piece of bacon peel, until at last he went into hiding below. And the young man with the thin volume of poetry stood in the stern watching England receding, looking, to Miss Winchelsea's eye, rather lonely and sad.

And then came Calais and tumultuous novelties, and the young man had not forgotten Miss Winchelsea's hold-all and the other little things. All three girls, though they had passed government examinations in French to any extent, were stricken with a dumb shame of their accents, and the young man was very useful. And he did not intrude. He put them in a comfortable carriage and raised his hat and went away. Miss Winchelsea thanked him in her best manner—a pleasing cultivated manner—and Fanny said he was "nice" almost before he was out of earshot. "I wonder what he can be," said Helen. "He's going to

Italy, because I noticed green tickets in his book." Miss Winchelsea almost told them of the poetry, and decided not to do so. And presently the carriage windows seized hold upon them and the young man was forgotten. It made them feel that they were doing an educated sort of thing to travel through a country whose commonest advertisements were in idiomatic French, and Miss Winchelsea made unpatriotic comparisons because there were weedy little sign-board advertisements by the rail side instead of the broad hoardings that deface the landscape in our land. But the north of France is really uninteresting country, and after a time Fanny reverted to Hare's *Walks* and Helen initiated lunch. Miss Winchelsea awoke out of a happy reverie; she had been trying to realise, she said, that she was actually going to Rome, but she perceived at Helen's suggestion that she was hungry, and they lunched out of their baskets very cheerfully. In the afternoon they were tired and silent until Helen made tea. Miss Winchelsea might have dozed, only she knew Fanny slept with her mouth open; and as their fellow passengers were two rather nice critical-looking ladies of uncertain age—who knew French well enough to talk it—she employed herself in keeping Fanny awake. The rhythm of the train became insistent, and the streaming landscape outside at last quite painful to the eye. Before their night's stoppage came they were already dreadfully tired of travelling.

The stoppage for the night was brightened by the appearance of the young man, and his manners were all that could be desired and his French quite serviceable. His coupons availed for the same hotel as theirs, and by chance as it seemed he sat next Miss Winchelsea at the *table d'hôte*. In spite of her enthusiasm for Rome, she had thought out some such possibility very thoroughly, and when he ventured to make a remark upon the tediousness of travelling—he let the soup and fish go by before he did this—she did not simply assent to his proposition, but responded with another. They were soon comparing their journeys, and Helen and Fanny were cruelly overlooked in the conversation. It was to be the same journey, they found; one day for the galleries at Florence—"from what I hear," said the young man, "it is barely enough,"—and the rest at Rome. He talked of Rome very pleasantly; he was evidently quite well read, and he quoted Horace about Soracte. Miss Winchelsea had "done" that book of Horace for her matriculation, and was delighted to cap his quotation. It gave a sort of tone to things, this incident—a touch of refinement to mere chatting. Fanny expressed a few emotions, and Helen interpolated a few sensible remarks, but the bulk of the talk on the girls' side naturally fell to Miss Winchelsea.

Before they reached Rome this young man was tacitly of their party. They did not know his name nor what he was, but it seemed he taught, and Miss Winchelsea had a shrewd idea he was an extension lecturer. At any rate he was something of that sort, something gentlemanly and refined without being opulent and impossible. She tried once or twice to ascertain whether he came from Oxford or Cambridge, but he missed her timid opportunities. She tried to get him to make remarks about those places to see if he would say "go up" to them instead of "go down"— she knew that was how you told a 'Varsity man. He used the word " 'Varsity"—not university—in quite the proper way.

They saw as much of Mr. Ruskin's Florence as their brief time permitted; the young man met them in the Pitti Gallery and went round with them, chatting brightly, and evidently very grateful for their recognition. He knew a great deal about art, and all four enjoyed the morning immensely. It was fine to go round recognising old favourites and finding new beauties, especially while so many people fumbled helplessly with Baedeker. Nor was he a bit of a prig, Miss Winchelsea said, and indeed she detested prigs. He had a distinct undertow of humour, and was funny, for example, without being vulgar, at the expense of the quaint work of Beato Angelico. He had a grave seriousness beneath it all, and was quick to seize the moral lessons of the pictures. Fanny went softly among these masterpieces; she admitted "she knew so little about them," and she confessed that to her they were "all beautiful." Fanny's "beautiful" inclined to be a little monotonous, Miss Winchelsea thought. She had been quite glad when the last sunny Alp had vanished, because of the staccato of Fanny's admiration. Helen said little, but Miss Winchelsea had found her a little wanting on the æsthetic side in the old days and was not surprised; sometimes she laughed at the young man's hesitating delicate little jests and sometimes she didn't, and sometimes she seemed quite lost to the art about them in the contemplation of the dresses of the other visitors.

At Rome the young man was with them intermittently. A rather "touristy" friend of his took him away at times. He complained comically to Miss Winchelsea. "I have only two short weeks in Rome," he said, "and my friend Leonard wants to spend a whole day at Tivoli looking at a waterfall."

"What is your friend Leonard?" asked Miss Winchelsea abruptly.

"He's the most enthusiastic pedestrian I ever met," the young man replied—amusingly, but a little unsatisfactorily, Miss Winchelsea thought.

They had some glorious times, and Fanny could not think what they would have done without him. Miss Winchelsea's interest and Fanny's enormous capacity for admiration were insatiable. They never flagged—through pictures and sculpture galleries, immense crowded churches, ruins and museums, Judas trees and prickly pears, wine carts and palaces, they admired their way unflinchingly. They never saw a stone pine nor a eucalyptus but they named and admired it; they never glimpsed Soracte but they exclaimed. Their common ways were made wonderful by imaginative play. "Here Cæsar may have walked," they would say. "Raphael may have seen Soracte from this very point." They happened on the tomb of Bibulus. "Old Bibulus," said the young man. "The oldest monument of Republican Rome!" said Miss Winchelsea.

"I'm dreadfully stupid," said Fanny, "but who *was* Bibulus?"

There was a curious little pause.

"Wasn't he the person who built the wall?" said Helen. The young man glanced quickly at her and laughed. "That was Balbus," he said. Helen reddened, but neither he nor Miss Winchelsea threw any light upon Fanny's ignorance about Bibulus.

Helen was more taciturn than the other three, but then she was always taciturn; and usually she took care of the tram tickets and things like that, or kept her eye on them if the young man took them, and told him where they were when he wanted them. Glorious times they had, these young people, in that pale brown cleanly city of memories that was once the world. Their only sorrow was the short-ness of the time. They said indeed that the electric trams and the '70 buildings, and that criminal advertise-ment that glares upon the Forum, outraged their æsthetic feelings unspeakably; but that was only part of the fun. And indeed Rome is such a wonderful place that at times it made Miss Winchelsea forget some of her most carefully prepared enthusiasms, and Helen, taken unawares, would suddenly admit the beauty of unexpected things. Yet Fanny and Helen would have liked a shop window or so in the English quarter if Miss Winchelsea's uncompromis-ing hostility to all other English visitors had not rendered that district impossible.

The intellectual and æsthetic fellowship of Miss Win-chelsea and the scholarly young man passed insensibly towards a deeper feeling. The exuberant Fanny did her best to keep pace with their recondite admiration by playing her "beautiful" with vigour, and saying "Oh! *let's* go," with enormous appetite whenever a new place of interest was mentioned. But Helen towards the end developed a certain want of sympathy, that disappointed Miss Win-chelsea a little. She refused to "see anything" in the face of Beatrice Cenci—Shelley's Beatrice Cenci!—in the Bar-berini gallery; and one day, when they were deploring the electric trams, she said rather snappishly that "people must get about somehow, and it's better than torturing horses up these horrid little hills." She spoke of the Seven Hills of Rome as "horrid little hills!"

And the day they went on the Palatine—though Miss Winchelsea did not know of this—she remarked suddenly to Fanny, "Don't hurry like that, my dear; *they* don't want us to overtake them. And we don't say the right things for them when we *do* get near."

"I wasn't trying to overtake them," said Fanny, slackening her excessive pace; "I wasn't indeed." And for a minute she was short of breath.

But Miss Winchelsea had come upon happiness. It was only when she came to look back across an intervening tragedy that she quite realised how happy she had been, pacing among the cypress-shadowed ruins, and exchanging the very highest class of information the human mind can possess, the most refined impressions it is possible to convey. Insensibly emotion crept into their intercourse, sunning itself openly and pleasantly at last when Helen's modernity was not too near. Insensibly their interest drifted from the wonderful associations about them to their more intimate and personal feelings. In a tentative way information was supplied; she spoke allusively of her school, of her examination successes, of her gladness that the days of "Cram" were over. He made it quite clear that he also was a teacher. They spoke of the greatness of their calling, of the necessity of sympathy to face its irksome details, of a certain loneliness they sometimes felt.

That was in the Colosseum, and it was as far as they got that day, because Helen returned with Fanny—she had taken her into the upper galleries. Yet the private dreams of Miss Winchelsea, already vivid and concrete enough, became now realistic in the highest degree. She figured that pleasant young man, lecturing in the most edifying way to his students, herself modestly prominent as his intellectual mate and helper; she figured a refined little home, with two bureaus, with white shelves of high-class books, and autotypes of the pictures of Rossetti and Burne-Jones, with Morris's wall papers and flowers in pots of beaten copper. Indeed she figured many things. On the Pincio the two had a few precious moments together, while Helen marched Fanny off to see the *muro Torto,* and he spoke at once plainly. He said he hoped their friendship was only beginning, that he already found her company very precious to him, that indeed it was more than that.

He became nervous, thrusting at his glasses with trembling fingers as though he fancied his emotions made them unstable. "I should of course," he said, "tell you things about myself. I know it is rather unusual my speaking to you like this. Only our meeting has been so accidental —or providential—and I am snatching at things. I came to Rome expecting a lonely tour . . . and I have been so very happy, so very happy. Quite recently I have found myself in a position—I have dared to think—— And——"

He glanced over his shoulder and stopped. He said "Demn!" quite distinctly—and she did not condemn him for that manly lapse into profanity. She looked and saw his friend Leonard advancing. He drew nearer; he raised his hat to Miss Winchelsea, and his smile was almost a grin. "I've been looking for you everywhere, Snooks," he said. "You promised to be on the Piazza steps half an hour ago."

Snooks! The name struck Miss Winchelsea like a blow in the face. She did not hear his reply. She thought afterwards that Leonard must have considered her the vaguest-minded person. To this day she is not sure whether she was introduced to Leonard or not, nor what she said to him. A sort of mental paralysis was upon her. Of all offensive surnames—Snooks!

Helen and Fanny were returning, there were civilities and the young men were receding. By a great effort she controlled herself to face the inquiring eyes of her friends. All that afternoon she lived the life of a heroine under the indescribable outrage of that name, chatting, observing, with "Snooks" gnawing at her heart. From the moment that it first rang upon her ears, the dream of her happiness was prostrate in the dust. All the refinement she had figured was ruined and defaced by that cognomen's inexorable vulgarity.

What was that refined little home to her now, spite of autotypes, Morris papers, and bureaus? Athwart it in

letters of fire ran an incredible inscription: "Mrs. Snooks."
That may seem a small thing to the reader, but consider
the delicate refinement of Miss Winchelsea's mind. Be
as refined as you can and then think of writing yourself
down: "S n o o k s." She conceived herself being addressed
as Mrs. Snooks by all the people she liked least, conceived
the patronymic touched with a vague quality of insult. She
figured a card of grey and silver bearing "Winchelsea"
triumphantly effaced by an arrow, Cupid's arrow, in favour
of "Snooks." Degrading confession of feminine weakness!
She imagined the terrible rejoicings of certain girl friends,
of certain grocer cousins from whom her growing refinement
had long since estranged her. How they would make it
sprawl across the envelope that would bring their sarcastic
congratulations. Would even his pleasant company com-
pensate her for that? "It is impossible," she muttered;
"impossible! *Snooks!*"

She was sorry for him, but not so sorry as she was for
herself. For him she had a touch of indignation. To be
so nice, so refined, while all the time he was "Snooks," to
hide under a pretentious gentility of demeanour the badge
sinister of his surname seemed a sort of treachery. To
put it in the language of sentimental science she felt he had
"led her on."

There were of course moments of terrible vacillation,
a period even when something almost like passion bid her
throw refinement to the winds. And there was something
in her, an unexpurgated vestige of vulgarity that made a
strenuous attempt at proving that Snooks was not so very
bad a name after all. Any hovering hesitation flew before
Fanny's manner, when Fanny came with an air of catastrophe to tell that she also knew the horror. Fanny's voice
fell to a whisper when she said *Snooks*. Miss Winchelsea
would not give him any answer when at last, in the Borghese,
she could have a minute with him; but she promised him
a note.

She handed him that note in the little book of poetry
he had lent her, the little book that had first drawn them
together. Her refusal was ambiguous, allusive. She could

no more tell him why she rejected him than she could have told a cripple of his hump. He too must feel something of the unspeakable quality of his name. Indeed he had avoided a dozen chances of telling it, she now perceived. So she spoke of "obstacles she could not reveal"—"reasons why the thing he spoke of was impossible." She addressed the note with a shiver, "E. K. Snooks."

Things were worse than she had dreaded; he asked her to explain. How *could* she explain? Those last two days in Rome were dreadful. She was haunted by his air of astonished perplexity. She knew she had given him intimate hopes, she had not the courage to examine her mind thoroughly for the extent of her encouragement. She knew he must think her the most changeable of beings. Now that she was in full retreat, she would not even perceive his hints of a possible correspondence. But in that matter he did a thing that seemed to her at once delicate and romantic. He made a go-between of Fanny. Fanny could not keep the secret, and came and told her that night under a transparent pretext of needed advice. "Mr. Snooks," said Fanny, "wants to write to me. Fancy! I had no idea. But should I let him?" They talked it over long and earnestly, and Miss Winchelsea was careful to keep the veil over her heart. She was already repenting his disregarded hints. Why should she not hear of him sometimes—painful though his name must be to her? Miss Winchelsea decided it might be permitted, and Fanny kissed her good-night with unusual emotion. After she had gone Miss Winchelsea sat for a long time at the window of her little room. It was moonlight, and down the street a man sang "Santa Lucia" with almost heart-dissolving tenderness. . . . She sat very still.

She breathed a word very softly to herself. The word was "*Snooks.*" Then she got up with a profound sigh, and went to bed. The next morning he said to her meaningly, "I shall hear of you through your friend."

Mr. Snooks saw them off from Rome with that pathetic interrogative perplexity still on his face, and if it had not been for Helen he would have retained Miss Winchelsea's hold-all in his hand as a sort of encyclopædic keepsake.

On their way back to England Miss Winchelsea on six separate occasions made Fanny promise to write to her the longest of long letters. Fanny, it seemed, would be quite near Mr. Snooks. Her new school—she was always going to new schools—would be only five miles from Steely Bank, and it was in the Steely Bank Polytechnic, and one or two first-class schools, that Mr. Snooks did his teaching. He might even see her at times. They could not talk much of him—she and Fanny always spoke of "him," never of Mr. Snooks—because Helen was apt to say unsympathetic things about him. Her nature had coarsened very much, Miss Winchelsea perceived, since the old Training College days; she had become hard and cynical. She thought he had a weak face, mistaking refinement for weakness as people of her stamp are apt to do, and when she heard his name was Snooks, she said she had expected something of the sort. Miss Winchelsea was careful to spare her own feelings after that, but Fanny was less circumspect.

The girls parted in London, and Miss Winchelsea returned with a new interest in life, to the Girls' High School in which she had been an increasingly valuable assistant for the last three years. Her new interest in life was Fanny as a correspondent, and to give her a lead she wrote her a lengthy descriptive letter within a fortnight of her return. Fanny answered very disappointingly. Fanny indeed had no literary gift, but it was new to Miss Winchelsea to find herself deploring the want of gifts in a friend. That letter was even criticised aloud in the safe solitude of Miss Winchelsea's study, and her criticism, spoken with great bitterness, was "Twaddle!" It was full of just the things Miss Winchelsea's letter had been full of, particulars of the school. And of Mr. Snooks, only this much: "I have had a letter from Mr. Snooks, and he has been over to see me on two Saturday afternoons running. He talked about Rome and you; we both talked about you. Your ears must have burnt, my dear. . . ."

Miss Winchelsea repressed a desire to demand more explicit information, and wrote the sweetest long letter again.

"Tell me all about yourself, dear. That journey has quite refreshed our ancient friendship, and I do so want to keep in touch with you." About Mr. Snooks she simply wrote on the fifth page that she was glad Fanny had seen him, and that if he *should* ask after her, she was to be remembered to him *very kindly* (underlined). And Fanny replied most obtusely in the key of that "ancient friendship," reminding Miss Winchelsea of a dozen foolish things of those old schoolgirl days at the training college, and saying not a word about Mr. Snooks!

For nearly a week Miss Winchelsea was so angry at the failure of Fanny as a go-between that she could not write to her. And then she wrote less effusively, and in her letter she asked point blank, "Have you seen Mr. Snooks?" Fanny's letter was unexpectedly satisfactory. "I *have* seen Mr. Snooks," she wrote, and having once named him she kept on about him; it was all Snooks—Snooks this and Snooks that. He was to give a public lecture, said Fanny, among other things. Yet Miss Winchelsea, after the first glow of gratification, still found this letter a little unsatisfactory. Fanny did not report Mr. Snooks as saying anything about Miss Winchelsea, nor as looking white and worn, as he ought to have been doing. And behold! before she had replied, came a second letter from Fanny on the same theme, quite a gushing letter, and covering six sheets with her loose feminine hand.

And about this second letter was a rather odd little thing that Miss Winchelsea only noticed as she re-read it the third time. Fanny's natural femininity had prevailed even against the round and clear traditions of the training college; she was one of those she-creatures born to make all her *m*'s and *n*'s and *u*'s and *r*'s and *e*'s alike, and to leave her *o*'s and *a*'s open and her *i*'s undotted. So that it was only after an elaborate comparison of word with word that Miss Winchelsea felt assured Mr. Snooks was not really "Mr. Snooks" at all! In Fanny's first letter of gush he was Mr. "Snooks," in her second the spelling was changed to Mr. "Senoks." Miss Winchelsea's hand positively trembled as she turned the sheet over—it meant so much to her. For it had already begun to seem to her that even the name of Mrs. Snooks might be avoided at too great a price, and suddenly—this possibility! She turned over the six sheets, all dappled with that critical name, and everywhere the first letter had the form of an e! For a time she walked the room with a hand pressed upon her heart.

She spent a whole day pondering this change, weighing a letter of inquiry that should be at once discreet and effectual, weighing too what action she should take after the answer came. She was resolved that if this altered spelling was anything more than a quaint fancy of Fanny's, she would write forthwith to Mr. Snooks. She had now reached a stage when the minor refinements of behaviour disappear. Her excuse remained uninvented but she had the subject of her letter clear in her mind, even to the hint that "circumstances in my life have changed very greatly since we talked together." But she never gave that hint. There came a third letter from that fitful correspondent Fanny. The first line proclaimed her "the happiest girl alive."

Miss Winchelsea crushed the letter in her hand—the rest unread—and sat with her face suddenly very still. She had received it just before morning school, and had opened it when the junior mathematicians were well under way. Presently she resumed reading with an appearance of great calm. But after the first sheet she went on reading the third without discovering the error: "told him frankly I did not like his name," the third sheet began. "He told me he did not like it himself—you know that sort of sudden frank way he has"—Miss Winchelsea did know. "So I said, 'Couldn't you change it?' He didn't see it at first. Well, you know, dear, he had told me what it really meant; it means Sevenoaks, only it has got down to Snooks—both Snooks and Noaks, dreadfully vulgar surnames though they be, are really worn forms of Sevenoaks. So I said— even I have my bright ideas at times—'if it got down from Sevenoaks to Snooks, why not get it back from Snooks to Sevenoaks?' And the long and the short of it is, dear, he couldn't refuse me, and he changed his spelling there and then to Senoks for the bills of the new lecture. And afterwards, when we are married, we shall put in the apostrophe and make it Se'noks. Wasn't it kind of him to mind that fancy of mine, when many men would have taken offence? But it is just like him all over; he is as kind as he is clever. Because he knew as well as I did that I would have had him in spite of it, had he been ten times Snooks. But he did it all the same."

The class was startled by the sound of paper being viciously torn, and looked up to see Miss Winchelsea white in the face, and with some very small pieces of paper clenched in one hand. For a few second they stared at her stare, and then her expression changed back to a more familiar one. "Has anyone finished number three?" she asked in an even tone. She remained calm after that. But impositions ruled high that day. And she spent two laborious evenings writing letters of various sorts to Fanny, before she found a decent congratulatory vein. Her reason struggled hopelessly against the persuasion that Fanny had behaved in an exceedingly treacherous manner.

One may be extremely refined and still capable of a very sore heart. Certainly Miss Winchelsea's heart was very sore. She had moods of sexual hostility, in which she generalised uncharitably about mankind. "He forgot himself with me," she said. "But Fanny is pink and pretty and soft and a fool—a very excellent match for a Man." And by way of a wedding present she sent Fanny a gracefully bound volume of poetry by George Meredith, and Fanny wrote back a grossly happy letter to say that it was "*all* beautiful." Miss Winchelsea hoped that some day Mr. Senoks might take up that slim book and think for a moment of the donor. Fanny wrote several times before and about her marriage, pursuing that fond legend of their "ancient friendship," and giving her happiness in the fullest detail. And Miss Winchelsea wrote to Helen for the first time after the Roman journey, saying nothing about the marriage, but expressing very cordial feelings.

They had been in Rome at Easter, and Fanny was married in the August vacation. She wrote a garrulous letter to Miss Winchelsea, describing her home-coming, and the astonishing arrangements of their "teeny weeny" little house. Mr. Se'noks was now beginning to assume a refinement in Miss Winchelsea's memory out of all proportion to the facts of the case, and she tried in vain to imagine his cultured greatness in a "teeny weeny" little house. "Am busy enamelling a cosy corner," said Fanny, sprawling to the end of her third sheet, "so excuse more." Miss Winchelsea answered in her best style, gently poking fun at Fanny's arrangements, and hoping intensely that Mr. Se'noks might see the letter. Only this hope enabled her to write at all, answering not only that letter but one in November and one at Christmas.

The two latter communications contained urgent invitations for her to come to Steely Bank on a visit during the Christmas holidays. She tried to think that *he* had told her to ask that, but it was too much like Fanny's opulent goodnature. She could not but believe that he must be sick of his blunder by this time; and she had more than a hope that he would presently write her a letter beginning "Dear Friend." Something subtly tragic in the separation was a great support to her, a sad misunderstanding. To have been jilted would have been intolerable. But he never wrote that letter beginning "Dear Friend."

For two years Miss Winchelsea could not go to Steely Bank, in spite of the reiterated invitations of Mrs. Sevenoaks —it became full Sevenoaks in the second year. Then one day near the Easter rest she felt lonely and without a soul to understand her in the world, and her mind ran once more on what is called Platonic friendship. Fanny was clearly happy and busy in her new sphere of domesticity, but no doubt *he* had his lonely hours. Did he ever think of those days in Rome—gone now beyond recalling. No one had understood her as he had done; no one in all the world. It would be a sort of melancholy pleasure to talk to him again, and what harm could it do? Why should she deny herself? That night she wrote a sonnet, all but the last two lines of the octave—which would not come, and the next day she composed a graceful little note to tell Fanny she was coming down.

And so she saw him again.

Even at the first encounter it was evident he had changed; he seemed stouter and less nervous, and it speedily appeared that his conversation had already lost much of its old delicacy. There even seemed a justification for Helen's discovery of weakness in his face—in certain lights it *was* weak. He seemed busy and pre-occupied about his affairs, and almost under the impression that Miss Winchelsea had come for the sake of Fanny. He discussed his dinner with Fanny in an intelligent way. They only had one good long talk together, and that came to nothing. He did not refer to Rome, and spent some time abusing a man who had stolen an idea he had had for a text-book. It did not seem a very wonderful idea to Miss Winchelsea. She discovered he had forgotten the names of more than half the painters whose work they had rejoiced over in Florence.

It was a sadly disappointing week, and Miss Winchelsea was glad when it came to an end. Under various excuses she avoided visiting them again. After a time the visitor's room was occupied by their two little boys, and Fanny's invitations ceased. The intimacy of her letters had long since faded away.

THE DREAM

A Dream of Armageddon

THE man with the white face entered the carriage at Rugby. He moved slowly in spite of the urgency of his porter, and even while he was still on the platform I noted how ill he seemed. He dropped into the corner over against me with a sigh, made an incomplete attempt to arrange his travelling shawl, and became motionless, with his eyes staring vacantly. Presently he was moved by a sense of my observation, looked up at me, and put out a spiritless hand for his newspaper. Then he glanced again in my direction.

I feigned to read. I feared I had unwittingly embarrassed him, and in a moment I was surprised to find him speaking.

"I beg your pardon?" said I.

"That book," he repeated, pointing a lean finger, "is about dreams."

"Obviously," I answered, for it was Fortnum-Roscoe's *Dream States*, and the title was on the cover.

He hung silent for a space as if he sought words. "Yes," he said at last, "but they tell you nothing."

I did not catch his meaning for a second.

"They don't know," he added.

I looked a little more attentively at his face.

"There are dreams," he said, "and dreams."

That sort of proposition I never dispute.

"I suppose——" he hesitated. "Do you ever dream? I mean vividly."

"I dream very little," I answered. "I doubt if I have three vivid dreams a year."

"Ah!" he said, and seemed for a moment to collect his thoughts.

"Your dreams don't mix with your memories?" he asked abruptly. "You don't find yourself in doubt; did this happen or did it not?"

"Hardly ever. Except just for a momentary hesitation now and then. I suppose few people do."

"Does *he* say——" he indicated the book.

"Says it happens at times and gives the usual explanation about intensity of impression and the like to account for its not happening as a rule. I suppose you know something of these theories——"

"Very little—except that they are wrong."

His emaciated hand played with the strap of the window for a time. I prepared to resume reading, and that seemed to precipitate his next remark. He leant forward almost as though he would touch me.

"Isn't there something called consecutive dreaming—that goes on night after night?"

"I believe there is. There are cases given in most books on mental trouble."

"Mental trouble! Yes. I dare say there are. It's the right place for them. But what I mean——" He looked at his bony knuckles. "Is that sort of thing always dreaming? Is it dreaming? Or is it something else? Mightn't it be something else?"

I should have snubbed his persistent conversation but for the drawn anxiety of his face. I remember now the look of his faded eyes and the lids red stained—perhaps you know that look.

"I'm not just arguing about a matter of opinion," he said. "The thing's killing me."

"Dreams?"

"If you call them dreams. Night after night. Vivid!—so vivid . . . this——" (he indicated the landscape that went streaming by the window) "seems unreal in comparison! I can scarcely remember who I am, what business I am on. . . ."

He paused. "Even now——"

"The dream is always the same—do you mean?" I asked.

"It's over."

"You mean?"

"I died."

"Died?"

"Smashed and killed, and now, so much of me as that dream was, is dead. Dead for ever. I dreamt I was another man, you know, living in a different part of the world and in a different time. I dreamt that night after night. Night after night I woke into that other life. Fresh scenes and fresh happenings—until I came upon the last——"

"When you died?"

"When I died."

"And since then——"

"No," he said. "Thank God! That was the end of the dream. . . ."

It was clear I was in for this dream. And after all, I had an hour before me, the light was fading fast, and Fortnum-Roscoe has a dreary way with him. "Living in a different time," I said: "do you mean in some different age?"

"Yes."

"Past?"

"No—to come—to come."

"The year three thousand, for example?"

"I don't know what year it was. I did when I was asleep, when I was dreaming, that is, but not now—not now that I am awake. There's a lot of things I have forgotten since I woke out of these dreams, though I knew them at the time when I was—I suppose it was dreaming. They called the year differently from our way of calling the year. . . . What *did* they call it?" He put his hand to his forehead. "No," said he, "I forget."

He sat smiling weakly. For a moment I feared he did not mean to tell me his dream. As a rule I hate people who tell their dreams, but this struck me differently. I proffered assistance even. "It began——" I suggested.

"It was vivid from the first. I seemed to wake up in it

suddenly. And it's curious that in these dreams I am speaking of I never remembered this life I am living now. It seemed as if the dream life was enough while it lasted. Perhaps—— But I will tell you how I find myself when I do my best to recall it all. I don't remember anything clearly until I found myself sitting in a sort of loggia looking out over the sea. I had been dozing, and suddenly I woke up—fresh and vivid—not a bit dreamlike—because the girl had stopped fanning me."

"The girl?"

"Yes, the girl. You must not interrupt or you will put me out."

He stopped abruptly. "You won't think I'm mad?" he said.

"No," I answered; "you've been dreaming. Tell me your dream."

"I woke up, I say, because the girl had stopped fanning me. I was not surprised to find myself there or anything of that sort, you understand. I did not feel I had fallen into it suddenly. I simply took it up at that point. Whatever memory I had of *this* life, this nineteenth-century life, faded as I woke, vanished like a dream. I knew all about myself, knew that my name was no longer Cooper but Hedon, and all about my position in the world. I've forgotten a lot since I woke—there's a want of connection—but it was all quite clear and matter of fact then."

He hesitated again, gripping the window strap, putting his face forward and looking up to me appealingly.

"This seems bosh to you?"

"No, no!" I cried. "Go on. Tell me what this loggia was like."

"It was not really a loggia—I don't know what to call it. It faced south. It was small. It was all in shadow except the semicircle above the balcony that showed the sky and sea and the corner where the girl stood. I was on a couch —it was a metal couch with light striped cushions—and the girl was leaning over the balcony with her back to me. The light of the sunrise fell on her ear and cheek. Her pretty white neck and the little curls that nestled there, and her white shoulder were in the sun, and all the grace of her body was in the cool blue shadow. She was dressed—how can I describe it? It was easy and flowing. And altogether there she stood, so that it came to me how beautiful and desirable she was, as though I had never seen her before. And when at last I sighed and raised myself upon my arm she turned her face to me——"

He stopped.

"I have lived three-and-fifty years in this world. I have had mother, sisters, friends, wife and daughters—all their faces, the play of their faces, I know. But the face of this girl—it is much more real to me. I can bring it back into memory so that I see it again—I could draw it or paint it. And after all——"

He stopped—but I said nothing.

"The face of a dream—the face of a dream. She was beautiful. Not that beauty which is terrible, cold, and worshipful, like the beauty of a saint; nor that beauty that stirs fierce passions; but a sort of radiation, sweet lips that softened into smiles, and grave grey eyes. And she moved gracefully, she seemed to have part with all pleasant and gracious things——"

He stopped, and his face was downcast and hidden. Then he looked up at me and went on, making no further attempt to disguise his absolute relief in the reality of his story.

"You see, I had thrown up my plans and ambitions, thrown up all I had ever worked for or desired for her sake. I had been a master man away there in the north, with influence and property and a great reputation, but none of it had seemed worth having beside her. I had come to the place, this city of sunny pleasures, with her, and left all those things to wreck and ruin just to save a remnant at least of my life. While I had been in love with her before I knew that she had any care for me, before I had imagined that she would dare—that we should dare, all my life had seemed vain and hollow, dust and ashes. It *was* dust and ashes. Night after night and through the long days I had longed and desired—my soul had beaten against the thing forbidden!

"But it is impossible for one man to tell another just these things. It's emotion, it's a tint, a light that comes and goes. Only while it's there, everything changes, everything. The thing is I came away and left them in their Crisis to do what they could."

"Left whom?" I asked, puzzled.

"The people up in the north there. You see—in this dream, anyhow—I had been a big man, the sort of man men come to trust in, to group themselves about. Millions of men who had never seen me were ready to do things and risk things because of their confidence in me. I had been playing that game for years, that big laborious game, that vague, monstrous political game amidst intrigues and betrayals, speech and agitation. It was a vast weltering world, and at last I had a sort of leadership against the Gang—you know it was called the Gang—a sort of compromise of scoundrelly projects and base ambitions and vast public emotional stupidities and catch-words—the Gang that kept the world noisy and blind year by year, and all the while that it was drifting, drifting towards infinite disaster. But I can't expect you to understand the shades and complications of the year—the year something or other ahead. I had it all—down to the smallest details—in my dream. I suppose I had been dreaming of it before I awoke, and the fading outline of some queer new development I had imagined still hung about me as I rubbed my eyes. It was some grubby affair that made me thank God for the sunlight. I sat up on the couch and remained looking at the woman and rejoicing —rejoicing that I had come away out of all that tumult and folly and violence before it was too late. After all, I thought, this is life—love and beauty, desire and delight, are they not worth all those dismal struggles for vague, gigantic ends. And I blamed myself for having ever sought to be a leader when I might have given my days to love. But then, thought I, if I had not spent my early days sternly and austerely, I might have wasted myself upon vain and worthless women, and at the thought all my being went out in love and tenderness to my dear mistress, my dear lady, who had come at last and compelled me—compelled me by her invincible charm for me—to lay that life aside.

"'You are worth it,' I said, speaking without intending her to hear; 'you are worth it, my dearest one; worth pride and praise and all things. Love! to have *you* is worth them all together.' And at the murmur of my voice she turned about.

"'Come and see,' she cried—I can hear her now—'come and see the sunrise upon Monte Solaro.'

"I remember how I sprang to my feet and joined her at the balcony. She put a white hand upon my shoulder and pointed towards great mases of limestone, flushing, as it were, into life. I looked. But first I noted the sunlight on her face caressing the lines of her cheeks and neck. How can I describe to you the scene we had before us? We were at Capri——"

"I have been there," I said. "I have clambered up Monte Solaro and drunk *vero Capri*—muddy stuff like cider—at the summit."

"Ah!" said the man with the white face; "then perhaps you can tell me—you will know if this was indeed Capri. For in this life I have never been there. Let me describe it. We were in a little room, one of a vast multitude of little rooms, very cool and sunny, hollowed out of the limestone of a sort of cape, very high above the sea. The whole island, you know, was one enormous hotel, complex beyond explaining, and on the other side there were miles of floating hotels, and huge floating stages to which the flying machines came. They called it a pleasure city. Of course, there was none of that in your time—rather, I should say, *is* none of that *now*. Of course. Now!—yes.

"Well, this room of ours was at the extremity of the cape, so that one could see east and west. Eastward was a great cliff—a thousand feet high perhaps—coldly grey except for one bright edge of gold, and beyond it the Isle of the Sirens, and a falling coast that faded and passed into the hot sunrise. And when one turned to the west, distinct and near was a little bay, a scimitar of beach still in shadow. And out of that shadow rose Solaro straight and tall, flushed and golden crested, like a beauty throned, and the white moon was floating behind her in the sky. And before us from east to west stretched the many-tinted sea all dotted with sailing boats.

"To the eastward, of course, these little boats were grey and very minute and clear, but to the westward they were little boats of gold—shining gold—almost like little flames. And just below us was a rock with an arch worn through it. The blue sea-water broke to green and foam all round the rock, and a galley came gliding out of the arch."

"I know that rock," I said. "I was nearly drowned there. It is called the Faraglioni."

"I *Faraglioni?* Yes, *she* called it that," answered the man with the white face. "There was some story—but that——"

He put his hand to his forehead again. "No," he said, "I forget that story.

"Well, that is the first thing I remember, the first dream I had, that shaded room and the beautiful air and sky and that dear lady of mine, with her shining arms and her graceful robe, and how we sat and talked in half whispers to one another. We talked in whispers not because there was anyone to hear, but because there was still such a freshness of mind between us that our thoughts were a little frightened, I think, to find themselves at last in words. And so they went softly.

"Presently we were hungry and we went from our apartment, going by a strange passage with a moving floor, until we came to the great breakfast room—there was a fountain and music. A pleasant and joyful place it was, with its sunlight and splashing, and the murmur of plucked strings. And we sat and ate and smiled at one another, and I would not heed a man who was watching me from a table near by.

"And afterwards we went on to the dancing-hall. But I cannot describe that hall. The place was enormous—larger than any building you have ever seen—and in one

place there was the old gate of Capri, caught into the
wall of a gallery high overhead. Light girders, stems and
threads of gold, burst from the pillars like fountains,
streamed like an Aurora across the roof and interlaced,
like—like conjuring tricks. All about the great circle for
the dancers there were beautiful figures, strange dragons,
and intricate and wonderful grotesques bearing lights. The
place was inundated with artificial light that shamed the
newborn day. And as we went through the throng the
people turned about and looked at us, for all through the
world my name and face were known, and how I had
suddenly thrown up pride and struggle to come to this place.
And they looked also at the lady beside me, though half
the story of how at last she had come to me was unknown
or mistold. And few of the men who were there, I know,
but judged me a happy man, in spite of all the shame
and dishonour that had come upon my name.

"The air was full of music, full of harmonious scents,
full of the rhythm of beautiful motions. Thousands of
beautiful people swarmed about the hall, crowded the galleries, sat in a myriad recesses; they were dressed in splendid
colours and crowned with flowers; thousands danced about
the great circle beneath the white images of the ancient
gods, and glorious processions of youths and maidens came
and went. We two danced, not the dreary monotonies of
your days—of this time, I mean—but dances that were
beautiful, intoxicating. And even now I can see my lady
dancing—dancing joyously. She danced, you know, with a
serious face; she danced with a serious dignity, and yet she
was smiling at me and caressing me—smiling and caressing
with her eyes.

"The music was different," he murmured. "It went—I
cannot describe it; but it was infinitely richer and more
varied than any music that has ever come to me awake.

"And then—it was when we had done dancing—a man
came to speak to me. He was a lean, resolute man, very
soberly clad for that place, and already I had marked his
face watching me in the breakfasting hall, and afterwards
as we went along the passage I had avoided his eye. But

now, as we sat in an alcove, smiling at the pleasure of all the people who went to and fro across the shining floor, he came and touched me, and spoke to me so that I was forced to listen. And he asked that he might speak to me for a while apart.

"'No,' I said. 'I have no secrets from this lady. What do you want to tell me?'

"He said it was a trivial matter, or at least a dry matter, for a lady to hear.

"'Perhaps for me to hear,' said I.

"He glanced at her, as though almost he would appeal to her. Then he asked me suddenly if I had heard of a great and avenging declaration that Evesham had made. Now, Evesham had always before been the man next to myself in the leadership of that great party in the north. He was a forcible, hard, and tactless man, and only I had been able to control and soften him. It was on his account even more than my own, I think, that the others had been so dismayed at my retreat. So this question about what he had done re-awakened my old interest in the life I had put aside just for a moment.

"'I have taken no heed of any news for many days,' I said. 'What has Evesham been saying?'

"And with that the man began, nothing loth, and I must confess even I was struck by Evesham's reckless folly in the wild and threatening words he had used. And this messenger they had sent to me not only told me of Evesham's speech, but went on to ask counsel and to point out what need they had of me. While he talked, my lady sat a little forward and watched his face and mine.

"My old habits of scheming and organising re-asserted themselves. I could even see myself suddenly returning to the north, and all the dramatic effect of it. All that this man said witnessed to the disorder of the party indeed, but not to its damage. I should go back stronger than I had come. And then I thought of my lady. You see— how can I tell you? There were certain peculiarities of our relationship—as things are I need not tell you about

that—which would render her presence with me impossible.
I should have had to leave her; indeed, I should have had
to renounce her clearly and openly, if I was to do all that
I could do in the north. And the man knew *that,* even as
he talked to her and me, knew it as well as she did, that
my steps to duty were—first, separation, then abandonment.
At the touch of that thought my dream of a return was
shattered. I turned on the man suddenly, as he was
imagining his eloquence was gaining ground with me.

"'What have I to do with these things now?' I said.
'I have done with them. Do you think I am coquetting
with your people in coming here?'

"'No,' he said; 'but——'

"'Why cannot you leave me alone. I have done with
these things. I have ceased to be anything but a private
man.'

"'Yes,' he answered. 'But have you thought?—this talk
of war, these reckless challenges, these wild aggressions——'

"I stood up.

"'No,' I cried. 'I won't hear you. I took count of all
those things, I weighed them—and I have come away.'

"He seemed to consider the possibility of persistence.
He looked from me to where the lady sat regarding us.

"'War,' he said, as if he were speaking to himself, and
then turned slowly from me and walked away.

"I stood, caught in the whirl of thoughts his appeal had
set going.

"I heard my lady's voice.

"'Dear,' she said; 'but if they have need of you——'

"She did not finish her sentence, she let it rest there. I
turned to her sweet face, and the balance of my mood swayed
and reeled.

"'They want me only to do the thing they dare not do
themselves,' I said. 'If they distrust Evesham they must
settle with him themselves.'

"She looked at me doubtfully.

"'But war——' she said.

"I saw a doubt on her face that I had seen before, a

doubt of herself and me, the first shadow of the discovery that, seen strongly and completely, must drive us apart for ever.

"Now I was an older mind than hers, and I could sway her to this belief or that.

" 'My dear one,' I said, 'you must not trouble over these things. There will be no war. Certainly there will be no war. The age of wars is past. Trust me to know the justice of this case. They have no right upon me, dearest, and no one has a right upon me. I have been free to choose my life, and I have chosen this.'

" 'But *war*——,' she said.

"I sat down beside her. I put an arm behind her and took her hand in mine. I set myself to drive that doubt away—I set myself to fill her mind with pleasant things again. I lied to her, and in lying to her I lied also to myself. And she was only too ready to believe me, only too ready to forget.

"Very soon the shadow had gone again, and we were hastening to our bathing-place in the Grotta del Bove Marino, where it was our custom to bathe every day. We swam and splashed one another, and in that buoyant water I seemed to become something lighter and stronger than a man. And at last we came out dripping and rejoicing and raced among the rocks. And then I put on a dry bathing-dress, and we sat to bask in the sun, and presently I nodded, resting my head against her knee, and she put her hand upon my hair and stroked it softly and I dozed. And behold! as it were with the snapping of the string of a violin, I was awakening, and I was in my own bed in Liverpool, in the life of to-day.

"Only for a time I could not believe that all these vivid moments had been no more than the substance of a dream.

"In truth, I could not believe it a dream for all the sobering reality of things about me. I bathed and dressed as it were by habit, and as I shaved I argued why I of all men should leave the woman I loved to go back to fantastic politics in the hard and strenuous north. Even if Evesham did force the world back to war, what was that to me? I was a man with the heart of a man, and why should I feel the responsibility of a deity for the way the world might go?

"You know that is not quite the way I think about affairs, about my real affairs. I am a solicitor, you know, with a point of view.

"The vision was so real, you must understand, so utterly unlike a dream that I kept perpetually recalling trivial irrelevant details; even the ornament of a book-cover that lay on my wife's sewing-machine in the breakfast-room recalled with the utmost vividness the gilt line that ran about the seat in the alcove where I had talked with the messenger from my deserted party. Have you ever heard of a dream that had a quality like that?"

"Like——?"

"So that afterwards you remembered details you had forgotten."

I thought. I had never noticed the point before, but he was right.

"Never," I said. "That is what you never seem to do with dreams."

"No," he answered. "But that is just what I did. I am a solicitor, you must understand, in Liverpool, and I could not help wondering what the clients and business people I found myself talking to in my office would think if I told them suddenly I was in love with a girl who would be born a couple of hundred years or so hence, and worried about the politics of my great-great-great-grandchildren. I was chiefly busy that day negotiating a ninety-nine-year building lease. It was a private builder in a hurry, and we wanted to tie him in every possible way. I had an interview with him, and he showed a certain want of temper that sent me to bed still irritated. That night I had no dream. Nor did I dream the next night, at least, to remember.

"Something of that intense reality of conviction vanished. I began to feel sure it *was* a dream. And then it came again.

"When the dream came again, nearly four days later, it was very different. I think it certain that four days had also elapsed in the dream. Many things had happened in the north, and the shadow of them was back again between us, and this time it was not so easily dispelled. I began I know with moody musings. Why, in spite of all, should I go back, go back for all the rest of my days to toil and stress, insults and perpetual dissatisfaction, simply to save hundreds of millions of common people, whom I did not love, whom too often I could do no other than despise, from the stress and anguish of war and infinite misrule? And after all I might fail. *They* all sought their own narrow ends, and why should not I—why should not I also live as a man? And out of such thoughts her voice summoned me, and I lifted my eyes.

"I found myself awake and walking. We had come out above the Pleasure City, we were near the summit of Monte Solaro and looking towards the bay. It was the late afternoon and very clear. Far away to the left Ischia hung in a golden haze between sea and sky, and Naples was coldly white against the hills, and before us was Vesuvius with a tall and slender streamer feathering at last towards the south, and the ruins of Torre Annunziata and Castellamare glittering and near."

I interrupted suddenly: "You have been to Capri, of course?"

"Only in this dream," he said, "only in this dream. All across the bay beyond Sorrento were the floating palaces of the Pleasure City moored and chained. And northward were the broad floating stages that received the aeroplanes. Aeroplanes fell out of the sky every afternoon, each bringing its thousands of pleasure-seekers from the uttermost parts of the earth to Capri and its delights. All these things, I say, stretched below.

"But we noticed them only incidentally because of an unusual sight that evening had to show. Five war aeroplanes that had long slumbered useless in the distant arsenals of the Rhinemouth were manœuvring now in the eastward sky. Evesham had astonished the world by producing them and others, and sending them to circle here and there. It was the threat material in the great game of bluff he was playing, and it had taken even me by surprise. He was one of those incredibly stupid energetic people who seem sent by heaven to create disasters. His energy to the first glance seemed so wonderfully like capacity! But he had no imagination, no invention, only a stupid, vast, driving force of will, and a mad faith in his stupid idiot 'luck' to pull him through. I remember how we stood out upon the headland watching the squadron circling far away, and how I weighed the full meaning of the sight, seeing clearly the way things must go. And then even it was not too late. I might have gone back, I think, and saved the world. The people of the north would follow me, I knew, granted only that in one thing I respected their moral standards. The east and south would trust me as they would trust no other northern man. And I knew I had only to put it to her and she would have let me go. . . . Not because she did not love me!

"Only I did not want to go; my will was all the other way about. I had so newly thrown off the incubus of responsibility: I was still so fresh a renegade from duty that the daylight clearness of what I *ought* to do had no power at all to touch my will. My will was to live, to gather pleasures and make my dear lady happy. But though this sense of vast neglected duties had no power to draw me, it could make me silent and preoccupied, it robbed the days I had spent of half their brightness and roused me into dark meditations in the silence of the night. And as I stood and watched Evesham's aeroplanes sweep to and fro—those birds of infinite ill omen—she stood beside me watching me, perceiving the trouble indeed, but not perceiving it clearly—her eyes questioning my face, her expression shaded with perplexity. Her face was grey because the sunset was fading out of the sky. It was no fault of hers that she held me. She had asked me to go from her, and again in the night time and with tears she had asked me to go.

"At last it was the sense of her that roused me from my mood. I turned upon her suddenly and challenged her to race down the mountain slopes. 'No,' she said, as if I jarred with her gravity; but I was resolved to end that gravity, and made her run—no one can be very grey and sad who is out of breath—and when she stumbled I ran with my hand beneath her arm. We ran down past a couple of men, who turned back staring in astonishment at my behaviour—they must have recognised my face. And half-way down the slope came a tumult in the air, clang-clank, clang-clank, and we stopped, and presently over the hill-crest those war things came flying one behind the other."

The man seemed hesitating on the verge of a description. "What were they like?" I asked.

"They had never fought," he said. "They were just like our ironclads are nowadays; they had never fought. No one knew what they might do, with excited men inside them; few even cared to speculate. They were great driving things shaped like spear-heads without a shaft, with a propeller in the place of the shaft."

"Steel?"

"Not steel."

"Aluminium?"

"No, no, nothing of that sort. An alloy that was very common—as common as brass, for example. It was called —let me see——" He squeezed his forehead with the fingers of one hand. "I am forgetting everything," he said.

"And they carried guns?"

"Little guns, firing high explosive shells. They fired the guns backwards, out of the base of the leaf, so to speak, and rammed with the beak. That was the theory, you know, but they had never been fought. No one could tell exactly what was going to happen. And meanwhile I suppose it was very fine to go whirling through the air like a flight of young swallows, swift and easy. I guess the captains tried not to think too clearly what the real thing would be like. And these flying war machines, you know, were only one sort of the endless war contrivances that had been invented and had fallen into abeyance during the long peace. There were all sorts of these things that people were routing out and furbishing up; infernal things, silly things; things that had never been tried; big engines, terrible explosives, great guns. You know the silly way of the ingenious sort of men who make these things; they turn 'em out as beavers build dams, and with no more sense of the rivers they're going to divert and the lands they're going to flood!

"As we went down the winding stepway to our hotel again, in the twilight, I foresaw it all: I saw how clearly and inevitably things were driving for war in Evesham's silly, violent hands, and I had some inkling of what war was bound to be under these new conditions. And even then, though I knew it was drawing near the limit of my opportunity, I could find no will to go back."

He sighed.

"That was my last chance.

"We didn't go into the city until the sky was full of stars, so we walked out upon the high terrace, to and fro, and—she counselled me to go back.

" 'My dearest,' she said, and her sweet face looked up to me, 'this is Death. This life you lead is Death. Go back to them, go back to your duty——'

"She began to weep, saying, between her sobs, and clinging to my arm as she said it, 'Go back—Go back.'

"Then suddenly she fell mute, and, glancing down at her face, I read in an instant the thing she had thought to do. It was one of those moments when one *sees*.

" 'No!' I said.

" 'No?' she asked, in surprise, and I think a little fearful at the answer to her thought.

" 'Nothing,' I said, 'shall send me back. Nothing! I have chosen. Love, I have chosen, and the world must go. Whatever happens I will live this life—I will live for *you!* It—nothing shall turn me aside; nothing, my dear one. Even if you died—even if you died——'

" 'Yes?' she murmured, softly.

" 'Then—I also would die.'

"And before she could speak again I began to talk, talking eloquently—as I *could* do in that life—talking to exalt love, to make the life we were living seem heroic and glorious; and the thing I was deserting something hard and enormously ignoble that it was a fine thing to set aside. I bent all my mind to throw that glamour upon it, seeking not only to convert her but myself to that. We talked, and she clung to me, torn too between all that she deemed noble and all that she knew was sweet. And at last I did make it heroic, made all the thickening disaster of the world only a sort of glorious setting to our unparalleled love, and we two poor foolish souls strutted there at last, clad in that splendid delusion, drunken rather with that glorious delusion, under the still stars.

"And so my moment passed.

"It was my last chance. Even as we went to and fro there, the leaders of the south and east were gathering their resolve, and the hot answer that shattered Evesham's bluffing for ever, took shape and waited. And all over Asia, and the ocean, and the South, the air and the wires were throbbing with their warnings to prepare—prepare.

"No one living, you know, knew what war was; no one could imagine, with all these new inventions, what horror war might bring. I believe most people still believed it would be a matter of bright uniforms and shouting charges and triumphs and flags and bands—in a time when half the world drew its food supply from regions ten thousand miles way——"

The man with the white face paused. I glanced at him, and his face was intent on the floor of the carriage. A little railway station, a string of loaded trucks, a signal-box, and the back of a cottage, shot by the carriage window, and a bridge passed with a clap of noise, echoing the tumult of the train.

"After that," he said, "I dreamt often. For three weeks of nights that dream was my life. And the worst of it was there were nights when I could not dream, when I lay tossing on a bed in *this* accursed life; and *there*—somewhere lost to me—things were happening—momentous, terrible things. . . . I lived at nights—my days, my waking days, this life I am living now, became a faded, far-away dream, a drab setting, the cover of the book."

He thought.

"I could tell you all, tell you every little thing in the dream, but as to what I did in the daytime—no. I could not tell—I do not remember. My memory—my memory has gone. The business of life slips from me———"

He leant forward, and pressed his hands upon his eyes. For a long time he said nothing.

"And then?" said I

"The war burst like a hurricane."

He stared before him at unspeakable things.

"And then?" I urged again.

"One touch of unreality," he said, in the low tone of a man who speaks to himself, "and they would have been nightmares. But they were not nightmares—they were not nightmares. *No!*"

He was silent for so long that it dawned upon me that there was a danger of losing the rest of the story. But he went on talking again in the same tone of questioning self-communion.

"What was there to do but flight? I had not thought the war would touch Capri—I had seemed to see Capri as being out of it all, as the contrast to it all; but two nights after the whole place was shouting and bawling, every woman almost and every other man wore a badge—Evesham's badge—and there was no music but a jangling war-song over and over again, and everywhere men enlisting, and in the dancing halls they were drilling. The whole island was awhirl with rumours; it was said, again and again, that fighting had begun. I had not expected this. I had seen so little of the life of pleasure that I had failed to reckon with this violence of the amateurs. And as for me, I was out of it. I was like a man who might have prevented the firing of a magazine. The time had gone. I was no one; the vainest stripling with a badge counted for more than I. The crowd jostled us and bawled in our ears; that accursed song deafened us; a woman shrieked at my lady because no badge was on her, and we two went back to our own place again, ruffled and insulted —my lady white and silent, and I aquiver with rage. So furious was I, I could have quarrelled with her if I could have found one shade of accusation in her eyes.

"All my magnificence had gone from me. I walked up and down our rock cell, and outside was the darkling sea and a light to the southward that flared and passed and came again.

" 'We must get out of this place,' I said over and over. 'I have made my choice, and I will have no hand in these troubles. I will have nothing of this war. We have taken our lives out of all these things. This is no refuge for us. Let us go.'

"And the next day we were already in flight from the war that covered the world.

"And all the rest was Flight—all the rest was Flight." He mused darkly.

"How much was there of it?"

He made no answer.

"How many days?"

His face was white and drawn and his hands were clenched. He took no heed of my curiosity.

I tried to draw him back to his story with questions.

"Where did you go?" I said.

"When?"

"When you left Capri."

"South-west," he said, and glanced at me for a second. "We went in a boat."

"But I should have thought an aeroplane?"

"They had been seized."

I questioned him no more. Presently I thought he was beginning again. He broke out in an argumentative monotone:

"But why should it be? If, indeed, this battle, this slaughter and stress *is* life, why have we this craving for pleasure and beauty? If there *is* no refuge, if there is no place of peace, and if all our dreams of quiet places are a folly and a snare, why have we such dreams? Surely it was no ignoble cravings, no base intentions, had brought us to this; it was Love had isolated us. Love had come to me with her eyes and robed in her beauty, more glorious than all else in life, in the very shape and colour of life, and summoned me away. I had silenced all the voices, I had answered all the questions—I had come to her. And suddenly there was nothing but War and Death!"

I had an inspiration. "After all," I said, "it could have been only a dream."

"A dream!" he cried, flaming upon me, "a dream—when, even now——"

For the first time he became animated. A faint flush crept into his cheek. He raised his open hand and clenched it, and dropped it to his knee. He spoke, looking away from me, and for all the rest of the time he looked away. "We are but phantoms," he said, "and the phantoms of phantoms, desires like cloud shadows and wills of straw that eddy in the wind; the days pass, use and wont carry us through as a train carries the shadow of its lights—so be it! But one thing is real and certain, one thing is no dreamstuff, but eternal and enduring. It is the centre of my life, and all other things about it are subordinate or altogether vain. I loved her, that woman of a dream. And she and I are dead together!

"A dream! How can it be a dream, when it has drenched a living life with unappeasable sorrow, when it makes all that I have lived for and cared for, worthless and unmeaning?

"Until that very moment when she was killed I believed we had still a chance of getting away," he said. "All through the night and morning that we sailed across the sea from Capri to Salerno, we talked of escape. We were full of hope, and it clung about us to the end, hope for the life together we should lead, out of it all, out of the battle and struggle. the wild and empty passions, the empty arbitrary 'thou shalt' and 'thou shalt not' of the world. We were uplifted, as though our quest was a holy thing, as though love for one another was a mission. . . .

"Even when from our boat we saw the fair face of that

great rock Capri—already scarred and gashed by the gun emplacements and hiding-places that were to make it a fastness—we reckoned nothing of the imminent slaughter, though the fury of preparation hung about in puffs and clouds of dust at a hundred points amidst the grey; but, indeed, I made a text of that and talked. There, you know, was the rock, still beautiful for all its scars, with its countless windows and arches and ways, tier upon tier, for a thousand feet, a vast carving of grey, broken by vine-clad terraces and lemon and orange groves and masses of agave and prickly pear, and puffs of almond blossom. And out under the archway that is built over the Marina Piccola other boats were coming; and as we came round the cape and within sight of the mainland, another string of boats came into view, driving before the wind towards the south-west. In a little while a multitude had come out, the remoter just specks of ultramarine in the shadow of the eastward cliff.

" 'It is love and reason,' I said, 'fleeing from all this madness of war.'

"And though we presently saw a squadron of aeroplanes flying across the southern sky we did not heed it. There it was—a line of dots in the sky—and then more, dotting the south-eastern horizon, and then still more, until all that quarter of the sky was stippled with blue specks. Now they were all thin little strokes of blue, and now one and now a multitude would heel and catch the sun and become short flashes of light. They came, rising and falling and growing larger like some huge flight of gulls or rooks or such-like birds, moving with a marvellous uniformity, and ever as they drew nearer they spread over a greater width of sky. The southward wing flung itself in an arrow-headed cloud athwart the sun. And then suddenly they swept round to the eastward and streamed eastward, growing smaller and smaller and clearer and clearer again until they vanished from the sky. And after that we noted to the northward and very high Evesham's fighting machines hanging high over Naples like an evening swarm of gnats.

"It seemed to have no more to do with us than a flight of birds.

"Even the mutter of guns far away in the south-east seemed to us to signify nothing. . . .

"Each day, each dream after that, we were still exalted, still seeking that refuge where we might live and love. Fatigue had come upon us, pain and many distresses. For though we were dusty and stained by our toilsome tramping, and half starved and with the horror of the dead men we had seen and the flight of the peasants—for very soon a gust of fighting swept up the peninsula—with these things haunting our minds it still resulted only in a deepening resolution to escape. Oh, but she was brave and patient! She who had never faced hardship and exposure had courage for herself—and me. We went to and fro seeking an outlet, over a country all commandeered and ransacked by the gathering hosts of war. Always we went on foot. At first there were other fugitives, but we did not mingle with them. Some escaped northward, some were caught in the torrent of peasantry that swept along the main roads, many gave themselves into the hands of the soldiery and were sent northward. Many of the men were impressed. But we kept away from these things; we had brought no money to bribe a passage north, and I feared for my lady at the hands of these conscript crowds. We had landed at Salerno, and we had been turned back from Cava, and we had tried to cross towards Taranto by a pass over Monte Alburno, but we had been driven back for want of food, and so we had come down among the marshes by Pæstum, where those great temples stand alone. I had some vague idea that by Pæstum it might be possible to find a boat or something, and take once more to sea. And there it was the battle overtook us.

"A sort of soul-blindness had me. Plainly I could see that we were being hemmed in; that the great net of that giant Warfare had us in its toils. Many times we had seen the levies that had come down from the north going to and fro, and had come upon them in the distance amidst the mountains making ways for the ammunition and preparing the mounting of the guns. Once we fancied they had fired at us, taking us for spies—at any rate a shot had gone shuddering over us. Several times we had hidden in woods from hovering aeroplanes.

"But all these things do not matter now, these nights of flight and pain. . . . We were in an open place near those great temples at Pæstum at last, on a blank stony place dotted with spiky bushes, empty and desolate and so flat that a grove of eucalyptus far away showed to the feet of its stems. How I can see it! My lady was sitting down under a bush resting a little, for she was very weak and weary, and I was standing up watching to see if I could tell the distance of the firing that came and went. They were still, you know, fighting far from each other, with those terrible new weapons that had never before been used: guns that would carry beyond sight, and aeroplanes that would do—— What *they* would do no man could foretell.

"I knew that we were between the two armies, and that they drew together. I knew we were in danger, and that we could not stop there and rest!

"Though all these things were in my mind, they were in the background. They seemed to be affairs beyond our concern. Chiefly, I was thinking of my lady. An aching distress filled me. For the first time she had owned herself beaten and had fallen a-weeping. Behind me I could hear her sobbing, but I would not turn round to her because I knew she had need of weeping, and had held herself so far and so long for me. It was well, I thought, that she would weep and rest and then we would toil on again, for I had no inkling of the thing that hung so near. Even now I can see her as she sat there, her lovely hair upon her shoulder, can mark again the deepening hollow of her cheek.

" 'If we had parted,' she said. 'if I had let you go.'

" 'No,' said I. 'Even now, I do not repent. I will not repent; I made my choice, and I will hold on to the end.'

"And then——

"Overhead in the sky flashed something and burst, and all about us I heard the bullets making a noise like a handful of peas suddenly thrown. They chipped the stones about us, and whirled fragments from the bricks and passed. . . ."

He put his hand to his mouth, and then moistened his lips.

"At the flash I had turned about. . . .

"You know—she stood up——

"She stood up, you know, and moved a step towards me——

"As though she wanted to reach me——

"And she had been shot through the heart."

He stopped and stared at me. I felt all that foolish incapacity an Englishman feels on such occasions. I met his eyes for a moment, and then stared out of the window. For a long space we kept silence. When at last I looked at him he was sitting back in his corner, his arms folded, and his teeth gnawing at his knuckles.

He bit his nail suddenly, and stared at it.

"I carried her," he said, "towards the temples, in my arms —as though it mattered. I don't know why. They seemed a sort of sanctuary, you know; they had lasted so long, I suppose.

"She must have died almost instantly. Only—I talked to her—all the way."

Silence again.

"I have seen those temples," I said abruptly, and indeed he had brought those still, sunlit arcades of worn sandstone very vividly before me.

"It was the brown one, the big brown one. I sat down on a fallen pillar and held her in my arms. . . . Silent after the first babble was over. And after a little while the lizards came out and ran about again, as though nothing unusual was going on, as though nothing had changed. . . . It was tremendously still there, the sun high and the shadows still; even the shadows of the weeds upon the entablature were still—in spite of the thudding and banging that went all about the sky.

"I seem to remember that the aeroplanes came up out of the south, and that the battle went away to the west. One aeroplane was struck, and overset and fell. I remember that—though it didn't interest me in the least. It didn't seem to signify. It was like a wounded gull, you know—

flapping for a time in the water. I could see it down the aisle of the temple—a black thing in the bright blue water.

"Three or four times shells burst about the beach, and then that ceased. Each time that happened all the lizards scuttled in and hid for a space. That was all the mischief done, except that once a stray bullet gashed the stone hard by—made just a fresh bright surface.

"As the shadows grew longer, the stillness seemed greater.

"The curious thing," he remarked, with the manner of a man who makes a trivial conversation, "is that I didn't *think* —I didn't think at all. I sat with her in my arms amidst the stones—in a sort of lethargy—stagnant.

"And I don't remember waking up. I don't remember dressing that day. I know I found myself in my office, with my letters all slit open in front of me, and how I was struck by the absurdity of being there, seeing that in reality I was sitting, stunned, in that Pæstum Temple with a dead woman in my arms. I read my letters like a machine. I have forgotten what they were about."

He stopped, and there was a long silence.

Suddenly I perceived that we were running down the incline from Chalk Farm to Euston. I started at this passing of time. I turned on him with a brutal question, in the tone of "Now or never."

"And did you dream again?"

"Yes."

He seemed to force himself to finish. His voice was very low.

"Once more, and as it were only for a few instants. I seemed to have suddenly awakened out of a great apathy, to have risen into a sitting position, and the body lay there on the stones beside me. A gaunt body. Not her, you know. So soon—it was not her. . . .

"I may have heard voices. I do not know. Only I knew clearly that men were coming into the solitude and that that was a last outrage.

"I stood up and walked through the temple, and then there came into sight—first one man with a yellow face, dressed in a uniform of dirty white, trimmed with blue, and then several, climbing to the crest of the old wall of the vanished city, and crouching there. They were little bright figures in the sunlight, and there they hung, weapon in hand, peering cautiously before them.

"And further away I saw others and then more at another point in the wall. It was a long lax line of men in open order.

"Presently the man I had first seen stood up and shouted a command, and his men came tumbling down the wall and into the high weeds towards the temple. He scrambled down with them and led them. He came facing towards me, and when he saw me he stopped.

"At first I had watched these men with a mere curiosity, but when I had seen they meant to come to the temple I was moved to forbid them. I shouted to the officer.

"'You must not come here,' I cried, 'I am here. I am here with my dead.'

"He stared, and then shouted a question back to me in some unknown tongue.

"I repeated what I had said.

"He shouted again, and I folded my arms and stood still. Presently he spoke to his men and came forward. He carried a drawn sword.

"I signed to him to keep away, but he continued to advance. I told him again very patiently and clearly: 'You must not come here. These are old temples and I am here with my dead.'

"Presently he was so close I could see his face clearly. It was a narrow face, with dull grey eyes, and a black moustache. He had a scar on his upper lip, and he was dirty and unshaven. He kept shouting unintelligible things, questions, perhaps, at me.

"I know now that he was afraid of me, but at the time that did not occur to me. As I tried to explain to him, he interrupted me in imperious tones, bidding me, I suppose, stand aside.

"He made to go past me, and I caught hold of him.

"I saw his face change at my grip.

"'You fool,' I cried. 'Don't you know? She is dead!'

"He started back. He looked at me with cruel eyes. I saw a sort of exultant resolve leap into them—delight. Then, suddenly, with a scowl, he swept his sword back—*so*—and thrust."

He stopped abruptly.

I became aware of a change in the rhythm of the train. The brakes lifted their voices and the carriage jarred and jerked. This present world insisted upon itself, became clamorous. I saw through the steamy window huge electric lights glaring down from tall masts upon a fog, saw rows of stationary empty carriages passing by; and then a signal-box, hoisting its constellation of green and red into the murky London twilight, marched after them. I looked again at his drawn features.

"He ran me through the heart. It was with a sort of astonishment—no fear, no pain—but just amazement, that I felt it pierce me, felt the sword drive home into my body. It didn't hurt, you know. It didn't hurt at all."

The yellow platform lights came into the field of view, passing first rapidly, then slowly, and at last stopping with a jerk. Dim shapes of men passed to and fro without.

"Euston!" cried a voice.

"Do you mean——?"

"There was no pain, no sting or smart. Amazement and then darkness sweeping over everything. The hot, brutal face before me, the face of the man who had killed me, seemed to recede. It swept out of existence——"

"Euston!" clamoured the voices outside; "Euston!"

The carriage door opened admitting a flood of sound, and a porter stood regarding us. The sounds of doors slamming, and the hoof-clatter of cab-horses, and behind these things the featureless remote roar of the London cobble-stones, came to my ears. A truckload of lighted lamps blazed along the platform.

"A darkness, a flood of darkness that opened and spread and blotted out all things."

"Any luggage, sir?" said the porter.

"And that was the end?" I asked.

He seemed to hesitate. Then, almost inaudibly, he answered, "*No.*"

"You mean?"

"I couldn't get to her. She was there on the other side of the temple—— And then——"

"Yes," I insisted. "Yes?"

"Nightmares," he cried; "nightmares indeed! My God! Great birds that fought and tore."

THE END

Printed in Great Britain

by Cox and Wyman Limited,

London, Fakenham and Reading